More Than Just a

Internet Resources

Step 1 Connect to NY Math Online glencoe.com

Step 2 Connect to online resources by using *QuickPass* codes. You can connect directly to the chapter you want.

NY3385c1

Enter this code with the appropriate chapter number.

For Students

Connect to the student edition *eBook* that contains all of the following online assets. You don't need to take your textbook home every night.

- Personal Tutor
- Extra Examples
- Self-Check Quizzes
- Multilingual eGlossary
- Concepts in Motion
- Chapter Test Practice
- Test Practice
- Study to Go

For Teachers

Connect to professional development content at glencoe.com and the *eBook Advance Tracker* at AdvanceTracker.com

For Parents

Connect to glencoe.com for access to the *eBook* and all the resources for students and teachers that are listed above.

Glencoe McGraw-Hill

New York
Math Connects

Concepts, Skills, and Problem Solving

Course 2

Authors
Day • Frey • Howard • Hutchens
Luchin • McClain • Molix-Bailey
Ott • Pelfrey • Price
Vielhaber • Willard

McGraw Hill Glencoe

About the Cover

Nothing beats the thrill of a giant water slide! Next time you are at a water park, think about math. The water slide is exciting because there is a big vertical drop in a small horizontal distance. In math, that is expressed as a ratio $\frac{rise}{run}$, which is called *slope*. You'll learn more about slope in Chapter 6.

Cover image photographed on location courtesy Busch Gardens/Adventure Island, Tampa, Florida. Credit Richard Hutchings.

 **Glencoe**

The *McGraw-Hill* Companies

Send all inquiries to:
Glencoe/McGraw-Hill
8787 Orion Place
Columbus, OH 43240-4027

ISBN: 978-0-07-888338-5
MHID: 0-07-888338-5

Printed in the United States of America.

1 2 3 4 5 6 7 8 9 10 027/055 16 15 14 13 12 11 10 09 08 07

Contents in Brief

Focal Points

The Curriculum Focal Points identify key mathematical ideas for this grade. They are not discrete topics or a checklist to be mastered; rather, they provide a framework for the majority of instruction at a particular grade level and the foundation for future mathematics study. The complete document may be viewed at www.nctm.org/focalpoints.

KEY

G7-FP1
Grade 7 Focal Point 1

G7-FP2
Grade 7 Focal Point 2

G7-FP3
Grade 7 Focal Point 3

G7-FP4C
Grade 7 Focal Point 4
Connection

G7-FP5C
Grade 7 Focal Point 5
Connection

G7-FP6C
Grade 7 Focal Point 6
Connection

G7-FP7C
Grade 7 Focal Point 7
Connection

G7-FP1   Number and Operations and Algebra and Geometry: **Developing an understanding of and applying proportionality, including similarity**

Students extend their work with ratios to develop an understanding of proportionality that they apply to solve single and multistep problems in numerous contexts. They use ratio and proportionality to solve a wide variety of percent problems, including problems involving discounts, interest, taxes, tips, and percent increase or decrease. They also solve problems about similar objects (including figures) by using scale factors that relate corresponding lengths of the objects or by using the fact that relationships of lengths within an object are preserved in similar objects. Students graph proportional relationships and identify the unit rate as the slope of the related line. They distinguish proportional relationships ($\frac{y}{x} = k$, or $y = kx$) from other relationships, including inverse proportionality ($xy = k$, or $y = \frac{k}{x}$).

G7-FP2   Measurement and Geometry and Algebra: **Developing an understanding of and using formulas to determine surface areas and volumes of three-dimensional shapes**

By decomposing two- and three-dimensional shapes into smaller, component shapes, students find surface areas and develop and justify formulas for the surface areas and volumes of prisms and cylinders. As students decompose prisms and cylinders by slicing them, they develop and understand formulas for their volumes (*Volume = Area of base × Height*). They apply these formulas in problem solving to determine volumes of prisms and cylinders. Students see that the formula for the area of a circle is plausible by decomposing a circle into a number of wedges and rearranging them into a shape that approximates a parallelogram. They select appropriate two- and three dimensional shapes to model real-world situations and solve a variety of problems (including multistep problems) involving surface areas, areas and circumferences of circles, and volumes of prisms and cylinders.

G7-FP3   Number and Operations and Algebra: **Developing an understanding of operations on all rational numbers and solving linear equations**

Students extend understandings of addition, subtraction, multiplication, and division, together with their properties, to all rational numbers, including negative integers. By applying properties of arithmetic and considering negative numbers in everyday contexts (e.g., situations of owing money or measuring elevations above and below sea level), students explain why the rules for adding, subtracting, multiplying, and dividing with negative numbers make sense. They use the arithmetic of rational numbers as they formulate and solve linear equations in one variable and use these equations to solve problems. Students make strategic choices of procedures to solve linear equations in one variable and implement them efficiently, understanding that when they use the properties of equality to express an equation in a new way, solutions that they obtain for the new equation also solve the original equation.

Connections to the Focal Points

G7-FP4C   **Measurement and Geometry:** Students connect their work on proportionality with their work on area and volume by investigating similar objects. They understand that if a scale factor describes how corresponding lengths in two similar objects are related, then the square of the scale factor describes how corresponding areas are related, and the cube of the scale factor describes how corresponding volumes are related. Students apply their work on proportionality to measurement in different contexts, including converting among different units of measurement to solve problems involving rates such as motion at a constant speed. They also apply proportionality when they work with the circumference, radius, and diameter of a circle; when they find the area of a sector of a circle; and when they make scale drawings.

G7-FP5C   **Number and Operations:** In grade 4, students used equivalent fractions to determine the decimal representations of fractions that they could represent with terminating decimals. Students now use division to express any fraction as a decimal, including fractions that they must represent with infinite decimals. They find this method useful when working with proportions, especially those involving percents. Students connect their work with dividing fractions to solving equations of the form $ax = b$, where a and b are fractions. Students continue to develop their understanding of multiplication and division and the structure of numbers by determining if a counting number greater than 1 is a prime, and if it is not, by factoring it into a product of primes.

G7-FP6C   **Data Analysis:** Students use proportions to make estimates relating to a population on the basis of a sample. They apply percentages to make and interpret histograms and circle graphs.

G7-FP7C   **Probability:** Students understand that when all outcomes of an experiment are equally likely, the theoretical probability of an event is the fraction of outcomes in which the event occurs. Students use theoretical probability and proportions to make approximate predictions.

Authors

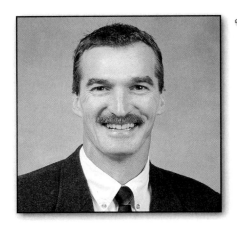

Roger Day, Ph.D.
Mathematics Department
  Chair
Pontiac Township High
  School
Pontiac, Illinois

Patricia Frey, Ed.D.
Math Coordinator at
  Westminster Community
  Charter School
Buffalo, New York

Arthur C. Howard
Mathematics Teacher
Houston Christian
  High School
Houston, Texas

**Deborah A. Hutchens,
  Ed.D.**
Principal
Chesapeake, Virginia

Beatrice Luchin
Mathematics Consultant
League City, Texas

Kay McClain, Ed.D.
Assistant Professor
Vanderbilt University
Nashville, Tennessee

Rhonda J. Molix-Bailey
Mathematics Consultant
Mathematics by Design
DeSoto, Texas

Jack M. Ott, Ph.D.
Distinguished Professor
  of Secondary Education
  Emeritus
University of South Carolina
Columbia, South Carolina

Ronald Pelfrey, Ed.D.
Mathematics Specialist
Appalachian Rural
  Systemic Initiative and
  Mathematics Consultant
Lexington, Kentucky

Jack Price, Ed.D.
Professor Emeritus
California State
  Polytechnic University
Pomona, California

Kathleen Vielhaber
Mathematics Consultant
St. Louis, Missouri

Teri Willard, Ed.D.
Assistant Professor
Department of Mathematics
Central Washington
  University
Ellensburg, Washington

Contributing Author

FOLDABLES **Dinah Zike**
Educational Consultant
Dinah-Might Activities, Inc.
San Antonio, Texas

WELCOME TO
New York Math Connects
COURSE 2

Master the New York State Standards in 3 Easy Steps

1 Practice the Standards Daily

- Each lesson addresses New York State Standards covered in that lesson.

- Questions aligned to the standards in a format like those on the New York State Mathematics Test provide you with ongoing opportunities to sharpen your test-taking skills.

> **NYS Core Curriculum**
>
> **7.PS.14** Determine information required to solve the problem
> **7.PS.15** Choose methods for obtaining required information
> *Also addresses 7.PS.12, 7.PS.13, 7.RP.3, 7.CM.3*

> **NYSMT PRACTICE**

② Practice the Standards throughout the Chapter

- Every chapter contains a completely worked-out New York State Mathematics Test Example to help you solve problems that are similar to those you might find on that test. NYSMT EXAMPLE

- Every chapter contains two full pages of NYSMT Practice with Test-Taking Tips.

 NYSMT **Practice**

③ Practice the Standards Before the Test

- If you've followed steps 1 and 2, you should be more than ready for the test. But just in case you want to make sure, use pages NY1–NY25 to practice questions that are organized by standard. Lesson references are included for you should you need a little refresher.

 GET READY FOR THE New York State Test

New York Reviewers

Each New York Reviewer gave feedback and suggestions for improving the effectiveness of the PreK–8 *Math Connects* program.

Paula Barnes
Mathematics Teacher
Minisink Valley CSD
Slate Hill, New York

Joanne DeMizio
Assistant Superintendent, Math and Science Curriculum
Archdiocese of New York
New York, New York

Roberta Grindle
Math and Language Arts Academic Intervention Service Provider
Cumberland Head Elementary School
Plattsburgh, New York

Consultants

Glencoe/McGraw-Hill wishes to thank the following professionals for their feedback. They were instrumental in providing valuable input toward the development of this program in these specific areas.

Mathematical Content

Viken Hovsepian
Professor of Mathematics
Rio Hondo College
Whittier, California

Grant A. Fraser, Ph.D.
Professor of Mathematics
California State University, Los Angeles
Los Angeles, California

Arthur K. Wayman, Ph.D.
Professor of Mathematics Emeritus
California State University, Long Beach
Long Beach, California

English Language Learners

Josefina V. Tinajero, Ph.D.
Dean, College of Education
The University of Texas at El Paso
El Paso, Texas

Gifted and Talented

Ed Zaccaro
Author and Consultant
Bellevue, Iowa

Graphing Calculator

Ruth M. Casey
National Mathematics Consultant
National Instructor, Teachers Teaching
  with Technology
Frankfort, Kentucky

Learning Disabilities

Kate Garnett, Ph.D.
Chairperson, Coordinator
  Learning Disabilities
School of Education
Department of Special Education
Hunter College, CUNY
New York, New York

Mathematical Fluency

Jason Mutford
Mathematics Instructor
Coxsackie-Athens Central School District
Coxsackie, New York

Pre-AP

Dixie Ross
Mathematics Teacher
Pflugerville High School
Pflugerville, Texas

Reading and Vocabulary

Douglas Fisher, Ph.D.
Professor of Language and Literacy
  Education
San Diego State University
San Diego, California

Lynn T. Havens
Director of Project CRISS
Kalispell, Montana

Reviewers

Each reviewer reviewed at least two chapters of the Student Edition, giving feedback and suggestions for improving the effectiveness of the mathematics instruction.

Sheila J. Allen
Mathematics Teacher
A.I. Root Middle School
Medina, Ohio

Paula Barnes
Mathematics Teacher
Minisink Valley CSD
Slate Hill, New York

Deborah Barnett
Mathematics Consultant
Lake Shore Public Schools
St. Clair Shores, Michigan

Laurel W. Blackburn
Teacher/Mathematics
    Department Chair
Hillcrest Middle School
Simpsonville, South Carolina

Drista Bowser
Mathematics Teacher
New Windsor Middle School
New Windsor, Maryland

Matthew Bowser
Teacher
Oil City Middle School
Oil City, Pennsylvania

Susan M. Brewer
Mathematics Teacher
Brunswick Middle School
Brunswick, Maryland

Patricia A. Bruzek
Mathematics Teacher
Glenn Westlake Middle School
Lombard, Illinois

Luanne Budd
Supervisor of Mathematics
Randolph Township
Randolph, New Jersey

Ella Violet Burch
Mathematics Teacher
Penns Grove High School
Carneys Point, New Jersey

Hailey Caldwell
7th Grade Mathematics Teacher
Greenville Middle Academy of
    Traditional Studies
Greenville, South Carolina

Linda K. Chandler
7th Grade Mathematics Teacher
Willard Middle School
Willard, Ohio

Debra M. Cline
7th Grade Mathematics Teacher
Thomas Jefferson Middle School
Winston-Salem, North Carolina

Randall G. Crites
Principal
Bunker R-3
Bunker, Missouri

Rose Dickinson
Science and Mathematics
    Teacher
Seneca Middle School
Clinton Township, Michigan

Joyce Wolfe Dodd
6th Grade Mathematics Teacher
Bryson Middle School
Simpsonville, South Carolina

John G. Doyle
Middle School Chairperson/
    Mathematics Teacher
Wyoming Valley West School
    District
Kingston, Pennsylvania

Katie England
Secondary Mathematics Resource
    Teacher
Carroll County Public Schools
Westminster, Maryland

Carol A. Fincannon
6th Grade Mathematics Teacher
Southwood Middle School
Anderson, South Carolina

Sally J. Fulmer
7th Grade Mathematics Teacher/
    Department Chair
C.E. Williams Middle School
Charleston, South Carolina

Marian K. Geist
Mathematics Teacher/Leadership
    Team
Baker Prairie Middle School
Canby, Oregon

Becky Gorniack
Middle School Mathematics
    Teacher
Fremont Middle School
Mundelein, Illinois

Donna Tutterow Hamilton
Curriculum Facilitator
Corriher Lipe Middle School
Landis, North Carolina

Danny Liebertz
8th Grade Mathematics
    Instructor
Fowler Middle School
Tigard, Oregon

Marie Merkel
Learning Support
North Pocono School District
Scranton, Pennsylvania

Tonda North
Algebra 1/8th Grade Mathematics
    Teacher
Indian Valley Middle School
Enon, Ohio

Natasha L.M. Nuttbrock
7th Grade Mathematics Teacher
Ferguson Middle School
Beavercreek, Ohio

Paul Penn
Curriculum Team Leader,
    Mathematics
Lima City Schools
Lima, Ohio

Casey Condran Plackett
7th Grade Mathematics Teacher
Kennedy Junior High School
Lisle, Illinois

E. Elaine Rafferty
Mathematics Consultant
Summerville, South Carolina

Edward M. Repko
Mathematics Teacher
Kilbourne Middle School
Worthington, Ohio

Alfreda Reynolds
Teacher
Charlotte-Mecklenburg School
    System
Charlotte, North Carolina

Alice Roberts
Mathematics Teacher
Oakdale Middle School
Ijamsville, Maryland

Jennifer L. Rodriguez
Mathematics Teacher
Glen Crest Middle School
Glen Ellyn, Illinois

Natalie Rohaley
6th Grade Mathematics
Riverside Middle School
Greer, South Carolina

Annika Lee Schilling
Mathematics and Science
    Teacher
Duniway Middle School
McMinnville, Oregon

Sherry Scott
Mathematics Teacher
E.A. Tighe School
Margate, New Jersey

Eli Shaheen
Mathematics Teacher/
    Department Chair
Plum Senior High School
Pittsburgh, Pennsylvania

Kelly Eady Shaw
7th Grade Mathematics Teacher
Rawlinson Road Middle School
Rock Hill, South Carolina

Evan J. Silver
Mathematics Teacher
Walkersville Middle School
Frederick, Maryland

Charlotte A. Thore
6th/7th Grade Mathematics
    Teacher
Northwest School of the Arts
Charlotte, North Carolina

Gene A. Tournoux
Mathematics Department Head
Shaker Heights High School
Shaker Heights, Ohio

Pamela J. Trainer
Mathematics Teacher
Roland-Grise Middle School
Wilmington, North Carolina

David A. Trez
Mathematics Teacher
Bloomfield Middle School
Bloomfield, New Jersey

Pauline D. Von Hoffer
Mathematics Teacher
Wentzville School District
Wentzville, Missouri

Kentucky Consultants

Jenn Crase
8th Grade Mathematics Teacher/
    Department Chair
South Oldham Middle School
Crestwood, Kentucky

Max DeBoer Lux
8th Grade Mathematics
Summit View Middle School
Independence, Kentucky

Jennifer Wells Phipps
Middle School Mathematics
    Teacher
Corbin Middle School
Corbin, Kentucky

Bea Torrence
Teacher/Mathematics Content
    Leader
Camp Ernst Middle School
Burlington, Kentucky

J. Ron Vanover
Advanced Placement Calculus
Boone County High School
Florence, Kentucky

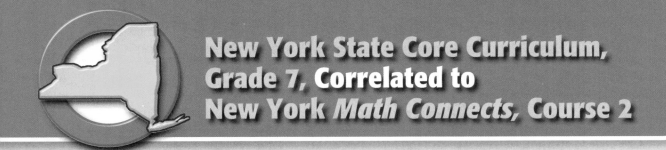

Lessons in which the standard is the primary focus are indicated in **bold**.

Process Strands and Performance Indicators		Lessons	Page References
Problem Solving Strand			
Students will build new mathematical knowledge through problem solving.			
7.PS.1	Use a variety of strategies to understand new mathematical content and to develop more efficient methods	Utilized throughout the text. For example, 1-5, 2-7, 9-6, 12-3	Utilized throughout the text. For example, 42–43, 112–113, 484–485, 646–647
7.PS.2	Construct appropriate extensions to problem situations	6-7	318–319
7.PS.3	Understand and demonstrate how written symbols represent mathematical ideas	1-6, 1-7	44–47, 49–52
Students will solve problems that arise in mathematics and in other contexts.			
7.PS.4	Observe patterns and formulate generalizations	1-9, Extend 1-9, **2-7**, 4-3, Extend 12-4	57–61, 62, **112–113**, 190–191, 654–655
7.PS.5	Make conjectures from generalizations	Utilized throughout the text. For example, Explore 3-7, Extend 10-8, Extend 11-6, Extend 12-4	Utilized throughout the text. For example, 162, 552, 600–601, 654–655
7.PS.6	Represent problem situations verbally, numerically, algebraically, and graphically	Utilized throughout the text. For example, 1-10, 6-7, Extend 7-8, 8-4, 11-6	Utilized throughout the text. For example, 63–67, 318–319, 383, 415–421, 596–599
Students will apply and adapt a variety of appropriate strategies to solve problems.			
7.PS.7	Understand that there is no one right way to solve mathematical problems but that different methods have advantages and disadvantages	Utilized throughout the text. For example, 1-5, 9-6, 10-5, 12-3	Utilized throughout the text. For example, 42–43, 484–485, 530–531, 646–647
7.PS.8	Understand how to break a complex problem into simpler parts or use a similar problem type to solve a problem	RSP, 4-3, **11-5**	150, 190–191, **594–595**
7.PS.9	Work backwards from a solution	**3-4**	**148–149**
7.PS.10	Use proportionality to model problems	6-2, 6-8, Extend 6-8, 10-7, LA	287–292, 320–327, 540–545, LA14–LA17
7.PS.11	Work in collaboration with others to solve problems	Explore 4-1, 6-7	180, 318–319
Students will monitor and reflect on the process of mathematical problem solving.			
7.PS.12	Interpret solutions within the given constraints of a problem	1-1, 7-5	25–29, 366–367
7.PS.13	Set expectations and limits for possible solutions	1-1	25–29
7.PS.14	Determine information required to solve the problem	1-1	25–29
7.PS.15	Choose methods for obtaining required information	1-1, 2-6, 3-5, 4-8, 5-1, 6-6, 7-1	25–29, 107–111, 151–155, 211–214, 230–235, 310–315, 344–348
7.PS.16	Justify solution methods through logical argument	Utilized throughout the text. For example, 5-6, 10-5	Utilized throughout the text. For example, 258–263, 530–531
7.PS.17	Evaluate the efficiency of different representations of a problem	Explore 3-2, 8-5	136–141, 424–425

LA = Looking Ahead to Next Year; CSB = Concepts and Skills Bank; RSP = Reading to Solve Problems

Process Strands and Performance Indicators		Lessons	Page References
Reasoning and Proof Strand			
Students will recognize reasoning and proof as fundamental aspects of mathematics.			
7.RP.1	Recognize that mathematical ideas can be supported by a variety of strategies	Utilized throughout the text. For example, 10-5, Explore 11-2	Utilized throughout the text. For example, 530–531, 578–582
Students will make and investigate mathematical conjectures.			
7.RP.2	Use mathematical strategies to reach a conclusion	Utilized throughout the text. For example, Extend 1-10, 10-5, 11-10	Utilized throughout the text. For example, 68–69, 530–531, 619–623
7.RP.3	Evaluate conjectures by distinguishing relevant from irrelevant information to reach a conclusion or make appropriate estimates	1-1	25–29
Students will develop and evaluate mathematical arguments and proofs.			
7.RP.4	Provide supportive arguments for conjectures	Utilized throughout the text. For example, 6-6, 8-1, 10-2, 11-2	Utilized throughout the text. For example, 310–315, 396–401, 514–517, 578–582
7.RP.5	Develop, verify, and explain an argument, using appropriate mathematical ideas and language	Utilized throughout the text. For example, RSP, 5-6, 10-5, 11-2	Utilized throughout the text. For example, 185, 262, 530–531, 578–562
Students will select and use various types of reasoning and methods of proof.			
7.RP.6	Support an argument by using a systematic approach to test more than one case	2-7, 10-5	112–113, 530–531
7.RP.7	Devise ways to verify results or use counterexamples to refute incorrect statements	Utilized throughout the text. For example, 6-6, 8-1, 12-2	Utilized throughout the text. For example, 310–315, 396–401, 640–645
7.RP.8	Apply inductive reasoning in making and supporting mathematical conjectures	2-7, **10-5**	112–113, **530–531**
Communication Strand			
Students will organize and consolidate their mathematical thinking through communication.			
7.CM.1	Provide a correct, complete, coherent, and clear rationale for thought process used in problem solving	Utilized throughout the text. For example, 5-1 Explore 7-1, 9-5	Utilized throughout the text. For example, 230–235, 344–348, 480–483
7.CM.2	Provide an organized argument which explains rationale for strategy selection	2-6, 3-5, 4-8, 5-1, 6-6, 7-1	107–111, 151–155, 211–214, 230–235, 310–315, 344–348
7.CM.3	Organize and accurately label work	1-1	25-29
Students will communicate their mathematical thinking coherently and clearly to peers, teachers, and others.			
7.CM.4	Share organized mathematical ideas through the manipulation of objects, numerical tables, drawings, pictures, charts, graphs, tables, diagrams, models and symbols in written and verbal form	Utilized throughout the text. For example, 1-6, Extend 1-9, Extend 8-6, Extend 12-4	44–47, 57–61, 426–431, 649–653
7.CM.5	Answer clarifying questions from others	Utilized throughout the text. For example, 3-6, 5-2, 8-1	156–161, 236–241, 396–401
Students will analyze and evaluate the mathematical thinking and strategies of others.			
7.CM.6	Analyze mathematical solutions shared by others	Utilized throughout the text. For example, 2-4, 4-7, 9-5, 11-3	Utilized throughout the text. For example, 95–99, 206–210, 480–483, 584–588

LA = Looking Ahead to Next Year; CSB = Concepts and Skills Bank; RSP = Reading to Solve Problems

Process Strands and Performance Indicators		Lessons	Page References
7.CM.7	Compare strategies used and solutions found by others in relation to their own work	Utilized throughout the text. For example, 1-4, 3-1, 5-2, 9-5	Utilized throughout the text. For example, 38–41, 128–133, 236–241, 480–483
7.CM.8	Formulate mathematical questions that elicit, extend, or challenge strategies, solutions, and/or conjectures of others	Utilized throughout the text. For example, 1-4, 3-1, 5-2, 7-3	38–41, 128–133, 236–241, 355–360
Students will use the language of mathematics to express mathematical ideas precisely.			
7.CM.9	Increase their use of mathematical vocabulary and language when communicating with others	Utilized throughout the text. For example, 3-7, 4-1, 7-1, 10-2	Utilized throughout the text. For example, 163–167, 181–184, 344–348, 514–517
7.CM.10	Use appropriate language, representations, and terminology when describing objects, relationships, mathematical solutions, and rationale	Utilized throughout the text. For example, 6-9, 8-2, 9-2, 11-3	Utilized throughout the text. For example, 328–332, 402–408, 465–470, 584–588
7.CM.11	Draw conclusions about mathematical ideas through decoding, comprehension, and interpretation of mathematical visuals, symbols, and technical writing	Utilized throughout the text. For example, RSP, 8-2, 10-3, 10-10, 11-9, 12-2	Utilized throughout the text. For example, 349, 402–408, 518–523, 558–562, 613–618, 640–645
Connections Strand			
Students will recognize and use connections among mathematical ideas.			
7.CN.1	Understand and make connections among multiple representations of the same mathematical idea	1-9, 1-10, Extend 1-10, Explore 3-7, 3-7, Extend 3-7	57–61, 63–67, 68–69, 162–168
7.CN.2	Recognize connections between subsets of mathematical ideas	1-9, 1-10, Extend 1-10	57–61, 63–67, 68–69
7.CN.3	Connect and apply a variety of strategies to solve problems	Utilized throughout the text. For example, 3-4, 4-3, 6-7, 8-5	Utilized throughout the text. For example, 148–149, 190–191, 318–319, 424–425
Students will understand how mathematical ideas interconnect and build on one another to produce a coherent whole.			
7.CN.4	Model situations mathematically, using representations to draw conclusions and formulate new situations	Utilized throughout the text. For example, Extend 1-9, Explore 11-3, 12-3	57–61, 584–588, 646–647
7.CN.5	Understand how concepts, procedures, and mathematical results in one area of mathematics can be used to solve problems in other areas of mathematics	Utilized throughout the text. For example, 2-3, 3-1, 5-2, 10-2	Utilized throughout the text. For example, 88–92, 128–133, 236–241, 514–517
Students will recognize and apply mathematics in contexts outside of mathematics.			
7.CN.6	Recognize and provide examples of the presence of mathematics in their daily lives	Utilized throughout the text. For example, 6-9, 9-5, 10-3, 11-1	Utilized throughout the text. For example, 328–332, 480–483, 518–523, 572–576
7.CN.7	Apply mathematical ideas to problem situations that develop outside of mathematics	Utilized throughout the text. For example, 3-7, 5-6, 8-2, 9-1	Utilized throughout the text. For example, 163–167, 258–263, 402–408, 460–464
7.CN.8	Investigate the presence of mathematics in careers and areas of interest	Utilized throughout the text. For example, 1-10, 4-6, 8-2, 9-8	Utilized throughout the text. For example, 63–67, 202–205, 402–408, 492–497
7.CN.9	Recognize and apply mathematics to other disciplines, areas of interest, and societal issues	Utilized throughout the text. For example, 3-3, 4-6, 7-2, 11-8	Utilized throughout the text. For example, 142–146, 202–205, 350–354, 608–612

LA = Looking Ahead to Next Year; CSB = Concepts and Skills Bank; RSP = Reading to Solve Problems

Process Strands and Performance Indicators		Lessons	Page References
Representation Strand			
Students will create and use representations to organize, record, and communicate mathematical ideas.			
7.R.1	Use physical objects, drawings, charts, tables, graphs, symbols, equations, or objects created using technology as representations	Utilized throughout the text. For example, Extend 3-7, Extend 6-8, Extend 8-6, 11-3	Utilized throughout the text. For example, 163–167, 327, 432–433, 584–588
7.R.2	Explain, describe, and defend mathematical ideas using representations	Utilized throughout the text. For example, 1-9, 3-1, 8-7	Utilized throughout the text. For example, 57–61, 128–133, 434–437
7.R.3	Recognize, compare, and use an array of representational forms	Explore 3-2, Explore 3-7, 3-7, Extend 3-7, 6-8, Extend 9-7	136–141, 162, 163–167, 168, 320–326, 491
7.R.4	Explain how different representations express the same relationship	Extend 1-10, Explore 3-7, 3-7, Extend 3-7	68–69, 162, 163–167, 168
7.R.5	Use standard and non-standard representations with accuracy and detail	1-10, Extend 1-10, Explore 3-7, 3-7, Extend 3-7	63–67, 68-69, 162, 163–167, 168
Students will select, apply, and translate among mathematical representations to solve problems.			
7.R.6	Use representations to explore problem situations	Utilized throughout the text. For example, 1-6, Explore 3-7, 3-7, 6-7, 8-5	Utilized throughout the text. For example, 44–47, 162, 163–167, 318–319, 424–425
7.R.7	Investigate relationships between different representations and their impact on a given problem	Explore 3-7, 3-7, Extend 3-7, 8-2, 8-9	162, 163–167, 168, 402–408, 444–449
7.R.8	Use representation as a tool for exploring and understanding mathematical ideas	Utilized throughout the text. For example, 6-8, 8-5, Extend 11-6	Utilized throughout the text. For example, 320–326, 424–425, 600–601
Students will use representations to model and interpret physical, social, and mathematical phenomena.			
7.R.9	Use mathematics to show and understand physical phenomena (e.g., make and interpret scale drawings of figures or scale models of objects)	Utilized throughout the text. For example, 3-6, 6-8, Extend 6-8 7-7, Extend 9-7	Utilized throughout the text. For example, 156–161, 327, 375–378, 491
7.R.10	Use mathematics to show and understand social phenomena (e.g., determine profit from sale of yearbooks)	Utilized throughout the text. For example, 7-8, 8-8, Extend 9-7, 10-3	Utilized throughout the text. For example, 379–382, 438–443, 491, 523
7.R.11	Use mathematics to show and understand mathematical phenomena (e.g., use tables, graphs, and equations to show a pattern underlying a function)	Utilized throughout the text. For example, Extend 1-9, 3-7, 5-6, 6-6	Utilized throughout the text. For example, 62, 163–167, 258–263, 310–315

LA = Looking Ahead to Next Year; CSB = Concepts and Skills Bank; RSP = Reading to Solve Problems

Content Strands and Performance Indicators		Lessons	Page References
Number Sense and Operations Strand			
Students will understand numbers, multiple ways of representing numbers, relationships among numbers, and number systems.			
Number Systems			
7.N.1	Distinguish between the various subsets of real numbers (counting/natural numbers, whole numbers, integers, rational numbers, and irrational numbers)	1-3, 2-1, 2-2, 12-1	34–37, 80–83, 84–87, 636–639
7.N.2	Recognize the difference between rational and irrational numbers (e.g., explore different approximations of π).	12-1	636–639
7.N.3	Place rational and irrational numbers (approximations) on a number line and justify the placement of the numbers	2-2, 4-9, 12-1	84–87, 215–220, 636–639
7.N.4	Develop the laws of exponents for multiplication and division	**LA**	**LA2–LA5**
7.N.5	Write numbers in scientific notation	**LA**	**LA2–LA5**
7.N.6	Translate numbers from scientific notation into standard form	**LA**	**LA2–LA5**
7.N.7	Compare numbers written in scientific notation	**LA**	**LA2–LA5**
Number Theory			
7.N.8	Find the common factors and greatest common factor of two or more numbers	**1-2, Explore 4-1**, 4-1, **4-2**, CSB	**30–33, 180**, 181–184, **186–189**, 734
7.N.9	Determine multiples and least common multiple of two or more numbers	**4-8**	**211–214**
7.N.10	Determine the prime factorization of a given number and write in exponential form	**4-1**	**181–184**
Students will understand meanings of operations and procedures, and how they relate to one another.			
Operations			
7.N.11	Simplify expressions using order of operations *Note: Expressions may include absolute value and/or integral exponents greater than 0.*	1-2, **1-4**, 1-6, 2-1	30–33, **38–41**, 44–47, 80–83
7.N.12	Add, subtract, multiply, and divide integers	**Explore 2-4, 2-4, Explore 2-5, 2-5, 2-6, 2-8**	**93–94, 95–99, 101–102, 103–106, 107–111, 114–118**
7.N.13	Add and subtract two integers (with and without the use of a number line)	**Explore 2-4, 2-4, Explore 2-5, 2-5**	**93–94, 95–99, 101–102, 103–106**
7.N.14	Develop a conceptual understanding of negative and zero exponents with a base of ten and relate to fractions and decimals (e.g., $10^{-2} = .01 = 1/100$)	1-2	30–33
7.N.15	Recognize and state the value of the square root of a perfect square (up to 225)	**1-3**	**34–37**
7.N.16	Determine the square root of non-perfect squares using a calculator	**12-1**	**636–639**
7.N.17	Classify irrational numbers as non-repeating/non-terminating decimals	4-5, 12-1	196–200, 636–639

LA = Looking Ahead to Next Year; CSB = Concepts and Skills Bank; RSP = Reading to Solve Problems

Content Strands and Performance Indicators		Lessons	Page References
Students will compute accurately and make reasonable estimates.			
Estimation			
7.N.18	Identify the two consecutive whole numbers between which the square root of a non-perfect square whole number less than 225 lies (with and without the use of a number line)	**12-1**	**636–639**
7.N.19	Justify the reasonableness of answers using estimation	7-3, **7-5**, 12-1	355–360, **366–367**, 636–639
Algebra Strand			
Students will represent and analyze algebraically a wide variety of problem solving situations.			
Variables and Expressions			
7.A.1	Translate two-step verbal expressions into algebraic expressions	**1-6**, 3-1	**44–47**, 128–133
Students will perform algebraic procedures accurately.			
Variables and Expressions			
7.A.2	Add and subtract monomials with exponents of one	CSB	742
7.A.3	Identify a polynomial as an algebraic expression containing one or more terms	CSB	742
Equations and Inequalities			
7.A.4	Solve multi-step equations by combining like terms, using the distributive property, or moving variables to one side of the equation	**3-5, LA**	**151–155, LA7–LA9**
7.A.5	Solve one-step inequalities (positive coefficients only) (See 7.G.10)	CSB	740–741
7.A.6	Evaluate formulas for given input values (surface area, rate, and density problems)	3-6, 6-2, Extend 11-6	156–161, 287–292, 600–601
Students will recognize, use, and represent algebraically patterns, relations, and functions.			
Patterns, Relations, and Functions			
7.A.7	Draw the graphic representation of a pattern from an equation or from a table of data	8-1, 8-3, 8-4, Extend 8-4	396–401, 410–414, 415–421, 422
7.A.8	Create algebraic patterns using charts/tables, graphs, equations, and expressions	1-6, **1-9**, Extend 1-9, 1-10, Extend 1-10, 2-7	44–47, **57–61**, 62, 63–67, 68–69, 112–113
7.A.9	Build a pattern to develop a rule for determining the sum of the interior angles of polygons	**Explore 10-6**	**532**
7.A.10	Write an equation to represent a function from a table of values	**1-10, Extend 1-10**	**63–67, 68–69**
Geometry Strand			
Students will use visualization and spatial reasoning to analyze characteristics and properties of geometric shapes.			
Shapes			
7.G.1	Calculate the radius or diameter, given the circumference or area of a circle	**Explore 11-3, 11-3**	**583, 584–588**
7.G.2	Calculate the volume of prisms and cylinders, using a given formula and a calculator	**11-9, 11-10**	**613–618, 619–623**

LA = Looking Ahead to Next Year; CSB = Concepts and Skills Bank; RSP = Reading to Solve Problems

Content Strands and Performance Indicators		Lessons	Page References
7.G.3	Identify the two-dimensional shapes that make up the faces and bases of three-dimensional shapes (prisms, cylinders, cones, and pyramids)	Extend 11-6, **11-7**, **11-8**, Extend 11-8, CSB	600–601, **603–606**, **608–612**, 745–746
7.G.4	Determine the surface area of prisms and cylinders, using a calculator and a variety of methods	**12-4**, **12-5**, CSB	**649–653**, **656–659**, 748–749
Students will identify and justify geometric relationships, formally and informally.			
Geometric Relationships			
7.G.5	Identify the right angle, hypotenuse, and legs of a right triangle	**12-2**	**640–645**
7.G.6	Explore the relationship between the lengths of the three sides of a right triangle to develop the Pythagorean Theorem	**12-2**	**640–645**
7.G.7	Find a missing angle when given angles of a quadrilateral	**10-6**	**533–538**
7.G.8	Use the Pythagorean Theorem to determine the unknown length of a side of a right triangle	**12-2**	**640–645**
7.G.9	Determine whether a given triangle is a right triangle by applying the Pythagorean Theorem and using a calculator	**LA**	**LA18–LA20**
Students will apply coordinate geometry to analyze problem solving situations.			
Coordinate Geometry			
7.G.10	Graph the solution set of an inequality (positive coefficients only) on a number line (See 7.A.5)	**CSB**	**740–741**
Measurement Strand			
Students will determine what can be measured and how, using appropriate methods and formulas.			
Units of Measurement			
7.M.1	Calculate distance using a map scale	**6-8**, Extend 6-8	**320–326**, 327
7.M.2	Convert capacities and volumes within a given system	**6-4**, **6-5**	**298–303**, **304–309**
7.M.3	Identify customary and metric units of mass	**6-4**, **6-5**	**298–303**, **304–309**
7.M.4	Convert mass within a given system	**6-4**, **6-5**	**298–303**, **304–309**
7.M.5	Calculate unit price using proportions	**6-2**	**287–292**
7.M.6	Compare unit prices	**6-2**	**287–292**
7.M.7	Convert money between different currencies with the use of an exchange rate table and a calculator	**CSB**	**739**
7.M.8	Draw central angles in a given circle using a protractor (circle graphs)	**10-3**	**518–523**
Tools and Methods			
7.M.9	Determine the tool and technique to measure with an appropriate level of precision: mass	6-3, 6-4	293–297, 298–303
Students will develop strategies for estimating measurements.			
Estimation			
7.M.10	Identify the relationships between relative error and magnitude when dealing with large numbers (e.g., money, population)	**CSB**	**750**

LA = Looking Ahead to Next Year; CSB = Concepts and Skills Bank; RSP = Reading to Solve Problems

Content Strands and Performance Indicators		Lessons	Page References
7.M.11	Estimate surface area	12-5	656–659
7.M.12	Determine personal references for customary /metric units of mass	6-4, 6-5	298–303, 304–309
7.M.13	Justify the reasonableness of the mass of an object	6-4, 6-5	298–303, 304–309
Statistics and Probability Strand			
Students will collect, organize, display, and analyze data.			
Collection of Data			
7.S.1	Identify and collect data using a variety of methods	8-1, 8-2, Extend 8-2, 8-3, 8-4, Extend 8-4, 8-8	396–401, 402–408, 409, 410–414, 415–421, 422, 438–443
Organization and Display of Data			
7.S.2	Display data in a circle graph	**Extend 8-4, 10-3**	**422, 518–523**
7.S.3	Convert raw data into double bar graphs and double line graphs	**Extend 8-6**	**432–433**
Analysis of Data			
7.S.4	Calculate the range for a given set of data	8-1	396–401
7.S.5	Select the appropriate measure of central tendency	**8-2**, Extend 8-2	**402–408**, 409
7.S.6	Read and interpret data represented graphically (pictograph, bar graph, histogram, line graph, double line/bar graphs or circle graph)	8-1, 8-2, Extend 8-2, 8-3, **8-4**, Extend 8-4, 8-6, 8-8, 10-3, LA	396–401, 402–408, 409, 410–414, **415–421**, 422, 426–431, 438–443, 518–523, LA21–LA25
Students will make predictions that are based upon data analysis.			
Predictions from Data			
7.S.7	Identify and explain misleading statistics and graphs	**8-9**	**444–449**
Students will understand and apply concepts of probability.			
Probability			
7.S.8	Interpret data to provide the basis for predictions and to establish experimental probabilities	**8-8**, 9-2	**438–443**, 465–470
7.S.9	Determine the validity of sampling methods to predict outcomes	**8-8**, 9-7, Extend 9-7	**438–443**, 486–490, 491
7.S.10	Predict the outcome of an experiment	8-8, 9-1, 9-7, Extend 9-7	438–443, 460–464, 486–490, 491
7.S.11	Design and conduct an experiment to test predictions	**Extend 9-7**	**491**
7.S.12	Compare actual results to predicted results	9-7, **Extend 9-7**	486–490, **491**

LA = Looking Ahead to Next Year; CSB = Concepts and Skills Bank; RSP = Reading to Solve Problems

Contents

Start Smart

Unit 1

Algebra and Functions

CHAPTER 1
Introduction to Algebra and Functions

NYSCC

Focal Points and Connections
See page iv for key.

G7-FP3 Number and Operations and Algebra

NYSMT PRACTICE

H.O.T. Problems
Higher Order Thinking

Table of Contents

CHAPTER 2 Integers

Focal Points and Connections
See page iv for key.

G7-FP3 Number and Operations and Algebra

Focal Points and Connections
See page iv for key.

G7-FP3 Number and Operations and Algebra

NYSMT PRACTICE

- Extended Response 175
- Multiple Choice 126, 131, 133, 141, 147, 155, 161, 167
- Short Response/Grid In 146, 175
- Worked Out Example 130

H.O.T. Problems
Higher Order Thinking

- Challenge 132, 141, 146, 155, 160, 161, 167
- Find the Error 132, 146
- Number Sense 160
- Open Ended 132, 160, 167
- Select the Operation 149
- Select a Technique 155
- Which One Doesn't Belong? 141

Unit 2

Number Sense: Fractions

CHAPTER 4 Fractions Decimals, and Percents

NYSCC

Focal Points and Connections
See page iv for key.

G7-FP3 Number and Operations and Algebra

G7-FP5C Number and Operations

NYSMT PRACTICE

- Extended Response 227
- Multiple Choice 184, 189, 195, 200, 205, 210, 214, 220
- Short Response/Grid In 214, 227
- Worked Out Example 217

H.O.T. *Problems*
Higher Order Thinking

- Challenge 184, 189, 195, 200, 205, 210, 214, 220
- Find the Error 195, 210
- Geometry 205
- Number Sense 191
- Open Ended 184, 194, 200, 210, 214
- Research 184
- Reasoning 213
- Select a Technique 214
- Which One Doesn't Belong? 205, 220

CHAPTER 5 Applying Fractions

Table of Contents

Unit 3

Algebra and Number Sense: Proportions and Percents

CHAPTER 6 — Ratios and Proportions

Focal Points and Connections
See page iv for key.

G7-FP1 Number and Operations and Algebra and Geometry

NYSMT PRACTICE

- Extended Response 339
- Multiple Choice 286, 289, 292, 297, 303, 309, 315, 326, 332
- Short Response/Grid In 297, 339
- Worked Out Example 288

H.O.T. Problems
Higher Order Thinking

- Challenge 286, 292, 302, 308, 314, 325
- Find the Error 286, 308
- Number Sense 292
- Open Ended 292, 297, 302, 305, 325
- Reasoning 302, 325
- Select a Technique 315
- Which One Doesn't Belong? 314

CHAPTER 7

Applying Percents

Table of Contents

Focal Points and Connections
See page iv for key.

G7-FP1 Number and Operations and Algebra and Geometry

NYSMT PRACTICE

H.O.T. Problems
Higher Order Thinking

Unit 4
Statistics, Data Analysis and Probability

CHAPTER 8 — Statistics: Analyzing Data

Focal Points and Connections
See page iv for key.

G7-FP6C Data Analysis

NYSMT PRACTICE

- Extended Response   457
- Multiple Choice   401, 405, 408, 414, 421, 431, 435, 443, 449
- Short Response/Grid In   421, 457
- Worked Out Example   404

H.O.T. Problems
Higher Order Thinking

- Challenge   400, 407, 413, 420, 430, 436, 443, 449
- Data Sense   420
- Find the Error   400, 413
- Open Ended   400, 407, 430, 436, 449
- Reasoning   400, 407
- Select a Technique   421
- Select a Tool   437
- Which One Doesn't Belong?   407, 437

CHAPTER 9 Probability

Table of Contents

Focal Points and Connections
See page iv for key.

G7-FP7C Probability

NYSMT PRACTICE

H.O.T. Problems
Higher Order Thinking

Unit 5
Geometry and Measurement

CHAPTER 10 Geometry: Polygons

Focal Points and Connections
See page iv for key.

G7-FP2 **Measurement and Geometry and Algebra**

NYSMT PRACTICE

H.O.T. Problems
Higher Order Thinking

CHAPTER 11

Measurement: Two-and Three-Dimensional Figures

Focal Points and Connections
See page iv for key.

G7-FP2 Measurement and Geometry and Algebra

G7-FP4C Measurement and Geometry

NYSMT PRACTICE

- Extended Response   633
- Multiple Choice   576, 582, 588, 593, 599, 606, 612, 618, 623
- Short Response/Grid In   582, 633
- Worked Out Example   590

H.O.T. Problems
Higher Order Thinking

- Challenge   576, 582, 588, 592, 599, 606, 611, 618, 622
- Find the Error   587, 592
- Number Sense   622
- Open Ended   576, 587, 606, 611, 622
- Reasoning   582, 606, 618
- Which One Doesn't Belong?   611

CHAPTER 12

Geometry and Measurement

Focal Points and Connections
See page iv for key.

G7-FP1 Number and Operations and Algebra and Geometry

Looking Ahead to Next Year

Student Handbook

Table of Contents

In March, you will take your Grade 7 New York State Mathematics Test. This test is divided into two books, Book 1 and Book 2. You will spend two days taking this test, one day for Book 1 and one day for Book 2. You will apply the concepts and skills that you have learned throughout the year in order to answer multiple-choice, short-response, and extended-response questions.

Book 1

This book contains multiple-choice questions only in which you will select the correct response from four answer choices. Your teacher will provide you with an answer sheet to fill in your answer choices.

Book 2

This book contains short-response and extended-response questions in which you will write an answer to an open-ended question. You are required to show your work to receive full credit. In some cases, you will be required to explain, in words, how you arrived at your response. You can write your responses directly in this test book.

How Can I Get Ready?

The New York State Test you will take in March covers the New York Core Curriculum for Grade 7 Mathematics. The following pages give you practice questions similar to those found on the test and include Grade 6 Post-March and Grade 7 Pre-March items. You can use these practice pages in the weeks before the test to determine if you are ready. If you are struggling with any of the items, lesson references are provided so that you can go back and review from the pages in your textbook.

Post March

1 The expression below represents the total cost in dollars, including postage p of a number n of theater tickets.

$$15n + p$$

Based on the expression above, what is the total cost of 5 tickets when postage is $4? (Lesson 1-6) **(6.A.2, 7.PS.3)**

A $19

B $34

C $64

D $79

2 What is the value of the expression below when $x = 4$ and $y = 1$?

$$3x^2 + 2y + 5$$

(Lesson 1-6) **(6.A.2, 7.R.3)**

A 19

B 55

C 65

D 151

3 Juan bought six packs of baseball cards. After the tax of $1 was added to the purchase, he paid $16. Which equation can be used to find the cost of one package of cards, c? (Lesson 3-1) **(6.A.3, 7.PS.9)**

A $6c + 1 = 16$

B $6c - 1 = 16$

C $c + 6 = 17$

D $6c - 1 = 17$

4 Janice had $115 in her savings account. She deposits $35 to her account each week. Which equation can be used to determine how many weeks, w, it will take for her balance to reach $465? (Lesson 3-1) **(6.A.3, 7.PS.6)**

A $115w - 35 = 465$

B $115w + 35 = 465$

C $35w - 115 = 465$

D $35w + 115 = 465$

5 Solve the equation below to find the value of n.

$$4n + 6 = 30$$

(Lesson 3-5) **(6.A.4, 7.PS.6)**

A $n = 4$

B $n = 6$

C $n = 9$

D $n = 24$

6 Canned soup is on sale at the grocery store at a price of $5 for 4 cans of soup. How much will it cost to buy 10 cans of soup at this price? (Lesson 6-6) **(6.A.5, 7.PS.10)**

A $10.00

B $12.50

C $17.50

D $20.00

7 An airplane travels 800 miles in two hours. If it maintains the same speed, how long will it take the airplane to travel the next 2,000 miles? (Lesson 6-2)
(6.A.5, 7.PS.10)

A 4 hours

B 5 hours

C 8 hours

D 10 hours

8 What are the coordinates of the point shown on the grid below? (Lesson 2-3)
(6.G.10, 7.R.1)

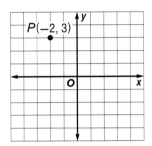

A (−2,−3)

B (−2, 3)

C (2, −3)

D (2, 3)

9 What is the area of the rectangle drawn on the coordinate grid below? (Lesson 11-1)
(6.G.11, 7.R.6)

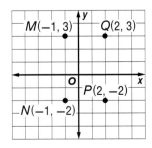

A 6 square units

B 12 square units

C 15 square units

D 24 square units

10 Alicia's garden is in the shape of a polygon and is shown on the grid below.

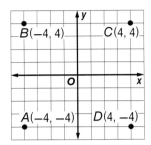

What is the area of Alicia's garden?
(Lesson 11-1) (6.G.11, 7.R.6)

A 16 square units

B 32 square units

C 48 square units

D 64 square units

11 Anne wants to find out which field trip would be most popular among the seventh-grade students in her school. Which group should she survey in order to get the most reliable results? (Lesson 11-1)
(6.S.1, 7.PS.14)

A all the students in her homeroom

B the first three students she sees

C all of her friends

D many students from several homerooms

12 Yolanda is working with a political candidate to take a survey of voters. If the survey is intended to predict the results of the next election, whom should Yolanda call? (Lesson 8-7) (6.S.1, 7.PS.15)

A some of her parents' friends

B a random selection of people

C people who voted for her candidate previously

D all of the students in her school

13 Lisa asked her classmates to tell her what type of pets they like best. She recorded her results in the table below.

Type of Pet	Tally
dog	卌 ‖
cat	卌 ‖‖
fish	‖‖
bird	‖
other	卌

How many of her classmates chose a dog as their favorite pet? (Lesson 8-1) **(6.S.2, 7.CM.6)**

A 5

B 6

C 7

D 8

14 Paolo randomly draws one of the numbers from 0 through 9. He then replaces the number and draws again. If Paolo lists all the possible outcomes of the two draws, how many results will be on the list? (Lesson 9-3) **(6.S.9, 7.PS.4)**

A 10

B 20

C 100

D 1,000

15 Solve the equation $7n + 1 = 29$ to find the value of n. (Lesson 3-5) **(6.A.4, 7.PS.6)**

A 3

B 4

C 5

D 6

16 Which type of graph *best* displays data that changes over time? (Lesson 8-6) **(6.S.4, 7.R.1)**

A line graph

B pictograph

C bar graph

D circle graph

17 The data set {3, 4, 8, 12, 13, 16, 22} can be sorted in the Venn diagram below.

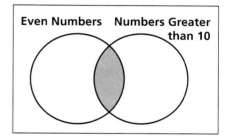

Which values from the data set will be located in the shaded portion of the diagram? (Lesson 4-2) **(6.S.3, 7.CN.2)**

A 3, 4, 8

B 4, 8, 12, 16, 22

C 12, 16, 22

D 12, 13, 16, 22

18 A bag contains four red marbles, four green marbles, and two yellow marbles. If marbles are not replaced in the bag after drawing them, what is the probability of drawing two red marbles in two draws? (Lesson 9-3) **(6.S.10, 7.RP.2)**

A $\frac{2}{15}$

B $\frac{1}{10}$

C $\frac{1}{5}$

D $\frac{2}{3}$

19 Lenny wants to order a sandwich at a deli. The table shows the choices of meat, cheese, and bread that are available.

Meat	Cheese	Bread
Turkey	American	White
Ham	Swiss	Whole wheat
Roast beef	Provolone	
	Cheddar	

If Lenny can choose one meat, one cheese, and one type of bread, how many different sandwiches are possible?
(Lesson 9-3) **(6.S.11, 7.CN.4)**

A 9
B 18
C 24
D 36

20 Mikaila has a bag with four different-colored marbles and a second bag with four tiles labeled with the letters A, B, C, and D. If Mikaila draws one object from each bag, how many possible combinations of two objects can she pick? (Lesson 9-8)
(6.S.11, 7.PS.4)

A 4
B 8
C 12
D 16

21 Look at the equation below.

$$8x + 32 = 48$$

Part A

Explain how the equation can be solved to find the value of x. (Lesson 3-5) **(6.A.4, 7.CM.1)**

Part B

Find the value of x.
Show your work.

22 What are the coordinates of the point shown on the grid below? (Lesson 2-3)
(6.G.10, 7.R.8)

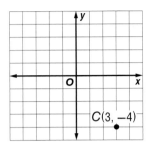

23 List all the possible outcomes if a number cube is rolled once and a letter is selected from the word *dog*. (Lesson 9-1)
(6.S.9, 7.R.2)

24 Marcia recorded how much total money she saved over a period of six months.

Month	Money Saved ($)
1	15
2	35
3	60
4	85
5	112
6	142

What type of graph is best to display Marcia's data? (Lesson 9-1) **(6.S.4, 7.R.2)**

25 A spinner with equal-size sections labeled A, B, C, and D is spun. What is the probability the spinner will land on a consonant? (Lesson 9-7) **(6.S.10, 7.PS.8)**

Write your answer as a percent, fraction, and a decimal.

26 Phyllis rolled a number cube 20 times and recorded these results: 1, 4, 5, 6, 6, 4, 3, 2, 1, 1, 5, 4, 2, 4, 5, 2, 3, 1, 6, 1

Make a frequency table, using tally marks to show her results. (Lesson 8-1) **(6.S.2, 7.R.5)**

27 Use the Venn diagram below to sort the data set {6, 36, 27, 5, 18, 9, 14, 2} into numbers divisible by 3 and numbers less than 20. (Lesson 4-2) **(6.S.3, 7.R.2)**

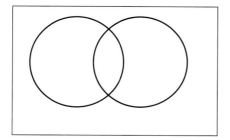

28 An ice cream shop sells ice cream in chocolate, vanilla, or strawberry flavors. The cones are either sugar or waffle. Customers can choose from hot fudge or caramel toppings. If a customer must choose one flavor, one type of cone, and one topping, how many different choices are possible? (Lesson 9-3) **(6.S.11, 7.CN.4)**

29 Stan recorded the ages of the 7th grade students in his school. He found that 10% of the students were 11 years old, 20% were 13 years old, and 70% were 12 years old. What type of graph is best to display Stan's data? (Lesson 9-1) **(6.S.4, 7.R.2)**

30 List all the possible outcomes if a coin is tossed three times. Use H for heads and T for tails. (Lesson 9-1) **(6.S.9, 7.R.2)**

31 Refer to Exercise 30. If a coin is tossed three times, what is the probability that heads will appear all three times? (Lesson 9-1) **(6.S.9, 7.R.2)**

32 The table below shows the plastic chips in a bag. Mary draws two chips out of the bag. She does not replace the first chip before she draws the second chip. (Lesson 9-7) **(6.S.10, 7.PS.8)**

Color	Number of chips
Red	4
Blue	3
Green	2
Yellow	1

Part A

What is the probability that Mary will draw a red chip on her first draw?

Part B

What is the probability that Mary will draw a red chip and then a yellow chip? Show your work.

Probability of yellow after red:

Probability of red, then yellow:

Number Sense and Operations

1 Which number below is not a whole number? (Lesson 2-1) (7.N.1, 7.CM.9)

A −9

B 0

C 3

D 19

2 Which response describes all the members of the set below? (Lesson 4-9) (7.N.1, 7.CM.10)

$$\left\{-77, \frac{5}{\sqrt{81}}, 8.266597, 0.52\overline{3}, 56\right\}$$

A irrational numbers

B rational numbers

C natural numbers

D whole numbers

3 Which of the numbers below is an irrational number? (Lesson 4-9) (7.N.2, 7.CM.10)

A −2.7

B $\frac{5}{93}$

C $\sqrt{18}$

D $\sqrt{49}$

4 Which number below is a rational number? (Lesson 4-9) (7.N.2, 7.CM.11)

A $\sqrt{56}$

B π

C 6.2481

D $\sqrt{99}$

5 Which point on the number line below represents the value of π? (Lesson 4-9) (7.N.3, 7.R.5)

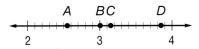

A point A

B point B

C point C

D point D

6 Evaluate the expression $6^8 \times 6^4$. (Lesson 1-2) (7.N.4, 7.CN.1)

A 6^4

B 6^{12}

C 6^{32}

D 36^{12}

7 Which of these numbers has the greatest value? (LA1) (7.N.7, 7.PS.1)

A 7^4

B 4^7

C 7^3

D 3^7

8 What is the greatest common factor of 16, 32, and 24? (Lesson 4-2) (7.N.8, 7.RP.2)

A 4

B 6

C 8

D 12

9 Evaluate the expression $3^6 \div 3^3$.
(Lesson 1-4) **(7.N.4, 7.R.2)**

  A 9

  B 27

  C 81

  D 216

10 The distance from Earth to the Sun is about 150,000,000 kilometers. What is this number written in scientific notation? (LA1) **(7.N.5, 7.CN.3)**

  A 1.5×10^7 kilometers

  B 15×10^7 kilometers

  C 1.5×10^8 kilometers

  D 1.5×10^9 kilometers

11 The population of Rochester is about 210,000 people. How is this number written in scientific notation? (LA1)
(7.N.5, 7.CN.6)

  A 2.1×10^5

  B 21×10^5

  C 21×10^4

  D 210×10^3

12 Light travels about 6.7×10^8 miles in one hour. How is this number written in standard form? (LA1) **(7.N.6, 7.R.9)**

  A 67,000,000

  B 670,000,000

  C 6,700,000,000

  D 67,000,000,000

13 Which of these numbers has the greatest value? (LA1) **(7.N.7, 7.PS.1)**

  A 7.65×10^7

  B 5.98×10^{-8}

  C 5.98×10^8

  D 1.22×10^9

14 Which of these numbers has the greatest value? (LA1) **(7.N.7, 7.PS.1)**

  A 9.62×10^7

  B -3.22×10^9

  C 8.0×10^8

  D 7.35×10^{-9}

15 What is the greatest common factor of 18, 24, and 30? (Lesson 4-2) **(7.N.8, 7.RP.2)**

  A 2

  B 3

  C 6

  D 120

16 What is the least common multiple of 15, 20, and 60? (Lesson 4-8) **(7.N.9, 7.RP.2)**

  A 60

  B 120

  C 180

  D 240

17 What is the prime factorization of 180 written in exponential form? (Lesson 4-1) **(7.N.10, 7.R.3)**

A $2^2 \times 3^2 \times 5$

B $2^2 \times 3 \times 5^2$

C $2^3 \times 3 \times 2^2$

D $2^3 \times 3^2 \times 5$

18 Simplify the expression below. (Lesson 2-1) **(7.N.11, 7.PS.8)**

$$(5^2 - 14) + |-2| \times 3$$

A 5

B 17

C 27

D 39

19 Betsy's camera can hold 428 pictures. On a school trip, she took 81 pictures the first day. She took twice as many pictures on the second day. On the third day, she took 68 pictures. How many more photos could Betsy take? (Lesson 1-4) **(7.N.12, 7.CN.7)**

A 107

B 117

C 198

D 213

20 The temperature on a certain winter night in Albany is $-10°$ Celsius. In the morning the temperature increases by $5°$ Celsius. What is the new temperature? (Lesson 2-4) **(7.N.12, 7.CN.6)**

A $-15°C$

B $-5°C$

C $5°C$

D $15°C$

21 Use the number line to find the value of the expression. (Lesson 2-5) **(7.N.13, 7.R.8)**

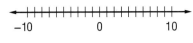

$$3 - (-4) = ?$$

A -7

B -1

C 1

D 7

22 Use the number line to find the value of the expression. (Lesson 2-5) **(7.N.13, 7.R.8)**

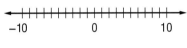

$$-2 - (-5) = ?$$

A -7

B -1

C 1

D 7

23 What is the value of 10^{-3}? (page 737) **(7.N.14, 7.CM.9)**

A $\dfrac{1}{1000}$

B $\dfrac{1}{100}$

C $\dfrac{1}{10}$

D -1000

24 A square classroom has a tile floor. It took 169 square meters of tile to cover the floor. How long is one side of the classroom? (Lesson 11-1) **(7.N.15, 7.CN.7)**

A 12 meters

B 13 meters

C 15 meters

D 169 meters

25 A square garden covers about 380 square feet. Using your calculator, find the approximate length of a side of the garden. (Lesson 11-1) **(7.N.16, 7.CN.7)**

A 18.6 feet

B   19.1 feet

C 19.5 feet

D 20.1 feet

26 Use your calculator to find the approximate value of the square root of 865. (Lesson 1-3) **(7.N.16, 7.RP.1)**

A 28.9 feet

B 29.4 feet

C 29.7 feet

D 32.1 feet

27 The number $\sqrt{23}$ is an irrational number. Which phrase below applies to $\sqrt{23}$ when it is written as a decimal number?
(Lesson 12-1) **(7.N.17, 7.R.7)**

A It terminates after many digits.

B The final digit repeats infinitely.

C The decimal does not terminate or repeat infinitely.

D It can be written as a ratio of two integers.

28 What is the value of 10^{-5}? (page 737)
(7.N.14, 7.CM.9)

A −50

B −0.00001

C 0.000001

D 0.00001

29 The value of $\sqrt{156}$ lies between what two whole numbers? (Lesson 2-1) **(7.N.18, 7.PS.12)**

A 12 and 13

B 13 and 14

C 74 and 75

D 75 and 76

30 Janice wants to leave approximately a 20% tip on a restaurant bill of $37.50. She calculates the tip to be $11.50. Use estimation to determine whether her answer is reasonable. (Lesson 7-5)
(7.N.19, 7.PS.13)

A Yes, the value is reasonable.

B No, the value is too low.

C No, the value is too high.

D No, the value should be more than $37.

31 Ling says that the square root of 200 is about 20. Use estimation to determine whether his answer is reasonable.
(Lesson 12-1) **(7.N.19, 7.PS.13)**

A yes, because $3100 = 10$

B no, because $12^2 = 144$

C no, because $20^2 = 200$

D no, because $14^2 = 196$

32 Which of the following is the closest approximation for $\sqrt{127}$? (Lesson 2-1)
(7.N.18, 7.PS.12)

A 10

B 11

C 12

D 13

33 This table shows geographic information about North America.

highest elevation	2.03×10^5 feet
area of largest lake	3.17×10^4 square miles

Write the number in standard form for the highest elevation and the area of the largest lake. (LA1) **(7.N.6, 7.R.9)**

34 Place the following numbers in order from least to greatest: (Lesson 4-9) **(7.N.7, 7.RP.2)**

$$9.6 \times 10^{-3}, -5.8 \times 10^4, 2.7 \times 10^1$$

35 Simplify this expression:
$$17 - 2^2 \div 2 + (4 \times 11)$$
Be sure to show your work. (Lesson 1-4) **(7.N.11, 7.PS.8)**

36 Mark has two pieces of string, measuring 36 inches and 54 inches. He wants to cut them into equal-length pieces so each piece will be a whole number of inches. (Lesson 4-2) **(7.N.8, 7.PS.8)**

Part A

In units of whole numbers, what lengths can Mark cut each string?

Part B

In units of whole numbers, what is the greatest length into which both pieces of string can be cut (the greatest common factor of the lengths)?

37 A carpenter is placing a wooden floor, using boards of two different lengths.

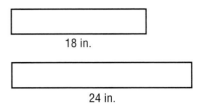

18 in.

24 in.

In inches, what is the shortest length that a row of all 18-inch boards and a row of all 24-inch-length boards will start and end at the same place (the least common multiple of the board lengths)? Show your work. (Lesson 4-8) **(7.N.9, 7.PS.6)**

38 For homework, Mrs. Lee required her students to find the prime factorization of 1,176. (Lesson 4-1) **(7.N.10, 7.PS.8)**

Part A

What is the prime factorization of 1,176?

Part B

How is the prime factorization of 1,176 written in exponential form?

39 Simplify this expression:
$$(4^2 - 2^3) + (3 \times 4 + 8)$$
Be sure to show your work. (Lesson 1-4) **(7.N.11, 7.PS.8)**

40 How many times larger is $5^3 \times 2^4$ than $5^2 + 2^3$? (Lesson 4-1) **(7.N.10, 7.PS.8)**

41 Multiply $-3.4 \times \frac{4}{5}$. Show your work.
(PS7, Lesson 5-5) **(7.N.12, 7.RP.2)**

42 Show the approximate location of $\sqrt{17} + (-4.1)$ on the number line.
(Lesson 4-9) **(7.N.13, 7.R.5)**

43 George read that a certain type of bacteria are about 10^{-4} centimeters in diameter. (PS8) **(7.N.14, 7.R.9)**

Part A

What is 10^{-4} expressed as a fraction?

Part B

What is 10^{-4} expressed as a decimal?

44 What is the length of each side of this square? (Lesson 11-1) **(7.N.15, 7.PS.9)**

Area = 225 cm²

45 Evaluate the expression $7^3 \times 3^2$.

Show your work. (Lesson 1-2) **(7.N.4, 7.CN.1)**

46 What is the square root of 196? (Lesson 1-3) **(7.N.16, 7.RP.2)**

47 Is the number $0.56\overline{12}$ a rational number or an irrational number? Explain your answer. (Lesson 12-1) **(7.N.17, 7.CM.10)**

48 The value of $\sqrt{113}$ between what two whole numbers? (Lesson 12-1) **(7.N.18, 7.PS.13)**

49 Sam works at a grocery store. The store received a shipment of 52 cases of canned vegetables. There are 24 cans in each case. Sam calculated that there are 1,248 cans of vegetables in the shipment. Use estimation to determine whether Sam's result is reasonable. (Lesson 2-6) **(7.N.19, 7.CN.6)**

50 What is the least common multiple of 7, 8, and 4? (Lesson 4-8) **(7.N.9, 7.RP.2)**

51 Is the square root of 625 a rational or irrational number? Explain your reasoning. (Lesson 4-9) **(7.N.2, 7.CM.11)**

Algebra

1 Kareem has already read 94 pages of a novel. If he reads 25 more pages every day, which expression shows how many pages he will have completed after x days? (Lesson 3-1) **(7.A.1, 7.CM.10)**

  A $25x + 94$

  B $25(x + 94)$

  C $25 - 94$

  D $94x + 25$

2 Joyce is decorating for an after-school party. She has a 300-foot-long roll of crepe paper. If she cuts off paper strips that are 27 feet long, which expression shows how much paper is left after she cuts off n strips? (Lesson 3-1) **(7.A.1, 7.R.6)**

  A $300n - 27$

  B $27n - 300$

  C $27 - 300n$

  D $300 - 27n$

3 Solve this inequality for x:

$$5x \geq 30$$

(CSB8) **(7.A.5, 7.CN.3)**

  A $x \geq 6$

  B $x \leq 6$

  C $x \geq 150$

  D $x \leq 150$

4 Wallace ran more than 12 miles in 3 hours. If he ran at a constant rate, what is the slowest he could have run? (Lesson 6-2) **(7.A.5, 7.PS.10)**

  A 3 miles per hour

  B 4 miles per hour

  C 12 miles per hour

  D 36 miles per hour

5 The formula for the density of a solid is $D = M/V$ where M = mass and V = volume. What is the density of a block weighing 10 kilograms with a volume of 5 cubic meters? (Lesson 1-6) **(7.A.6, 7.CN.7)**

  A 15 kg/m^3

  B 10.5 kg/m^3

  C 2 kg/m^3

  D 0.5kg/m^3

6 The expression below represents the total cost in dollars of renting a certain number of movies. If n represents the number of movies, what is the total cost to rent 3 movies? (Lesson 1-6) **(7.A.6, 7.CN.7)**

$$4n + 3$$

  A $9

  B $12

  C $15

  D $46

7 A club charges $20 each year for dues. There is also a charge of $3 for supplies for each party. Write an expression that shows the total cost of a one-year membership and participation in n parties. (Lesson 3-1) **(7.A.1, 7.R.1)**

8 Solve this inequality for b:

$$3b \geq 9$$

(CSB8) **(7.A.5, 7.RP.2)**

9 The formula for the surface area of a rectangular prism is the following: Surface area $= 2wl + 2lh + 2wh$ What is the surface area of a box that measures 8 cm × 6 cm × 3 cm? (Lesson 12-4) **(7.A.6, 7.PS.8)**

Geometry

1 The circumference of the circle below is 31.4 centimeters. What is the radius? (Lesson 11-3) **(7.G.1, 7.PS.9)**

[not drawn to scale]

Circumference = $2\pi r$

A 3.1 cm

B 5.0 cm

C 10.0 cm

D 31.4 cm

2 Use your calculator to find the volume of the triangular prism shown below. (Lesson 11-9) **(7.G.2, 7.PS.1)**

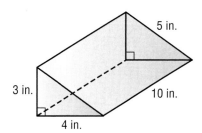

[not drawn to scale]

Volume = $\frac{1}{2}wh \times l$

A 60 cubic inches

B 80 cubic inches

C 100 cubic inches

D 120 cubic inches

3 What is the circumference of a circle whose radius is 7 centimeters? Round to the nearest hundredth. (Lesson 11-3) **(7.G.1, 7.PS.9)**

A 21.99 cm **B** 31.42 cm

C 43.98 cm **D** 153.94 cm

4 The figure below is a square pyramid. How many triangular faces does it have? (CSB13) **(7.G.3, 7.R.3)**

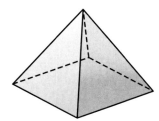

A 2

B 3

C 4

D 5

5 Which response *best* describes the faces of a rectangular prism? (Lesson 12-4) **(7.G.3, 7.PS.4)**

A 6 squares

B 6 rectangles

C 2 pairs of identical rectangles

D 3 pairs of identical rectangles

6 Find the volume of a box with dimensions 12in., 15in., and 14in. (Lesson 11-9) **(7.G.2, 7.PS.1)**

A 168 in²

B 180 in²

C 210 in²

D 2,520 in²

New York State Test

7 Use a calculator to find the surface area of the cylindrical tank shown below. (Lesson 12-5) **(7.G.4, 7.R.6)**

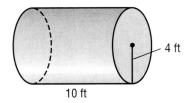

4 ft

10 ft

[not drawn to scale]

A 125 square feet
B 251 square feet
C 352 square feet
D 704 square feet

8 Three of the angles of a quadrilateral each measure 75°. What is the measure of the fourth angle of the quadrilateral? (Lesson 10-6) **(7.G.7, 7.PS.9)**

A 75°
B 90°
C 125°
D 135°

9 What is the surface area of a cylinder that is 2.25 m tall with a 0.75 m radius? (Lesson 12-5) **(7.G.4, 7.R.6)**

A 10.1 m²
B 12.1 m²
C 14.1 m²
D 16.1 m²

10 What is the measure of the angle labeled x in the illustration below? (Lesson 10-6) **(7.G.7, 7.R.1)**

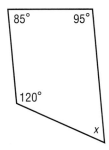

85° 95°

120°

x

A 45°
B 60°
C 90°
D 120°

11 Which inequality is graphed on the number line? (CSB8) **(7.G.10, 7.R.3)**

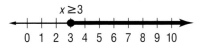

$x \geq 3$

0 1 2 3 4 5 6 7 8 9 10

A $2x > 6$
B $2x \geq 6$
C $3x > 6$
D $3x \geq 6$

12 Which describes the inequality graphed on a number line? (CSB8) **(7.G.10, 7.R.3)**

$$3x < 12$$

A open dot on 4, arrow pointing to the left
B closed dot on 4, arrow pointing to the left
C open dot on 4, arrow pointing to the right
D closed dot on 4, arrow pointing to the right

13 A circular pond in a public garden has a circumference of 12π meters.

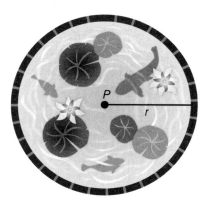

$$\boxed{\text{Circumference} = 2\pi r}$$

Part A: What is the radius of the pond? Show your work. (Lesson 11-3) **(7.G.1, 7.CN.6)**

Part B: What is the diameter of the circle?

14 A can of soup has the dimensions shown below. (Lesson 11-10) **(7.G.2, 7.PS.6)**

[not drawn to scale]

What is the volume of the soup can to the nearest cubic centimeter?

$$\boxed{\text{Volume of a cylinder} = \pi r^2 h}$$

15 What length of paper is needed to wrap a $12 \times 10 \times 8$-inch gift box if you are using 16-inch wide paper? (Lesson 12-4) **(7.G.4, 7.CN.6)**

16 Margie is making a gift box with the dimensions shown below to sell at a craft fair. How much fabric will she need to cover the entire surface area of the box? (Lesson 12-4) **(7.G.4, 7.CN.6)**

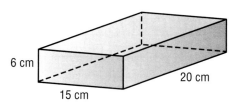

[not drawn to scale]

17 Jennie uses the inequality below to determine how many 8-inch pieces of wood she can cut from a 96-inch board. (CSB8) **(7.G.10, 7.R.5)**

$$8x \geq 96$$

Part A

Solve the inequality for x.

Show your work.

Part B

Plot the inequality on the number line.

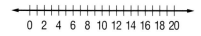

18 What shapes make up a rectangular pyramid? (CSB 13) **(7.G.3, 7.R.3)**

Measurement

1 How many liters of water are in a bottle with a label that says it contains 750 milliliters? (Lesson 6-5) (7.M.2, 7.PS.10)

| 1 liter = 1,000 milliliters |

A 0.75 liter

B 7.5 liters

C 75 liters

D 750 liters

2 Which of the following is a *not* a metric unit of mass ? (Lesson 6-5) (7.M.3, 7.CM.10)

A kilogram

B ounce

C gram

D milligram

3 Which of the following is not a customary unit of length? (Lesson 6-5) (7.M.3, 7.CM.10)

A meter

B foot

C inch

D yard

4 Which of the following is a metric unit of mass? (Lesson 6-5) (7.M.3, 7.CM.10)

A pound

B liter

C kilogram

D quart

5 Wanda bought a box of cereal that contains 2,000 grams of cereal. How many kilograms of cereal does the box contain? (Lesson 6-5) (7.M.4, 7.PS.10)

| 1 kilogram = 1,000 grams |

A 1 kilogram

B 1.5 kilograms

C 2 kilograms

D 2.5 kilograms

6 Use your protractor to solve this problem. What angle is formed by the two radii of the circle? (PS11) (7.M.3, 7.R.8)

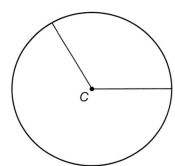

A 60°

B 90°

C 105°

D 120°

7 How many milliliters are in a 2-liter bottle of soda? (Lesson 6-5) (7.M.2, 7.PS.10)

A 2 milliliters

B 20 milliliters

C 200 milliliters

D 2000 milliliters

8 What would be the most accurate method to determine the mass of a bicycle? (Lessons 6-3, 6-4) (**7.M.9, 7.PS.15**)

  A Weigh it on a scale with a range of 1 to 1,000 grams.

  B Weigh it on a scale with range of 1 to 100 kilograms.

  C Measure the volume of the bicycle with a ruler and multiply by π.

  D Measure the surface area of the bicycle with a ruler and multiply by π.

9 What is the most accurate method to determine the mass of a dime? (Lessons 6-3, 6-4) (**7.M.9, 7.PS.15**)

  A Weigh it on a scale with a range of 1 to 1,000 grams.

  B Weigh it on a scale with a range of 1 to 100 kilograms.

  C Measure the circumference with a ruler and multiply by π.

  D Measure the radius with a ruler and multiply by π.

10 In 2006, the census bureau estimated the female population of New York City to be 4,229,829 ±705. What is the relative error of the estimate? (CSB13) (**7.M.10, 7.CN.7**)

  A 0.000017

  B 0.00017

  C 0.0017

  D 0.017

11 Lewis is painting the outside of the box below for a stage set for a school play. Which of the following is the most accurate estimate of the surface area he will paint? (Lesson 12-4) (**7.M.11, 7.PS.12**)

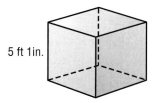

5 ft 1in.

[not drawn to scale]

  A 200 square feet

  B 150 square feet

  C 125 square feet

  D 100 square feet

12 Which of the following has a mass closest to 100 grams? (Lesson 6-5) (**7.M.12, 7.CN.6**

  A a mouse

  B a needle

  C a bicycle

  D a horse

13 Which of the following has a mass closest to 4 kg? (Lesson 6-5) (**7.M.12, 7.CN.6**)

  A an elephant

  B a person 12 years old

  C a marble

  D a cat

14 Jason wants to know how a mass of 500 grams feels. Which of the following objects should he pick up? (Lesson 6-5) (7.M.12, 7.CN.6)

   A a sheet of paper

   B a small textbook

   C a backpack with four textbooks

   D a bookcase with 25 books

15 Jack says his father's pickup truck weighs about 50 kilograms. Which of the following is a reasonable evaluation of Jack's estimate? (Lesson 6-5) (7.M.13, 7.PS.13)

   A Jack's estimate is a good one because a truck's mass is about the same as 50 textbooks.

   B Jack's estimate is a good one because a truck's mass is about the same as 20 bowling balls.

   C Jack's estimate is not a good one because a truck's mass is less than that of 10 backpacks.

   D Jack's estimate is not a good one because a truck's mass is more than 50 watermelons.

16 Kara has 54 inches of fabric. How many feet of fabric does she have? (Lesson 6-4) (7.M.2, 7.PS.10)

   A 648

   B 27

   C 9

   D 4.5

17 Yolanda unpacked a case of bottled water and placed them on a shelf in the cafeteria. (Lesson 6-4) (7.M.2, 7.PS.10)

Part A

If each bottle contains one pint of water and there are 16 bottles in a case, how many quarts of water does she unpack from the case? Show your work.

$$\boxed{\text{1 quart = 2 pints}}$$

Part B

If a recipe calls for one cup per serving, how many servings can Yolanda pour from the case? Show your work.

18 How many 100-gram servings can Wanda prepare from a 2-kilogram package of trail mix? (Lesson 6-5) (7.M.4, 7.CN.7)

$$\boxed{\text{1 pint = 2 cups}}$$

19 A nail is 4 centimeters long.

Part A

How many millimeters long is the nail? (Lesson 6-5) (7.M.3, 7.CM.10)

Part B

How many nails would fit lengthwise in a meter?

20 A truck used 6.3 gallons of gasoline to travel 107 miles. How many gallons of gasoline would it need to travel an additional 250 miles? (Lesson 6-4) (7.M.2, 7.PS.10)

   A 8.4 gal

   B 14.7 gal

   C 18.9 gal

   D 21.0 gal

21 Copy the circle below onto your paper and then use your protractor to draw two radii of the circle that form a 90° angle. (PS11) **(7.M.3, 7.R.1)**

22 Describe an appropriate technique for measuring the mass of a rock about the size of a baseball within 10 grams.
(Lessons 6-3, 6-4) **(7.M.9, 7.PS.15)**

23 A farmer packs tomatoes in boxes that weigh 1.4 kilograms when empty. The average tomato weighs 0.2 kilogram and the total weight of a box filled with tomatoes is 11 kilograms. How many tomatoes are packed in each box?
(Lesson 6-5) **(7.M.2, 7.PS.10)**

24 About how much paper is needed to make a label that covers only the sides of a soup can 5 inches tall with a diameter of 2 inches? (Lesson 12-4) **(7.M.11, 7.PS.12)**

25 A company manufactures a part for a scientific instrument shaped like the rectangular prism shown below. A thin layer of gold must cover the entire surface. Estimate the surface area of the part by rounding to the nearest centimeter. Show your work. (Lesson 12-4) **(7.M.11, 7.PS.7)**

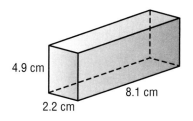

4.9 cm

8.1 cm

2.2 cm

26 What is a reasonable estimate of the mass of a textbook? (Lesson 6-5) **(7.M.13, 7.PS.16)**

How can you estimate the mass of a textbook without measuring?

27 A scale drawing of a football field was made using a scale of 1 inch = 20 yards. What is the length, in yards, of the football field if it is 6 inches on the scale drawing? (Lesson 3-6) **(7.M.3, 7.CM.10)**

28 Ms. Leigh wants to organize the desks in study hall into a square. If she has 64 desks, how many should be in each row?
(Lesson 3-6) **(7.M.3, 7.CM.10)**

Statistics and Probability

1 You are conducting a survey to find out which of several activities your classmates want to do in a school program. Which method of collecting data would give you the best results? (Lesson 8-8) **(7.S.1, 7.CN.7)**

  A Ask your friends what they prefer.

  B Survey the entire class.

  C Ask your teacher which activity is best.

  D Ask students at random from the whole school.

2 The best data about the preference between two grocery stores in your town would come from a survey of which group of people? (Lesson 8-8) **(7.S.1, 7.CN.9)**

  A a group of your parents' friends

  B a group of people shopping at one of the two stores

  C people on a sidewalk that is not near either store

  D people in a different state

3 Which would be the best way to predict the outcome of elections for student president? (Lesson 8-8) **(7.S.1, 7.PS.15)**

  A Have the candidates count people who promise to vote for them.

  B Have teachers predict whom their class will vote for.

  C Conduct an anonymous poll of all the students.

  D Check which candidate gets the most applause after his or her speech.

4 Mike collected data on the eye color of students in his class. He recorded the data in the chart below.

Eye Color	
Color	**Number of Students**
brown	⊮⊮
blue	⊪⊪
green	⊫
other	⊫

Which circle graph most accurately represents the data shown? (Lesson 8-4) **(7.S.2, 7.R.1)**

A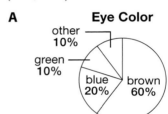

B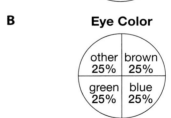

C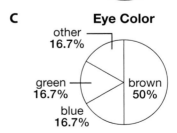

D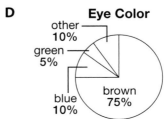

5 Lee used the data in the newspaper to make a table of high and low temperatures for 5 days.

Day	Mon	Tues	Wed	Thurs	Fri
High	72°F	78°F	81°F	83°F	85°F
Low	62°F	58°F	65°F	65°F	61°F

Which double line graph best presents the data? (Lesson 8-6) (**7.S.3, 7.CM.3**)

A **Daily High and Low Temperatures**

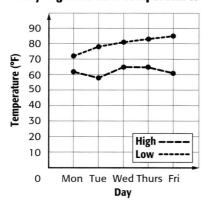

B **Daily High and Low Temperatures**

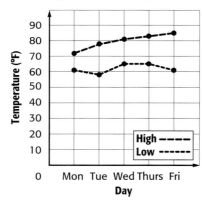

C **Daily High and Low Temperatures**

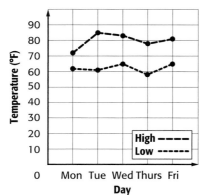

D **Daily High and Low Temperatures**

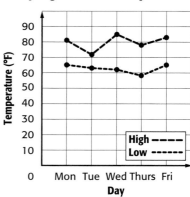

6 Mickey recorded the time he spent practicing his violin during the week. On Monday, he practiced for 23 minutes, Tuesday for 45 minutes, Wednesday for 31 minutes, Thursday for 48 minutes, and Friday for 42 minutes. What is the range for times he spent practicing this week? (Lesson 8-2) (**7.S.4, 7.CM.10**)

A 23 minutes

B 25 minutes

C 26 minutes

D 48 minutes

7 Angela received the following scores on her math quizzes during the marking period:

10, 9, 9, 6, 8, 9, 10, 10, 10

What is the range of her scores? (Lesson 8-2)
(**7.S.4, 7.CN.6**)

A 4

B 5

C 9

D 10

8 Which measure of central tendency will give the greatest emphasis to the most common number in the data? (Lesson 8-2) **(7.S.5, 7.PS.7)**

{45, 45, 44, 45, 13, 44, 45, 45, 45}

A mean

B median

C mode

D range

9 The line graph below shows the number of cars passing a particular intersection for the first ten days of a month.

Cars Passing Intersection

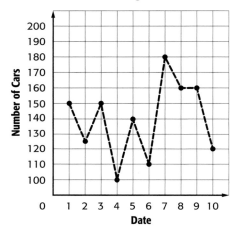

How many cars pass the intersection on the 5th of the month? (Lesson 8-1) **(7.S.6, 7.R.1)**

A 120

B 130

C 140

D 150

10 Which measure of central tendency will give the greatest emphasis to the outlier in the data? (Lesson 8-2) **(7.S.5, 7.PS.7)**

{103, 110, 109, 25, 105, 107, 100}

A mean

B median

C mode

D range

11 Which of the following statistics is most likely to be misleading? (Lesson 8-9) **(7.S.7, 7.PS.12)**

A Candidate A is ahead in the election campaign (based on a scientific poll).

B Everyone loves our store (based on a survey of our customers).

C Bungee jumping is dangerous (based on a report by experts).

D Town Z has a population of 12,521 people (based on the U.S. census).

12 Using some letter tiles from a word game, Anna draws a random tile, records the letter, and replaces the tile. Her results for 100 draws are shown below.

A-35, B-5, C-25, D-5, E-30

What is the experimental probability that her next draw will be a letter E? (Lesson 9-7) **(7.S.8, 7.PS.4)**

A $\frac{1}{26}$

B $\frac{3}{100}$

C $\frac{1}{5}$

D $\frac{3}{10}$

13 The number of minutes Casey spent exercising is listed below.

22, 25, 30, 35, 18, 40

What is the range for these minutes? (Lesson 8-2) **(7.S.4, 7.CN.6)**

A 18

B 22

C 32

D 40

14 Nancy asked three students about their lunch preference. Two chose the sandwich and one chose pizza. Nancy concluded that the sandwich was twice as popular as the pizza. Why was her data misleading? (Lesson 8-9) **(7.S.9, 7.PS.15)**

  A Most students would choose sandwiches.

  B The preference might be different on another day.

  C There were only two choices.

  D The sample size was too small.

15 Which data sample would give the most accurate prediction of the high temperature in Albany on May 1st? (Lesson 8-7) **(7.S.9, 7.PS.12)**

  A the high temperature on that date in Albany for the past 20 years

  B the high temperature in Albany two months before

  C the high temperature in several American cities the previous day

  D the high temperature in Albany each day for the last 6 months

16 Miguel has 6 T-shirts and 3 sweatshirts in his closet. He randomly picks one of these articles of clothing. How likely is it that he will pick a T-shirt? (Lesson 9-1) **(7.S.10, 7.PS.4)**

  A very unlikely

  B unlikely

  C likely

  D very likely

17 Leslie is rolling a number cube 300 times. What is the most accurate prediction of the number of times she will roll an even number? (Lesson 8-7) **(7.S.10, 7.PS.4)**

  A 50

  B 100

  C 150

  D 175

18 Which comparison would most accurately test the prediction that more people drive cars than pickup trucks? (Lesson 8-8) **(7.S.11, 7.PS.15)**

  A counting cars / trucks at a car lot

  B counting cars / trucks at a busy intersection

  C counting cars / trucks at a construction site

  D counting cars / trucks driven by your friends' parents

19 LeeAnn predicted that she would hit 80% of her soccer penalty kicks. She made 30 goals in 40 attempts. How did LeeAnn's prediction compare to her actual results? (Lesson 8-7) **(7.S.12, 7.PS.1)**

  A prediction = results

  B prediction < results

  C prediction > results

  D Her prediction cannot be compared to her actual results.

20 Refer to the table and graph below.

(Lesson 8-4) **(7.S.2, 7.R.3)**

Instruments played by students in class

Instrument	Number of students
Piano	
Woodwind	8
Brass	
Percussion	2

Instruments Played by Students in Class

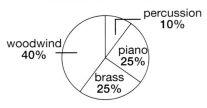

Part A

If 20 students were surveyed, find the number of students who play the piano or brass.

Part B

What fraction of the students play percussion?

21 Myra surveyed 60 students about their plans for spring break. She displayed the results on this circle graph.

Spring Break Plans

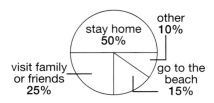

How many of the students said they plan to visit family or friends? (Lesson 8-4)

(7.S.6, 7.PS.16)

22 Refer to the table and graph below. On the graph, which bar represents 2008?

(Lesson 8-6) **(7.S.3, 7.R.1)**

Number of Sales per Month

Total sales in month	2007	2008
January	200	225
February	150	200
March	175	225
April	250	300

Number of Sales Per Month

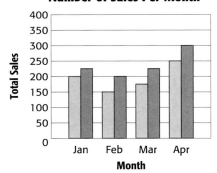

23 Explain why this graph of the scores on a test is misleading. (Lesson 8-9) **(7.S.7, 7.PS.16)**

Scores on Test

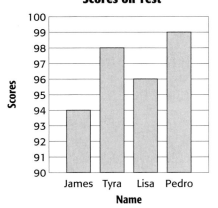

24 Juan tossed a coin 500 times. He recorded the results and found that the coin showed heads 240 times and tails the rest of the tosses. What is his experimental probability of tossing tails? Show your work. (Lesson 9-7) **(7.S.8, 7.PS.10)**

25 A bag contains 6 red marbles, 4 blue marbles, and 5 green marbles. Predict the number of times a red marble will be drawn in 100 draws if the marble is replaced after each draw. Show your work. (Lesson 8-8) **(7.S.10, 7.PS.12)**

26 Mandy asked the students in her school which is their favorite sport to play. The table below shows the results.

FAVORITE SPORT

Sport	Number of Students
Basketball	14
Football	10
Soccer	7
Volleyball	9

What is the experimental probability that the next student Mandy asks will say soccer? (Lesson 9-7) **(7.S.8, 7PS.10)**

27 How could you collect data to predict the results of an election for class president? (Lesson 8-8) **(7.S.11, 7.CN.7)**

28 The high temperature of Lakewood last week is shown in the table below. (Lesson 8-2) **(7.S.4, 7.CN.6)**

TEMPERATURE

Day	Temperature (°F)
Mon	85°
Tue	70°
Wed	83°
Th	77°
Fri	82°

Part A

What is the range of the temperatures?

Part B

How is the range affected if the low temperature is removed?

29 Joel rolled a number cube 50 times. He landed on 3 nine times. What is his experimental probability of rolling a 3? (Lesson 9-7) **(7.S.8, 7.PS.10)**

To the Student

As you gear up to study mathematics, you are probably wondering, "What will I learn this year?" You will focus on these three areas.

- **Number and Operations, Algebra, and Geometry:** Develop and apply an understanding of proportions, including similarity.
- **Measurement and Geometry:** Find surface areas and volumes of three-dimensional shapes.
- **Number and Operations and Algebra:** Solve linear equations and deepen an understanding of operations on all rational numbers.

Along the way, you'll learn more about problem solving, how to use the tools and language of mathematics, and how to THINK mathematically.

How to Use Your Math Book

Have you ever been in class and not understood all of what was being presented? Or, you understood everything in class, but got stuck on how to solve some of the homework problems? Don't worry. You can find answers in your math book!

- **Read** the **MAIN IDEA** at the beginning of the lesson.

- **Find** the **New Vocabulary** words, **highlighted in yellow**, and read their definitions.

- **Review** the **EXAMPLE** problems, solved step-by-step, to remind you of the day's material.

- **Refer** to the **HOMEWORK HELP** boxes that show you which examples may help with your homework problems.

- **Go to** **NY Math Online** where you can find extra examples to coach you through difficult problems.

- **Review** the notes you've taken on your **FOLDABLES**™.

- **Find** the answers to odd-numbered problems in the back of the book. Use them to see if you are solving the problems correctly.

Scavenger Hunt

Let's Get Started

Use the Scavenger Hunt below to learn where things are located in each chapter.

1. What is the title of Chapter 1?

2. How can you tell what you'll learn in Lesson 1-1?

3. Sometimes you may ask, "When am I ever going to use this?" Name a situation that uses the concepts from Lesson 1-2.

4. In the margin of Lesson 1-2, there is a Vocabulary Link. What can you learn from that feature?

5. What is the key concept presented in Lesson 1-3?

6. How many examples are presented in Lesson 1-3?

7. What is the title of the feature in Lesson 1-3 that tells you how to read square roots?

8. What is the Web address where you could find extra examples?

9. Suppose you're doing your homework on page 40 and you get stuck on Exercise 18. Where could you find help?

10. What problem-solving strategy is presented in the Problem-Solving Investigation in Lesson 1-5?

11. List the new vocabulary words that are presented in Lesson 1-7.

12. What is the web address that would allow you to take a self-check quiz to be sure you understand the lesson?

13. There is a Real-World Career mentioned in Lesson 1-10. What is it?

14. On what pages will you find the Study Guide and Review for Chapter 1?

15. Suppose you can't figure out how to do Exercise 25 in the Study Guide and Review on page 72. Where could you find help?

New York

Start Smart

Let's Review!

Lesson 1

A Plan for Problem Solving

Wonderous Waterfalls

Niagara Falls are massive waterfalls located on the Niagara River on the border between Ontario, Canada, and New York state. Niagara Falls is famous for its breath-taking beauty and power and has been a popular tourist site for over a century. Facts about Niagara Falls are shown in the table.

	American Falls (including Bridal Veil Falls)	Horseshoe Falls
Height (ft)	176	167
Width (ft)	1,060	2,600
Volume of water (gal/sec)	150,000	600,000

Source: Niagara Falls Live

How much wider are the Horseshoe Falls than the American Falls?

You can use the four-step problem-solving plan to solve many kinds of problems. The four steps are Understand, Plan, Solve, and Check.

Understand

- **Read the problem carefully.**
- **What facts do you know?**
- **What do you need to find?**

You know the widths of both waterfalls. You need to find how much wider the Horseshoe Falls are than the American Falls.

Plan

- **How do the facts relate to each other?**
- **Plan a strategy to solve the problem.**

The width of the American Falls is 1,060 feet, and the width of the Horseshoe Falls is 2,600 feet. To find how much wider the Horseshoe Falls are than the American Falls, subtract the width of the American Falls from the width of the Horseshoe Falls.

Solve

- **Use your plan to solve the problem.**

$$
\begin{array}{r}
\overset{5\ 10}{2{,}6\cancel{0}0} \\
-\ 1{,}060 \\
\hline
1{,}540
\end{array}
\quad
\begin{array}{l}
\text{width of the Horseshoe Falls} \\
\text{width of the American Falls}
\end{array}
$$

So, the Horseshoe Falls are 1,540 feet wider than the American Falls.

Check

- **Look back at the problem.**
- **Does your answer make sense?**
- **If not, solve the problem another way.**

Estimate. Round 1,060 to 1,100.

$2,600 - 1,100 = 1,500$.

Since 1,500 is close to 1,540, the answer is reasonable.

CHECK Your Understanding

1. What is the volume of water flowing over the American Falls per minute? the Horseshoe Falls?

2. **WRITING IN MATH** The American Falls have a clear drop of 70 feet before reaching a collection of fallen rocks. Explain how you would use the four-step plan to determine the height of the fallen rocks at the bottom of the American Falls.

Lesson 2

Number and Operations

Batter Up

Baseball is one of America's most popular sports. Originating in 1901, Major League Baseball now consists of 30 teams throughout the United States and Canada. New York is the home of two Major League Baseball teams, the New York Mets and the New York Yankees. The top ten batting averages for both teams for a recent season are shown in the table below.

New York Batting Averages			
Mets		Yankees	
Player	Average	Player	Average
Beltran	0.275	Abreu	0.330
Chavez	0.306	Cano	0.342
Delgado	0.265	Cabrera	0.280
Franco	0.273	Damon	0.285
Green	0.257	Jeter	0.343
Lo Duca	0.318	Matsui	0.302
Nady	0.264	Posada	0.277
Reyes	0.300	Rodriguez	0.290
Wright	0.311	Sheffield	0.298
Valentin	0.271	Williams	0.281

Source: ESPN

 **What You Know**

Comparing and Ordering Decimals

When comparing and ordering decimals, first line up the decimal points. Then compare the digits in each place-value position.

Use the information in the table on page 6 to answer each question.

1. Which of the Mets players listed in the table had the greatest batting average? the least?

2. Which of the Yankees players listed in the table had the greatest batting average? the least?

3. List the top 10 players from both teams combined in order from least to greatest batting average.

✓ **CHECK** **What You Know**

Subtracting Decimals ·············

When subtracting decimals, first line up the decimal points. Then subtract as with whole numbers and place the decimal point in the answer.

Use the table to answer each question.

4. Of all fields combined, how much greater was the greatest average ticket price than the least average ticket price?

5. How much greater is the average ticket price at Wrigley Field than at Shea Stadium?

6. How much less is the average ticket price at Minute Maid Park than at Jacob's Field?

7. **WRITING IN MATH** Use the table on page 6 to write and solve a real-world problem about the batting averages.

Baseball Field	Average Ticket Price ($)
Busch Stadium	29.78
Fenway Park	46.46
Jacob's Field	21.54
Minute Maid Park	26.66
Safeco Field	24.01
Shea Stadium	23.92
Wrigley Field	34.30
Yankee Stadium	24.86

Source: Team Marketing Report

Algebra

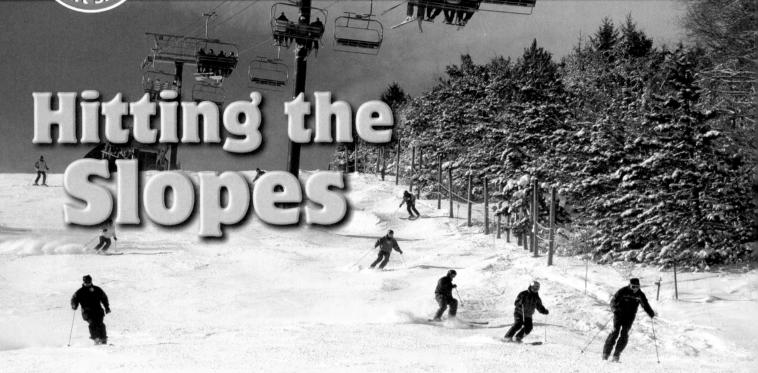

Hitting the Slopes

Belleayre Mountain, located just hours from New York City, is New York's Winter Snow Park. It was a big success when it opened in 1949, and after millions of dollars worth of improvements and expansions, now serves as a model in the ski industry. The weekend costs of full-day lift passes at Belleayre Mountain are shown in the table.

Full-Day Lift Fees for Weekends	
Age	Fee ($)
Toddler	5
Youth (6–12)	32
Junior (13–17)	38
College Student	38
Adult	48
Senior	38

Source: Belleayre Mountain

Variables and Expressions . . . .

Use the table on page 8 to answer the following questions.

1. Suppose an adult was accompanying several juniors for a day of skiing. Write an addition expression to represent the cost for any number of juniors and one adult.

2. Jill has a coupon for $4 off the price of each youth admission. Write a subtraction expression to represent the cost for any number of youths to ski using the coupon.

3. A two-day junior lift package for a weekend is t more than the full-day junior lift fee. Write an addition expression that could be used to find the cost of a two-day junior lift package for three juniors.

4. The Cesta family decides to go skiing at Belleayre Mountain. Write an expression for the cost for toddlers, youths, and adults.

5. The uphill lift capacity at Belleayre Mountain is t trips per hour. Suppose the lift carries two people at a time. Write an expression that represents the number of hours it would take for the lift to carry 50,000 skiers.

6. Refer to Exercise 5. Suppose the uphill lift carries four people at a time. Write a new expression that represents the number of hours it would take for the lift to carry 50,000 skiers.

7. **WRITING IN MATH** Write a real-world problem to represent the expression $3a - 5$.

Geometry

Lighting the Way

Lighthouses provide navigation for ships. Since New York is so rich in waterways, there are over 80 lighthouses in the state. Many different geometric angles can be seen in their designs. The photo above shows several angles in the 30 Mile Point Lighthouse on Lake Ontario in New York.

 **CHECK** What You Know   **Estimating Angle Measures** . . . .

Use the photo of the 30 Mile Point Lighthouse to estimate the measure of each of the following angles.

1. the angle outlined in pink

2. the angle outlined in green

3. the angle outlined in red

4. the angle outlined in yellow

Use a protractor and the photo below of the Esopus lighthouse to measure each of the following angles.

5. the angle outlined in green

6. the angle outlined in blue

7. the angle outlined in yellow

8. the angle outlined in red

9. Use the Internet or another source to find a photo of another New York lighthouse that illustrates geometric angles. Estimate the measure of several angles in the photo. Then use a protractor to measure the angles. Compare with your estimates.

10. **WRITING IN MATH** Write a few sentences explaining how to use a protractor to find the measure of an angle.

Measurement

The Sky's the Limit

New York City is famous for its breath-taking skyline, including the Empire State Building and numerous other skyscrapers. The heights of several New York City high-rise buildings are shown in the table below.

Building	Height (ft)
American International	952
Bloomberg Tower	806
Carnegie Hall Tower	757
Chrysler Building	1,046
Empire State Building	1,250
Rockefeller Center	850
Time Warner Center	750
Trump World Tower	861

Source: Emporis Buildings

Standard units of measurement are most commonly used in the United States. They include inch, foot, yard, and mile.

 **CHECK Your Understanding**

Understanding Units ....

1. How many inches are in one foot? How many feet are in one yard?

2. How many inches tall is the Time Warner Center? How many yards tall is it?

3. About how many yards tall is the Rockefeller Center?

4. How many yards taller is the Chrysler Building than the Bloomberg Tower?

5. Is 7 inches, 7 feet, or 7 yards a more reasonable estimate for the height of the front doors of a skyscraper?

6. Estimate how many times taller the Empire State Building is than the Trump World Tower.

7. **WRITING IN MATH** Use the Internet or another source to find the height of another building in New York. Write a problem involving standard units using the data that you found.

Statistics and Probability

Wonderful Wheat

Next to dairy products, wheat and wheat products are one of the most valuable agriculture exports for New York state, over 85 million dollars. The table below shows the number of acres of wheat harvested for each New York county in a recent year.

Number of Acres of Wheat Harvested by New York Counties							
800	1,500	9,100	11,800	10,900	7,100	7,600	10,100
7,600	1,100	3,800	2,300	3,100	7,000	700	1,700
4,700	300	400	1,400	300	1,700	700	900
10,100	3,400	13,000	11,200	8,000	8,200	11,600	1,300
4,100	6,100	1,500	2,400	6,300	900		

Source: United States Department of Agriculture

These data can be arranged in a frequency table. A frequency table shows the number of pieces of data that fall within the given intervals.

Frequency Tables · · · · · · · · · · · · · · · · · ·

1. Complete the frequency table using the data about the number of acres of wheat harvested by New York counties.

Number of Acres of Wheat Harvested by New York Counties		
Number of Acres	Tally	Frequency
0–1,999		
2,000–3,999		
4,000–5,999		
6,000–7,999		
8,000–9,999		
10,000–11,999		
12,000–13,999		

2. Which interval has the greatest number of counties?

3. Which interval has the least number of counties?

4. How many counties harvested 10,000 acres of wheat or more?

5. How many counties harvested less than 6,000 acres of wheat?

6. **WRITING IN MATH** Use the Internet to research another of New York's crops. In a report, explain how you could use a frequency table to make a conclusion about the crop.

New York Data File

The following pages contain data that you'll use throughout the book.

Eastern Bluebird

Length: 5.5–7 in. (14–18 cm)

Wingspread: 11–13 in. (28–33 cm)

Life Span: 6 years

Eggs: 3–6 pale blue eggs

Incubation: 12–14 days

Source: Enchanted Learning

Buffalo, New York

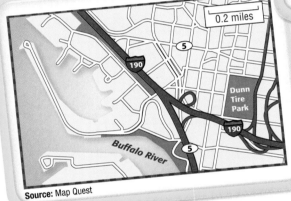

Source: Map Quest

Temperature

Average High Temperatures for Ithaca, New York			
Month	**Temperature (°F)**	**Month**	**Temperature (°F)**
Jan.	32	July	82
Feb.	33	Aug.	80
Mar.	42	Sept.	72
Apr.	56	Oct.	61
May	67	Nov.	48
June	77	Dec.	35

Source: Weatherbase

Maple Oatmeal Chocolate Chip Cookies

Maple Oatmeal Chocolate Chip Cookies

Ingredients

$\frac{1}{2}$ cup butter

1 cup maple syrup

1 large egg

1 teaspoon vanilla extract

$1\frac{3}{4}$ cups flour

2 cups rolled oats

$\frac{1}{2}$ teaspoon baking soda

$\frac{1}{8}$ teaspoon salt

1 cup miniature chocolate chips

$\frac{1}{2}$ cup walnuts (optional)

Source: Maple Weekend

Apple Nutrition Facts

Nutrition Facts

Serving Size 1 medium apple (154g / 5.5 oz.)

Amount Per Serving

Calories 80	Calories from Fat 0

	% Daily Value*
	0%
Total Fat 0g	0%
Saturated Fat 0g	0%
Cholesterol 0mg	0%
Sodium 0mg	5%
Potassium 170mg	7%
Total Carbohydrate 22g	20%
Dietary Fiber 5g	
Sugars 16g	
Protein 0g	

Source: New York Apple Country

NBA Basketball

Number of Points Scored by New York Knicks in Each Game, 2006–2007					
118	92	95	93	**109**	94
92	96	**102**	**100**	118	90
89	**101**	95	85	**101**	100
100	**98**	102	**115**	90	**94**
96	100	**97**	**111**	**103**	77
151	86	80	100	99	**111**
106	110	**102**	98	100	**108**
83	107	**116**	105	**99**	87
94	**102**	102	**107**	101	**100**
84	**95**	92	**99**	94	**106**
104	99	**90**	94	90	**92**
77	86	68	89	**97**	103
94	90	94	**118**	83	69
86	105	95	**94**		

*Data in boldface type indicate winning games for the Knicks.

Source: New York Post

Sugar Maple Tree

The Sugar Maple is a prominent tree in the hardwood forests of eastern North America.

Tree Height: 50–80 ft (15–24 m)

Leaves: 8–15 cm long and equally wide

Source: National Parks Service

The Cyclone

Amusement Park: Astroland, Brooklyn, New York

Type: wood twister

First Season: 1927

Height: 85 ft (25.9 m)

Speed: 60 mph

Track Length: 2,640 ft (804.7 m)

Vehicles: 3

Riders Per Vehicle: 24

Ride Time: 1.5 min

Source: Coasterbuzz

Kodak Tower, Rochester

NFL Football

New York Jets Quarterbacks, 2007

Jersey Number	Height	Age	Experience (yr)
11	6 ft 2 in.	24	2
8	6 ft 1 in.	28	7
10	6 ft 3 in.	31	8

Metropolitan Museum of Art

General Admission	
Adult	$20
Senior (65 and over)	$15
Student	$10
Child (under 12)	Free
Memberships	
Met Net	$60
Associate	$50
Friend	$275
Individual	$95
Family/Dual	$190
Sustaining	$500
Contributing	$1,200
Donor	$1,800
Sponsor	$4,000
Patron	$8,000
Patron Circle	$12,000
President's Circle	$20,000

Source: The Metropolitan Museum of Art

New York State Capitol Building in Albany

Darien Lake State Park, Darien Center

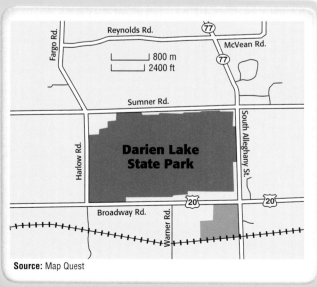

Source: Map Quest

Agriculture

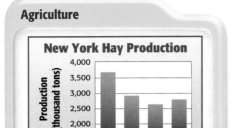

New York Hay Production

Year	Production (thousand tons)
2003	~3,650
2004	~2,900
2005	~2,600
2006	~2,750

Source: United States Department of Agriculture

New York Aquarium

Summer Hours:
Mon. – Fri. 10 A.M. – 6 P.M.
Weekends/Holidays 10 A.M. – 7 P.M.

Aquarium Admission:
Adult: $12.00
Child (3–12): $8.00
Senior (65 and over): $8.00
Child (2 and under): free

Deep Sea 3D:
All ages: $6.00

Combo Pass: (Aquarium plus Deep Sea 3D)
Adult: $16.00
Child (3–12): $12.00
Senior (65 and over): $12.00

Source: New York Aquarium

Water Safari Enchanted Forest, Old Forge

Water Ride	Description	Minimum Height Requirement (in.)
The Amazon	1,100-foot family tube ride	36
Black River	500-foot double tube slide takes 30 seconds	42
Cascade Falls	500 feet long	36
Killermanjaro	drop 280 feet over 30 mph	42
Raging Rapids	700 feet	42
Tidal Wave Pool	$\frac{1}{4}$ acre pool $4\frac{1}{2}$ feet deep waves up to 4 feet tall	none

Source: Water Safari

LaGuardia International Airport

LaGuardia International Airport Runways		
Surface	**Length (ft)**	**Length (m)**
Asphalt/concrete	7,001	2,134
Asphalt/concrete	7,003	2,135

Unit 1
Algebra and Functions

Focus
Use appropriate operations to solve problems and linear equations.

CHAPTER 1
Introduction to Algebra and Functions

BIG Idea Represent relationships in numerical, verbal, geometric and symbolic form.

CHAPTER 2
Integers

BIG Idea Know the properties of, and compute with, integers.

CHAPTER 3
Algebra: Linear Equations and Functions

BIG Idea Solve linear equations in one variable.

Problem Solving in Social Studies

Real-World Unit Project

Stand Up and Be Counted! Every 10 years, the U.S. Census takes a count of the U.S. population. How does the U.S. Census affect the number of members in the House of Representatives from each state? You're on a mission to find out! Along the way, you will create a map of the United States, make a line plot, and write a paragraph about these changes. Don't forget to bring your math tool kit. This adventure will appeal to your "census."

NY Math Online Log on to glencoe.com to begin.

CHAPTER 1

Introduction to Algebra and Functions

New York State Core Curriculum

7.A.8 Create algebraic patterns using charts/tables, graphs, equations, and expressions

Key Vocabulary

algebra (p. 44)

defining the variable (p. 50)

evaluate (p. 31)

numerical expression (p. 38)

🌐 Real-World Link

Parks Admission to the Kentucky Horse Park in Lexington, Kentucky, costs $15 for each adult and $8 for each child. You can use the four-step problem-solving plan to determine the cost of admission for a family of 2 adults and 3 children.

Introduction to Algebra and Functions Make this Foldable to help you organize your notes. Begin with eleven sheets of notebook paper.

① **Staple** the eleven sheets together to form a booklet.

② **Cut** tabs. Make each one 2 lines longer than the one before it.

③ **Write** the chapter title on the cover and label each tab with the lesson number.

GET READY for Chapter 1

Diagnose Readiness You have two options for checking Prerequisite Skills.

Option 2

NY Math Online Take the Online Readiness Quiz at glencoe.com.

Option 1

Take the Quick Quiz below. Refer to the Quick Review for help.

QUICK Quiz

Add. (Prior Grade)

1. $89.3 + 16.5$
2. $7.9 + 32.45$
3. $54.25 + 6.39$
4. $10.8 + 2.6$

5. **TECHNOLOGY** Patrick bought a personal electronic organizer for $59.99 and a carrying case for $12.95. What was his total cost, not including tax? (Prior Grade)

Subtract. (Prior Grade)

6. $24.6 - 13.3$
7. $9.1 - 6.6$
8. $30.55 - 2.86$
9. $17.4 - 11.2$

Multiply. (Prior Grade)

10. 4×7.7
11. 9.8×3
12. 2.7×6.3
13. 8.5×1.2

Divide. (Prior Grade)

14. $37.49 \div 4.6$
15. $14.31 \div 2.7$
16. $6.16 \div 5.6$
17. $11.15 \div 2.5$

18. **PIZZA** Four friends decided to split the cost of a pizza evenly. The total cost was $25.48. How much does each friend need to pay? (Prior Grade)

QUICK Review

Example 1 Find $17.89 + 43.2$.

$$\begin{array}{r} 17.89 \\ + 43.20 \\ \hline 61.09 \end{array}$$

Line up the decimal points.
Annex a zero.

Example 2 Find $37.45 - 8.52$.

$$\begin{array}{r} 37.45 \\ - 8.52 \\ \hline 28.93 \end{array}$$

Line up the decimal points.

Example 3 Find 1.7×3.5.

$$\begin{array}{r} 1.7 \\ \times 3.5 \\ \hline 5.95 \end{array}$$

← 1 decimal place
← + 1 decimal place
← 2 decimal places

Example 4 Find $24.6 \div 2.5$.

$2.5\overline{)24.6} \rightarrow 25.\overline{)246.}$ Multiply both numbers by the same power of 10.

$$\begin{array}{r} 9.84 \\ 25\overline{)246.00} \\ -225 \\ \hline 210 \\ -200 \\ \hline 100 \\ -100 \\ \hline 0 \end{array}$$

Annex zeros.

Divide as with whole numbers.

READING to SOLVE PROBLEMS

Making Sense

When you solve a word problem, the first thing to do is to read the problem carefully. The last thing to do is to see whether your answer makes sense. Sometimes a picture or diagram can help.

> Kelly lives 5 miles from school. This is 4 times as far as Miguel lives from school. How far does Miguel live from school?

NYSCC 7.CM.4 Share **organized mathematical ideas through the manipulation of** objects, numerical tables, drawings, **pictures**, charts, graphs, tables, **diagrams**, models and symbols in written and verbal form

If you look just at the key words in the problem, it might seem that 4 *times* 5 would give the solution.

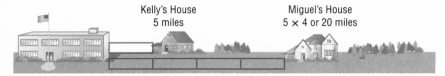

Kelly's House
5 miles

Miguel's House
5 × 4 or 20 miles

But the important question is, "Does this solution make sense?" In this case, the solution does *not* make sense because Kelly lives farther away. This problem is solved by dividing.

Miguel's House
5 ÷ 4 or 1.25 miles

Kelly's House
5 miles

So, Miguel lives 1.25 miles away from school.

PRACTICE

For Exercises 1 and 2, choose the model that illustrates each problem. Explain your reasoning. Then solve.

1. Jennifer has saved $210 to purchase an MP3 player. She needs $299 to buy it. How much more money does she need?

Model A

| 299 |
| 210 |

Model B

| 210 | 299 |

2. The school cafeteria sold 465 lunches on Thursday. They expect to sell 75 more lunches on Friday because they serve pizza that day. How many lunches do they expect to sell on Friday?

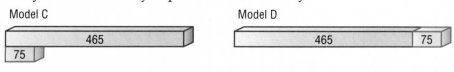

Model C

| 465 |
| 75 |

Model D

| 465 | 75 |

1-1 A Plan for Problem Solving

MAIN IDEA

Solve problems using the four-step plan.

NYS Core Curriculum

7.PS.14 Determine information required to solve the problem
7.PS.15 Choose methods for obtaining required information
Also addresses 7.PS.12, 7.PS.13, 7.RP.3, 7.CM.3

NY Math Online

glencoe.com

• Extra Examples
• Personal Tutor
• Self-Check Quiz

▷ GET READY for the Lesson

ANALYZE GRAPHS The graph shows the countries with the most world championship motocross wins. What is the total number of wins for these five countries?

1. Do you have all of the information necessary to solve this problem?

2. Explain how you would solve this problem. Then solve it.

3. Does your answer make sense? Explain.

4. What can you do if your first attempt at solving the problem does not work?

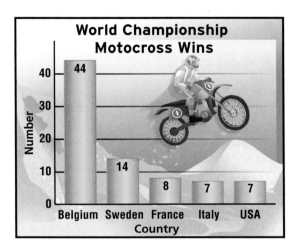

In mathematics, there is a *four-step plan* you can use to help you solve any problem.

Understand
• Read the problem carefully.
• What information is given?
• What do you need to find out?
• Is enough information given?
• Is there any extra information?

Plan
• How do the facts relate to each other?
• Select a strategy for solving the problem. There may be several that you can use.
• Estimate the answer.

Solve
• Use your plan to solve the problem.
• If your plan does not work, revise it or make a new plan.
• What is the solution?

Check
• Does your answer fit the facts given in the problem?
• Is your answer reasonable compared to your estimate?
• If not, make a new plan and start again.

EXAMPLE Use the Four-Step Plan

1 **TELEVISION** There were about 268 million TVs in the U.S. in 2007. This amount increases by 4 million each year after 2007. In what year will there be at least 300 million TVs?

Understand   *What are you trying to find?*
In what year will there be at least 300 million TVs in the U.S.?

What information do you need to solve the problem?
You know how many TVs there were in 2007. Also, the number increases by 4 million each year.

Plan   Find the number of TVs needed to reach 300 million. Then divide this number by 4 to find the number of years that will pass before the total reaches 300 million TVs.

Solve   The change in the number of TVs from 268 million to 300 million is $300 - 268$ or 32 million TVs. Dividing the difference by 4, you get $32 \div 4$ or 8.

You can also use the *make a table* strategy.

Year	'07	'08	'09	'10	'11	'12	'13	'14	'15
Number (millions)	268	272	276	280	284	288	292	296	300

+4 +4 +4 +4 +4 +4 +4 +4

So, there will be at least 300 million TVs in the U.S. in the year 2015.

Check   8 years × 4 million = 32 million
268 million + 32 million = 300 million ✔

 CHECK Your Progress

a. **WHALES** A baby blue whale gains about 200 pounds each day. About how many pounds does a baby blue whale gain per hour?

Problems can be solved using different operations or strategies.

Problem-Solving Strategies   Concept Summary

guess and check	use a graph
look for a pattern	work backward
make an organized list	eliminate possibilities
draw a diagram	estimate reasonable answers
act it out	use logical reasoning
solve a simpler problem	make a model

EXAMPLE Use a Strategy in the Four-Step Plan

2 **GEOMETRY** A *diagonal* connects two nonconsecutive vertices in a figure, as shown at the right. Find how many diagonals a figure with 7 sides would have.

3 sides   4 sides   5 sides
0 diagonals   2 diagonals   5 diagonals

Understand You know the number of diagonals for figures with three, four, and five sides.

Plan You can look for a pattern by organizing the information in a table. Then continue the pattern until you find the diagonals for an object with 7 sides.

Solve

Sides	3	4	5	6	7
Diagonals	0	2	5	9	14

+2 +3 +4 +5

So, a 7-sided figure would have 14 diagonals.

Check Check your answer by making a drawing.

CHECK Your Progress

b. GEOMETRY Numbers that can be represented by a triangular arrangement of dots are called *triangular numbers.* The first five triangular numbers are shown below. Write a sequence formed by the first eight triangular numbers. Write a rule for generating the sequence.

   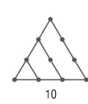 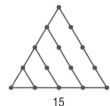

1   3   6   10   15

CHECK Your Understanding

Use the four-step plan to solve each problem.

Example 1
(p. 26)

1. ANALYZE TABLES The table lists the sizes of six of the largest lakes in North Carolina. About how many times as large is High Rock Lake than Hyco Lake?

Example 2
(p. 27)

2. ALGEBRA What are the next two numbers in the pattern below?

1, 1, 2, 6, 24, ▦, ▦

Lake	Size (acres)
Lake Mattamuskeet	40,000
Falls Lake	12,000
Hyco Lake	3,750
Lake Gaston	20,000
Lake James	6,500
High Rock Lake	15,000

HOMEWORK HELP	
For Exercises	**See Examples**
3–6	1
7–10	2

Use the four-step plan to solve each problem.

3. **BIRDS** Most hummingbirds flap their wings about 50 times a second. How many times can a hummingbird flap its wings in one minute?

4. **PLANETS** Jupiter is about 3 times the size of Neptune. If the diameter of Jupiter is 88,736 miles, estimate the diameter of Neptune.

5. **FIELD TRIPS** To attend a field trip to a museum, each student will have to pay $6.00 for transportation and $5.75 for admission. If there are 65 students attending the field trip, how much money will their teacher need to collect?

6. **CANOE RENTALS** A state park took in $12,000 in canoe rentals during March. June rentals are expected to double that amount. If canoes rent for $40, how many canoe rentals are expected in June?

7. **GEOMETRY** What are the next two figures in the pattern?

8. **ALGEBRA** What are the next two numbers in the pattern below?

9, 27, 81, 243, 729, ■ , ■

ANALYZE TABLES For Exercises 9 and 10, use the commuter train schedule shown.

A commuter train departs from a train station and travels to the city each day. The schedule shows the first five departure and arrival times.

9. How often does the commuter train arrive in the city?

10. What is the latest time that passengers can depart from the train station if they need to arrive in the city no later than noon?

Commuter Train Schedule	
Departure	**Arrival**
6:30 A.M.	6:50 A.M.
7:15 A.M.	7:35 A.M.
8:00 A.M.	8:20 A.M.
8:45 A.M.	9:05 A.M.
9:30 A.M.	9:50 A.M.

11. **HOMEWORK** Angel has guitar practice at 7:00 P.M. He has homework in math, science, and history that will take him 30 minutes each to complete. He also has to allow 20 minutes for dinner. What is the latest time Angel can start his homework?

12. **ESTIMATION** Terry opened a savings account in December with $132 and saved $27 each month beginning in January. Estimate the value of Terry's account in July. Then calculate the amount and evaluate the reasonableness of your estimate.

13. **FIND THE DATA** Refer to the Data File on pages 16–19 of your book. Choose some data and write a real-world problem in which you would use the four-step plan to solve the problem.

14. **ANALYZE TABLES** The sizes of Earth's oceans in millions of square kilometers are shown in the table. If the combined size of Earth's oceans is 367 million square kilometers, what is the size of the Pacific Ocean?

Earth's Oceans	
Ocean	Size (million km²)
Arctic	45
Atlantic	77
Indian	69
Pacific	▓
Southern	20

Source: *The World Factbook*

15. **MONEY** Meli wants to buy a pair of rollerblades that cost $140.75. So far, she has saved $56.25. If she saves $6.50 every week, in how many weeks will she be able to purchase the rollerblades?

NYSCC • NYSMT
Extra Practice, pp. 668, 704

H.O.T. Problems

16. **CHALLENGE** Use the digits 5, 6, 7, and 8 to form two 2-digit numbers so that their product is as great as possible. Use each digit only once.

17. **OPEN ENDED** Create a real-world problem that can be solved by adding 79 and 42 and then multiplying the result by 3.

18. **WRITING IN MATH** Explain why it is important to plan before solving a problem.

NYSMT PRACTICE 7.PS.14, 7.PS.15

19. Sheryl has $2 to spend at the school store. Based on the choices below, which three items from the table could Sheryl purchase?

Item	Cost
Folder	$1.50
Pencil	$0.20
Pen	$0.50
Ruler	$1.75
Highlighter	$0.40

A folder, pencil, pen

B folder, highlighter, pencil

C pencil, pen, highlighter

D ruler, highlighter, pencil

20. Mr. Brooks went on a business trip. The trip was 380 miles, and the average price of gasoline was $3.15 per gallon. What information is needed to find the amount Mr. Brooks spent on gasoline for the trip?

F Number of times Mr. Brooks stopped to fill his tank with gasoline

G Number of miles the car can travel using one gallon of gasoline

H Number of hours the trip took

J Average number of miles Mr. Brooks drove per day

▷ **GET READY** for the Next Lesson

PREREQUISITE SKILL Multiply.

21. 10×10

22. $3 \times 3 \times 3$

23. $5 \times 5 \times 5 \times 5$

24. $2 \times 2 \times 2 \times 2 \times 2$

1-2 Powers and Exponents

MAIN IDEA

Use powers and exponents.

NYS Core Curriculum

7.N.8 Find the common factors and greatest common factor of two or more numbers
Also addresses 7.N.11, 7.N.14

New Vocabulary

factors
exponent
base
powers
squared
cubed
evaluate
standard form
exponential form

NY Math Online

glencoe.com

• Extra Examples
• Personal Tutor
• Self-Check Quiz

▷ GET READY for the Lesson

TEXT MESSENGING Suppose you text message one of your friends. That friend then text messages two friends after one minute. The pattern continues.

1. How is doubling shown in the table?

2. How many text messages will be sent after 4 minutes?

3. What is the relationship between the number of 2s and the number of minutes?

Minutes	Number of Text Messages	
0	1	= 1
1	1 × 2	= 2
2	2 × 2	= 4
3	2 × 2 × 2	= 8

Two or more numbers that are multiplied together to form a product are called **factors**. When the same factor is used, you may use an exponent to simplify the notation. The **exponent** tells how many times the base is used as a factor. The common factor is called the **base**.

$$16 = 2 \cdot 2 \cdot 2 \cdot 2 = 2^4 \longleftarrow \text{exponent}$$

base

Numbers expressed using exponents are called **powers**.

Powers	Words
5^2	five to the second power or five **squared**
4^3	four to the third power or four **cubed**
2^4	two to the fourth power

EXAMPLES   Write Powers as Products

Write each power as a product of the same factor.

1 7^5

Seven is used as a factor five times.

$7^5 = 7 \cdot 7 \cdot 7 \cdot 7 \cdot 7$

2 3^2

Three is used as a factor twice.

$3^2 = 3 \cdot 3$

✓ CHECK Your Progress

Write each power as a product of the same factor.

a. 6^4 b. 1^3 c. 9^5

You can **evaluate**, or find the value of, powers by multiplying the factors. Numbers written without exponents are in **standard form**.

 **EXAMPLES** Write Powers in Standard Form

**Vocabulary Link** . . . .
Evaluate
Everyday Use to find what something is worth
Math Use find the value of

Evaluate each expression.

3 2^5

$2^5 = 2 \cdot 2 \cdot 2 \cdot 2 \cdot 2$ 2 is used as a factor 5 times.

     $= 32$ Multiply.

4 4^3

$4^3 = 4 \cdot 4 \cdot 4$ 4 is used as a factor 3 times.

     $= 64$ Multiply.

CHECK Your Progress

Evaluate each expression.

d. 10^2 **e.** 7^3 **f.** 5^4

Numbers written with exponents are in **exponential form**.

 **EXAMPLE** Write Numbers in Exponential Form

5 Write $3 \cdot 3 \cdot 3 \cdot 3$ in exponential form.

3 is the base. It is used as a factor 4 times. So, the exponent is 4.

$3 \cdot 3 \cdot 3 \cdot 3 = 3^4$

CHECK Your Progress

Write each product in exponential form.

g. $5 \cdot 5 \cdot 5$ **h.** $12 \cdot 12 \cdot 12 \cdot 12 \cdot 12 \cdot 12$

CHECK Your Understanding

Examples 1, 2
(p. 30)

Write each power as a product of the same factor.

1. 9^3 **2.** 3^4 **3.** 8^5

Examples 3, 4
(p. 31)

Evaluate each expression.

4. 2^4 **5.** 7^2 **6.** 10^3

7. POPULATION There are approximately 5^{10} people living in North Carolina. About how many people is this?

Example 5
(p. 31)

Write each product in exponential form.

8. $5 \cdot 5 \cdot 5 \cdot 5 \cdot 5 \cdot 5$ **9.** $1 \cdot 1 \cdot 1 \cdot 1$ **10.** $4 \cdot 4 \cdot 4 \cdot 4 \cdot 4$

Lesson 1-2 Powers and Exponents   **31**

Practice and Problem Solving

HOMEWORK HELP

For Exercises	See Examples
11–16	1, 2
17–24	3, 4
25–28	5

Write each power as a product of the same factor.

11. 1^5 **12.** 4^2 **13.** 3^8

14. 8^6 **15.** 9^3 **16.** 10^4

Evaluate each expression.

17. 2^6 **18.** 4^3 **19.** 7^4

20. 4^6 **21.** 1^{10} **22.** 10^1

23. BIKING In a recent year, the number of 12- to 17-year-olds that went off-road biking was 10^6. Write this number in standard form.

24. TRAINS The Maglev train in China is the fastest passenger train in the world. Its average speed is 3^5 miles per hour. Write this speed in standard form.

Write each product in exponential form.

25. $3 \cdot 3$ **26.** $7 \cdot 7 \cdot 7 \cdot 7$

27. $1 \cdot 1 \cdot 1 \cdot 1 \cdot 1 \cdot 1 \cdot 1 \cdot 1 \; 1^8$ **28.** $6 \cdot 6 \cdot 6 \cdot 6 \cdot 6$

Write each power as a product of the same factor.

29. *four to the fifth power*   **30.** *nine squared*

Evaluate each expression.

31. *six to the fourth power*   **32.** *6 cubed*

GEOMETRY For Exercises 33 and 34, use the puzzle cube below.

33. Suppose the puzzle cube is made entirely of unit cubes. Find the number of unit cubes in the puzzle. Write your answer using exponents.

34. Why do you think the expression 3^3 is sometimes read as *3 cubed*?

35. NUMBERS Write $5 \cdot 5 \cdot 5 \cdot 5 \cdot 4 \cdot 4 \cdot 4$ in exponential form.

36. COMPUTERS A. gigabyte is a measure of computer data storage capacity. One gigabyte stores 2^{30} bytes of data. Use a calculator to find the number in standard form that represents two gigabytes.

Order the following powers from least to greatest.

37. $6^5, 1^{14}, 4^{10}, 17^3$ **38.** $2^8, 15^2, 6^3, 3^5$ **39.** $5^3, 4^6, 2^{11}, 7^2$

NYSCC • NYSMT
Extra Practice, pp. 668, 704

40. OPEN ENDED Select a number between 1,000 and 2,000 that can be expressed as a power.

41. CHALLENGE Write two different powers that have the same value.

42. Which One Doesn't Belong? Identify the number that does not belong with the other three. Explain your reasoning.

| 121 | 361 | 576 | 1,000 |

43. **WRITING IN MATH** Analyze the number pattern shown at the right. Then write a convincing argument as to the value of 2^0. Based on your argument, what do you think will be the value of 2^{-1}?

$2^4 = 16$
$2^3 = 8$
$2^2 = 4$
$2^1 = 2$
$2^0 = ?$

NYSMT PRACTICE 7.N.8

44. Which model represents 6^3?

A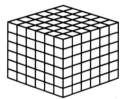

B

C
6
12

D

Spiral Review

45. FOOTBALL The graph shows the number of wins the Pittsburgh Steelers had from 2003–2006. How many more wins did the Steelers have in 2004 than 2006? (Lesson 1-1)

46. COOKING Ms. Jackson is serving fried turkey at 5:00 P.M. The 12-pound turkey has to cook 3 minutes for every pound, and then cool for at least 45 minutes. What is the latest time she can start frying? (Lesson 1-1)

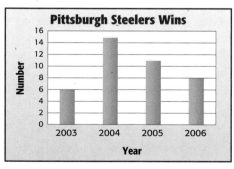

Pittsburgh Steelers Wins

Source: National Football League

▷ **GET READY for the Next Lesson**

PREREQUISITE SKILL Multiply.

47. $2 \cdot 2$ **48.** $3 \cdot 3$ **49.** $5 \cdot 5$ **50.** $7 \cdot 7$

1-3 Squares and Square Roots

MAIN IDEA

Find squares of numbers and square roots of perfect squares.

NYS Core Curriculum

7.N.15 Recognize and state the value of the square root of a perfect square (up to 225)

New Vocabulary

square
perfect squares
square root
radical sign

NY Math Online

glencoe.com

- Extra Examples
- Personal Tutor
- Self-Check Quiz

▷ MINI Lab

A square with an area of 36 square units is shown.

1. Using tiles, try to construct squares with areas of 4, 9, and 16 square units.
2. Try to construct squares with areas 12, 18, and 20 square units.
3. Which of the areas form squares?
4. What is the relationship between the lengths of the sides and the areas of these squares?
5. Using your square tiles, create a square that has an area of 49 square units. What are the lengths of the sides of the square?

The area of the square at the right is $5 \cdot 5$ or 25 square units. The product of a number and itself is the **square** of that number. So, the square of 5 is 25.

5 units   25 units²
5 units

EXAMPLES   Find Squares of Numbers

① Find the square of 3.

$3 \cdot 3 = 9$ Multiply 3 by itself.

9 units² 3 units
3 units

② Find the square of 28.

METHOD 1	Use paper and pencil.
28	Multiply 28 by itself.
× 28	
224	
+ 560	Annex a zero.
784	

METHOD 2	Use a calculator.
28 $\boxed{x^2}$ $\boxed{\text{ENTER}}$ 784	

✓ CHECK Your Progress

Find the square of each number.

a. 8   b. 12   c. 23

Numbers like 9, 16, and 225 are called square numbers or **perfect squares** because they are squares of whole numbers.

The factors multiplied to form perfect squares are called **square roots**. A **radical sign**, $\sqrt{}$, is the symbol used to indicate a square root of a number.

Reading Math

Square Roots Read $\sqrt{16} = 4$ as *the square root of 16 is 4.*

Square Root	Key Concept

Words A square root of a number is one of its two equal factors.

Examples   **Numbers**   **Algebra**

$4 \cdot 4 = 16$, so $\sqrt{16} = 4$.   If $x \cdot x$ or $x^2 = y$, then $\sqrt{y} = x$.

EXAMPLES Find Square Roots

 Find $\sqrt{81}$.

$9 \cdot 9 = 81$, so $\sqrt{81} = 9$.   What number times itself is 81?

 Find $\sqrt{225}$.

[2nd] [$\sqrt{}$] 225 [ENTER] 15

So, $\sqrt{225} = 15$.

CHECK Your Progress

Find each square root.

d. $\sqrt{64}$ e. $\sqrt{289}$

Real-World EXAMPLE

5 **SPORTS** The infield of a baseball field is a square with an area of 8,100 square feet. What are the dimensions of the infield?

The infield is a square. By finding the square root of the area, 8,100, you find the length of one side of the infield.

$90 \cdot 90 = 8,100$, so $\sqrt{8,100} = 90$.

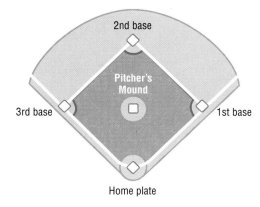

The length of one side of the infield is 90 feet. So, the dimensions of the infield are 90 feet by 90 feet.

CHECK Your Progress

f. **SPORTS** The largest ring in amateur boxing is a square with an area of 400 square feet. What are the dimensions of the ring?

Examples 1, 2
(p. 34)

Find the square of each number.

1. 6   2. 10   3. 17   4. 30

Examples 3, 4
(p. 35)

Find each square root.

5. $\sqrt{9}$ 6. $\sqrt{36}$ 7. $\sqrt{121}$ 8. $\sqrt{169}$

Example 5
(p. 35)

9. **ROAD SIGNS** Historic Route 66 from Chicago to Los Angeles is known as the Main Street of America. If the area of a Route 66 sign measures 576 square inches and the sign is a square, what are the dimensions of the sign?

Practice and Problem Solving

HOMEWORK HELP	
For Exercises	See Examples
10–17	1, 2
18–25	3, 4
26–27	5

Find the square of each number.

10. 4   11. 1   12. 7   13. 11
14. 16   15. 20   16. 18   17. 34

Find each square root.

18. $\sqrt{4}$ 19. $\sqrt{16}$ 20. $\sqrt{49}$ 21. $\sqrt{100}$
22. $\sqrt{144}$ 23. $\sqrt{256}$ 24. $\sqrt{529}$ 25. $\sqrt{625}$

26. **MEASUREMENT** Emma's bedroom is shaped like a square. What are the dimensions of the room if the area of the floor is 196 square feet?

27. **SPORTS** For the floor exercise, gymnasts perform their tumbling skills on a mat that has an area of 1,600 square feet. How much room does a gymnast have to run along one side of the mat?

28. What is the square of 12?   29. Find the square of 19.

30. **GARDENING** A square garden has an area of 225 square feet. How much fencing will a gardener need to buy in order to place fencing around the garden?

GEOGRAPHY For Exercises 31–33, refer to the squares in the diagram. They represent the approximate areas of Florida, North Carolina, and Pennsylvania.

31. What is the area of North Carolina in square miles?

32. How much larger is Florida than Pennsylvania?

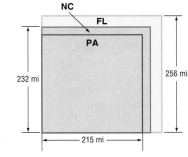

33. The water areas of Florida, North Carolina, and Pennsylvania are 11,881 square miles; 5,041 square miles; and 1,225 square miles, respectively. Make a similar diagram comparing the water areas of these states.

NYSCC • NYSMT

Extra Practice, pp. 668, 704

34. MEASUREMENT A chessboard has an area of 324 square inches. There is a 1-inch border around the 64 squares on the board. What is the length of one side of the region containing the small squares?

35. MEASUREMENT The area of a square that is 7 meters by 7 meters is how much greater than the area of a square containing 8 square meters? Explain.

H.O.T. Problems

36. OPEN ENDED Write a number whose square is between 100 and 150.

CHALLENGE For Exercises 37 and 38, use the diagram shown.

37. Could the area of the dog's pen be made larger using the same amount of fencing? Explain.

38. Describe the largest pen area possible using the same amount of fencing. How do the perimeter and area compare to the original pen?

6 ft

14 ft

39. **WRITING IN MATH** Explain why raising a number to the second power is called *squaring* the number.

NYSMT PRACTICE 7.N.15

40. Which model represents the square of 4?

A   C

B   D

41. Which measure can be the area of a square if the measure of the side length is a whole number?

F 836 sq ft

G 949 sq ft

H 1,100 sq ft

J 1,225 sq ft

Spiral Review

Write each power as a product of the same factor. (Lesson 1-2)

42. 3^4 **43.** 8^5 **44.** 7^2 **45.** 2^6

46. SHIPPING Jocelyn spent a total of $24 to ship 4 packages. If the packages are equal in size and weight, how much did it cost to ship each package? (Lesson 1-1)

▷ **GET READY for the Next Lesson**

PREREQUISITE SKILL Add, subtract, multiply, or divide.

47. $13 + 8$ **48.** $10 - 6$ **49.** 5×6 **50.** $36 \div 4$

Order of Operations

MAIN IDEA

Evaluate expressions using the order of operations.

NYS Core Curriculum

7.N.11 Simplify expressions using order of operations *Note: Expressions may include absolute value and/or integral exponents greater than 0* *Also addresses 7.CM.7, 7.CM.8*

New Vocabulary

numerical expression
order of operations

NY Math Online

glencoe.com

• Extra Examples
• Personal Tutor
• Self-Check Quiz
• Reading in the Content Area

▷ **GET READY for the Lesson**

SPORTS The Kent City football team made one 6-point touchdown and four 3-point field goals in its last game. Megan and Dexter each use an expression to find the total number of points the team scored.

Megan	Dexter
$6 + 4 \cdot 3 = 6 + 12$	$(6 + 4) \cdot 3 = 10 \cdot 3$
$= 18$	$= 30$
The team scored 18 points.	The team scored 30 points.

1. List the differences between their calculations.

2. Whose calculations are correct?

3. Make a conjecture about what should be the first step in simplifying $6 + 4 \cdot 3$.

The expression $6 + 4 \cdot 3$ is a **numerical expression**. To evaluate expressions, use the **order of operations**. These rules ensure that numerical expressions have only one value.

Order of Operations   Key Concept

1. Evaluate the expressions inside grouping symbols.
2. Evaluate all powers.
3. Multiply and divide in order from left to right.
4. Add and subtract in order from left to right.

EXAMPLES Use Order of Operations

1 Evaluate $5 + (12 - 3)$. Justify each step.

$5 + (12 - 3) = 5 + 9$ Subtract first, since $12 - 3$ is in parentheses.
$\qquad\qquad = 14$ Add 5 and 9.

2 Evaluate $8 - 3 \cdot 2 + 7$. Justify each step.

$8 - 3 \cdot 2 + 7 = 8 - 6 + 7$ Multiply 3 and 2.
$\qquad\qquad = 2 + 7$ Subtract 6 from 8.
$\qquad\qquad = 9$ Add 2 and 7.

CHECK Your Progress

Evaluate each expression. Justify each step.

a. $39 \div (9 + 4)$ **b.** $10 + 8 \div 2 - 6$

EXAMPLE Use Order of Operations

3 Evaluate $5 \cdot 3^2 - 7$. Justify each step.

$$5 \cdot 3^2 - 7 = 5 \cdot 9 - 7 \qquad \text{Find the value of } 3^2.$$
$$= 45 - 7 \qquad \text{Multiply 5 and 9.}$$
$$= 38 \qquad \text{Subtract 7 from 45.}$$

✓ **CHECK Your Progress**

c. 3×10^4

d. $(5 - 1)^3 \div 4$

In addition to using the symbols $\times$ and $\cdot$, multiplication can be indicated by using parentheses. For example, $2(3 + 5)$ means $2 \times (3 + 5)$.

EXAMPLE Use Order of Operations

4 Evaluate $14 + 3(7 - 2)$. Justify each step.

$$14 + 3(7 - 2) = 14 + 3(5) \qquad \text{Subtract 2 from 7.}$$
$$= 14 + 15 \qquad \text{Multiply 3 and 5.}$$
$$= 29 \qquad \text{Add 14 and 15.}$$

✓ **CHECK Your Progress**

e. $20 - 2(4 - 1) \cdot 3$

f. $6 + 8 \div 2 + 2(3 - 1)$

Real-World EXAMPLE

5 **MONEY** Julian orders crepe paper, balloons, and favors for the school dance. What is the total cost?

Item	Quantity	Unit Cost
crepe paper	3 rolls	$2
favors	2 boxes	$7
balloons	4 boxes	$5

Words	cost of 3 rolls of crepe paper	+	cost of 4 boxes of balloons	+	cost of 2 boxes of favors
Expression	3×2	+	4×5	+	2×7

$$3 \times 2 + 4 \times 5 + 2 \times 7 = 6 + 20 + 14 \qquad \text{Multiply from left to right.}$$
$$= 40 \qquad \text{Add.}$$

The total cost is $40.

✓ **CHECK Your Progress**

g. What is the total cost of twelve rolls of crepe paper, three boxes of balloons, and three boxes of favors?

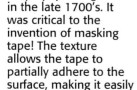

Evaluate each expression. Justify each step.

Examples 1, 2
(p. 38)

1. $8 + (5 - 2)$

2. $25 \div (9 - 4)$

3. $14 - 2 \cdot 6 + 9$

4. $8 \cdot 5 - 4 \cdot 3$

Examples 3, 4
(p. 39)

5. 4×10^2

6. $45 \div (4 - 1)^2$

7. $17 + 2(6 - 3) - 3 \times 4$

8. $22 - 3(8 - 2) + 12 \div 4$

Example 5
(p. 39)

9. **COINS** Isabelle has 3 nickels, 2 quarters, 2 dimes, and 7 pennies. Write an expression that can be used to find how much money Isabelle has altogether. How much money does Isabelle have?

Practice and Problem Solving

For Exercises	**See Examples**
10–17	1, 2
18–23	3
24–27	4
28, 29	5

HOMEWORK HELP

Evaluate each expression. Justify each step.

10. $(1 + 8) \times 3$

11. $10 - (3 + 4)$

12. $(25 \div 5) + 8$

13. $(11 - 2) \div 9$

14. $3 \cdot 2 + 14 \div 7$

15. $4 \div 2 - 1 + 7$

16. $12 + 6 \div 3 - 4$

17. $18 - 3 \cdot 6 + 5$

18. 6×10^2

19. 3×10^4

20. $5 \times 4^3 + 2$

21. $8 \times 7^2 - 6$

22. $8 \div 2 \times 6 + 6^2$

23. $9^2 - 14 \div 7 \cdot 3$

24. $(17 + 3) \div (4 + 1)$

25. $(6 + 5) \cdot (8 - 6)$

26. $6 + 2(4 - 1) + 4 \times 9$

27. $3(4 + 7) - 5 \cdot 4 \div 2$

For Exercises 28 and 29, write an expression for each situation. Then evaluate to find the solution.

28. **MP3 PLAYERS** Reina is buying an MP3 player, a case, three packs of batteries, and six songs. What is the total cost?

29. **BOOKS** Ian goes to the library's used book sale. Paperback books are $0.25, and hardback books are $0.50. If Ian buys 3 paperback books and 5 hardback books, how much does he spend?

Item	Quantity	Unit Cost
MP3 player	1	$200
case	1	$30
pack of batteries	3	$4
songs	6	$2

Evaluate each expression. Justify each step.

30. $(2 + 10)^2 \div 4$

31. $(3^3 + 8) - (10 - 6)^2$

32. $3 \cdot 4(5.2 + 3.8) + 2.7$

33. $7 \times 9 - (4 - 3.2) + 1.8$

34. **MONEY** Suppose that your family orders 2 pizzas, 2 orders of garlic bread, and 1 order of BBQ wings from Mario's Pizza Shop. Write an expression to find the amount of change you would receive from $30. Then evaluate the expression.

Mario's Pizza Shop	
Item	**Cost**
14" pizza	$8
garlic bread	$2
BBQ wings	$4

NYSCC • NYSMT
Extra Practice, pp. 669, 704

H.O.T. Problems

35. **FIND THE ERROR** Phoung and Peggy are evaluating $16 - 24 \div 6 \cdot 2$. Who is correct? Explain your reasoning.

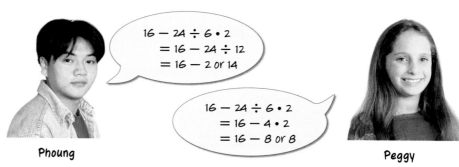

$16 - 24 \div 6 \cdot 2$
$= 16 - 24 \div 12$
$= 16 - 2 \text{ or } 14$

Phoung

$16 - 24 \div 6 \cdot 2$
$= 16 - 4 \cdot 2$
$= 16 - 8 \text{ or } 8$

Peggy

36. **CHALLENGE** Insert parentheses to make $72 \div 9 + 27 - 2 = 0$ a true statement.

37. **WRITING IN MATH** Write a real-world problem in which you would need to use the order of operations or a scientific calculator to solve it.

NYSMT PRACTICE 7.N.11

38. Simplify $3^2 + 9 \div 3 + 3$.

A 3   C 15

B 9   D 18

39. Grace has 2 boxes that contain 24 straws each and 3 boxes that contain 15 cups each. Which expression *cannot* be used to find the total number of items she has?

F $2(24) + 3(15)$

G $3 \times 15 + 2 \times 24$

H $5 \times (24 + 15)$

J $15 + 15 + 15 + 24 + 24$

40. The steps Alana took to evaluate the expression $4y + 4 \div 4$ when $y = 7$ are shown below.

$4y + 4 \div 4$ when $y = 7$
$4 \times 7 = 28$
$28 + 4 = 32$
$32 \div 4 = 8$

What should Alana have done differently in order to evaluate the expression correctly?

A divided $(28 + 4)$ by (28×4)

B divided $(28 + 4)$ by $(28 + 4)$

C added $(4 \div 4)$ to 28

D added 4 to $(28 \div 4)$

Spiral Review

Find each square root. (Lesson 1-3)

41. $\sqrt{64}$ 42. $\sqrt{2,025}$ 43. $\sqrt{784}$

44. **INTERNET** Each day, Internet users perform 2^5 million searches using a popular search engine. How many searches is this? (Lesson 1-2)

GET READY for the Next Lesson

45. **PREREQUISITE SKILL** A Chinese checkerboard has 121 holes. How many holes can be found on eight Chinese checkerboards? (Lesson 1-1)

1-5 Problem-Solving Investigation

MAIN IDEA: Solve problems using the guess and check strategy.

NYSCC **7.PS.1** Use a variety of strategies to understand new mathematical content and to develop more efficient methods
7.PS.7 Understand that there is no one right way to solve mathematical problems but that different methods have advantages and disadvantages

P.S.I. TEAM +

e-Mail: GUESS AND CHECK

TREVOR: My soccer team held a car wash to help pay for a trip to a tournament. We charged $5 for a car and $7 for an SUV. During the first hour, we washed 10 vehicles and earned $58.

YOUR MISSION: Use guess and check to find how many of each type of vehicle were washed.

Understand	You know car washes are $5 for cars and $7 for SUVs. Ten vehicles were washed for $58.
Plan	Make a guess and check it. Adjust the guess until you get the correct answer.
Solve	Make a guess.
	5 cars and 5 SUVs $\qquad$ $5(5) + 7(5) = \$60$ too high
	Adjust the number of SUVs downward.
	5 cars and 4 SUVs $\qquad$ $5(5) + 7(4) = \$49$ too low
	Adjust the number of cars upward.
	6 cars and 4 SUVs $\qquad$ $5(6) + 7(4) = \$58$ correct ✔
	So, 6 cars and 4 SUVs were washed.
Check	Six cars cost $30, and four SUVs cost $28. Since $\$30 + \$28 = \$58$, the guess is correct.

Analyze The Strategy

1. Explain why you should keep a careful record of each of your guesses.

2. **WRITING IN MATH** Write a problem that could be solved by guess and check. Then write the steps you would take to find the solution to your problem.

Mixed Problem Solving

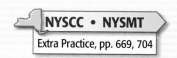
Use the *guess and check* strategy to solve Exercises 3–6.

3. **TICKET SALES** The total ticket sales for the school basketball game were $1,625. Adult tickets were $7, and student tickets were $3. Twice as many students bought tickets as adults. How many adult and student tickets were sold?

4. **NUMBERS** A number is multiplied by 6. Then 4 is added to the product. The result is 82. What is the number?

5. **ANALYZE TABLES** Camila is transferring her home videos onto a DVD. Suppose the DVD holds 60 minutes. Which videos should Camila select to have the maximum time on the DVD without going over?

Video	Time
birthday	25 min 15 s
family picnic	18 min 10 s
holiday	15 min 20 s
vacation	19 min 20 s

6. **MONEY** Susan has $1.60 in change in her purse. If she has an equal number of nickels, dimes, and quarters, how many of each does she have?

Use any strategy to solve Exercises 7–13. Some strategies are shown below.

PROBLEM-SOLVING STRATEGIES
• Guess and check.
• Find a pattern.

7. **BRIDGES** The total length of wire used in the cables supporting the Golden Gate Bridge in San Francisco is about 80,000 miles. This is 5,300 miles longer than three times the distance around Earth at the Equator. What is the distance around Earth at the Equator?

8. **GEOMETRY** What are the next two figures in the pattern?

9. **ALGEBRA** What are the next two numbers in the pattern?

16, 32, 64, 128, 256, ▒ , ▒

10. **FRUIT** Mason places 4 apples and 3 oranges into each fruit masket he makes. If he has used 24 apples and 18 oranges, how many fruit baskets has he made?

11. **ANALYZE TABLES** The table gives the average snowfall, in inches, for Valdez, Alaska, for the months of October through April.

Month	Snowfall
October	11.6
November	40.3
December	73.0
January	65.8
February	59.4
March	52.0
April	22.7

Source: National Climatic Data Center

How many inches total of snowfall could a resident of Valdez expect to receive from October to April?

12. **ROLLER COASTERS** The Jackrabbit roller coaster can handle 1,056 passengers per hour. The coaster has 8 vehicles. If each vehicle carries 4 passengers, how many runs are made in one hour?

13. **NUMBERS** Della is thinking of 3 numbers from 1 through 9 with a product of 36. Find the numbers.

1-6 Algebra: Variables and Expressions

MAIN IDEA

Evaluate simple algebraic expressions.

NYS Core Curriculum

7.N.11 Simplify expressions using order of operations
Note: Expressions may include absolute value and/or integral exponents greater than 0
7.A.1 Translate two-step verbal expressions into algebraic expressions
Also addresses 7.PS.3, 7.A.8, 7.CM.4, 7.R.6

New Vocabulary

variable
algebra
algebraic expression
coefficient

NY Math Online

glencoe.com

• Concepts In Motion
• Extra Examples
• Personal Tutor
• Self-Check Quiz

▷ **MINI Lab**

A pattern of squares is shown.

1. Draw the next three figures in the pattern.
2. Find the number of squares in each figure and record your data in a table like the one shown below. The first three are completed for you.

Figure	1	2	3	4	5	6
Number of Squares	3	4	5	▦	▦	▦

3. Without drawing the figure, determine how many squares would be in the 10th figure. Check by making a drawing.
4. Find a relationship between the figure and its number of squares.

In the Mini Lab, you found that the number of squares in the figure is two more than the figure number. You can use a placeholder, or variable, to represent the number of squares. A **variable** is a symbol that represents an unknown quantity.

$$\text{figure number} \longrightarrow \boldsymbol{n + 2}$$
$$\underset{\longrightarrow \ \text{number of squares}}{}$$

The branch of mathematics that involves expressions with variables is called **algebra**. The expression $n + 2$ is called an **algebraic expression** because it contains variables, numbers, and at least one operation.

EXAMPLE Evaluate an Algebraic Expression

① Evaluate $n + 3$ if $n = 4$.

$n + 3 = 4 + 3$ Replace n with 4.
$\quad\ = 7$ Add 4 and 3.

✓ **CHECK Your Progress**

Evaluate each expression if $c = 8$ and $d = 5$.

a. $c - 3$ b. $15 - c$ c. $c + d$

In algebra, the multiplication sign is often omitted.

6d
↑
6 times *d*

9st
↑
9 times *s* **times** *t*

mn
↑
m **times** *n*

The numerical factor of a multiplication expression that contains a variable is called a **coefficient**. So, 6 is the coefficient of 6*d*.

EXAMPLES Evaluate Expressions

2 **Evaluate** $8w - 2v$ **if** $w = 5$ **and** $v = 3$.

$$8w - 2v = 8(5) - 2(3)$$ Replace *w* with 5 and *v* with 3.

$$= 40 - 6$$ Do all multiplications first.

$$= 34$$ Subtract 6 from 40.

3 **Evaluate** $4y^2 + 2$ **if** $y = 3$.

$$4y^2 + 2 = 4(3)^2 + 2$$ Replace *y* with 3.

$$= 4(9) + 2$$ Evaluate the power.

$$= 38$$ Multiply, then add.

CHECK Your Progress

d. $9a - 6b$ **e.** $\dfrac{ab}{2}$ **f.** $2a^2 + 5$

The fraction bar is a grouping symbol. Evaluate the expressions in the numerator and denominator separately before dividing.

Real-World EXAMPLE

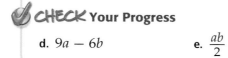

4 **HEALTH** Use the formula at the left to find Latrina's minimum training heart rate if she is 15 years old.

$$\frac{3(220 - a)}{5} = \frac{3(220 - 15)}{5}$$ Replace *a* with 15.

$$= \frac{3(205)}{5}$$ Subtract 15 from 220.

$$= \frac{615}{5}$$ Multiply 3 and 205.

$$= 123$$ Divide 615 by 5.

Latrina's minimum training heart rate is 123 beats per minute.

CHECK Your Progress

g. **MEASUREMENT** To find the area of a triangle, you can use the formula $\dfrac{bh}{2}$, where *h* is the height and *b* is the base. What is the area in square inches of a triangle with a height of 6 inches and base of 8 inches?

Real-World Link . . . . .
Athletic trainers use the formula $\dfrac{3(220 - a)}{5}$, where *a* is a person's age, to find their minimum training heart rate.
Source: CMPMedica Ltd.

Example 1
(p. 44)

Evaluate each expression if $a = 3$ and $b = 5$.

1. $a + 7$ **2.** $8 - b$ **3.** $b - a$

4. HEALTH The standard formula for finding your maximum heart rate is $220 - a$, where a represents a person's age in years. What is your maximum heart rate?

Examples 2–4
(p. 45)

Evaluate each expression if $m = 2$, $n = 6$, and $p = 4$.

5. $6n - p$ **6.** $7m - 2n$ **7.** $3m + 4p$

8. $n^2 + 5$ **9.** $15 - m^3$ **10.** $3p^2 - n$

11. $\dfrac{mn}{4}$ **12.** $\dfrac{3n}{9}$ **13.** $\dfrac{5n + m}{8}$

Practice and Problem Solving

HOMEWORK HELP	
For Exercises	**See Examples**
14–29	1–3
30–31	4

Evaluate each expression if $d = 8$, $e = 3$, $f = 4$, and $g = 1$.

14. $d + 9$ **15.** $10 - e$ **16.** $4f + 1$ **17.** $8g - 3$

18. $f - e$ **19.** $d + f$ **20.** $10g - 6$ **21.** $8 + 5d$

22. $\dfrac{d}{5}$ **23.** $\dfrac{16}{f}$ **24.** $\dfrac{5d - 25}{5}$ **25.** $\dfrac{(5 + g)^2}{2}$

26. $6f^2$ **27.** $4e^2$ **28.** $d^2 + 7$ **29.** $e^2 - 4$

30. BOWLING The expression $5n + 2$ can be used to find the total cost in dollars of bowling where n is the number of games bowled. How much will it cost Vincent to bowl 3 games?

31. HEALTH The expression $\dfrac{w}{30}$, where w is a person's weight in pounds, is used to find the approximate number of quarts of blood in the person's body. How many quarts of blood does a 120-pound person have?

Evaluate each expression if $x = 3.2$, $y = 6.1$, and $z = 0.2$.

32. $x + y - z$ **33.** $14.6 - (x + y + z)$ **34.** $xz + y^2$

35. CAR RENTAL A car rental company charges $19.99 per day and $0.17 per mile to rent a car. Write an expression that gives the total cost in dollars to rent a car for d days and m miles.

36. MUSIC A Web site charges $0.99 to download a song onto an MP3 player and $12.49 to download an entire album. Write an expression that gives the total cost in dollars to download a albums and s songs.

37. SCIENCE The expression $\frac{32t^2}{2}$ gives the falling distance of an object in feet after t seconds. How far would a bungee jumper fall 2 seconds after jumping?

38. GEOMETRY To find the total number of diagonals for any given polygon, you can use the expression $\frac{n(n-3)}{2}$, where n is the number of sides of the polygon. What is the total number of diagonals for a 10-sided polygon?

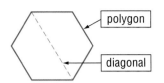

NYSCC • NYSMT
Extra Practice, pp. 669, 704

H.O.T. Problems

39. OPEN ENDED Write an algebraic expression with the variable x that has a value of 3 when evaluated.

40. CHALLENGE Name values of x and y so that the value of $5x + 3$ is greater than the value of $2y + 14$.

41. WRITING IN MATH Tell whether the statement below is *sometimes*, *always*, or *never* true. Justify your reasoning.

The expressions $x - 3$ and $y - 3$ represent the same value.

NYSMT PRACTICE 7.A.1

42. Which expression could be used to find the cost of buying b books at $7.95 each and m magazines at $4.95 each?

A $7.95b + 4.95m$

B $7.95b - 4.95m$

C $12.9(b + m)$

D $12.9(bm)$

43. Tonya has x quarters, y dimes, and z nickels in her pocket. Which of the following expressions gives the total amount of change she has in her pocket?

F $\$0.25x + \$0.05y + \$0.10z$

G $\$0.25x + \$0.10y + \$0.05z$

H $\$0.05x + \$0.25y + \$0.10z$

J $\$0.10x + \$0.05y + \$0.25z$

Spiral Review

44. SHOPPING A grocery store sells hot dog buns in packages of 8 and 12. How many 8-packs and 12-packs could you buy if you needed 44 hot dog buns? Use the *guess and check* strategy. (Lesson 1-5)

Evaluate each expression. (Lesson 1-4)

45. $6(5) - 2$

46. $9 + 9 \div 3$

47. $4 \cdot 2(8 - 1)$

48. $(17 + 3) \div 5$

49. Find $\sqrt{361}$. (Lesson 1-3)

▷ **GET READY** for the Next Lesson

PREREQUISITE SKILL Determine whether each sentence is *true* or *false*. (Lesson 1-4)

50. $15 - 2(3) = 9$

51. $20 \div 5 \times 4 = 1$

52. $4^2 + 6 \cdot 7 = 154$

1. **MULTIPLE CHOICE** A cycling club is planning an 1,800-mile trip. The cyclers average 15 miles per hour. What additional information is needed to determine the number of days it will take them to complete the trip? (Lesson 1-1)

   A The number of cyclists in the club

   B The number of miles of rough terrain

   C The number of hours they plan to cycle each day

   D Their average speed per minute

Write each power as a product of the same factor. (Lesson 1-2)

2. 4^5

3. 9^6

4. **OCEANS** The world's largest ocean, the Pacific Ocean, covers approximately 4^3 million square miles. Write this area in standard form. (Lesson 1-2)

5. **ZOOS** The Lincoln Park Zoo in Illinois is $2 \cdot 2 \cdot 2 \cdot 2 \cdot 2 \cdot 2 \cdot 2$ years old. Write this age in exponential form. (Lesson 1-2)

6. **MULTIPLE CHOICE** The model below represents $\sqrt{49} = 7$.

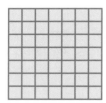

   Which arrangement of small squares can be used to model a large square that represents $\sqrt{324}$? (Lesson 1-3)

   F 9 rows of 36 squares

   G 18 rows of 18 squares

   H 12 rows of 27 squares

   J 6 rows of 54 squares

Find the square of each number. (Lesson 1-3)

7. 4

8. 12

Find each square root. (Lesson 1-3)

9. $\sqrt{64}$

10. $\sqrt{289}$

11. **LANDSCAPING** A bag of lawn fertilizer covers 2,500 square feet. Describe the largest square that one bag of fertilizer could cover. (Lesson 1-3)

Evaluate each expression. (Lesson 1-4)

12. $25 - (3^2 + 2 \times 5)$

13. $\dfrac{2(7 - 3)}{2^2}$

14. **MEASUREMENT** The perimeter of a rectangle is 42 inches, and its area is 104 square inches. Find the dimensions of the rectangle. Use the *guess and check* strategy. (Lesson 1-5)

15. **MULTIPLE CHOICE** Ana buys some baseball bats at $35 each and some baseball gloves at $48 each. Which expression could be used to find the total cost of the sports items? (Lesson 1-6)

   A $35b \cdot 48g$

   B $\dfrac{35b}{48g}$

   C $35b + 48g$

   D $48g - 35b$

Evaluate each expression if $x = 12$, $y = 4$, and $z = 8$. (Lesson 1-6)

16. $x - 5$

17. $3y + 10z$

18. $\dfrac{yz}{2}$

19. $\dfrac{(y + 8)^2}{x}$

20. **HEALTH** A nurse can use the expression $110 + \dfrac{A}{2}$, where A is a person's age, to estimate a person's normal systolic blood pressure. Estimate the normal systolic blood pressure for a 16-year-old. (Lesson 1-6)

Algebra: Equations

1-7

MAIN IDEA

Write and solve equations using mental math.

NYS Core Curriculum

7.PS.3 Understand and demonstrate how written symbols represent mathematical ideas

New Vocabulary

equation
solution
solving an equation
defining the variable

NY Math Online

glencoe.com

• Extra Examples
• Personal Tutor
• Self-Check Quiz

▷ **GET READY for the Lesson**

VOLLEYBALL The table shows the number of wins for six women's college volleyball teams.

1. Suppose each team played 34 games. How many losses did each team have?

2. Write a rule to describe how you found the number of losses.

3. Let w represent the number of wins and ℓ represent the number of losses. Rewrite your rule using numbers, variables, and an equals sign.

Women's College Volleyball		
Team	**Wins**	**Losses**
Bowling Green State University	28	■
Kent State University	13	■
Ohio University	28	■
University of Akron	7	■
University at Buffalo	14	■
Miami University	13	■

Source: Mid-American Conference

An **equation** is a sentence that contains two expressions separated by an equals sign, $=$. The equals sign tells you that the expression on the left is equivalent to the expression on the right.

$$7 = 8 - 1 \qquad 3(4) = 12 \qquad 17 = 13 + 2 + 2$$

An equation that contains a variable is neither true nor false until the variable is replaced with a number. A **solution** of an equation is a numerical value for the variable that makes the sentence true.

The process of finding a solution is called **solving an equation**. Some equations can be solved using mental math.

EXAMPLE **Solve an Equation Mentally**

① Solve $18 = 14 + t$ mentally.

$18 = 14 + t$ Write the equation.

$18 = 14 + 4$ You know that $14 + 4$ is 18.

$18 = 18$ Simplify.

So, $t = 4$. The solution is 4.

 CHECK Your Progress

Solve each equation mentally.

a. $p - 5 = 20$ **b.** $8 = y \div 3$ **c.** $7h = 56$

2 Each day, Sierra cycles 3 miles on a bicycle trail. The equation $3d = 36$ represents how many days it will take her to cycle 36 miles. How many days *d* will it take her to cycle 36 miles?

A 10   **B** 12   **C** 15   **D** 20

Read the Item

Solve $3d = 36$ to find how many days it will take to cycle 36 miles.

Solve the Item

$3d = 36$ Write the equation.

$3 \cdot 12 = 36$ You know that 3 • 12 is 36.

Therefore, $d = 12$. The answer is B.

✓ CHECK Your Progress

d. Jordan has 16 video games. This is 3 less than the number Casey has. To find how many video games Casey has, the equation $v - 3 = 16$ can be used. How many video games *v* does Casey have?

     F 13   **G** 15   **H** 18   **J** 19

Choosing a variable to represent an unknown quantity is called **defining the variable**.

Real-World EXAMPLE

3 **WHALES** Each winter, Humpback whales migrate 1,500 miles to the Indian Ocean. However, one whale migrated 5,000 miles in one season. How many miles farther than normal did this whale travel?

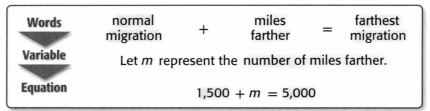

Words	normal migration	+	miles farther	=	farthest migration
Variable	Let *m* represent the number of miles farther.				
Equation	$1{,}500 + m = 5{,}000$				

$1{,}500 + m = 5{,}000$ Write the equation.

$1{,}500 + 3{,}500 = 5{,}000$ Replace *m* with 3,500 to make the equation true.

So, $m = 3{,}500$. The whale went 3,500 miles farther than normal.

✓ CHECK Your Progress

e. Aaron buys a movie rental, popcorn, and a soft drink for a total cost of $6.25. What is the cost of the popcorn if the movie rental and soft drink cost $4.70 together?

Example 1
(p. 49)

Solve each equation mentally.

1. $75 = w + 72$
2. $y - 18 = 20$
3. $\dfrac{r}{9} = 6$

Example 2
(p. 50)

4. **MULTIPLE CHOICE** Daniel scored 7 points in a football game. Together, he and Judah scored 28 points. Solve the equation $7 + p = 28$ to find how many points p Judah scored.

    A 14   **B** 21   **C** 23   **D** 35

Example 3
(p. 50)

5. **MONEY** Jessica buys a notebook and a pack of pencils for a total of $3.50. What is the cost of the notebook if the pack of pencils costs $1.25?

Practice and Problem Solving

HOMEWORK HELP

For Exercises	See Examples
6–17	1
18–19 33–34	2
20–21	3

Solve each equation mentally.

6. $b + 7 = 13$
7. $8 + x = 15$
8. $y - 14 = 20$
9. $a - 18 = 10$
10. $25 - n = 19$
11. $x + 17 = 63$
12. $77 = 7t$
13. $3d = 99$
14. $n = \dfrac{30}{6}$
15. $16 = \dfrac{u}{4}$
16. $20 = y \div 5$
17. $84 \div z = 12$

18. **MONEY** Rosa charges $9 per hour of baby-sitting. Solve the equation $9h = 63$ to find how many hours h Rosa needs to baby-sit to earn $63.

19. **SNACKS** A box initially contained 25 snack bars. There are 14 snack bars remaining. Solve the equation $25 - x = 14$ to find how many snack bars x were eaten.

For Exercises 20 and 21, define a variable. Then write and solve an equation.

20. **BASKETBALL** During one game of his rookie year, LeBron James scored 41 of the Cleveland Cavaliers' 107 points. How many points did the rest of the team score?

21. **EXERCISE** On Monday and Tuesday, Derrick walked a total of 6.3 miles. If he walked 2.5 miles on Tuesday, how many miles did he walk on Monday?

Solve each equation mentally.

22. $1.5 + j = 10.0$
23. $1.2 = m - 4.2$
24. $n - 1.4 = 3.5$
25. $13.4 - h = 9.0$
26. $9.9 + r = 24.2$
27. $w + 15.8 = 17.0$

28. **CATS** The table shows the average weight of lions. Write and solve an addition equation to find how much more male lions weigh than female lions.

Lions	Weight (lb)
Female	275
Male	372

NYSCC • NYSMT

Extra Practice, pp. 670, 704

29. **FOOD** The total cost of a chicken sandwich and a drink is $6.25. The drink costs $1.75. Write and solve an equation that can be used to find how much the chicken sandwich is alone.

H.O.T. Problems

30. **CHALLENGE** Find the values of a and b if $0 \cdot a = b$. Explain your reasoning.

31. **FIND THE ERROR** Justin and Antonio each solved $w - 35 = 70$. Whose solution is correct? Explain your reasoning.

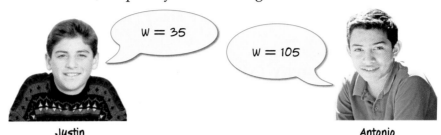

Justin   Antonio

32. **WRITING IN MATH** Explain what it means to solve an equation.

NYSMT PRACTICE 7.PS.3

33. The diagram shows the distance from Madison to Hudson and from Lawrence to Hudson. Which equation can be used to find how many more miles x Lawrence is from Madison?

Lawrence   Madison   Hudson

├─── 36 mi ───┤

├──── 58 mi ────┤

A $58 = x + 36$ **C** $36 \cdot 58 = x$

B $58 = \frac{x}{36}$ **D** $x - 36 = 58$

34. **SHORT RESPONSE** What value of h makes the following equation true?

$$h \div 4 = 32$$

35. Solve $u + 8 = 15$.

F 23   **H** 8

G 22   **J** 7

Spiral Review

36. **ALGEBRA** Evaluate $3a + b^2$ if $a = 2$ and $b = 3$. (Lesson 1-6)

Evaluate each expression. (Lesson 1-4)

37. $11 \cdot 6 \div 3 + 9$ 38. $5 \cdot 13 - 6^2$ 39. $1 + 2(8 - 5)^2$

40. **FARMING** A farmer planted 389 acres of land with 78,967 corn plants. How many plants were planted per acre? (Lesson 1-1)

▷ **GET READY for the Next Lesson**

PREREQUISITE SKILL **Multiply.** (Lesson 1-4)

41. $2 \cdot (4 + 10)$ 42. $(9 \cdot 1) \cdot 8$ 43. $(5 \cdot 3)(5 \cdot 2)$ 44. $(6 + 8) \cdot 12$

1-8 Algebra: Properties

MAIN IDEA

Use Commutative, Associative, Identity, and Distributive properties to solve problems.

NYS Core Curriculum

Reinforcement of 6.N.2 Define and identify the commutative and associative properties of addition and multiplication
Reinforcement of 6.N.3 Define and identify the distributive property of multiplication over addition

New Vocabulary

equivalent expressions
properties

NY Math Online

glencoe.com
• Extra Examples
• Personal Tutor
• Self-Check Quiz

▷ **GET READY** for the Lesson

MUSEUMS The admission costs for the Louisville Science Center are shown.

Louisville Science Center Admission	
Admission	$12
IMAX Movie	$8

Source: Louisville Science Center

1. Find the total cost of admission and a movie ticket for a 4-person family.

2. Describe the method you used to find the total cost.

Here are two ways to find the total cost.

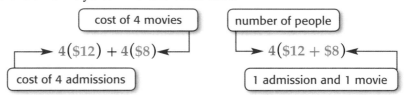

The expressions 4($12) + 4($8) and 4($12 + $8) are **equivalent expressions** because they have the same value, $80. This shows how the **Distributive Property** combines addition and multiplication.

Distributive Property   Key Concept

Words	To multiply a sum by a number, multiply each addend of the sum by the number outside the parentheses.

Examples	**Numbers**	**Algebra**
	$3(4 + 6) = 3(4) + 3(6)$	$a(b + c) = a(b) + a(c)$
	$5(7) + 5(3) = 5(7 + 3)$	$a(b) + a(c) = a(b + c)$

EXAMPLES Write Sentences as Equations

Use the Distributive Property to rewrite each expression. Then evaluate it.

1 $5(3 + 2)$

$5(3 + 2) = 5(3) + 5(2)$
$\qquad = 15 + 10$ Multiply.
$\qquad = 25$ Add.

2 $3(7) + 3(4)$

$3(7) + 3(4) = 3(7 + 4)$
$\qquad = 3(11)$ Add.
$\qquad = 33$ Multiply.

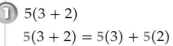 **CHECK** Your Progress

a. $6(1 + 4)$

b. $6(9) + 6(3)$

Real-World EXAMPLE

 TOUR DE FRANCE The Tour de France is a cycling race through France that lasts 22 days. If a cyclist averages 90 miles per day, about how far does he travel?

Use the Distributive Property to multiply 90×22 mentally.

$90(22) = 90(20 + 2)$	Rewrite 22 as $20 + 2$.
$= 90(20) + 90(2)$	Distributive Property
$= 1{,}800 + 180$	Multiply.
$= 1{,}980$	Add.

The cyclist travels about 1,980 miles.

✓ CHECK Your Progress

c. Jennifer saved $120 each month for five months. How much did she save in all? Explain your reasoning.

Properties are statements that are true for all numbers.

Real Number Properties   **Concept Summary**

Commutative Properties	The order in which two numbers are added or multiplied does not change their sum or product. $$a + b = b + a \qquad a \cdot b = b \cdot a$$
Associative Properties	The way in which three numbers are grouped when they are added or multiplied does not change their sum or product. $$a + (b + c) = (a + b) + c \qquad a \cdot (b \cdot c) = (a \cdot b) \cdot c$$
Identity Properties	The sum of an addend and 0 is the addend. The product of a factor and 1 is the factor. $$a + 0 = a \qquad a \cdot 1 = a$$

EXAMPLE Use Properties to Evaluate Expressions

④ Find $4 \cdot 12 \cdot 25$ mentally. Justify each step.

$4 \cdot 12 \cdot 25 = 4 \cdot 25 \cdot 12$	Commutative Property of Multiplication
$= (4 \cdot 25) \cdot 12$	Associative Property of Multiplication
$= 100 \cdot 12$ or $1{,}200$	Multiply 100 and 12 mentally.

✓ CHECK Your Progress

Find each of the following. Justify each step.

d. $40 \cdot (7 \cdot 5)$ e. $(89 + 15) + 1$

Examples 1, 2
(p. 53)

Use the Distributive Property to rewrite each expression. Then evaluate it.

1. $7(4 + 3)$ 2. $5(6 + 2)$ 3. $3(9) + 3(6)$ 4. $6(17) + 6(3)$

Example 3
(p. 54)

5. **MENTAL MATH** Admission to a baseball game is $12, and a hot dog costs $5. Use the Distributive Property to mentally find the total cost for 4 tickets and 4 hot dogs. Explain your reasoning.

6. **MENTAL MATH** A cheetah can run 65 miles per hour at maximum speed. At this rate, how far could a cheetah run in 2 hours? Use the Distributive Property to multiply mentally. Explain your reasoning.

Example 4
(p. 54)

Find each expression mentally. Justify each step.

7. $44 + (23 + 16)$ 8. $50 \cdot (33 \cdot 2)$

Practice and Problem Solving

For Exercises	See Examples
9–12	1, 2
13–22	4
23, 24	3

HOMEWORK HELP

Use the Distributive Property to rewrite each expression. Then evaluate it.

9. $2(6 + 7)$ 10. $5(8 + 9)$ 11. $4(3) + 4(8)$ 12. $7(3) + 7(6)$

Find each expression mentally. Justify each step.

13. $(8 + 27) + 52$ 14. $(13 + 31) + 17$

15. $91 + (15 + 9)$ 16. $85 + (46 + 15)$

17. $(4 \cdot 18) \cdot 25$ 18. $(5 \cdot 3) \cdot 8$

19. $15 \cdot (8 \cdot 2)$ 20. $2 \cdot (16 \cdot 50)$

21. $5 \cdot (30 \cdot 12)$ 22. $20 \cdot (48 \cdot 5)$

MENTAL MATH For Exercises 23 and 24, use the Distributive Property to multiply mentally. Explain your reasoning.

23. **TRAVEL** Each year about 27 million people visit Paris, France. About how many people will visit Paris over a five-year period?

24. **ROLLER COASTERS** One ride on a roller coaster lasts 108 seconds. How long will it take to ride this coaster three times?

The Distributive Property also can be applied to subtraction. Use the Distributive Property to rewrite each expression. Then evaluate it.

25. $7(9) - 7(3)$ 26. $12(8) - 12(6)$ 27. $9(7) - 9(3)$ 28. $6(12) - 6(5)$

ALGEBRA Use one or more properties to rewrite each expression as an equivalent expression that does not use parentheses.

29. $(y + 1) + 4$ 30. $2 + (x + 4)$ 31. $4(8b)$ 32. $(3a)2$

33. $2(x + 3)$ 34. $4(2 + b)$ 35. $6(c + 1)$ 36. $3(f + 4) + 2f$

MILEAGE For Exercises 37 and 38, use the table that shows the driving distance between certain cities in Pennsylvania.

37. Write a sentence that compares the mileage from Pittsburgh to Johnstown to Allentown, and the mileage from Allentown to Johnstown to Pittsburgh.

From	To	Driving Distance (mi)
Pittsburgh	Johnstown	55
Johnstown	Allentown	184

NYSCC • NYSMT
Extra Practice, pp. 670, 704

38. Name the property that is illustrated by this sentence.

H.O.T. Problems

39. **OPEN ENDED** Write an equation that illustrates the Associative Property of Addition.

40. **NUMBER SENSE** Analyze the statement $(18 + 35) \times 4 = 18 + 35 \times 4$. Then tell whether the statement is *true* or *false*. Explain your reasoning.

41. **CHALLENGE** A *counterexample* is an example showing that a statement is not true. Provide a counterexample to the following statement.

Division of whole numbers is associative.

42. **WRITING IN MATH** Write about a real-world situation that can be solved using the Distributive Property. Then use it to solve the problem.

NYSMT PRACTICE 6.N.3

43. Which expression can be written as $6(9 + 8)$?

    A $8 \cdot 6 + 8 \cdot 9$

    B $6 \cdot 9 + 6 \cdot 8$

    C $6 \cdot 9 \cdot 6 \cdot 8$

    D $6 + 9 \cdot 6 + 8$

44. Jared deposited $5 into his savings account. Six months later, his account balance had doubled. If his old balance was b dollars, which of the following would be equivalent to his new balance of $2(b + 5)$ dollars?

    F $2b + 5$ H $b + 10$

    G $2b + 7$ J $2b + 10$

Spiral Review

Name the number that is the solution of the given equation. (Lesson 1-7)

45. $7.3 = t - 4$; 10.3, 11.3, 12.3

46. $35.5 = 5n$; 5.1, 7.1, 9.1

47. **CATS** It is believed that a cat ages 5 human years for every calendar year. This situation can be represented by the expression $5y$ where y is the age of the cat in calendar years. Find the human age of a cat that has lived for 15 calendar years. (Lesson 1-6)

48. Evaluate $(14 - 9)^4$. (Lesson 1-4)

▷ **GET READY for the Next Lesson**

PREREQUISITE SKILL Find the next number in each pattern.

49. 2, 4, 6, 8, ▉

50. 10, 21, 32, 43, ▉

51. 1.4, 2.2, 3.0, 3.8, ▉

Algebra: Arithmetic Sequences

1-9

MINI Lab

Use centimeter cubes to make the three figures shown.

Figure 1   Figure 2   Figure 3

1. How many centimeter cubes are used to make each figure?

2. What pattern do you see? Describe it in words.

3. Suppose this pattern continues. Copy and complete the table to find the number of cubes needed to make each figure.

Figure	1	2	3	4	5	6	7	8
Cubes Needed	4	8	12					

4. How many cubes would you need to make the 10th figure? Explain your reasoning.

A **sequence** is an ordered list of numbers. Each number in a sequence is called a **term**. In an **arithmetic sequence**, each term is found by adding the same number to the previous term. An example of an arithmetic sequence is shown.

$$8, \ 11, \ 14, \ 17, \ 20, \ \ldots$$
$$+3 \ \ +3 \ \ +3 \ \ +3$$

Each term is found by adding 3 to the previous term.

MAIN IDEA

Describe the relationships and extend terms in arithmetic sequences.

NYS Core Curriculum

7.PS.4 Observe patterns and formulate generalizations
7.A.8 Create algebraic patterns using charts/tables, graphs, equations, and expressions *Also addresses 7.CN.4*

New Vocabulary

sequence
term
arithmetic sequence

NY Math Online

glencoe.com
• Concepts In Motion
• Extra Examples
• Personal Tutor
• Self-Check Quiz

EXAMPLE   Describe and Extend Sequences

① Describe the relationship between the terms in the arithmetic sequence 8, 13, 18, 23, … Then write the next three terms in the sequence.

$$8, \ 13, \ 18, \ 23, \ \ldots$$
$$+5 \ \ +5 \ \ +5$$

Each term is found by adding 5 to the previous term. Continue the pattern to find the next three terms.

$$23 + 5 = 28 \qquad 28 + 5 = 33 \qquad 33 + 5 = 38$$

The next three terms are 28, 33, and 38.

✓ CHECK Your Progress

Describe the relationship between the terms in each arithmetic sequence. Then write the next three terms in the sequence.

a. 0, 13, 26, 39, …

b. 4, 7, 10, 13 …

Arithmetic sequences can also involve decimals.

 EXAMPLE Describe and Extend Sequences

2 Describe the relationship between the terms in the arithmetic sequence 0.4, 0.6, 0.8, 1.0, ... Then write the next three terms in the sequence.

$$0.4, \ 0.6, \ 0.8, \ 1.0, \ ...$$
$$+0.2 \ +0.2 \ +0.2$$

Each term is found by adding 0.2 to the previous term. Continue the pattern to find the next three terms.

$$1.0 + 0.2 = 1.2 \qquad 1.2 + 0.2 = 1.4 \qquad 1.4 + 0.2 = 1.6$$

The next three terms are 1.2, 1.4, and 1.6.

 CHECK Your Progress

Describe the relationship between the terms in each arithmetic sequence. Then write the next three terms in the sequence.

c. 1.0, 1.3, 1.6, 1.9, ...   d. 2.5, 3.0, 3.5, 4.0, ...

In a sequence, each term has a specific position within the sequence. Consider the sequence 2, 4, 6, 8, 10, ...

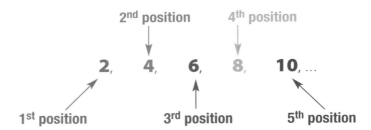

The table below shows the position of each term in this sequence. Notice that as the position number increases by 1, the value of the term increases by 2.

Position	Operation	Value of Term
1	$1 \cdot 2 = 2$	2
2	$2 \cdot 2 = 4$	4
3	$3 \cdot 2 = 6$	6
4	$4 \cdot 2 = 8$	8
5	$5 \cdot 2 = 10$	10

You can also write an algebraic expression to represent the relationship between any term in a sequence and its position in the sequence. In this case, if n represents the position in the sequence, the value of the term is $2n$.

3 **GREETING CARDS** The homemade greeting cards that Meredith makes are sold in boxes at a local gift store. Each week, the store sells five more boxes.

Week 1   Week 2   Week 3

If this pattern continues, what algebraic expression can be used to help her find the total number of boxes sold at the end of the 100th week? Use the expression to find the total.

Make a table to display the sequence.

Position	Operation	Value of Term
1	$1 \cdot 5$	5
2	$2 \cdot 5$	10
3	$3 \cdot 5$	15
n	$n \cdot 5$	$5n$

Each term is 5 times its position number. So, the expression is $5n$.

$5n$ Write the expression.

$5(100) = 500$ Replace n with 100.

So, at the end of 100 weeks, 500 boxes will have been sold.

> **Study Tip**
>
> **Arithmetic Sequences**
> When looking for a pattern between the position number and each term in the sequence, it is often helpful to make a table.

CHECK Your Progress

e. **GEOMETRY** If the pattern continues, what algebraic expression can be used to find the number of circles used in the 50th figure? How many circles will be in the 50th figure?

Figure 1   Figure 2   Figure 3

CHECK Your Understanding

Examples 1, 2
(pp. 57–58)

Describe the relationship between the terms in each arithmetic sequence. Then write the next three terms in each sequence.

1. 0, 9, 18, 27, …

2. 4, 9, 14, 19, …

3. 1, 1.1, 1.2, 1.3, …

4. 5, 5.4, 5.8, 6.2, …

Example 3
(p. 59)

5. **PLANTS** The table shows the height of a certain plant each month after being planted. If this pattern continues, what algebraic expression can be used to find the height of the plant at the end of twelve months? Find the plant's height after 12 months.

Month	Height (in.)
1	3
2	6
3	9
4	12

Describe the relationship between the terms in each arithmetic sequence. Then write the next three terms in each sequence.

6. 0, 7, 14, 21, …   **7.** 1, 7, 13, 19, …   **8.** 26, 34, 42, 50, …

9. 19, 31, 43, 55, …   **10.** 6, 16, 26, 36, …   **11.** 33, 38, 43, 48, …

12. 0.1, 0.4, 0.7, 1.0, …   **13.** 2.4, 3.2, 4.0, 4.8, …   **14.** 2.0, 3.1, 4.2, 5.3, …

15. 4.5, 6.0, 7.5, 9.0, …   **16.** 1.2, 3.2, 5.2, 7.2, …   **17.** 4.6, 8.6, 12.6, 16.6, …

18. COLLECTIONS Hannah is starting a doll collection. Each year, she buys 6 dolls. Suppose she continues this pattern. What algebraic expression can be used to find the number of dolls in her collection after any number of years? How many dolls will Hannah have after 25 years?

19. EXERCISE The table shows the number of laps that Jorge swims each week. Jorge's goal is to continue this pace. What algebraic expression can be used to find the total number of laps he will swim after any given number of weeks? How many laps will Jorge swim after 6 weeks?

Week	Number of Laps
1	7
2	14
3	21
4	28

Describe the relationship between the terms in each arithmetic sequence. Then write the next three terms in each sequence.

20. 18, 33, 48, 63, …   **21.** 20, 45, 70, 95, …   **22.** 38, 61, 84, 107, …

In a *geometric sequence*, each term is found by multiplying the previous term by the same number. Write the next three terms of each geometric sequence.

23. 1, 4, 16, 64, …   **24.** 2, 6, 18, 54, …   **25.** 4, 12, 36, 108, …

26. GEOMETRY Kendra is stacking boxes of tissues for a store display. Each minute, she stacks another layer of boxes. If the pattern continues, how many boxes will be displayed after 45 minutes?

1 Minute

2 Minutes

3 Minutes

NUMBER SENSE Find the 100th number in each sequence.

27. 12, 24, 36, 48, …   **28.** 14, 28, 42, 56, …

29. 0, 50, 100, 150, …   **30.** 0, 75, 150, 225, …

31. RESEARCH The Fibonacci sequence is one of the most well-known sequences in mathematics. Use the Internet or another source to write a paragraph about the Fibonacci sequence.

NYSCC • NYSMT
Extra Practice, pp. 670, 704

CHALLENGE Not all sequences are arithmetic. But, there is still a pattern. Describe the relationship between the terms in each sequence. Then write the next three terms in the sequence.

32. 1, 2, 4, 7, 11, …

33. 0, 2, 6, 12, 20, …

34. OPEN ENDED Write five terms of an arithmetic sequence and describe the rule for finding the terms.

35. SELECT A TOOL Suppose you want to begin saving $15 each month. Which of the following tools would you use to determine the amount you will have saved after 2 years? Justify your selection(s). Then use the tool(s) to solve the problem.

paper/pencil	real object	technology

36. WRITING IN MATH Janice earns $6.50 per hour running errands for her neighbor. Explain how the hourly earnings form an arithmetic sequence.

NYSMT PRACTICE   7.PS.4, 7.A.8

37. Which sequence follows the rule $3n - 2$, where n represents the position of a term in the sequence?

A 21, 18, 15, 12, 9, …

B 3, 6, 9, 12, 15, …

C 1, 7, 10, 13, 16, …

D 1, 4, 7, 10, 13, …

38. Which expression can be used to find the nth term in this sequence?

Position	1	2	3	4	5	n^{th}
Value of Term	2	5	10	17	26	

F $n^2 + 1$

G $2n + 1$

H $n + 1$

J $2n^2 + 2$

Spiral Review

Find each expression mentally. Justify each step. (Lesson 1-8)

39. $(23 + 18) + 7$

40. $5 \cdot (12 \cdot 20)$

Solve each equation mentally. (Lesson 1-7)

41. $f - 26 = 3$

42. $\frac{a}{4} = 8$

43. $30 + y = 50$

44. SCIENCE At normal temperatures, sound travels through water at a rate of $5 \cdot 10^3$ feet per second. Write this rate in standard form. (Lesson 1-2)

▷ GET READY for the Next Lesson

PREREQUISITE SKILL Find the value of each expression. (Lesson 1-6)

45. $2x$ if $x = 4$

46. $d - 5$ if $d = 8$

47. $3m - 3$ if $m = 2$

Algebra Lab
Exploring Sequences

MAIN IDEA

Explore patterns in sequences of geometric figures.

NYS Core Curriculum

**7.PS.4 Observe patterns and formulate generalizations
7.A.8 Create algebraic patterns using charts/tables, graphs, equations, and expressions** *Also addresses 7.R.11*

ACTIVITY

STEP 1 Use toothpicks to build the figures below.

Figure 1   Figure 2   Figure 3

STEP 2 Make a table like the one shown and record the figure number and number of toothpicks used in each figure.

Figure Number	Number of Toothpicks
1	4
2	
3	

STEP 3 Construct the next figure in this pattern. Record your results.

STEP 4 Repeat Step 3 until you have found the next four figures in the pattern.

ANALYZE THE RESULTS

1. How many additional toothpicks were used each time to form the next figure in the pattern? Where is this pattern found in the table?

2. Based on your answer to Exercise 1, how many toothpicks would be in Figure 0 of this pattern?

3. Remove one toothpick from your pattern so that Figure 1 is made up of just three toothpicks as shown. Then create a table showing the number of toothpicks that would be in the first 7 figures by continuing the same pattern as above.

Figure 1

4. How many toothpicks would there be in Figure n of this new pattern?

5. How could you adapt the expression you wrote in Exercise 4 to find the number of toothpicks in Figure n of the original pattern?

6. **MAKE A PREDICTION** How many toothpicks would there be in Figure 10 of the original pattern? Explain your reasoning. Then check your answer by constructing the figure.

7. Find the number of toothpicks in Figure n of the pattern below, and predict the number of toothpicks in Figure 12. Justify your answer.

Figure 1   Figure 2   Figure 3

1-10

Algebra: Equations and Functions

MAIN IDEA

Make function tables and write equations.

NYS Core Curriculum

7.A.10 Write an equation to represent a function from a table of values *Also addresses 7.A.8, 7PS.6, 7.CN.8, 7.R.5*

New Vocabulary

function
function rule
function table
domain
range

NY Math Online

glencoe.com

• Concepts In Motion
• Extra Examples
• Personal Tutor
• Self-Check Quiz

▷ **GET READY** for the Lesson

MAGAZINES Suppose you can buy magazines for $4 each.

1. Copy and complete the table to find the cost of 2, 3, and 4 magazines.

2. Describe the pattern in the table between the cost and the number of magazines.

Number	Multiply by 4	Cost ($)
1	4 × 1	4
2		
3		
4		

A relationship that assigns exactly one *output* value for each *input* value is called a **function**. In a function, you start with an input number, perform one or more operations on it, and get an output number. The operation performed on the input is given by the **function rule**.

Input → **Function Rule** → Output

You can organize the input numbers, output numbers, and the function rule in a **function table**. The set of input values is called the **domain**, and the set of output values is called the **range**.

EXAMPLE Make a Function Table

① **MONEY** Javier saves $20 each month. Make a function table to show his savings after 1, 2, 3, and 4 months. Then identify the domain and range.

The domain is {1, 2, 3, 4}, and the range is {20, 40, 60, 80}.

Input	Function Rule	Output
Number of Months	Multiply by 20	Total Savings ($)
1	20 × 1	20
2	20 × 2	40
3	20 × 3	60
4	20 × 4	80

✓ **CHECK Your Progress**

a. Suppose a student movie ticket costs $3. Make a function table that shows the total cost for 1, 2, 3, and 4 tickets. Then identify the domain and range.

Functions are often written as equations with two variables—one to represent the input and one to represent the output. Here's an equation for the situation in Example 1.

Function rule: multiply by 20

$$20x = y$$

Input: number of months Output: total savings

Real-World EXAMPLES

2 **ANIMALS** An armadillo sleeps 19 hours each day. Write an equation using two variables to show the relationship between the number of hours *h* an armadillo sleeps in *d* days.

Input	Function Rule	Output
Number of Days (*d*)	Multiply by 19	Number of Hours Slept (*h*)
1	1 × 19	19
2	2 × 19	38
3	3 × 19	57
d	*d* × 19	19*d*

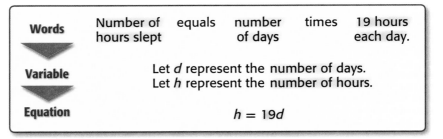

| Words | Number of hours slept | equals | number of days | times | 19 hours each day. |

Variable   Let *d* represent the number of days.
Let *h* represent the number of hours.

Equation $h = 19d$

3 How many hours does an armadillo sleep in 4 days?

$h = 19d$ Write the equation.

$h = 19(4)$ Replace *d* with 4.

$h = 76$ Multiply.

An armadillo sleeps 76 hours in 4 days.

Real-World Career . . . .
How Does a Botanist Use Math? A botanist gathers and studies plant statistics to solve problems and draw conclusions about various plants.

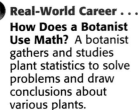

NY Math Online

For more information, go to glencoe.com.

CHECK Your Progress

BOTANIST A botanist discovers that a certain species of bamboo grows 4 inches each hour.

b. Write an equation using two variables to show the relationship between the growth *g* in inches of this bamboo plant in *h* hours.

c. Use your equation to explain how to find the growth in inches of this species of bamboo after 6 hours.

Example 1
(p. 63)

Copy and complete each function table. Then identify the domain and range.

1. $y = 3x$

x	3x	y
1	3 • 1	3
2	3 • 2	
3	3 • 3	
4		

2. $y = 4x$

x	4x	y
0	4 • 0	
1	4 • 1	
2		
3		

3. **MUSIC** Jonas downloads 8 songs each month onto his digital music player. Make a function table that shows the total number of songs downloaded after 1, 2, 3, and 4 months. Then identify the domain and range.

Examples 2, 3
(p. 64)

SPORTS For Exercises 4 and 5, use the following information.

The top speed reached by a race car is 231 miles per hour.

4. Write an equation using two variables to show the relationship between the number of miles m that a race car can travel in h hours.

5. Use your equation to explain how to find the distance in miles the race car will travel in 3 hours.

Practice and Problem Solving

HOMEWORK HELP	
For Exercises	**See Examples**
6–10	1
11–14	2, 3

Copy and complete each function table. Then identify the domain and range.

6. $y = 2x$

x	2x	y
0	2 • 0	0
1	2 • 1	
2		
3		

7. $y = 6x$

x	6x	y
1		
2		
3		
4		

8. $y = 9x$

x	9x	y
1		
2		
3		
4		

Make a function table for each situation. Then identify the domain and range.

9. **PIZZA** A pizza shops sells 25 pizzas each hour. Find the number of pizzas sold after 1, 2, 3, and 4 hours.

10. **TYPING** Suppose you can type 60 words per minute. What is the total number of words typed after 5, 10, 15, and 20 minutes?

CELL PHONES For Exercises 11 and 12, use the following information.

A cell phone provider charges a customer $40 for each month of service.

11. Write an equation using two variables to show the relationship between the total amount charged c, after m months of cell phone service.

12. Use your equation to explain how to find the total cost for 6 months of cell phone service.

Real-World Link · · · ·
Crickets are among the 800,000 different types of insects in the world.

·· **INSECTS** For Exercises 13 and 14, use the following information.

A cricket will chirp approximately 35 times per minute when the outside temperature is 72°F.

13. Write an equation using two variables to show the relationship between the total number of times a cricket will chirp t, after m minutes at this temperature.

14. Use your equation to explain how to find the number of times a cricket will have chirped after 15 minutes at this temperature.

Copy and complete each function table. Then identify the domain and range.

15. $y = x - 1$

x	x − 1	y
1		
2		
3		
4		

16. $y = x + 5$

x	x + 5	y
1		
2		
3		
4		

17. $y = x + 0.25$

x	x + 0.25	y
0		
1		
2		
3		

18. $y = x - 1.5$

x	x − 1.5	y
2		
3		
4		
5		

MEASUREMENT For Exercises 19 and 20, use the following information.

The formula for the area of a rectangle with length 6 units is $A = 6w$.

19. Make a function table that shows the area in square units of a rectangle with a width of 2, 3, 4, and 5 units.

20. Study the pattern in your table. Explain how the area of a rectangle with a length of 6 units changes when the width is increased by 1 unit.

ANALYZE TABLES For Exercises 21–23, use the table that shows the approximate velocity of certain planets as they orbit the Sun.

21. Write an equation to show the relationship between the total number of miles m Jupiter travels in s seconds as it orbits the Sun.

22. What equation can be used to show the total number of miles Earth travels?

23. Use your equation to explain how to find the number of miles Jupiter and Earth each travel in 1 minute.

Orbital Velocity Around Sun	
Planet	**Velocity (mi/s)**
Mercury	30
Earth	19
Jupiter	8
Saturn	6
Neptune	5

NYSCC • NYSMT
Extra Practice, pp. 671, 704

CHALLENGE Write an equation for the function shown in each table.

24.

x	y
1	3
2	4
3	5
4	6

25.

x	y
2	6
4	12
6	18
8	24

26.

x	y
1	3
2	5
3	7
4	9

27. **OPEN ENDED** Write about a real-world situation that can be represented by the equation $y = 3x$.

28. **WRITING IN MATH** Explain the relationship among an *input*, an *output*, and a *function rule*.

NYSMT PRACTICE 7.A.10

29. The table shows the number of hand-painted T-shirts Mi-Ling can make after a given number of days.

Number of Days (x)	Total Number of T-Shirts (y)
1	6
2	12
3	18
4	24

Which function rule represents the data?

A $y = 4x$ **C** $y = 6x$

B $y = 5x$ **D** $y = 12x$

30. Cristina needs to have 50 posters printed to advertise a community book fair. The printing company charges $3 to print each poster. Which table represents this situation?

F

Posters	Cost ($)
3	3
6	6
9	9
p	p

H

Posters	Cost ($)
1	3
2	6
3	9
p	$3 + p$

G

Posters	Cost ($)
1	3
2	6
3	9
p	$3p$

J

Posters	Cost ($)
3	1
6	2
9	3
p	$p \div 3$

Spiral Review

31. **ALGEBRA** Write the next three terms of the sequence $27, 36, 45, 54, \ldots$ (Lesson 1-9)

Use the Distributive Property to rewrite each expression. Then evaluate it. (Lesson 1-8)

32. $5(9 + 7)$ 33. $(12 + 4)4$ 34. $8(7) - 8(2)$ 35. $10(6) - 10(5)$

36. **ALLOWANCE** If Karen receives a weekly allowance of $8, about how much money in all will she receive in two years? (Lesson 1-1)

Graphing Calculator Lab
Functions and Tables

You can use a graphing calculator to represent functions.

ACTIVITY

① **GROCERIES** A grocery store has 12-ounce bottles of sports drink on sale for $1.80 each, with no limit on how many you can buy. In addition, you can use a coupon for $1 off one bottle. Make a table showing the cost for 3, 4, 5, 6, and 7 bottles of this drink.

STEP 1 Write an equation to show the relationship between the number of bottles purchased x and their cost y.

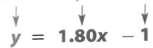

Cost is $1.80 per bottle less $1.

$$y = 1.80x - 1$$

STEP 2

Press Y= on your calculator. Then enter the function into Y_1 by pressing 1.80 X,T,θ,n − 1 ENTER.

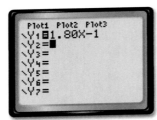

STEP 3

Next, set up a table of x- and y-values. Press 2nd [TBLSET] to display the table setup screen. Then press ⬇ ⬇ ➡ ENTER to highlight Indpnt: Ask.

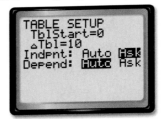

STEP 4 Access the table by pressing 2nd [TABLE]. Then key in each number of bottles, pressing ENTER after each entry.

ANALYZE THE RESULTS

1. Analyze the table to determine how many bottles you can buy for $10. Explain your reasoning.

2. **MAKE A CONJECTURE** Notice that you can purchase 5 bottles for the whole dollar amount of $8. How many bottles will you be able to purchase for $9, the next whole dollar amount? Use the calculator to test your conjecture.

ACTIVITY

2 **CAMPING** Out-There Campground charges each group a camping fee of $20 plus $4.25 per person per night. Roughing-It Campground charges $6.25 per person per night. Make a table showing the one-night fee for 2, 3, 4, 5, and 6 people to camp at each campground.

STEP 1 Write an equation to show the relationship between the number of people x and the one-night fee y for them to camp at each campground.

Out-There Campground

Fee is $20 plus $4.25 per person.

$$y \quad = \quad 20 \quad + \quad 4.25x$$

Roughing-It Campground

Fee is $6.25 per person.

$$y \quad = \quad 6.25x$$

STEP 2 Enter the function for the Out-There Campground into Y_1 and the function for the Roughing-It Campground into Y_2.

STEP 3 Next, set up a table of x- and y-values as in Activity 1.

STEP 4 Then access the table and key in each number of people. Notice that the calculator follows the order of operations multiplying each x-value by 4.25 first and then adding 20.

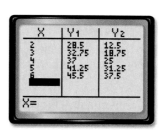

Reading Math

The phrase *$4.25 per person* means *$4.25 for each person.*

ANALYZE THE RESULTS

3. For 2, 3, 4, 5, and 6 people, which campground charges the greater total nightly cost to camp?

4. **MAKE A CONJECTURE** Will the total nightly cost to camp at each campground ever be the same? If so, for what number of people?

5. Use the graphing calculator to test your conjecture from Exercise 4. Were you correct? If not, use the graphing calculator to guess and check until you find the correct number of people.

6. If all other aspects of these two campgrounds are equal, write a recommendation as to which campground a group of n people should choose based on your cost analysis.

FOLDABLES Study Organizer

GET READY to Study

Be sure the following Big Ideas are noted in your Foldable.

Introduction to Algebra and Functions

BIG Ideas

Squares and Square Roots (Lesson 1-3)
• The square of a number is the product of a number and itself.
• A square root of a number is one of its two equal factors.

Order of Operations (Lesson 1-4)
• Do all operations within grouping symbols first. Evaluate all powers before other operations. Multiply and divide in order from left to right. Add and subtract in order from left to right.

Properties (Lesson 1-8)
• Distributive Property
  $5(2 + 4) = 5 \cdot 2 + 5 \cdot 4$
  $(3 + 2)4 = 3 \cdot 4 + 2 \cdot 4$
• Commutative Property
  $3 + 2 = 2 + 3$
  $7 \cdot 4 = 4 \cdot 7$
• Associative Property
  $6 + (3 + 8) = (6 + 3) + 8$
  $5 \cdot (2 \cdot 3) = (5 \cdot 2) \cdot 3$
• Identity Property
  $4 + 0 = 4$
  $4 \cdot 1 = 4$

Functions (Lesson 1-10)
• A function is a relationship that assigns exactly one *output* value for each *input* value.
• In a function, the function rule gives the operation to perform on the input.

Key Vocabulary

algebra (p. 44)
algebraic expression (p. 44)
arithmetic sequence (p. 57)
base (p. 30)
coefficient (p. 45)
defining the variable (p. 50)
domain (p. 63)
equation (p. 49)
equivalent expressions (p. 53)
evaluate (p. 31)
exponent (p. 30)
factors (p. 30)
function (p. 63)

function rule (p. 63)
numerical expression (p. 38)
order of operations (p. 38)
perfect square (p. 34)
powers (p. 30)
radical sign (p. 35)
range (p. 63)
sequence (p. 57)
solution (p. 49)
square (p. 34)
square root (p. 35)
term (p. 57)
variable (p. 44)

Vocabulary Check

State whether each sentence is *true* or *false*. If *false*, replace the underlined word or number to make a true sentence.

1. <u>Numerical expressions</u> have the same value.

2. Two or more numbers that are multiplied together are called <u>powers</u>.

3. The <u>range</u> of a function is the set of input values.

4. A function assigns exactly <u>two</u> *output* values for each *input* value.

5. An <u>equation</u> is a sentence that contains an equals sign.

6. A <u>sequence</u> is an ordered list of numbers.

7. The product of a number and itself is the <u>square root</u> of the number.

Lesson-by-Lesson Review

 1-1 **A Plan for Problem Solving** (pp. 25–29)

7.PS.14, 7.PS.15

Use the four-step plan to solve each problem.

8. **PHONE CALLS** When Tamik calls home from college, she talks ten minutes per call for 3 calls each week. How many minutes does she use in a 15-week semester?

9. **RUNNING** Darren runs at a rate of 6 feet per second, and Kim runs at a rate of 7 feet per second. If they both start a race at the same time, how far apart are they after one minute?

10. **WORK** Alan was paid $9 per hour and earned $128.25. How many hours did he work?

Example 1 One quart of paint covers 40 square feet of wall space. Brock uses 5 quarts of paint to cover his walls. How many square feet did Brock paint?

Understand	Brock uses 5 quarts of paint, each covering 40 square feet.
Plan	Multiply 40 by 5.
Solve	$40 \cdot 5 = 200$ Brock painted 200 square feet.
Check	$200 \div 5 = 40$, so the answer is reasonable.

 1-2 **Powers and Exponents** (pp. 30–33)

7.N.8

Write each power as a product of the same factor.

11. 3^4

12. 9^6

13. 5^1

14. 7^5

15. Write *5 to the fourth power* as a product of the same factor.

Evaluate each expression.

16. 3^5

17. 7^9

18. 2^8

19. 18^2

20. 10^4

21. 100^1

22. Write $15 \cdot 15 \cdot 15$ in exponential form.

23. **PATHS** At the edge of a forest, there are two paths. At the end of each path, there are two additional paths. If at the end of each of those paths there are two more paths, how many paths are there at the end?

Example 2 Write 2^3 as a product of the same factor.

The base is 2. The exponent 3 means that 2 is used as a factor 3 times.

$2^3 = 2 \cdot 2 \cdot 2$

Example 3 Evaluate 4^5.

The base is 4. The exponent 5 means that 4 is used as a factor 5 times.

$4^5 = 4 \cdot 4 \cdot 4 \cdot 4 \cdot 4$
$\quad\;\; = 1,024$

1-3 **Squares and Square Roots** (pp. 34–37)

7.N.15

Find the square of each number.

24. 4   25. 13

Find each square root.

26. $\sqrt{81}$ 27. $\sqrt{324}$

28. **MEASUREMENT** The area of a certain kind of ceramic tile is 25 square inches. What is the length of one side?

Example 4 Find the square of 15.

$15 \cdot 15 = 225$ Multiply 15 by itself.

Example 5 Find the square root of 441.

$21 \cdot 21 = 441$, so $\sqrt{441} = 21$.

1-4 **Order of Operations** (pp. 38–41)

7.N.11

Evaluate each expression.

29. $24 - 8 + 3^2$ 30. $48 \div 6 + 2 \cdot 5$

31. $9 + 3(7 - 5)^3$ 32. $15 + 9 \div 3 - 7$

33. **SEATING** In planning for a ceremony, 36 guests need to be seated with 4 guests per table. An additional 12 guests need to be seated with 3 guests per table. Write an expression to determine how many tables are needed. Then evaluate the expression.

Example 6 Evaluate $24 - (8 \div 4)^4$.

$24 - (8 \div 4)^4 = 24 - 2^4$ Divide 8 by 4.

$\qquad\qquad\qquad = 24 - 16$ Find the value of 2^4.

$\qquad\qquad\qquad = 8$ Subtract.

1-5 **PSI: Guess and Check** (pp. 42–43)

7.PS.1,
7.PS.7

Solve. Use the *guess and check* strategy.

34. **TRAVEL** Lucinda is driving away from Redding at 50 miles per hour. When she is 100 miles away, Tom leaves Redding, driving at 60 miles per hour in the same direction. After how many hours will Tom pass Lucinda?

35. **FARMING** A farmer sells a bushel of soybeans for $5 and a bushel of corn for $3. If he hopes to earn $164 and plans to sell 40 bushels in all, how many bushels of soybeans does he need to sell?

Example 7 Find two numbers with a product of 30 and a difference of 13.

Make a guess, and check to see if it is correct. Then adjust the guess until it is correct.

5 and 6 $5 \cdot 6 = 30$ and $6 - 5 = 1$
            incorrect

3 and 10 $3 \cdot 10 = 30$ and $10 - 3 = 7$
            incorrect

2 and 15 $2 \cdot 15 = 30$ and $15 - 2 = 13$
            correct

The two numbers are 2 and 15.

Mixed Problem Solving
For mixed problem-solving practice,
see page 704.

1-6 **Algebra: Variables and Expressions** (pp. 44–47)

7.N.11,
7.A.1

Evaluate each expression if $a = 10$, $b = 4$, and $c = 8$.

36. $(a - b)^2$

37. $ab \div c$

38. $3b^2 + c$

39. $\dfrac{(b + c)^2}{3}$

40. **CLOTHING** The cost of buying h hats and s shirts is given by the expression $\$5.75h + \$8.95s$. Find the cost of purchasing 3 hats and 5 shirts.

Example 8 Evaluate $2m^2 - 5n$ if $m = 4$ and $n = 3$.

$2m^2 - 5n = 2(4)^2 - 5(3)$	Replace m with 4 and n with 3.
$= 2(16) - 5(3)$	Find the value of 4^2.
$= 32 - 15$	Multiply.
$= 17$	Subtract.

1-7 **Algebra: Equations** (pp. 49–52)

7.PS.3

Solve each equation mentally.

41. $h + 9 = 17$

42. $31 - y = 8$

43. $\dfrac{t}{9} = 12$

44. $100 = 20g$

45. **COUNTY FAIRS** Five friends wish to ride the Ferris wheel, which requires 3 tickets per person. The group has a total of 9 tickets. Write and solve an equation to find the number of additional tickets needed for everyone to ride the Ferris wheel.

Example 9 Solve $14 = 5 + x$ mentally.

$14 = 5 + x$	Write the equation.
$14 = 5 + 9$	You know that $5 + 9 = 14$.
$14 = 14$	Simplify.

The solution is 9.

1-8 **Algebra: Properties** (pp. 53–56)

6.N.2,
6.N.3

Find each expression mentally. Justify each step.

46. $(25 \cdot 15) \cdot 4$

47. $14 + (38 + 16)$

48. $8 \cdot (11 \cdot 5)$

49. **ROSES** Wesley sold roses in his neighborhood for $\$2$ a rose. He sold 15 roses on Monday and 12 roses on Tuesday. Use the Distributive Property to mentally find the total amount Wesley earned. Explain your reasoning.

Example 10 Find $8 + (17 + 22)$ mentally. Justify each step.

$8 + (17 + 22)$

$= 8 + (22 + 17)$	Commutative Property of Addition
$= (8 + 22) + 17$	Associative Property of Addition
$= 30 + 17$ or 47	Add 30 and 17 mentally.

1-9 Algebra: Arithmetic Sequences (pp. 57–61)

7.PS.4, 7.A.8

Describe the relationship between the terms in each arithmetic sequence. Then find the next three terms in each sequence.

50. 3, 9, 15, 21, 27, …

51. 2.6, 3.4, 4.2, 5, 5.8, …

52. 0, 7, 14, 21, 28, …

MONEY For Exercises 53 and 54, use the following information.

Tanya collected $4.50 for the first car washed at a band fund-raiser. After the second and third cars were washed, the donations totaled $9 and $13.50, respectively.

53. If this donation pattern continues, what algebraic expression can be used to find the amount of money earned for any number of cars washed?

54. How much money will be collected after a total of 8 cars have been washed?

Example 11 At the end of day 1, Sierra read 25 pages of a novel. By the end of days 2 and 3, she read a total of 50 and 75 pages, respectively. If the pattern continues, what expression will give the total number of pages read after any number of days?

Make a table to display the sequence.

Position	Operation	Value of Term
1	1 • 25	25
2	2 • 25	50
3	3 • 25	75
n	n • 25	$25n$

Each term is 25 times its position number. So, the expression is $25n$.

1-10 Algebra: Equations and Functions (pp. 63–67)

7.A.10

Copy and complete the function table. Then identify the domain and range.

55. $y = 4x$

x	4x	y
5		
6		
7		
8		

56. NAME TAGS Charmaine can make 32 name tags per hour. Make a function table that shows the number of name tags she can make in 3, 4, 5, and 6 hours.

Example 12 Create and complete a function table for $y = 3x$. Then identify the domain and range.

Select any four values for the input x.

x	3x	y
3	3(3)	9
4	3(4)	12
5	3(5)	15
6	3(6)	18

The domain is {3, 4, 5, 6}.
The range is {9, 12, 15, 18}.

1. **PIZZA** Ms. Carter manages a pizza parlor. The average daily cost is $40, plus $52 to pay each employee. It also costs $2 to make each pizza. If 42 pizzas were made one day, requiring the work of 7 employees, what was her total cost that day?

Write each power as a product of the same factor. Then evaluate the expression.

2. 3^5

3. 15^4

4. **MEASUREMENT** Gregory wants to stain the 15-foot-by-15-foot deck in his backyard. One can of stain covers 200 square feet of surface. Is one can of stain enough to cover his entire deck? Explain your reasoning.

Find each square root.

5. $\sqrt{121}$

6. $\sqrt{900}$

7. **MULTIPLE CHOICE** What is the value of $8 + (12 \div 3)^3 - 5 \times 9$?

A 603

B 135

C 27

D 19

8. **ANIMALS** Sally has 6 pets, some dogs and some birds. Her animals have a total of 16 legs. How many of each pet does Sally have?

Evaluate each expression if $x = 12$, $y = 5$, and $z = 3$.

9. $x - 9$

10. $8y$

11. $(y - z)^3$

12. $\dfrac{xz}{y + 13}$

Solve each equation mentally.

13. $9 + m = 16$

14. $d - 14 = 37$

15. $32 = \dfrac{96}{t}$

16. $6x = 126$

17. **SAVINGS** Deb is saving $54 per month to buy a new camera. Use the Distributive Property to mentally find how much she has saved after 7 months. Explain.

Find each expression mentally. Justify each step.

18. $13 + (34 + 17)$

19. $50 \cdot (17 \cdot 2)$

20. **MULTIPLE CHOICE** The table shows the number of hours Teodoro spent studying for his biology test over four days. If the pattern continues, how many hours will Teodoro study on Sunday?

Day	Study Time (hours)
Monday	0.5
Tuesday	0.75
Wednesday	1.0
Thursday	1.25

F 1.5 hours

G 1.75 hours

H 2.0 hours

J 2.5 hours

Describe the relationship between the terms in each arithmetic sequence. Then write the next three terms in the sequence.

21. 7, 16, 25, 34, …

22. 59, 72, 85, 98, …

23. **TRAVEL** Beth drove at the rate of 65 miles per hour for several hours. Make a function table that shows her distance traveled after 2, 3, 4, and 5 hours. Then identify the domain and range.

MONEY For Exercises 24 and 25, use the following information.

Anthony earns extra money after school doing yard work for his neighbors. He charges $12 for each lawn he mows.

24. Write an equation using two variables to show the relationship between the number of lawns mowed m and number of dollars earned d.

25. Then find the number of dollars earned if he mows 14 lawns.

PART 1   Multiple Choice

Read each question. Then fill in the correct answer on the answer document provided by your teacher or on a sheet of paper.

1. A store owner bought some paperback books and then sold them for $4.50 each. He sold 35 books on Monday and 52 books on Tuesday. What piece of information is needed to find the amount of profit made from sales on Monday and Tuesday?

   A Number of books sold on Wednesday

   B Number of hardback books sold on Monday and Tuesday

   C Total number of paperback books sold

   D How much the owner paid for each of the paperback books

2. The table shows the number of milkshakes sold at an ice cream shop each day last week.

Day of Week	Number of Milkshakes
Sunday	31
Monday	9
Tuesday	11
Wednesday	15
Thursday	18
Friday	24
Saturday	28

   Which statement does *not* support the data?

   F There were almost three times as many milkshakes sold on Sunday as on Tuesday.

   G There were half as many milkshakes sold on Monday as on Thursday.

   H There were 11 more milkshakes sold on Tuesday than on Saturday.

   J The total number of milkshakes sold during the week was 136.

3. Which description shows the relationship between the value of a term and n, its position in the sequence?

Position	1	2	3	4	5	n
Value of Term	3	6	9	12	15	

   A Add 2 to n.

   B Divide n by 3.

   C Multiply n by 3.

   D Subtract n from 2.

TEST-TAKING TIP

Question 3 Have students eliminate unlikely answer choices. Since the value of each term is greater than its position, eliminate answer choices B and D.

4. Andrew spent $\frac{1}{2}$ of his Saturday earnings on a pair of jeans and $\frac{1}{2}$ of the remaining amount on a DVD. After he spent $7.40 on lunch, he had $6.10 left. How much did Andrew earn on Saturday?

   F $13.50

   G $27

   H $54

   J $108

5. Lemisha drove an average of 50 miles per hour on Sunday, 55 miles per hour on Monday, and 53 miles per hour on Tuesday. If s represents the number of hours she drove on Sunday, m represents the number of hours she drove on Monday, and t represents the number of hours she drove on Tuesday, which of the following expressions gives the total distance Lemisha traveled?

   A $50s + 53m + 55t$

   B $55s + 50m + 53t$

   C $50s + 55m + 53t$

   D $53s + 55m + 50t$

6. Mrs. Albert drove 850 miles and the average price of gasoline was $2.50 per gallon. What information is needed to find the amount Mrs. Albert spent on gasoline for the trip?

   F Number of hours the trip took

   G Number of miles per hour traveled

   H Average number of miles the car traveled per gallon of gasoline

   J Average number of miles Mrs. Albert drove per day

7. Mr. Thompson wants to estimate the total amount he spends on insurance and fuel for his car each month. Insurance costs about $300 per month, and he expects to drive an average of 150 miles per week. What else does he need to estimate his monthly expenses?

   A The cost of fuel and the one-way distance to work

   B The cost of fuel and the number of miles per gallon his car gets

   C The cost of fuel and his weekly pay

   D The gallons of fuel needed per week

8. Jeremy bought 3 hamburgers at $1.99 each, 2 orders of onion rings at $0.89 each, and 4 soft drinks at $1.25 each. He paid 6.75% tax on the whole order. What other information is necessary to find Jeremy's correct change?

   F Total cost of the order

   G Amount he paid in tax

   H Reason for buying the food

   J Amount he gave the cashier

PART 2 Short Response/Grid In

Record your answers on the answer sheet provided by your teacher or on a sheet of paper.

9. Emily bought 2.5 pounds of salami for $1.99 per pound. About how much did she pay?

10. How do you correctly evaluate the expression $4 \times (5 + 4) - 27$?

11. What value of t makes the following equation true?
$$t \div 6 = 48$$

12. Use the Distributive Property to rewrite $4(3 + 5)$.

PART 3 Extended Response

Record your answers on the answer sheet provided by your teacher or on a sheet of paper. Show your work.

13. **GEOMETRY** The first and fifth terms of a toothpick sequence are shown below.

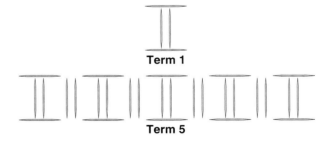

Term 1

Term 5

   a. What might the third term look like?

   b. Write a rule that connects the term number and the number of toothpicks in your sequence.

NEED EXTRA HELP?												
If You Missed Question...	1	2	3	4	5	6	7	8	9	10	11	12	13
Go to Lesson...	1-1	1-1	1-9	1-1	1-6	1-1	1-1	1-1	1-1	1-4	1-6	1-8	1-9
NYS Core Curriculum	7.PS.14	7.PS.14	7.A.8	7.N.11	7.A.1	7.PS.14	7.PS.14	7.PS.14	7.PS.14	7.N.11	7.N.11	6.N.3	7.A.8

CHAPTER 2

Integers

New York State Core Curriculum

7.N.12 Add subtract, multiply, and divide integers

Key Vocabulary

graph (p. 80)

integer (p. 80)

negative integer (p. 80)

positive integer (p. 80)

🌐 Real-World Link

Sports In miniature golf, a score above par can be written as a positive integer and a score below par can be written as a negative integer.

FOLDABLES®
Study Organizer

Integers Make this Foldable to help you organize your notes. Begin with two sheets of $8\frac{1}{2}''$ by 11" paper.

❶ **Fold** one sheet in half from top to bottom. Cut along fold from edges to margin.

❷ **Fold** the other sheet in half from top to bottom. Cut along fold between margins.

❸ **Insert** first sheet through second sheet and align folds.

❹ **Label** each inside page with a lesson number and title.

2-1
Integers and Absolute Value

GET READY for Chapter 2

Diagnose Readiness You have two options for checking Prerequisite Skills.

Option 1

Option 2

NY Math Online ➤ Take the Online Readiness Quiz at glencoe.com.

Take the Quick Check below. Refer to the Quick Review for help.

QUICK **Quiz**	QUICK **Review**
Replace each ● with < or > to make a true sentence. (Prior Grade) 1. 1,458 ● 1,548　　2. 36 ● 34 3. 1.02 ● 1.20　　4. 76.7 ● 77.6 5. **COINS** Philippe has $5.17 in coins and Garrett has $5.71 in coins. Who has the greater amount? (Prior Grade)	**Example 1** **Replace the ● with < or > to make a true sentence.** **3.14 ● 3.41** 3.14　Line up the decimal points. 3.41　Starting at the left, compare ↑　　the digits in each place-value position. The digits in the tenths place are not the same. Since 1 tenth < 4 tenths, 3.14 < 3.41.
Evaluate each expression if $a = 7$, $b = 2$, and $c = 11$. (Prior Grade) 6. $a + 8$　　　7. $a + b + c$ 8. $c - b$　　　9. $a - b + 4$ 10. **TEMPERATURE** At 8 A.M., it was 63°F. By noon, the temperature had risen 9 degrees Fahrenheit. What was the temperature at noon? (Prior Grade)	**Example 2** **Evaluate the expression $11 - a + b$ if $a = 2$ and $b = 8$.** $11 - a + b = 11 - 2 + 8$　Replace a with 2 and b with 8. $= 9 + 8$　Subtract 2 from 11. $= 17$　Add 9 and 8.
Evaluate each expression if $m = 9$ and $n = 4$. (Prior Grade) 11. $6mn$　　　12. $n \div 2 - 1$ 13. $m + 5 \times n$　　14. $m^2 \div (n + 5)$ 15. **PLANES** The distance in miles that an airplane travels is given by rt where r is the rate of travel and t is the time. Find the distance an airplane traveled if $t = 4$ hours and $r = 475$ miles per hour. (Prior Grade)	**Example 3** **Evaluate the expression $n^2 \div 16 + m$ if $m = 3$ and $n = 8$.** $n^2 \div 16 + m = 8^2 \div 16 + 3$　Replace m with 3 and n with 8. $= 64 \div 16 + 3$　Evaluate 8^2. $= 4 + 3$　Divide 64 by 16. $= 7$　Add 4 and 3.

Integers and Absolute Value

MAIN IDEA

Read and write integers, and find the absolute value of a number.

NYS Core Curriculum

7.N.11 Simplify expressions using order of operations
Note: Expressions may include absolute value and/or integral exponents greater than 0
7.N.1 Distinguish between the various subsets of real numbers (counting/natural numbers, whole numbers, integers, rational numbers, and irrational numbers)

New Vocabulary

integer
negative integer
positive integer
graph
absolute value

NY Math Online

glencoe.com

• Extra Examples
• Personal Tutor
• Self-Check Quiz

▶ **GET READY for the Lesson**

SKATEBOARDING The bottom of a skateboarding ramp is 8 feet below streetlevel. A value of −8 represents 8 feet *below* street level.

1. What does a value of −10 represent?

2. The top deck of the ramp is 5 feet *above* street level. How can you represent 5 feet *above* street level?

Numbers like 5 and −8 are called integers. An **integer** is any number from the set {..., −4, −3, −2, −1, 0, 1, 2, 3, 4, ...} where ... means *continues without end*.

Negative integers are integers less than zero. They are written with a − sign.

Positive integers are integers greater than zero. They can be written with or without a + sign.

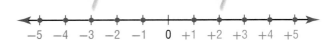

−5 −4 −3 −2 −1 0 +1 +2 +3 +4 +5

Zero is neither negative nor positive.

Real-World EXAMPLES

WEATHER Write an integer for each situation.

1. **an average temperature of 5 degrees below normal**

Because it represents *below* normal, the integer is −5.

2. **an average rainfall of 5 inches above normal**

Because it represents *above* normal, the integer is +5 or 5.

✓ **CHECK Your Progress**

Write an integer for each situation.

a. 6 degrees above normal   b. 2 inches below normal

Integers can be graphed on a number line. To **graph** a point on the number line, draw a point on the line at its location.

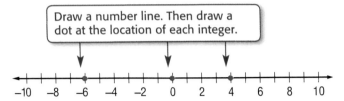 **EXAMPLE** Graph Integers

3 Graph the set of integers {4, −6, 0} on a number line.

Draw a number line. Then draw a dot at the location of each integer.

✓ **CHECK Your Progress**

Graph each set of integers on a number line.

c. {−2, 8, −7}   d. {−4, 10, −3, 7}

On the number line below, notice that −5 and 5 are each 5 units from 0, even though they are on opposite sides of 0. Numbers that are the same distance from zero on a number line have the same **absolute value**.

Reading Math

Absolute Value

|−5| *absolute value of negative five*

Absolute Value	**Key Concept**
Words	The absolute value of a number is the distance between the number and zero on a number line.

5 units   5 units

−5 −4 −3 −2 −1 0 1 2 3 4 5

Examples |−5| = 5 |5| = 5

EXAMPLES Evaluate Expressions

Evaluate each expression.

4 |−4|

On the number line, the point −4 is 4 units from 0.

So, |−4| = 4.

4 units

−4 −3 −2 −1 0 1 2 3 4

Study Tip

Order of Operations
The absolute value bars are considered to be a grouping symbol. When evaluating |−5| − |2|, evaluate the absolute values before subtracting.

5 |−5| − |2|

|−5| − |2| = 5 − 2 |−5| = 5, |2| = 2

So, |−5| − |2| = 3.

✓ **CHECK Your Progress**

Evaluate each expression.

e. |8| f. 2 + |−3| g. |−6| − 5

Examples 1, 2
(p. 80)

Write an integer for each situation.

1. a loss of 11 yards

2. 6°F below zero

3. a deposit of $16

4. 250 meters above sea level

5. **FOOTBALL** The quarterback lost 15 yards on one play. Write an integer to represent the number of yards lost.

Example 3
(p. 81)

Graph each set of integers on a number line.

6. $\{11, -5, -8\}$

7. $\{2, -1, -9, 1\}$

Examples 4, 5
(p. 81)

Evaluate each expression.

8. $|-9|$

9. $1 + |7|$

10. $|-1| - |-6|$

Practice and Problem Solving

HOMEWORK HELP	
For Exercises	**See Examples**
11–20	1, 2
21–24	3
25–30	4, 5

Write an integer for each situation.

11. a profit of $9

12. a bank withdrawal of $50

13. 53°C below zero

14. 7 inches more than normal

15. 2 feet below flood level

16. 160 feet above sea level

17. an elevator goes up 12 floors

18. no gains or losses on first down

19. **GOLF** In golf, scores are often written in relationship to *par*, the average score for a round at a certain course. Write an integer to represent a score that is 7 under par.

20. **PETS** Javier's pet guinea pig gained 8 ounces in one month. Write an integer to describe the amount of weight his pet gained.

Graph each set of integers on a number line.

21. $\{0, 1, -3\}$

22. $\{3, -7, 6\}$

23. $\{-5, -1, 10, -9\}$

24. $\{-2, -4, -6, -8\}$

Evaluate each expression.

25. $|10|$

26. $|-12|$

27. $|-7| - 5$

28. $7 + |4|$

29. $|-9| + |-5|$

30. $|8| - |-10|$

31. $|-10| \div 2 \times |5|$

32. $12 - |-8| + 7$

33. $|27| \div 3 - |-4|$

34. **SCUBA DIVING** One diver descended 10 feet, and another ascended 8 feet. Which situation has the greater absolute value? Explain.

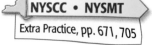

35. **SCIENCE** If you rub a balloon through your hair, you can make the balloon stick to a wall. Suppose there are 17 positive charges on the wall and 25 negative charges on the balloon. Write an integer for each charge.

Extra Practice, pp. 671, 705

36. REASONING If $|x| = 3$, what is the value of x?

37. CHALLENGE Determine whether the following statement is *true* or *false*. If *false*, give a counterexample.

The absolute value of every integer is positive.

38. WRITING IN MATH Write a real-world situation that uses negative integers. Explain what the negative integer means in that situation.

NYSMT PRACTICE 7.N.11, 7.N.1

39. Which point has a coordinate with the greatest absolute value?

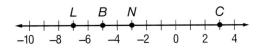

A Point B

B Point C

C Point L

D Point N

40. Which statement about these real-world situations is *not* true?

F A $100 check deposited in a bank can be represented by $+100$.

G A loss of 15 yards in a football game can be represented by -15.

H A temperature of 20 below zero can be represented by -20.

J A submarine diving 300 feet under water can be represented by $+300$.

Spiral Review

Copy and complete each function table. Identify the domain and range. (Lesson 1-10)

41. $y = x - 4$

x	x − 4	y
4		
5		
6		
7		

42. $y = 9x$

x	9x	y
0		
1		
2		
3		

43. $y = 5x + 1$

x	5x + 1	y
1		
2		
3		
4		

44. GEOMETRY The table shows the side length and perimeter of several equilateral triangles. Write an expression that describes the perimeter if x represents the side length. (Lesson 1-9)

Side Length (in.)	2	3	4	5	6
Perimeter (in.)	6	9	12	15	18

▷ **GET READY for the Next Lesson**

PREREQUISITE SKILL Replace each ● with < or > to make a true sentence.

45. 16 ● 6

46. 101 ● 111

47. 87.3 ● 83.7

48. 1,051 ● 1,015

2-2

Comparing and Ordering Integers

MAIN IDEA

Compare and order integers.

NYS Core Curriculum

7.N.3 Place rational and irrational numbers (approximations) on a number line and justify the placement of the numbers
7.N.1 Distinguish between the various subsets of real numbers (counting/natural numbers, whole numbers, **integers**, rational numbers, and irrational numbers)

NY Math Online

glencoe.com

• Concepts In Motion
• Extra Examples
• Personal Tutor
• Self-Check Quiz
• Reading in the Content Area

▷ **GET READY** for the Lesson

TOYS The timeline shows when some toys were invented.

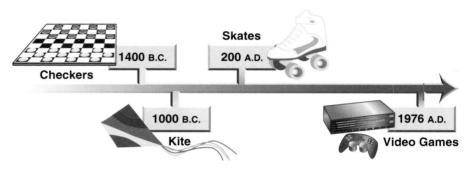

1. The yo-yo was invented around 500 B.C. Was it invented before or after the kite?

2. Modern chess was invented around 600 A.D. Between which two toys was this invented?

When two numbers are graphed on a number line, the number to the left is always less than the number to the right. The number to the right is always greater than the number to the left.

Compare Integers		**Key Concept**
Model	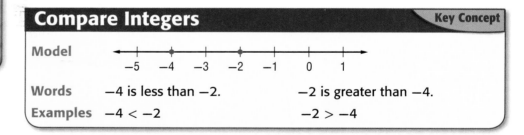	
Words	−4 is less than −2.	−2 is greater than −4.
Examples	−4 < −2	−2 > −4

EXAMPLE Compare Two Integers

1️⃣ **Replace the ⬤ with < or > to make −5 ⬤ −3 a true sentence.**

Graph each integer on a number line.

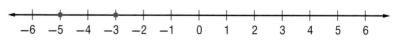

Since −5 is to the left of −3, −5 < −3.

☑ **CHECK Your Progress**

Replace each ⬤ with < or > to make a true sentence.

a. −8 ⬤ −4   b. 5 ⬤ −1   c. −10 ⬤ −13

2 The elevations, in feet, for the lowest points in California, Oklahoma, Louisiana, and Kentucky are listed. Which list shows the elevation in order from highest to lowest?

State	Elevation (ft)
California	−282
Oklahoma	289
Louisiana	−8
Kentucky	257

A 289, −282, 257, −8

B −8, 257, −282, 289

C −282, −8, 257, 289

D 289, 257, −8, −282

Test-Taking Tip

Eliminating Answer Choices
If you are unsure of the correct answer, eliminate the choices you know are incorrect. Then consider the remaining choices. You can eliminate choices B and C since those lists begin with a negative number.

Read the Item

To order the integers, graph them on a number line.

Solve the Item

Order the integers by reading from right to left: 289, 257, −8, −282. So, the answer is D.

 CHECK Your Progress

d. A newspaper reporter lists the third round scores of the top five finishers in a golf tournament. Which list shows these scores from least to greatest?

F 5, 2, 0, −1, −3   **H** −3, −1, 0, 2, 5

G −1, −3, 0, 2, 5   **J** 0, −1, 2, −3, 5

CHECK Your Understanding

Example 1
(p. 84)

Replace each ● with < or > to make a true sentence.

1. −4 ● −6   2. −2 ● 8   3. 0 ● −10

Example 2
(p. 85)

Order the integers in each set from least to greatest.

4. $\{-13, 9, -2, 0, 4\}$ 5. $\{12, -16, -10, 19, -18\}$

6. **MULTIPLE CHOICE** The lowest temperatures in Hawaii, Illinois, Minnesota, and South Carolina are listed. Which list shows these temperatures in order from coldest to warmest?

A −19, −36, −60, 12   **C** −60, −36, −19, 12

B 12, −19, −36, −60   **D** −60, −19, 12, −36

Replace each ● with < or > to make a true sentence.

7. −7 ● −3   **8.** −21 ● −12   **9.** −6 ● −11   **10.** −15 ● −33

11. 17 ● −20   **12.** 4 ● −4   **13.** −5 ● 17   **14.** −12 ● 8

Order the integers in each set from least to greatest.

15. {−8, 11, 6, −5, −3}   **16.** {7, −2, 14, −9, 2}

17. {5, −6, −7, −4, 1, 3}   **18.** {−12, 15, 8, −15, −23, 10}

19. ANALYZE TABLES The ocean floor is divided into five zones according to how deep sunlight penetrates. Order the zones from closest to the surface to nearest to the ocean floor.

Zone	Beginning Ocean Depth
Abyssal	−4,000 m
Hadal	−6,000 m
Midnight	−1,000 m
Sunlight	0 m
Twilight	−200 m

20. STOCK MARKET Kevin's dad owns stock in five companies. The change in the stock value for each company was as follows: Company A, +12; Company B, −5; Company C, −25; Company D, +18; Company E, −10. Order the companies from the worst performing to best performing.

Replace each ● with <, >, or = to make a true sentence.

21. −13 ● |−14| **22.** |36| ● −37   **23.** −12 ● |12| **24.** |−29| ● |92|

FOOTBALL For Exercises 25 and 26, use the information at the right. It shows the yardage gained each play for five plays.

Play	1	2	3	4	5
Yardage	10	−2	5	−5	20

25. Order the yardage from least to greatest.

26. Which run is the middle, or *median*, yardage?

27. WEATHER The wind chill index was invented in 1939 by Paul Siple, a polar explorer and authority on Antarctica. Wind chill causes the air to feel colder on human skin. Which feels colder: a temperature of 10° with a 15-mile-per-hour wind or a temperature of 5° with a 10 mile per hour wind?

WIND CHILL

Wind (mph)	Temperature (°F)				
	15	10	5	0	−5
5	7	1	−5	−11	−16
10	3	−4	−10	−16	−22
15	0	−7	−13	−19	−26
20	−2	−9	−15	−23	−29

Determine whether each sentence is *true* or *false*. If *false*, change one number to make the sentence true.

28. −8 > 5   **29.** −7 < 0   **30.** |5| < −6   **31.** 10 > |−8|

H.O.T. Problems

32. **NUMBER SENSE** If 0 is the greatest integer in a set of five integers, what can you conclude about the other four integers?

33. **CHALLENGE** What is the greatest integer value of n such that $n < 0$?

34. **WRITING IN MATH** Develop a method for ordering a set of negative integers from least to greatest without the aid of a number line. Explain your method and use it to order the set $\{-5, -8, -1, -3\}$.

NYSMT PRACTICE 7.N.1

35. On a certain game show, contestants receive positive numbers of points for correct responses and negative numbers of points for incorrect responses. Which list gives the points a contestant received during one round of the game in order from highest to lowest?

   A $-200, -400, -1000, 200, 600$

   B $600, -1000, -400, -200, 200$

   C $600, 200, -200, -400, -1000$

   D $-1000, -400, -200, 600, 200$

36. Which statement about the values shown is *not* true?

State	Low Temperature (°F)
AR	-29
GA	-17
MS	-19
VA	-30
TX	-23

Source: *The World Almanac*

   F Virginia's record low is less than the record low for Arkansas.

   G Arkansas' record low is less than the record low for Georgia.

   H Mississippi's record low is greater than the record low for Texas.

   J Texas' record low is less than the record low for Arkansas.

Spiral Review

Write an integer for each situation. (Lesson 2-1)

37. 9°C below zero

38. a gain of 20 feet

HOBBIES For Exercises 39 and 40, use the following information. (Lesson 1-10)

Sophia estimates that she knits 6 rows of an afghan each hour.

39. Write an equation using two variables to represent the total number of rows r completed by Sophia after time t.

40. How many rows will Sophia have completed after 4 hours?

▷ **GET READY for the Next Lesson**

PREREQUISITE SKILL Graph each point on a vertical number line that goes from −10 on the bottom to 10 at the top. (Lesson 2-1)

41. -3

42. 0

43. 4

44. -7

The Coordinate Plane

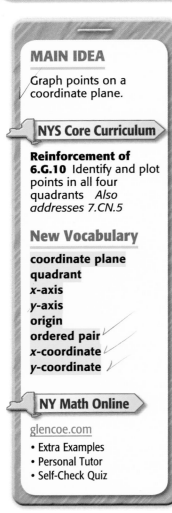

MAIN IDEA

Graph points on a coordinate plane.

NYS Core Curriculum

Reinforcement of 6.G.10 Identify and plot points in all four quadrants *Also addresses 7.CN.5*

New Vocabulary

coordinate plane
quadrant
x-axis
y-axis
origin
ordered pair
x-coordinate
y-coordinate

NY Math Online

glencoe.com

• Extra Examples
• Personal Tutor
• Self-Check Quiz

▶ **GET READY** for the Lesson

GPS A GPS, or global positioning system, is a satellite based navigation system. A GPS map of Raleigh, North Carolina, is shown.

1. Suppose Mr. Diaz starts at Shaw University and drives 2 blocks north. Name the street he will cross.

2. Using the words *north, south, east,* and *west,* write directions to go from Chavis Park to Moore Square.

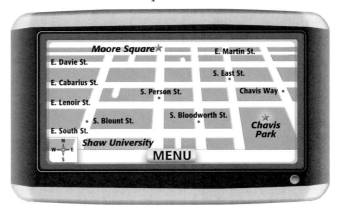

On a GPS, towns and streets are often located on a grid. In mathematics, we use a grid called a coordinate plane, to locate points. A **coordinate plane** is formed when two number lines intersect. The number lines separate the coordinate plane into four regions called **quadrants**.

Coordinate Plane **Key Concept**

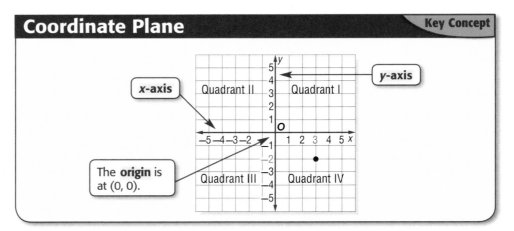

An **ordered pair** is a pair of numbers, such as (3, −2), used to locate a point in the coordinate plane.

x-coordinate corresponds to a number on the *x*-axis. ╱ **(3, −2)** ╲ *y*-coordinate corresponds to a number on the *y*-axis.

When locating an ordered pair, moving *right* or *up* on a coordinate plane is in the *positive* direction. Moving *left* or *down* is in the *negative* direction.

 EXAMPLE **Naming Points Using Ordered Pairs**

① Write the ordered pair that corresponds to point D. Then state the quadrant in which the point is located.

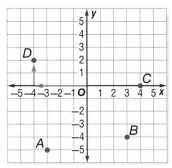

- Start at the origin.

- Move left on the x-axis to find the x-coordinate of point D, which is −4.

- Move up to find the y-coordinate, which is 2.

So, point D corresponds to the ordered pair (−4, 2). Point D is located in Quadrant II.

✓ **CHECK** Your Progress

Write the ordered pair that corresponds to each point. Then state the quadrant or axis on which the point is located.

a. A   **b.** B   **c.** C

 EXAMPLE **Graph an Ordered Pair**

② Graph and label point K at (2, −5).

Reading Math

Scale When no numbers are shown on the x- or y-axis, you can assume that each square is 1 unit long on each side.

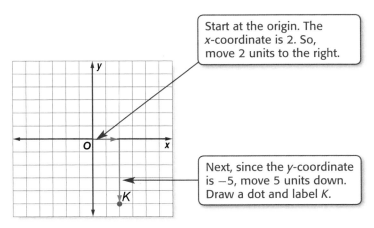

Start at the origin. The x-coordinate is 2. So, move 2 units to the right.

Next, since the y-coordinate is −5, move 5 units down. Draw a dot and label K.

 ✓ **CHECK** Your Progress

On graph paper, draw a coordinate plane. Then graph and label each point.

d. L(−4, 2)   **e.** M(−5, −3)   **f.** N(0, 1)

Real-World EXAMPLES

 ③ AQUARIUMS A map can be divided into a coordinate plane where the *x*-coordinate represents how far to move right or left and the *y*-coordinate represents how far to move up or down. What exhibit is located at (6, 5)?

New York Aquarium, Bronx, NY

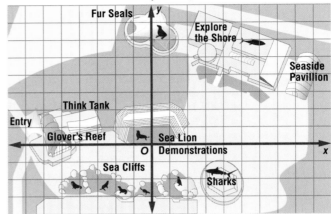

Real-World Link . . . .
New York Aquarium has a marine mammal exhibit that includes sea otters. Large male sea otters can grow to be over 4 feet in length.

Start at the origin. Move 6 units to the right and then 5 units up. Explore the Shore is located at (6, 5).

④ In which quadrant is the Shark Exhibit located?

The Shark Exhibit is located in Quadrant IV.

CHECK Your Progress

For Exercises g and h, use the map above.

g. Find the ordered pair that represents the location of the Think Tank.

h. What is located at the origin?

CHECK Your Understanding

Example 1
(p. 89)

Write the ordered pair corresponding to each point graphed at the right. Then state the quadrant or axis on which each point is located.

1. *P* 2. *Q*

3. *R* 4. *S*

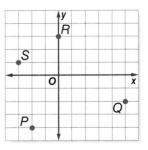

Example 2
(p. 89)

On graph paper, draw a coordinate plane. Then graph and label each point.

5. *T*(2, 3) 6. *U*(−4, 6)

7. *V*(−5, 0) 8. *W*(1, −2)

Examples 3, 4
(p. 90)

GEOGRAPHY For Exercises 9 and 10, use the map in Example 3 above.

9. What exhibit is located at (0, −3)?

10. In which quadrant is the Seaside Pavillion located?

Write the ordered pair corresponding to each point graphed at the right. Then state the quadrant or axis on which each point is located.

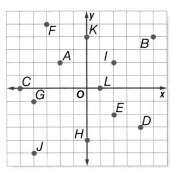

11. A	**12.** B	**13.** C
14. D	**15.** E	**16.** F
17. G	**18.** H	**19.** I
20. J	**21.** K	**22.** L

On graph paper, draw a coordinate plane. Then graph and label each point.

23. $M(5, 6)$	**24.** $N(-2, 10)$	**25.** $P(7, -8)$	**26.** $Q(3, 0)$
27. $R(-1, -7)$	**28.** $S(8, 1)$	**29.** $T(-3, 7)$	**30.** $U(5, -2)$
31. $V(0, 6)$	**32.** $W(-5, -7)$	**33.** $X(-4, 0)$	**34.** $Y(0, -5)$

GEOGRAPHY For Exercises 35–38, use the world map.

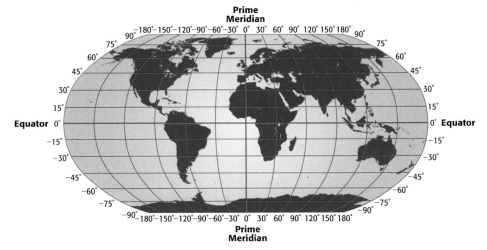

35. The world map can be divided into a coordinate plane where (x, y) represents (degrees longitude, degrees latitude). In what continent is the point (30° longitude, −15° latitude) located?

36. Which of the continents is located entirely in Quadrant II?

37. In what continent is the point (−90° longitude, 0° latitude) located?

38. Name a continent on the map that is located entirely in Quadrant I.

On graph paper, draw a coordinate plane. Then graph and label each point.

39. $X(1.5, 3.5)$ **40.** $Y\left(3\frac{1}{4}, 2\frac{1}{2}\right)$ **41.** $Z\left(2, 1\frac{2}{3}\right)$

42. **GEOMETRY** Graph four points on a coordinate plane so that they form a square when connected. Identify the ordered pairs.

43. **RESEARCH** Use the Internet or other resources to explain why the coordinate plane is sometimes called the Cartesian plane.

Determine whether each statement is *sometimes*, *always*, or *never* true. Explain or give a counterexample to support your answer.

44. Both *x*- and *y*-coordinates of a point in Quadrant III are negative.

NYSCC • NYSMT
Extra Practice, pp. 672, 705

45. The *y*-coordinate of a point that lies on the *y*-axis is negative.

46. The *y*-coordinate of a point in Quadrant II is negative.

H.O.T. Problems

47. **OPEN ENDED** Create a display that shows how to determine in what quadrant a point is located without graphing. Then provide an example that demonstrates how your graphic is used.

48. **CHALLENGE** Find the possible locations for any ordered pair with *x*- and *y*-coordinates always having the same sign. Explain.

49. **WRITING IN MATH** Explain why the location of point $A(1, -2)$ is different than the location of point $B(-2, 1)$.

NYSMT PRACTICE 6.G.10

50. Which of the following points lie within the triangle graphed at the right?

    A $A(-4, -1)$
    B $B(1, 3)$
    C $C(-1, 2)$
    D $D(2, -2)$

51. What are the coordinates of the point that shows the location of the lunch room on the map?

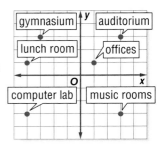

    F $(4, -1)$
    G $(-4, 1)$
    H $(1, 4)$
    J $(1, -4)$

Spiral Review

Replace each ● with <, >, or = to make a true sentence. (Lesson 2-2)

52. -8 ● -3

53. 26 ● -30

54. 14 ● $|-15|$

55. -40 ● $|40|$

56. Find the absolute value of -101. (Lesson 2-1)

57. **RUNNING** Salvador is training for a marathon. He runs 5 miles each day on weekdays and 8 miles each day on the weekends. How many miles does Salvador run in one week? (Lesson 1-1)

▷ **GET READY** for the Next Lesson

PREREQUISITE SKILL Add.

58. $138 + 246$

59. $814 + 512$

60. $2,653 + 4,817$

61. $6,003 + 5,734$

Algebra Lab
Adding Integers

You can use positive and negative counters to model the addition of integers. The counter ⊕ represents 1, and the counter ⊖ represents −1. Remember that addition means *combining* two sets.

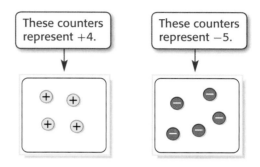

These counters represent +4.

These counters represent −5.

ACTIVITY

1. **Use counters to find −3 + (−6).**

Combine a set of 3 negative counters and a set of 6 negative counters.

Find the total number of counters.

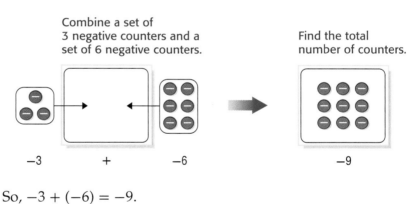

−3   +   −6   −9

So, −3 + (−6) = −9.

CHECK Your Progress

Use counters or a drawing to find each sum.

a. $5 + 6$
b. $-3 + (-5)$
c. $-5 + (-4)$
d. $7 + 3$
e. $-2 + (-5)$
f. $-8 + (-6)$

The following two properties are important when modeling operations with integers.

• When one positive counter is paired with one negative counter, the result is called a **zero pair.** The value of a zero pair is 0.

• You can add or remove zero pairs from a mat because adding or removing zero does not change the value of the counters on the mat.

Use counters to find each sum.

2 −4 + 2

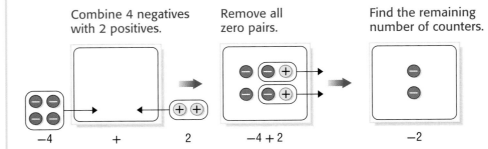

Combine 4 negatives with 2 positives.

Remove all zero pairs.

Find the remaining number of counters.

−4   +   2   −4 + 2   −2

So, −4 + 2 = −2.

3 5 + (−3)

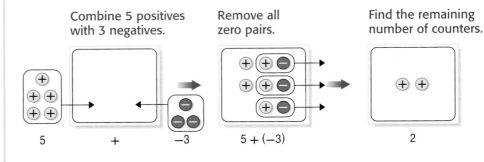

Combine 5 positives with 3 negatives.

Remove all zero pairs.

Find the remaining number of counters.

5   +   −3   5 + (−3)   2

So, 5 + (−3) = 2.

✓ CHECK Your Progress

Use counters or a drawing to find each sum.

g. −6 + 5   h. 3 + (−6)   i. −2 + 7

j. 8 + (−3)   k. −9 + 1   l. −4 + 10

ANALYZE THE RESULTS

1. Write two addition sentences where the sum is positive. In each sentence, one addend should be positive and the other negative.

2. Write two addition sentences where the sum is negative. In each sentence, one addend should be positive and the other negative.

3. **MAKE A CONJECTURE** What is a rule you can use to determine how to find the sum of two integers with the same sign? two integers with different signs?

2-4 Adding Integers

MAIN IDEA

Add integers.

NYS Core Curriculum

7.N.12 Add, subtract, multiply, and divide **integers** **7.N.13 Add** and subtract **two integers (with and without the use of a number line)** *Also addresses 7.CM.6*

New Vocabulary

opposites
additive inverse

NY Math Online

glencoe.com

• Extra Examples
• Personal Tutor
• Self-Check Quiz

▷ **GET READY** for the Lesson

SCIENCE Atoms are made of negative charges (electrons) and positive charges (protons). The helium atom shown has a total of 2 electrons and 2 protons.

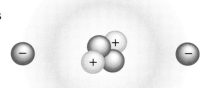

1. Represent the electrons in an atom of helium with an integer.

2. Represent the protons in an atom of helium with an integer.

3. Each proton-electron pair has a value of 0. What is the total charge of an atom of helium?

Combining protons and electrons in an atom is similar to adding integers.

EXAMPLE Add Integers with the Same Sign

1 **Find $-3 + (-2)$.**

Use a number line.

• Start at 0.
• Move 3 units left to show -3.
• From there, move 2 units left to show -2.

So, $-3 + (-2) = -5$.

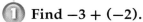

☑ **CHECK Your Progress**

a. $-5 + (-7)$ b. $-10 + (-4)$

These and other examples suggest the following rule.

Add Integers with the Same Sign	Key Concept
Words	To add integers with the same sign, add their absolute values. The sum is: • positive if both integers are positive. • negative if both integers are negative.
Examples	$7 + 4 = 11$ $-7 + (-4) = -11$

EXAMPLE Add Integers with the Same Sign

② **Find −26 + (−17).**

−26 + (−17) = −43   Both integers are negative, so the sum is negative.

✓**CHECK Your Progress**

c. −14 + (−16)   d. 23 + 38

Vocabulary Link · · · ·
Opposite

Everyday Use something that is across from or is facing the other way, as in running the opposite way

Math Use two numbers that are the same distance from 0, but on opposite sides of 0 on the number line

The integers 5 and −5 are called **opposites** because they are the same distance from 0, but on opposite sides of 0. Two integers that are opposites are also called **additive inverses**.

Additive Inverse Property	**Key Concept**
Words	The sum of any number and its additive inverse is 0.
Examples	5 + (−5) = 0

Number lines can also help you add integers with different signs.

EXAMPLES Add Integers with Different Signs

③ **Find 5 + (−3).**

Use a number line.

• Start at zero.
• Move 5 units right.
• Then move 3 units left.

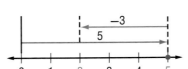

So, 5 + (−3) = 2.

④ **Find −3 + 2.**

Use a number line.

• Start at zero.
• Move 3 units left.
• Then move 2 units right.

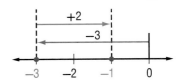

So, −3 + 2 = −1.

✓**CHECK Your Progress**

e. 6 + (−7)   f. −15 + 19

Study Tip

Look Back You can review **absolute value** in Lesson 2-1.

Add Integers with Different Signs	**Key Concept**
Words	To add integers with different signs, subtract their absolute values. The sum is:
	• positive if the positive integer's absolute value is greater.
	• negative if the negative integer's absolute value is greater.
Examples	9 + (−4) = 5 −9 + 4 = −5

5 Find $7 + (-1)$.

$7 + (-1) = 6$ Subtract absolute values; $7 - 1 = 6$. Since 7 has the greater absolute value, the sum is positive.

6 Find $-8 + 3$.

$-8 + 3 = -5$ Subtract absolute values; $8 - 3 = 5$. Since -8 has the greater absolute value, the sum is negative.

Study Tip

Properties Using the Commutative, Associative, and Additive Inverse Properties allows the calculation to be as simple as possible.

---

7 Find $2 + (-15) + (-2)$.

$$\begin{aligned}
2 + (-15) + (-2) &= 2 + (-2) + (-15) && \text{Commutative Property (+)}\\
&= [2 + (-2)] + (-15) && \text{Associative Property (+)}\\
&= 0 + (-15) && \text{Additive Inverse Property}\\
&= -15 && \text{Additive Identity Property}
\end{aligned}$$

✔ **CHECK Your Progress**

g. $10 + (-12)$ **h.** $-13 + 18$ **i.** $(-14) + (-6) + 6$

Real-World EXAMPLE

8 **ROLLER COASTERS** The graphic shows the change in height at several points on a roller coaster. Write an addition sentence to find the height at point D in relation to point A.

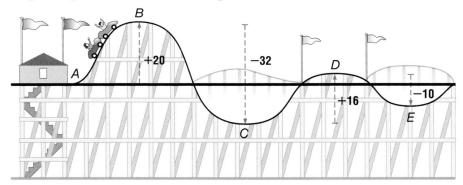

$$\begin{aligned}
20 + (-32) + 16 &= 20 + 16 + (-32) && \text{Commutative Property (+)}\\
&= 36 + (-32) && 20 + 16 = 36\\
&= 4 && \text{Subtract absolute values. Since 36}\\
& && \text{has the greater absolute value, the}\\
& && \text{sum is positive.}
\end{aligned}$$

The result is a positive integer. So, point D is 4 feet higher than point A.

✔ **CHECK Your Progress**

j. **WEATHER** The temperature is $-3°F$. An hour later, it drops $6°$ and 2 hours later, it rises $4°$. Write an addition sentence to describe this situation. Then find the sum and explain its meaning.

CHECK Your Understanding

Examples 1–6
(pp. 95–97)

Add.

1. $-6 + (-8)$
2. $4 + 5$
3. $-3 + 10$
4. $-15 + 8$
5. $7 + (-11)$
6. $14 + (-6)$

Example 7
(p. 97)

7. $-17 + 20 + (-3)$
8. $15 + 9 + (-9)$

Example 8
(p. 97)

9. **MONEY** Camilia owes her brother $25, so she gives her brother the $18 she earned dog-sitting for the neighbors. Write an addition sentence to describe this situation. Then find the sum and explain its meaning.

Practice and Problem Solving

HOMEWORK HELP	
For Exercises	See Examples
10–13	1, 2
14–21	3–6
22–27	7
28–31	8

Add.

10. $-22 + (-16)$
11. $-10 + (-15)$
12. $6 + 10$
13. $17 + 11$
14. $18 + (-5)$
15. $13 + (-19)$
16. $13 + (-7)$
17. $7 + (-20)$
18. $-19 + 24$
19. $-12 + 10$
20. $-30 + 16$
21. $-9 + 11$
22. $21 + (-21) + (-4)$
23. $-8 + (-4) + 12$
24. $-34 + 25 + (-25)$
25. $-16 + 16 + 22$
26. $25 + 3 + (-25)$
27. $7 + (-19) + (-7)$

Write an addition expression to describe each situation. Then find each sum and explain its meaning.

28. **SCUBA DIVING** Lena was scuba diving 14 meters below the surface of the water. She saw a nurse shark 3 meters above her.

29. **PELICANS** A pelican starts at 60 feet above sea level. It descends 60 feet to catch a fish.

30. **BANKING** Stephanie has $152 in the bank. She withdraws $20. Then she deposits $84.

31. **FOOTBALL** A quarterback is sacked for a loss of 5 yards. On the next play, his team receives a penalty and loses 15 more yards. Then the team gains 12 yards on the third play.

32. **MONEY** Josephine is saving money for a new bike and has already saved $17. Write the integers she should use to represent each entry.

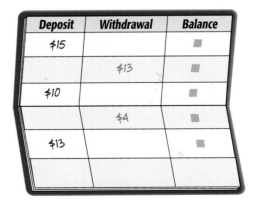

Deposit	Withdrawal	Balance
$15		■
	$13	■
$10		■
	$4	■
$13		■

ALGEBRA Evaluate each expression if $x = -10$, $y = 7$, and $z = -8$.

33. $x + 14$
34. $z + (-5)$
35. $x + y$
36. $x + z$

37. FIND THE DATA Refer to the Data File on pages 16–19 of your book. Choose some data and write a real-world problem in which you would add a positive and a negative integer. Then find the sum and explain its meaning.

NYSCC • NYSMT

Extra Practice, pp. 672, 705

H.O.T. Problems

38. FIND THE ERROR Beth and Jordan are finding $-12 + 15$. Who is correct? Explain your reasoning.

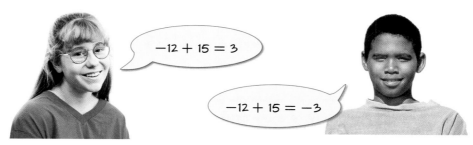

$-12 + 15 = 3$

$-12 + 15 = -3$

Beth   Jordan

CHALLENGE Simplify.

39. $8 + (-8) + a$ **40.** $x + (-5) + 1$ **41.** $-9 + m + (-6)$ **42.** $-1 + n + 7$

43. **WRITING IN MATH** Explain how you know whether a sum is positive, negative, or zero without actually adding.

NYSMT PRACTICE 7.N.13, 7.N.12

44. SHORT RESPONSE Find $-8 + (-11)$.

45. Find $-8 + 7 + (-3)$.

   A -18 **C** 2

   B -4 **D** 18

46. At 8 A.M., the temperature was 3°F below zero. By 1 P.M., the temperature rose 14°F and by 10 P.M. dropped 12°F. What was the temperature at 10 P.M.?

   F 5°F above zero

   G 5°F below zero

   H 1°F above zero

   J 1°F below zero

Spiral Review

Write the ordered pair for each point graphed at the right. Then name the quadrant or axis on which each point is located. (Lesson 2-3)

47. J **48.** K **49.** L **50.** M

51. Order 6, -3, 0, 4, -8, 1, and -4 from least to greatest. (Lesson 2-2)

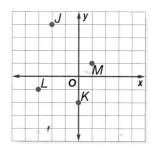

▷ **GET READY for the Next Lesson**

PREREQUISITE SKILL Subtract.

52. $287 - 125$ **53.** $420 - 317$ **54.** $5,684 - 2,419$ **55.** $7,000 - 3,891$

CHAPTER 2

Mid-Chapter Quiz

Lessons 2-1 through 2-4

NYS Core Curriculum
11: 7.N.1, 20: 6.G.10,
25: 7.N.12

Write an integer for each situation. (Lesson 2-1)

1. dropped 45 feet

2. a bank deposit of $100

3. gained 8 pounds

4. lost a $5 bill

5. **OCEANS** The deepest point in the world is the Mariana Trench in the Western Pacific Ocean at a depth of 35,840 feet below sea level. Write this depth as an integer. (Lesson 2-1)

Evaluate each expression. (Lesson 2-1)

6. $|-16|$

7. $|24|$

8. $|-9| - |3|$

9. $|-13| + |-1|$

10. **ANALYZE TABLES** The table shows the record low temperatures for January and February in Lincoln, Nebraska.

Month	Temperature (°F)
January	−33
February	−27

Source: University of Nebraska, Lincoln

Which month had the coldest temperature? (Lesson 2-2)

11. **MULTIPLE CHOICE** The local news records the following changes in average daily temperature for the past week: 4°, −7°, −3°, 2°, 9°, −8°, 1°. Which list shows the temperatures from least to greatest? (Lesson 2-2)

A 9°, 4°, 2°, 1°, −3°, −7°, −8°

B −7°, −8°, 1°, −3°, 2°, 4°, 9°

C −8°, −7°, −3°, 1°, 2°, 4°, 9°

D −8°, −7°, 1°, 2°, 3°, 4°, 9°

Replace each ● with <, >, or = to make a true sentence. (Lesson 2-2)

12. $-4 \,●\, 4$

13. $-8 \,●\, -11$

14. $|-14| \,●\, |3|$

15. $|-12| \,●\, |12|$

On graph paper, draw a coordinate plane. Then graph and label each point. (Lesson 2-3)

16. $D(4, -3)$

17. $E(-1, 2)$

18. $F(0, -5)$

19. $G(-3, 0)$

20. **MULTIPLE CHOICE** Which line contains the ordered pair $(-1, 4)$? (Lesson 2-3)

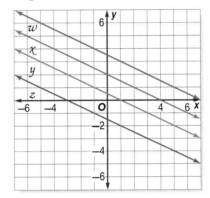

F line w

H line y

G line x

J line z

Add. (Lesson 2-4)

21. $3 + 4 + (-3)$

22. $7 + (-11)$

23. $-5 + (-6)$

24. $8 + (-1) + 1$

25. **MULTIPLE CHOICE** Kendra deposited $78 into her savings account. Two weeks later, she deposited a check for $50 into her account and withdrew $27. Which of the following expressions represents the amount of money left in her account? (Lesson 2-4)

A $78 + (−$50) + (−$27)

B $78 + (−$50) + $27

C $78 + $50 + (−$27)

D $78 + $50 + $27

Algebra Lab
Subtracting Integers

MAIN IDEA

Use counters to model the subtraction of integers.

NYS Core Curriculum

7.N.12 Add, **subtract**, multiply, and divide **integers**   **7.N.13** Add and **subtract two integers** (with and **without the use of a number line**)

You can also use counters to model subtraction of integers. Remember one meaning of subtraction is to *take away*.

ACTIVITY

Use counters to find each difference.

1 5 − 2

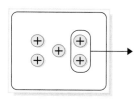

Place 5 positive counters on the mat. Remove 2 positive counters.

So, 5 − 2 = 3.

2 4 − (−3)

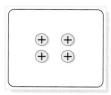

Place 4 positive counters on the mat. Remove 3 negative counters. However, there are 0 negative counters.

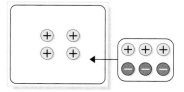

Add 3 zero pairs to the set.

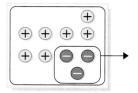

Now you can remove 3 negative counters. Find the remaining number of counters.

So, 4 − (−3) = 7.

CHECK Your Progress

Use counters or a drawing to find each difference.

a. 7 − 6   **b.** 5 − (−3)   **c.** 6 − (−3)   **d.** 5 − 8

Use counters to find each difference.

3 −6 − (−3)

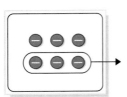

Place 6 negative counters on the mat. Remove 3 negative counters.

So, −6 − (−3) = −3.

4 −5 − 1

Place 5 negative counters on the mat. Remove 1 positive counter. However, there are 0 positive counters.

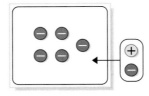

Add 1 zero pair to the set.

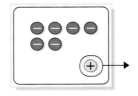

Now you can remove 1 positive counter. Find the remaining number of counters.

So, −5 − 1 = −6.

Reading Math

Minuends, Subtrahends, and Differences In the subtraction sentence −5 − 1 = −6, −5 is the *minuend*, 1 is the *subtrahend*, and −6 is the *difference*.

 CHECK Your Progress

Use counters or a drawing to find each difference.

e. −6 − (−3) **f.** −7 − 3 **g.** −5 − (−7)

ANALYZE THE RESULTS

1. Write two subtraction sentences where the difference is positive. Use a combination of positive and negative integers.

2. Write two subtraction sentences where the difference is negative. Use a combination of positive and negative integers.

3. **MAKE A CONJECTURE** Write a rule that will help you determine the sign of the difference of two integers.

2-5 Subtracting Integers

MAIN IDEA

Subtract integers.

NYS Core Curriculum

7.N.12 Add, **subtract**, multiply, and divide **integers 7.N.13** Add and **subtract two integers** (with and **without the use of a number line**)

NY Math Online

glencoe.com

- Concepts In Motion
- Extra Examples
- Personal Tutor
- Self-Check Quiz

MINI Lab

You can use a number line to model a subtraction problem.

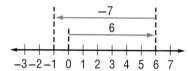

1. Write a related addition sentence for the subtraction sentence.

Use a number line to find each difference. Write an equivalent addition sentence for each.

2. $1 - 5$ **3.** $-2 - 1$ **4.** $-3 - 4$ **5.** $0 - 5$

When you subtract 7, the result is the same as adding its opposite, -7.

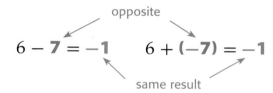

This and other examples suggest the following rule.

Subtract Integers	Key Concept
Words	To subtract an integer, add its opposite.
Examples	$4 - 9 = 4 + (-9) = -5$ $7 - (-10) = 7 + (10) = 17$

EXAMPLES Subtract Positive Integers

1 **Find $8 - 13$.**

$8 - 13 = 8 + (-13)$ To subtract 13, add -13.

$\qquad\quad = -5$ Simplify.

2 **Find $-10 - 7$.**

$-10 - 7 = -10 + (-7)$ To subtract 7, add -7.

$\qquad\qquad = -17$ Simplify.

CHECK Your Progress

a. $6 - 12$ **b.** $-20 - 15$ **c.** $-22 - 26$

Subtract Negative Integers

3 Find $1 - (-2)$.

$1 - (-2) = 1 + 2$ To subtract -2, add 2.

$\qquad\qquad = 3$ Simplify.

4 Find $-10 - (-7)$.

$-10 - (-7) = -10 + 7$ To subtract -7, add 7.

$\qquad\qquad = -3$ Simplify.

✓ CHECK Your Progress

d. $4 - (-12)$ **e.** $-15 - (-5)$ **f.** $18 - (-6)$

EXAMPLE Evaluate an Expression

5 **ALGEBRA** Evaluate $x - y$ if $x = -6$ and $y = -5$.

$x - y = -6 - (-5)$ Replace x with -6 and y with -5.

$\qquad = -6 + (5)$ To subtract -5, add 5.

$\qquad = -1$ Simplify.

✓ CHECK Your Progress

Evaluate each expression if $a = 5$, $b = -8$, and $c = -9$.

g. $b - 10$ **h.** $a - b$ **i.** $c - a$

Real-World EXAMPLE

6 **SPACE** The temperatures on the Moon vary from $-173°C$ to $127°C$. Find the difference between the maximum and minimum temperatures.

To find the difference in temperatures, subtract the lower temperature from the higher temperature.

Estimate $100 + 200 = 300$

$127 - (-173) = 127 + 173$ To subtract -173, add 173.

$\qquad\qquad = 300$ Simplify.

So, the difference between the temperatures is 300°C.

Real-World Link . . . .
The mean surface temperature on the Moon during the day is 107°C.
Source: Views of the Solar System

✓ CHECK Your Progress

j. **GEOGRAPHY** The Dead Sea's deepest part is 799 meters below sea level. A plateau to the east of the Dead Sea rises to about 1,340 meters above sea level. What is the difference between the top of the plateau and the deepest part of the Dead Sea?

Examples 1, 2
(p. 103)

Subtract.

1. $14 - 17$

2. $10 - 30$

3. $-4 - 8$

4. $-2 - 23$

Examples 3, 4
(p. 104)

5. $14 - (-10)$

6. $5 - (-16)$

7. $-3 - (-1)$

8. $-11 - (-9)$

Example 5
(p. 104)

ALGEBRA Evaluate each expression if $p = 8$, $q = -14$, and $r = -6$.

9. $r - 15$

10. $q - r$

11. $p - q$

Example 6
(p. 104)

12. **EARTH SCIENCE** The sea-surface temperatures range from $-2°C$ to $31°C$. Find the difference between the maximum and minimum temperatures.

Practice and Problem Solving

Subtract.

For Exercises	See Examples
13–16, 21–24	1, 2
17–20, 25–28	3, 4
29–36	5
37–40	6

HOMEWORK HELP

13. $0 - 10$

14. $13 - 17$

15. $-9 - 5$

16. $-8 - 9$

17. $4 - (-19)$

18. $27 - (-8)$

19. $-11 - (-42)$

20. $-27 - (-19)$

21. $12 - 26$

22. $31 - 48$

23. $-25 - 5$

24. $-44 - 41$

25. $52 - (-52)$

26. $15 - (-14)$

27. $-27 - (-33)$

28. $-18 - (-20)$

ALGEBRA Evaluate each expression if $f = -6$, $g = 7$, and $h = 9$.

29. $g - 7$

30. $f - 6$

31. $-h - (-9)$

32. $f - g$

33. $h - f$

34. $g - h$

35. $5 - f$

36. $4 - (-g)$

ANALYZE TABLES For Exercises 37–40, use the information below.

State	California	Georgia	Louisiana	New Mexico	Texas
Lowest Elevation (ft)	−282	0	−8	2,842	0
Highest Elevation (ft)	14,494	4,784	535	13,161	8,749

37. What is the difference between the highest elevation in Texas and the lowest elevation in Louisiana?

38. Find the difference between the lowest elevation in New Mexico and the lowest elevation in California.

39. Find the difference between the highest elevation in Georgia and the lowest elevation in California.

40. What is the difference between the lowest elevations in Texas and Louisiana?

NYSCC • NYSMT

Extra Practice, pp. 672, 705

ALGEBRA Evaluate each expression if $h = -12$, $j = 4$, and $k = 15$.

41. $-j + h - k$

42. $|h - j|$

43. $k - j - h$

44. OPEN ENDED Write a subtraction sentence using integers. Then, write the equivalent addition sentence, and explain how to find the sum.

45. FIND THE ERROR Alicia and Mei are finding $-15 - (-18)$. Who is correct? Explain your reasoning.

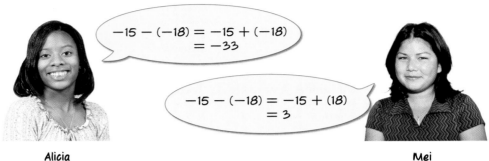

$$-15 - (-18) = -15 + (-18)$$
$$= -33$$

$$-15 - (-18) = -15 + (18)$$
$$= 3$$

Alicia   Mei

46. CHALLENGE *True* or *False*? When n is a negative integer, $n - n = 0$.

47. WRITING IN MATH Explain how additive inverses are used in subtraction.

NYSMT PRACTICE 7.N.12, 7.N.13

48. Which sentence about integers is *not* always true?

A positive − positive = positive

B positive + positive = positive

C negative + negative = negative

D positive − negative = positive

49. Morgan drove from Los Angeles (elevation 330 feet) to Death Valley (elevation −282 feet). What is the difference in elevation between Los Angeles and Death Valley?

F 48 feet **H** 582 feet

G 148 feet **J** 612 feet

Spiral Review

Add. (Lesson 2-4)

50. $10 + (-3)$ **51.** $-2 + (-9)$ **52.** $-7 + (-6)$ **53.** $-18 + 4$

54. In which quadrant does the ordered pair $(5, -6)$ lie? (Lesson 2-3)

55. NUMBERS A number times 2 is added to 7. The result is 23. What is the number? Use the *guess and check* strategy. (Lesson 1-5)

GET READY for the Next Lesson

Add. (Lesson 2-4)

56. $-6 + (-6) + (-6) + (-6)$ **57.** $-11 + (-11) + (-11)$

58. $-2 + (-2) + (-2) + (-2)$ **59.** $-8 + (-8) + (-8)$

2-6 Multiplying Integers

MAIN IDEA

Multiply integers.

NYS Core Curriculum

7.N.12 Add, subtract, **multiply**, and divide **integers** *Also addresses 7.PS.15, 7.CM.2*

NY Math Online

glencoe.com

• Extra Examples
• Personal Tutor
• Self-Check Quiz

▷ **MINI Lab**

Counters can be used to multiply integers.

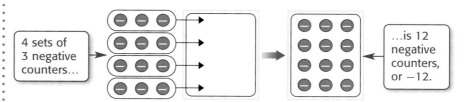

4 sets of 3 negative counters…

…is 12 negative counters, or −12.

1. Write a multiplication sentence that describes the model above.

Find each product using counters or a drawing.

2. 3(−2)   **3.** 4(−3)   **4.** 1(−7)   **5.** 5(−2)

Remember that multiplication is the same as repeated addition.

$4(-3) = (-3) + (-3) + (-3) + (-3)$ −3 is used as an addend four times.

$ = -12$

By the Commutative Property of Multiplication, $4(-3) = -3(4)$.

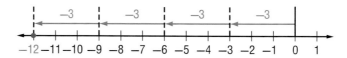

Multiply Integers with Different Signs	Key Concept
Words	The product of two integers with different signs is negative.
Examples	$6(-4) = -24$ $-5(7) = -35$

EXAMPLES Multiply Integers with Different Signs

① **Find 3(−5).**

$3(-5) = -15$ The integers have different signs. The product is negative.

② **Find −6(8).**

$-6(8) = -48$ The integers have different signs. The product is negative.

 **Your Progress**

a. 9 (−2)   **b.** −7(4)

The product of two positive integers is positive. You can use a pattern to find the sign of the product of two negative integers.

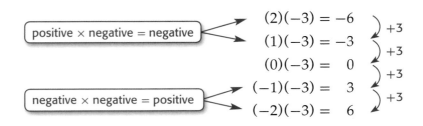

Study Tip

Multiplying by Zero
The Multiplicative Property of Zero states that when any number is multiplied by zero, the product is zero.

Each product is 3 more than the previous product. This pattern can also be shown on a number line.

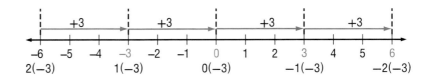

These and other examples suggest the following rule.

Multiply Integers with Same Sign   Key Concept

Words   The product of two integers with the same sign is positive.

Examples $2(6) = 12$ $-10(-6) = 60$

EXAMPLES   Multiply Integers with the Same Sign

3 Find $-11(-9)$.

$-11(-9) = 99$ The integers have the same sign. The product is positive.

Study Tip

Look Back
You can review **exponents** in Lesson 1-2.

4 Find $(-4)^2$.

$(-4)^2 = (-4)(-4)$ There are two factors of -4.

$\quad\quad = 16$ The product is positive.

5 Find $-3(-4)(-2)$.

$-3(-4)(-2) = [-3(-4)](-2)$ Associative Property

$\quad\quad\quad = 12(-2)$ $-3(-4) = 12$

$\quad\quad\quad = -24$ $12(-2) = -24$

CHECK Your Progress

c. $-12(-4)$ d. $(-5)^2$ e. $-7(-5)(-3)$

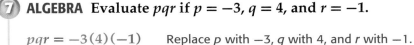

Real-World EXAMPLE

6 **SUBMERSIBLES** A submersible is diving from the surface of the water at a rate of 90 feet per minute. What is the depth of the submersible after 7 minutes?

If the submersible descends 90 feet per minute, then after 7 minutes, the vessel will be at $7(-90)$ or -630 feet. Thus, the submersible will descend to 630 feet below the surface.

CHECK Your Progress

f. **MONEY** Mr. Simon's bank automatically deducts a $4 monthly maintenance fee from his savings account. What integer represents a change in his savings account from one year of fees?

Negative numbers are often used when evaluating algebraic expressions.

EXAMPLE   Evaluate Expressions

7 **ALGEBRA** Evaluate pqr if $p = -3$, $q = 4$, and $r = -1$.

$$
\begin{aligned}
pqr &= -3(4)(-1) &&\text{Replace } p \text{ with } -3, q \text{ with } 4, \text{ and } r \text{ with } -1.\\
&= (-12)(-1) &&\text{Multiply } -3 \text{ and } 4.\\
&= 12 &&\text{Multiply } -12 \text{ and } -1.
\end{aligned}
$$

CHECK Your Progress

g. Evaluate xyz if $x = -7$, $y = -4$, and $z = 2$.

CHECK Your Understanding

Examples 1, 2
(p. 107)

Multiply.

1. $6(-10)$ **2.** $11(-4)$ **3.** $-2(14)$ **4.** $-8(5)$

Examples 3–5
(p. 108)

Multiply.

5. $-15(-3)$ **6.** $-7(-9)$ **7.** $(-8)^2$

8. $(-3)^3$ **9.** $-1(-3)(-4)$ **10.** $2(4)(5)$

Example 6
(p. 109)

11. **MONEY** Tamera owns 100 shares of a certain stock. Suppose the price of the stock drops by $3 per share. Write a multiplication expression to find the change in Tamera's investment. Explain the answer.

Example 7
(p. 109)

ALGEBRA Evaluate each expression if $f = -1$, $g = 7$, and $h = -10$.

12. $5f$ **13.** fgh

Practice and Problem Solving

HOMEWORK HELP

For Exercises	See Examples
14–19, 28	1, 2
20–27, 29	3–5
30–37	7
38–39	6

Multiply.

14. $8(-12)$

15. $11(-20)$

16. $-15(4)$

17. $-7(10)$

18. $-7(11)$

19. $25(-2)$

20. $-20(-8)$

21. $-16(-5)$

22. $(-6)^2$

23. $(-5)^3$

24. $(-4)^3$

25. $(-9)^2$

26. $-4(-2)(-8)$

27. $-9(-1)(-5)$

28. Find the product of 10 and -10.

29. Find -7 squared.

ALGEBRA Evaluate each expression if $w = 4$, $x = -8$, $y = 5$, and $z = -3$.

30. $-4w$

31. $3x$

32. xy

33. xz

34. $7wz$

35. $-2wx$

36. xyz

37. wyx

Write a multiplication expression to represent each situation. Then find each product and explain its meaning.

38. **ECOLOGY** Wave erosion causes a certain coastline to recede at a rate of 3 centimeters each year. This occurs uninterrupted for a period of 8 years.

39. **EXERCISE** Ethan burns 650 Calories when he runs for 1 hour. Suppose he runs 5 hours in one week.

ALGEBRA Evaluate each expression if $a = -6$, $b = -4$, $c = 3$, and $d = 9$.

40. $-3a^2$

41. $-cd^2$

42. $-2a + b$

43. $b^2 - 4ac$

44. **BANKING** Tamika's aunt writes a check for $150 each month for her car loan. She writes another check for $300 twice a year to pay for car insurance. Write an expression involving multiplication and addition to describe how these expenses affect her checking account balance on a yearly basis. Then evaluate the expression and explain its meaning.

45. **FIND THE DATA** Refer to the Data File on pages 16–19 of your book. Choose some data and write a real-world problem in which you would multiply integers.

GEOMETRY For Exercises 46–48, use the graph at the right.

46. Name the ordered pairs for A, B, and C. Multiply each x- and y-coordinate by -1 to get three new ordered pairs.

47. Graph the ordered pairs and connect them to form a new triangle. Describe its position with respect to the original triangle.

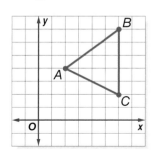

48. In which quadrant would a new triangle lie if only the y-coordinates of the original triangle are multiplied by -1?

NYSCC • NYSMT

Extra Practice, pp. 673, 705

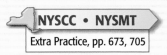
Use the *look for a pattern* strategy to solve Exercises 4–6.

4. **DISPLAYS** A display of cereal boxes is stacked as shown below.

If the display contains 7 rows of boxes and the top three rows are shown, how many boxes are in the display?

5. **MONEY** Peter is saving money to buy an MP3 player. After one month, he has $50. After 2 months, he has $85. After 3 months, he has $120. After 4 months, he has $155. He plans to keep saving at the same rate. How long will it take Peter to save enough money to buy an MP3 player that costs $295?

6. **INSECTS** The table shows how many times a cricket chirps at different temperatures. About how many times will a cricket chirp when the temperature is 60°F?

Outside Temperature (°F)	Chirps per Minute
85°F	180
80°F	160
75°F	140
70°F	120

Use any strategy to solve Exercises 7–14. Some strategies are shown below.

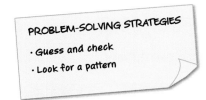

PROBLEM-SOLVING STRATEGIES
· Guess and check
· Look for a pattern

7. **COINS** Adelina has six coins that total $0.86. What are the coins?

8. **ELEVATION** The lowest point in Mexico is Laguna Salada with an elevation of −10 meters. The highest point in Mexico is Volcan Pico de Orizaba with an elevation of 5,700 meters. What is the difference in these elevations?

9. **MONEY** While on vacation, Edmundo sent postcards and letters to his friends. He spent $3.42 on postage. A stamp for a letter costs 41¢, and a stamp for a postcard costs 26¢. How many postcards and letters did he send?

10. **GEOMETRY** What is the next figure in the pattern shown?

11. **POPULATION** The total land area of North Carolina is about 48,711 square miles. If an average of 183 persons were living in each square mile of North Carolina in 2007, what was the population of North Carolina in 2007?

12. **GOLF** Allie's golf scores for the first five holes are given in the table. What is her total score after the first five holes?

Hole	Score
1	0
2	1
3	−1
4	−2
5	3

13. **FLOWERS** A sunflower grows to be about 252 centimeters tall in 3 months. What is the average rate of growth per month?

14. **NUMBERS** Determine the next three numbers in the pattern below.
48, 42, 36, 30, 24, …

2-8 Dividing Integers

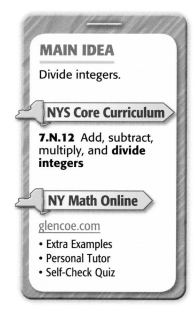

MAIN IDEA

Divide integers.

NYS Core Curriculum

7.N.12 Add, subtract, multiply, and **divide integers**

NY Math Online

glencoe.com

• Extra Examples
• Personal Tutor
• Self-Check Quiz

MINI Lab

You can use counters to model division of integers. Follow these steps to find $-8 \div 2$.

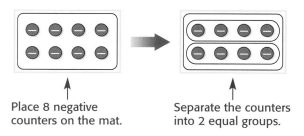

Place 8 negative counters on the mat.

Separate the counters into 2 equal groups.

There are 4 negative counters in each group. So, $-8 \div 2 = -4$.

Find each quotient using counters or a drawing.

1. $-6 \div 2$ **2.** $-12 \div 3$

Division of numbers is related to multiplication. When finding the quotient of two integers, you can use a related multiplication sentence.

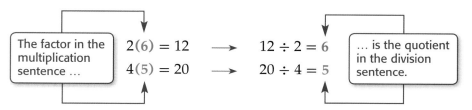

The factor in the multiplication sentence …

$2(6) = 12 \longrightarrow 12 \div 2 = 6$

$4(5) = 20 \longrightarrow 20 \div 4 = 5$

… is the quotient in the division sentence.

Since multiplication and division sentences are related, you can use them to find the quotient of integers with different signs.

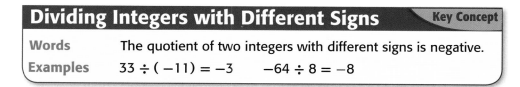

different signs

$2(-6) = -12 \longrightarrow -12 \div 2 = -6$

$-2(-6) = 12 \longrightarrow 12 \div (-2) = -6$

negative quotient

These related sentences lead to the following rule.

Dividing Integers with Different Signs	**Key Concept**
Words	The quotient of two integers with different signs is negative.
Examples	$33 \div (-11) = -3$ $-64 \div 8 = -8$

Practice and Problem Solving

HOMEWORK HELP

For Exercises	See Examples
10–13, 16–19	1, 2
14–15, 20–23	3
24–31	4
32–33	5

Divide.

10. $50 \div (-5)$

11. $56 \div (-8)$

12. $-18 \div 9$

13. $-36 \div 4$

14. $-15 \div (-3)$

15. $-100 \div (-10)$

16. $\dfrac{22}{-2}$

17. $\dfrac{84}{-12}$

18. $\dfrac{-26}{13}$

19. $\dfrac{-27}{3}$

20. $\dfrac{-21}{-7}$

21. $\dfrac{-54}{-6}$

22. Divide -200 by -100.

23. Find the quotient of -65 and -13.

ALGEBRA Evaluate each expression if $r = 12$, $s = -4$, and $t = -6$.

24. $-12 \div r$

25. $72 \div t$

26. $r \div s$

27. $rs \div 16$

28. $\dfrac{t - r}{3}$

29. $\dfrac{8 - r}{-2}$

30. $\dfrac{s + t}{5}$

31. $\dfrac{t + 9}{-3}$

32. **MONEY** Last year, Mr. Engle's total income was $52,000, while his total expenses were $53,800. Use the expression $\dfrac{I - E}{12}$, where I represents total income and E represents total expenses, to find the average difference between his income and expenses each month.

33. **SCIENCE** The boiling point of water is affected by changes in elevation. Use the expression $\dfrac{-2A}{1{,}000}$, where A represents the altitude in feet, to find the number of degrees Fahrenheit the boiling point of water changes at an altitude of 5,000 feet.

ALGEBRA Evaluate each expression if $d = -9$, $f = 36$, and $g = -6$.

34. $\dfrac{-f}{d}$

35. $\dfrac{12 - (-f)}{-g}$

36. $\dfrac{f^2}{d^2}$

37. $g^2 \div f$

38. **PLANETS** The temperature on Mars ranges widely from $-207°F$ at the winter pole to almost $80°F$ on the dayside during the summer. Use the expression $\dfrac{-207 + 80}{2}$ to find the average of the temperature extremes on Mars.

39. **ANALYZE GRAPHS** The *mean* of a set of data is the sum of the data divided by the number of items in the data set. The graph shows the approximate depths where certain fish are found in the Caribbean. What is the mean depth of the fish shown?

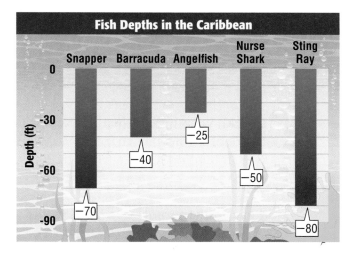

Fish Depths in the Caribbean

NYSCC • NYSMT

Extra Practice, pp. 673, 705

40. **OPEN ENDED** Write a division sentence with a quotient of −12.

41. **Which One Doesn't Belong?** Identify the expression that does not belong with the other three. Explain your reasoning.

| −66 ÷ 11 | −32 ÷ (−4) | 16 ÷ (−4) | −48 ÷ 4 |

42. **PATTERNS** Find the next two numbers in the pattern 729, −243, 81, −27, 9, … . Explain your reasoning.

43. **CHALLENGE** Order from least to greatest all of the numbers by which −20 is divisible.

44. **WRITING IN MATH** Evaluate $-2 \cdot (2^2 + 2) \div 2^2$. Justify each step in the process.

NYSMT PRACTICE 7.N.12

45. Find 18 ÷ (−3).

   A −6

   B $\dfrac{-1}{6}$

   C 6

   D 15

46. On December 24, 1924, the temperature in Fairfield, Montana, fell from 63°F at noon to −21°F at midnight. What was the average temperature change per hour?

   F −3.5°F

   G −7°F

   H −42°F

   J −84°F

Spiral Review

47. **GEOMETRY** What is the next figure in the pattern shown at the right? (Lesson 2-7)

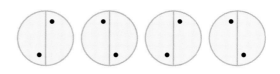

Multiply. (Lesson 2-6)

48. 14(−2)　　　　49. −20(−3)　　　　50. −5(7)　　　　51. (−9)²

52. Find 6 − (−12). (Lesson 2-5)

53. **DIVING** Valentina jumped into 10 feet of water and touched the bottom of the pool before she surfaced. Write an integer to describe where Valentina was in relation to the surface of the water when she touched the bottom of the pool. (Lesson 2-1)

Find each square root. (Lesson 1-3)

54. $\sqrt{324}$　　　　　55. $\sqrt{900}$　　　　　56. $\sqrt{196}$

FOLDABLES® Study Organizer ▶ GET READY to Study

Be sure the following Big Ideas are noted in your Foldable.

> **2-1**
> Integers and
> ○ Absolute Value

BIG Ideas

Absolute Value (Lesson 2-1)
• The absolute value of a number is the distance the number is from zero on a number line.

Comparing and Ordering Integers (Lesson 2-2)
• When two numbers are graphed on a number line, the number to the left is always less than the number to the right.

Graphing Points (Lesson 2-3)
• On a coordinate plane, the horizontal number line is the *x*-axis and the vertical number line is the *y*-axis. The origin is at (0, 0) and is the point where the number lines intersect. The *x*-axis and *y*-axis separate the plane into four quadrants.

Integer Operations (Lessons 2-4, 2-5, 2-6, 2-8)
• To add integers with the same sign, add their absolute value. The sum is positive if both integers are positive and negative if both integers are negative.

• The sum of any number and its additive inverse is 0.

• To add integers with different signs, subtract their absolute values. The sum is positive if the positive integer's absolute value is greater and negative if the negative integer's absolute value is greater.

• To subtract an integer, add its opposite.

• The product or quotient of two integers with different signs is negative.

• The product or quotient of two integers with the same sign is positive.

⬦ Key Vocabulary

absolute value (p. 81)	origin (p. 88)
additive inverse (p. 96)	positive integer (p. 80)
coordinate plane (p. 88)	quadrant (p. 88)
graph (p. 80)	*x*-axis (p. 88)
integer (p. 80)	*x*-coordinate (p. 88)
negative integer (p. 80)	*y*-axis (p. 88)
opposites (p. 96)	*y*-coordinate (p. 88)
ordered pair (p. 88)	

⬦ Vocabulary Check

State whether each sentence is *true* or *false*. If *false*, replace the underlined word or number to make a true sentence.

1. Integers less than zero are <u>positive</u> integers.

2. The <u>origin</u> is the point where the *x*-axis and *y*-axis intersect.

3. The <u>absolute value</u> of 7 is −7.

4. The sum of two negative integers is <u>positive</u>.

5. The <u>*x*-coordinate</u> of the ordered pair (2, −3) is −3.

6. Two integers that are opposites are also called <u>additive inverses</u>.

7. The product of a positive and a negative integer is <u>negative</u>.

8. The *x*-axis and the *y*-axis separate the plane into four <u>coordinates</u>.

9. The quotient of two negative integers is <u>negative</u>.

Lesson-by-Lesson Review

2-1

7.N.1,
7.N.11

Integers and Absolute Value (pp. 80–83)

Write an integer for each situation.

10. a loss of $150

11. 350 feet above sea level

12. a gain of 8 yards

13. 12°F below 0

Evaluate each expression.

14. $|100|$

15. $|-32|$

16. $|-16| + |9|$

17. **JUICE** Mavis drank 48 milliliters of apple juice before replacing the carton in the refrigerator. Write an integer that shows the change in the volume of juice in the carton.

Example 1 Write an integer for 8 feet below sea level.

Since this situation represents an elevation *below* sea level, −8 represents the situation.

Example 2 Evaluate $|-10|$.

On the number line, the graph of −10 is 10 units from 0.

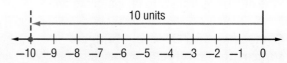

So, $|-10| = 10$.

2-2

7.N.1,
7.N.3

Comparing and Ordering Integers (pp. 84–87)

Replace each ● with < or > to make a true sentence.

18. −3 ● −9   19. 8 ● −12

20. −3 ● 3   21. $|-10|$ ● $|-13|$

22. 25 ● $|8|$ 23. 0 ● $|-4|$

Order each set of integers from least to greatest.

24. $\{-3, 8, -10, 0, 5, -12, 9\}$

25. $\{-21, 19, -23, 14, -32, 25\}$

26. $\{-17, -18, 18, 15, -16, 16\}$

27. **WEATHER** The high temperatures in degrees Celsius for ten cities were 0, 10, −5, 12, 25, −6, 20, −10, 5 and 2. Order these temperatures from least to greatest.

Example 3 Replace ● with < or > to make −4 ● −7 a true sentence.

Graph each integer on a number line.

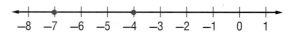

Since −4 is to the right of −7, −4 > −7.

Example 4 Order the integers −4, −3, 5, 3, 0, −2 from least to greatest.

Graph the integers on a number line.

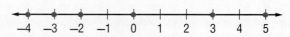

Order the integers by reading from left to right: −4, −3, −2, 0, 3, 5.

2-3

The Coordinate Plane (pp. 88–92)

6.G.10

On graph paper, draw a coordinate plane.
Then graph and label each point.

28. $E(1, -4)$

29. $F(-4, 2)$

30. $G(-2, -3)$

31. $H(4, 0)$

32. **ROUTES** Starting at the school, Pilar walked 1 block east and 3 blocks south. From there, she walked 5 blocks west and 4 blocks north to the park. If the school represents the origin, what is the ordered pair for the park?

Example 5 Graph and label the point $S(3, -1)$.

Draw a coordinate plane. Move 3 units to the right. Then move 1 unit down. Draw a dot and label it $S(3, -1)$.

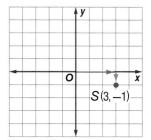

2-4

Adding Integers (pp. 95–99)

7.N.13,
7.N.12

Add.

33. $-6 + 8$ 34. $-4 + (-9)$

35. $7 + (-12)$ 36. $-18 + 18$

37. **HIKING** Samuel hiked 75 feet up a mountain. He then hiked 22 feet higher. Then, he descended 8 feet, and finally climbed up another 34 feet. What is Samuel's final elevation?

Example 6 Find $-4 + 3$.

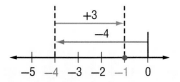

So, $-4 + 3 = -1$.

2-5

Subtracting Integers (pp. 103–106)

7.N.12,
7.N.13

Subtract.

38. $-5 - 8$ 39. $3 - 6$

40. $5 - (-2)$ 41. $-4 - (-8)$

42. **GOLF** Owen shot 2 under par while his friend Nathan shot 3 above par. By how many shots was Owen's score better than Nathan's?

Example 7 Find $-3 - 9$.

$-3 - 9 = -3 + (-9)$ To subtract 9, add −9.

$\quad\quad = -12$ Simplify.

2-6 **Multiplying Integers** (pp. 107–111)

7.N.12

Multiply.

43. $-4(3)$
44. $8(-6)$
45. $-5(-7)$
46. $-2(40)$

ALGEBRA Evaluate each expression if $a = -4$, $b = -7$, and $c = 5$.

47. ab
48. $-3c$
49. bc
50. abc

Example 8 Find $-4(3)$.

$-4(3) = -12$ The integers have different signs. The product is negative.

Example 9 Evaluate xyz if $x = -6$, $y = 11$, and $z = -10$.

xyz
$= (-6)(11)(-10)$ $x = -6, y = 11, z = -10$.
$= (-66)(-10)$ Multiply -6 and 11.
$= 660$ Multiply -66 and -10.

2-7 **PSI: Look for a Pattern** (pp. 112–113)

7.PS.4

Solve. Look for a pattern.

51. **HEALTH** The average person blinks 12 times per minute. At this rate, how many times does the average person blink in one day?

52. **SALARY** Suki gets a job that pays $31,000 per year. She is promised a $2,200 raise each year. At this rate, what will her salary be in 7 years?

53. **DOGS** A kennel determined that they need 144 feet of fencing to board 2 dogs, 216 feet to board 3 dogs, and 288 feet to board 4 dogs. If this pattern continues, how many feet of fencing is needed to board 8 dogs?

Example 10 A theater has 18 seats in the first row, 24 seats in the second row, 30 seats in the third row, and so on. If this pattern continues, how many seats are in the sixth row?

Begin with 18 seats and add 6 seats for each additional row. So, there are 48 seats in the sixth row.

Row	Number of Seats
1	18
2	24
3	30
4	36
5	42
6	48

2-8 **Dividing Integers** (pp. 114–118)

7.N.12

Divide.

54. $-45 \div (-9)$
55. $36 \div (-12)$
56. $-12 \div 6$
57. $-81 \div (-9)$

Example 11 Find $-72 \div (-9)$.

$-72 \div (-9) = 8$ The integers have the same sign. The quotient is positive.

1. **WEATHER** Adam is recording the change in the outside air temperature for a science project. At 8:00 A.M., the high temperature was 42°F. By noon, the outside temperature had fallen 11°F. By mid-afternoon, the outside air temperature had fallen 12°F and by evening, it had fallen an additional 5°F. Write an integer that describes the final change in temperature.

Evaluate each expression.

2. $|-3|$

3. $|-18| - |6|$

Replace each ● with <, >, or = to make a true sentence.

4. -3 ● -9

5. $|9|$ ● $|-12|$

6. The Iowa Hawkeyes recorded the following yardage in six plays: 9, −2, 5, 0, 12, and −7. Order these integers from least to greatest.

7. **MULTIPLE CHOICE** Which of the following coordinates lie within the rectangle graphed below?

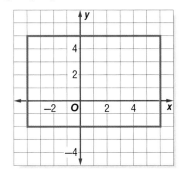

A $(5, 6)$ C $(-5, 1)$

B $(0, -3)$ D $(-3, 0)$

8. **DEBT** Amanda owes her brother $24. If she plans to pay him back an equal amount from her piggy bank each day for six days, describe the change in the amount of money in her piggy bank each day.

Write the ordered pair for each point graphed. Then name the quadrant in which each point is located.

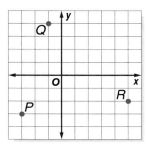

9. P 10. Q 11. R

Add, subtract, multiply, or divide.

12. $12 + (-9)$ 13. $-3 - 4$

14. $-7 - (-20)$ 15. $-7(-3)$

16. $5(-11)$ 17. $-36 \div (-9)$

18. $-15 + (-7)$ 19. $8 + (-6) + (-4)$

20. $-9 - 7$ 21. $-13 + 7$

22. **MULTIPLE CHOICE** Kendrick created a 6-week schedule for practicing the piano. If the pattern continues, how many hours will he practice during the sixth week?

The table shows the number of hours he practiced in the first three weeks.

Week	1	2	3
Hours	4	7	10

F   15 hours   H   19 hours

G   18 hours   J   22 hours

Evaluate each expression if $a = -5$, $b = 4$, and $c = -12$.

23. $ac \div b$ 24. $\dfrac{a - b}{3}$

25. **STOCKS** The value of a stock decreased $4 each week for a period of six weeks. Describe the change in the value of the stock at the end of the six-week period.

PART 1   Multiple Choice

Read each question. Then fill in the correct answer on the answer document provided by your teacher or on a sheet of paper.

1. The daily low temperatures for Cleveland, Ohio, over the last five days were 15°F, −2°F, 8°F, −6°F, and 5°F. Which expression can be used to find the average daily low temperature during the last five days?

   A $(15 + 2 + 8 + 6 + 5) \div 5$

   B $15 + 2 + 8 + 6 + 5 \div 5$

   C $[15 + (-2) + 8 + (-6) + 5] \div 5$

   D $15 + (-2) + 8 + (-6) + 5 \div 5$

TEST-TAKING TIP

Question 1 Check every answer choice of a multiple-choice question. Each time you find an incorrect answer, cross it off so you remember that you've eliminated it.

2. Marcia runs r miles on Mondays, Tuesdays, and Thursdays. She bicycles for b miles on Wednesday and Saturdays. If she rests on Fridays and Sundays, which equation represents the total number of miles M she exercises each week?

   F $M = 3r + 2b$

   G $M = r + b$

   H $M = 2r + 3b$

   J $M = 5(r + b)$

3. Simplify the expression below.

   $$3 + 6(10 - 7) - 3^2$$

   A 0

   B 12

   C 18

   D 74

4. At 8 A.M., the temperature was 13°F below zero. By 1 P.M., the temperature rose 22°F and by 6 P.M. dropped 14°F. What was the temperature at 6 P.M.?

   F 5°F above zero

   G 5°F below zero

   H 21°F above zero

   J 21°F below zero

5. Sue typically spends between $175 and $250 each month on clothes. Which of the following is the best estimate for the amount she spends in 6 months?

   A From $600 to $1,200

   B From $900 to $1,300

   C From $1,050 to $1,500

   D From $1,200 to $1,500

6. On their first play, a football team gained 17 yards. On their next play, they lost 22 yards. On their third play, they gained 14 yards. Which expression represents the total number of yards gained after the third play?

   F $17 + (-22) + 14$

   G $17 + 22 + (-14)$

   H $-17 + (-22) + (-14)$

   J $-17 + 22 + 14$

7. The top four runners of a race were Alicia, Kyle, Drew, and Juanita. Drew finished before Juanita. Alicia finished after both boys, but before Juanita. What information is needed to determine the order of the runners from first to fourth?

   A Did Kyle finish before or after Drew?

   B Did Alicia finish before or after Juanita?

   C Did Drew finish before or after Alicia?

   D Did Kyle finish before or after Alicia?

8. The lowest point in Japan is Hachiro-gata (elevation −4 meters), and the highest point is Mount Fuji (elevation 3,776 meters). What is the difference in elevation between Mount Fuji and Hachiro-gata?

   F 3,780 meters

   G 3,772 meters

   H 3,080 meters

   J 944 meters

9. Which of the following relationships is best represented by the data in the table?

x	y
1	36
2	72
3	108
4	144
5	180

   A Conversion of feet to inches

   B Conversion of inches to yards

   C Conversion of feet to yards

   D Conversion of yards to inches

10. A storeowner has n employees and pays each employee $440 per week. If the owner also pays d weekly to rent the building and w for utilities, which equation below represents the total E of these weekly expenses?

    F $E = 440w + n + w$

    G $E = 440n + dw$

    H $E = 440n + d + w$

    J $E = 440(n + d + w)$

PART 2   Short Response/Grid In

Record your answers on the answer sheet provided by your teacher or on a sheet of paper.

11. Nick spends a total of 75 hours per week at work and at the gym. He goes to the gym from 6:45 A.M. to 8:45 A.M., Monday through Friday. Write an equation that can be used to find t, the maximum number of hours Nick works at his job each week.

12. Find $-8 - 17$.

PART 3   Extended Response

Record your answers on the answer sheet provided by your teacher or a sheet of paper. Show your work.

13. A rectangle and a square are graphed on a coordinate plane. Use the graph below to answer the questions.

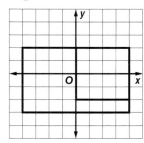

   a. Pick an ordered pair both objects have in common.

   b. Find an ordered pair that is inside the rectangle, but not the square.

   c. How could you increase the size of the square so that it still lies within the rectangle? What are the four coordinates?

   d. Draw a graph with the new square inside the rectangle.

NEED EXTRA HELP?													
If You Missed Question...	1	2	3	4	5	6	7	8	9	10	11	12	13
Go to Lesson...	1–6	1–7	1–4	2–4	1–1	2–4	1–1	2–5	2–7	1–6	1–7	2–5	2–3
NYS Core Curriculum	7.A.1	7.P3.3	7.N.11	7.N.12	7.PS.14	7.N.12	7.PS.14	7.N.12	7.PS.4	7.A.1	7.PS.3	7.N.12	6.G.10

CHAPTER 3

Algebra: Linear Equations and Functions

New York State Core Curriculum

7.A.1 Translate two-step verbal expressions into algebraic expressions

7.A.4 Solve multi-step equations by combining like terms, using the distributive property, or moving variables to one side of the equation

Key Vocabulary

formula (p. 144)

linear equation (p. 164)

two-step equation (p. 151)

work backward strategy (p. 148)

🌐 Real-World Link

Segways The Segway's top speed is 12.5 miles per hour—two to three times faster than walking. You can use the equation $d = 12.5t$ to find the distance d you can travel in t hours.

FOLDABLES®
Study Organizer

Algebra: Linear Equations and Functions Make this Foldable to help you organize your notes. Begin with a sheet of 11" by 17" paper.

① **Fold** the short sides toward the middle.

② **Fold** the top to the bottom.

③ **Open.** Cut along the second fold to make four tabs.

④ **Label** each of the tabs as shown.

Expressions · Equations · Perimeter and Area · Functions

GET READY for Chapter 3

Diagnose Readiness You have two options for checking Prerequisite Skills.

Option 2

NY Math Online Take the Online Readiness Quiz at glencoe.com.

Option 1

Take the Quick Quiz below. Refer to the Quick Review for help.

QUICK Quiz

Name the number that is the solution of the given equation. (Lesson 1-6)

1. $a + 15 = 19$; 4, 5, 6

2. $11k = 77$; 6, 7, 8

3. $x + 9 = -2$; 7, −11, 11

Graph each point on a coordinate plane. (Lesson 2-3)

4. $(-4, 3)$ 5. $(-2, -1)$

6. **HIKING** Keith hiked 4 miles north and 2 miles west from the campground before he rested. If the origin represents the campground, graph Keith's resting point. (Lesson 2-3)

Add. (Lesson 2-4)

7. $-3 + (-5)$ 8. $-8 + 3$

9. $9 + (-5)$ 10. $-10 + 15$

Subtract. (Lesson 2-5)

11. $-5 - 6$ 12. $8 - 10$

13. $8 - (-6)$ 14. $-3 - (-1)$

Divide. (Lesson 2-8)

15. $-6 \div (-3)$ 16. $-12 \div 3$

17. $10 \div (-5)$ 18. $-24 \div (-4)$

QUICK Review

Example 1 Name the number that is the solution of $24 \div a = 3$; 7, 8, or 9.

$24 \div a = 3$	Write the equation.
$24 \div 7 = 3$? No.	Substitute $a = 7$.
$24 \div 8 = 3$? Yes.	Substitute $a = 8$.
$24 \div 9 = 3$? No.	Substitute $a = 9$.

Example 2 Graph the point $(-1, 3)$ on a coordinate plane.

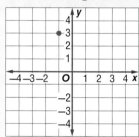

The first number in an ordered pair tells you to move left or right from the origin. The second number tells you to move up or down.

Example 3 Find $-4 + (-2)$.

$-4 + (-2) = -6$ Since −4 and −2 are both negative, add their absolute values. The sum is negative also.

Example 4 Find $9 - (-7)$.

$9 - (-7) = 9 + (7)$ Subtracting −7 is the same as adding 7.

$= 16$ Add.

Example 5 Find $-16 \div 2$.

$-16 \div 2 = -8$ Since −16 and 2 have opposite signs, their quotient is negative.

3-1 Writing Expressions and Equations

MAIN IDEA

Write verbal phrases and sentences as simple algebraic expressions and equations.

NYS Core Curriculum

Preparation for 7.A.1
Translate two-step verbal expressions into algebraic expressions *Also addresses 7.CM.8*

NY Math Online

glencoe.com

• Extra Examples
• Personal Tutor
• Self-Check Quiz

▷ GET READY for the Lesson

PLANETS Earth has only one moon, but other planets have many moons. For example, Uranus has 21 moons, and Saturn has 10 more moons than Uranus.

1. What operation would you use to find how many moons Saturn has? Explain.

2. Jupiter has about three times as many moons as Uranus. What operation would you use to find how many moons Jupiter has?

Words and phrases in problems often suggest addition, subtraction, multiplication, and division. Here are some examples.

Addition and Subtraction		Multiplication and Division	
sum	difference	each	divide
more than	less than	product	quotient
increased by	less	multiplied	per
in all	decreased by	twice	separate

EXAMPLE Write a Phrase as an Expression

1. Write the phrase *five dollars more than Jennifer earned* as an algebraic expression.

Words	five dollars more than Jennifer earned.
▼	
Variable	Let *d* represent the number of dollars Jennifer earned.
▼	
Expression	$d + 5$

 **CHECK Your Progress**

Write the phrase as an algebraic expression.

a. 3 more runs than the Pirates scored

Remember, an equation is a sentence in mathematics that contains an equals sign. When you write a verbal sentence as an equation, you can use the equals sign (=) for the words *equals* or *is*.

EXAMPLES Write Sentences as Equations

Write each sentence as an algebraic equation.

② Six less than a number is 20.

Six less than a number is 20.

Let n represent the number.

$n - 6 = 20$

③ Three times Jack's age equals 12.

Three times Jack's age equals 12.

Let a represent Jack's age.

$3a = 12$

Reading Math

Less Than You can write *six more than a number* as either $6 + n$ or $n + 6$. But *six less than a number* can only be written as $n - 6$.

CHECK Your Progress

Write each sentence as an algebraic equation.

b. Seven more than a number is 15.

c. Five times the number of students is 250.

Real-World EXAMPLE

④ WATERFALLS The tallest waterfall in the United States is **Yosemite Falls** in California with a height of about **739 meters**. This height is **617 meters taller than Raven Cliff Falls**. What is the height of Raven Cliff Falls? Write an equation that models this situation.

Words	*Yosemite Falls* is 617 meters taller than *Raven Cliff Falls*.
Variable	Let h represent the height of *Raven Cliff Falls*.
Equation	739 = 617 + h

The equation is $739 = 617 + h$.

CHECK Your Progress

d. **ANIMALS** North American cougars are about 1.5 times as long as cougars found in the tropical jungles of Central America. If North American cougars are about 75 inches long, how long is the tropical cougar? Write an equation that models this situation.

Real-World Link . . . .
The tallest waterfall in South Carolina is Raven Cliff Falls, located in Caesars Head State Park.

Source: South Carolina Department of Parks, Recreation, and Tourism

5 **Which problem situation matches the equation $x - 5.83 = 3.17$?**

A Tyler ran 3.17 kilometers. His friend ran the same distance 5.83 seconds faster than Tyler. What is x, the time in seconds that Tyler ran?

B Lynn and Heather measured the length of worms in science class. Lynn's worm was 5.83 centimeters long, and Heather's worm was 3.17 centimeters long. What is x, the average length of the worms?

C Keisha's lunch cost $5.83. She received $3.17 in change when she paid the bill. What is x, the amount of money she gave the cashier?

D Mr. Carlos paid $3.17 for a notebook that originally cost $5.83. What is x, the amount of money that Mr. Carlos saved?

Test-Taking Tip

Vocabulary Terms
Before taking a standardized test, review the meaning of vocabulary terms such as *average*.

Read the Item

You need to find which problem situation matches the equation $x - 5.83 = 3.17$.

Solve the Item

- You can eliminate A because you cannot add or subtract different units of measure.

- You can eliminate B because to find an average you add and then divide.

- Act out C. If you gave the cashier x dollars and your lunch cost $5.83, you would subtract to find your change, $3.17. This is the correct answer.

- Check D, just to be sure. To find the amount Mr. Carlos saved, you would calculate $5.83 - 3.17$, not $x - 5.83$.

The solution is C.

✓**CHECK Your Progress**

e. **Which problem situation matches the equation $4y = 6.76$?**

F Mrs. Thomas bought 4 gallons of gas. Her total cost was $6.76. What is y, the cost of one gallon of gas?

G Jordan bought 4 CDs that were on sale for $6.76 each. What is y, the total cost of the CDs?

H The width of a rectangle is 4 meters. The length is 6.76 meters more than the width. What is y, the length of the rectangle?

J The average yearly rainfall is 6.76 inches. What is y, the amount of rainfall you might expect in 4 years?

Example 1
(p. 128)

Write each phrase as an algebraic expression.

1. a number increased by eight

2. ten dollars more than Grace has

Examples 2, 3
(p. 129)

Write each sentence as an algebraic equation.

3. Nine less than a number equals 24.

4. Two points less than his score is 4.

5. Twice the number of miles is 18.

6. One half the regular price is $13.

Example 4
(p. 129)

7. **ALGEBRA** The median age of people living in Arizona is 1 year younger than the median age of people living in the United States. Use this information and the information at the right to write an equation to find the median age in the United States.

Median Age
Arizona   34.3
United States   ?

Example 5
(p. 130)

8. **MULTIPLE CHOICE** Which problem situation matches the equation $x - 15 = 46$?

A The original price of a jacket is $46. The sale price is $15 less. What is x, the sale price of the jacket?

B Mark had several baseball cards. He sold 15 of the cards and had 46 left. What is x, the amount of cards Mark had to start with?

C Sonja scored 46 points in last week's basketball game. Talisa scored 15 points less. What is x, the amount of points Talisa scored?

D Katie earned $15 babysitting this week. Last week she earned $46. What is x, her average earnings for the two weeks?

Practice and Problem Solving

HOMEWORK HELP

For Exercises	See Examples
9–16	1
17–22	2, 3
23–24	4
41	5

Write each phrase as an algebraic expression.

9. fifteen increased by t

10. five years older than Luis

11. a number decreased by ten

12. three feet less than the length

13. the product of r and 8

14. twice as many oranges

15. Emily's age divided by 3

16. the quotient of a number and -12

Write each sentence as an algebraic equation.

17. The sum of a number and four is equal to -8.

18. Two more than the number of frogs is 4.

19. The product of a number and five is -20.

20. Ten times the number of students is 280.

21. Ten inches less than her height is 26.

22. Five less than a number is 31.

For Exercises 23 and 24, write an equation that models each situation.

23. **ANIMALS** A giraffe is 3.5 meters taller than a camel. If a giraffe is 5.5 meters tall, how tall is a camel?

24. **FOOTBALL** Carson Palmer led the National Football League with 32 touchdown passes in a season. This was twice as many touchdown passes as Donovan McNabb had. Find the number of touchdown passes for McNabb.

MEASUREMENT For Exercises 25–28, describe the relationship that exists between the length and width of each rectangle.

25. The width is x, and the length is $4x$.

26. The length is $x + 3$, and the width is x.

27. The length is x, and the width is $x - 5$.

28. The length is x, and the width is $0.5x$.

Write each phrase as an algebraic expression.

29. 2 more than twice as many bikes

30. nine CDs less than three times the number of CDs Margaret owns

31. 43 dollars off the price of each admission, which is then multiplied by 3 admissions

32. the quotient of a number w and (-8), which is then increased by 7

33. the square of a number k which is then multiplied by 13

34. the sum of a number p and 0.4 which is then decreased by the fifth power of the same number

ANALYZE TABLES For Exercises 35 and 36, use the table.

The table shows the average lifespan of several types of pets. Let y represent the average lifespan of a gerbil.

35. Which lifespan can be represented by $3y$?

36. Write an expression to represent the lifespan of a cat.

Pets	
Type	**Lifespan (years)**
American toad	15
cat	25
dog	22
gerbil	5
rabbit	9

NYSCC • NYSMT
Extra Practice, pp. 674, 706

H.O.T. Problems

37. **OPEN ENDED** Write a verbal sentence for the equation $n - 3 = 6$.

38. **FIND THE ERROR** Sancho and Candace are writing an algebraic expression for the phrase *5 less than a number*. Who is correct? Explain.

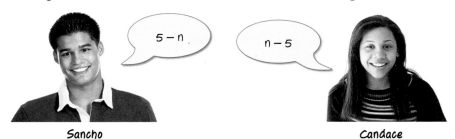

Sancho

Candace

39. CHALLENGE If x is an odd number, how would you represent the odd number immediately following it? preceding it?

40. **WRITING IN MATH** Analyze the meaning of the expressions $a + 5$, $a - 3$, $2a$, and $\frac{a}{2}$ if a represents someone's age.

NYSMT PRACTICE 7.A.1

41. Asha had some change in her purse. After her brother gave her $0.79, Asha had $2.24 altogether. Which equation can she use to find the original amount of money m she had in her purse?

A $2.24 = m - 0.79$

B $m = 2.24 \times 0.79$

C $m + 0.79 = 2.24$

D $m + 2.24 = 0.79$

42. Which algebraic equation best describes the total distance D traveled in miles after a 6-hour period, if r represents the rate of travel in miles per hour?

F $D = 6 + r$

G $D = \frac{r}{6}$

H $D = 6r$

J $D = \frac{6}{r}$

Spiral Review

Divide (Lesson 2-8)

43. $-42 \div 6$

44. $36 \div (-3)$

45. $-45 \div (-3)$

46. MONEY Jordan withdraws $14 per week from his savings account for a period of 7 weeks. Write a multiplication expression to represent this situation. Then find the product and explain its meaning. (Lesson 2-7)

Evaluate each expression. (Lesson 1-4)

47. $3 + 7 \cdot 4 - 6$

48. $8(16 - 5) - 6$

49. $75 \div 3 + 6(5 - 1)$

ANALYZE DATA For Exercises 50–52, use the table that shows the cost of two different plans for downloading music. (Lesson 1-1)

50. Suppose you download 12 songs in one month. Find the cost per song using Plan B.

51. Which plan is less expensive for downloading 9 songs in one month?

52. When is it less expensive to use Plan B instead of Plan A?

Music Downloads	
Plan A	$0.99 per song
Plan B	$15.99 per month for unlimited downloads

GET READY for the Next Lesson

PREREQUISITE SKILL Find each sum. (Lesson 2-4)

53. $-8 + (-3)$

54. $-10 + 9$

55. $12 + (-20)$

56. $-15 + 15$

Algebra Lab
Solving Equations Using Models

MAIN IDEA

Solve equations using models.

NYS Core Curriculum

7.PS.17 Evaluate the efficiency of different representations of a problem
7.R.3 Recognize, compare, and use an array of representational forms

In Chapter 2, you used counters to add, subtract, multiply, and divide integers. Integers can also be modeled using algebra tiles. The table shows how these two types of models are related.

Type of Model	Variable x	Integer 1	Integer -1
Cups and Counters		$+$	$-$
Algebra Tiles	x	1	-1

You can use either type of model to solve equations.

ACTIVITY

1 Solve $x + 2 = 5$ using cups and counters or a drawing.

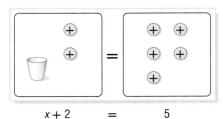

$$x + 2 \quad = \quad 5$$

Model the equation.

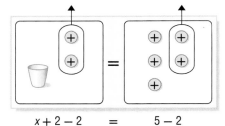

$$x + 2 - 2 \quad = \quad 5 - 2$$

Remove the same number of counters from each side of the mat until the cup is by itself on one side.

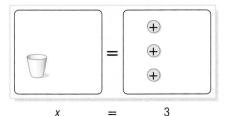

$$x \quad = \quad 3$$

The number of counters remaining on the right side of the mat represents the value of x.

Therefore, $x = 3$. Since $3 + 2 = 5$, the solution is correct.

✓ CHECK Your Progress

Solve each equation using cups and counters or a drawing.

a. $x + 4 = 4$ **b.** $5 = x + 4$ **c.** $4 = 1 + x$ **d.** $2 = 2 + x$

Review Vocabulary

zero pair a number paired with its opposite; Example: 2 and −2. (Explore 2-4)

You can add or subtract a zero pair from either side of an equation without changing its value, because the value of a zero pair is zero.

ACTIVITY

(2) Solve $x + 2 = -1$ using models.

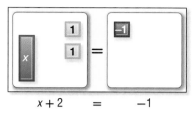

$x + 2 \quad = \quad -1$

Model the equation.

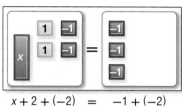

$x + 2 + (-2) \quad = \quad -1 + (-2)$

Add 2 negative tiles to the left side of the mat and add 2 negative tiles to the right side of the mat.

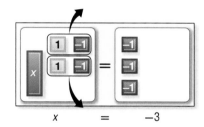

$x \quad = \quad -3$

Remove all of the zero pairs from the left side. There are 3 negative tiles on the right side of the mat.

Therefore, $x = -3$. Since $-3 + 2 = -1$, the solution is correct.

✓CHECK Your Progress

Solve each equation using models or a drawing.

e. $-2 = x + 1$ **f.** $x - 3 = -2$ **g.** $x - 1 = -3$ **h.** $4 = x - 2$

ANALYZE THE RESULTS

Explain how to solve each equation using models or a drawing.

1.

$x + 1 \quad = \quad 3$

2.
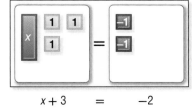

$x + 3 \quad = \quad -2$

3. **MAKE A CONJECTURE** Write a rule that you can use to solve an equation like $x + 3 = 2$ without using models or a drawing.

Solving Addition and Subtraction Equations

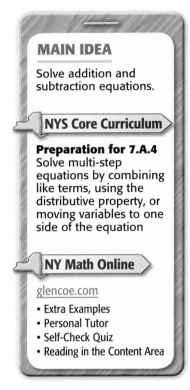

MAIN IDEA

Solve addition and subtraction equations.

NYS Core Curriculum

Preparation for 7.A.4
Solve multi-step equations by combining like terms, using the distributive property, or moving variables to one side of the equation

NY Math Online

glencoe.com

• Extra Examples
• Personal Tutor
• Self-Check Quiz
• Reading in the Content Area

▷ **GET READY** for the Lesson

VIDEO GAMES Max had some video games, and then he bought two more games. Now he has six games.

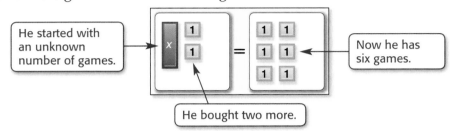

He started with an unknown number of games.

Now he has six games.

He bought two more.

1. What does x represent in the figure?
2. What addition equation is shown in the figure?
3. Explain how to solve the equation.
4. How many games did Max have in the beginning?

You can solve the equation $x + 2 = 6$ by *removing*, or subtracting, the same number of positive tiles from each side of the mat. You can also subtract 2 from each side of the equation. The variable is now by itself on one side of the equation.

Use Models

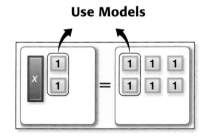

Use Symbols

$$\begin{aligned} x + 2 &= 6 \\ -2 &= -2 \\ \hline x &= 4 \end{aligned}$$

Subtracting 2 from each side of an equation illustrates the Subtraction Property of Equality.

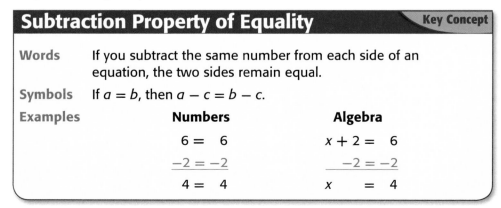

Subtraction Property of Equality   **Key Concept**

Words   If you subtract the same number from each side of an equation, the two sides remain equal.

Symbols   If $a = b$, then $a - c = b - c$.

Examples

Numbers

$$\begin{aligned} 6 &= 6 \\ -2 &= -2 \\ \hline 4 &= 4 \end{aligned}$$

Algebra

$$\begin{aligned} x + 2 &= 6 \\ -2 &= -2 \\ \hline x &= 4 \end{aligned}$$

EXAMPLES Solve Addition Equations

1 Solve $x + 5 = 8$. Check your solution.

$$
\begin{array}{llr}
x + 5 = & 8 & \text{Write the equation.} \\
\underline{-5 = -5} & & \text{Subtract 5 from each side.} \\
x \quad = & 3 & \text{Simplify.}
\end{array}
$$

Study Tip

Solutions Notice that your new equation, $x = 3$, has the same solution as the original equation, $x + 5 = 8$.

Check $\qquad x + 5 = 8 \qquad$ Write the original equation.
$\qquad\qquad\quad 3 + 5 \overset{?}{=} 8 \qquad$ Replace x with 3.
$\qquad\qquad\qquad\quad 8 = 8 \checkmark \quad$ The sentence is true.

The solution is 3.

2 Solve $x + 6 = 4$. Check your solution.

$$
\begin{array}{llr}
x + 6 = & 4 & \text{Write the equation.} \\
\underline{-6 = -6} & & \text{Subtract 6 from each side.} \\
x \quad = & -2 & \text{Simplify.}
\end{array}
$$

The solution is -2. $\qquad$ Check the solution.

 CHECK Your Progress

Solve each equation. Check your solution.

a. $y + 6 = 9$ $\qquad$ b. $x + 3 = 1$ $\qquad$ c. $-3 = a + 4$

 Real-World EXAMPLE

3 **MARINE BIOLOGY** Clownfish and angelfish are popular tropical fish. An angelfish can grow to be 12 inches long. If an angelfish is 8.5 inches longer than a clownfish, how long is a clownfish?

Words	An angelfish	is	8.5 inches longer than	a clownfish.
Variable	Let c represent the length of the clownfish.			
Equation	12	=	8.5 +	c

$$
\begin{array}{llr}
12 \quad = & 8.5 + c & \text{Write the equation.} \\
\underline{-8.5 = -8.5} & & \text{Subtract 8.5 from each side.} \\
3.5 \quad = & c & \text{Simplify.}
\end{array}
$$

A clownfish is 3.5 inches long.

 **Real-World Career** . . . .
How Does a Marine Biologist Use Math?
A marine biologist uses math to analyze data about marine plants, animals, and organisms.

 **NY Math Online**

For more information, go to: glencoe.com.

 CHECK Your Progress

d. **WEATHER** The highest recorded temperature in Warsaw, Missouri, is 118°F. This is 158° greater than the lowest recorded temperature. Write and solve an equation to find the lowest recorded temperature.

Vocabulary Link · · · · ·
Inverse
Everyday Use
something that is
opposite
Math Use undo

· · · Similarly, you can use inverse operations and the Addition Property of
    Equality to solve equations like $x - 2 = 1$.

Addition Property of Equality   **Key Concept**

Words   If you add the same number to each side of an equation, the
              two sides remain equal.

Symbols   If $a = b$, then $a + c = b + c$.

Examples   **Numbers**   **Algebra**

$$\begin{array}{rcl} 5 &=& 5 \\ +\,3 &=& +\,3 \\ \hline 8 &=& 8 \end{array}$$ $$\begin{array}{rcl} x - 2 &=& 4 \\ +\,2 &=& +\,2 \\ \hline x &=& 6 \end{array}$$

 EXAMPLE   Solve a Subtraction Equation

4 Solve $x - 2 = 1$. Check your solution.

$$\begin{array}{rcl} x - 2 &=& 1 \\ +\,2 &=& +\,2 \\ \hline x &=& 3 \end{array}$$ Write the equation.
                                         Add 2 to each side.
                                         Simplify.

Check the solution. Since $3 - 2 = 1$, the solution is 3.

 CHECK Your Progress

e. $y - 3 = 4$ **f.** $r - 4 = -2$ **g.** $q - 8 = -9$

Real-World EXAMPLE

5 **SHOPPING** A pair of shoes costs $25. This is $14 **less than** the cost of
a pair of jeans. **Find the cost of the jeans.**

Study Tip

Check for Reasonableness
Ask yourself which costs
more: the shoes or the
jeans. Then check your
answer. Does it show that
the jeans cost more than
the shoes?

Words	Shoes	are	$14 less than	jeans
Variable		Let j represent the cost of jeans.		
Equation	25	=	j	$-$ 14

$$\begin{array}{rcl} 25 &=& j - 14 \\ +\,14 &=& +\,14 \\ \hline 39 &=& j \end{array}$$ Write the equation.
                                          Add 14 to each side.
                                          Simplify.

The jeans cost $39.

 CHECK Your Progress

h. **ANIMALS** The average lifespan of a tiger is 22 years. This is
13 years less than a lion. Write and solve an equation to find the
lifespan of a lion.

Examples 1, 2
(p. 137)

Solve each equation. Check your solution.

1. $n + 6 = 8$ **2.** $7 = y + 2$

3. $m + 5 = 3$ **4.** $-2 = a + 6$

Example 3
(p. 137)

5. FLYING Orville and Wilbur Wright made the first airplane flights in 1903. Wilbur's flight was 364 feet. This was 120 feet longer than Orville's flight. Write and solve an equation to find the length of Orville's flight.

Example 4
(p. 138)

Solve each equation. Check your solution.

6. $x - 5 = 6$ **7.** $-1 = c - 6$

Example 5
(p. 138)

8. PRESIDENTS John F. Kennedy was the youngest president to be inaugurated. He was 43 years old. This was 26 years younger than the oldest president to be inaugurated—Ronald Reagan. Write and solve an equation to find how old Reagan was when he was inaugurated.

Practice and Problem Solving

Solve each equation. Check your solution.

HOMEWORK HELP	
For Exercises	**See Examples**
9–12	1
13–16	2
17–20	4
21–24	3, 5

9. $a + 3 = 10$ **10.** $y + 5 = 11$ **11.** $9 = r + 2$

12. $14 = s + 7$ **13.** $x + 8 = 5$ **14.** $y + 15 = 11$

15. $r + 6 = -3$ **16.** $k + 3 = -9$ **17.** $s - 8 = 9$

18. $w - 7 = 11$ **19.** $-1 = q - 8$ **20.** $-2 = p - 13$

For Exercises 21–24, write an equation. Then solve the equation.

21. MUSIC Last week Tiffany practiced her bassoon a total of 7 hours. This was 2 hours more than she practiced the previous week. How many hours did Tiffany practice the previous week?

22. CIVICS In the 2004 presidential election, Ohio had 20 electoral votes. This is 14 votes less than Texas had. How many electoral votes did Texas have in 2004?

23. AGES Zack is 15 years old. This is 3 years younger than his brother Tyler. How old is Tyler?

24. BASKETBALL The Miami Heat scored 79 points in a recent game. This was 13 points less than the Chicago Bulls score. How many points did the Chicago Bulls score?

Solve each equation. Check your solution.

25. $34 + r = 95$

26. $64 + y = 84$

27. $-23 = x - 18$

28. $-59 = m - 11$

29. $-18 + c = -30$

30. $-34 = t + 9$

31. $a - 3.5 = 14.9$

32. $x - 2.8 = 9.5$

33. $r - 8.5 = -2.1$

34. $z - 9.4 = -3.6$

35. $n + 1.4 = 0.72$

36. $b + 2.25 = 1$

For Exercises 37–42, write an equation. Then solve the equation.

37. **MONEY** Suppose you have d dollars. After you pay your sister the $5 you owe her, you have $18 left. How much money did you have at the beginning?

38. **MONEY** Suppose you have saved $38. How much more do you need to save to buy a small television that costs $65?

39. **GEOMETRY** The sum of the measures of the angles of a triangle is 180°. Find the missing measure.

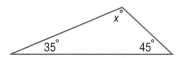

40. **VOLCANOES** Alaska, Hawaii, and Washington have active volcanoes. Alaska has 43, Hawaii has 5, and Washington has v. If they have 52 active volcanoes in all, how many volcanoes does Washington have?

41. **GOLF** The table shows Cristie Kerr's scores for four rounds of the 2007 U.S. Women's Open. Her total score was -5 (5 under par). What was her score for the third round?

Round	Score
First	0
Second	+1
Third	s
Fourth	-1

42. **BUSINESS** At the end of the day, the closing price of XYZ Stock was $62.87 per share. This was $0.62 less than the opening price. Find the opening price.

ANALYZE TABLES For Exercises 43–45, use the table.

Tallest Wooden Roller Coasters	Height (feet)	Drop (feet)	Speed (mph)
Son of Beast	218	214	s
El Toro	181	176	70
The Rattler	180	d	65
Colossos	h	159	75
Voyage	163	154	67

Source: Coaster Grotto

43. The difference in speeds of Son of Beast and The Rattler is 13 miles per hour. If Son of Beast has the greater speed, write and solve a subtraction equation to find its speed.

44. The Rattler has a drop that is 52 feet less than El Toro. Write and solve an addition equation to find the height of The Rattler.

45. Colossos is 13 feet taller than Voyage. Write and solve a subtraction equation to find the height of Colossos.

NYSCC • NYSMT
Extra Practice, pp. 674, 706

46. **Which One Doesn't Belong?** Identify the equation that does not have the same solution as the other three. Explain your reasoning.

| $x - 1 = -4$ | $b + 5 = -8$ | $11 + y = 8$ | $-6 + a = -9$ |

47. **CHALLENGE** Suppose $x + y = 11$ and the value of x increases by 2. If their sum remains the same, what must happen to the value of y?

48. **WRITING IN MATH** Write a problem about a real-world situation that can be represented by the equation $p - 25 = 50$.

NYSMT PRACTICE > 7.A.4

49. The Oriental Pearl Tower in Shanghai, China, is 1,535 feet tall. It is 280 feet shorter than the Canadian National Tower in Toronto, Canada. Which equation can be used to find the height of the Canadian National Tower?

   A $1,535 + h = 280$

   B $h = 1,535 - 280$

   C $1,535 = h - 280$

   D $280 - h = 1,535$

50. Which of the following statements is true concerning the equation $x + 3 = 7$?

   F To find the value of x, add 3 to each side.

   G To find the value of x, add 7 to each side.

   H To find the value of x, find the sum of 3 and 7.

   J To find the value of x, subtract 3 from each side.

Spiral Review

51. **SCIENCE** The boiling point of water is 180° higher than its freezing point. If p represents the freezing point, write an expression that represents the boiling point of water. (Lesson 3-1)

52. **ALGEBRA** Evaluate the expression $xy \div (-4)$ if $x = 12$ and $y = -2$. (Lesson 2-8)

53. **ALGEBRA** The table shows the number of pages of a novel Ferguson read each hour. If the pattern continues, how many pages will Ferguson read during the 8th hour? (Lesson 2-7)

Hour	Number of Pages Read
1	11
2	13
3	16
4	20
5	25

▷ **GET READY** for the Next Lesson

PREREQUISITE SKILL Find each quotient.

54. $15.6 \div 13$

55. $8.84 \div 3.4$

56. $75.25 \div 0.25$

57. $0.76 \div 0.5$

Solving Multiplication Equations

MAIN IDEA

Solve multiplication equations.

NYS Core Curriculum

Preparation for 7.A.4 Solve multi-step equations by combining like terms, using the distributive property, or moving variables to one side of the equation

New Vocabulary

formula

NY Math Online

glencoe.com
- Concepts In Motion
- Extra Examples
- Personal Tutor
- Self-Check Quiz

▷ MINI Lab

MONEY Suppose three friends order an appetizer of nachos that costs $6. They agree to split the cost equally. The figure below illustrates the multiplication equation $3x = 6$, where x represents the amount each friend pays.

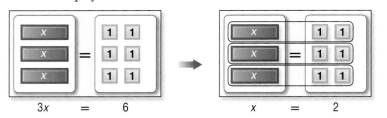

Each x is matched with $2.

Each friend pays $2. The solution of $3x = 6$ is 2.

Solve each equation using models or a drawing.

1.

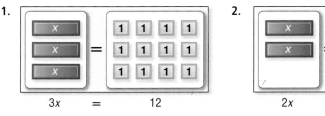

$$3x = 12$$

2.

$$2x = -8$$

3. $4x = 20$ 4. $8 = 2x$ 5. $3x = -9$

6. What operation did you use to find each solution?

7. How can you use the coefficient of x to solve $8x = 40$?

Equations like $3x = 6$ are called multiplication equations because the expression $3x$ means 3 *times the value of x*. So, you can use the Division Property of Equality to solve multiplication equations.

Division Property of Equality		Key Concept
Words	If you divide each side of an equation by the same nonzero number, the two sides remain equal.	
Symbols	If $a = b$ and $c \neq 0$, then $\frac{a}{c} = \frac{b}{c}$.	
Examples	**Numbers**	**Algebra**
	$8 = 8$	$2x = -6$
	$\frac{8}{2} = \frac{8}{2}$	$\frac{2x}{2} = \frac{-6}{2}$
	$4 = 4$	$x = -3$

Review Vocabulary

coefficient the numerical factor for a multiplication expression; *Example:* the coefficient of *x* in the expression 4*x* is 4.
(Lesson 1-4)

EXAMPLES Solve Multiplication Equations

1 **Solve 20 = 4x. Check your solution.**

$20 = 4x$	Write the equation.
$\dfrac{20}{4} = \dfrac{4x}{4}$	Divide each side of the equation by 4.
$5 = x$	$20 \div 4 = 5$

The solution is 5.   Check the solution.

2 **Solve −8y = 24. Check your solution.**

$-8y = 24$	Write the equation.
$\dfrac{-8y}{-8} = \dfrac{24}{-8}$	Divide each side by −8.
$y = -3$	$24 \div (-8) = -3$

The solution is −3.   Check the solution.

✓ CHECK Your Progress

Solve each equation. Check your solution.

a. $30 = 6x$ **b.** $-6a = 36$ **c.** $-9d = -72$

Many real-world situations increase at a constant rate. These can be represented by multiplication equations.

Real-World EXAMPLE

3 **TEXT MESSAGING** It costs $0.10 to send a text message. You can spend a total of $5.00. How many text messages can you send?

Words	Total	is equal to	cost of each message	times	number of messages.
Variable	Let *m* represent the number of messages you can send.				
Equation	5.00	=	0.10	•	m

$5.00 = 0.10m$	Write the equation.
$\dfrac{5.00}{0.10} = \dfrac{0.10m}{0.10}$	Divide each side by 0.10.
$50 = m$	$5.00 \div 0.10 = 50$

At $0.10 per message, you can send 50 text messages for $5.00.

Real-World Link . . . . Over 60% of teenagers' text messages are sent from their homes—even when a landline is available.
Source: Xerox

✓ CHECK Your Progress

d. TRAVEL Mrs. Acosta's car can travel an average of 24 miles on each gallon of gasoline. Write and solve an equation to find how many gallons of gasoline she will need for a trip of 348 miles.

A **formula** is an equation that shows the relationship among certain quantities. One of the most common formulas is the equation $d = rt$, which gives the relationship among distance d, rate r, and time t.

Real-World EXAMPLE

Reading Math

Speed Another name for *rate* is *speed*.

④ **ANIMALS** The tortoise is one of the slowest land animals, reaching an average top speed of about 0.25 mile per hour. At this speed, how long will it take a tortoise to travel 1.5 miles?

You are asked to find the time t it will take to travel a distance d of 1.5 miles at a rate r of 0.25 mile per hour.

METHOD 1 Substitute, then solve.

$d = rt$	Write the equation.
$1.5 = 0.25t$	Replace d with 1.5 and r with 0.25.
$\dfrac{1.5}{0.25} = \dfrac{0.25t}{0.25}$	Divide each side by 0.25.
$6 = t$	$1.5 \div 0.25 = 6$

METHOD 2 Solve, then substitute.

$d = rt$	Write the equation.
$\dfrac{d}{r} = \dfrac{rt}{r}$	Divide each side by r to solve the equation for t.
$\dfrac{d}{r} = t$	Simplify.
$\dfrac{1.5}{0.25} = t$	Replace d with 1.5 and r with 0.25.
$6 = t$	$1.5 \div 0.25 = 6$

It would take a tortoise 6 hours to travel 1.5 miles.

✔ CHOOSE Your Method

    e. SCIENCE A sound wave travels a distance of 700 meters in 2.5 seconds. Find the average speed of the sound wave.

✔ CHECK Your Understanding

Examples 1, 2
(p. 143)

Solve each equation. Check your solution.

1. $6c = 18$

2. $15 = 3z$

3. $-8x = 24$

4. $-9r = -36$

Example 3
(p. 143)

5. **WORKING** Antonia earns $6 per hour helping her grandmother. How many hours does she need to work to earn $48?

Example 4
(p. 144)

6. **SWIMMING** A shark can swim at an average speed of about 25 miles per hour. At this rate, how long will it take a shark to swim 60 miles?

Solve each equation. Check your solution.

7. $7a = 49$

8. $9e = 27$

9. $2x = -6$

10. $3y = -21$

11. $35 = 5v$

12. $72 = 12r$

13. $-4j = 36$

14. $-12y = 60$

15. $-4s = -16$

16. $-6z = -36$

17. $48 = -6r$

18. $-28 = -7f$

For Exercises 19–22, write an equation. Then solve the equation.

19. **MONEY** Brandy wants to buy a digital camera that costs $300. If she saves $15 each week, in how many weeks will she have enough money for the camera?

20. **COMPUTERS** The width of a computer monitor is 1.25 times as long as its height. Find the height of the computer monitor at the right.

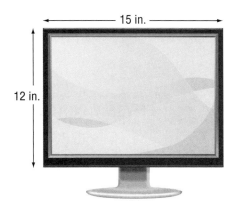

21. **SPEED** A racecar can travel at a rate of 205 miles per hour. At this rate, how long would it take to travel 615 miles?

22. **INSECTS** A dragonfly, the fastest insect, can fly a distance of 50 feet in about 2 seconds. Find a dragonfly's average speed in feet per second.

Solve each equation. Check your solution.

23. $0.4x = 9.2$

24. $0.9y = 13.5$

25. $5.4 = 0.3p$

26. $9.72 = 1.8a$

27. $3.9y = 18.33$

28. $2.6b = 2.08$

ANALYZE TABLES For Exercises 29 and 30, use the following information.
The table shows women's championship record holders for several track events.

29. Without calculating explain whether Evelyn Ashford or Sanya Richards has the faster average speed.

30. Find the average speed of each athlete in meters per second. Round to the nearest hundredth.

Name	Race (m)	Time (s)
Marion Jones	100	10.72
Evelyn Ashford	200	21.88
Sanya Richards	400	49.27

Source: USA Outdoor Track & Field

31. **HURRICANES** A category 3 hurricane reaches speeds up to 20.88 kilometers per hour. The distance from Cuba to Key West is 145 kilometers. Write and solve a multiplication equation to find how long it would take a category 3 hurricane to travel from Cuba to Key West.

32. **WATER** A case of water bottles costs $9.48. If there are 12 water bottles in the case, find the cost per bottle. Then find the decrease in cost per bottle if the cost of a case is reduced to $8.64.

33. **FIND THE ERROR** Steve and Becky are solving $-6x = 72$. Who is correct? Explain.

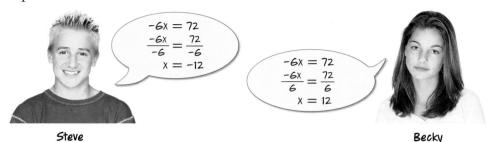

Steve

$$-6x = 72$$
$$\frac{-6x}{-6} = \frac{72}{-6}$$
$$x = -12$$

Becky

$$-6x = 72$$
$$\frac{-6x}{6} = \frac{72}{6}$$
$$x = 12$$

34. **CHALLENGE** Solve $3|x| = 12$. Explain your reasoning.

WRITING IN MATH Write a real-world problem that could be represented by each equation.

35. $2x = 16$

36. $3x = 75$

37. $4x = -8$

NYSMT PRACTICE 7.A.4

38. A football player can run 20 yards in 3.4 seconds. Which equation could be used to find y, the number of yards the football player can run in a second?

A $20y = 3.4$

B $3.4 - y = 20$

C $3.4y = 20$

D $20 + y = 3.4$

39. **SHORT RESPONSE** Use the formula $A = bh$ to find the base in inches of a rhombus with a height of 7 inches and an area of 56 square inches.

Spiral Review

ALGEBRA Solve each equation. Check your solution. (Lesson 3-2)

40. $y + 8 = -2$

41. $x - 7 = -2$

42. $20 = z + 23$

43. **ALGEBRA** Write an algebraic expression for the phrase *the product of -3 and y.* (Lesson 3-1)

44. **MONTHS** A lunar month, the time from one new moon to the next, is 29.5 days. How many days longer is our calendar year of 365 days than 12 lunar months? (Lesson 1-1)

▷ **GET READY for the Next Lesson**

PREREQUISITE SKILL Draw the next two figures in the pattern. (Lesson 2-7)

45.

Write each sentence as an algebraic equation.
(Lesson 3-1)

1. The product of a number and 3 is −16.

2. 10 less than a number is 45.

3. **CLIMBING** A rock climber is at an altitude of a feet before she climbs up another 80 feet. Write an expression for her new altitude. (Lesson 3-1)

4. **MULTIPLE CHOICE** Stephanie has 5 dollars more than Necie. If Necie has d dollars, which expression represents the number of dollars Stephanie has? (Lesson 3-1)

   A $d - 5$ **C** $5 - d$

   B $d + 5$ **D** $5d$

Solve each equation. Check your solution.
(Lesson 3-2)

5. $21 + m = 33$ 6. $a - 5 = -12$

7. $p + 1.7 = -9.8$ 8. $56 = k - (-33)$

9. **GEOMETRY** The sum of the measures of the angles of a triangle is 180°. Write and solve an equation to find the missing measure m. (Lesson 3-2)

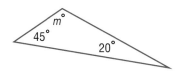

10. **MULTIPLE CHOICE** Trevor's test score was 5 points lower than Ursalina's test score. If Ursalina scored 85 on the test, which equation would give Trevor's score d when solved? (Lesson 3-2)

   F $85 = d + 5$

   G $d - 5 = 85$

   H $80 = d + 5$

   J $d - 5 = 80$

11. **PETS** Cameron has 11 adult Fantail goldfish. This is 7 fewer Fantail goldfish than his friend Julia has. Write and solve a subtraction equation to determine the number of Fantail goldfish g that Julia has. (Lesson 3-2)

12. **MEASUREMENT** The Grand Canyon has a maximum depth of almost 5,280 feet. An average four-story apartment building has a height of 66 feet. Write and solve a multiplication equation to determine the number of apartment buildings b, stacked on top of each other, that would fill the depth of the Grand Canyon. (Lesson 3-3)

Solve each equation. Check your solution.
(Lesson 3-3)

13. $5f = -75$ 14. $-1.6w = 4.8$

15. $63 = 7y$ 16. $-28 = -2d$

17. $3.7g = -4.44$ 18. $2.25 = 1.5b$

19. **MULTIPLE CHOICE** Michelann drove 44 miles per hour and covered a distance of 154 miles. Which equation accurately describes this situation if h represents the number of hours Michelann drove? (Lesson 3-3)

   A $154 = 44 + h$

   B $44h = 154$

   C $154 = 44 \div h$

   D $h - 44 = 154$

20. **LAWN SERVICE** Trey estimates he will earn $470 next summer cutting lawns in his neighborhood. This amount is 2.5 times the amount a he earned this summer. Write and solve a multiplication equation to find how much Trey earned this summer. (Lesson 3-3)

MAIN IDEA: Solve problems using the work backward strategy.

 NYSCC > **7.PS.9** Work backwards from a solution

P.S.I. TEAM +

e-Mail: WORK BACKWARD

MIGUEL: Yesterday, I earned extra money by doing yardwork for my neighbor. Then I spent $5.50 at the convenience store and four times that amount at the bookstore. Now I have $7.75 left.

YOUR MISSION: Work backward to find how much money Miguel had before he went to the convenience store and the bookstore.

Understand	You know he has $7.75 left. You need to find the amount he started with.	
Plan	Start with the end result and work backward.	
Solve	He has $7.75 left.	
	Undo the four times $5.50 spent at the bookstore. Since $5.50 × 4 is $22, add $7.75 and $22.	$7.75 + 22.00 $29.75
	Undo the $5.50 spent at the convenience store. Add $5.50 and $29.75.	+ $5.50 $35.25
	So, Miguel had $35.25 to start with.	
Check	Assume Miguel started with $35.25. After going to the convenience store, he had $35.25 − $5.50 or $29.75. He spent four times the amount he spent at the convenience store at the bookstore. So, he had $29.75 − 4($5.50) or $7.75 left. So, $35.25 is correct. ✓	

Analyze The Strategy

1. Explain when you would use the work backward strategy to solve a problem.

2. Describe how to solve a problem by working backward.

3. **WRITING IN MATH** Write a problem that could be solved by working backward. Then write the steps you would take to find the solution to your problem.

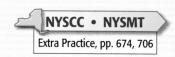
Use the *work backward* strategy to solve Exercises 4–7.

4. **MONEY** Marisa spent $8 on a movie ticket. Then she spent $5 on popcorn and one half of what was left on a drink. She has $2 left. How much did she have initially?

5. **NUMBER THEORY** A number is multiplied by −3. Then 6 is subtracted from the product. After adding −7, the result is −25. What is the number?

6. **TIME** Timothy's morning schedule is shown. At what time does Timothy wake up?

Timothy's Schedule	
Activity	**Time**
Wakes up	■
Get ready for school − 45 min	■
Walk to school − 25 min	9:00 A.M.

7. **LOGIC** A small box has 4 tennis balls inside it. There are 6 of these small boxes inside a medium box. There are 8 medium boxes inside each large box, and there are 100 large boxes shipped in a large truck. How many tennis balls are on the truck?

Use any strategy to solve Exercises 8–15. Some strategies are shown below.

PROBLEM-SOLVING STRATEGIES
· Guess and check.
· Look for a pattern.
· Work backward.

8. **GEOGRAPHY** The land area of North Dakota is 68,976 square miles. This is about 7 times the land area of Vermont. Estimate the land area of Vermont.

9. **AGE** Brie is two years older than her sister Kiana. Kiana is 4 years older than their brother Jeron, who is 8 years younger than their brother Trey. Trey is 16 years old. How old is Brie?

10. **ELEVATION** New Orleans, Lousiana, has an elevation of −8 feet related to sea level. Death Valley, California, is 274 feet lower than New Orleans. What is the elevation of Death Valley?

11. **GEOMETRY** Draw the sixth figure in the pattern shown.

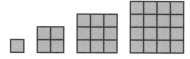

12. **WATERFALLS** Angel Falls in Venezuela, the highest waterfall in the world, is 3,212 feet high. It is 87 feet higher than 2.5 times the height of the Empire State Building. Find the height of the Empire State Building.

13. **AIRCRAFT** An aircraft carrier travels about 6 inches per gallon of fuel. Raquel's car travels about 28 miles per gallon of fuel. If there are 5,280 feet in one mile, how many more inches per gallon would Raquel's car get than an aircraft carrier?

14. **SCHOOL SUPPLIES** Alexandra wishes to buy 5 pens, 1 ruler, and 7 folders to start the school year. The prices are shown in the table.

Item	Cost
Pens	$2.09
Ruler	$0.99
Folder	$1.19

If there is no tax, is $20 enough to pay for Alexandra's school supplies? Explain your reasoning.

15. **MONEY** Antonio has saved $25 in cash to spend at the arcade. If he has 10 bills, how many of each kind of bill does he have?

READING to SOLVE PROBLEMS

Simplify the Problem

Have you ever tried to solve a long word problem and didn't know where to start? Try to rewrite the problem using only the most important words. Here's an example.

STEP 1 Read the problem and identify the important words and numbers.

CELL PHONES There is a wide range of cell phone plans available for students. With Janelle's plan, she pays $15 per month for 200 minutes, plus $0.10 per minute once she talks for more than 200 minutes. Suppose Janelle can spend $20 each month for her cell phone. How many minutes can she talk?

STEP 2 Simplify the problem. Keep all of the important words and numbers, but use fewer of them.

The total monthly cost is the $15 for 200 minutes plus $0.10 times the number of minutes over 200. How many minutes can she talk for $20?

STEP 3 Simplify it again. Use a variable for the unknown.

The cost of m minutes at $0.10 per minute plus $15 is $20.

PRACTICE

Use the method above to simplify each problem.

1. **MONEY** Akira is saving money to buy a scooter that costs $125. He has already saved $80 and plans to save an additional $5 each week. In how many weeks will he have enough money for the scooter?

2. **SHOPPING** Online shopping is a popular way to buy books. Cheryl wants to order several books that cost $7 each. In addition, she will pay a shipping fee of $8. How many books can she order with $43?

3. **TEMPERATURE** The current temperature is 40°. It is expected to rise 5° each hour for the next several hours. In how many hours will the temperature be 60°?

4. **MONEY** Joaquin wants to buy some DVDs that are each on sale for $10 plus a CD that costs $15. How many DVDs can he buy if he has $75 to spend?

Solving Two-Step Equations

MAIN IDEA

Solve two-step equations.

NYS Core Curriculum

7.A.4 Solve multi-step equations by combining like terms, using the distributive property, **or moving variables to one side of the equation** *Also addresses 7.PS.15, 7.CM.2*

New Vocabulary

two-step equation

NY Math Online

glencoe.com

• Concepts In Motion
• Extra Examples
• Personal Tutor
• Self-Check Quiz

▷ **MINI Lab**

MONEY A florist charges $2 for each balloon in an arrangement and a $3 delivery fee. You have $9 to spend. The model illustrates the equation $2x + 3 = 9$, where x represents the number of balloons.

To solve $2x + 3 = 9$, remove three 1-tiles from each side of the mat. Then divide the remaining tiles into two equal groups. The solution of $2x + 3 = 9$ is 3.

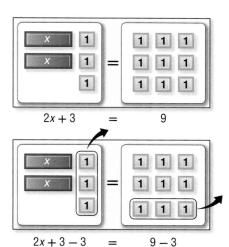

$2x + 3 = 9$

$2x + 3 - 3 = 9 - 3$

$x = 3$

Solve each equation by using models or a drawing.

1. $2x + 1 = 5$

2. $3x + 2 = 8$

3. $2 = 5x + 2$

A **two-step equation** has two different operations. To solve a two-step equation, undo the operations in reverse order of the order of operations. You can review the order of operations in Lesson 1-4.

EXAMPLE Solve a Two-Step Equation

① **Solve $2x + 3 = 9$. Check your solution.**

$2x + 3 =$		9	Write the equation.
$-3 =$		-3	Undo the addition first by subtracting 3 from each side.
$2x$	$=$	6	
$\dfrac{2x}{2}$	$=$	$\dfrac{6}{2}$	Next, undo the multiplication by dividing each side by 2.
x	$=$	3	Simplify.

Check the solution. Since $2(3) + 3 = 9$, the solution is 3.

✓ **CHECK Your Progress**

Solve each equation. Check your solution.

a. $2x + 4 = 10$ b. $3x + 1 = 7$ c. $5 = 2 + 3x$

Study Tip

Order of Operations
Multiplication comes before addition in the order of operations. To undo these operations, reverse the order. So, undo the addition first by subtracting. Then, undo the multiplication by dividing.

EXAMPLES Solve Two-Step Equations

2 Solve $3x + 2 = 23$. Check your solution.

$3x + 2 =$	23		Write the equation.
$-2 = -2$			Undo the addition first by subtracting 2 from each side.
$3x =$	21		
$\dfrac{3x}{3} =$	$\dfrac{21}{3}$		Divide each side by 3.
$x =$	7		Simplify.

Check $\quad 3x + 2 = 23 \qquad$ Write the original equation.

$\qquad\qquad 3(7) + 2 \overset{?}{=} 23 \qquad$ Replace x with 7.

$\qquad\qquad 21 + 2 \overset{?}{=} 23 \qquad$ Simplify.

$\qquad\qquad\qquad 23 = 23 \checkmark \quad$ The sentence is true.

The solution is 7.

3 Solve $-2y - 7 = 3$. Check your solution.

$-2y - 7 =$	3	Write the equation.
$+7 = +7$		Undo the subtraction first by adding 7 to each side.
$-2y =$	10	
$\dfrac{-2y}{-2} =$	$\dfrac{10}{-2}$	Divide each side by −2.
$y =$	-5	Simplify.

The solution is −5. Check the solution.

Study Tip

Equations
Remember, solutions of the new equation are also solutions of the original equation.

4 Solve $4 + 5r = -11$. Check your solution.

$4 + 5r =$	-11	Write the equation.
$-4 = -4$		Undo the addition of 4 first by subtracting 4 from each side.
$5r =$	-15	
$\dfrac{5r}{5} =$	$-\dfrac{15}{5}$	Divide each side by 5.
$r =$	-3	Simplify.

The solution is −3. Check the solution.

✓ CHECK Your Progress

Solve each equation. Check your solution.

d. $4x + 5 = 13$ **e.** $8y + 15 = 71$ **f.** $-3n - 8 = 7$

g. $-5s + 8 = -2$ **h.** $1 + 2y = -3$ **i.** $-2 + 6w = 10$

Solving Two-Step Equations   Key Concept

To solve a two-step equation like $3x + 4 = 16$ or $2x - 1 = -3$:

Step 1 Undo the addition or subtraction first.

Step 2 Then undo the multiplication or division.

Some real-world situations start with a given amount and increase at a certain rate.

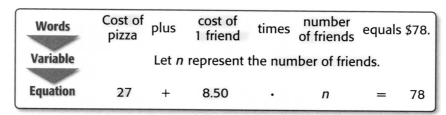

Real-World EXAMPLE

5 **MOVIES** Aisha wants to have her birthday party at the movies. It costs **$27 for pizza** and **$8.50 per friend** for the movie tickets. Since it is Aisha's birthday, she does not have to pay for her movie ticket. How many friends can Aisha have at her party if she has $78 to spend?

Words	Cost of pizza	plus	cost of 1 friend	times	number of friends	equals $78.
Variable	Let n represent the number of friends.					
Equation	27	+	8.50	·	n	= 78

$$27 + 8.50n = 78 \qquad \text{Write the equation.}$$
$$\underline{-27 \qquad\qquad = -27} \qquad \text{Subtract 27 from each side.}$$
$$8.50n = 51$$
$$\frac{8.50n}{8.50} = \frac{51}{8.50} \qquad \text{Divide each side by 8.50.}$$
$$n = 6 \qquad 51 \div 8.50 = 6$$

Check $\quad 27 + 8.50n = 78 \qquad$ Write the original equation.
$\qquad 27 + 8.50(6) \stackrel{?}{=} 78 \qquad$ Replace n with 6.
$\qquad\qquad 27 + 51 \stackrel{?}{=} 78 \qquad$ Simplify.
$\qquad\qquad\qquad 78 = 78 \ \checkmark \qquad$ The sentence is true.

Aisha can have 6 friends at her party.

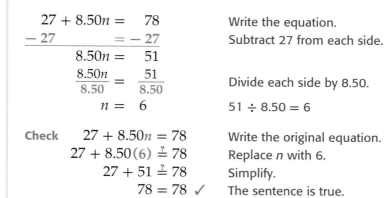

Real-World Link . . . .
Most teenagers see more than 7 movies a year.
Source: Gallup News Service

**CHECK Your Progress**

j. **FITNESS** A fitness club is having a special offer where you pay $22 to join plus a $16 monthly fee. You have $150 to spend. Write and solve an equation to find how many months you can use the fitness club.

**Your Understanding**

Examples 1–4
(pp. 151–152)

Solve each equation. Check your solution.

1. $3x + 1 = 7$

2. $4h - 6 = 22$

3. $-6r + 1 = -17$

4. $-3y - 5 = 10$

5. $13 = 1 + 4s$

6. $-7 = 1 + 2n$

Example 5
(p. 153)

7. **MONEY** Syreeta wants to buy some CDs, each costing $14, and a DVD that cost $23. She has $65 to spend. Write and solve an equation to find how many CDs she can buy.

Solve each equation. Check your solution.

HOMEWORK HELP	
For Exercises	See Examples
8–11	1, 2
12–15	3
16–19	4
20–21	5

8. $3x + 1 = 10$

9. $5x + 4 = 19$

10. $2t + 7 = -1$

11. $6m + 1 = -23$

12. $-4w - 4 = 8$

13. $-7y + 3 = -25$

14. $-8s + 1 = 33$

15. $-2x + 5 = -13$

16. $3 + 8n = -5$

17. $5 + 4d = 37$

18. $14 + 2p = 8$

19. $25 + 2y = 47$

For Exercises 20 and 21, write an equation. Then solve the equation.

20. **BICYCLES** Cristiano is saving money to buy a new bike that costs $189. He has saved $99 so far. He plans on saving $10 each week. In how many weeks will Cristiano have enough money to buy a new bike?

21. **PETTING ZOOS** It cost $10 to enter a petting zoo. Each cup of food to feed the animals is $2. If you have $14, how many cups of food can you buy?

Solve each equation. Check your solution.

22. $2r - 3.1 = 1.7$

23. $4t + 3.5 = 12.5$

24. $16b - 6.5 = 9.5$

25. $5w + 9.2 = 19.7$

26. $16 = 0.5r - 8$

27. $0.2n + 3 = 8.6$

For Exercises 28 and 29, write an equation. Then solve the equation.

28. **CELL PHONES** A cell phone company charges a monthly fee of $39.99 for unlimited *off-peak* minutes on the nights and weekends but $0.45 for each *peak* minute during the weekday. If Brad's monthly cell phone bill was $62.49, for how many *peak* minutes did he get charged?

29. **PLANTS** In ideal conditions, bamboo can grow 47.6 inches each day. At this rate, how many days will it take a bamboo shoot that is 8 inches tall to reach a height of 80 feet?

TEMPERATURE For Exercises 30 and 31, use the following information and the table.

Temperature is usually measured on the Fahrenheit scale (°F) or the Celsius scale (°C). Use the formula $F = 1.8C + 32$ to convert from one scale to the other.

Alaska Record Low Temperatures (°F) by Month	
January	−80
April	−50
July	16
October	−48

30. Convert the temperature for Alaska's record low in July to Celsius. Round to the nearest degree.

31. Hawaii's record low temperature is −11°C. Find the difference in degrees Fahrenheit between Hawaii's record low temperature and the record low temperature for Alaska in January.

NYSCC • NYSMT
Extra Practice, pp. 675, 706

32. CHALLENGE Refer to Exercises 30 and 31. Is there a temperature at which the number of Celsius degrees is the same as the number of Fahrenheit degrees? If so, find it. If not, explain why not.

33. CHALLENGE Suppose your school is selling magazine subscriptions. Each subscription costs $20. The company pays the school half of the total sales in dollars. The school must also pay a one-time fee of $18. What is the fewest number of subscriptions that can be sold to earn a profit of $200?

34. SELECT A TECHNIQUE Bianca rented a car for a flat fee of $19.99 plus $0.26 per mile. Which of the following techniques might she use to determine the approximate number of miles she can drive for $50? Justify your selection(s). Then use the technique(s) to solve the problem.

Mental Math	Number Sense	Estimation

35. WRITING IN MATH Write a real-world problem that would be represented by the equation $2x + 5 = 15$.

NYSMT PRACTICE 7.A.4

36. A rental car company charges $30 a day plus $0.05 a mile. Which expression could be used to find the cost of renting a car for m miles?

A $30.05m$

B $30m + 0.05m$

C $30 + 0.05m$

D $30m + 0.05$

37. The Rodriguez family went on a vacation. They started with $1,875. If they spent $140 each day, which expression represents how much money they had after d days?

F $1,735d$

G $1,875 - 140d$

H $140d$

J $1,875 + 140d$

Spiral Review

38. SCHEDULES Jaime needs to be at the bus stop by 7:10 A.M. If it takes her 7 minutes to walk to the bus stop and 40 minutes to get ready in the morning, what is the latest time that she can set her alarm in order to be at the bus stop 5 minutes earlier than she needs to be? (Lesson 3-4)

ALGEBRA Solve each equation. Check your solution. (Lessons 3-2 and 3-3)

39. $4f = 28$ **40.** $-3y = -15$ **41.** $p - 14 = 27$ **42.** $-11 = n + 2$

43. HIKING Two people are hiking in the Grand Canyon. One is 987 feet below the rim and the other is 1,200 feet below the rim. Find the vertical distance between them (Lesson 2-5)

▷ GET READY for the Next Lesson

PREREQUISITE SKILL Multiply or divide.

44. 2.5×20 **45.** 3.5×4 **46.** $4,200 \div 2.1$ **47.** $104 \div 6.5$

Measurement: Perimeter and Area

MAIN IDEA

Find the perimeters and areas of figures.

NYS Core Curriculum

7.A.6 Evaluate formulas for given input values (surface area, rate, and density problems) *Also addresses 7.CM.5*

New Vocabulary

perimeter
area

NY Math Online

glencoe.com
• Extra Examples
• Personal Tutor
• Self-Check Quiz

▷ **GET READY** for the Lesson

MEASUREMENT At the end of gym class, Mrs. Dalton has the students run around the perimeter of the gym.

1. If the students run around the gym 5 times, how far would they run?

2. Explain how you can use both multiplication and addition to find the distance.

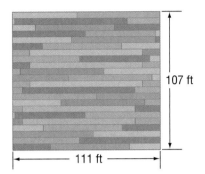

The distance around a geometric figure is called the **perimeter**. To find the perimeter of a rectangle, you can use these formulas.

Perimeter of a Rectangle		Key Concept
Words	The perimeter *P* of a rectangle is twice the sum of the length ℓ and width *w*.	**Model**
Symbols	$P = \ell + \ell + w + w$ $P = 2\ell + 2w \text{ or } 2(\ell + w)$	

EXAMPLE Find the Perimeter of a Rectangle

1. **Find the perimeter of the rectangle shown at the right.**

   4 cm

   15 cm

   $P = 2\ell + 2w$ Perimeter of a rectangle
   $P = 2(15) + 2(4)$ Replace ℓ with 15 and *w* with 4.
   $P = 30 + 8$ Multiply.
   $P = 38$ Add.

   The perimeter is 38 centimeters.

✓ **CHECK Your Progress**

a. Find the perimeter of a rectangle whose length is 14.5 inches and width is 12.5 inches.

GARDENS Elan is designing a rectangular garden. He wants the width to be 8 feet. He also wants to put a fence around the garden. If he has 40 feet of fencing, what is the greatest length the garden can be?

$P = 2\ell + 2w$	Perimeter of a rectangle
$40 = 2\ell + 2(8)$	Replace P with 40 and w with 8.
$40 = 2\ell + 16$	Multiply.
$\underline{-16 = \quad -16}$	Subtract 16 from each side.
$24 = 2\ell$	Simplify.
$12 = \ell$	Divide each side by 2.

The greatest length the garden can be is 12 feet.

✓ **CHECK Your Progress**

b. FRAMES Angela bought a frame for a photo of her friends. The width of the frame is 8 inches. If the distance around the frame is 36 inches, what is the length of the frame?

The distance *around* a rectangle is its perimeter. The measure of the surface *enclosed* by a rectangle is its **area**.

Area of a Rectangle

Key Concept

Words	The area A of a rectangle is the product of the length ℓ and width w.	**Model**	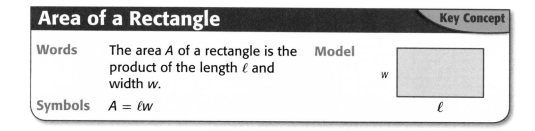
Symbols	$A = \ell w$		

Study Tip

Area Units
When finding area, the units are also multiplied. So, area is given in *square* units. Consider a rectangle 2 ft by 3 ft.

2 ft
3 ft

$A = 2\,ft \cdot 3\,ft$
$A = (2 \cdot 3)(ft \cdot ft)$
$A = 6\,ft^2$

EXAMPLE Find the Area of a Rectangle

③ **TOYS** Find the area of the top of the wooden train table shown at the right.

$A = \ell w$	Area of a rectangle
$A = 49 \cdot 35$	Replace ℓ with 49 and w with 35.
$A = 1{,}715$	Multiply.

The area is 1,715 square inches.

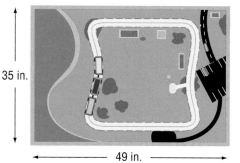
35 in.
49 in.

✓ **CHECK Your Progress**

c. VIDEO GAMES Find the perimeter and area of the top of a video game console that measures 18 inches long and 15 inches wide.

EXAMPLE Use Area to Find a Missing Side

4 The area of a rectangle is 53.94 square feet. If the width is 8.7 feet, find the length.

METHOD 1 Substitute, then solve.

$A = \ell w$	Write the equation.
$53.94 = \ell(8.7)$	Replace A with 53.94 and w with 8.7.
$\dfrac{53.94}{8.7} = \dfrac{\ell(8.7)}{8.7}$	Divide each side by 8.7.
$\ell = 6.2$	Simplify.

Study Tip

Check for Reasonableness
You know that
$53.94 \approx 54$ and that
$8.77 \approx 9$. Since
$54 \div 9 = 6$, the answer
is reasonable.

METHOD 2 Solve, then substitute.

$A = \ell w$	Write the equation.
$\dfrac{A}{w} = \dfrac{\ell w}{w}$	Divide each side by w.
$\dfrac{A}{w} = \ell$	Simplify.
$\dfrac{53.94}{8.7} = \ell$	Replace A with 53.94 and w with 8.7.
$\ell = 6.2$	Simplify.

So, the length of the rectangle is 6.2 feet.

 CHOOSE Your Method

d. What is the width of a rectangle that has an area of 135 square meters and a length of 9 meters?

CHECK Your Understanding

Example 1
(p. 156)

Find the perimeter of each rectangle.

1.
4 yd
5 yd

2.
4.5 cm
1.9 cm

Example 2
(p. 157)

3. PHOTOGRAPHY A photograph is 5 inches wide. The perimeter of the photograph is 24 inches. What is the length of the photograph?

Example 3
(p. 157)

Find the area of each rectangle.

4.
1 m
3.8 m

5.
5 ft
5.25 ft

Example 4
(p. 158)

6. MEASUREMENT The area and length of a rectangle are 30 square feet and 6 feet, respectively. What is the width of the rectangle?

HOMEWORK HELP

For Exercises	See Examples
7–12	1
13–14	2
15–20	3
21–22	4

Find the perimeter of each rectangle.

7.

12 ft

6 ft

8.
18 in.

23 in.

9.

2 mm

5.4 mm

10.
3.8 cm

2.4 cm

11. $\ell = 1.5$ ft, $w = 4$ ft

12. $\ell = 5.75$ ft, $w = 8$ ft

13. **SEWING** The fringe used to outline a placemat is 60 inches. If the width of the placemat is 12 inches, what is the measure of its length?

14. **GAMES** Orlando marks off a rectangular section of grass to use as a playing field. He knows that he needs to mark off 111 feet around the border of the playing field. If the field is 25.5 feet long, what is its width?

Find the area of each rectangle.

15.

13 ft

6 ft

16.

3 yd

11 yd

17.
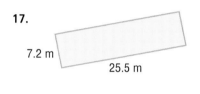
7.2 m

25.5 m

18.

16.5 cm

8.4 cm

19. $\ell = 3.25$ in., $w = 2$ in.

20. $\ell = 4.5$ ft, $w = 10.6$ ft

21. **QUILTING** Daniela's grandmother is making a quilt that is 7 squares wide. If she needs 35 squares total, how many squares long is the length?

22. **PAINTING** A rectangular mural is painted on a wall. If the mural is 12 feet wide and covers 86.4 square feet, what is the height of the mural?

Find the missing measure.

23. $P = 115.6$ ft, $w = 24.8$ ft

24. $A = 189.28$ cm^2, $w = 16.9$ cm

ANALYZE TABLES For Exercises 25 and 26, use the table shown.

25. How much greater is the area of a soccer field for a 14-year-old than for a 10-year-old?

26. An *acre* equals 4,840 square yards. How many acres are there in a field for a 12-year-old? Round to the nearest hundredth.

Youth Soccer Fields		
Age (years)	Length (yards)	Width (yards)
10	70	40
12	80	50
14	100	60

For Exercises 27–30, determine whether the problem involves perimeter, area, or both. Then solve.

27. **HIKING** Ramón walked along a rectangular hiking path. Initially, he walked 3 miles north before he turned east. If he walked a total of 14 miles, how many miles did he walk east before he turned south?

28. **BORDERS** Kaitlyn's bedroom is shaped like a rectangle with rectangular walls. She is putting a wallpaper border along the top of the two longer walls and one of the shorter walls. If the length of the room is 13 feet and the width is 9.8 feet, how many feet of border does she need?

29. **DECKS** Armando built a rectangular deck in his backyard. The deck takes up 168.75 square feet of space. If the deck's width is 12.8 feet, what is its approximate length?

30. **FENCING** Kina plans to fence her rectangular backyard on three sides. Her backyard measures 48 feet in length. She will not fence one of the shorter sides. If the area of her backyard is 1,752 square feet, how many feet of fencing is required?

31. **VOLLEYBALL** For safety, regulation courts have a "free zone" added to the width of each side of the court. If the free zone on each side is 3 meters wide, use the information to the left to find the area of the entire volleyball court including the free zone.

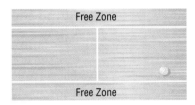

32. **GEOMETRY** Use the diagram at the right to write formulas for the perimeter P and area A of a square.

33. **FIND THE DATA** Refer to the Data File on pages 16–19. Choose some data and write a real-world problem in which you would find the perimeter and area of a rectangle.

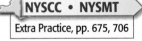
NYSCC • NYSMT
Extra Practice, pp. 675, 706

H.O.T. Problems

34. **OPEN ENDED** Draw and label three different rectangles that have an area of 24 square centimeters.

CHALLENGE For Exercises 35–38, find each equivalent measurement. Provide a diagram to justify your answers. The diagram for Exercise 35 is shown.

35. $1 \text{ yd}^2 = \blacksquare \text{ ft}^2$

36. $4 \text{ yd}^2 = \blacksquare \text{ ft}^2$

37. $1 \text{ ft}^2 = \blacksquare \text{ in}^2$

38. $2 \text{ ft}^2 = \blacksquare \text{ in}^2$

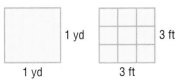

NUMBER SENSE For Exercises 39 and 40, describe the effect on the perimeter and area in each of the following situations.

39. The width of a rectangle is doubled.

40. The length of a side of a square is doubled.

41. **CHALLENGE** A rectangle has width w. Its length is one unit more than 3 times its width. Write an expression that represents the perimeter of the rectangle.

42. **WRITING IN MATH** Decide whether the statement is *true* or *false*. Explain your reasoning and provide examples.

> *Of all rectangles with a perimeter of 24 square inches, the one with the greatest area is a square.*

NYSMT PRACTICE 7.A.6

43. Oakland Garden Center created a design plan for the Nelson family's rock garden. The shaded areas will hold flowers and the rest of the garden will be rock.

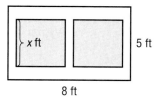

8 ft

If each shaded area is a square, which expression represents the area of the garden that will be rock?

A $(40 - 2x^2)$ ft^2 **C** $(40 + x)$ ft^2

B $(40 - x)$ ft^2 **D** $(40 + x^2)$ ft^2

44. The rectangle below has width 4.75 feet and perimeter P feet.

4.75 ft

ℓ ft

Which of the following could be used to find the length of the rectangle?

F $P = 4.75 + \dfrac{\ell}{2}$

G $P = 4.75 - \ell$

H $P = 9.5 + 2\ell$

J $P = 9.5 - 2\ell$

Spiral Review

Solve each equation. Check your solution. (Lesson 3-5)

45. $5d + 12 = 2$
46. $13 - f = 7$
47. $10 = 2g + 3$
48. $6 = 3 - 3h$

49. **ALGEBRA** Anna was charged $11.25 for returning a DVD 5 days late. Write and solve an equation to find how much the video store charges per day for late fees. (Lesson 3-3)

Multiply. (Lesson 2-6)

50. $14(-5)$
51. $(-3)^3$
52. $-10(2)(-8)$

53. **AGE** The sum of Denise's and Javier's ages is 26 years. If Denise is 4 years older than Javier, find Javier's age. Use the *guess and check* strategy. (Lesson 1-5)

GET READY for the Next Lesson

PREREQUISITE SKILL Graph and label each point on a coordinate plane. (Lesson 2-3)

54. $(-4, 2)$
55. $(3, -1)$
56. $(-3, -4)$
57. $(2, 0)$

Measurement Lab
Representing Relationships

MAIN IDEA

Graph data to demonstrate the relationship between the dimensions and the perimeter of a rectangle.

NYS Core Curriculum

7.R.4 Explain how different representations express the same relationship *Also addresses 7.R.3, 7.R.5, 7.R.7*

In this lab, you will investigate the relationships between the dimensions and the perimeter of a rectangle.

ACTIVITY

STEP 1 Use 10 chenille stems, 24 centimeters in length, to form 10 rectangles with different dimensions.

STEP 2 Measure and record the width and length of each rectangle to the nearest centimeter in a table like the one at the right.

Width (cm)	Length (cm)

ANALYZE THE RESULTS

1. What rectangle measure does 24 centimeters represent?

2. Find the sum of the width and length for each of your rectangles. Write a sentence that describes the relationship between this sum and the measure of the length of the stem for each rectangle. Then write a rule that describes this relationship for a rectangle with a width w and length ℓ.

3. In this activity, if a rectangle has a length of 4.5 centimeters, what is its width? Explain your reasoning. Write a rule that can be used to find w when ℓ is known for any rectangle in this Activity.

4. **GRAPH THE DATA** Graph the data in your table on a coordinate plane like the one at the right.

5. Describe what the ordered pair (w, ℓ) represents. Describe how these points appear on the graph.

6. Use your graph to find the width of a rectangle with a length of 7 centimeters. Explain your method.

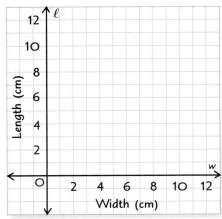

7. **MAKE A CONJECTURE** If the length of each chenille stem was 20 centimeters, how would this affect the data in your table? the rule you wrote in Exercise 3? the appearance of your graph?

Functions and Graphs

▷ **GET READY** for the Lesson

MONEY The Westerville Marching Band is going on a year-end trip to an amusement park. Each band member must pay an admission price of $15. In the table, this is represented by 15m.

Total Cost of Admission		
Number of Members	15m	Total Cost ($)
1	15(1)	15
2	15(2)	30
3	15(3)	
4		
5		
6		

1. Copy and complete the function table for the total cost of admission.

2. Graph the ordered pairs (number of members, total cost).

3. Describe how the points appear on the graph.

If you are given a function, ordered pairs in the form (input, output), or (x, y), provide useful information about that function. These ordered pairs can then be graphed on a coordinate plane and form part of the graph of the function. The graph of the function consists of the points in the coordinate plane that correspond to *all* the ordered pairs of the form (input, output).

Real-World EXAMPLE

1 **TEMPERATURE** The table shows temperatures in Celsius and the corresponding temperatures in Fahrenheit. Make a graph of the data to show the relationship between Celsius and Fahrenheit.

The ordered pairs (5, 41), (10, 50), (15, 59), (20, 68), (25, 77), and (30, 86) represent this function. Graph the ordered pairs.

Celsius (input)	Fahrenheit (output)
5	41
10	50
15	59
20	68
25	77
30	86

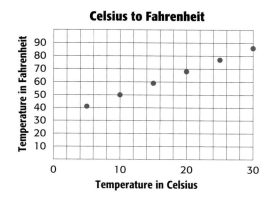

Celsius to Fahrenheit

 **CHECK** Your Progress

a. **MUSIC** The table shows the money remaining on a $75 gift certificate after a certain number of CDs are bought. Make a graph to show how the number of CDs bought and the remaining balance are related.

$75 Music Gift Certificate	
Number of CDs	**Balance ($)**
1	63
2	51
3	39
4	27
5	15

The solution of an equation with two variables consists of two numbers, one for each variable, that make the equation true. The solution is usually written as an ordered pair (x, y).

EXAMPLE **Graph Solutions of Linear Equations**

 Graph $y = 2x + 1$.

Select any four values for the input x. We chose 2, 1, 0, and −1. Substitute these values for x to find the output y.

x	$2x + 1$	y	(x, y)
2	$2(2) + 1$	5	$(2, 5)$
1	$2(1) + 1$	3	$(1, 3)$
0	$2(0) + 1$	1	$(0, 1)$
−1	$2(-1) + 1$	−1	$(-1, -1)$

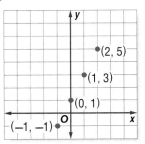

The four inputs correspond to the solutions $(2, 5)$, $(1, 3)$, $(0, 1)$, and $(-1, -1)$. By graphing these ordered pairs, you can create the graph of $y = 2x + 1$.

 CHECK Your Progress **Graph each equation.**

b. $y = x - 3$ c. $y = -3x$ d. $y = -3x + 2$

Notice that all four points in the graph lie on the same straight line. Draw a line through the points to graph *all* solutions of the equation $y = 2x + 1$. Note that the point $(3, 7)$ is also on this line.

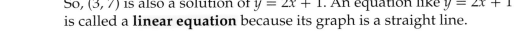

$y = 2x + 1$ Write the equation.
$7 \stackrel{?}{=} 2(3) + 1$ Replace x with 3 and y with 7.
$7 = 7$ ✓   This sentence is true.

So, $(3, 7)$ is also a solution of $y = 2x + 1$. An equation like $y = 2x + 1$ is called a **linear equation** because its graph is a straight line.

Real-World EXAMPLE

3 **SWIMMING** Michael Phelps swims the 400-meter individual medley at an average speed of 100 meters per minute. The equation $d = 100t$ describes the distance d that he can swim in t minutes at this speed. Represent the function by a graph.

Step 1 Select any four values for t. Select only positive numbers because t represents time. Make a function table.

t	100t	d	(t, d)
1	100(1)	100	(1, 100)
2	100(2)	200	(2, 200)
3	100(3)	300	(3, 300)
4	100(4)	400	(4, 400)

Step 2 Graph the ordered pairs and draw a line through the points.

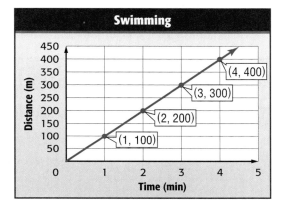

CHECK Your Progress

e. JOBS Sandi makes $6 an hour babysitting. The equation $m = 6h$ describes how much money m she earns babysitting for h hours. Represent this function by a graph.

Representing Functions   Key Concept

Words   There are 12 inches in one foot.

Table

Feet	Inches
1	12
2	24
3	36
4	48

Graph

Convert Feet to Inches

(graph with Inches on vertical axis from 10 to 60, Feet on horizontal axis from 0 to 4)

Equation $f = 12n$, where f represents the number of feet and n represents the number of inches.

Example 1
(p. 163)

Graph the function represented by each table.

1.

Total Cost of Baseballs	
Baseball	Total Cost ($)
1	4
2	8
3	12
4	16

2.

Convert Minutes to Seconds	
Minutes	Seconds
1	60
2	120
3	180
4	240

Example 2
(p. 164)

Graph each equation.

3. $y = x - 1$

4. $y = -1x$

5. $y = -2x + 3$

Example 3
(p. 165)

6. **MEASUREMENT** The perimeter of a square is 4 times greater than the length of one of its sides. The equation $p = 4s$ describes the perimeter p of a square with sides s units long. Represent this function by a graph.

▶ Practice and Problem Solving

HOMEWORK HELP

For Exercises	See Examples
7–8	1
9–14	2
15–16	3

Graph the function represented by each table.

7.

Total Phone Bill	
Time (min)	Total (¢)
1	8
2	16
3	24
4	32

8.

Calories in Fruit Cups	
Servings	Total Calories
1	70
3	210
5	350
7	490

Graph each equation.

9. $y = x + 1$

10. $y = x + 3$

11. $y = x$

12. $y = -2x$

13. $y = 2x + 3$

14. $y = 3x - 1$

For Exercises 15 and 16, represent each function by a graph.

15. **CARS** A car averages 36 miles per gallon of gasoline. The function $m = 36g$ represents the miles m driven using g gallons of gasoline.

16. **FITNESS** A health club charges $35 a month for membership fees. The equation $c = 35m$ describes the total charge c for m months of membership.

Graph each equation.

17. $y = 0.25x$

18. $y = x + 0.5$

19. $y = 0.5x - 1$

20. **SHOPPING** You buy a DVD for $14 and CDs for $9 each. The equation $t = 14 + 9c$ represents the total amount t that you spend if you buy 1 DVD and c CDs. Represent this function by a graph.

NYSCC • NYSMT
Extra Practice, pp. 676, 706

H.O.T. Problems

21. **OPEN ENDED** Draw the graph of a linear function. Name three ordered pairs in the function.

CHALLENGE For Exercises 22 and 23, let x represent the first number and let y represent the second number. Draw a graph of each function.

22. The second number is three more than the first number.

23. The second number is the product of -3 and the first number.

24. **WRITING IN MATH** Describe how you use a function table to create the graph of a function.

NYSMT PRACTICE 8.A.4

25. The graph shows the relationship between the number of hours Jennifer spent jogging and the total number of miles she jogged. Which table best represents the data in the graph?

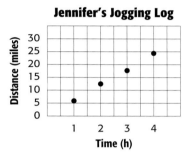

Jennifer's Jogging Log

A
Time (h)	Distance (mi)
6	4
12	3
18	2
24	1

B
Time (h)	Distance (mi)
2	6
3	12
4	18
5	24

C
Time (h)	Distance (mi)
1	6
2	12
3	18
4	24

D
Time (h)	Distance (mi)
4	6
3	6
2	6
1	6

Spiral Review

26. **MEASUREMENT** The area and width of a rug are 323 square inches and 17 inches respectively. What is the perimeter of the rug? (Lesson 3-6)

Solve each equation. Check your solution. (Lesson 3-5)

27. $4y + 19 = 7$ 28. $10x + 2 = 32$ 29. $48 - 8j = 16$ 30. $14 = 2 - 6d$

31. Evaluate $|5| + |-10|$. (Lesson 2-1)

Problem Solving in Social Studies   🌐 **Real-World Unit Project**

Stand Up and Be Counted! It's time to complete your project. Use the data you have gathered about how the U.S. Census affects the House of Representatives to prepare a poster. Be sure to include a map, frequency table, and paragraph discussing the changes in the House of Representatives.

NY Math Online Unit Project at glencoe.com

Graphing Calculator Lab
Graphing Relationships

MAIN IDEA

Use technology to graph relationships involving conversions of measurement.

NYS Core Curriculum

Preparation for 8.A.4 Create a graph given a description or an expression for a situation involving a linear or nonlinear relationship *Also addresses 7.R.3, 7.R.4, 7.R.5, 7.R.7*

You can use a graphing calculator to graph relationships.

ACTIVITY

1. **MEASUREMENT** Use the table at the right to write a function that relates the number of yards x to the number of feet y. Then graph your function.

Yards (x)	Feet (y)
1	3
2	6
3	9
4	12

STEP 1 By examining the table, you can see that the number of feet is 3 times the number of yards. Write a function.

The number of feet   is   3 times   the number of yards.
$$y \quad = \quad 3 \quad\quad x$$

STEP 2 Press $\boxed{Y=}$ and enter the function $y = 3x$ into Y₁.

STEP 3
Adjust your viewing window. Press $\boxed{\text{WINDOW}}$ and change the values to reflect the range of values in the table.

STEP 4
Finally, graph the function by pressing $\boxed{\text{Graph}}$.

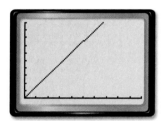

ANALYZE THE RESULTS

1. Test the function above using one of the values from the table and the CALC feature on your calculator. Press $\boxed{\text{2nd}}$ [CALC] 1 and then enter an x-value of 3. What y-value is displayed? What do each of these values represent and how are they represented on the graph?

2. Use your graph to convert 7 yards into feet. Explain your method.

3. **MAKE A CONJECTURE** Write a function that could be used to convert feet into yards. What is an appropriate window for a graph of this function? Graph and test your function.

4. Use your function from Exercise 3 to convert 16 feet into yards.

5. Write a function that could be used to convert 36 ounces to pounds. Indicate an appropriate window, then use a graph of the function to convert 36 ounces to pounds. (*Hint:* 1 pound = 16 ounces)

FOLDABLES®
Study Organizer

GET READY to Study

Be sure the following Big Ideas are noted in your Foldable.

BIG Ideas

Solving Equations (Lessons 3-2, 3-3, and 3-5)
• If you add or subtract the same number from each side of an equation, the two sides remain equal.

• If you divide each side of an equation by the same nonzero number, the two sides remain equal.

• To solve a two–step equation like $3x + 4 = 19$ or $2x - 1 = -5$:
  Step 1 Undo the addition or subtraction first.
  Step 2 Then undo the multiplication or division.

Perimeter and Area Formulas (Lesson 3-6)
• The perimeter P of a rectangle is twice the sum of the length ℓ and width w.

• The area A of a rectangle is the product of the length ℓ and width w.

Linear Functions (Lesson 3-7)
• The graph of a linear function is a straight line. Ordered pairs in the form (x, y) can be used to represent a function and graphed on the coordinate plane as part of the graph of the function.

Key Vocabulary

formula (p. 144)

linear equation (p. 164)

two–step equation (p. 151)

work backward strategy (p. 148)

Vocabulary Check

State whether each sentence is *true* or *false*. If *false*, replace the underlined word or number to make a true sentence.

1. The expression $\frac{1}{3}y$ means <u>one third of y</u>.

2. The words *more than* sometimes suggest the operation of <u>multiplication</u>.

3. The formula <u>$d = rt$</u> gives the distance d traveled at a rate of r for t units of time.

4. The algebraic expression representing the words *six less than m* is <u>$6 - m$</u>.

5. Use the <u>work backward</u> strategy when you are given a final result and asked to find an earlier amount.

6. The word *each* sometimes suggests the operation of <u>division</u>.

7. In solving the equation $4x + 3 = 15$, first <u>divide each side by 4</u>.

8. The solution to the equation $p + 4.4 = 11.6$ is <u>7.2</u>.

9. The process of solving a <u>two-step equation</u> uses the work backward strategy.

10. The expression $5x$ means <u>5 more than x</u>.

11. To find the distance around a rectangle, use the formula for its <u>area</u>.

12. The word *per* sometimes suggests the operation of <u>subtraction</u>.

Lesson-by-Lesson Review

3-1 Writing Expressions and Equations (pp. 128–133)

7.A.1

Write each phrase as an algebraic expression.

13. the sum of a number and 5

14. six inches less than her height

15. twice as many apples

Write each sentence as an algebraic equation.

16. Ten years older than Theresa's age is 23.

17. Four less than a number is 19.

18. The quotient of 56 and a number is 14.

19. **AMUSEMENT PARKS** This year, admission to a popular amusement park is $8.75 more than the previous year's admission fee. Write an expression describing the cost of this year's admission.

Example 1 Write the phrase as an algebraic expression.

four times the price

Let p represent the price.
The algebraic expression is $4p$.

Example 2 Write the sentence as an algebraic equation.

Six less than the number of cookies is 24.

Let c represent the number of cookies.
The equation is $c - 6 = 24$.

3-2 Solving Addition and Subtraction Equations (pp. 136–141)

7.A.4

Solve each equation. Check your solution.

20. $x + 5 = 8$

21. $r + 8 = 2$

22. $p + 9 = -4$

23. $s - 8 = 15$

24. $n - 1 = -3$

25. $w - 9 = 28$

26. **COOKIES** Marjorie baked some chocolate chip cookies for her and her sister. Her sister ate 6 of these cookies. If there were 18 cookies left, write and solve an equation to find how many cookies c Marjorie baked.

Example 3 Solve $x + 6 = 4$.

$$x + 6 = 4$$
$$\underline{-6 = -6} \quad \text{Subtract 6 from each side.}$$
$$x \quad = -2$$

Example 4 Solve $y - 3 = -2$.

$$y - 3 = -2$$
$$\underline{+3 = +3} \quad \text{Add 3 to each side.}$$
$$y \quad = 1$$

Mixed Problem Solving
For mixed problem-solving practice,
see page 706.

3-3

Solving Multiplication Equations (pp. 142–146)

7.A.4

Solve each equation. Check your solution.

27. $7c = 28$ 28. $-8w = 72$

29. $10y = -90$ 30. $-12r = -36$

31. **MONEY** Matt borrowed $98 from his father. He plans to repay his father at $14 per week. Write and solve an equation to find the number of weeks w required to pay back his father.

Example 5   Solve $-4b = 32$.

$$-4b = 32$$

$$\frac{-4b}{-4} = \frac{32}{-4} \qquad \text{Divide each side by } -4.$$

$$b = -8$$

3-4

PSI: Work Backward (pp. 148–149)

7.PS.9

Solve. Use the *work backward* strategy.

32. **BASEBALL** Last baseball season, Nelson had four less than twice the number of hits Marcus had. Nelson had 48 hits. How many hits did Marcus have last season?

33. **CREDIT CARDS** Alicia paid off $119 of her credit card balance and made an additional $62.75 in purchases. If she now owes $90.45, what was her starting balance?

Example 6   A number is divided by 2. Then 4 is added to the quotient. After subtracting 3, the result is 18. What is the number?

Start with the final value and work backward with each resulting value until you arrive at the starting value.

$18 + 3 = 21$ Undo subtracting 3.

$21 - 4 = 17$ Undo adding 4.

$17 \cdot 2 = 34$ Undo dividing by 2.

The number is 34.

3-5

Solving Two-Step Equations (pp. 151–155)

7.A.4

Solve each equation. Check your solution.

34. $3y - 12 = 6$ 35. $6x - 4 = 20$

36. $2x + 5 = 3$ 37. $5m + 6 = -4$

38. $10c - 8 = 90$ 39. $3r - 20 = -5$

40. **ALGEBRA** Ten more than five times a number is 25. Find the number.

Example 7   Solve $3p - 4 = 8$.

$$3p - 4 = \quad 8$$

$$\underline{+4 = +4} \qquad \text{Add 4 to each side.}$$

$$3p \quad = \quad 12$$

$$\frac{3p}{3} = \frac{12}{3} \qquad \text{Divide each side by 3.}$$

$$p = 4$$

3-6 **Measurement: Perimeter and Area** (pp. 156–161)

7.A.6

Find the perimeter and area of each rectangle.

41.

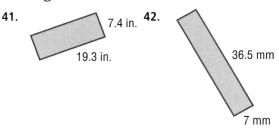

7.4 in.

19.3 in.

42.

36.5 mm

7 mm

Find the missing measure.

43. $P = 56$ mi, $\ell = 21$ mi

44. $A = 10.26$ ft^2, $w = 2.7$ ft

45. $A = 272$ yd^2, $\ell = 17$ yd

46. $P = 14.2$ cm, $w = 2.6$ cm

47. **CARPET** In order to carpet her rectangular living room, Flora needs 192 square feet of carpet. If the length of her living room is 16 feet, find the width.

Example 8 The perimeter of a rectangle is 38 meters. If the length is 7 meters, find the width.

$\begin{aligned} P &= 2\ell + 2w & \text{Perimeter formula} \\ 38 &= 2(7) + 2w & \text{Replace } P \text{ with 38 and } \ell \text{ with 7.} \\ 38 &= 14 + 2w & \text{Simplify.} \\ 24 &= 2w & \text{Subtract 14 from each side.} \\ \frac{24}{2} &= \frac{2w}{2} & \text{Divide each side by 2.} \\ 12 &= w & \text{Simplify.} \end{aligned}$

The width is 12 meters.

Example 9 The area of a rectangle is 285 square inches. If the width is 15 inches, find the length.

$\begin{aligned} A &= \ell w & \text{Area formula} \\ 285 &= \ell(15) & \text{Replace } A \text{ with 285 and } w \text{ with 15.} \\ \frac{285}{15} &= \frac{15\ell}{15} & \text{Divide each side by 15.} \\ 19 &= \ell & \text{Simplify.} \end{aligned}$

The length is 19 inches.

3-7 **Functions and Graphs** (pp. 163–167)

8.A.4

Graph each equation.

48. $y = x + 5$

49. $y = x - 4$

50. $y = 2x$

51. $y = -1x$

52. $y = 3x + 2$

53. $y = -2x + 3$

MONEY For Exercises 54–56, use the following information.

Clara earns $9 per hour mowing lawns.

54. Make a table that shows her total earnings for 2, 4, 6, and 8 hours.

55. Write an equation in which x represents the number of hours and y represents Clara's total earnings.

56. Graph the equation.

Example 10 Graph $y = x + 3$.

Select four values for x. Substitute these values for x to find values for y.

x	$x + 3$	y
-1	$-1 + 3$	2
0	$0 + 3$	3
1	$1 + 3$	4
2	$2 + 3$	5

Four solutions are $(-1, 2)$, $(0, 3)$, $(1, 4)$, and $(2, 5)$. The graph is shown below.

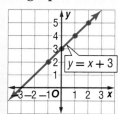

Write each phrase as an algebraic expression or equation.

1. $5 less than Tomasita has

2. 4 years older than Hana

3. 56 inches is 9 inches shorter than Jacob's height.

4. Twice the distance from the park to the post office is 5 miles.

5. **FLOWERS** The number of tulips in Paula's garden is 8 less than the number of marigolds. If there are 16 tulips, write and solve an equation to determine the number of marigolds m.

6. **MULTIPLE CHOICE** If you divide a number by 8 and subtract 11 from the result, the answer is 4. Which equation matches this relationship?

   A $\dfrac{n-11}{8} = 4$ C $\dfrac{n}{8} = 11 - 4$

   B $4 = \dfrac{n}{8} - 11$ D $4 = 11 - \dfrac{n}{8}$

ANALYZE TABLES For Exercises 7–9, use the table below. It shows how Jared's age and his sister Emily's are related.

Jared's age (yr)	1	2	3	4	5
Emily's age (yr)	7	8	9	10	11

7. Write an equation in which x represents Jared's age, and y represents Emily's age.

8. Graph the equation.

9. Predict how old Emily will be when Jared is 10 years old.

10. **TOURISM** The Statue of Liberty is 151 feet tall. It is 835 feet shorter than the Eiffel Tower. Write and solve an equation to find the height of the Eiffel Tower.

Solve each equation. Check your solution.

11. $x + 5 = -8$ 12. $y - 11 = 15$

13. $9z = -81$ 14. $-6k + 4 = -38$

15. $3z - 7 = 17$ 16. $2g - 9 = -5$

17. **PIZZA** Chris and Joe shared a pizza. Chris ate two more than twice as many pieces as Joe, who ate 3 pieces. If there were 3 pieces left, how many pieces were there initially? Use the *work backward* strategy.

18. **MULTIPLE CHOICE** A giant cake in the shape of and decorated as an American flag was 60 feet in length. If it took A square feet of icing to cover the top of the cake, which of the following would represent the cake's perimeter?

   F $P = 120 + 2 \cdot \dfrac{A}{60}$ H $P = 60 + \dfrac{A}{60}$

   G $P = 120 + 2A$ J $P = 60 + 2A$

Find the perimeter and area of each rectangle.

19. 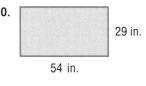 8.9 cm

13.2 cm

20. 29 in.

54 in.

Graph each equation.

21. $y = x + 1$

22. $y = 2x$

23. $y = 2x - 3$

24. $y = -x + 1$

25. **MOVIES** A student ticket to the movies costs $6. The equation $c = 6t$ describes the total cost c for t tickets. Make a function table that shows the total cost for 1, 2, 3, and 4 tickets and then graph the equation.

PART 1 Multiple Choice

Read each question. Then fill in the correct answer on the answer sheet provided by your teacher or on a sheet of paper.

1. During a bike-a-thon, Shalonda cycled at a constant rate. The table shows the distance she covered in half-hour intervals.

Time (h)	Distance (mi)
$\frac{1}{2}$	6
1	12
$1\frac{1}{2}$	18
2	24

Which of the following equations represents the distance d Shalonda covered after h hours?

A $d = 6 + h$

B $d = 6h$

C $d = 12 + h$

D $d = 12h$

2. Which line contains the ordered pair $(-2, 4)$?

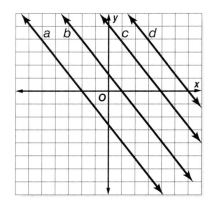

F line a **H** line b

G line c **J** line d

3. The table gives the value of several terms and their positions in a sequence.

Position	3	4	5	6	7	n
Value of Term	13	17	21	25	29	

Which description shows the relationship between a term and n, its position in the sequence?

A Add 4 to n.

B Multiply n by 5 and add 1.

C Multiply n by 3 and add 2.

D Multiply n by 4 and add 1.

4. Mr. McDowell owes $1,750 on his car loan and pays off $185 each month towards the loan balance. Which expression represents how much money in dollars he still owes after x months?

F $1,750x$

G $1,750x + 185$

H $1,750x - 185$

J $1,750 - 185x$

5. Which of the following problems can be solved by solving the equation $x - 9 = 15$?

A Allison is nine years younger than her sister Pam. Allison is 15 years old. What is x, Pam's age?

B David's portion of the bill is $9 more than Sam's portion of the bill. If Sam pays $9, find x, the amount in dollars that David pays.

C The sum of two numbers is 15. If one of the numbers is 9, what is x, the other number?

D Pedro owns 15 CDs. If he gave 9 of them to a friend, what is x, the number of CDs he has left?

6. The table below shows values for x and corresponding values for y.

x	y
18	2
27	3
9	1
36	4

Which of the following represents the relationship between x and y?

F $y = x + 16$

G $y = 9x$

H $y = \frac{1}{9}x$

J $y = x + 9$

TEST-TAKING TIP

Question 7 Read the question carefully to check that you answered the question that was asked. In question 7, you are asked to pick which coordinates lie within the triangle, not to identify a vertex.

7. Which of the following coordinates lies within the triangle graphed below?

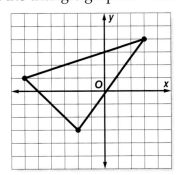

A $(3, 4)$

B $(2, 0)$

C $(-3, 6)$

D $(-1, -1)$

PART 2 **Short Response/Grid In**

Record your answers on the answer sheet provided by your teacher or on a sheet of paper.

8. Write an expression that can be used to find the maximum number of 3-foot pieces of lumber than can be cut from a 12-foot length of lumber.

9. Tonya bought p pounds of sand. Each pound of sand costs $0.40. How many pounds of sand could Tonya buy with $3.20?

PART 3 **Extended Response**

Record your answers on the answer sheet provided by your teacher or on a sheet of paper. Show your work.

10. The distances traveled by cars traveling at 40 miles per hour and at 60 miles per hour are given in the table.

Time (hours)	Distance (miles) at 40 mph	Distance (miles) at 60 mph
0	0	0
1	40	60
2	80	120
3	120	180
4	160	240

a. Graph the ordered pairs (time, distance) for 40 miles per hour.

b. Graph the ordered pairs (time, distance) for 60 miles per hour.

c. Predict where the ordered pairs for 50 miles per hour would be graphed. Explain how you know.

NEED EXTRA HELP?										
If You Missed Question...	1	2	3	4	5	6	7	8	9	10
Go to Lesson...	3-7	3-7	1-9	3-1	3-1	3-7	3-7	2-8	1-7	3-7
NYS Core Curriculum	8.A.4	8.A.4	7.A.8	7.A.1	7.A.1	8.A.4	8.A.4	7.N.12	7.PS.3	8.A.4

Unit 2

Number Sense: Fractions

Focus
Represent and use numbers in a variety of equivalent forms and apply addition, subtraction, multiplication, and division of fractions.

CHAPTER 4
Fractions, Decimals, and Percents

BIG Idea Represent and use numbers in a variety of forms.

CHAPTER 5
Applying Fractions

BIG Idea Extend understandings of operations. Add, subtract, multiply, and divide to solve fraction problems.

Problem Solving in Geography

A Traveling We Will Go You're about to embark on a journey to your favorite vacation spot in the United States. In your role as a travel agent, you will plan a vacation for you and your family. You will calculate the total cost including transportation, lodging, and tourist attractions. So bring your sense of adventure and get ready to set off on your trip!

NY Math Online Log on to glencoe.com to begin.

CHAPTER 4

Fractions, Decimals, and Percents

 New York State Core Curriculum

Reinforcement of 6.N.11 Read, write, and identify percents of a whole (0% to 100%)

7.N.8 Find the common factors and greatest common factor of two or more numbers

Key Vocabulary

equivalent fractions (p. 192)

percent (p. 202)

ratio (p. 202)

simplest form (p. 192)

🌐 **Real-World Link**

Reptiles North Carolina's state reptile is the Eastern Box Turtle. Adults range in size from $4\frac{1}{2}$ inches to $5\frac{9}{10}$ inches. You can write these fractions as 4.5 and 5.9, respectively.

FOLDABLES®
Study Organizer

Fractions, Decimals, and Percents Make this Foldable to help you organize your notes. Begin with five sheets of $8\frac{1}{2}" \times 11"$ paper.

❶ **Stack** five sheets of paper $\frac{3}{4}$ inch apart.

❷ **Roll** up bottom edges so that all tabs are the same size.

❸ **Crease** and staple along the fold.

❹ **Write** the chapter title on the front. Label each tab with a lesson number and title.

GET READY for Chapter 4

Diagnose Readiness You have two options for checking Prerequisite Skills.

Option 2

NY Math Online Take the Online Readiness Quiz at glencoe.com.

Option 1

Take the Quick Quiz below. Refer to the Quick Review for help.

QUICK Quiz

State which decimal is greater. (Prior Grade)

1. 0.6, 0.61   **2.** 1.25, 1.52

3. 0.33, 0.13   **4.** 1.08, 10.8

5. LUNCH Kirsten spent $4.21 on lunch while Almanzo spent $4.12. Who spent the greater amount? (Lesson 3-9)

QUICK Review

Example 1

State which decimal is greater, 7.4 or 7.04.

7.4
7.04
↑
7.4 is greater.

Line up the decimal points and compare place value. The 4 in the tenths place is greater than the 0 in the tenths place.

Use divisibility rules to determine whether each number is divisible by 2, 3, 5, 6, or 10. (Prior Grade)

6. 125   **7.** 78   **8.** 37

9. MUFFINS Without calculating, determine whether 51 banana nut muffins can be evenly distributed among 3 persons. Explain. (Prior Grade)

Example 2

Use divisibility rules to determine whether 84 is divisible by 2, 3, 5, 6, or 10.

2: Yes, the ones digit, 4, is divisible by 2.
3: Yes, the sum of the digits, 12, is divisible by 3.
5: No, the ones digit is neither 0 nor 5.
6: Yes, the number is divisible by both 2 and 3.
10: No, the ones digit is not 0.

Divide. (Prior Grade)

10. $12 \div 6$ **11.** $18 \div 3$

12. $2 \div 5$ **13.** $3 \div 4$

Example 3

Find $1 \div 5$.

$$\begin{array}{r} 0.2 \\ 5\overline{)1.0} \\ -10 \\ \hline 0 \end{array}$$

Divide 1 by 5 until there is a remainder of 0 or a repeating pattern.

Write each power as a product of the same factor. (Lesson 1-2)

14. 2^3 **15.** 5^5

16. 7^2 **17.** 9^4

Example 4

Write 4^3 as a product of the same factor.

$4^3 = 4 \times 4 \times 4$

Math Lab
Exploring Factors

MAIN IDEA

Discover factors of whole numbers.

NYS Core Curriculum

7.N.8 Find the common factors and greatest common factor of two or more numbers
7.PS.11 Work in collaboration with others to solve problems

The students in Mrs. Faccinto's homeroom have lockers numbered 1–30, located down a long hallway. One day the class did an experiment.

ACTIVITY

STEP 1 The first student, Student 1, walked down the hall and opened every locker.

STEP 2 Student 2 closed Locker 2 and every second locker after it.

STEP 3 Student 3 closed Locker 3 and changed the state of every third locker after it. This means that if the locker was open, Student 3 closed it; if the locker was closed, Student 3 opened it.

STEP 4 Student 4 changed the state of every fourth locker, starting with Locker 4. The student continued this pattern until all 30 students had a turn.

ANALYZE THE RESULTS

1. Which lockers were open after Student 30 took a turn? What do the numbers on the open lockers have in common?

2. Explain why the lockers you listed in Exercise 1 were open after Student 30 took a turn.

3. Suppose there were 100 lockers. Which lockers would be open after Student 100 took a turn?

4. **CHALLENGE** Which lockers were touched the greatest number of times?

5. What are the fewest lockers and students needed for 31 lockers to be open at the end of the experiment?

Prime Factorization

MAIN IDEA

Find the prime factorization of a composite number.

NYS Core Curriculum

7.N.10 Determine the prime factorization of a given number and write in exponential form 7.N.8 Find the common factors and greatest common factor of two or more numbers

New Vocabulary

prime number
composite number
prime factorization
factor tree

NY Math Online

glencoe.com

• Concepts In Motion
• Extra Examples
• Personal Tutor
• Self-Check Quiz

▷ **MINI Lab**

There is only one way that 2 can be expressed as the product of whole numbers. The geoboard shows that there is only one way that two squares can form a rectangle.

1. Using your geoboard, make as many different rectangles as possible containing 3, 4, 5, 6, 7, 8, 9, and 10 squares.

2. Which numbers of squares can be made into only one rectangle? Into more than one rectangle?

The rectangles in the Mini Lab illustrate prime and composite numbers.

A **prime number** is a whole number greater than 1 that has exactly two factors, 1 and itself.

A **composite number** is a whole number greater than 1 that has more than two factors.

Whole Numbers	Factors
2	1, 2
3	1, 3
5	1, 5
7	1, 7
4	1, 2, 4
6	1, 2, 3, 6
8	1, 2, 4, 8
9	1, 3, 9
10	1, 2, 5, 10
0	many
1	1

The numbers 0 and 1 are neither prime nor composite.

EXAMPLES **Identify Numbers as Prime or Composite**

Determine whether each number is *prime* or *composite*.

1 **17**

The number 17 has only two factors, 1 and 17, so it is prime.

2 **12**

The number 12 has six factors: 1, 2, 3, 4, 6, and 12. So, it is composite.

✓ **CHECK Your Progress**

Determine whether each number is *prime* or *composite*.

a. 11   b. 15   c. 24

Review Vocabulary

factor two or more numbers that are multiplied together to form a product; *Example*: 2 and 3 are factors of 6.
(Lesson 1-2)

Every composite number can be written as a product of prime numbers. This product is the **prime factorization** of the number. You can use a **factor tree** to find the prime factorization. The following two factor trees show the prime factorization of 60.

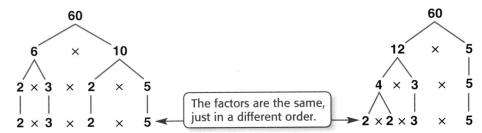

The factors are the same, just in a different order.

The prime factorization of 60 is $2 \times 2 \times 3 \times 5$, or $2^2 \times 3 \times 5$.

EXAMPLE Find the Prime Factorization

3 Find the prime factorization of 24.

METHOD 1

Use a factor tree.

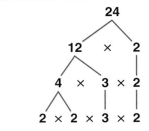

METHOD 2

Divide by prime numbers.

$2)\overline{24}$
$2)\overline{12}$
$2)\underline{\ 6}$
$\quad 3$

The divisors are 2, 2, 2, and 3.

The prime factorization of 24 is $2 \times 2 \times 3 \times 2$ or $2^3 \times 3$.

CHOOSE Your Method

Find the prime factorization.

d. 18   e. 28   f. 16

Algebraic expressions like $6ab$ can also be factored as the product of prime numbers and variables.

EXAMPLE Factor an Algebraic Expression

4 ALGEBRA Factor $6ab$.

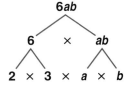

$6ab = 2 \cdot 3 \cdot a \cdot b$

CHECK Your Progress

g. **ALGEBRA** Factor $18xy$.

✓ CHECK Your Understanding

Examples 1, 2
(p. 181)

Determine whether each number is *prime* or *composite*.

1. 7
2. 50
3. 67

4. **GARDENING** Louisa has 72 flowers to plant in rows in a flower bed. How many different ways can she plant the flowers? Justify your answer.

Example 3
(p. 182)

Find the prime factorization of each number.

5. 34
6. 30
7. 12

Example 4
(p. 182)

ALGEBRA Factor each expression.

8. $10ac$
9. $16x^2$
10. $11g^3$

Practice and Problem Solving

HOMEWORK HELP	
For Exercises	See Examples
11–18	1, 2
19–28	3
29–34	4

Determine whether each number is *prime* or *composite*.

11. 22
12. 44
13. 13
14. 39
15. 81
16. 31
17. 97
18. 43

Find the prime factorization of each number.

19. 96
20. 42
21. 99
22. 64
23. 210
24. 180
25. 126
26. 375

27. **LOBSTERS** Lobsters can live up to 50 years. What is this amount expressed as a product of primes?

28. **DOGS** Greyhounds can jump a distance of 27 feet. Write this distance as a product of primes.

ALGEBRA Factor each expression.

29. $15mn$
30. $20pq$
31. $34jkm$
32. $49y^2$
33. $52gh^2$
34. $48a^2b^2$

Replace each ■ with prime factors to make a true sentence.

35. $2^3 \cdot \blacksquare \cdot 11 = 616$
36. $2 \cdot \blacksquare \cdot 5^2 = 450$
37. $3 \cdot 2^4 \cdot \blacksquare = 1{,}200$
38. $2^2 \cdot \blacksquare \cdot 3 = 1{,}500$

39. **RIVERS** The Colorado River is 1,450 miles long. Write 1,450 as a product of primes.

ALGEBRA For Exercises 40 and 41, determine whether the value of each expression is *prime* or *composite* if $a = 1$ and $b = 5$.

40. $3a + 6b$
41. $7b - 4a$

CONTESTS For Exercises 42 and 43, use the following information.

Sandcastle Day at Cannon Beach, Oregon, is one of the largest sandcastle contests on the west coast. Entrants from each age division are given lots measuring 81 square feet, 225 square feet, or 441 square feet, on which to build their castles.

42. Find the prime factorization of 81, 225, and 441.

NYSCC • NYSMT

Extra Practice, pp. 676, 707

43. Use the prime factors to determine two possible dimensions for each plot.

H.O.T. Problems

44. **RESEARCH** Use the Internet or another source to make a Sieve of Eratosthenes to determine the prime numbers up to 100.

45. **CHALLENGE** This whole number is between 30 and 40. It has only two prime factors whose sum is 5. What is the number?

46. **OPEN ENDED** Primes that differ by two are called *twin primes*. For example, 59 and 61 are twin primes. Give three examples of twin primes that are less than 50.

47. **WRITING IN MATH** Suppose n represents a whole number. Is $2n$ prime or composite? Explain.

NYSMT PRACTICE 7.N.8

48. Which number is a prime factor of both 63 and 140?

   A 2   **C** 5

   B 3   **D** 7

49. Which of the following numbers is *not* a prime number?

   F 2   **H** 16

   G 11   **J** 31

Spiral Review

50. **ALGEBRA** Graph $y = 3x$. (Lesson 3-7)

51. **MEASUREMENT** Find the perimeter and area of a rectangle with a length of 13 feet and width of 5 feet. (Lesson 3-6)

Add. (Lesson 2-4)

52. $6 + (-4)$ 53. $-13 + 9$ 54. $25 + (-26)$ 55. $-5 + 5$

▷ **GET READY** for the Next Lesson

PREREQUISITE SKILL State whether each number is divisible by **2, 3, 5, 6, 9, or 10.** (Page 668)

56. 24   57. 70   58. 120   59. 99

READING to SOLVE PROBLEMS

Everyday Meaning

The key to understanding word problems is to understand the meaning of the mathematical terms in the problem. Many words used in mathematics are also used in everyday language.

For example, you will use the terms *factor* and *multiple* in this chapter. Here are two sentences that show their everyday meanings.

- Weather was a *factor* in their decision to postpone the picnic.
- The star quarterback won *multiple* post-season awards.

The table below shows how the everyday meaning is connected to the mathematical meaning.

NYSCC ▷ 7.RP.5 Develop, verify, and explain an argument, using appropiate mathematical ideas and language

Term	Everyday Meaning	Mathematical Meaning	Connection
factor from the Latin *factor,* meaning doer	something that actively contributes to a decision or result	2 and 3 are *factors* of 6.	A *factor* helps to make a decision, and in mathematics, factors "make up" a product.
multiple from the Latin *multi-,* meaning many, and *plex,* meaning fold	consisting of more than one or shared by many	The *multiples* of 2 are 0, 2, 4, 6, ...	*Multiple* means many, and in mathematics, a number has infinitely many multiples.

PRACTICE

1. Make a list of other words that have the prefixes *fact-* or *multi-*. Determine what the words in each list have in common.

2. **WRITING IN MATH** Write your own rule for remembering the difference between *factor* and *multiple*.

RESEARCH Use a dictionary to find the everyday meanings of *least, greatest,* and *common.* Then use the definitions to determine how to find each number. Do not solve.

3. the greatest common factor of 10 and 15

4. the least common multiple of 2 and 3.

Greatest Common Factor

▷ GET READY for the Lesson

VENN DIAGRAM The Venn diagram shows the prime factors of 12 and 18.

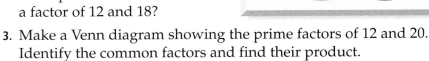

1. Which factors are in the overlapping section? What does this mean?

2. Is the product of 2 and 3 also a factor of 12 and 18?

3. Make a Venn diagram showing the prime factors of 12 and 20. Identify the common factors and find their product.

As shown above, **Venn diagrams** use overlapping circles to show how common elements among sets of numbers or objects are related. They can also show common factors. The greatest of the common factors of two or more numbers is the **greatest common factor, or GCF**.

EXAMPLE   Find the Greatest Common Factor

① **Find the GCF of 18 and 48.**

METHOD 1   List the factors of the numbers.

factors of 18: 1, 2, 3, 6, 9, 18   List the factors
factors of 48: 1, 2, 3, 4, 6, 8, 12, 16, 24, 48   of 18 and 48.

The common factors of 18 and 48 are 1, 2, 3, and 6.
So, the greatest common factor or GCF is 6.

METHOD 2   Use prime factorization.

Write the prime factorization. Circle the common prime factors.

$18 = 2 \times 3 \times 3$ Write the prime factorizations
$48 = 2 \times 2 \times 2 \times 2 \times 3$ of 18 and 48.

The greatest common factor or GCF is 2×3 or 6.

☑ **CHOOSE** Your Method

Find the GCF of each pair of numbers.

a. 8, 10   b. 6, 12   c. 10, 17

Greatest Common Factor
The greatest common factor is also called the *greatest common divisor* because it is the greatest number that divides evenly into the given numbers.

EXAMPLE Find the GCF of Three Numbers

2 Find the GCF of 12, 24, and 60.

Write the prime factorization. Circle the common prime factors.

$12 = 2 \times 2 \times 3$
$24 = 2 \times 2 \times 2 \times 3$ Write the prime factorization of 12, 24, and 60.
$60 = 2 \times 2 \times 3 \times 5$

The common prime factors are 2, 2, and 3. So, the GCF is $2 \times 2 \times 3$, or 12.

✓ **CHECK Your Progress**

Find the GCF of each set of numbers.

d. 30, 45, 75   **e.** 42, 70, 84

Real-World EXAMPLES

3 **SCHOOL SPIRIT** The cheerleaders are making spirit ribbons. Blue ribbon comes in a 24 inch spool, red ribbon comes in a 30 inch spool, and gold ribbon comes in a 36 inch spool. The cheerleaders want to cut strips of equal length and use the entire spool of each ribbon. What is the length of the longest piece of ribbon that can be cut from each spool?

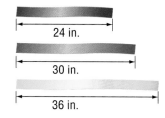

24 in.

30 in.

36 in.

The length of the longest ribbon that can be cut from each spool is the GCF of the three lengths.

$24 = 2 \times 2 \times 2 \times 3$
$30 = 2 \qquad \times 3 \times 5$ Write the prime factorization of 24, 30, and 36.
$36 = 2 \times 2 \qquad \times 3 \times 3$

The GCF of 24, 30, and 36 is 2×3 or 6. So, the ribbons should be 6 inches long.

4 How many spirit ribbons can be made if the ribbons are cut into 6-inch pieces?

There is a total of $24 + 30 + 36$, or 90 inches of ribbon. So, $90 \div 6$, or 15 spirit ribbons can be made.

Study Tip

Prime Numbers
The GCF of a group of prime numbers is 1.

✓ **CHECK Your Progress**

f. **CARPENTRY** Mr. Glover wants to make shelves for his garage using an 18-foot board and a 36-foot board. He will cut the boards to make shelves of the same length and wants to use all of both boards. Find the longest possible length of each shelf. How many shelves can he make?

Examples 1, 2
(pp. 186–187)

Find the GCF of each set of numbers.

1. 18, 30
2. 45, 60
3. 20, 50
4. 6, 8, 12
5. 8, 20, 40
6. 18, 42, 60

JOBS For Exercises 7 and 8, use the following information.
A store manager wants to display the inventory of three styles of bicycle helmets in rows with the same number of each style in each row.

Examples 3, 4
(p. 187)

7. Find the greatest number of helmets that can be placed in each row.

8. How many rows of each helmet are there?

Bike Helmets	
Style	Inventory
Sport	36
Road	72
Mountain	45

Practice and Problem Solving

Find the GCF of each set of numbers.

HOMEWORK HELP	
For Exercises	See Examples
9–16	1
17–20	2
21–24	3, 4

9. 12, 78
10. 40, 50
11. 20, 45
12. 32, 48
13. 24, 48
14. 45, 75
15. 56, 96
16. 40, 125
17. 18, 24, 30
18. 36, 60, 84
19. 35, 49, 84
20. 36, 50, 130

BANDS For Exercises 21 and 22, use the following information.
In a marching band, there are 64 woodwinds, 88 brass, and 16 percussion players. When they march in a parade, there is the same number of students in each row.

21. Find the greatest number of students in each row.

22. How many rows of each group are there?

COMMUNITY SERVICE For Exercises 23 and 24, use the following information.
You want to make care packages for a local shelter. You have 18 toothbrushes, 30 combs, and 12 bars of soap. Each package has the same number of each item.

23. What is the greatest number of care packages you can make using all the items?

24. How many of each item are in each package?

Find the GCF of each set of numbers.

25. 25¢, $1.50, 75¢, $3.00
26. 6 feet, 15 feet, 21 feet, 9 feet

ALGEBRA Find the greatest common divisor of each set of expressions.

27. $24a, 6a$
28. $30mn, 40mn$
29. $15xy, 55y$

Find two numbers whose greatest common divisor is the given number.

30. 9
31. 12
32. 15
33. 30

GEOMETRY For Exercises 34 and 35, use the following information.

Jeremy is building rectangular prisms using one-inch cubes. He is planning to build three prisms, the first with 96 blue cubes, the second with 240 red cubes, and the third with 200 yellow cubes. All of prisms must be the same height, but not necessarily the same length and width.

34. What is the maximum height of each prism Jeremy can build?

NYSCC • NYSMT
Extra Practice, pp. 676, 707

35. What are the dimensions of all three prisms?

H.O.T. Problems **CHALLENGE** Determine whether each statement is *sometimes, always,* or *never* true.

36. The GCF of two numbers is greater than both numbers.

37. If two numbers have no common prime factors, the GCF is 1.

38. The GCF of two numbers is one of the numbers.

39. **WRITING IN MATH** Using the words *factor* and *greatest common factor*, explain the relationship between the numbers 4, 12, and 24.

NYSMT PRACTICE   7.N.8

40. Student Council earned $26 selling bottled water, $32 selling oranges, and $28 selling energy bars. If all items cost the same, what is the greatest possible price per item?

   A $2   **C** $7

   B $4   **D** $8

41. Which set of numbers has the greatest GCF?

   F 4, 5, 20

   G 18, 36

   H 18, 36, 45

   J 23, 29

Spiral Review

42. What is the prime factorization of 75? (Lesson 4-1)

ALGEBRA Graph each equation. (Lesson 3-7)

43. $y = -x$ **44.** $y = x + 3$ **45.** $y = 2x - 1$

46. ALGEBRA Solve the equation $-7y + 18 = 39$. Check your solution. (Lesson 3-5)

▷ **GET READY** for the Next Lesson

47. PREREQUISITE SKILL Serena received a gift card to download music from the Internet. She downloaded 3 songs on Monday, 5 songs on Tuesday, and one half of what was left on Wednesday. She has 6 songs left. How many songs were initially on the gift card? Use the *work backward* strategy. (Lesson 3-4)

4-3 Problem-Solving Investigation

MAIN IDEA: Solve problems by making an organized list.

NYSCC **7.PS.4** Observe patterns and formulate generalizations **7.PS.8** Understand how to break a complex problem into simpler parts or use a similar problem type to solve a problem

P.S.I. TEAM +

e-Mail: MAKE AN ORGANIZED LIST

NIKKI: I am ordering a pizza. The crust choices are thin or hand-tossed. The meat choices are pepperoni or sausage. The vegetable choices are olives, mushrooms, or banana peppers.

YOUR MISSION: Nikki chooses from one crust, one meat, and one vegetable choice. Make an organized list to find how many different types of pizza Nikki can order.

Understand	You know the crust, meat, and vegetable options for the pizza. You need to find all the possible pizza combinations that can be made.
Plan	Make an organized list of all the possible combinations. Use T for thin, H for hand-tossed, P for pepperoni, S for sausage, O for olives, M for mushrooms, and B for banana peppers.
Solve	**Choosing thin crust:** T P O T P M T P B T S O T S M T S B **Choosing hand-tossed crust:** H P O H P M H P B H S O H S M H S B There are 12 different combinations of pizza that can be ordered.
Check	Draw a tree diagram to check the result.

Analyze The Strategy

1. Explain why making an organized list was a useful strategy in solving this problem.

2. **WRITING IN MATH** Write a problem that can be solved by making an organized list. Then explain how to solve the problem using this strategy.

For Exercises 3–6, solve each problem by making an organized list.

3. **SHOPPING** Charmaine went to the store and bought a yellow shirt, a blue shirt, and a red shirt. She also bought a pair of jeans and a pair of khaki dress pants. How many different outfits can be made using one shirt and one pair of pants?

4. **WORK** The following four numbers are used for employee identification numbers at a small company: 0, 1, 2, and 3. How many different employee identification numbers can be made using each digit once?

5. **PHOTOS** Joshua, Diego, and Audri stand side-by-side for a photo. How many different ways can the three friends stand next to each other?

6. **CELLPHONES** How many phone numbers are possible for one area code if the first three numbers are 268, in that order, and the last four numbers are 0, 9, 7, 1 in any order?

Use any strategy to solve Exercises 7–14. Some strategies are shown below.

PROBLEM-SOLVING STRATEGIES
· Guess and check.
· Work backward.
· Make an organized list.

7. **ALGEBRA** Consecutive odd numbers are numbers like 1, 3, 5, and 7. Find two consecutive odd numbers whose sum is 56 and whose product is 783.

8. **FOOD** The table shows the choices for ordering a deli sandwich. How many different subs can be ordered if you choose only one kind of bread and one kind of meat?

The Sandwich Shop	
Bread	White, Wheat, Whole Grain
Meat	Ham, Turkey, Roast Beef
Cheese	American, Swiss
Dressing	Italian, Ranch

9. **DVDS** Paul rented 2 times as many DVDs as Angelina last month. Angelina rented 4 fewer than Bret, but 4 more than Jill. Bret rented 9 DVDs. How many DVDs did each person rent?

10. **CLOTHES** Jeffrey owns 3 shirts, 2 pairs of pants, and 2 pairs of shoes. How many different outfits can he create?

11. **SNOWFALL** A total of 17 inches of snow fell in a 72-hour period. In the last 24 hours, 6 inches fell, and in the previous 24 hours, 4 inches fell. How many inches fell in the first 24 hours?

12. **GAS MILEAGE** Mrs. Acosta travels 44 miles in her car and uses 2 gallons of gas. If her gas mileage continues at the same rate, how many gallons of gas would she use to travel 528 miles?

13. **SCIENCE** Hydrothermal vents are similar to geysers, but are found on the ocean floor. A hydrothermal vent chimney can grow at an average rate of 9 meters in 18 months. What is the average rate of growth per month?

14. **CRAFTS** Paul makes three different sizes of birdhouses. He can paint each style in red, brown, or yellow. How many different birdhouses can Paul make?

4-4 Simplifying Fractions

MAIN IDEA

Write fractions in simplest form.

NYS Core Curriculum

Reinforcement of 6.N.21 Find multiple representations of rational numbers (fractions, decimals, and percents 0 to 100) *Also addresses 7.N.8, 7.CN.8*

New Vocabulary

equivalent fractions
simplest form

NY Math Online

glencoe.com
• Extra Examples
• Personal Tutor
• Self-Check Quiz

▷ MINI Lab

On grid paper, draw the two figures shown. Shade 4 out of the 10 squares in one figure. Shade 2 out of the 5 rectangles in the other.

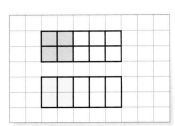

1. Write a fraction to describe each figure:

$$\frac{\text{number of shaded parts}}{\text{total number of parts}}$$

2. Based on the figures, what can you conclude about the fractions?

Equivalent fractions have the same value. A fraction is in **simplest form** when the GCF of the numerator and denominator is 1.

EXAMPLES Write a Fraction in Simplest Form

① Write $\frac{6}{24}$ in simplest form.

METHOD 1 Divide by common factors.

$$\frac{6}{24} = \frac{6 \div 2}{24 \div 2} = \frac{3}{12}$$ 2 is a common factor of 6 and 24, so divide by 2.

$$\frac{3}{12} = \frac{3 \div 3}{12 \div 3} = \frac{1}{4}$$ 3 is a common factor of 3 and 12, so divide by 3.

The fraction $\frac{1}{4}$ is in simplest form since 1 and 4 have no common factors greater than 1.

METHOD 2 Divide by the GCF.

First, find the GCF of the numerator and denominator.

factors of 6: 1, 2, 3, 6

factors of 24: 1, 2, 3, 4, 6, 8, 12, 24   The GCF of 6 and 24 is 6.

Then, divide the numerator and denominator by the GCF, 6.

$$\frac{6}{24} = \frac{6 \div 6}{24 \div 6} = \frac{1}{4}$$ Divide the numerator and denominator by the GCF, 6.

So, $\frac{6}{24}$ written in simplest form is $\frac{1}{4}$.

2 Write $\frac{36}{45}$ in simplest form.

First, find the GCF of the numerator and denominator.

factors of 36: 1, 2, 3, 4, 6, **9**, 12, 18, 36

factors of 45: 1, 3, 5, **9**, 15, 45

The GCF of 36 and 45 is 9.

Then, divide the numerator and denominator by the GCF, 9.

$$\frac{36}{45} = \frac{36 \div 9}{45 \div 9} = \frac{4}{5}$$

Divide the numerator and denominator by the GCF, 9.

So, $\frac{36}{45}$ written in simplest form is $\frac{4}{5}$.

Study Tip

More Than One Way
To write in simplest form, you can also divide by common factors.

$$\frac{36}{45} = \frac{36 \div 3}{45 \div 3} = \frac{12}{15}$$

$$\frac{12}{15} = \frac{12 \div 3}{15 \div 3} = \frac{4}{5}$$

So, $\frac{36}{45} = \frac{4}{5}$.

✓ **CHOOSE Your Method**

Write each fraction in simplest form.

a. $\frac{7}{28}$
b. $\frac{8}{20}$
c. $\frac{27}{36}$

Real-World EXAMPLE

3 **SCIENCE** Lauren measured a grasshopper for her science project. Each line on her ruler represents $\frac{1}{16}$ inch. Find the length of the grasshopper in simplest form.

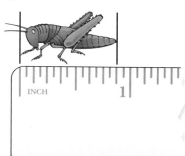

$$\frac{14}{16} = \frac{\overset{1}{\cancel{2}} \cdot 7}{\underset{}{\cancel{2}} \cdot 2 \cdot 2 \cdot 2}$$

$$= \frac{7}{8}$$

The slashes mean that part of the numerator and part of the denominator are both divided by the same number. For example, $2 \div 2 = 1$.

So, $\frac{14}{16}$ written in simplest form is $\frac{7}{8}$.

✓ **CHECK Your Progress**

d. **CARPENTRY** A carpenter measured a shelf and found it to be $\frac{10}{16}$ inch thick. Find the simplified fraction.

e. **CARPENTRY** A carpenter measured a board to be $\frac{8}{16}$ inch thick. Find the simplified fraction.

Real-World Career· · · ·
How Does a Carpenter Use Math?
Carpenters use fractions when they measure and cut boards.

NY Math Online

For more information visit: glencoe.com.

Examples 1, 2
(pp. 192–193)

Write each fraction in simplest form.

1. $\frac{3}{9}$　　　2. $\frac{4}{18}$　　　3. $\frac{10}{25}$　　　4. $\frac{36}{40}$

Example 3
(p. 193)

5. **ALLOWANCE** Mary received $15 for her weekly allowance. She spent $10 at the movie theater with her friends. What fraction of the money, in simplest form, was spent at the theater?

Practice and Problem Solving

HOMEWORK HELP	
For Exercises	See Examples
6–17	1, 2
18–19	3

Write each fraction in simplest form.

6. $\frac{9}{12}$　　　7. $\frac{25}{35}$　　　8. $\frac{16}{32}$　　　9. $\frac{14}{20}$

10. $\frac{10}{20}$　　11. $\frac{12}{21}$　　12. $\frac{15}{25}$　　13. $\frac{24}{28}$

14. $\frac{48}{64}$　　15. $\frac{32}{32}$　　16. $\frac{20}{80}$　　17. $\frac{45}{54}$

18. **PRESIDENTS** Of the 43 U.S. presidents, 15 were elected to serve two terms. What fraction of the U.S. presidents, in simplest form, was elected to serve two terms?

19. **TV SHOWS** A television station has 28 new TV shows scheduled to air this week. What fraction of the television shows, in simplest form, are 30-minute programs?

WYTB Programming	
30–minute	60–minute
20	8

Write each fraction in simplest form.

20. $\frac{45}{100}$　　21. $\frac{60}{150}$　　22. $\frac{16}{120}$　　23. $\frac{35}{175}$

24. **TIME** Fifteen minutes is what part of one hour?

25. **MEASUREMENT** Nine inches is what part of one foot?

26. **CALENDAR** Four days is what part of the month of April?

27. **SLEEP** Marcel spends 8 hours each day sleeping. What fraction of a week, written in simplest form, does Marcel spend sleeping?

28. **MONEY** Each week, Lorenzo receives a $10 allowance. What fraction of his yearly allowance, in simplest form, does he receive each week?

29. **FIND THE DATA** Refer to the Data File on pages 16–19 of your book. Choose some data and write a real-world problem in which you would simplify fractions.

NYSCC • NYSMT
Extra Practice, pp. 677, 707

30. **OPEN ENDED** Select a fraction in simplest form. Then, write two fractions that are equivalent to it.

31. CHALLENGE Both the numerator and denominator of a fraction are even. Is the fraction in simplest form? Explain your reasoning.

32. FIND THE ERROR Nhu and Booker both wrote $\frac{16}{36}$ in simplest form. Who is correct? Explain.

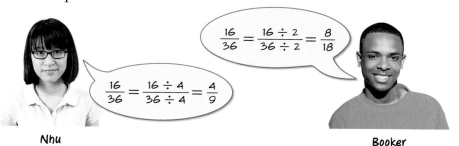

$$\frac{16}{36} = \frac{16 \div 2}{36 \div 2} = \frac{8}{18}$$

$$\frac{16}{36} = \frac{16 \div 4}{36 \div 4} = \frac{4}{9}$$

Nhu

Booker

33. WRITING IN MATH Explain how to determine whether a fraction is in simplest form.

NYSMT PRACTICE 6.N.21

34. It takes Benito 12 minutes to walk to school. What fraction represents the part of an hour it takes Benito to walk to school?

A $\frac{12}{1}$ C $\frac{5}{30}$

B $\frac{4}{15}$ D $\frac{1}{5}$

35. What fraction of a foot is 2 inches?

F $\frac{1}{6}$ H $\frac{1}{3}$

G $\frac{1}{4}$ J $\frac{1}{2}$

Spiral Review

36. SANDWICHES A deli offers sandwiches with ham, turkey, or roast beef with American, provolone, Swiss, or mozzarella cheese. How many different types of sandwiches can be made if you choose one meat and one cheese? Use the *make an organized list* strategy. (Lesson 4-3)

Find the GCF of each set of numbers. (Lesson 4-2)

37. 27, 36   **38.** 16, 28   **39.** 20, 50, 65

40. ANALYZE GRAPHS Refer to the graph. At these rates, about how much longer would it take a blue shark to swim 280 miles than it would a sailfish? Use the formula $d = rt$. Justify your answer. (Lesson 3-3)

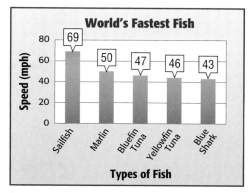

World's Fastest Fish

Source: *Top Ten of Everything*

▷ **GET READY** for the Next Lesson

PREREQUISITE SKILL Divide. (Page 676)

41. $2\overline{)1.0}$ **42.** $4\overline{)1.00}$

43. $10\overline{)7.0}$ **44.** $8\overline{)3.000}$

Fractions and Decimals

▶ **GET READY** for the Lesson

NASCAR The table shows the winning speeds for a 10-year-period at the Daytona 500.

1. What fraction of the speeds are between 130 and 145 miles per hour?

2. Express this fraction using words and then as a decimal.

3. What fraction of the speeds are between 145 and 165 miles per hour? Express this fraction using words and then as a decimal.

Daytona 500		
Year	Winner	Speed (mph)
1998	D. Earnhardt	172.712
1999	J. Gordon	148.295
2000	D. Jarrett	155.669
2001	M. Waltrip	161.783
2002	W. Burton	142.971
2003	M. Waltrip	133.870
2004	D. Earnhardt Jr.	156.345
2005	J. Gordon	135.173
2006	J. Johnson	142.667
2007	K. Harvick	149.335

Source: ESPN Sports Almanac

Our decimal system is based on powers of 10. So, if the denominator of a fraction is a power of 10, you can use place value to write the fraction as a decimal. For example, to write $\frac{7}{10}$ as a decimal, place a 7 in the tenths place.

Words	**Fraction**	**Decimal**
seven tenths	$\frac{7}{10}$	0.7

If the denominator of a fraction is a *factor* of 10, 100, 1,000, or any higher power of ten, you can use mental math and place value.

EXAMPLES Use Mental Math

Write each fraction or mixed number as a decimal.

1 $\frac{7}{20}$

THINK  $\frac{7}{20} = \frac{35}{100}$

So, $\frac{7}{20} = 0.35$.

2 $5\frac{3}{4}$

$5\frac{3}{4} = 5 + \frac{3}{4}$ Think of it as a sum.

$= 5 + 0.75$ You know that $\frac{3}{4} = 0.75$.

$= 5.75$ Add mentally.

So, $5\frac{3}{4} = 5.75$.

✓ **CHECK Your Progress**

a. $\frac{3}{10}$ b. $\frac{3}{25}$ c. $6\frac{1}{2}$

Any fraction can be written as a decimal by dividing its numerator by its denominator. Division ends when the remainder is zero.

EXAMPLES Use Division

3 Write $\frac{3}{8}$ as a decimal.

$$
\begin{array}{r}
0.375 \\
8\overline{)3.000} \\
-24 \\
\hline
60 \\
-56 \\
\hline
40 \\
-40 \\
\hline
0
\end{array}
$$

Divide 3 by 8.

Division ends when the remainder is 0.

So, $\frac{3}{8} = 0.375$.

4 Write $\frac{1}{40}$ as a decimal.

$$
\begin{array}{r}
0.025 \\
40\overline{)1.000} \\
-80 \\
\hline
200 \\
-200 \\
\hline
0
\end{array}
$$

Divide 1 by 40.

So, $\frac{1}{40} = 0.025$.

CHECK Your Progress

Write each fraction or mixed number as a decimal.

d. $\frac{7}{8}$ e. $2\frac{1}{8}$ f. $7\frac{9}{20}$

Vocabulary Link . . . . .
Terminate

Everyday Use coming to an end, as in terminate a game

Math Use a decimal whose digits end

In Examples 1–4, the decimals 0.35, 5.75, 0.375, and 0.025 are called terminating decimals. A **terminating decimal** is a decimal whose digits end.

Repeating decimals have a pattern in their digits that repeats forever. Consider $\frac{1}{3}$.

$$
\begin{array}{r}
0.333\ldots \\
3\overline{)1.000} \\
-9 \\
\hline
10 \\
-9 \\
\hline
10 \\
-9 \\
\hline
1
\end{array}
$$

The number 3 repeats. The repetition of 3 is represented by three dots.

You can use **bar notation** to indicate that a number pattern repeats indefinitely. A bar is written only over the digits that repeat.

$0.33333\ldots = 0.\overline{3}$ $0.121212\ldots = 0.\overline{12}$ $11.3858585\ldots = 11.3\overline{85}$

Write Fractions as Repeating Decimals

⑤ Write $\frac{7}{9}$ as a decimal.

$$
\begin{array}{r}
0.777\ldots \\
9\overline{)7.000} \\
-63 \\
\hline
70 \\
-63 \\
\hline
70 \\
-63 \\
\hline
7
\end{array}
$$

Divide 7 by 9.

> Notice that the remainder will never be zero. That is, the division never ends.

So, $\frac{7}{9} = 0.777\ldots$ or $0.\overline{7}$.

✓ **CHECK Your Progress**

Write each fraction or mixed number as a decimal. Use bar notation if the decimal is a repeating decimal.

g. $\frac{2}{3}$ h. $\frac{3}{11}$ i. $8\frac{1}{3}$

Every terminating decimal can be written as a fraction with a denominator of 10, 100, 1,000, or a higher power of ten. Place the digits that come after the decimal point in the numerator. Use the place value of the final digit as the denominator.

numerator

$$0.25 = \frac{25}{100}$$

hundredths place

Real-World Link . . . . :
The recommended water temperature for goldfish is 65–72°F.
Source: Animal-World

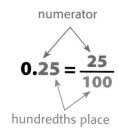 **Real-World EXAMPLE** **Use a Power of 10**

⑥ **FISH** Use the table to find what fraction of the fish in an aquarium are goldfish. Write in simplest form.

$0.15 = \frac{15}{100}$ The final digit, 5, is in the hundredths place.

$ = \frac{3}{20}$ Simplify.

Fish	Amount
Guppy	0.25
Angel Fish	0.4
Goldfish	0.15
Molly	0.2

✓ **CHECK Your Progress**

Determine the fraction of the aquarium made up by each fish. Write the answer in simplest form.

j. molly   k. guppy   l. angel fish

Examples 1–5
(pp. 196–198)

Write each fraction or mixed number as a decimal. Use bar notation if the decimal is a repeating decimal.

1. $\frac{2}{5}$

2. $\frac{9}{10}$

3. $7\frac{1}{2}$

4. $4\frac{3}{20}$

5. $\frac{1}{8}$

6. $3\frac{5}{8}$

7. $\frac{5}{9}$

8. $1\frac{5}{6}$

Example 6
(p. 198)

Write each decimal as a fraction or mixed number in simplest form.

9. 0.22

10. 0.1

11. 4.6

12. **HOCKEY** During a hockey game, an ice resurfacer travels 0.75 mile during each ice resurfacing. What fraction represents this distance?

Practice and Problem Solving

HOMEWORK HELP	
For Exercises	See Examples
13–16	1, 2
17–22	3, 4
23–28	5
29–36	6

Write each fraction or mixed number as a decimal. Use bar notation if the decimal is a repeating decimal.

13. $\frac{4}{5}$

14. $\frac{1}{2}$

15. $4\frac{4}{25}$

16. $7\frac{1}{20}$

17. $\frac{5}{16}$

18. $\frac{3}{16}$

19. $\frac{33}{50}$

20. $\frac{17}{40}$

21. $5\frac{7}{8}$

22. $9\frac{3}{8}$

23. $\frac{4}{9}$

24. $\frac{8}{9}$

25. $\frac{1}{6}$

26. $\frac{8}{11}$

27. $5\frac{1}{3}$

28. $2\frac{6}{11}$

Write each decimal as a fraction or mixed number in simplest form.

29. 0.2

30. 0.9

31. 0.55

32. 0.34

33. 5.96

34. 2.66

35. **INSECTS** The maximum length of a praying mantis is 30.5 centimeters. What mixed number represents this length?

36. **GROCERIES** Suppose you buy a 1.25 pound package of ham for $4.99. What fraction of a pound did you buy?

37. **FIND THE DATA** Refer to the Data File on page 16–19 of your book. Choose some data and write a real-world problem in which you would write a percent as a decimal.

Write each of the following as an integer over a whole number.

38. −13

39. $7\frac{1}{3}$

40. −0.028

41. −3.2

42. **MUSIC** Nicolás practiced playing the cello for 2 hours and 18 minutes. Write the time Nicolás spent practicing as a decimal.

43. SOFTBALL The batting average of a softball player is the number of hits divided by the number of at-bats. If Felisa had 50 hits in 175 at-bats and Harmony had 42 hits in 160 at-bats, who had the better batting average? Justify your answer.

H.O.T. Problems

44. OPEN ENDED Write a fraction that is equivalent to a terminating decimal between 0.5 and 0.75.

45. CHALLENGE The value of pi (π) is 3.1415926… . The mathematician Archimedes believed that π was between $3\frac{1}{7}$ and $3\frac{10}{71}$. Was Archimedes correct? Explain your reasoning.

46. WRITING IN MATH Fractions with denominators of 2, 4, 8, 16, and 32 produce terminating decimals. Fractions with denominators of 6, 12, 18, and 24 produce repeating decimals. What causes the difference? Explain.

NYSMT PRACTICE 7.N.17

47. Which decimal represents the shaded region of the model?

A 0.666   **C** 0.667

B $0.\overline{6}$ **D** $0.66\overline{7}$

48. Based on the information given in the table, what fraction represents $0.\overline{8}$?

F $\frac{4}{5}$

G $\frac{80}{99}$

H $\frac{5}{6}$

J $\frac{8}{9}$

Decimal	Fraction
$0.\overline{3}$	$\frac{3}{9}$
$0.\overline{4}$	$\frac{4}{9}$
$0.\overline{5}$	$\frac{5}{9}$
$0.\overline{6}$	$\frac{6}{9}$

Spiral Review

Write each fraction in simplest form. (Lesson 4-4)

49. $\frac{10}{24}$ **50.** $\frac{39}{81}$ **51.** $\frac{28}{98}$ **52.** $\frac{51}{68}$

53. PIZZA How many different pizzas can Alfonso order if he can choose thick, thin, or deep dish crust and one topping from either pepperoni, sausage, or mushrooms? Use the *make an organized list* strategy. (Lesson 4-3)

GET READY for the Next Lesson

PREREQUISITE SKILL Write a fraction for the number of shaded squares to the total number of squares.

54.   **55.**   **56.** 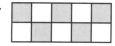   **57.**

Determine whether each number is *prime* or *composite*. (Lesson 4-1)

1. 24   **2.** 61   **3.** 2

4. AGE Kevin just turned 13 years old. How old will he be the next time his age is a prime number? (Lesson 4-1)

Find the prime factorization of each number. (Lesson 4-1)

5. 30   **6.** 120

Factor each expression. (Lesson 4-1)

7. $14x^2y$ **8.** $50mn$

Find the GCF of each set of numbers. (Lesson 4-2)

9. 16, 40   **10.** 65, 100

11. MULTIPLE CHOICE Lakeesha wants to cut a 14-inch by 21-inch poster board into equal-size squares for an art project. She does not want to waste any of the poster board, and she wants the largest squares possible. What is the length of the side of the largest squares she can cut? (Lesson 4-2)

  A 14   **C** 6

  B 7   **D** 2

12. MULTIPLE CHOICE Lynne cannot remember her password to check the messages on her cell phone. She knows that it is a three-digit number consisting of the numbers 1, 4, and 7, but she cannot remember the order. Which list shows all the different possibilities for her password? (Lesson 4-3)

  F 147, 174, 417, 714, 741

  G 147, 174, 417, 471, 714, 741

  H 417, 471, 714, 741

  J 147, 174, 74, 417, 17, 471, 714, 741

13. PIANO Evelina spends 40 minutes practicing the flute each afternoon after school. What part of one hour does she spend practicing? Write as a fraction in simplest form. (Lesson 4-4)

Write each fraction in simplest form. (Lesson 4-4)

14. $\dfrac{20}{36}$

15. $\dfrac{45}{60}$

16. $\dfrac{63}{108}$

17. $\dfrac{60}{72}$

18. VOTING In the 2004 presidential election, Kentucky had 8 out of 538 total electoral votes. Write Kentucky's portion of the electoral votes as a fraction in simplest form. (Lesson 4-4)

Write each fraction or mixed number as a decimal. Use bar notation if the decimal is a repeating decimal. (Lesson 4-5)

19. $\dfrac{7}{8}$ **20.** $\dfrac{2}{9}$ **21.** $3\dfrac{13}{20}$

Write each decimal as a fraction in simplest form. (Lesson 4-5)

22. 0.6   **23.** 0.48   **24.** 7.02

25. ANIMALS The maximum height of an Asian elephant is 9.8 feet. What mixed number represents this height? (Lesson 4-5)

▶ **GET READY for the Lesson**

In a recent survey, students were asked to choose their favorite sport to play. The results are shown in the table.

Sport	Number of Students
Basketball	3 out of 20
Football	3 out of 25
Gymnastics	1 out of 20
Swimming	9 out of 100

1. For each sport, shade a 10 × 10 grid that represents the number of students that chose the sport.

2. What fraction of the students chose swimming?

A **ratio** is a comparison of two quantities by division. Ratios like 9 out of 100 can also be written as 9:100 or $\frac{9}{100}$. When a ratio compares a number to 100, it can be written as a **percent.**

Percent   Key Concept

Words   A percent is a part to whole ratio that compares a number to 100.

Example

Symbols $\frac{n}{100} = n\%$

9 out of 100 = 9%

EXAMPLES   Write Ratios as Percents

Write each ratio as a percent.

 Annie answered 90 out of 100 questions correctly.

$\frac{90}{100} = 90\%$

Annie answered 90% of the questions correctly.

90%

2. On average, 50.5 out of 100 students own a pet.

$\frac{50.5}{100} = 50.5\%$

On average, 50.5% of the students own a pet.

Reading Math

Percent *Percent* means per *hundred* or *hundredths.* The symbol % means percent.

✓ **CHECK Your Progress**

a. 45 out of 100 cars sold   b. $3.30:$100 spent on soft drinks

Fractions and percents are ratios that can represent the same number. You can write a fraction as a percent by finding an equivalent fraction with a denominator of 100.

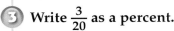 **EXAMPLE** **Write a Fraction as a Percent**

3 Write $\frac{3}{20}$ as a percent.

First, find an equivalent fraction with a denominator of 100.
Then write the fraction as a percent.

$$\frac{3}{20} = \frac{3 \times 5}{20 \times 5} = \frac{15}{100} \text{ or } 15\%$$

So, $\frac{3}{20} = 15\%$.

$\frac{3}{20}$ = $\frac{15}{100}$

✓ **CHECK Your Progress**

Write each fraction as a percent.

c. $\frac{17}{20}$ d. $\frac{3}{5}$ e. $\frac{2}{25}$

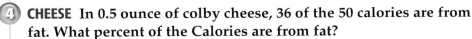 **Real-World EXAMPLE**

4 **CHEESE** In 0.5 ounce of colby cheese, 36 of the 50 calories are from fat. What percent of the Calories are from fat?

$\frac{36}{50} = \frac{72}{100}$ Write an equivalent fraction with a denominator of 100.

$= 72\%$ $\frac{72}{100} = 72\%$

About 72% of the Calories are from fat.

✓ **CHECK Your Progress**

f. **CHEESE** In a small piece of mozzarella cheese, 17 of the 25 Calories come from fat. What percent of the Calories come from fat?

 **EXAMPLE** **Write a Percent as a Fraction**

5 Write 48% as a fraction in simplest form.

$48\% = \frac{48}{100}$ Definition of percent

$= \frac{12}{25}$ Simplify.

✓ **CHECK Your Progress**

Write each percent as a fraction in simplest form.

g. 40%   h. 6%   i. 24%

Examples 1, 2
(p. 202)

Write each ratio as a percent.

1. 57:100 insects are spiders

2. $29.20 per $100

Example 3
(p. 203)

Write each fraction as a percent.

3. $\frac{1}{4}$

4. $\frac{6}{10}$

5. $\frac{17}{20}$

Example 4
(p. 203)

6. **TECHNOLOGY** Tansy used $\frac{2}{5}$ of the memory available on her flash drive. What percent of the memory did she use?

Example 5
(p. 203)

Write each percent as a fraction in simplest form.

7. 90%

8. 75%

9. 22%

Practice and Problem Solving

HOMEWORK HELP	
For Exercises	**See Examples**
10–15	1, 2
16–23	3
24–25	4
26–33	5

Write each ratio as a percent.

10. 87 out of 100 books read

11. 42 per 100 teenagers

12. 12.2 out of 100 points earned

13. 99.9:100 miles driven

14. $11\frac{3}{4}$ out of 100 feet

15. $66\frac{2}{3}$:100 yards run

Write each fraction as a percent.

16. $\frac{7}{10}$

17. $\frac{16}{20}$

18. $\frac{15}{25}$

19. $\frac{13}{50}$

20. $\frac{1}{5}$

21. $\frac{3}{5}$

22. $\frac{19}{20}$

23. $\frac{10}{10}$

24. **PETS** Twenty out of every 25 households own one dog. What percent of households own one dog?

25. **SPORTS** If 15 out of every 50 teens like to ski, what percent of teens like to ski?

Write each percent as a fraction in simplest form.

26. 45%

27. 30%

28. 62%

29. 88%

30. 68%

31. 13%

32. 2%

33. 300%

Replace each ● with >, <, or =.

34. $\frac{1}{4}$ ● 25%

35. $\frac{9}{20}$ ● 55%

36. 78% ● $\frac{3}{5}$

37. 38% ● $\frac{19}{50}$

38. 12% ● $1\frac{1}{5}$

39. $2\frac{2}{5}$ ● 24%

NYSCC • NYSMT

Extra Practice, pp. 678, 708

40. **VOLUNTEERING** A Girl Scouts event is expecting 40 out of 50 people to volunteer at a charity auction. So far 30 volunteers have arrived at the auction. What percent of the volunteers have not yet arrived?

41. GEOMETRY What percent of the larger rectangle shown is *not* shaded?

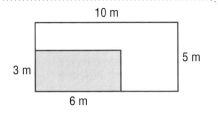

10 m

5 m

3 m

6 m

42. CHALLENGE Apply what you know about percents, fractions, and decimals to write $12\frac{1}{2}\%$ as a fraction. Justify your answer.

43. Which One Doesn't Belong? Identify the ratio that does not have the same value as the other three. Explain your reasoning.

$\frac{5}{20}$	5%	$\frac{1}{4}$	50 out of 200

44. **WRITING IN** **MATH** Explain how you know which model represents 36%. Then explain why that model also represents $\frac{9}{25}$.

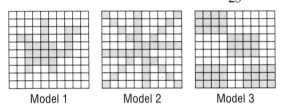

Model 1   Model 2   Model 3

NYSMT PRACTICE 6.N.11

45. What percent of the model is shaded?

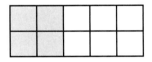

A 25%   **C** 40%

B 30%   **D** 60%

46. Cats spend about $\frac{3}{10}$ of their time awake grooming themselves. Which number is equivalent to $\frac{3}{10}$?

F 3%   **H** 33.3%

G 30%   **J** 300%

Spiral Review

Write each decimal as a fraction or mixed number in simplest form. (Lesson 4-5)

47. 0.6   **48.** 0.15   **49.** 2.8

50. TIME The drive to a football game took 56 minutes. Twenty-four minutes of the time was spent stopped in traffic. What fraction of the drive, in simplest form, was spent in stopped traffic? (Lesson 4-4)

Solve each equation. (Lesson 3-2)

51. $x + 7 = 10$ **52.** $m - 2 = 8$ **53.** $12 + a = 16$

▷ **GET READY** for the Next Lesson

PREREQUISITE SKILL **Multiply or divide.** (Pages 675, 676)

54. 16.2×10 **55.** 0.71×100 **56.** $14.4 \div 100$ **57.** $791 \div 1,000$

Percents and Decimals

▷ **GET READY** for the Lesson

SCHOOL UNIFORMS The graph shows students' favorite school uniform colors.

1. Write the percent of students who chose black as a fraction.

2. Write the fraction as a decimal.

3. What do you notice about the percent and decimal form for students who chose black?

Favorite School Uniform Colors

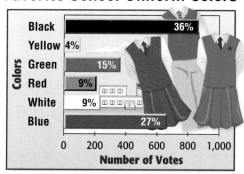

Source: *Kids USA Survey*

Any percent can be written as a fraction with the percent as the numerator and 100 as the denominator. Use this fact to help write percents as decimals.

EXAMPLE Write Percents as Decimals

1 Write 7% as a decimal.

METHOD 1 Write the percent as a fraction.

$7\% = \dfrac{7}{100}$ Write the percent as a fraction.

$ = 0.07$ Write the fraction as a decimal.

METHOD 2 Divide mentally.

The % symbol means to divide by 100. So, remove the % symbol and divide by 100.

$7\% = .07$ Remove the % symbol and divide by 100. Add placeholder zero.

$ = 0.07$ Add leading zero.

So, 7% = 0.07.

Study Tip

Division When dividing by 100, the decimal point moves two places left.

✓ **CHOOSE** Your Method

Write each percent as a decimal.

a. 8%   b. 54%   c. 85.2%

2 **GEOGRAPHY** Alaska is the largest state, making up about $16\frac{1}{10}\%$ of the land area of the United States. Write this amount as a decimal.

$16\frac{1}{10}\% = 16.1\%$ Write $\frac{1}{10}$ as 0.1.

$= 16.1$ Remove the % symbol and divide by 100.

$= 0.161$ Add leading zero.

So, $16\frac{1}{10}\% = 0.161$.

✓ CHECK Your Progress

d. **GEOGRAPHY** About $6\frac{4}{5}\%$ of the total area of the United States is water. What decimal represents this amount?

e. **GEOGRAPHY** The total area of Australia is about $38\frac{11}{20}\%$ of the total area of North America. Write the amount as a decimal.

EXAMPLE **Write Decimals as Percents**

3 Write 0.4 as a percent.

METHOD 1 **Write the decimal as fraction.**

$0.4 = \frac{4}{10}$ or $\frac{40}{100}$ Write the decimal as a fraction.

$= 40\%$ Write the fraction as a percent.

METHOD 2 **Multiply mentally.**

Multiply by 100 and add the % symbol.

$0.4 = 0.40$ Multiply by 100. Add a placeholder zero.

$= 40\%$ Add the % symbol.

So, $0.4 = 40\%$.

Study Tip

Multiplication When multiplying by 100, the decimal point moves two places right.

✓ CHOOSE Your Method

Write each decimal as a percent.

f. 0.5   g. 0.34   h. 0.98

④ **SCIENCE** About 0.875 of an iceberg's mass is underwater. What percent of an iceberg's mass is underwater?

$$0.875 = 0.\underset{\smile}{875} \qquad \text{Multiply by 100.}$$
$$= 87.5\% \qquad \text{Add the \% symbol.}$$

So, 87.5% of an iceberg's mass is underwater.

Real-World Link . . . . .
Of the world's icebergs, 93% are found surrounding the Antarctic.
Source: JPL Polar Oceanography Group

✓ **CHECK** Your Progress

i. **EXERCISE** Each day, Rico and his dog walk 0.625 mile. What percent of a mile do they walk?

Percent as a Decimal   Key Concept

To write a percent as a decimal, divide the percent by 100 and remove the percent symbol.

$$25\% = .\underset{\smile}{25} \quad \text{or} \quad 0.25$$

Decimal as a Percent   Key Concept

To write a decimal as a percent, multiply the percent by 100 and add the percent symbol.

$$0.58 = 0.\underset{\smile}{58} = 58\%$$

✓ CHECK Your Understanding

Example 1
(p. 206)

Write each percent as a decimal.

1. 68%   **2.** 5%   **3.** 27.6%

Example 2
(p. 207)

4. SPENDING A family spends about $33\frac{2}{5}\%$ of their annual income on housing. What decimal represents the amount spent on housing?

Example 3
(p. 207)

Write each decimal as a percent.

5. 0.09   **6.** 0.3   **7.** 0.73

Example 4
(p. 208)

8. BASKETBALL The table shows the top five WNBA players with the highest free throw averages. What percent of the time does Sue Bird make a free throw?

Player	Average
Eva Nemcova	0.897
Seimone Augustus	0.897
Elena Tornikidou	0.882
Sue Bird	0.878
Cynthia Cooper	0.871

Source: Women's National Basketball Association

HOMEWORK HELP	
For Exercises	**See Examples**
9–20	1
21–22	2
23–34	3
35–37	4

Write each percent as a decimal.

9. 27%   **10.** 70%   **11.** 6%   **12.** 4%

13. 18.5%   **14.** 56.4%   **15.** 2.2%   **16.** 3.8%

17. $27\frac{7}{10}\%$ **18.** $15\frac{1}{2}\%$ **19.** $30\frac{1}{4}\%$ **20.** $46\frac{2}{5}\%$

21. BONES An adult human body has $68\frac{3}{5}\%$ of the number of bones it had at birth. What decimal represents this amount?

22. VIDEO GAMES Brian reaches the sixth level of a video game $92\frac{3}{4}\%$ of the time he plays. What decimal represents this percent?

Write each decimal as a percent.

23. 0.7   **24.** 0.6   **25.** 5.8   **26.** 8.2

27. 0.95   **28.** 0.08   **29.** 0.17   **30.** 0.78

31. 0.675   **32.** 0.145   **33.** 0.012   **34.** 0.7025

BASEBALL The table shows the top five Major League Baseball players with the highest batting averages in a recent year. Express each player's batting average as a percent.

35. Joe Mauer

36. Derek Jeter

37. Miguel Cabrera

Player	Average
Joe Mauer	0.347
Freddy Sanchez	0.344
Derek Jeter	0.344
Robinson Cano	0.342
Miguel Cabrera	0.339

Source: Major League Baseball

Replace each ● with >, <, or = to make a true sentence.

38. 0.25% ● 0.125   **39.** 0.76 ● 76.5%   **40.** 500% ● 50

41. 99% ● 0.985   **42.** 0.325 ● 30%   **43.** 56% ● 0.5625

44. SPORTS A tennis player won 0.805 of the matches she played. What percent of the matches did she lose?

ANALYZE TABLES For Exercises 45–48, use the table and the information given.

Model airplane measurements are based on a scale of the life-size original. For example, a scale of $\frac{1}{72}$ is about 1% of the size of the original. Percents are rounded to the nearest thousandth.

45. Find the decimal equivalents for each scale.

46. Which scale is the smallest?

47. About how long is a model of an actual 88.5-foot C-119 boxcar plane using a $\frac{1}{72}$ scale?

48. Which scale is used if a model of an 88.5-foot C-119 boxcar plane is about 33.1875 inches long?

Scale	Percent Equivalent
$\frac{1}{72}$	1.389%
$\frac{1}{24}$	4.167%
$\frac{1}{48}$	2.083%
$\frac{1}{20}$	5%
$\frac{1}{32}$	3.125%

NYSCC • NYSMT

Extra Practice, pp. 678, 708

49. OPEN ENDED Write any decimal between 0 and 1. Then write it as a fraction in simplest form and as a percent.

50. FIND THE ERROR Emilio and Janelle both wrote 0.992 as a percent. Who is correct? Explain.

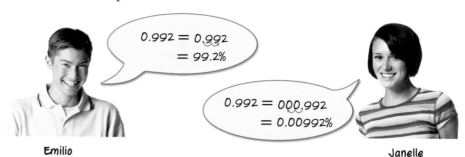

$0.992 = 0.992$
$= 99.2\%$

$0.992 = 000.992$
$= 0.00992\%$

Emilio

Janelle

CHALLENGE Write each fraction as a percent.

51. $\dfrac{3}{8}$

52. $\dfrac{1}{40}$

53. $\dfrac{1}{32}$

54. WRITING IN MATH Write a word problem about a real-world situation in which you would change a decimal to a percent.

NYSMT PRACTICE 6.N.11

55. It is estimated that 13.9% of the population of Texas was born outside the United States. Which number is *not* equivalent to 13.9%?

A $\dfrac{139}{1,000}$ **C** 0.139

B $\dfrac{13.9}{100}$ **D** 1.39

56. Which of the following is ordered from least to greatest?

F 0.42, $\dfrac{2}{5}$, 50%, $\dfrac{3}{4}$

G $\dfrac{2}{5}$, 0.42, 50%, $\dfrac{3}{4}$

H $\dfrac{3}{4}$, $\dfrac{2}{5}$, 0.42%, 50%

J $\dfrac{3}{4}$, 0.42, $\dfrac{2}{5}$, 50%

Spiral Review

Write each ratio as a percent. (Lesson 4-6)

57. 72 out of 100 animals   **58.** $9.90:$100   **59.** 3.1 out of 100 households

60. Write $9\dfrac{3}{8}$ as a decimal. (Lesson 4-5)

61. AIRPLANES Write an integer that represents an airplane descending 125 feet. (Lesson 2-1)

62. MONEY Marina earned $187.50 by working 30 hours. If she works 35 hours at this rate, how much will she earn? (Lesson 1-1)

▷ **GET READY** for the Next Lesson

PREREQUISITE SKILL Write the prime factorization of each number. (Lesson 4-1)

63. 50   **64.** 32   **65.** 76   **66.** 105

Least Common Multiple

▷ **MINI Lab**

Use cubes to build the first row of each prism as shown.

1. Add a second row to each prism. Record the total number of cubes used in a table like the one shown below.

Prism A

Number of Rows	1	2	3	4
Cubes in Prism A	4	■	■	■
Cubes in Prism B	6	■	■	■

Prism B

2. Add rows until each prism has four rows.

3. Describe two prisms that have the same number of cubes.

4. If you keep adding rows, will the two prisms have the same number of cubes again?

A **multiple** is the product of a number and any whole number. The **least common multiple**, or **LCM**, of two or more numbers is the least of their common multiples, excluding zero.

EXAMPLES Find the LCM

① **Find the LCM of 6 and 10.**

METHOD 1 List the nonzero multiples.

List the multiples of 6 until you come to a number that is also a multiple of 10.

multiples of 6: 6, 12, 18, 24, 30, …

multiples of 10: 10, 20, 30, …

Notice that 30 is also a multiple of 10. The LCM of 6 and 10 is 30.

METHOD 2 Use prime factorization.

$6 = 2 \cdot 3$
$10 = 2 \cdot 5$

The prime factors of 6 and 10 are 2, 3, and 5.

The LCM is the least product that contains the prime factors of each number. So, the LCM of 6 and 10 is $2 \cdot 3 \cdot 5$ or 30.

Vocabulary Link · · · · ·

Multiply
Everyday Use to find the product

Multiple
Math Use the product of a number and any whole number

2 **Find the LCM of 45 and 75.**

Use Method 2. Find the prime factorization of each number.

$45 = 3 \cdot 3 \cdot 5$ or $3^2 \cdot 5$

$75 = 3 \cdot 5 \cdot 5$ or $3 \cdot 5^2$

> The prime factors of 45 and 75 are 3 and 5. Write the prime factorization using exponents.

The LCM is the product of the prime factors 3 and 5, with each one raised to the *highest* power it occurs in *either* prime factorization. The LCM of 45 and 75 is $3^2 \cdot 5^2$, which is 225.

✓ **CHOOSE Your Method**

Find the LCM of each set of numbers.

a. 3, 12   b. 10, 12   c. 25, 30

Real-World Link . . . .
Each day, about 700,000 people in the U.S. celebrate their birthday.

Real-World EXAMPLE

3 **PARTY** Ling needs to buy paper plates, napkins, and cups for a party. Plates come in packages of 12, napkins come in packages of 16, and cups come in packages of 8. What is the least number of packages she will have to buy if she wants to have the same number of plates, napkins, and cups?

First find the LCM of 8, 12, and 16.

$8 = 2 \cdot 2 \cdot 2$ or 2^3

$12 = 2 \cdot 2 \cdot 3$ or $2^2 \cdot 3$

$16 = 2 \cdot 2 \cdot 2 \cdot 2$ or 2^4

> The prime factors of 8, 12, and 16 are 2 and 3. Write the prime factorization using exponents.

The LCM of 8, 12, and 16 is $2^4 \cdot 3$, which is 48.

To find the number of packages of each Ling needs to buy, divide 48 by the amount in each package.

cups: $48 \div 8$ or 6 packages

plates: $48 \div 12$ or 4 packages

napkins: $48 \div 16$ or 3 packages

So, Ling will need to buy 6 packages of cups, 4 packages of plates, and 3 packages of napkins.

✓ **CHECK Your Progress**

d. **VEHICLES** Mr. Hernandez changes his car's oil every 3 months, rotates the tires every 6 months, and replaces the air filter once a year. If he completed all three tasks in April, what will be the next month he again completes all three tasks?

CHECK Your Understanding

Examples 1–3
(pp. 211–212)

Find the LCM of each set of numbers.

1. 4, 14
2. 6, 7
3. 12, 15
4. 21, 35
5. 3, 5, 12
6. 6, 14, 21

Example 3
(p. 212)

7. **GOVERNMENT** The number of years per term for a U.S. President, senator, and representative is shown. Suppose a senator was elected in the presidential election year 2008. In what year will he or she campaign again during a presidential election year?

Elected Office	Term (yr)
President	4
Senator	6
Representative	2

Practice and Problem Solving

Find the LCM for each set of numbers.

HOMEWORK HELP	
For Exercises	**See Examples**
8–13, 20	1, 2
14–19, 21	3

8. 6, 8
9. 8, 18
10. 12, 16
11. 24, 36
12. 11, 12
13. 45, 63
14. 2, 3, 5
15. 6, 8, 9
16. 8, 12, 16
17. 12, 15, 28
18. 22, 33, 44
19. 12, 16, 36

20. **CHORES** Hernando walks his dog every two days. He gives his dog a bath once a week. Today, Hernando walked his dog and then gave her a bath. How many days will pass before he does both chores on the same day?

21. **TEXT MESSAGING** Three friends use text messaging to notify their parents of their whereabouts. If all three contact their parents at 3:00 P.M., at what time will all three contact their parents again at the same time?

Friend	Time Interval
Linda	every 30 min
Brandon	every 45 min
Edward	every 60 min

Find the LCM of each set.

22. $3.00, $14.00
23. 10¢, 25¢, 5¢
24. 9 inches, 2 feet

Write two numbers whose LCM is the given number.

25. 35
26. 56
27. 70
28. 30

29. **SNACKS** Alvin's mom needs to buy snacks for soccer practice. Juice boxes come in packages of 10. Oatmeal snack bars come in packages of 8. She wants to have the same number of juice boxes and snack bars, what is the least number of packages of each snack that she will have to buy?

30. **REASONING** The LCM of two consecutive positive numbers is greater than 200 and is a multiple of 7. What are the least possible numbers?

NYSCC • NYSMT
Extra Practice, pp. 678, 707

H.O.T. Problems

31. **CHALLENGE** Two numbers have a GCF of $3 \cdot 5$. Their LCM is $2^2 \cdot 3 \cdot 5$. If one of the numbers is $3 \cdot 5$, what is the other number?

32. **SELECT A TECHNIQUE** The schedule for each of three trains is shown. Suppose a train from each line leaves Clark Street at 11:35 A.M. Which of the following technique(s) might you use to determine the next time all three trains will be leaving at the same time? Justify your selection(s). Then use the technique to solve the problem.

Clark Street Train Station	
Train	**Leaves Station**
Red-line	every 14 minutes
Blue-line	every 16 minutes
Brown-line	every 8 minutes

mental math	number sense	estimation

33. **OPEN ENDED** Write three numbers that have an LCM of 30.

34. **WRITING IN MATH** Describe the relationship between 4, 20, and 5 using the words *factor* and *multiple*.

NYSMT PRACTICE 7.N.9

35. Which rule describes the common multiples of 12 and 18, where n represents the counting numbers?

   A $12n$

   B $18n$

   C $36n$

   D $216n$

36. **SHORT RESPONSE** Wil swims every third day, runs every fourth day, and lifts weights every fifth day. If Wil does all three activities today, how many days will pass before he does all three activities on the same day again?

Spiral Review

Write each percent as a decimal. (Lesson 4-7)

37. 55%   38. 26.4%   39. $\frac{1}{4}$%   40. 2%

41. **DIAMONDS** Sixty-eight percent of engagement rings have a diamond that is round in shape. Write this percent as a fraction in simplest form. (Lesson 4-6)

42. **ALGEBRA** Solve $3x = 18$. (Lesson 3-3)

43. **ALGEBRA** Rose swam 7 laps more than twice the number of laps her sister swam. Write an algebraic expression to represent this situation. (Lesson 3-1)

▷ **GET READY** for the Next Lesson

PREREQUISITE SKILL Replace each ● with $<$, $>$ or $=$ to make a true sentence. (Page 670)

44. 6.85 ● 5.68   45. 2.34 ● 2.43   46. 6.9 ● 5.99

Comparing and Ordering Rational Numbers

MAIN IDEA

Compare and order fractions, decimals, and percents.

NYS Core Curriculum

7.N.3 Place rational and irrational numbers (approximations) on a number line and justify the placement of the numbers

New Vocabulary

rational numbers
common denominator
least common denominator (LCD)

NY Math Online

glencoe.com

• Extra Examples
• Personal Tutor
• Self-Check Quiz

▷ **MINI Lab**

In Chapter 2, you used a number line to compare integers. You can also use a number line to compare positive and negative fractions. The number line shows that $-\frac{1}{8} < \frac{3}{8}$.

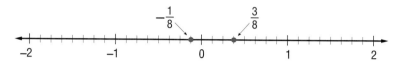

Graph each pair of numbers on a number line. Then determine which number is less.

1. $-\frac{7}{8}, -\frac{3}{8}$

2. $-\frac{5}{8}, -1\frac{1}{8}$

3. $-\frac{13}{8}, -\frac{3}{8}$

4. $-1\frac{7}{8}, -1\frac{5}{8}$

5. $-\frac{1}{2}, -\frac{3}{4}$

6. $1\frac{1}{4}, -1\frac{1}{4}$

The different types of numbers you have been using are all examples of rational numbers. A **rational number** is a number that can be expressed as a fraction. Fractions, terminating and repeating decimals, percents, and integers are all rational numbers. The points corresponding to rational numbers begin to "fill in" the number line.

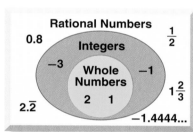

EXAMPLE Compare Rational Numbers

① **Replace the ● with <, >, or = to make $-1\frac{5}{6}$ ● $-1\frac{1}{6}$ a true sentence.**

Graph each rational number on a number line. Mark off equal size increments of $\frac{1}{6}$ between -2 and -1.

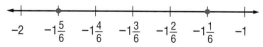

The number line shows that $-1\frac{5}{6} < -1\frac{1}{6}$.

✓ **CHECK Your Progress**

a. **Replace the ● with <, >, or = to make $-5\frac{5}{9}$ ● $-5\frac{1}{9}$ a true sentence.**

A **common denominator** is a common multiple of the denominators of two or more fractions. The **least common denominator** or **LCD** is the LCM of the denominators. You can use the LCD to compare fractions.

EXAMPLE Compare Rational Numbers

2 Replace the ● with <, >, or = to make $\frac{7}{12}$ ● $\frac{8}{18}$ a true sentence.

$12 = 2^2 \cdot 3$ and $18 = 2 \cdot 3^2$. So, the LCM is $2^2 \cdot 3^2$ or 36. The LCD of the denominators 12 and 18 is 36.

$$\frac{7}{12} = \frac{7 \times 3}{12 \times 3} \qquad\qquad \frac{8}{18} = \frac{8 \times 2}{18 \times 2}$$

$$= \frac{21}{36} \qquad\qquad\qquad = \frac{16}{36}$$

Since $\frac{21}{36} > \frac{16}{36}$, then $\frac{7}{12} > \frac{8}{18}$.

CHECK Your Progress

Replace each ● with <, >, or = to make a true sentence.

b. $\frac{5}{6}$ ● $\frac{7}{9}$ c. $\frac{1}{5}$ ● $\frac{7}{50}$ d. $-\frac{9}{16}$ ● $-\frac{7}{10}$

You can also compare fractions by writing each fraction as a decimal and then comparing the decimals.

Real-World EXAMPLE

3 **ROLLER SHOES** In Mr. Huang's math class, 6 out of 32 students own roller shoes. In Mrs. Trevino's math class, 5 out of 29 students own roller shoes. In which class do a greater fraction of students own roller shoes?

Since the denominators are large, write $\frac{6}{32}$ and $\frac{5}{29}$ as decimals and then compare.

$6 \div 32 = 0.1875$ $5 \div 29 \approx 0.1724$ Divide.

Since $0.1875 > 0.1724$, then $\frac{6}{32} > \frac{5}{29}$.

So, a greater fraction of students in Mr. Huang's class own roller shoes.

Real-World Link . . . .
The first roller shoe was introduced in 2001.
Source: Associated Content, Inc.

CHECK Your Progress

e. **BOWLING** Twelve out of 32 students in second period class like to bowl. In fifth period class, 12 out of 29 students like to bowl. In which class do a greater fraction of the students like to bowl?

These fraction-decimal-percent equivalents are used frequently.

Fractions-Decimals-Percents

$\frac{1}{4} = 0.25 = 25\%$	$\frac{1}{5} = 0.2 = 20\%$	$\frac{1}{8} = 0.125 = 12.5\%$	$\frac{1}{10} = 0.1 = 10\%$
$\frac{1}{2} = 0.5 = 50\%$	$\frac{2}{5} = 0.4 = 40\%$	$\frac{3}{8} = 0.375 = 37.5\%$	$\frac{3}{10} = 0.3 = 30\%$
$\frac{3}{4} = 0.75 = 75\%$	$\frac{3}{5} = 0.6 = 60\%$	$\frac{1}{3} = 0.\overline{3} = 33.\overline{3}\%$	$\frac{7}{10} = 0.7 = 70\%$
$1 = 1.00 = 100\%$	$\frac{4}{5} = 0.8 = 80\%$	$\frac{2}{3} = 0.\overline{6} = 66.\overline{6}\%$	$\frac{9}{10} = 0.9 = 90\%$

The Greek letter π (pi) represents the nonterminating and nonrepeating number whose first few digits are 3.1415926.... This number is not rational. You will learn more about *irrational numbers* in Chapter 12.

NYSMT EXAMPLE

4 Which list shows the numbers 3.44, π, 3.14, and $3.\overline{4}$ in order from least to greatest?

A π, 3.14, 3.44, $3.\overline{4}$

C 3.14, π, $3.\overline{4}$, 3.44

B π, 3.14, $3.\overline{4}$, 3.44

D 3.14, π, 3.44, $3.\overline{4}$

Read the Item

Compare the digits using place value.

Solve the Item

Line up the decimal points and compare using place value.

3.14**0**	Annex a zero.	3.44**0**	Annex a zero.
3.14**1**5926...	$\pi \approx 3.1415926...$	3.44**4**...	$3.\overline{4} = 3.444...$
Since $0 < 1$, $3.14 < \pi$.		Since $0 < 4$, $3.44 < 3.\overline{4}$.	

So, the order of the numbers from least to greatest is 3.14, π, 3.44, and $3.\overline{4}...$. The answer is D.

CHECK Your Progress

f. The amount of rain received on four consecutive days was 0.3 inch, $\frac{3}{5}$ inch, 0.75 inch, and $\frac{2}{3}$ inch. Which list shows the amounts from least to greatest?

F 0.3 in., $\frac{2}{3}$ in., $\frac{3}{5}$ in., 0.75 in.

H 0.75 in., $\frac{2}{3}$ in., $\frac{3}{5}$ in., 0.3 in.

G 0.3 in., $\frac{3}{5}$ in., $\frac{2}{3}$ in., 0.75 in.

J $\frac{3}{5}$ in., $\frac{2}{3}$ in., 0.3 in., 0.75 in.

Examples 1–2
(pp. 215–216)

Replace each ⬤ with <, >, or = to make a true sentence. Use a number line if necessary.

1. $-\dfrac{4}{9}$ ⬤ $-\dfrac{7}{9}$

2. $-1\dfrac{3}{4}$ ⬤ $-1\dfrac{6}{8}$

3. $\dfrac{3}{8}$ ⬤ $\dfrac{6}{15}$

4. $2\dfrac{4}{5}$ ⬤ $2\dfrac{7}{8}$

Example 3
(p. 216)

5. **SOCCER** The table shows the average saves for two soccer goalies. Who has the better average, Elliot or Shanna? Explain.

Name	Average
Elliot	3 saves out of 4
Shanna	7 saves out of 11

6. **SCHOOL** On her first quiz in social studies, Majorie answered 23 out of 25 questions correctly. On her second quiz, she answered 27 out of 30 questions correctly. On which quiz did Majorie have the greater score?

Example 4
(p. 217)

7. **MULTIPLE CHOICE** The lengths of four insects are 0.02-inch, $\dfrac{1}{8}$-inch, 0.1-inch, and $\dfrac{2}{3}$-inch. Which list shows the lengths in inches from least to greatest?

A $0.1, 0.02, \dfrac{1}{8}, \dfrac{2}{3}$

B $\dfrac{1}{8}, 0.02, 0.1, \dfrac{2}{3}$

C $0.02, 0.1, \dfrac{1}{8}, \dfrac{2}{3}$

D $\dfrac{2}{3}, 0.02, 0.1, \dfrac{1}{8}$

Practice and Problem Solving

HOMEWORK HELP	
For Exercises	**See Examples**
8–19	1, 2
20–25, 49	3
26–31, 48	4

Replace each ⬤ with <, >, or = to make a true sentence. Use a number line if necessary.

8. $-\dfrac{3}{5}$ ⬤ $-\dfrac{4}{5}$

9. $-\dfrac{5}{7}$ ⬤ $-\dfrac{2}{7}$

10. $-7\dfrac{5}{8}$ ⬤ $-7\dfrac{1}{8}$

11. $-3\dfrac{2}{3}$ ⬤ $-3\dfrac{4}{6}$

12. $\dfrac{7}{10}$ ⬤ $\dfrac{2}{3}$

13. $\dfrac{4}{7}$ ⬤ $\dfrac{5}{8}$

14. $\dfrac{2}{3}$ ⬤ $\dfrac{10}{15}$

15. $-\dfrac{17}{24}$ ⬤ $-\dfrac{11}{12}$

16. $2\dfrac{3}{4}$ ⬤ $2\dfrac{2}{3}$

17. $6\dfrac{2}{3}$ ⬤ $6\dfrac{1}{2}$

18. $5\dfrac{5}{7}$ ⬤ $5\dfrac{11}{14}$

19. $3\dfrac{11}{16}$ ⬤ $3\dfrac{7}{8}$

20. 40% ⬤ 112 out of 25

21. 3 out of 5 ⬤ 59%

22. 0.82 ⬤ 5 out of 6

23. 9 out of 20 ⬤ 0.45

24. **MONEY** The table shows how much copper is in each type of coin. Which coin contains the greatest amount of copper?

Coin	Amount of Copper
Dime	$\dfrac{12}{16}$
Nickel	$\dfrac{3}{4}$
Penny	$\dfrac{1}{400}$
Quarter	$\dfrac{23}{25}$

25. **BASKETBALL** Gracia and Jim were shooting free throws. Gracia made 4 out of 15 free throws. Jim *missed* the free throw 6 out of 16 times. Who made the free throw a greater fraction of the time?

Order each set of numbers from least to greatest.

26. 0.23, 19%, $\frac{1}{5}$

27. $\frac{8}{10}$, 81%, 0.805

28. $-0.615, -\frac{5}{8}, -0.62$

29. $-1.4, -1\frac{1}{25}, -1.25$

30. $7.49, 7\frac{49}{50}, 7.5$

31. $3\frac{4}{7}, 3\frac{3}{5}, 3.47$

MEASUREMENT Replace each ● with <, >, or = to make a true sentence.

32. $\frac{5}{8}$ yard ● $\frac{1}{16}$ yard

33. 0.25 pound ● $\frac{2}{9}$ pound

34. $2\frac{5}{6}$ hours ● 2.8 hours

35. $1\frac{7}{12}$ gallons ● $1\frac{5}{8}$ gallons

MEASUREMENT Order each of the following from least to greatest.

36. 4.4 miles, $4\frac{3}{8}$ miles, $4\frac{5}{12}$ miles

37. 6.5 cups, $6\frac{1}{3}$ cups, 6 cups

38. 1.2 laps, 2 laps, $\frac{1}{2}$ lap

39. $\frac{1}{5}$ gram, 5 grams, 1.5 grams

ANIMALS For Exercises 40–42, use the table that shows the lengths of the smallest mammals.

Animal	Length (ft)
Eastern Chipmunk	$\frac{1}{3}$
Kitti's Hog-Nosed Bat	$0.8\overline{3}$
European Mole	$\frac{5}{12}$
Masked Shrew	$\frac{1}{6}$
Spiny Pocket Mouse	0.25

Source: *Scholastic Book of World Records*

Real-World Link . . . . .
The Olympic gold medals are actually made out of 92.5% silver, with the gold medal covered in 6 grams of pure gold.

40. Which animal is the smallest mammal?

41. Which animal is smaller than the European mole but larger than the spiny pocket mouse?

42. Order the animals from greatest to least size.

SOFTBALL For Exercises 43 and 44, use the following table which shows the at-bats, hits, and home run statistics for four players on the 2004 Olympics U.S. Women's softball team.

Player	At-Bats	Hits	Home Runs
Crystal Bustos	26	9	5
Kelly Krestschman	21	7	1
Stacey Nuveman	16	5	2
Natasha Watley	30	12	0

Source: Olympic Movement

43. Write the ratio of hits to at-bats as a decimal to the nearest thousandth for each player. Who had the greatest batting average during the Olympic games?

NYSCC • NYSMT
Extra Practice, pp. 679, 707

44. Write the ratio of home runs to at-bats as a decimal for each player. Who had the greatest home run average during the Olympic games?

45. **Which One Doesn't Belong?** Identify the ratio that does not have the same value as the other three. Explain your reasoning.

| 12 out of 15 | 0.08 | 80% | $\frac{4}{5}$ |

46. **CHALLENGE** Explain how you know which number, $1\frac{15}{16}$, $\frac{17}{8}$, or $\frac{63}{32}$, is nearest to 2.

47. **WRITING IN MATH** Write a word problem about a real-world situation in which you would compare rational numbers. Then solve the problem.

NYSMT PRACTICE 7.N.3

48. Which point shows the location of $\frac{7}{2}$ on the number line?

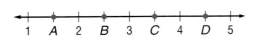

A point A
B point B
C point C
D point D

49. Which list of numbers is ordered from least to greatest?

F $\frac{1}{4}$, $4\frac{1}{4}$, 0.4, 0.04

G 0.04, 0.4, $4\frac{1}{4}$, $\frac{1}{4}$

H 0.04, $\frac{1}{4}$, 0.4, $4\frac{1}{4}$

J 0.4, $\frac{1}{4}$, 0.04, $4\frac{1}{4}$

50. Which of the following fractions is closest to 0?

A $-\frac{3}{4}$ C $\frac{7}{12}$

B $-\frac{2}{3}$ D $\frac{5}{8}$

Spiral Review

Find the LCM of each set of numbers. (Lesson 4-8)

51. 14, 21
52. 3, 13
53. 12, 16

SALES TAX The table shows the sales tax rate for the states shown. Write each sales tax rate as a decimal. (Lesson 4-7)

54. Kentucky
55. Illinois
56. North Carolina

State	Sales Tax
Illinois	6.25%
Kentucky	6%
North Carolina	4.25%
South Carolina	5%

Source: Federation of Tax Administrators

Find the GCF of each set of numbers. (Lesson 4-2)

57. 18, 72
58. 40, 12
59. 72, 20

ALGEBRA Solve each equation. (Lesson 3-5)

60. $4x + 3 = 15$
61. $2n - 5 = 19$
62. $-8 = -3d + 1$

FOLDABLES Study Organizer

GET READY to Study

Be sure the following Big Ideas are noted in your Foldable.

Fractions, Decimals, and Percents

4-1 Prime Factorization
4-2 Greatest Common Factors
4-3 Make an Organized List
4-4 Simplifying Fractions
4-5 Fractions and Decimals
4-6 Fractions and Percents
4-7 Percents and Decimals
4-8 Least Common Multiple
4-9 Comparing and Ordering Rational Numbers

BIG Ideas

Greatest Common Factor (Lesson 4-2)
• The greatest common factor or GCF is the greatest of the common factors of two or more numbers.

Fractions, Decimals, and Percents
(Lessons 4-4 to 4-7)
• A fraction is in simplest form when the GCF of the numerator and denominator is 1.

• A terminating decimal is a decimal whose digits end. Repeating decimals have a pattern in their digits that repeats forever.

• A percent is a part to whole ratio that compares a number to 100.

• To write a percent as a decimal, divide the percent by 100 and remove the percent symbol.

• To write a decimal as a percent, multiply the percent by 100 and add the percent symbol.

Least Common Multiple (Lesson 4-8)
• The least common multiple or LCM of two or more numbers is the least of their common multiples.

Rational Numbers (Lesson 4-9)
• A rational number is one that can be expressed as a fraction.

Key Vocabulary

bar notation (p. 197)

common denominator (p. 216)

composite number (p. 181)

equivalent fractions (p. 192)

factor tree (p. 182)

greatest common factor (GCF) (p. 186)

least common denominator (p. 216)

least common multiple (LCM) (p. 211)

multiple (p. 211)

percent (p. 202)

prime factorization (p. 182)

prime number (p. 181)

ratio (p. 202)

rational number (p. 215)

repeating decimal (p. 197)

simplest form (p. 192)

terminating decimal (p. 197)

Venn diagram (p. 186)

Vocabulary Check

State whether each sentence is *true* or *false*. If *false*, replace the underlined word or number to make a true sentence.

1. A ratio is a comparison of two numbers by <u>multiplication</u>.

2. A <u>rational number</u> is a whole number greater than 1 that has exactly two factors, 1 and itself.

3. 1.875 is an example of a <u>terminating decimal</u>.

4. A common denominator for the fractions $\frac{2}{3}$ and $\frac{1}{4}$ is <u>12</u>.

5. The <u>greatest common factor</u> of 3 and 5 is 15.

6. A ratio that compares a number to <u>100</u> is a percent.

7. The fractions $\frac{9}{21}$ and $\frac{3}{7}$ are <u>equivalent fractions</u>.

Lesson-by-Lesson Review

4-1 **Prime Factorization** (pp. 181–184)

7.N.8,
7.N.10

Find the prime factorization of each number.

8. 54   **9.** 128   **10.** 68   **11.** 95

12. ALGEBRA Factor $36x^2yz^3$.

13. PLANTS The palm tree *raffia* has leaves up to 65 feet long. Write this length as a product of primes.

Example 1 Find the prime factorization of 18.

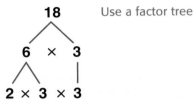

Use a factor tree

The prime factorization of 18 is 2×3^2.

4-2 **Greatest Common Factor** (p. 186–189)

7.N.8

Find the GCF of each set of numbers.

14. 18, 27   **15.** 30, 72

16. 28, 70, 98   **17.** 42, 63, 105

18. ALGEBRA Find the GCF of $18w$ and $54w^2y$.

19. CLOTHING Maria spent a total of $24 on earrings, $36 on shirts, and $48 on shorts. If each item cost the same amount, what is the greatest possible price per item?

Example 2 Find the GCF of 24 and 56.

First, make a list of all the factors of 24 and 56.

factors of 24: 1, 2, 3, 4, 6, 8, 12, 24

factors of 56: 1, 2, 4, 7, 8, 14, 28, 56

common factors: 1, 2, 4, 8

The GCF of 24 and 56 is 8.

4-3 **PSI: Make an Organized List** (pp. 190–191)

7.PS.4,
7.PS.8

Solve by making an organized list.

20. SEATING In how many ways can four friends sit in a row at the movies?

21. TELEPHONES A phone company offers 5 different types of long-distance plans and 3 different caller features (call waiting, caller ID, and call forward). How many different kinds of plans can be set up that include a long-distance service and a caller feature?

Example 3 In how many ways can the letters A, B, and C be arranged?

The possible outcomes are written below.

ABC   ACB
BAC   BCA
CAB   CBA

There are 6 different ways to arrange the letters.

Mixed Problem Solving
For mixed problem-solving practice,
see page 707.

4-4 · **Simplifying Fractions** (pp. 192–195)

7.N.1

Write each fraction in simplest form.

22. $\frac{12}{15}$ 23. $\frac{35}{60}$ 24. $\frac{11}{121}$

25. $\frac{14}{63}$ 26. $\frac{37}{45}$ 27. $\frac{55}{110}$

28. **CATS** The average household cat sleeps 18 hours a day. Write a fraction in simplest form comparing the number of hours a household cat sleeps to the number of hours in a day.

Example 4 Write $\frac{24}{32}$ in simplest form.

Find the GCF of the numerator and denominator.

24 = 1, 2, 3, 4, 6, **8**, 12, 24
32 = 1, 2, 4, **8**, 16, 32

$\frac{24}{32} = \frac{24 \div 8}{32 \div 8} = \frac{3}{4}$ Divide the numerator and denominator by the GCF.

4-5 · **Fractions and Decimals** (pp. 196–200)

7.N.17

Write each fraction or mixed number as a decimal. Use bar notation if the decimal is a repeating decimal.

29. $\frac{3}{4}$ 30. $\frac{7}{8}$ 31. $\frac{5}{9}$

32. $4\frac{1}{3}$ 33. $6\frac{2}{5}$ 34. $1\frac{6}{7}$

Write each decimal as a fraction in simplest form.

35. 0.7   36. 0.44   37. 0.05

38. 0.18   39. 0.54   40. 0.08

41. **RUNNING** Jeremy ran a mile in 5 minutes and 8 seconds. Write this time in minutes as a decimal.

Example 5 Write $\frac{3}{8}$ as a decimal.

$$\begin{array}{r} 0.375 \\ 8)\overline{3.000} \\ -24 \\ \hline 60 \\ -56 \\ \hline 40 \\ -40 \\ \hline 0 \end{array}$$

So, $\frac{3}{8} = 0.375$.

Example 6 Write 0.64 as a fraction.

$0.64 = \frac{64}{100}$ Write as a fraction with a denominator of 100.

$= \frac{16}{25}$ Simplify.

4-6 · **Fractions and Percents** (pp. 202–205)

6.N.11

Write each fraction as a percent.

42. $\frac{32}{100}$ 43. $\frac{11}{25}$ 44. $\frac{47}{50}$ 45. $\frac{8}{20}$

Write each percent as a fraction in simplest form.

46. 68%   47. 95%   48. 42%   49. 16%

50. **LUNCH** In Mrs. Soulise's class, 56% of the students buy their lunch. Write this percent as a fraction in simplest form.

Example 7 Write $\frac{27}{50}$ as a percent.

$\frac{27}{50} = \frac{54}{100}$ Write an equivalent fraction with a denominator of 100.

$= 54\%$ Definition of percent

Example 8 Write 96% as a fraction.

$96\% = \frac{96}{100}$ Definition of percent

$= \frac{24}{25}$ Simplify.

4-7 **Percents and Decimals** (pp. 206–210)

6.N.11

Write each percent as a decimal.

51. 48%   **52.** 7%

53. 12.5%   **54.** $75\frac{1}{4}$%

Write each decimal as a percent.

55. 0.61   **56.** 0.055   **57.** 0.19   **58.** 0.999

59. FOOD A serving of oatmeal contains 3 grams of fiber. This is 12% of the recommended daily allowance. Write this percent as a decimal.

Example 9   Write 35% as a decimal.

$35\% = \dfrac{35}{100}$ Write the percent as a fraction.

    $= 0.35$ Write the fraction as a decimal.

Example 10   Write 0.625 as a percent.

$0.625 = 0.625$ Multiply by 100.

     $= 62.5\%$ Add the % symbol.

4-8 **Least Common Multiple** (pp. 211–214)

7.N.9

Find the LCM of each set of numbers.

60. 9, 15   **61.** 4, 8

62. 16, 24   **63.** 3, 8, 12

64. 4, 9, 12   **65.** 15, 24, 30

66. BREAKFAST At a bakery, muffins come in dozens and individual serving containers of orange juice come in packs of 8. If Avery needs to have the same amount of muffins as orange juice containers, what is the least possible number of sets of each he needs to buy?

Example 11   Find the LCM of 8 and 36.

Write each prime factorization.

$8 = 2 \times 2 \times 2 = 2^3$

$36 = 2 \times 2 \times 3 \times 3 = 2^2 \times 3^2$

LCM: $2^3 \times 3^2 = 72$

The LCM of 8 and 36 is 72.

4-9 **Comparing and Ordering Rational Numbers** (pp. 215–220)

7.N.3

Replace each ● with <, >, or = to make a true sentence.

67. $\dfrac{3}{8}$ ● $\dfrac{2}{3}$ **68.** -0.45 ● $-\dfrac{9}{20}$

69. $\dfrac{8}{9}$ ● 85%   **70.** $-3\dfrac{3}{4}$ ● $-3\dfrac{5}{8}$

71. SCHOOL Michael received a $\dfrac{26}{30}$ on his English quiz and received 81% on his biology test. In which class did he receive the higher score?

Example 12   Replace ● with <,>, or = to make $\dfrac{3}{5}$ ● $\dfrac{5}{8}$ a true sentence.

Find equivalent fractions. The LCD is 40.

$\dfrac{3}{5} = \dfrac{3 \times 8}{5 \times 8} = \dfrac{24}{40}$ $\dfrac{5}{8} = \dfrac{5 \times 5}{8 \times 5} = \dfrac{25}{40}$

Since $\dfrac{24}{40} < \dfrac{25}{40}$, then $\dfrac{3}{5} < \dfrac{5}{8}$.

1. Find the prime factorization of 72.

2. Find the GCF of 24 and 40.

3. **SCHEDULES** Farijah registered for French, Pre-Algebra, Life Science, English, and Social Studies. French is only offered first period, Pre-Algebra is only offered fifth period, and she must have lunch fourth period. How many different schedules can she create out of a six period day? Use the *make an organized list* strategy.

Write each fraction in simplest form.

4. $\frac{24}{60}$

5. $\frac{64}{72}$

Write each fraction, mixed number, or percent as a decimal. Use bar notation if the decimal is a repeating decimal.

6. $\frac{7}{9}$

7. $4\frac{5}{8}$

8. 91%

9. **COINS** The United States Mint released a new quarter every ten weeks from 1999 to 2008 commemorating the 50 states. By the end of 2006, 40 state coins had been released. What percent of the coins is this?

Write each decimal or percent as a fraction in simplest form.

10. 0.84

11. 0.006

12. 42%

13. **MULTIPLE CHOICE** Which of the following is equivalent to the decimal 0.087?

  A 0.87%
  B 8.7%
  C 87%
  D 870%

Write each fraction or decimal as a percent.

14. $\frac{15}{25}$

15. 0.26

16. 0.135

17. **FLOORING** Mr. Daniels is putting new floor tiles in his bathroom. He has already tiled 34 square feet of the floor measuring 5 feet by 10 feet. What percent of the floor has he tiled?

18. **MULTIPLE CHOICE** What percent of the figure below is unshaded?

  F 15%
  G 30%
  H 40%
  J 60%

Find the LCM of each set of numbers.

19. 18, 42

20. 4, 5, 12

21. **PRACTICE** Rico has track practice every 3 days. He has saxophone practice every 4 days. If Rico has both track and saxophone practice today, after how many days will Rico have both track and saxophone practice again?

Replace each ● with <, >, or = to make a true sentence.

22. $-\frac{3}{5}$ ● $-\frac{5}{9}$

23. $4\frac{7}{12}$ ● $4\frac{6}{8}$

24. $\frac{13}{20}$ ● 65%

25. **BASKETBALL** To make it past the first round of tryouts for the basketball team, Paul must make at least 35% of his free-throw attempts. During the first round of tryouts he makes 17 out of 40 attempts. Did Paul make it to the next round of tryouts? Explain your reasoning.

PART 1   Multiple Choice

Read each question. Then fill in the correct answer on the answer sheet provided by your teacher or on a sheet of paper.

1. A large school system estimates that 0.706 of its students will take the bus to school throughout the school year. Which number is greater than 0.706?

   A $\dfrac{706}{1,000}$

   B $-1\dfrac{6}{7}$

   C $\dfrac{76}{100}$

   D -7.06

2. Debra is working on three different art projects. She has completed $\dfrac{1}{4}$, $\dfrac{3}{8}$, and $\dfrac{1}{2}$ of these projects, respectively. Which list shows the percent of work completed on these projects from least to greatest?

   F 37.5%, 50%, 25%

   G 50%, 37.5%, 25%

   H 25%, 37.5%, 50%

   J 25%, 50%, 87.5%

3. Which of the following is the prime factored form of the lowest common denominator of $\dfrac{1}{6}$ and $\dfrac{3}{8}$?

   A $2^2 \times 3 \times 5$ C 2×6

   B $2^3 \times 5$ D $2^3 \times 3$

TEST-TAKING TIP

Question 3 Eliminate any answer choices that you know are incorrect. Since the LCD, 24, does not have a factor of 5, you can eliminate answer choices A and B.

4. Solve the equation $x + 7 = -3$. What is the value of x?

   F 4   H -4

   G 3   J -10

5. At a wedding reception, the number of seats s is equal to 8 times the number of tables t. Which equation matches this situation?

   A $s = 8 + t$

   B $t = 8 \cdot s$

   C $s = 8 \cdot t$

   D $t = 8 - t$

6. Which problem situation matches the equation below?

$$x + 12 = 35$$

   F The difference between two numbers is 35. One of the numbers is 12. What is x, the other number?

   G Laura is 12 years younger than her brother. If Laura is 35 years old, find her brother's age x.

   H The sum of a number, x, and 12 is 35. What is the value of x?

   J Karen had $35. If she received $12, what is x, the total amount she now has?

7. Which of the following is true when evaluating the expression $3 \cdot 4^2 - 12 \div 6$?

   A Multiply 3 by 4 first since multiplication comes before subtraction.

   B Evaluate 4^2 first since it is a power.

   C Divide 12 by 6 first since division comes before multiplication.

   D Multiply 3 by 4 first since all operations occur in order from left to right.

8. Which sequence follows the rule $2n + 5$, where n represents the position of a term in the sequence?

   F   3, 5, 7, 9, 11, ...   H   7, 9, 11, 13, 15, ...

   G   6, 8, 10, 12, 14, ...   J   8, 12, 16, 20, 24, ...

9. Nicholas used the Distributive Property to evaluate the expression $5(12 + 7)$ mentally. Which of the following is a correct use of the Distributive Property to evaluate this expression?

   A $5(12 + 7) = 5(12) + 5(7) = 60 + 35$ or 95

   B $5(12 + 7) = 5(12) + 7 = 60 + 7$ or 67

   C $5(12 + 7) = 12 + 5(7) = 12 + 35$ or 47

   D $5(12 + 7) = 5 + 60 + 5 + 7 = 65 + 12$ or 77

10. Which of the following relationships is represented by the data in the table?

x	y
1	5,280
2	10,560
3	15,840
4	21,120
5	26,400

   F   conversion of miles to feet

   G   conversion of inches to yards

   H   conversion of feet to miles

   J   conversion of yards to inches

11. If $g = 4$, $m = 3$, and $n = 6$, then $\frac{mn + 2}{g} + 1$ is equivalent to which of the following?

   A   6   C   3

   B   5   D   2

PART 2 Short Response/Grid In

Record your answers on the answer sheet provided by your teacher or on a sheet of paper.

12. Write 7.2% as a decimal.

13. Jeremy expects 8 out of the 10 friends he invited to come to his party. What percent of his friends does he expect to come?

PART 3 Extended Response

Record your answers on the answer sheet provided by your teacher or on a sheet of paper. Show your work.

14. The prime factorization of 24 is $2 \times 2 \times 2 \times 3$. The table lists each unique prime factor and the products of all possible unique combinations of two, three, and four prime factors.

Unique Prime Factors	2, 3
Products of Two Factors	$2 \times 2, 2 \times 3$
Products of Three Factors	$2 \times 2 \times 2, 2 \times 2 \times 3$
Product of Four Factors	$2 \times 2 \times 2 \times 3$

   a. Find each product.

   b. What do the products have in common?

   c. What other numbers are factors of 24?

   d. Explain how you can use the prime factors of a number to find all of its factors. Test your conjecture by finding the factors of 60.

NEED EXTRA HELP?														
If You Missed Question...	1	2	3	4	5	6	7	8	9	10	11	12	13	14
Go to Lesson...	4-9	4-9	4-2	3-2	3-1	3-1	1-4	1-9	1-8	2-6	1-4	4-7	4-6	4-9
NYS Core Curriculum	7.N.3	7.N.3	7.N.8	7.A.4	7.A.1	7.A.1	7.N.11	7.A.8	6.N.2	7.N.12	7.N.11	6.N.11	6.N.11	7.N.3

CHAPTER 5

Applying Fractions

New York State Core Curriculum

Reinforcement of 6.N.18
Add, subtract, multiply, and divide mixed numbers with unlike denominators

Key Vocabulary

compatible numbers (p. 232)
like fractions (p. 236)
reciprocal (p. 258)
unlike fractions (p. 237)

Real-World Link

Baking The measurements found on measuring cups and spoons are written as fractions. You will use fractions to find how much of each ingredient is needed when you make part of a whole recipe.

Applying Fractions Make this Foldable to help you organize your notes. Begin with a plain sheet of 11" by 17" paper, four index cards, and glue.

① **Fold** the paper in half widthwise.

② **Open** and fold along the length about $2\frac{1}{2}$" from the bottom.

③ **Glue** the edge on each side to form two pockets.

④ **Label** the pockets *Fractions* and *Mixed Numbers*, respectively. Place two index cards in each pocket.

GET READY for Chapter 5

Diagnose Readiness You have two options for checking Prerequisite Skills.

Option 1

Take the Quick Quiz below. Refer to the Quick Review for help.

Option 2

NY Math Online Take the Online Readiness Quiz at glencoe.com.

QUICK Quiz

Find the LCD of each pair of fractions. (Lesson 4-8)

1. $\frac{5}{7}, \frac{3}{5}$ 2. $\frac{1}{2}, \frac{4}{9}$

3. $\frac{8}{15}, \frac{1}{6}$ 4. $\frac{3}{4}, \frac{7}{10}$

Multiply or divide. (Prior Grade)

5. 1.8×12 6. $99 \div 12$

7. $83 \div 100$ 8. 4.6×0.3

9. **MEASUREMENT** How many 1.6-meter sections of rope can be cut from a length of rope 6.4 meters? (Prior Grade)

10. **COINS** Manuel owes each of 8 friends $0.35. How much does he owe in all? (Prior Grade)

Complete to show equivalent mixed numbers. (Prior Grade)

11. $3\frac{1}{5} = 2\frac{\blacksquare}{5}$ 12. $9\frac{2}{3} = \blacksquare\frac{5}{3}$

13. $6\frac{1}{4} = 5\frac{\blacksquare}{4}$ 14. $8\frac{6}{7} = 7\frac{\blacksquare}{7}$

15. **RECIPES** A recipe calls for $4\frac{2}{3}$ cups of flour. This is equivalent to 3 cups of flour plus an additional how many cups of flour? (Prior Grade)

QUICK Review

Example 1

Find the LCD of $\frac{5}{6}$ and $\frac{3}{10}$.
The LCD is the LCM of the denominators, 6 and 10, or 30.

Example 2

Find $7.8 \div 0.25$.

$$\begin{array}{r} 31.2 \\ 0.25\overline{)7.80\,0} \\ \underline{-7\,5} \\ 30 \\ \underline{-25} \\ 50 \\ \underline{-50} \\ 0 \end{array}$$

Move the decimal point 2 places to the right and divide as with whole numbers.

Example 3

Complete $4\frac{2}{9} = \blacksquare\frac{11}{9}$ to show equivalent mixed numbers.

$4\frac{2}{9} = 3 + 1\frac{2}{9}$
$\phantom{4\frac{2}{9}} = 3 + \frac{9}{9} + \frac{2}{9}$
$\phantom{4\frac{2}{9}} = 3 + \frac{11}{9}$
$\phantom{4\frac{2}{9}} = 3\frac{11}{9}$

Estimating with Fractions

▶ **GET READY** for the Lesson

MAMMALS The table below lists the average length for a few mammals.

1. Graph $9\frac{1}{4}$ on a number line. To the nearest whole number, how long is an American Bison?

2. Graph $3\frac{3}{4}$ on a number line. To the nearest whole number, how long is a dingo?

3. About how much longer is the American bison than a dingo?

Mammal	Length (ft)
Brown Bear	$6\frac{1}{2}$
American Bison	$9\frac{1}{4}$
Opossum	$2\frac{1}{2}$
Dingo	$3\frac{3}{4}$

To estimate the sum, difference, product, or quotient of mixed numbers, round the mixed numbers to the nearest whole number.

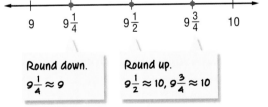

Round down.
$9\frac{1}{4} \approx 9$

Round up.
$9\frac{1}{2} \approx 10, 9\frac{3}{4} \approx 10$

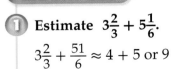 **Estimate with Mixed Numbers**

① **Estimate** $3\frac{2}{3} + 5\frac{1}{6}$.

$3\frac{2}{3} + \frac{51}{6} \approx 4 + 5$ or 9

The sum is *about* 9.

② **Estimate** $6\frac{2}{5} \times 1\frac{7}{8}$.

$6\frac{2}{5} \times 1\frac{7}{8} \approx 6 \times 2$ or 12

The product is *about* 12.

✓ **CHECK Your Progress**

Estimate.

a. $2\frac{1}{5} + 3\frac{1}{2}$

b. $4\frac{3}{8} \times 5\frac{1}{4}$

c. $8\frac{7}{9} \div 2\frac{3}{4}$

To estimate the sum, difference, product, or quotient of fractions, round each fraction to 0, $\frac{1}{2}$, or 1, whichever is closest. Number lines and fraction models, like the ones shown below, can help you decide how to round.

Fractions Close to 0	Fractions Close to $\frac{1}{2}$	Fractions Close to 1
0 $\frac{1}{6}$ 1	0 $\frac{1}{2}$ $\frac{5}{8}$ 1	0 $\frac{9}{10}$ 1
$\frac{1}{7}$	$\frac{4}{9}$	$\frac{5}{6}$
The numerator is much smaller than the denominator.	The numerator is about half of the denominator.	The numerator is almost as large as the denominator.

EXAMPLES **Estimate with Fractions**

3 Estimate $\frac{1}{8} + \frac{2}{3}$.

1 is much smaller than 8, so $\frac{1}{8} \approx 0$.

2 is close to half of 3, so $\frac{2}{3} \approx \frac{1}{2}$.

$\frac{1}{8} + \frac{2}{3} \approx 0 + \frac{1}{2} = \frac{1}{2}$ The sum is *about* $\frac{1}{2}$.

Study Tip

Estimating with Fractions If one of the fractions is a mixed number, such as $3\frac{5}{8} + \frac{2}{3}$, round the mixed number to the nearest whole number and the fraction to the nearest half. $3\frac{5}{8} + \frac{2}{3} \approx 4 + \frac{1}{2}$ or about $4\frac{1}{2}$.

4 Estimate $\frac{6}{7} - \frac{7}{10}$.

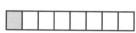

6 is almost as large as 7, so $\frac{6}{7} \approx 1$.

7 is about half of 10, so $\frac{7}{10} \approx \frac{1}{2}$.

$\frac{6}{7} - \frac{7}{10} \approx 1 - \frac{1}{2} = \frac{1}{2}$ The difference is *about* $\frac{1}{2}$.

5 Estimate $\frac{8}{9} \div \frac{5}{6}$.

$\frac{8}{9} \div \frac{5}{6} \approx 1 \div 1 = 1$ $\frac{8}{9} \approx 1$ and $\frac{5}{6} \approx 1$.

The quotient is *about* 1.

CHECK Your Progress

Estimate.

d. $\frac{1}{7} + \frac{3}{5}$ e. $\frac{7}{8} - \frac{5}{9}$ f. $\frac{3}{5} \times \frac{11}{12}$ g. $\frac{7}{8} \div \frac{2}{5}$

Compatible numbers, or numbers that are easy to compute mentally, can also be used to estimate.

EXAMPLES Use Compatible Numbers

Estimate using compatible numbers.

⑥ $\frac{1}{3} \cdot 14$

THINK What is $\frac{1}{3}$ of 14?

$\frac{1}{3} \cdot 14 \approx \frac{1}{3} \cdot 15$ or 5

Round 14 to 15, since 15 is divisible by 3.
$\frac{1}{3}$ of 15 is $15 \div 3$ or 5.

⑦ $9\frac{7}{8} \div 4\frac{1}{5}$

$9\frac{7}{8} \div 4\frac{1}{5} \approx 10 \div 4\frac{1}{5}$ Round $9\frac{7}{8}$ to 10.

$\approx 10 \div 5$ or 2   Round $4\frac{1}{5}$ to 5, since 10 is divisible by 5.

Study Tip

Compatible Numbers
When dividing mixed numbers, round so that the dividend is a multiple of the divisor.

✓ **CHECK Your Progress**

Estimate using compatible numbers.

h. $\frac{1}{4} \cdot 21$ i. $\frac{1}{3} \cdot 17$ j. $12 \div 6\frac{2}{3}$

Real-World EXAMPLE

Real-World Link.....
The monster truck shown is $15\frac{1}{2}$ feet tall and weighs 28,000 pounds.
Source: Monster Trucks UK

⑧ **MONSTER TRUCKS** The height of the wheels on the monster truck at the left is about $\frac{2}{3}$ of the total height of the truck. Estimate the height of the wheels.

Words	Wheel height is $\frac{2}{3}$ of the truck height.
Variable	Let x represent the wheel height.
Equation	$x = \frac{2}{3} \cdot 15\frac{1}{2}$

$x \approx \frac{2}{3} \cdot 15$ Round $15\frac{1}{2}$ to 15, since 15 is divisible by 3.

$x \approx 10$ $\frac{1}{3}$ of 15 is 5, so $\frac{2}{3}$ of 15 is $2 \cdot 5$ or 10.

The wheels are about 10 feet high.

✓ **CHECK Your Progress**

k. **MEASUREMENT** The area of a rectangle is $19\frac{3}{4}$ square feet. The width of the rectangle is $5\frac{1}{4}$ feet. What is the approximate length of the rectangle?

Examples 1–5
(p. 230–231)

Estimate.

1. $8\frac{3}{8} + 1\frac{4}{5}$
2. $2\frac{5}{6} - 1\frac{1}{8}$
3. $5\frac{5}{7} \cdot 2\frac{7}{8}$
4. $9\frac{2}{7} \div 2\frac{2}{3}$

5. $\frac{1}{6} + \frac{2}{5}$
6. $\frac{6}{7} - \frac{1}{5}$
7. $\frac{5}{8} \cdot \frac{8}{9}$
8. $\frac{4}{5} \div \frac{6}{7}$

Examples 6, 7
(p. 232)

Estimate using compatible numbers.

9. $\frac{1}{4} \cdot 15$
10. $21\frac{5}{6} \div 9\frac{3}{4}$

Example 8
(p. 232)

11. **BIRDS** A seagull's wingspan is about $\frac{2}{3}$ of a bald eagle's wingspan. The eagle's wingspan is shown at the right. Estimate the wingspan of a seagull.

$\longmapsto 6\frac{2}{5}$ ft $\longrightarrow$

Practice and Problem Solving

HOMEWORK HELP	
For Exercises	**See Examples**
12–19	1, 2
20–29	3–5
30–35	6–8

Estimate.

12. $3\frac{3}{4} + 4\frac{5}{6}$
13. $1\frac{1}{8} + 5\frac{11}{12}$
14. $5\frac{1}{3} - 3\frac{1}{6}$
15. $4\frac{2}{5} - 1\frac{1}{2}$

16. $2\frac{2}{3} \cdot 6\frac{1}{3}$
17. $1\frac{4}{5} \cdot 3\frac{1}{4}$
18. $6\frac{1}{8} \div 1\frac{2}{3}$
19. $8\frac{1}{2} \div 2\frac{5}{8}$

20. $\frac{3}{4} + \frac{3}{8}$
21. $\frac{5}{8} + \frac{3}{7}$
22. $\frac{5}{9} - \frac{1}{6}$
23. $\frac{3}{4} - \frac{3}{5}$

24. $\frac{1}{8} \cdot \frac{3}{4}$
25. $\frac{4}{9} \cdot \frac{11}{12}$
26. $\frac{4}{5} \div \frac{7}{8}$
27. $\frac{1}{10} \div \frac{5}{6}$

28. **COOKING** Joaquim wants to make the macaroni and cheese shown at the right, but he has only about $1\frac{3}{4}$ cups of macaroni. About how much more macaroni does he need?

Macaroni & Cheese

3 tbsp butter
$2\frac{1}{2}$ c uncooked macaroni
1 tbsp salt
$\frac{1}{4}$ tbsp pepper
1 qt milk
$\frac{1}{2}$ lb cheese

29. **MEASUREMENT** Isabella is sewing a trim that is $1\frac{1}{8}$ inches wide on the bottom of a skirt that is $15\frac{7}{8}$ inches long. Approximately how long will the skirt be?

Estimate using compatible numbers.

30. $\frac{1}{4} \cdot 39$
31. $\frac{1}{6} \cdot 37$
32. $23\frac{2}{9} \div 3$
33. $25\frac{3}{10} \div 5\frac{2}{3}$

34. **MONEY** Arleta has $22. She uses $\frac{1}{3}$ of her money to buy a pair of earrings. About how much money did she spend on the earrings?

35. **SNACKS** A cereal company has 24 pounds of granola to package in bags that contain $1\frac{3}{4}$ pounds of granola. About how many bags will they have?

36. FIND THE DATA Refer to the Data File on pages 16–19. Choose some data and write a real-world problem in which you would estimate with fractions.

37. SPORTS Paquito and Jeff are on a basketball team. The table shows the approximate fraction of the team's points that each of them scored in a game. If the team scored a total of 72 points, about how many did Paquito and Jeff score together?

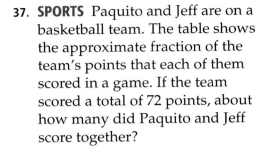

Player's Names	Fraction of Total Points Scored
Paquito	$\frac{3}{8}$
Jeff	$\frac{1}{6}$

38. RESEARCH Research the statistics of any basketball team. How can you use fractions to analyze the statistics?

39. COOKING Kathryn baked the sheet of brownies shown. She wants to cut it into brownies that are about 2 inches square. How many brownies will there be?

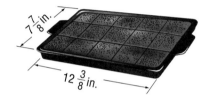

$7\frac{7}{8}$ in.

$12\frac{3}{8}$ in.

ANALYZE TABLES For Exercises 40–43, use the following information and the table shown.

The adult human skeleton is made up of 206 bones. The table shows the approximate fraction of the bones that each body part(s) makes up.

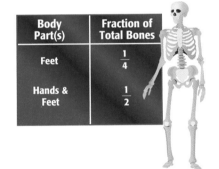

Body Part(s)	Fraction of Total Bones
Feet	$\frac{1}{4}$
Hands & Feet	$\frac{1}{2}$

40. About how many bones are in the feet?

41. About how many bones are in both hands and feet?

42. About how many bones are in one hand?

43. The length of your thighbone is equal to $\frac{1}{4}$ of your height. About how many inches long is your thighbone?

NYSCC • NYSMT

Extra Practice, pp. 679, 708

Real-World Link . . . . .
The femur or thigh bone is the longest in length, largest in volume, and strongest bone of the human body.

H.O.T. Problems

44. CHALLENGE In a division expression, the divisor is rounded up and the dividend is rounded down. How does the new quotient compare to the original quotient? Explain.

45. OPEN ENDED Select two fractions whose estimated difference and product is $\frac{1}{2}$. Justify your selection.

46. NUMBER SENSE Decide which of the following have sums that are less than 1. Explain.

a. $\frac{1}{3} + \frac{2}{5}$

b. $\frac{7}{8} + \frac{1}{2}$

c. $\frac{5}{6} + \frac{2}{3}$

d. $\frac{1}{7} + \frac{3}{9}$

47. **SELECT A TECHNIQUE** To make the crust for a peach cobbler, Dion needs $3\frac{1}{4}$ cups of flour, $1\frac{2}{3}$ cups of sugar, and $1\frac{2}{3}$ cups of hot water. He needs to mix all of these in a large bowl. The largest bowl he can find holds 6 cups. Which of the following techniques might Dion use to determine whether he can use this bowl to mix the ingredients? Justify your selection(s). Then use the technique(s) to solve the problem.

mental math	number sense	estimation

48. **WRITING IN MATH** Explain when estimation would *not* be the best method for solving a problem. Then give an example.

49. **SHORT RESPONSE** A chef has $15\frac{2}{3}$ cups of penne pasta and $22\frac{1}{4}$ cups of rigatoni pasta. About how much pasta is there altogether?

50. On a full tank of gasoline, a certain car can travel 360 miles. The needle on its gasoline gauge is shown. Without refueling, which is the best estimate of how far the car can travel?

   A 150 miles
   B 180 miles
   C 240 miles
   D 329 miles

Spiral Review

Replace each ● with <, >, or = to make a true sentence. (Lesson 4-9)

51. $2\frac{7}{8}$ ● 2.75

52. $\frac{-1}{3}$ ● $\frac{-7}{3}$

53. $\frac{5}{7}$ ● $\frac{4}{5}$

54. $3\frac{6}{11}$ ● $3\frac{9}{14}$

55. **SHOPPING** A store sells a 3-pack of beaded necklaces and a 5-pack of beaded bracelets. How many packages of each must you buy so that you have the same number of necklaces and bracelets? (Lesson 4-8)

Write each decimal as a percent. (Lesson 4-7)

56. 0.56

57. 0.375

58. 0.07

59. 0.019

GET READY for the Next Lesson

PREREQUISITE SKILL Find the LCD of each pair of fractions. (Lesson 4-9)

60. $\frac{3}{4}, \frac{5}{12}$

61. $\frac{1}{2}, \frac{7}{10}$

62. $\frac{1}{6}, \frac{1}{8}$

63. $\frac{4}{5}, \frac{2}{3}$

Adding and Subtracting Fractions

MAIN IDEA

Add and subtract fractions.

NYS Core Curriculum

Reinforcement of 6.N.16 Add and subtract fractions with unlike denominators *Also addresses 7.CM.5, 7.CM.8*

New Vocabulary

like fractions
unlike fractions

NY Math Online

glencoe.com

• Extra Examples
• Personal Tutor
• Self-Check Quiz
• Reading in the Content Area

▷ **GET READY** for the Lesson

INSTANT MESSENGER Sean surveyed ten classmates to find which abbreviation they use most when they instant message.

1. What fraction uses L8R? BRB?

2. What fraction uses either L8R or BRB?

Abbreviation	Number
L8R	5
LOL	3
BRB	2

Fractions that have the same denominators are called **like fractions**.

Add and Subtract Like Fractions   Key Concept

Words   To add or subtract like fractions, add or subtract the numerators and write the result over the denominator.

Examples   **Numbers**   **Algebra**

$$\frac{5}{10} + \frac{2}{10} = \frac{5+2}{10} \text{ or } \frac{7}{10}$$ $$\frac{a}{c} + \frac{b}{c} = \frac{a+b}{c}, \text{ where } c \neq 0$$

$$\frac{11}{12} - \frac{4}{12} = \frac{11-4}{12} \text{ or } \frac{7}{12}$$ $$\frac{a}{c} - \frac{b}{c} = \frac{a-b}{c}, \text{ where } c \neq 0$$

EXAMPLES   Add and Subtract Like Fractions

1 Add $\frac{5}{9} + \frac{2}{9}$. Write in simplest form.

$$\frac{5}{9} + \frac{2}{9} = \frac{5+2}{9}$$ Add the numerators.

$$= \frac{7}{9}$$ Write the sum over the denominator.

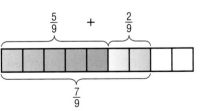

2 Subtract $\frac{9}{10} - \frac{1}{10}$. Write in simplest form.

$$\frac{9}{10} - \frac{1}{10} = \frac{9-1}{10}$$ Subtract the numerators.

$$= \frac{8}{10}$$ Write the difference over the denominator.

$$= \frac{4}{5}$$ Simplify.

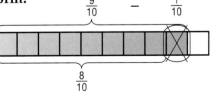

✓ **CHECK** Your Progress

a. $\frac{1}{6} + \frac{3}{6}$

b. $\frac{3}{7} - \frac{1}{7}$

To add or subtract **unlike fractions**, or fractions with different denominators, rename the fractions using the LCD. Then add or subtract as with like fractions.

EXAMPLES Add and Subtract Unlike Fractions

3 Add $\frac{1}{2} + \frac{1}{6}$. Write in simplest form.   Estimate $\frac{1}{2} + 0 = \frac{1}{2}$

METHOD 1 Use a model.

$$\frac{1}{2}$$
$$+\frac{1}{6}$$
$$\overline{\frac{4}{6} \text{ or } \frac{2}{3}}$$

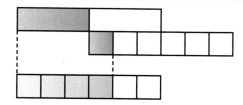

Study Tip

Renaming Fractions
To rename a fraction, multiply both the numerator and the denominator of the original fraction by the same number. By doing so, the renamed fraction has the same value as the original fraction.

METHOD 2 Use the LCD.

The least common denominator (LCD) of $\frac{1}{2}$ and $\frac{1}{6}$ is 6.

Rename using the LCD, 6.   Add.

$$\frac{1}{2} \rightarrow \frac{1 \times 3}{2 \times 3} = \frac{3}{6} \rightarrow \frac{3}{6}$$
$$+\frac{1}{6} \rightarrow \frac{1 \times 1}{6 \times 1} = +\frac{1}{6} \rightarrow +\frac{1}{6}$$
$$\overline{\frac{4}{6} \text{ or } \frac{2}{3}}$$

So, $\frac{1}{2} + \frac{1}{6} = \frac{2}{3}$. **Check for Reasonableness** $\frac{2}{3} \approx \frac{1}{2}$ ✔

4 Subtract $\frac{11}{12} - \frac{3}{8}$. Write in simplest form. Estimate $1 - \frac{1}{2} = \frac{1}{2}$

Since $12 = 2^2 \cdot 3$ and $8 = 2^3$, the LCM of 12 and 8 is $2^3 \cdot 3$ or 24. Rename each fraction using a denominator of 24. Then subtract.

Think: $12 \times 2 = 24$, so $\frac{11 \times 2}{12 \times 2}$ or $\frac{22}{24}$.

Think: $8 \times 3 = 24$, so $\frac{3}{8} = \frac{3 \times 3}{8 \times 3}$ or $\frac{9}{24}$.

$$\frac{11}{12} - \frac{3}{8} = \frac{11 \times 2}{12 \times 2} - \frac{3 \times 3}{8 \times 3}.$$ The LCD of $\frac{11}{12}$ and $\frac{3}{8}$ is 24.

$$= \frac{22}{24} - \frac{9}{24}$$ Rename the fractions using LCD, 24.

$$= \frac{13}{24}$$ Subtract the fractions.

Check for Reasonableness $\frac{13}{24} \approx \frac{1}{2}$ ✔

✓ CHOOSE Your Method

c. $\frac{8}{9} - \frac{2}{3}$ d. $\frac{5}{6} - \frac{3}{8}$ e. $\frac{7}{8} + \frac{3}{4}$

Healthy Lunch

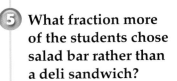

SURVEYS In a recent survey, students were asked what they would choose for a healthy lunch. The results are shown in the graph.

⑤ What fraction more of the students chose salad bar rather than a deli sandwich?

$$\frac{9}{20} - \frac{3}{25} = \frac{9 \times 5}{20 \times 5} - \frac{3 \times 4}{25 \times 4}$$ The LCD of $\frac{9}{20}$ and $\frac{3}{25}$ is 100.

$$= \frac{45}{100} - \frac{12}{100}$$ Rename the fractions using the LCD.

$$= \frac{33}{100}$$ Subtract the numerators.

So, $\frac{33}{100}$ more students chose salad bar rather than a deli sandwich.

⑥ What fraction of students chose pizza or chicken nuggets?

$$\frac{3}{20} + \frac{1}{10} = \frac{3}{20} + \frac{2}{20}$$ Rename.

$$= \frac{5}{20}$$ Add.

$$= \frac{1}{4}$$ Simplify.

So, $\frac{1}{4}$ of the students chose pizza or chicken nuggets combined.

 CHECK Your Progress

f. SURVEYS What fraction more of the students chose a deli sandwich rather than a burger and fries?

CHECK Your Understanding

Examples 1–4
(pp. 236–237)

Add or subtract. Write in simplest form.

1. $\frac{4}{9} + \frac{2}{9}$ 2. $\frac{5}{6} + \frac{4}{9}$ 3. $\frac{3}{8} - \frac{1}{8}$ 4. $\frac{4}{5} - \frac{2}{5}$

5. $\frac{1}{6} + \frac{3}{8}$ 6. $\frac{2}{3} + \frac{5}{6}$ 7. $\frac{5}{6} - \frac{7}{12}$ 8. $\frac{3}{4} - \frac{1}{3}$

Examples 5, 6
(p. 238)

For Exercises 9 and 10, choose an operation to solve each problem. Explain your reasoning. Then solve the problem.

9. **MEASUREMENT** Cassandra cuts $\frac{5}{16}$ inch off the top of a photo and $\frac{3}{8}$ inch off the bottom. How much smaller is the total height of the photo now?

10. **CHORES** A bucket was $\frac{7}{8}$ full with soapy water. After washing the car, the bucket was only $\frac{1}{4}$ full. What part of the water was used?

HOMEWORK HELP	
For Exercises	See Examples
11–14	1, 2
15–22	3, 4
23–26	5, 6

Add or subtract. Write in simplest form.

11. $\dfrac{3}{7} + \dfrac{1}{7}$

12. $\dfrac{5}{8} + \dfrac{7}{8}$

13. $\dfrac{5}{6} - \dfrac{1}{6}$

14. $\dfrac{7}{10} - \dfrac{3}{10}$

15. $\dfrac{1}{15} + \dfrac{3}{5}$

16. $\dfrac{7}{12} + \dfrac{7}{10}$

17. $\dfrac{5}{8} + \dfrac{11}{12}$

18. $\dfrac{7}{9} + \dfrac{5}{6}$

19. $\dfrac{7}{9} - \dfrac{1}{3}$

20. $\dfrac{4}{5} - \dfrac{1}{6}$

21. $\dfrac{4}{9} - \dfrac{2}{15}$

22. $\dfrac{3}{10} - \dfrac{1}{4}$

For Exercises 23–26, choose an operation to solve each problem. Explain your reasoning. Then solve the problem.

23. **MEASUREMENT** Ebony is building a shelf to hold the two boxes shown. What is the smallest width she should make the shelf?

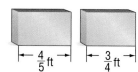

24. **WEATHER** Using the information under the photo, find the difference of the average precipitation for Boise in February and November.

25. **MEASUREMENT** Makayla bought $\dfrac{1}{4}$ pound of ham and $\dfrac{5}{8}$ pound of turkey. How much more turkey did she buy?

26. **ANIMALS** The three-toed sloth can travel $\dfrac{3}{20}$ miles per hour while a giant tortoise can travel $\dfrac{17}{100}$ miles per hour. How much faster, in miles per hour, is the giant tortoise?

Real-World Link . . . . .
The average precipitation for February and November for Boise, Idaho, is $\dfrac{4}{10}$ and $\dfrac{7}{10}$ inches, respectively.
Source: The Weather Channel

Simplify.

27. $\dfrac{1}{7} + \dfrac{1}{2} + \dfrac{5}{28}$

28. $\dfrac{1}{4} + \dfrac{5}{6} + \dfrac{7}{12}$

29. $\dfrac{1}{6} + \left(\dfrac{2}{3} - \dfrac{1}{4}\right)$

30. $\dfrac{5}{6} - \left(\dfrac{1}{2} + \dfrac{1}{3}\right)$

31. $1 + \dfrac{1}{4}$

32. $1 - \dfrac{5}{8}$

33. $2 + \dfrac{2}{3}$

34. $3 - \dfrac{1}{6}$

35. **MONEY** Chellise saves $\dfrac{1}{5}$ of her allowance and spends $\dfrac{2}{3}$ of her allowance at the mall. What fraction of her allowance remains?

36. **ANALYZE TABLES** Pepita and Francisco each spend an equal amount of time on homework. The table shows the fraction of their time they spend on each subject. Determine the missing fraction for each student.

Homework	Fraction of Time	
	Pepita	Francisco
Math	▨	$\dfrac{1}{2}$
English	$\dfrac{2}{3}$	▨
Science	$\dfrac{1}{6}$	$\dfrac{3}{8}$

ALGEBRA Evaluate each expression if $a = \dfrac{3}{4}$ and $b = \dfrac{5}{6}$.

37. $\dfrac{1}{2} + a$

38. $b - \dfrac{7}{10}$

39. $b - a$

40. $a + b$

41. **BOOK REPORTS** Four students were scheduled to give book reports in a 1-hour class period. After the first report, $\frac{2}{3}$ hour remained. If the next two students' reports took $\frac{1}{6}$ hour and $\frac{1}{4}$ hour, respectively, what fraction of the hour remained after the final students' report? Justify your answer.

42. **MEASUREMENT** Mrs. Escalante was riding a bicycle on a bike path. After riding $\frac{2}{3}$ of a mile, she discovered that she still needed to travel $\frac{3}{4}$ of a mile to reach the end of the path. How long is the bike path?

43. **CELL PHONES** One hundred sixty cell phone owners were surveyed. What fraction of owners prefers using their cell phone for text messaging or taking pictures?

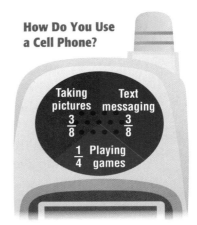

How Do You Use a Cell Phone?

Taking pictures $\frac{3}{8}$ Text messaging $\frac{3}{8}$

$\frac{1}{4}$ Playing games

44. **MEASUREMENT** LaTasha and Eric are jogging on a track. LaTasha jogs $\frac{1}{4}$ of a mile and then stops. Eric jogs $\frac{5}{8}$ of a mile, stops and then turns around and jogs $\frac{1}{2}$ of a mile. Who is farther ahead on the track? How much farther?

NYSCC • NYSMT
Extra Practice, pp. 679, 708.

H.O.T. Problems

45. **CHALLENGE** Fractions, such as $\frac{1}{2}$ or $\frac{1}{3}$, whose numerators are 1, are called *unit fractions*. Describe a method you can use to add two unit fractions mentally. Explain your reasoning and use your method to find $\frac{1}{99} + \frac{1}{100}$.

46. **OPEN ENDED** Provide a counterexample to the following statement.

    The sum of three fractions with odd numerators is never $\frac{1}{2}$.

47. **FIND THE ERROR** Meagan and Lourdes are finding $\frac{1}{4} + \frac{3}{5}$. Who is correct? Explain.

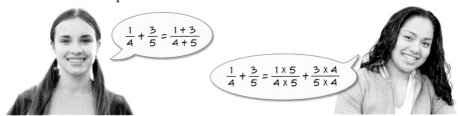

$\frac{1}{4} + \frac{3}{5} = \frac{1+3}{4+5}$

$\frac{1}{4} + \frac{3}{5} = \frac{1 \times 5}{4 \times 5} + \frac{3 \times 4}{5 \times 4}$

Meagan

Lourdes

48. **WRITING IN MATH** To make a cake, Felicia needs 1 cup of flour but she only has a $\frac{2}{3}$-measuring cup and a $\frac{3}{4}$-measuring cup. Which method will bring her closest to having the amount of flour she needs? Explain.

    a. Fill the $\frac{2}{3}$-measuring cup twice.   b. Fill the $\frac{2}{3}$-measuring cup once.

    c. Fill the $\frac{3}{4}$-measuring cup twice.   d. Fill the $\frac{3}{4}$-measuring cup once.

49. The table gives the number of hours Orlando spent at football practice for one week.

Day	Time (hours)
Monday	$1\frac{1}{2}$
Tuesday	2
Wednesday	$2\frac{1}{3}$
Thursday	$1\frac{5}{6}$
Friday	$2\frac{1}{2}$
Saturday	$1\frac{3}{4}$

How many more hours did he practice over the last three days than he did over the first three days?

A $\frac{1}{4}$ h

B $\frac{1}{2}$ h

C $\frac{2}{3}$ h

D $\frac{3}{4}$ h

50. Which of the following is the prime factored form of the lowest common denominator of $\frac{7}{12} + \frac{11}{18}$?

F 2×3

G 2×3^2

H $2^2 \times 3^2$

J $2^3 \times 3$

51. Find $\frac{5}{6} - \frac{1}{8}$.

A $\frac{4}{7}$

B $\frac{3}{8}$

C $\frac{7}{12}$

D $\frac{17}{24}$

Spiral Review

Estimate. (Lesson 5-1)

52. $\frac{6}{7} - \frac{5}{12}$

53. $4\frac{1}{9} + 3\frac{3}{4}$

54. $16\frac{2}{3} \div 8\frac{1}{5}$

55. $5\frac{4}{5} \cdot 3\frac{1}{3}$

56. WEATHER The table shows about how much rain falls in Albuquerque and Denver. Which city has the greater fraction of inches of rain per day? Explain. (Lesson 4-9)

City	Amount of Rain (in.)	Number of Days
Albuquerque, NM	9	60
Denver, CO	15	90

Source: The Weather Channel

57. Write 0.248 as a percent. (Lesson 4-7)

ALGEBRA Find each sum if $a = -3$ and $b = 2$. (Lessons 2-4 and 2-5)

58. $a + b$

59. $a - b$

60. $b - a$

▷ **GET READY for the Next Lesson**

PREREQUISITE SKILL Complete.

61. $5\frac{2}{3} = 5 + \blacksquare$

62. $1 = \frac{\blacksquare}{9}$

63. $1 = \frac{\blacksquare}{5}$

64. $\blacksquare = 4 + \frac{3}{8}$

Adding and Subtracting Mixed Numbers

MAIN IDEA

Add and subtract mixed numbers.

NYS Core Curriculum

Reinforcement of 6.N.18 Add, subtract, multiply, and divide mixed numbers with unlike denominators *Also addresses 7.CN.8*

NY Math Online

glencoe.com

- Extra Examples
- Personal Tutor
- Self-Check Quiz

▷ **GET READY** for the Lesson

BABIES The birth weights of several babies in the hospital nursery are shown.

Birth Weight (pounds)	
Jackson	$8\frac{1}{8}$
Nicolás	$7\frac{15}{16}$
Rebekah	$6\frac{13}{16}$
Mia	$5\frac{7}{8}$

1. Write an expression to find how much more Nicolás weighs than Mia.
2. Rename the fractions using the LCD.
3. Find the difference of the fractional parts of the mixed numbers.
4. Find the difference of the whole numbers.
5. **MAKE A CONJECTURE** Explain how to find $7\frac{15}{16} - 5\frac{7}{8}$. Then use your conjecture to find the difference.

To add or subtract mixed numbers, first add or subtract the fractions. If necessary, rename them using the LCD. Then add or subtract the whole numbers and simplify if necessary.

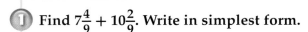 **EXAMPLES** Add and Subtract Mixed Numbers

① Find $7\frac{4}{9} + 10\frac{2}{9}$. Write in simplest form.

Estimate $7 + 10 = 17$

$$7\frac{4}{9}$$

Add the whole numbers and fractions separately.

$$+ 10\frac{2}{9}$$

$$17\frac{6}{9} \text{ or } 17\frac{2}{3}$$ Simplify.

Check for Reasonableness $17\frac{2}{3} \approx 17$ ✔

✓ **CHECK Your Progress**

a. $6\frac{1}{8} + 2\frac{5}{8}$ b. $5\frac{1}{5} + 2\frac{3}{10}$ c. $1\frac{5}{9} + 4\frac{1}{6}$

2 Find $8\frac{5}{6} - 2\frac{1}{3}$. Write in simplest form.

Estimate $9 - 2 = 7$

$$
\begin{array}{rcl}
8\frac{5}{6} & \rightarrow & 8\frac{5}{6} \\
-2\frac{1}{3} & \rightarrow & -2\frac{2}{6} \\
\hline
& & 6\frac{3}{6} \text{ or } 6\frac{1}{2}
\end{array}
$$

Rename the fraction using the LCD. Then subtract.

Simplify.

Check for Reasonableness $6\frac{1}{2} \approx 7$ ✔

CHECK Your Progress

Subtract. Write in simplest form.

d. $5\frac{4}{5} - 1\frac{3}{10}$ e. $13\frac{7}{8} - 9\frac{3}{4}$ f. $8\frac{2}{3} - 2\frac{1}{2}$

g. $7\frac{3}{4} - 4\frac{1}{3}$ h. $11\frac{5}{6} - 3\frac{1}{8}$ i. $9\frac{4}{7} - 5\frac{1}{2}$

Sometimes when you subtract mixed numbers, the fraction in the first mixed number is less than the fraction in the second mixed number. In this case, rename the first fraction as an improper fraction in order to subtract.

EXAMPLES **Rename Mixed Numbers to Subtract**

3 Find $2\frac{1}{3} - 1\frac{2}{3}$.

Estimate $2 - 1\frac{1}{2} = \frac{1}{2}$

Since $\frac{1}{3}$ is less than $\frac{2}{3}$, rename $2\frac{1}{3}$ before subtracting.

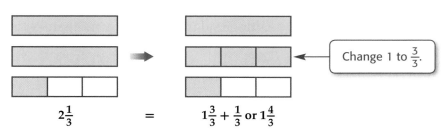

Change 1 to $\frac{3}{3}$.

$2\frac{1}{3}$ = $1\frac{3}{3} + \frac{1}{3}$ or $1\frac{4}{3}$

$$
\begin{array}{rcl}
2\frac{1}{3} & \rightarrow & 1\frac{4}{3} \\
-1\frac{2}{3} & \rightarrow & -1\frac{2}{3} \\
\hline
& & \frac{2}{3}
\end{array}
$$

Rename $2\frac{1}{3}$ as $1\frac{4}{3}$.

Subtract the whole numbers and then the fractions.

Check for Reasonableness $\frac{2}{3} \approx \frac{1}{2}$ ✔

4 Find $8 - 3\frac{3}{4}$. **Estimate** $8 - 4 = 4$

Using the denominator of the fraction in the subtrahend, $8 = 8\frac{0}{4}$.

Since $\frac{0}{4}$ is less than $\frac{3}{4}$, rename 8 before subtracting.

$$
\begin{array}{rl}
8 \quad \rightarrow & 7\frac{4}{4} \qquad \text{Rename 8 as } 7 + \frac{4}{4} \text{ or } 7\frac{4}{4}. \\
\underline{-3\frac{3}{4}} \quad \rightarrow & \underline{-3\frac{3}{4}} \qquad \text{Subtract.} \\
& 4\frac{1}{4} \qquad \text{Check for Reasonableness } 4\frac{1}{4} \approx 4 \; \checkmark
\end{array}
$$

CHECK Your Progress

j. $11\frac{2}{5} - 2\frac{3}{5}$ **k.** $5\frac{3}{8} - 4\frac{11}{12}$ **l.** $7 - 1\frac{1}{2}$

Real-World Career . . . .
How Does an Urban Planner Use Math?
An urban planner uses math to measure and draw site plans for future development.

NY Math Online

For more information, go to glencoe.com.

Real-World EXAMPLE

5 **MEASUREMENT** An urban planner is designing a skateboard park. What will be the length of the park and the parking lot combined?

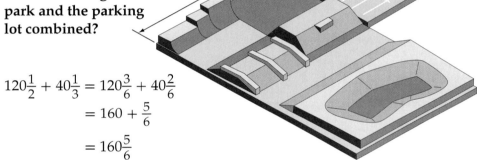

$40\frac{1}{3}$ ft

$120\frac{1}{2}$ ft

$$
\begin{aligned}
120\frac{1}{2} + 40\frac{1}{3} &= 120\frac{3}{6} + 40\frac{2}{6} \\
&= 160 + \frac{5}{6} \\
&= 160\frac{5}{6}
\end{aligned}
$$

The total length is $160\frac{5}{6}$ feet.

CHECK Your Progress

m. MEASUREMENT Jermaine walked $1\frac{5}{8}$ miles on Saturday and $2\frac{1}{2}$ miles on Sunday. How many more miles did he walk on Sunday?

CHECK Your Understanding

Examples 1–4
(pp. 242–244)

Add or subtract. Write in simplest form.

1. $1\frac{5}{7} + 8\frac{1}{7}$ **2.** $8\frac{1}{2} + 3\frac{4}{5}$ **3.** $7\frac{5}{6} - 3\frac{1}{6}$ **4.** $9\frac{4}{5} - 2\frac{3}{4}$

5. $3\frac{1}{4} - 1\frac{3}{4}$ **6.** $5\frac{2}{3} - 2\frac{3}{5}$ **7.** $11 - 6\frac{3}{8}$ **8.** $16 - 5\frac{5}{6}$

Example 5
(p. 244)

9. CARS A hybrid car's gas tank can hold $11\frac{9}{10}$ gallons of gasoline. It contains $8\frac{3}{4}$ gallons of gasoline. How much more gasoline is needed to fill the tank?

HOMEWORK HELP

For Exercises	See Examples
10–17	1, 2
18–23	3
24–25	4
26–29	5

Add or subtract. Write in simplest form.

10. $2\frac{1}{9} + 7\frac{4}{9}$

11. $3\frac{2}{7} + 4\frac{3}{7}$

12. $10\frac{4}{5} - 2\frac{1}{5}$

13. $8\frac{6}{7} - 6\frac{5}{7}$

14. $9\frac{4}{5} - 2\frac{3}{10}$

15. $11\frac{3}{4} - 4\frac{1}{3}$

16. $8\frac{5}{12} + 11\frac{1}{4}$

17. $8\frac{3}{8} + 10\frac{1}{3}$

18. $9\frac{1}{5} - 2\frac{3}{5}$

19. $6\frac{1}{4} - 2\frac{3}{4}$

20. $6\frac{3}{5} - 1\frac{2}{3}$

21. $4\frac{3}{10} - 1\frac{3}{4}$

22. $14\frac{1}{6} - 7\frac{1}{3}$

23. $12\frac{1}{2} - 6\frac{5}{8}$

24. $8 - 3\frac{2}{3}$

25. $13 - 5\frac{5}{6}$

For Exercises 26–29, choose an operation to solve each problem. Explain your reasoning. Then solve the problem.

26. **HIKING** If Sara and Maggie hiked both of the trails listed in the table, how far did they hike altogether?

Trail	Length (mi)
Woodland Park	$3\frac{2}{3}$
Mill Creek Way	$2\frac{5}{6}$

27. **JEWELRY** Margarite made the jewelry shown at the right. If the necklace is $10\frac{5}{8}$ inches longer than the bracelet, how long is the necklace that Margarite made?

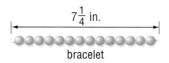

$7\frac{1}{4}$ in.

bracelet

necklace

28. **GARDENS** The length of Kasey's garden is $4\frac{5}{8}$ feet. Find the width of Kasey's garden if it is $2\frac{7}{8}$ feet shorter than the length.

29. **HAIRSTYLES** Before Alameda got her haircut, the length of her hair was $9\frac{3}{4}$ inches. After her haircut, the length was $6\frac{1}{2}$ inches. How many inches did she have cut?

Add or subtract. Write in simplest form.

30. $10 - 3\frac{5}{11}$

31. $24 - 8\frac{3}{4}$

32. $6\frac{1}{6} + 1\frac{2}{3} + 5\frac{5}{9}$

33. $3\frac{1}{4} + 2\frac{5}{6} - 4\frac{1}{3}$

34. **TIME** Karen wakes up at 6:00 A.M. It takes her $1\frac{1}{4}$ hours to shower, get dressed, and comb her hair. It takes her $\frac{1}{2}$ hour to eat breakfast, brush her teeth, and make her bed. At what time will she be ready for school?

MEASUREMENT Find the perimeter of each figure.

35.

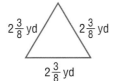

$2\frac{3}{8}$ yd $2\frac{3}{8}$ yd

$2\frac{3}{8}$ yd

36.

$5\frac{1}{3}$ in.

$3\frac{1}{6}$ in. $4\frac{2}{3}$ in.

$4\frac{5}{6}$ in.

NYSCC • NYSMT
Extra Practice, pp. 680, 708

37. **NUMBER SENSE** Which of the following techniques could be used to determine whether $6\frac{3}{4} + \frac{4}{5}$ is *greater than, less than,* or *equal to* $2\frac{1}{9} + 6\frac{7}{8}$? Justify your selection(s). Then use the technique(s) to solve the problem.

number sense	mental math	estimation

38. **CHALLENGE** A string is cut in half. One of the halves is thrown away. One fifth of the remaining half is cut away and the piece left is 8 feet long. How long was the string initially? Justify your answer.

39. **WRITING IN MATH** The fence of a rectangular garden is constructed from 12 feet of fencing wire. Suppose that one side of the garden is $2\frac{5}{12}$ feet long. Explain how to find the length of the other side.

NYSMT PRACTICE 6.N.18

40. The distance from home plate to the pitcher's mound is 60 feet 6 inches and from home plate to second base is 127 feet $3\frac{3}{8}$ inches. Find the distance from the pitcher's mound to second base.

   A 68 ft $3\frac{1}{4}$ in.

   B 67 ft $8\frac{3}{4}$ in.

   C 67 ft $2\frac{5}{8}$ in.

   D 66 ft $9\frac{3}{8}$ in.

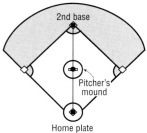

41. A recipe for party mix calls for $4\frac{3}{4}$ cups of cereal. The amount of peanuts needed is $1\frac{2}{3}$ cups less than the amount of cereal needed. How many cups of peanuts and cereal are needed?

   F $3\frac{1}{12}$ cups

   G $6\frac{1}{2}$ cups

   H $7\frac{5}{6}$ cups

   J $8\frac{1}{2}$ cups

Spiral Review

42. **SCHOOL** Kai did $\frac{1}{5}$ of her homework in class and $\frac{1}{3}$ more of it on the bus. What fraction of homework does she still need to do? (Lesson 5-2)

Estimate. (Lesson 5-1)

43. $\frac{8}{9} \div \frac{9}{10}$

44. $3\frac{1}{2} + 6\frac{2}{3}$

45. $8\frac{4}{5} \times 7\frac{1}{9}$

46. $4\frac{2}{9} - 1\frac{1}{4}$

47. **MEASUREMENT** To carpet a living room with a length of 17 feet, 255 square feet of carpet is needed. Find the width of the living room. (Lesson 3-6)

▷ **GET READY for the Next Lesson**

48. **PREREQUISITE SKILL** Andre needs to be at the train station by 5:30 P.M. It takes him $\frac{1}{3}$ hour to pack and $1\frac{1}{4}$ hours to get to the station. Find the latest time he should begin packing. Use the *work backward* strategy. (Lesson 3-4)

1. **MONEY** Latisha spends $\frac{3}{4}$ of her money on a birthday present for her brother. If she has $33, estimate the amount she spends on her brother's present. (Lesson 5-1)

Estimate. (Lesson 5-1)

2. $5\frac{1}{9} + 1\frac{7}{8}$

3. $13\frac{1}{2} \div 7\frac{2}{9}$

4. $\frac{11}{20} - \frac{5}{8}$

5. $4\frac{2}{3} \times 1\frac{3}{4}$

6. $7\frac{3}{4} \div 1\frac{4}{5}$

7. $\frac{8}{9} + 2\frac{13}{15}$

8. **MULTIPLE CHOICE** Mrs. Ortega is making 5 batches of muffins for the school bake sale. Each batch uses $2\frac{1}{4}$ cups sugar and $1\frac{1}{2}$ cups milk. Which is the best estimate of the total amount of sugar and milk Mrs. Ortega uses for the muffins? (Lesson 5-1)

   A less than 15 cups

   B between 15 and 20 cups

   C between 20 and 25 cups

   D more than 25 cups

Add or subtract. Write in simplest form.
(Lesson 5-2)

9. $\frac{11}{15} - \frac{1}{15}$

10. $\frac{4}{7} - \frac{3}{14}$

11. $\frac{1}{2} + \frac{2}{9}$

12. $\frac{5}{8} + \frac{3}{4}$

13. **SCIENCE** $\frac{39}{50}$ of Earth's atmosphere is made up of nitrogen while only $\frac{21}{100}$ is made up of oxygen. What fraction of Earth's atmosphere is either nitrogen or oxygen? (Lesson 5-2)

Add or subtract. Write in simplest form.
(Lesson 5-3)

14. $8\frac{3}{4} - 2\frac{5}{12}$

15. $5\frac{1}{6} - 1\frac{1}{3}$

16. $2\frac{5}{9} + 1\frac{2}{3}$

17. $2\frac{3}{5} + 6\frac{13}{15}$

18. **MULTIPLE CHOICE** The table shows the weight of a newborn infant for the first year. (Lesson 5-3)

Month	Weight (lb)
0	$7\frac{1}{4}$
3	$12\frac{1}{2}$
6	$16\frac{5}{8}$
9	$19\frac{4}{5}$
12	$23\frac{3}{20}$

During which three-month period was the infant's weight gain the greatest?

F 0–3 months   H 6–9 months

G 3–6 months   J 9–12 months

19. **MEASUREMENT** How much does a $50\frac{1}{4}$-pound suitcase weigh after $3\frac{7}{8}$ pounds is removed? (Lesson 5-3)

20. **MULTIPLE CHOICE** The table gives the average annual snowfall for several U.S. cities. (Lesson 5-3)

City	Average Snowfall (in.)
Anchorage, AK	$70\frac{4}{5}$
Mount Washington, NH	$259\frac{9}{10}$
Buffalo, NY	$93\frac{3}{5}$
Birmingham, AL	$1\frac{1}{2}$

Source: Fact Monster

On average, how many more inches of snow does Mount Washington, New Hampshire, receive than Anchorage, Alaska?

A $330\frac{7}{10}$ in.   C $166\frac{3}{10}$ in.

B $189\frac{1}{10}$ in.   D $92\frac{1}{10}$ in.

5-4 Problem-Solving Investigation

MAIN IDEA: Solve problems by eliminating possibilities.

NYSCC **7.PS.1** Use a variety of strategies to understand new mathematical content and to develop more efficient methods

P.S.I. TEAM +

e-Mail: ELIMINATE POSSIBILITIES

MADISON: I am making school pennants to decorate the cafeteria. I use $1\frac{1}{4}$ yards of fabric for each pennant.

YOUR MISSION: Eliminate possibilities to find the greatest number of pennants Madison can make with 12 yards of fabric. Is it 6, 9, or 12?

Understand	You know she has 12 yards of fabric. Each pennant uses $1\frac{1}{4}$ yards of fabric.
Plan	Eliminate the answers that are not reasonable.
Solve	Madison needs more than 1 yard of fabric for each pennant. So, she needs more than 12 yards for 12 pennants. Eliminate this choice. Now check the choice of 9 pennants. • $1\frac{1}{4} + 1\frac{1}{4} + 1\frac{1}{4} + 1\frac{1}{4} = 5$. So, Madison can make 4 pennants with 5 yards of fabric. Therefore, she can make 8 pennants out of 10 yards of fabric. • She can also make 1 more pennant with the remaining 2 yards. So, Madison can make $8 + 1$ or 9 pennants.
Check	Making 6 pennants would take $1\frac{1}{4} + 1\frac{1}{4} + 1\frac{1}{4} + 1\frac{1}{4} + 1\frac{1}{4} + 1\frac{1}{4} = 7\frac{1}{2}$ yards. This is not the greatest number she can make. So, making 6 pennants is *not* reasonable.

Analyze The Strategy

1. Describe different ways that you can eliminate possibilities when solving problems.

2. Explain how the strategy of eliminating possibilities is useful for taking multiple-choice tests.

3. **WRITING IN MATH** Write a problem that could be solved by eliminating possibilities.

Mixed Problem Solving

Eliminate possibilities to solve Exercises 4–6.

4. **TRAINS** A train passes through an intersection at the rate of 3 cars per 30 seconds. Assume that it takes 5 minutes for the train to completely pass through the intersection. How many cars does the train have altogether?

   A 6 cars   C 30 cars

   B 15 cars   D 45 cars

5. **PIZZA** A pizza shop used 100 pounds of pizza dough to make 125 pizzas. If a large pizza requires 1 pound of dough and a medium pizza requires $\frac{1}{2}$ pound, how many large- and medium-sized pizzas were made?

   F 40 large, 85 medium

   G 65 large, 60 medium

   H 55 large, 70 medium

   J 75 large, 50 medium

6. **PILLOWS** Pat is making pillows out of fabric. He uses $\frac{3}{4}$ yard of fabric for each pillow. What is the greatest number of pillows Pat can make with 9 yards of fabric: 9, 12, or 15?

Use any strategy to solve Exercises 7–14. Some strategies are shown below.

PROBLEM-SOLVING STRATEGIES
- Look for a pattern.
- Work backward.
- Make an organized list.

7. **MEASUREMENT** The diagram shows a shelf that holds CDs. Each shelf is $\frac{1}{8}$ inch thick, and the distance between shelves is as shown. How much space is available on each layer of the shelf for a CD?

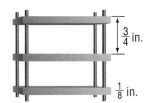

$\frac{3}{4}$ in.

$\frac{1}{8}$ in.

8. **BRIDGES** A covered bridge has a maximum capacity of 48,000 pounds. If an average school bus weighs 10,000 pounds, about how many school buses could a covered bridge hold?

9. **GEOMETRY** Draw the next two figures in the pattern.

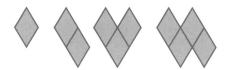

10. **SLEEP** A 9-month-old infant needs about 14 hours of sleep each day while a teenager needs about 10 hours of sleep each day. How much more sleep does a 9-month-old need than a teenager? Write as a fraction of a day.

11. **PRECIPITATION** In Olympia, Washington, the average annual precipitation is $50\frac{3}{5}$ inches. Is $\frac{1}{49}$ inch, 1 inch, or 14 inches the best estimate for the average precipitation per day?

12. **MONEY** Kristen has $15 to go to the movies. Her ticket costs $7.25, drinks are $3.50, popcorn is $5.75, and pretzels are $4.25. Which two items can Kristen get from the concession stand?

13. **PIZZA** Sebastian ate $\frac{2}{5}$ of a pizza while his sister ate $\frac{1}{3}$ of the same pizza. The remainder was stored in the refrigerator. What fraction of the pizza was stored in the refrigerator?

14. **GRADES** Jerome had an average of 88 on his first three science tests. His score on the second and third tests were 92 and 87. What was his score on the first test?

Math Lab
Multiplying Fractions

MAIN IDEA

Use area models to
multiply fractions and
mixed numbers.

NYS Core Curriculum

**Reinforcement of
6.N.17** Multiply and
divide fractions with
unlike denominators

NY Math Online

glencoe.com
• Concepts In Motion

Just as the product of 3×4 is the number of square units in a rectangle, the product of two fractions can be shown using area models.

ACTIVITY

1 Find $\frac{3}{4} \times \frac{2}{3}$ using a geoboard.

The first factor is 3 *fourths* and the second factor is 2 *thirds*.

STEP 1 Use one geoband to show fourths and another to show thirds on the geoboard.

STEP 2 Use geobands to form a rectangle. Place one geoband on the peg to show 3 fourths and another on the peg to show 2 thirds.

STEP 3 Connect the geobands to show a small rectangle.

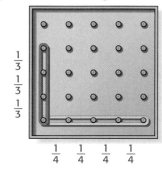 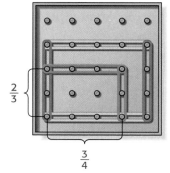

The area of the small square is 6 square units. The area of the large rectangle is 12 square units. So, $\frac{3}{4} \times \frac{2}{3} = \frac{6}{12}$ or $\frac{1}{2}$.

CHECK Your Progress

Find each product using a geoboard.

a. $\frac{1}{4} \times \frac{1}{3}$ b. $\frac{1}{2} \times \frac{1}{2}$ c. $\frac{3}{4} \times \frac{1}{2}$ d. $\frac{2}{3} \times \frac{1}{4}$

ACTIVITY

2 Find $2 \times \frac{1}{4}$ using an area model.

STEP 1 To represent 2 or $\frac{2}{1}$, draw 2 large rectangles, side by side. Divide each rectangle horizontally into fourths. Color both large rectangles blue.

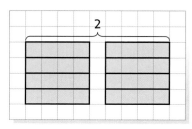

STEP 2 Color 1 fourth of each large rectangle yellow.

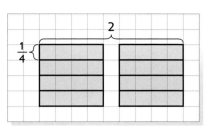

The fraction that compares the number of green sections, 2, to the number of sections in one rectangle, 4, is $\frac{2}{4}$ or $\frac{1}{2}$.

So, $2 \times \frac{1}{4} = \frac{1}{2}$.

✓ **CHECK Your Progress**

Find each product using a model.

e. $3 \times \frac{2}{3}$ f. $2 \times \frac{2}{5}$ g. $4 \times \frac{1}{2}$ h. $3 \times \frac{3}{4}$

ACTIVITY

3 Find $1\frac{2}{3} \times \frac{1}{2}$ using a model.

STEP 1 Draw 2 rectangles divided vertically into thirds and horizontally into halves. Color $1\frac{2}{3}$ of the squares blue.

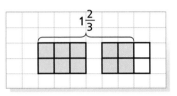

STEP 2 Color $\frac{1}{2}$ of the squares yellow. Then count the small squares that are green.

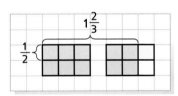

Since the green area is $\frac{3}{6}$ of the first rectangle and $\frac{2}{6}$ of the second rectangle, the total area shaded green is $\frac{3}{6} + \frac{2}{6}$ or $\frac{5}{6}$. So, $1\frac{2}{3} \times \frac{1}{2} = \frac{5}{6}$.

✓ **CHECK Your Progress**

Find each product using a model.

i. $1\frac{1}{4} \times \frac{1}{5}$ j. $2\frac{1}{2} \times \frac{3}{4}$ k. $1\frac{2}{3} \times \frac{1}{3}$

ANALYZE THE RESULTS

1. Analyze Exercises a–k. What is the relationship between the numerators of the factors and of the product? between the denominators of the factors and of the product?

2. **MAKE A CONJECTURE** Write a rule you can use to multiply two fractions.

Explore 5-5 Math Lab: Multiplying Fractions **251**

Study Tip

Shading
Yellow and blue make green. So, the green sections have been shaded twice and represent the product.

Multiplying Fractions and Mixed Numbers

MAIN IDEA

Multiply fractions and mixed numbers.

NYS Core Curriculum

Reinforcement of 6.N.18 Add, subtract, multiply, and divide mixed numbers with unlike denominators

NY Math Online

glencoe.com
• Extra Examples
• Personal Tutor
• Self-Check Quiz

▷ **GET READY** for the Lesson

LUNCH Two thirds of the students at the lunch table ordered a hamburger for lunch. One half of those students ordered cheese on their hamburgers.

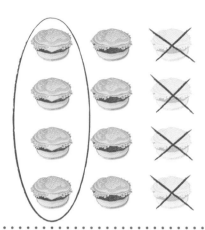

1. What fraction of the students at the lunch table ordered a cheeseburger?

2. How are the numerators and denominators of $\frac{2}{3}$ and $\frac{1}{2}$ related to the fraction in Exercise 1?

Multiply Fractions
Key Concept

Words	To multiply fractions, multiply the numerators and multiply the denominators.

Examples

Numbers

$$\frac{1}{2} \times \frac{2}{3} = \frac{1 \times 2}{2 \times 3} \text{ or } \frac{2}{6}$$

Algebra

$$\frac{a}{b} \cdot \frac{c}{d} = \frac{a \cdot c}{b \cdot d} \text{ or } \frac{ac}{bd}, \text{ where } b, d \neq 0$$

EXAMPLES Multiply Fractions

Multiply. Write in simplest form.

① $\frac{1}{2} \times \frac{1}{3}$

$\frac{1}{2} \times \frac{1}{3} = \frac{1 \times 1}{2 \times 3}$ ⟵ Multiply the numerators.
⟵ Multiply the denominators.

$= \frac{1}{6}$ Simplify.

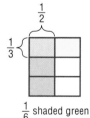

$\frac{1}{6}$ shaded green

② $2 \times \frac{3}{4}$

$2 \times \frac{3}{4} = \frac{2}{1} \times \frac{3}{4}$ Write 2 as $\frac{2}{1}$.

$= \frac{2 \times 3}{1 \times 4}$ ⟵ Multiply the numerators.
⟵ Multiply the denominators.

$= \frac{6}{4}$ or $1\frac{1}{2}$ Simplify.

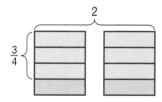

$\frac{6}{4}$ or $1\frac{1}{2}$ shaded green

Study Tip

Whole Numbers
When multiplying a fraction by a whole number, write the whole number as a fraction with a denominator of 1.

✓ **CHECK** Your Progress

a. $\frac{3}{5} \times \frac{1}{2}$ b. $\frac{1}{3} \times \frac{3}{4}$ c. $\frac{2}{3} \times 4$

If the numerator and denominator of either fraction have common factors, you can simplify before multiplying.

Review Vocabulary

GCF the greatest of the common factors of two or more numbers; *Example:* the GCF of 8 and 12 is 4.
(Lesson 4-2)

EXAMPLE Simplify Before Multiplying

3 Find $\frac{2}{7} \times \frac{3}{8}$. Write in simplest form.

$$\frac{2}{7} \times \frac{3}{8} = \frac{\overset{1}{\cancel{2}}}{7} \times \frac{3}{\underset{4}{\cancel{8}}} \qquad \text{Divide 2 and 8 by their GCF, 2.}$$

$$= \frac{1 \times 3}{7 \times 4} \text{ or } \frac{3}{28} \qquad \text{Multiply.}$$

✓ **CHECK Your Progress**

Multiply. Write in simplest form.

d. $\frac{1}{3} \times \frac{3}{7}$ e. $\frac{4}{9} \times \frac{1}{8}$ f. $\frac{5}{6} \times \frac{3}{5}$

EXAMPLE Multiply Mixed Numbers

4 Find $\frac{1}{2} \times 4\frac{2}{5}$. Write in simplest form. **Estimate** $\frac{1}{2} \times 4 = 2$

Study Tip

Simplifying
If you forget to simplify before multiplying, you can always simplify the final answer. However, it is usually easier to simplify before multiplying.

METHOD 1 Rename the mixed number.

$$\frac{1}{2} \times 4\frac{2}{5} = \frac{1}{\underset{1}{\cancel{2}}} \times \frac{\overset{11}{\cancel{22}}}{5} \qquad \begin{array}{l}\text{Rename } 4\frac{2}{5} \text{ as an improper fraction, } \frac{22}{5}.\\ \text{Divide 2 and 22 by their GCF, 2.}\end{array}$$

$$= \frac{1 \times 11}{1 \times 5} \qquad \text{Multiply.}$$

$$= \frac{11}{5} \text{ or } 2\frac{1}{5} \qquad \text{Simplify.}$$

METHOD 2 Use mental math.

The mixed number $4\frac{2}{5}$ is equal to $4 + \frac{2}{5}$.

So, $\frac{1}{2} \times 4\frac{2}{5} = \frac{1}{2}\left(4 + \frac{2}{5}\right)$. Use the Distributive Property to multiply, then add mentally.

$$\frac{1}{2}\left(4 + \frac{2}{5}\right) = 2 + \frac{1}{5} \qquad \textbf{THINK} \text{ Half of 4 is 2 and half of 2 fifths is 1 fifth.}$$

$$= 2\frac{1}{5} \qquad \text{Rewrite the sum as a mixed number.}$$

So, $\frac{1}{2} \times 4\frac{2}{5} = 2\frac{1}{5}$. **Check for Reasonableness** $2\frac{1}{5} \approx 2$ ✔

✓ **CHOOSE Your Method**

Multiply. Write in simplest form.

g. $\frac{1}{4} \times 8\frac{4}{9}$ h. $5\frac{1}{3} \times 3$ i. $1\frac{7}{8} \times 2\frac{2}{5}$

5 **SLEEP** Humans sleep about $\frac{1}{3}$ of each day. If each year is equal to $365\frac{1}{4}$ days, determine the number of days in a year the average human sleeps.

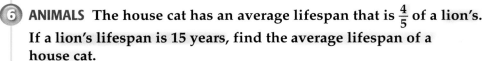

Words	Humans sleep about $\frac{1}{3}$ of $365\frac{1}{4}$ days.
Variable	Let d represent the number of days a human sleeps.
Equation	$d = \frac{1}{3} \cdot 365\frac{1}{4}$

$d = \frac{1}{3} \cdot 365\frac{1}{4}$ Write the equation.

$d = \frac{1}{3} \cdot \frac{1{,}461}{4}$ Rename the mixed number as an improper fraction.

$d = \frac{1}{\underset{1}{3}} \cdot \frac{\overset{487}{1{,}461}}{4}$ Divide 3 and 1,461 by their GCF, 3.

$d = \frac{487}{4}$ or $121\frac{3}{4}$ Multiply. Then rename as a mixed number.

The average human sleeps $121\frac{3}{4}$ days each year.

6 **ANIMALS** The house cat has an average lifespan that is $\frac{4}{5}$ of a lion's. If a lion's lifespan is 15 years, find the average lifespan of a house cat.

Words	The lifespan of a house cat is $\frac{4}{5}$ of that of the lion.
Variable	Let c represent the lifespan of a house cat.
Equation	$c = \frac{4}{5} \cdot 15$

$c = \frac{4}{5} \cdot 15$ Write the equation.

$c = \frac{4}{5} \cdot \frac{15}{1}$ Write the whole number 15 as an improper fraction.

$c = \frac{4}{\underset{1}{5}} \cdot \frac{\overset{3}{15}}{1}$ Divide 5 and 15 by their GCF, 5.

$c = \frac{12}{1}$ or 12   Multiply, then simplify.

The average lifespan of a house cat is 12 years.

Real-World Link . . . . .
The average group of
lions, called a pride,
consists of about 15
lions with about $\frac{2}{3}$ of
the pride being female.
Source: *African Wildlife Foundation*

✓CHECK Your Progress

j. **COOKING** Sofia wishes to make $\frac{1}{2}$ of a recipe. If the original recipe calls for $3\frac{3}{4}$ cups of flour, how many cups should she use?

Examples 1–4
(pp. 252–253)

Multiply. Write in simplest form.

1. $\frac{2}{3} \times \frac{1}{3}$

2. $2 \times \frac{2}{5}$

3. $\frac{1}{6} \times 4$

4. $\frac{1}{4} \times \frac{8}{9}$

5. $2\frac{1}{4} \times \frac{2}{3}$

6. $1\frac{5}{6} \times 3\frac{3}{5}$

Examples 5, 6
(p. 254)

7. **WEIGHT** The weight of an object on Mars is about $\frac{2}{5}$ its weight on Earth. How much would an 80-pound dog weigh on Mars?

Practice and Problem Solving

HOMEWORK HELP	
For Exercises	**See Examples**
8–11	1, 2
12–19	3, 4
20–23	5, 6

Multiply. Write in simplest form.

8. $\frac{3}{4} \times \frac{1}{8}$

9. $\frac{2}{5} \times \frac{2}{3}$

10. $9 \times \frac{1}{2}$

11. $\frac{4}{5} \times 6$

12. $\frac{1}{5} \times \frac{5}{6}$

13. $\frac{4}{9} \times \frac{1}{4}$

14. $\frac{2}{3} \times \frac{1}{4}$

15. $\frac{1}{12} \times \frac{3}{5}$

16. $\frac{4}{7} \times \frac{7}{8}$

17. $\frac{2}{5} \times \frac{15}{16}$

18. $\frac{3}{8} \times \frac{10}{27}$

19. $\frac{9}{10} \times \frac{5}{6}$

20. **DVDs** Each DVD storage case is about $\frac{1}{5}$ inch thick. What will be the height of 12 cases sold together in plastic wrapping?

21. **PIZZA** Mark left $\frac{3}{8}$ of a pizza in the refrigerator. On Friday, he ate $\frac{1}{2}$ of what was left of the pizza. What fraction of the entire pizza did he eat on Friday?

22. **MEASUREMENT** The width of a vegetable garden is $\frac{1}{3}$ times its length. If the length of the garden is $7\frac{3}{4}$ feet, what is the width?

23. **RECIPES** A recipe to make one batch of blueberry muffins calls for $4\frac{2}{3}$ cups of flour. How many cups of flour are needed to make 3 batches of blueberry muffins?

Multiply. Write in simplest form.

24. $4\frac{2}{3} \times \frac{4}{7}$

25. $\frac{5}{8} \times 2\frac{1}{2}$

26. $14 \times 1\frac{1}{7}$

27. $3\frac{3}{4} \times 8$

28. $9 \times 4\frac{2}{3}$

29. $4 \times 7\frac{5}{6}$

30. $3\frac{1}{4} \times 2\frac{2}{3}$

31. $5\frac{1}{3} \times 3\frac{3}{4}$

32. **MEASUREMENT** The width of the fish tank is $\frac{2}{5}$ of its length. What is the width of the fish tank?

33. **BICYCLING** Philip rode his bicycle at $9\frac{2}{5}$ miles per hour. If he rode for $\frac{3}{4}$ of an hour, how many miles did he cover?

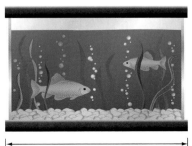

30 in.

Evaluate each verbal expression.

34. one half of five eighths

35. four sevenths of two thirds

36. nine tenths of one fourth

37. one third of eleven sixteenths

MEASUREMENT Find the perimeter and area of each rectangle.

38.

$3\frac{1}{2}$ ft

$5\frac{1}{4}$ ft

39.

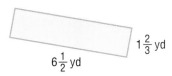

$1\frac{2}{3}$ yd

$6\frac{1}{2}$ yd

40. POOLS A community swimming pool is $90\frac{2}{5}$ feet long and $55\frac{1}{2}$ feet wide. If Natalie swims the perimeter of the pool four times, what is the total number of feet she will swim? Explain how you solved the problem.

MEASUREMENT For Exercises 41–44, use measurement conversions.

41. Find $\frac{1}{2}$ of $\frac{1}{4}$ of a gallon.

42. What is $\frac{1}{60}$ of $\frac{1}{24}$ of a day?

43. Find $\frac{1}{100}$ of $\frac{1}{1,000}$ of a kilometer.

44. What is $\frac{1}{12}$ of $\frac{1}{3}$ of a yard?

ALGEBRA Evaluate each expression if $a = 4$, $b = 2\frac{1}{2}$, and $c = 5\frac{3}{4}$.

45. $a \times b + c$

46. $b \times c - a$

47. $2bc$

Real-World Link . . . .
There are an estimated 5 million in-ground swimming pools in the U.S.
Source: *Pool & Spa Service Industry News*

48. TELEVISION One evening, $\frac{2}{3}$ of the students in Rick's class watched television, and $\frac{3}{8}$ of those students watched a reality show, of which $\frac{1}{4}$ taped the show. What fraction of the students in Rick's class watched and taped a reality TV show?

49. FOOD Alano wants to make one and a half recipes of the pasta salad recipe shown at the right. How much of each ingredient will Alano need? Explain how you solved the problem.

50. FIND THE DATA Refer to the Data File on pages 16–19. Choose some data and write a real-world problem in which you would multiply fractions.

Pasta Salad Recipe	
Ingredient	**Amount**
broccoli	$1\frac{1}{4}$ c
cooked pasta	$3\frac{3}{4}$ c
salad dressing	$\frac{2}{3}$ c
cheese	$1\frac{1}{3}$ c

Write and evaluate a multiplication expression to represent each model. Explain how the models show the multiplication process.

51.

52.

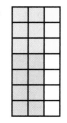

H.O.T. Problems

53. **CHALLENGE** Two improper fractions are multiplied. Is the product *sometimes, always,* or *never* less than 1? Explain your reasoning.

54. **OPEN ENDED** Write a word problem that involves finding the product of $\frac{3}{4}$ and $\frac{1}{8}$.

55. **WRITING IN MATH** Refer to Example 2. Explain how the model represents the meaning of the multiplication process.

NYSMT PRACTICE 6.N.18

56. Of the dolls in Marjorie's doll collection, $\frac{1}{5}$ have red hair. Of these, $\frac{3}{4}$ have green eyes. What fraction of Marjorie's doll collection has both red hair and green eyes?

A $\frac{2}{9}$ C $\frac{4}{9}$

B $\frac{3}{20}$ D $\frac{19}{20}$

57. Which description gives the relationship between a term and n, its position in the sequence?

Position	1	2	3	4	5	n
Value of Term	$\frac{1}{4}$	$\frac{1}{2}$	$\frac{3}{4}$	1	$1\frac{1}{4}$	

F Subtract 4 from n.

G Add $\frac{1}{4}$ to n.

H Multiply n by $\frac{1}{4}$.

J Divide n by $\frac{1}{4}$.

Spiral Review

58. **MEASUREMENT** Find which room dimensions would give an area of $125\frac{3}{8}$ square feet. Use the *eliminate possibilities* strategy. (Lesson 5-4)

A $11\frac{1}{2}$ feet by $10\frac{3}{8}$ feet   C $13\frac{5}{8}$ feet by 9 feet

B $10\frac{7}{8}$ feet by $12\frac{1}{4}$ feet   D $14\frac{3}{4}$ feet by $8\frac{1}{2}$ feet

59. **MEASUREMENT** How much longer is a $2\frac{1}{2}$-inch-long piece of string than a $\frac{2}{5}$-inch-long piece of string? (Lesson 5-3)

Replace each ● with <, >, or = to make a true sentence. (Lesson 4-9)

60. $\frac{5}{12}$ ● $\frac{2}{5}$ 61. $\frac{3}{16}$ ● $\frac{1}{8}$ 62. $3\frac{7}{6}$ ● $3\frac{6}{5}$

63. **PHONES** A long-distance telephone company charges a flat monthly fee of \$4.95 and \$0.06 per minute on all long-distance calls. Write and solve an equation to find the number of monthly minutes spent talking long-distance if the bill total was \$22.95. (Lesson 3-5)

▷ **GET READY for the Next Lesson**

PREREQUISITE SKILL Solve each equation mentally. (Lesson 1-7)

64. $x + 2 = 8$ 65. $9 + m = 12$ 66. $7 - w = 2$

5-6 Algebra: Solving Equations

MAIN IDEA

Solve equations with rational number solutions.

NYS Core Curriculum

Preparation for 7.A.4
Solve multi-step equations by combining like terms, using the distributive property, or moving variables to one side of the equation

New Vocabulary

multiplicative inverse
reciprocal

NY Math Online

glencoe.com
• Extra Examples
• Personal Tutor
• Self-Check Quiz

▷ **GET READY** for the Lesson

HOMEWORK Shawnda spends $\frac{1}{2}$ hour doing homework after school. Then she spends another $\frac{1}{2}$ hour doing homework before bed.

1. Write a multiplication expression to find how much time Shawnda spends doing homework. Then find the product.

2. Copy and complete the table below.

$\frac{3}{2} \times \frac{2}{3} = $ ■	$\frac{1}{5} \times $ ■ $ = 1$	$\frac{5}{6} \times \frac{6}{5} = $ ■	$\frac{7}{8} \times \frac{8}{7} = $ ■
■ $\times \frac{5}{7} = 1$	$\frac{2}{6} \times \frac{6}{2} = $ ■	$\frac{7}{1} \times $ ■ $ = 1$	■ $\times 8 = 1$

3. What is true about the numerators and denominators in the fractions in Exercise 2?

Two numbers with a product of 1 are called **multiplicative inverses**, or **reciprocals**.

Inverse Property of Multiplication		**Key Concept**
Words	The product of a number and its multiplicative inverse is 1.	
Examples	**Numbers**	**Algebra**
	$\frac{3}{4} \times \frac{4}{3} = 1$	$\frac{a}{b} \cdot \frac{b}{a} = 1$, for $a, b \neq 0$

EXAMPLES Find Multiplicative Inverses

① Find the multiplicative inverse of $\frac{2}{5}$.

$\frac{2}{5} \cdot \frac{5}{2} = 1$ Multiply $\frac{2}{5}$ by $\frac{5}{2}$ to get the product 1.

The multiplicative inverse of $\frac{2}{5}$ is $\frac{5}{2}$, or $2\frac{1}{2}$.

② Find the multiplicative inverse of $2\frac{1}{3}$.

$2\frac{1}{3} = \frac{7}{3}$ Rename the mixed number as an improper fraction.

$\frac{7}{3} \cdot \frac{3}{7} = 1$ Multiply $\frac{7}{3}$ by $\frac{3}{7}$ to get the product 1.

The multiplicative inverse of $2\frac{1}{3}$ is $\frac{3}{7}$.

✔ **CHECK Your Progress**

a. $\frac{5}{6}$ b. $1\frac{1}{2}$ c. 8 d. $\frac{4}{3}$

In Chapter 3, you learned to solve equations using the Addition, Subtraction, and Division Properties of Equality. You can also solve equations by multiplying each side by the same number. This is called the **Multiplication Property of Equality**.

Multiplication Property of Equality Key Concept

Words	If you multiply each side of an equation by the same nonzero number, the two sides remain equal.

Examples	**Numbers**	**Algebra**

	$5 = 5$	$\dfrac{x}{2} = -3$	$\dfrac{2}{3}x = 4$
	$5 \cdot 2 = 5 \cdot 2$	$\dfrac{x}{2}(2) = -3(2)$	$\dfrac{3}{2} \cdot \dfrac{2}{3}x = \dfrac{3}{2} \cdot 4$
	$10 = 10$	$x = -6$	$x = 6$

Study Tip

Fractions
The fraction bar indicates division. So, $\dfrac{x}{2}$ means x divided by 2.

EXAMPLES Solve a Division Equation

③ Solve $7 = \dfrac{n}{4}$. Check your solution.

$7 = \dfrac{n}{4}$	Write the equation.
$7 \cdot 4 = \dfrac{n}{4} \cdot 4$	Multiply each side of the equation by 4.
$28 = n$	Simplify.

Check	$7 = \dfrac{n}{4}$	Write the original equation.
	$7 \stackrel{?}{=} \dfrac{28}{4}$	Replace n with 28.
	$7 = 7$ ✔	Is this sentence true?

④ Solve $\dfrac{d}{3.5} = 4.2$.

$\dfrac{d}{3.5} = 4.2$	Write the equation.
$\dfrac{d}{3.5} \cdot 3.5 = 4.2 \cdot 3.5$	Multiply each side by 3.5.
$d = 14.7$	Simplify.

The solution is 14.7.

Check	$\dfrac{d}{3.5} = 4.2$	Write the original equation.
	$\dfrac{14.7}{3.5} \stackrel{?}{=} 4.2$	Replace d with 14.7.
	$4.2 = 4.2$ ✔	Is this sentence true?

CHECK Your Progress

Solve each equation. Check your solution.

e. $6 = \dfrac{m}{8}$ f. $\dfrac{p}{2.8} = 1.5$ g. $\dfrac{k}{4.7} = 2.3$

EXAMPLE Solve a Multiplication Equation

⑤ Solve $\frac{3}{4}x = \frac{12}{20}$.

$$\frac{3}{4}x = \frac{12}{20}$$ Write the equation.

$$\left(\frac{4}{3}\right) \cdot \frac{3}{4}x = \left(\frac{4}{3}\right) \cdot \frac{12}{20}$$ Multiply each side by the reciprocal of $\frac{3}{4}$, $\frac{4}{3}$.

$$\overset{1}{\underset{3}{\cancel{\frac{4}{3}}}} \cdot \overset{1}{\underset{1}{\cancel{\frac{3}{4}}}}x = \overset{1}{\underset{1}{\cancel{\frac{4}{3}}}} \cdot \overset{4}{\underset{5}{\cancel{\frac{12}{20}}}}$$ Divide by common factors.

$$x = \frac{4}{5}$$ Simplify.

Study Tip

Fractions as Coefficients
The expression $\frac{3}{4}x$ can be read as $\frac{3}{4}$ of x, $\frac{3}{4}$ multiplied by x, $3x$ divided by 4, or $\frac{x}{4}$ multiplied by 3.

✓ **CHECK Your Progress**

Solve each equation. Check your solution.

h. $\frac{1}{2}x = 8$ i. $\frac{3}{4}x = 9$ j. $\frac{7}{8}x = \frac{21}{64}$

NYSMT EXAMPLE

⑥ Valerie needs $\frac{2}{3}$ yard of fabric to make each hat for the school play. How many hats can she make with 6 yards of fabric?

A 12 C 8

B 9 D 4

Read the Item

Each hat needs $\frac{2}{3}$ yard of fabric. Given the number of hats, you would multiply by $\frac{2}{3}$ to find the number of yards of fabric needed.

Solve the Item

Write and solve a multiplication equation.

$$\frac{2}{3}n = 6$$ Write the equation.

$$\left(\frac{3}{2}\right) \cdot \frac{2}{3}n = \left(\frac{3}{2}\right) \cdot 6$$ Multiply each side by $\frac{3}{2}$.

$$n = 9$$ Simplify.

So, the answer is B.

Test-Taking Tip

Verify Your Answer
It is a good idea to verify your answer by checking the other answer choices. By doing so, you can greatly reduce your chances of making an error.

✓ **CHECK Your Progress**

k. Wilson has 9 pounds of trail mix. How many $\frac{3}{4}$-pound bags of trail mix can he make?

F 3 H 9

G 6 J 12

260 Chapter 5 Applying Fractions

Examples 1, 2
(p. 258)

Find the multiplicative inverse of each number.

1. $\dfrac{8}{5}$　　　　2. $\dfrac{2}{9}$　　　　3. $5\dfrac{4}{5}$　　　　4. 9

Examples 3–5
(p. 259–260)

Solve each equation. Check your solution.

5. $\dfrac{k}{16} = 2$　　　　6. $4 = \dfrac{y}{3}$　　　　7. $\dfrac{b}{8.2} = 2.5$

8. $0.5 = \dfrac{h}{3.6}$　　　　9. $\dfrac{3}{8}a = \dfrac{12}{40}$　　　　10. $6 = \dfrac{4}{7}x$

Example 5
(p. 260)

11. **FRUIT** Three fourths of the fruit in a refrigerator are apples. There are 24 apples in the refrigerator. The number of pieces of fruit is given by the equation $\dfrac{3}{4}f = 24$. How many pieces of fruit are in the refrigerator?

Example 6
(p. 260)

12. **MULTIPLE CHOICE** Dillon deposited $\dfrac{3}{4}$ of his paycheck into the bank. The deposit slip shows how much he deposited. What was the amount of his paycheck?

A $15　　　　C $60

B $33.75　　　　D $75

▶ **Practice and Problem Solving**

HOMEWORK HELP

For Exercises	See Examples
13–20	1, 2
21–26 33–34	3, 4
27–32	5
51–52	6

Find the multiplicative inverse of each number.

13. $\dfrac{5}{6}$　　　　14. $\dfrac{11}{2}$　　　　15. $\dfrac{1}{6}$　　　　16. $\dfrac{1}{10}$

17. 3　　　　18. 14　　　　19. $5\dfrac{1}{8}$　　　　20. $6\dfrac{2}{3}$

Solve each equation. Check your solution.

21. $\dfrac{x}{12} = 3$　　　　22. $28 = \dfrac{d}{4}$　　　　23. $\dfrac{b}{2.4} = 6$

24. $5 = \dfrac{w}{4.9}$　　　　25. $0.8 = \dfrac{h}{3.6}$　　　　26. $\dfrac{m}{4.6} = 2.8$

27. $\dfrac{2}{5}t = \dfrac{12}{25}$　　　　28. $\dfrac{24}{16} = \dfrac{3}{4}a$　　　　29. $\dfrac{7}{8}k = \dfrac{5}{6}$

30. $\dfrac{2}{3} = \dfrac{8}{3}b$　　　　31. $\dfrac{1}{2}g = 3\dfrac{1}{3}$　　　　32. $\dfrac{3}{5}c = 6\dfrac{1}{4}$

33. **DISTANCE** The distance d Toya travels in her car while driving 60 miles per hour for 3.25 hours is given by the equation $\dfrac{d}{3.25} = 60$. How far did she travel?

34. **ANIMALS** An adult Fitch ferret weighs about 1.8 kilograms. To find its weight in pounds p, you can use the equation $\dfrac{p}{1.8} = 2.2$. How many pounds does an adult Fitch ferret weigh?

Solve each equation. Check your solution.

35. $\dfrac{a}{-5} = 15$

36. $-8 = \dfrac{r}{-2}$

37. $34.5 = \dfrac{5}{6}m$

38. $\dfrac{5}{7}x = -1.5$

39. $\dfrac{1}{4}t = \dfrac{3}{8}$

40. $\dfrac{3}{8}m = 1\dfrac{1}{2}$

For Exercises 41–46, define a variable and write an equation. Then solve.

41. **CAVES** The self-guided Mammoth Cave Discovery Tour includes an elevation change of 140 feet. This is $\dfrac{7}{15}$ of the elevation change on the Wild Cave Tour. What is the elevation change on the Wild Cave Tour?

42. **MUSEUMS** Twenty-four students brought their permission slips to attend the class field trip to the local art museum. If this represented $\dfrac{4}{5}$ of the class, how many students are in the class?

43. **MEASUREMENT** If one serving of cooked rice is $\dfrac{3}{4}$ cup, how many servings will $16\dfrac{1}{2}$ cups of rice yield?

44. **HIKING** After Alana hiked $2\dfrac{5}{8}$ miles along a hiking trail, she realized that she was only $\dfrac{3}{4}$ of the way to the end of the trail. How long is the trail?

45. **SLEEP** The average person spends $\dfrac{1}{3}$ of his life asleep. According to this, if a person has spent 26 years asleep, how old is he?

46. **ANALYZE TABLES** Tierra recorded the distance she ran each day last week. If she ran $\dfrac{5}{6}$ of her weekly running goal, what was her running goal?

NYSCC • NYSMT
Extra Practice, pp. 681, 708

H.O.T. Problems

47. **REASONING** Complete the statement: If $8 = \dfrac{m}{4}$, then $m - 12 = \blacksquare$. Explain your reasoning.

Distance Ran in One Week	
Day	**Distance (mi)**
Monday	$1\dfrac{3}{4}$
Wednesday	2
Friday	$1\dfrac{1}{2}$
Saturday	$2\dfrac{1}{4}$

48. **Which One Doesn't Belong?** Identify the pair of numbers that does not belong with the other three. Explain.

| $\dfrac{9}{6}, \dfrac{6}{9}$ | $4, \dfrac{1}{4}$ | $\dfrac{3}{5}, 5$ | $\dfrac{2}{7}, \dfrac{7}{2}$ |

49. **CHALLENGE** The formula for the area of a trapezoid is $A = \dfrac{1}{2}h(b_1 + b_2)$, where b_1 and b_2 are both bases and h is the height. Find the value of h in terms of A, b_1, and b_2. Justify your answer.

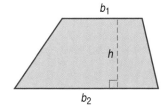

50. **WRITING IN MATH** Explain the Multiplication Property of Equality. Then give an example of an equation in which you would use this property to solve the equation.

51. Audrey drove 200 miles in 3.5 hours. Which equation can you use to find the rate r at which Audrey was traveling?

 A $200 = 3.5r$

 B $200 \cdot 3.5 = r$

 C $\dfrac{r}{3.5} = 200$

 D $200r = 3.5$

52. The table shows the results of a survey.

Music Preference	
Type	**Fraction of Students**
pop	$\dfrac{5}{8}$
jazz	$\dfrac{1}{8}$
rap	$\dfrac{1}{4}$

If there are 420 students surveyed, which equation can be used to find the number of students s who prefer rap?

 F $\dfrac{1}{4}s = 420$ H $s + \dfrac{1}{4} = 420$

 G $s = \dfrac{1}{4} \cdot 420$ J $420 + s = \dfrac{1}{4}$

Spiral Review

Multiply. Write in simplest form. (Lesson 5-5)

53. $\dfrac{3}{8} \times \dfrac{4}{9}$ 54. $1\dfrac{1}{2} \times 6$ 55. $2\dfrac{2}{5} \times \dfrac{1}{6}$ 56. $1\dfrac{1}{2} \times 1\dfrac{7}{9}$

57. **COOKING** Lawana had $4\dfrac{2}{3}$ cups of chopped walnuts. She used $1\dfrac{1}{4}$ cups in a recipe. How many cups of chopped walnuts are left? (Lesson 5-3)

Write each percent as a decimal. (Lesson 4-7)

58. 25% 59. 8% 60. 25.6% 61. 123%

62. **MEASUREMENT** A farmer has a rectangular pumpkin field with a perimeter of 3,800 feet. If the width of the pumpkin field is 800 feet, what is the length of the field? (Lesson 3-6)

ALGEBRA Solve each equation. Check your solution. (Lesson 3-2)

63. $7 = x + 8$ 64. $k - 3 = -14$ 65. $-2 = m + 6$

66. **ALGEBRA** The table below shows the time needed to complete 4 art projects. If the pattern continues, how much time is needed to complete the fifth art project? (Lesson 2-7)

Project	1	2	3	4	5
Time (min)	8	25	42	59	

▶ GET READY for the Next Lesson

PREREQUISITE SKILL Estimate. (Lesson 5-1)

67. $18\dfrac{1}{6} \div 3$ 68. $24\dfrac{3}{8} \div 11\dfrac{7}{9}$ 69. $\dfrac{2}{11} \div \dfrac{11}{12}$ 70. $\dfrac{9}{10} \div \dfrac{6}{7}$

READING to SOLVE PROBLEMS

Meaning of Division

You know that one meaning of division is to *put objects into equal groups*. But there are other meanings too. Look for these meanings when you're solving a word problem.

NYSCC 7.CM.11
Draw conclusions about mathematical ideas through decoding, comprehension, and interpretation of mathematical visuals, symbols, and technical writing

● **To share**

Zach and his friend are going to share 3 apples equally. How many apples will each boy have?

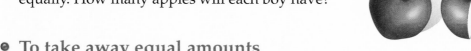

● **To take away equal amounts**

Isabel is making bookmarks from a piece of ribbon. Each bookmark is 6.5 centimeters long. How many bookmarks can she make from a piece of ribbon that is 27 centimeters long?

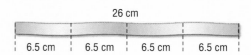

26 cm

6.5 cm   6.5 cm   6.5 cm   6.5 cm

● **To find how many times greater**

The Nile River, the longest river on Earth, is 4,160 miles long. The Rio Grande River is 1,900 miles long. About how many times longer is the Nile than the Rio Grande?

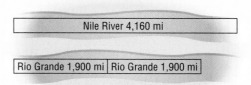

Nile River 4,160 mi

Rio Grande 1,900 mi | Rio Grande 1,900 mi

PRACTICE

1. Solve each problem above.

Identify the meaning of division shown in each problem. Then solve the problem.

2. A landscape architect wants to make a border along one side of a garden using bricks that are 0.25 meter long. If the garden is 11.25 meters long, how many bricks does she need?

3. The Jackson family wants to buy a flat-screen television that costs $1,200. They plan to pay in six equal payments. What will be the amount of each payment?

4. A full-grown blue whale can weigh 150 tons. An adult African elephant weighs about 5 tons. How many times greater does a blue whale weigh than an African elephant?

5. Each story in an office building is about 4 meters tall. The Eiffel Tower in Paris, France, is 300 meters tall. How many stories tall is the Eiffel Tower?

5-7 Dividing Fractions and Mixed Numbers

MAIN IDEA

Divide fractions and mixed numbers.

NYS Core Curriculum

Reinforcement of 6.N.18 Add, subtract, multiply, and divide mixed numbers with unlike denominators

NY Math Online

glencoe.com

• Concepts In Motion
• Extra Examples
• Personal Tutor
• Self-Check Quiz

▷ **MINI Lab**

Cut two paper plates into four equal pieces each to show $2 \div \frac{1}{4}$.

1. How many $\frac{1}{4}$'s are in 2 plates?

2. How would you model $3 \div \frac{1}{2}$?

3. What is true about $3 \div \frac{1}{2}$ and 3×2?

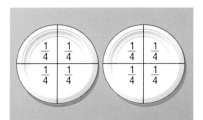

Dividing 8 by 2 gives the same result as multiplying 8 by $\frac{1}{2}$, which is the reciprocal of 2. In the same way, dividing 4 by $\frac{1}{3}$ is the same as multiplying 4 by the reciprocal of $\frac{1}{3}$, or 3.

Is this pattern true for any division expression?

Consider $\frac{7}{8} \div \frac{3}{4}$, which can be rewritten as $\dfrac{\frac{7}{8}}{\frac{3}{4}}$.

$$\dfrac{\frac{7}{8}}{\frac{3}{4}} = \dfrac{\frac{7}{8} \times \frac{4}{3}}{\frac{3}{4} \times \frac{4}{3}}$$

Multiply the numerator and denominator by the reciprocal of $\frac{3}{4}$, which is $\frac{4}{3}$.

$$= \dfrac{\frac{7}{8} \times \frac{4}{3}}{1} \qquad \frac{3}{4} \times \frac{4}{3} = 1$$

$$= \frac{7}{8} \times \frac{4}{3}$$

So, $\frac{7}{8} \div \frac{3}{4} = \frac{7}{8} \times \frac{4}{3}$. These examples suggest the following rule.

Divide by Fractions	**Key Concept**
Words	To divide by a fraction, multiply by its multiplicative inverse, or reciprocal.
Examples	**Numbers** $\qquad$ **Algebra**
	$\frac{7}{8} \div \frac{3}{4} = \frac{7}{8} \cdot \frac{4}{3}$ $\qquad$ $\frac{a}{b} \div \frac{c}{d} = \frac{a}{b} \cdot \frac{d}{c}$, where $b, c, d \neq 0$

EXAMPLE Divide by a Fraction

① Find $\frac{3}{4} \div \frac{1}{2}$. Write in simplest form.

Estimate $1 \div \frac{1}{2} =$

Think: How many groups of $\frac{1}{2}$ are in 1? $1 \div \frac{1}{2} = 2$

$\frac{3}{4} \div \frac{1}{2} = \frac{3}{4} \cdot \frac{2}{1}$ Multiply by the reciprocal of $\frac{1}{2}$, which is $\frac{2}{1}$.

$\qquad = \frac{3}{\overset{}{\underset{2}{4}}} \cdot \frac{\overset{1}{2}}{1}$ Divide 4 and 2 by their GCF, 2.

$\qquad = \frac{3}{2}$ or $1\frac{1}{2}$ Multiply.

Check for Reasonableness $1\frac{1}{2} \approx 2$ ✔

CHECK Your Progress

Divide. Write in simplest form.

a. $\frac{3}{4} \div \frac{1}{4}$ b. $\frac{4}{5} \div \frac{8}{9}$ c. $\frac{5}{6} \div \frac{2}{3}$

To divide by a mixed number, first rename the mixed number as an improper fraction. Then multiply the first fraction by the reciprocal, or multiplicative inverse, of the second fraction.

EXAMPLE Divide by Mixed Numbers

② Find $\frac{2}{3} \div 3\frac{1}{3}$. Write in simplest form.

Estimate $\frac{1}{2} \div 3 = \frac{1}{2} \times \frac{1}{3}$ or $\frac{1}{6}$

$\frac{2}{3} \div 3\frac{1}{3} = \frac{2}{3} \div \frac{10}{3}$ Rename $3\frac{1}{3}$ as an improper fraction.

$\qquad = \frac{2}{3} \cdot \frac{3}{10}$ Multiply by the reciprocal of $\frac{10}{3}$, which is $\frac{3}{10}$.

$\qquad = \frac{\overset{1}{2}}{\underset{1}{3}} \cdot \frac{\overset{1}{3}}{\underset{5}{10}}$ Divide out common factors.

$\qquad = \frac{1}{5}$ Multiply.

Check for Reasonableness $\frac{1}{5}$ is close to $\frac{1}{6}$. ✔

CHECK Your Progress

Divide. Write in simplest form.

d. $5 \div 1\frac{1}{3}$ e. $-\frac{3}{4} \div 1\frac{1}{2}$ f. $2\frac{1}{3} \div 5$

g. **NUTS** In planning for a party, $5\frac{1}{4}$ pounds of cashews will be divided into $\frac{3}{4}$-pound bags. How many such bags can be made?

Study Tip

Dividing by a Whole Number
Remember that a whole number can be written as a fraction with a 1 in the denominator.

So, $2\frac{1}{3} \div 5$ can be rewritten as $2\frac{1}{3} \div \frac{5}{1}$.

3 **WOODWORKING** Students in a woodworking class are making butterfly houses. The side pieces of the house need to be $8\frac{1}{4}$ inches long. How many side pieces can be cut from a board measuring $49\frac{1}{2}$ inches long?

To find how many side pieces can be cut, divide $49\frac{1}{2}$ by $8\frac{1}{4}$.

Estimate Use compatible numbers. $48 \div 8 = 6$

$$49\frac{1}{2} \div 8\frac{1}{4} = \frac{99}{2} \div \frac{33}{4} \qquad \text{Rename the mixed numbers as improper fractions.}$$

$$= \frac{99}{2} \cdot \frac{4}{33} \qquad \text{Multiply by the reciprocal of } \frac{33}{4}, \text{ which is } \frac{4}{33}.$$

$$= \frac{\overset{3}{\cancel{99}}}{\underset{1}{\cancel{2}}} \cdot \frac{\overset{2}{\cancel{4}}}{\underset{1}{\cancel{33}}} \qquad \text{Divide out common factors.}$$

$$= \frac{6}{1} \text{ or } 6 \qquad \text{Multiply.}$$

So, 6 side pieces can be cut.

Check for Reasonableness The answer matches the estimate. ✔

CHECK Your Progress

h. **FOOD** Suppose a small box of cereal contains $12\frac{2}{3}$ cups of cereal. How many $1\frac{1}{3}$-cup servings are in the box?

i. **MEASUREMENT** The area of a rectangular bedroom is $146\frac{7}{8}$ square feet. If the width of the bedroom is $11\frac{3}{4}$ feet, find the length.

CHECK Your Understanding

Examples 1–3
(pp. 266–267)

Divide. Write in simplest form.

1. $\frac{1}{8} \div \frac{1}{3}$ 2. $\frac{3}{5} \div \frac{1}{4}$ 3. $3 \div \frac{6}{7}$ 4. $\frac{3}{4} \div 6$

5. $\frac{1}{2} \div 7\frac{1}{2}$ 6. $\frac{4}{7} \div 1\frac{2}{7}$ 7. $5\frac{3}{5} \div 4\frac{2}{3}$ 8. $6\frac{1}{2} \div 3\frac{5}{7}$

Example 2
(p. 266)

9. **FOOD** Deandre has 7 apples, and each apple is divided evenly into eighths. How many apple slices does Deandre have?

Example 3
(p. 267)

10. **WALKING** On Saturday, Lindsay walked $3\frac{1}{2}$ miles in $1\frac{2}{5}$ hours. What was her walking pace, in miles per hour?

HOMEWORK HELP	
For Exercises	See Examples
11–14	1
15–32	2, 3

Divide. Write in simplest form.

11. $\dfrac{3}{8} \div \dfrac{6}{7}$ 12. $\dfrac{5}{9} \div \dfrac{5}{6}$ 13. $\dfrac{2}{3} \div \dfrac{1}{2}$ 14. $\dfrac{7}{8} \div \dfrac{3}{4}$

15. $6 \div \dfrac{1}{2}$ 16. $\dfrac{4}{9} \div 2$ 17. $2\dfrac{2}{3} \div 4$ 18. $5 \div 1\dfrac{1}{3}$

19. **FOOD** Mason has 8 cups of popcorn kernels to divide into $\dfrac{2}{3}$-cup portions. How many portions will there be?

20. **MOVIES** Cheryl is organizing her movie collection. If each movie case is $\dfrac{3}{4}$ inch wide, how many movies can fit on a shelf 5 feet wide?

Divide. Write in simplest form.

21. $\dfrac{2}{3} \div 2\dfrac{1}{2}$ 22. $\dfrac{8}{9} \div 5\dfrac{1}{3}$ 23. $4\dfrac{1}{2} \div 6\dfrac{3}{4}$ 24. $5\dfrac{2}{7} \div 2\dfrac{1}{7}$

25. $3\dfrac{4}{5} \div 1\dfrac{1}{3}$ 26. $9\dfrac{1}{2} \div 2\dfrac{5}{6}$ 27. $5\dfrac{1}{5} \div \dfrac{2}{3}$ 28. $6\dfrac{7}{8} \div \dfrac{3}{4}$

29. **ICE CREAM** Vinh bought $4\dfrac{1}{2}$ gallons of ice cream to serve at his birthday party. If a pint is $\dfrac{1}{8}$ of a gallon, how many pint-sized servings can be made?

30. **BEVERAGES** William has $8\dfrac{1}{4}$ cups of fruit juice. If he divides the juice into $\dfrac{3}{4}$-cup servings, how many servings will he have?

BIRDS For Exercises 31 and 32, use the table that gives information about several types of birds of prey found at the Woodland Park Zoo in Seattle, Washington.

Bird	Maximum Weight (lb)
Golden Eagle	$13\dfrac{9}{10}$
Northern Bald Eagle	$9\dfrac{9}{10}$
Red-Tailed Hawk	$3\dfrac{1}{2}$

Source: Woodland Park Zoo

31. How many times as heavy is the Golden Eagle as the Red-tailed Hawk?

32. How many times as heavy is the Golden Eagle as the Northern Bald Eagle?

Draw a model of each verbal expression and then evaluate the expression. Explain how the model shows the division process.

Real-World Link · · · · ·
Red-tailed hawks are large, stocky birds with long, broad wings and short, broad tails. Females are larger than males and can weigh up to $3\dfrac{1}{2}$ pounds.
Source: Woodland Park Zoo

33. one half divided by two fifths

34. five eighths divided by one fourth

35. one and three eighths divided by one half

36. two and one sixth divided by two thirds

37. **PIZZA** A concession stand sells three types of pizza. The diagram shows how much pizza of each type is left after the concession stand was open for one hour. If the pizza is sold in slices that are $\dfrac{1}{8}$ of a whole pizza, how many more slices can be sold?

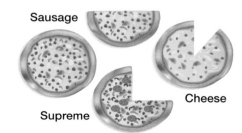

Sausage

Cheese

Supreme

ALGEBRA Evaluate each expression if $g = \frac{1}{6}$, $h = \frac{1}{2}$, and $j = 3\frac{2}{3}$.

38. $j \div h$ 39. $g \div j$ 40. $3g \div h$ 41. $h \div \left(\frac{1}{2}j\right)$

42. **SHOPPING** A supermarket sells pretzels in $\frac{3}{4}$-ounce snack-sized bags or $12\frac{1}{2}$-ounce regular-sized bags. How many times larger is the regular-sized bag than the snack-sized bag?

43. **MEASUREMENT** A recipe calls for $2\frac{2}{3}$ cups of brown sugar and $\frac{2}{3}$ cup of confectioner's sugar. How many times greater is the number of cups of brown sugar in the recipe than of confectioner's sugar?

SCHOOL For Exercises 44 and 45, use the table that shows the number of hours students spend studying each week during the school year.

44. How many times greater was the number of students who spent over 10 hours each week studying than those who spent only 1–2 hours each week studying?

45. How many times greater was the number of students who spent 3 or more hours each week studying than those who spent less than 3 hours each week studying?

Weekly Study Hours	
Hours	**Fraction of Students**
none	$\frac{1}{50}$
1–2	$\frac{2}{25}$
3–5	$\frac{11}{50}$
6–7	$\frac{17}{100}$
8–10	$\frac{1}{5}$
Over 10	$\frac{1}{5}$
Not sure	$\frac{3}{25}$

Source: *Time Magazine*

46. **SCHOOL SUPPLIES** Tara bought a dozen folders. She took $\frac{1}{3}$ of the dozen and then divided the remaining folders equally among her four friends. What fraction of the dozen did each of her four friends receive and how many folders was this per person?

47. **WEATHER** A meteorologist has issued a thunderstorm warning. So far, the storm has traveled 35 miles in $\frac{1}{2}$ hour. If it is currently 5:00 P.M., and the storm is 105 miles away from you, at what time will the storm reach you? Explain how you solved the problem.

NYSCC • NYSMT

Extra Practice, pp. 681, 708

H.O.T. Problems

48. **CHALLENGE** If $\frac{5}{6}$ is divided by a certain fraction $\frac{a}{b}$, the result is $\frac{1}{4}$. What is the fraction $\frac{a}{b}$?

49. **SELECT A TOOL** Reynaldo cut a rope to make the running knot shown. The rope used to make the knot was $1\frac{1}{2}$ feet long and was $\frac{3}{4}$ of the original rope length. Which of the following tools could be used to determine the rope's original length? Justify your selection(s). Then use your tool(s) to find the length.

paper and pencil	model	calculator	real object

50. **FIND THE ERROR** Evan and José are finding $\frac{4}{5} \div \frac{6}{7}$. Who is correct? Explain your reasoning.

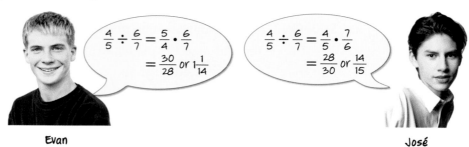

Evan

$$\frac{4}{5} \div \frac{6}{7} = \frac{5}{4} \cdot \frac{6}{7}$$
$$= \frac{30}{28} \text{ or } 1\frac{1}{14}$$

$$\frac{4}{5} \div \frac{6}{7} = \frac{4}{5} \cdot \frac{7}{6}$$
$$= \frac{28}{30} \text{ or } \frac{14}{15}$$

José

51. **WRITING IN MATH** If you divide a proper fraction by another proper fraction, is it possible to get a mixed number as an answer? Explain your reasoning.

NYSMT PRACTICE 6.N.18

52. The Corbett family owns 300 acres of land that they plan to rent to people for their horses. How many $7\frac{1}{2}$-acre lots can they make using the 300 acres?

 A 21 C $40\frac{1}{2}$

 B 40 D $292\frac{1}{2}$

53. How many $1\frac{1}{8}$-pound boxes of peanuts can be made using $6\frac{3}{4}$ pounds of peanuts?

 F 4 H 6

 G 5 J 7

Spiral Review

Find the multiplicative inverse of each number. (Lesson 5-6)

54. $\frac{6}{7}$

55. $\frac{4}{13}$

56. 8

57. $5\frac{1}{4}$

58. Find $\frac{1}{10} \times \frac{5}{8}$. Write in simplest form. (Lesson 5-5)

59. **MEASUREMENT** Find the length of a rectangular flower bed if the perimeter is 12 feet and the width is 1.5 feet. (Lesson 3-6)

60. **ANIMALS** An elephant herd can move 50 miles in a day. At this rate, about how many miles can an elephant herd move each hour? (Lesson 1-1)

Problem Solving in Geography Real-World Unit Project

A Traveling We Will Go It's time to complete your project. Use the data you have gathered about schedules and costs to prepare a travel brochure for your vacation destination. Be sure to include a paragraph describing why you chose your vacation spot.

NY Math Online Unit Project at glencoe.com

FOLDABLES® Study Organizer

GET READY to Study

Be sure the following Big Ideas are noted in your Foldable.

BIG Ideas

Estimating with Fractions (Lesson 5-1)

• When the numerator is much smaller than the denominator, round the fraction to 0.

• When the numerator is about half of the denominator, round the fraction to $\frac{1}{2}$.

• When the numerator is almost as large as the denominator, round the fraction to 1.

Adding and Subtracting Fractions

(Lessons 5-2 and 5-3)

• To add or subtract like fractions, add or subtract the numerators and write the result over the denominator.

• To add or subtract unlike fractions, rename the fractions using the LCD. Then add or subtract as with like fractions.

• To add or subtract mixed numbers, first add or subtract the fractions. If necessary, rename them using the LCD. Then add or subtract the whole numbers and simplify if necessary.

Multiplying and Dividing Fractions

(Lessons 5-5 and 5-7)

• To multiply fractions, multiply the numerators and multiply the denominators.

• The product of a number and its multiplicative inverse is 1.

• To divide by a fraction, multiply by its multiplicative inverse, or reciprocal.

Solving Equations (Lesson 5-6)

If you multiply each side of an equation by the same nonzero number, the two sides remain equal.

Key Vocabulary

compatible numbers (p. 232)

like fractions (p. 236)

multiplicative inverse (p. 258)

reciprocal (p. 258)

unlike fractions (p. 237)

Vocabulary Check

Choose the correct term or number to complete each sentence.

1. To add like fractions, add the (numerators, denominators).

2. The symbol $\approx$ means (*approximately, exactly*) *equal to*.

3. When dividing by a fraction, multiply by its (value, reciprocal).

4. When estimating, if the numerator of a fraction is much smaller than the denominator, round the fraction to 0, $\frac{1}{2}$.

5. Fractions with different denominators are called (like, unlike) fractions.

6. The multiplicative inverse of $\frac{5}{6}$ is $\left(\frac{6}{5}, -\frac{5}{6}\right)$.

7. The mixed number $2\frac{4}{7}$ can be renamed as $\left(2\frac{7}{7}, 1\frac{11}{7}\right)$.

8. When multiplying fractions, multiply the numerators and (multiply, keep) the denominators.

9. The reciprocal of $\frac{1}{3}$ is $(-3, 3)$.

10. The fractions $\frac{4}{16}$ and $\frac{2}{4}$ are (like, unlike) fractions.

11. Another word for multiplicative inverse is (reciprocal, denominator).

12. The fraction $\frac{x}{2}$ can be read *x multiplied by* $\left(2, \frac{1}{2}\right)$.

Lesson-by-Lesson Review

  Estimating with Fractions (pp. 230–235)

7.PS.15, 7.CM.2

Estimate.

13. $2\frac{9}{10} \div 1\frac{1}{8}$ **14.** $6\frac{2}{9} - 5\frac{1}{7}$

15. $\frac{13}{15} \times \frac{1}{5}$ **16.** $\frac{1}{2} + \frac{3}{8}$

17. $\frac{1}{2} \cdot 25$ **18.** $15\frac{6}{7} \div 7\frac{1}{3}$

19. MEASUREMENT Gina wishes to carpet her living room. It has a length of $18\frac{5}{8}$ feet and the width of her living room is $9\frac{1}{2}$ feet. About how many square feet of carpet would be needed for her living room?

20. FOOTBALL Jamil practiced football for $1\frac{3}{4}$ hours on Saturday and $2\frac{2}{3}$ hours on Sunday. About how many more hours did he practice on Sunday than on Saturday?

Example 1 Estimate $5\frac{1}{12} + 2\frac{5}{6}$.

1 is much smaller than 12, so $5\frac{1}{12} \approx 5$.

5 is almost as large as 6, so $2\frac{5}{6} \approx 3$.

$5\frac{1}{12} + 2\frac{5}{6} \approx 5 + 3$ or 8

The sum is *about* 8.

Example 2 Estimate $\frac{7}{8} - \frac{4}{7}$.

7 is almost as large as 8, so $\frac{7}{8} \approx 1$.

4 is about half of 7, so $\frac{4}{7} \approx \frac{1}{2}$.

$\frac{7}{8} - \frac{4}{7} \approx 1 - \frac{1}{2}$ or $\frac{1}{2}$

The difference is *about* $\frac{1}{2}$.

  Adding and Subtracting Fractions (pp. 236–241)

6.N.16

Add or subtract. Write in simplest form.

21. $\frac{2}{6} - \frac{1}{6}$ **22.** $\frac{3}{7} + \frac{9}{14}$

23. $\frac{1}{9} + \frac{5}{9}$ **24.** $\frac{9}{10} - \frac{3}{10}$

25. $\frac{5}{8} - \frac{5}{12}$ **26.** $\frac{3}{4} + \frac{7}{20}$

27. RAIN At 8 A.M., Della's rain gauge read $\frac{1}{8}$ inch. By 4 P.M., the gauge read $\frac{3}{4}$ inch. How much rain fell between 8 A.M. and 4 P.M.?

28. PIZZA Owen ate $\frac{1}{8}$ of a pizza Tuesday night. The next day, he ate an additional $\frac{1}{2}$ of the pizza. What fraction of the pizza has he eaten?

Example 3 Find $\frac{1}{6} + \frac{2}{3}$.

Use a model.

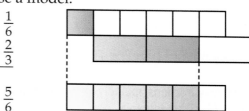

$\begin{array}{c}\frac{1}{6}\\+\frac{2}{3}\\\hline\\\frac{5}{6}\end{array}$

Example 4 Find $\frac{3}{10} - \frac{1}{4}$.

$\frac{3}{10} - \frac{1}{4} = \frac{6}{20} - \frac{5}{20}$ Rename the fractions using the LCD, 20.

$\qquad\qquad = \frac{1}{20}$ Subtract the numerators.

5-3 **Adding and Subtracting Mixed Numbers** (pp. 242–246)

6.N.18

Add or subtract. Write in simplest form.

29. $3\frac{2}{15} + 6\frac{9}{15}$

30. $4\frac{1}{3} - 2\frac{2}{3}$

31. $8\frac{2}{7} + 1\frac{6}{7}$

32. $7\frac{11}{12} - 4\frac{3}{12}$

33. $7\frac{3}{5} - 5\frac{1}{3}$

34. $5\frac{3}{4} + 1\frac{1}{6}$

35. $3\frac{5}{8} + 11\frac{1}{2}$

36. $4\frac{3}{10} - 2\frac{4}{5}$

37. **BABYSITTING** Lucas watched his little sister for $2\frac{1}{2}$ hours on Friday, $3\frac{2}{3}$ hours on Saturday, and $1\frac{3}{4}$ hours on Sunday. How many hours did Lucas watch his little sister?

Example 5 Find $5\frac{2}{3} + 3\frac{1}{2}$.

$5\frac{2}{3} + 3\frac{1}{2} = 5\frac{4}{6} + 3\frac{3}{6}$ Rename the fractions.

$= 8\frac{7}{6}$ Add the whole numbers and add the fractions.

$= 9\frac{1}{6}$ $8\frac{7}{6} = 8 + 1\frac{1}{6}$ or $9\frac{1}{6}$

Example 6 Find $4\frac{1}{5} - 2\frac{3}{5}$.

$4\frac{1}{5} - 2\frac{3}{5} = 3\frac{6}{5} - 2\frac{3}{5}$ Rename $4\frac{1}{5}$ as $3\frac{6}{5}$.

$= 1\frac{3}{5}$ Subtract the whole numbers and subtract the fractions.

5-4 **PSI: Eliminate Possibilities** (pp. 248–249)

7.PS.1

Solve by eliminating possibilities.

38. **SCHOOL** It takes Beth 15 minutes to walk to school, $\frac{1}{2}$ mile away. What is her walking pace?

A 2 miles per hour

B 1 mile per hour

C 7.5 miles per hour

D 30 miles per hour

39. **COOKING** Which of the following would yield a larger batch of bagels?

F Multiply a recipe by $\frac{1}{2}$.

G Divide a recipe by $\frac{1}{2}$.

H Multiply a recipe by $\frac{3}{4}$.

J Divide a recipe by 3.

Example 7 A group of friends went to a theme park. Six of the friends rode the Ferris wheel. If this was $\frac{2}{3}$ of the group, how many friends were in the group?

A 3

B 6

C 9

D 12

Since 6 friends rode the Ferris wheel and this was $\frac{2}{3}$ of the total number of friends in the group, the number of friends in the group must be greater than 6. So, eliminate choices A and B.

If there were 12 friends in the group, the 6 that rode the Ferris wheel would represent $\frac{1}{2}$ of the group. So, eliminate choice D.

Choice C is the only remaining possibility. Since 6 out of 9 is $\frac{6}{9}$ or $\frac{2}{3}$, C is correct.

Multiplying Fractions and Mixed Numbers (pp. 252–257)

6.N.18

Multiply. Write in simplest form.

40. $\frac{3}{5} \times \frac{2}{7}$ 41. $\frac{5}{12} \times \frac{4}{9}$

42. $\frac{3}{5} \times \frac{10}{21}$ 43. $4 \times \frac{13}{20}$

44. $2\frac{1}{3} \times \frac{3}{4}$ 45. $4\frac{1}{2} \times 2\frac{1}{12}$

46. **FOOD** An average slice of American cheese is about $\frac{1}{8}$ inch thick. What is the height of a package containing 20 slices?

Example 8   Find $\frac{5}{9} \times \frac{2}{3}$.

$\frac{5}{9} \times \frac{2}{3} = \frac{5 \times 2}{9 \times 3}$ Multiply the numerators and multiply the denominators.

$= \frac{10}{27}$ Simplify.

Example 9   Find $3\frac{1}{2} \times 2\frac{3}{4}$.

$3\frac{1}{2} \times 2\frac{3}{4} = \frac{7}{2} \times \frac{11}{4}$ Rename $3\frac{1}{2}$ and $2\frac{3}{4}$.

$= \frac{7 \times 11}{2 \times 4}$ Multiply the numerators and multiply the denominators.

$= \frac{77}{8}$ or $9\frac{5}{8}$ Simplify.

Algebra: Solving Equations (pp. 258–263)

7.A.4

Find the multiplicative inverse of each number.

47. $\frac{7}{12}$ 48. 5   49. $3\frac{1}{3}$

Solve each equation. Check your solution.

50. $8 = \frac{w}{2}$ 51. $\frac{4}{5}b = 12$

52. $-7.6 = \frac{n}{3}$ 53. $\frac{x}{0.3} = 2.5$

54. **BOOKS** Of the books on a shelf, $\frac{2}{3}$ are mysteries. If there are 10 mystery books, how many books are on the shelf?

Example 10   Find the multiplicative inverse of $\frac{9}{5}$.

$\frac{9}{5} \cdot \frac{5}{9} = 1$ The product of $\frac{9}{5}$ and $\frac{5}{9}$ is 1.

The multiplicative inverse of $\frac{9}{5}$ is $\frac{5}{9}$.

Example 11   Solve $\frac{3}{4}g = 2$.

$\frac{3}{4}g = 2$ Write the equation.

$\frac{4}{3} \cdot \frac{3}{4}g = \frac{4}{3} \cdot 2$ Multiply each side by the reciprocal of $\frac{3}{4}$.

$g = \frac{8}{3}$ or $2\frac{2}{3}$ Simplify.

Dividing Fractions and Mixed Numbers (pp. 265–270)

6.N.18

Divide. Write in simplest form.

55. $\frac{3}{5} \div \frac{6}{7}$ 56. $4 \div \frac{2}{3}$

57. $2\frac{3}{4} \div \frac{5}{6}$ 58. $-\frac{2}{5} \div 3$

59. $4\frac{3}{10} \div 2\frac{1}{5}$ 60. $-\frac{2}{7} \div \frac{8}{21}$

61. **MEASUREMENT** How many $\frac{1}{8}$-inch lengths are in $6\frac{3}{4}$ inches?

Example 12   Find $2\frac{4}{5} \div \frac{7}{10}$.

$2\frac{4}{5} \div \frac{7}{10} = \frac{14}{5} \div \frac{7}{10}$ Rename $2\frac{4}{5}$.

$= \frac{\overset{2}{\cancel{14}}}{\underset{1}{\cancel{5}}} \cdot \frac{\overset{2}{\cancel{10}}}{\underset{1}{\cancel{7}}}$ Multiply by the reciprocal of $\frac{7}{10}$.

$= \frac{4}{1}$ or 4   Simplify.

Estimate.

1. $5\frac{7}{9} - 1\frac{2}{13}$

2. $3\frac{1}{12} + 6\frac{5}{7}$

3. $\frac{3}{7} \times \frac{13}{15}$

4. $5\frac{2}{3} \div 1\frac{4}{5}$

5. **BAKING** A restaurant uses $2\frac{3}{4}$ pounds of flour to make a batch of dinner rolls. About how many pounds of flour are needed if 3 batches of dinner rolls are to be made?

Add, subtract, multiply, or divide. Write in simplest form.

6. $\frac{4}{15} + \frac{8}{15}$

7. $\frac{7}{10} - \frac{1}{6}$

8. $\frac{5}{8} \times \frac{2}{5}$

9. $6 \times \frac{8}{21}$

10. $4\frac{5}{12} - 2\frac{1}{12}$

11. $6\frac{7}{9} + 3\frac{5}{12}$

12. $8\frac{4}{7} - 1\frac{5}{14}$

13. $4\frac{5}{6} \times 1\frac{2}{3}$

14. $\frac{8}{9} \div 5\frac{1}{3}$

15. $\frac{1}{6} \div 5$

16. **MULTIPLE CHOICE** Seth drove $5\frac{3}{4}$ miles to the bank, $6\frac{1}{3}$ miles to the post office, and $4\frac{5}{6}$ miles to the park. What is the total distance Seth drove?

 A $15\frac{9}{13}$ miles

 B $\frac{7}{12}$ miles

 C $\frac{11}{12}$ miles

 D $16\frac{11}{12}$ miles

17. **SPORTS** Tyler's football practice lasted $2\frac{1}{2}$ hours. If $\frac{1}{4}$ of the time was spent catching passes, how many hours were spent catching passes?

18. **MEASUREMENT** The floor of a moving van is $11\frac{1}{3}$ feet long and $7\frac{5}{12}$ feet wide. Find the area of the moving van floor.

19. **MULTIPLE CHOICE** For his birthday, Keith received a check from his grandmother. Of this amount, the table shows how he spent or saved the money.

Fraction of Check	How Spent or Saved
$\frac{2}{5}$	spent on baseball cards
$\frac{1}{4}$	spent on a CD
$\frac{7}{20}$	deposited into savings account

Two weeks later, he withdrew $\frac{2}{3}$ of the amount he had deposited into his savings account. What fraction of the original check did he withdraw from his savings account?

 F $\frac{2}{3}$

 G $\frac{9}{23}$

 H $\frac{7}{30}$

 J $\frac{7}{40}$

20. **MEASUREMENT** An ounce is $\frac{1}{16}$ of a pound. How many ounces are in $8\frac{3}{4}$ pounds?

ALGEBRA Solve each equation. Check your solution.

21. $\frac{y}{3.7} = 8.1$

22. $6 = \frac{2}{5}m$

23. $\frac{3}{4} = \frac{5}{8}x$

24. $3\frac{1}{6} = \frac{2}{3}p$

25. **MULTIPLE CHOICE** Maria is making a mural that is $9\frac{2}{3}$ feet long. She wants to divide the mural into sections that are each $\frac{5}{8}$ foot. Which equation can be used to find n, the number of sections in Maria's mural?

 A $\frac{5}{8}n = 9\frac{2}{3}$

 B $9\frac{2}{3}n = \frac{5}{8}$

 C $\frac{5}{8} + n = 9\frac{2}{3}$

 D $n - \frac{5}{8} = 9\frac{2}{3}$

PART 1 Multiple Choice

Read each question. Then fill in the correct answer on the answer sheet provided by your teacher or on a sheet of paper.

1. Mrs. Brown needs to make two different desserts for a dinner party. The first recipe requires $2\frac{1}{4}$ cups of flour, and the second recipe requires $\frac{3}{4}$ cup less than the first. Which equation can be used to find n, the number of cups of flour needed for the second recipe?

   A $n = 2\frac{1}{4} - \frac{3}{4}$　　　**C** $n = 2\frac{1}{4} + \frac{3}{4}$

   B $n = 2\frac{1}{4} \cdot \frac{3}{4}$　　　**D** $n = 2\frac{1}{4} \div \frac{3}{4}$

2. Which of the following is true concerning the least common multiple of 6 and 9?

   F It is greater than the least common multiple of 8 and 12.

   G It is greater than the least common multiple of 5 and 15.

   H It is less than the least common multiple of 4 and 6.

   J It is less than the least common multiple of 3 and 4.

3. Kyle's hockey team has 6 sixth graders, 9 seventh graders, and 5 eighth graders. Which statement below is true?

   A One fourth of the team members are sixth graders.

   B More than half of the team members are seventh graders.

   C 25% of the team members are eighth graders.

   D 30% of the team members are seventh graders.

4. The fraction $\frac{5}{6}$ is found between which pair of fractions on a number line?

   F $\frac{1}{4}$ and $\frac{5}{8}$

   G $\frac{1}{3}$ and $\frac{4}{9}$

   H $\frac{11}{12}$ and $\frac{31}{36}$

   J $\frac{7}{12}$ and $\frac{17}{18}$

5. The table shows the distance Kelly swam over a four-day period. What was the total distance, in miles, Kelly swam?

Kelly's Swimming	
Day	**Distance (miles)**
Monday	1.5
Tuesday	$2\frac{3}{4}$
Wednesday	2.3
Thursday	$3\frac{1}{2}$

   A 10.5 miles　　　**C** $10\frac{1}{20}$ miles

   B $10\frac{1}{4}$ miles　　　**D** 9 miles

6. Which of the following gives the correct meaning of the expression $\frac{5}{8} \div \frac{1}{3}$?

   F $\frac{5}{8} \div \frac{1}{3} = \frac{8}{5} \times \frac{3}{1}$

   G $\frac{5}{8} \div \frac{1}{3} = \frac{5}{8} \times \frac{3}{1}$

   H $\frac{5}{8} \div \frac{1}{3} = \frac{5+1}{8+3}$

   J $\frac{5}{8} \div \frac{1}{3} = \frac{5}{8} \times \frac{1}{3}$

7. What is the value of the expression $(3 + 4)^2 \div 7 - 2 \times 6$?

A −9   C 30

B −5   D 1

8. Which line contains the ordered pair $(−1, 2)$?

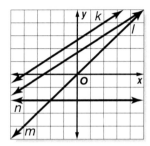

F Line k

G Line l

H Line m

J Line n

9. A pizza shop tried 45 new types of pizza during the past year and 20% of them became popular. Which best represents the fraction of pizzas that did *not* become popular?

A $\frac{1}{5}$ C $\frac{3}{5}$

B $\frac{4}{9}$ D $\frac{4}{5}$

TEST-TAKING TIP

Question 9 Be sure to pay attention to emphasized words, or words that are italicized. In Question 9, you are asked to find which answer is *not* correct.

PART 2 Short Response/Grid In

Record your answers on the answer sheet provided by your teacher or on a sheet of paper.

10. Write an equation using two variables to show the relationship between the position x and the value of a term y.

Position (x)	1	2	3	4	5	n
Value of Term (y)	5	9	13	17	21	

11. Evan runs $2\frac{3}{8}$ miles each week. He runs $\frac{3}{4}$ mile on Mondays and $\frac{3}{4}$ mile on Tuesdays. How far does he run, in miles, if the only other day he runs each week is Thursday?

12. Find $15 \div (−5)$.

PART 3 Extended Response

Record your answers on the answer sheet provided by your teacher or on a sheet of paper. Show your work.

13. A box of laundry detergent contains 35 cups. It takes $1\frac{1}{4}$ cups per load of laundry.

a. Write an equation to represent how many loads ℓ you can wash with one box.

b. How many loads can you wash with one box?

c. How many loads can you wash with 3 boxes?

NEED EXTRA HELP?													
If You Missed Question...	1	2	3	4	5	6	7	8	9	10	11	12	13
Go to Lesson...	5-2	4-8	4-6	4-9	4-5	5-7	1-4	3-7	4-6	1-10	5-2	2-8	5-7
NYS Core Curriculum	6.N.16	7.N.9	6.N.11	7.N.3	7.N.17	6.N.18	7.N.11	8.A.4	6.N.11	7.A.10	6.N.16	7.N.12	6.N.18

Unit 3

Algebra and Number Sense: Proportions and Percents

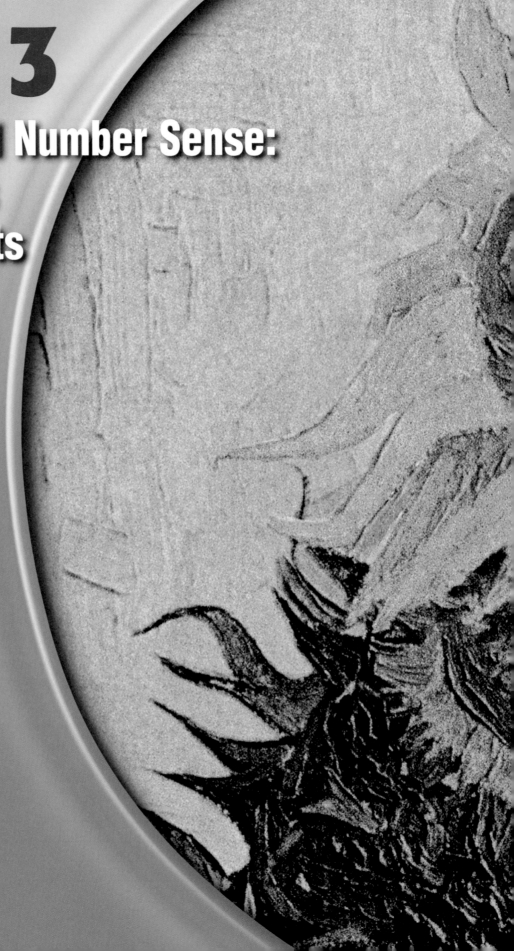

Focus

Develop an understanding of and apply proportionality.

CHAPTER 6
Ratios and Proportions

BIG Idea Use ratio and proportionality to solve problems, including those with tables and graphs.

CHAPTER 7
Applying Percents

BIG Idea Solve percent problems using ratios and proportions.

Problem Solving in Social Studies

Real-World Unit Project

It's Golden! On this adventure, you'll learn about the Golden Ratio in art and nature. Along the way, you'll research how artists use the Golden Ratio to create masterpieces. Also, you'll learn how the Golden Ratio occurs in nature as well. Our adventure will begin soon, so pack your art supplies and math tool kit. This is one golden adventure!

NY Math Online ▶ Log on to glencoe.com to begin.

CHAPTER 6

Ratios and Proportions

New York State Core Curriculum

7.PS.10 Use proportionality to model problems
7.M.5 Calculate unit price using proportions

Key Vocabulary

rate (p. 287)
ratio (p. 282)
proportional (p. 310)

🌐 Real-World Link

Statues A bronze replica of the Statue of Liberty can be found in Paris, France. The replica has the ratio 1 : 4 with the Statue of Liberty that stands in New York Harbor.

FOLDABLES®
Study Organizer

Ratios and Proportions Make this Foldable to help you organize your notes. Begin with a sheet of notebook paper.

1 **Fold** lengthwise to the holes.

2 **Cut** along the top line and then make equal cuts to form 7 tabs.

3 **Label** the major topics as shown.

Ratios
Rates
Rate of Change and Slope
Customary/ Metric Units
Proportions
Scale
Fractions, Decimals, and Percents

GET READY for Chapter 6

Diagnose Readiness You have two options for checking Prerequisite Skills.

Option 2

NY Math Online Take the Online Readiness Quiz at glencoe.com.

Option 1

Take the Quick Quiz below. Refer to the Quick Review for help.

QUICK Quiz

Evaluate each expression. Round to the nearest tenth if necessary. (Lesson 1-4)

1. $100 \times 25 \div 52$ 2. $10 \div 4 \times 31$

3. $\dfrac{63 \times 4}{34}$ 4. $\dfrac{2 \times 100}{68}$

Write each fraction in simplest form.
(Lesson 4-4)

5. $\dfrac{9}{45}$ 6. $\dfrac{16}{24}$ 7. $\dfrac{38}{46}$

8. **AGES** Mikhail is 14 years old. His father is 49 years old. What fraction, in simplest form, of his father's age is Mikhail? (Lesson 4-4)

Write each decimal as a fraction in simplest form. (Lesson 4-5)

9. 0.78 10. 0.320 11. 0.06

12. **SAVINGS** Belinda has saved 0.92 of the cost of a new bicycle. What fraction, in simplest form, represents her savings? (Lesson 4-5)

Multiply. (Lesson 1-2)

13. 4.5×10^2 14. 1.78×10^3

15. 0.22×10^4 16. 0.03×10^5

QUICK Review

Example 1 Evaluate $15 \times 32 \div 40$.

$15 \times 32 \div 40 = 480 \div 40$ Multiply 15 by 32.
$= 12$ Divide.

Example 2 Write $\dfrac{16}{44}$ in simplest form.

$$\overset{\div\, 4}{\underset{\div\, 4}{\dfrac{16}{44} = \dfrac{4}{11}}}$$ Divide the numerator and denominator by their GCF, 4.

Example 3 Write 0.62 as a fraction in simplest form.

$0.62 = \dfrac{62}{100}$ 0.62 is sixty-two hundredths.

$= \dfrac{31}{50}$ Divide the numerator and denominator by their GCF, 2.

Example 4 Find 3.9×10^3.

$3.9 \times 10^3 = 3.900$ Move the decimal point 3 places to the right. Annex two zeros.

$= 3,900$

Ratios

MAIN IDEA

Write ratios as fractions in simplest form and determine whether two ratios are equivalent.

NYS Core Curriculum

Preparation for 7.M.5 Calculate unit price using proportions
Preparation for 7.M.6 Compare unit prices

New Vocabulary

ratio
equivalent ratios

NY Math Online

glencoe.com

• Extra Examples
• Personal Tutor
• Self-Check Quiz
• Reading in the Content Area

▶ **GET READY** for the Lesson

SCHOOL The student-teacher ratio of a school compares the total number of students to the total number of teachers.

Middle School	Students	Teachers
Prairie Lake	396	22
Green Brier	510	30

1. Write the student-teacher ratio of Prairie Lake Middle School as a fraction. Then write this fraction with a denominator of 1.

2. Can you determine which school has the lower student-teacher ratio by examining just the number of teachers at each school? just the number of students at each school? Explain.

Ratios   **Key Concept**

Words   A **ratio** is a comparison of two quantities by division.

Examples   **Numbers**   **Algebra**

3 to 4   3:4 $\frac{3}{4}$ a to b $a:b$ $\frac{a}{b}$

Ratios can express part to part, part to whole, or whole to part relationships and are often written as fractions in simplest form.

EXAMPLE   Write Ratios in Simplest Form

① **GRILLING** Seasonings are often added to meat prior to grilling. Using the recipe, write a ratio comparing the amount of garlic powder to the amount of dried oregano as a fraction in simplest form.

Recipe: Greek Style Seasonings

4 tsp. garlic powder
6 tsp. dried oregano
2 tsp. pepper

garlic powder $\dfrac{4 \text{ tsp}}{6 \text{ tsp}} = \dfrac{\overset{2}{\cancel{4 \text{ tsp}}}}{\underset{3}{\cancel{6 \text{ tsp}}}}$ or $\dfrac{2}{3}$
dried oregano

The ratio of garlic powder to dried oregano is $\frac{2}{3}$, 2:3, or 2 to 3. That is, for every 2 units of garlic powder there are 3 units of dried oregano.

✓ **CHECK** Your Progress

Use the recipe to write each ratio as a fraction in simplest form.

a. pepper:garlic powder   b. oregano:pepper

Ratios that express the same relationship between two quantities are called **equivalent ratios**. Equivalent ratios have the same value.

Study Tip

Writing Ratios
Ratios greater than 1 are expressed as improper fractions and not as mixed numbers.

EXAMPLE Identify Equivalent Ratios

2 Determine whether the ratios 250 miles in 4 hours and 500 miles in 8 hours are equivalent.

METHOD 1   Compare the ratios written in simplest form.

$$250 \text{ miles} : 4 \text{ hours} = \frac{250 \div 2}{4 \div 2} \text{ or } \frac{125}{2}$$ Divide the numerator and denominator by the GCF, 2

$$500 \text{ miles} : 8 \text{ hours} = \frac{500 \div 4}{8 \div 4} \text{ or } \frac{125}{2}$$ Divide the numerator and denominator by the GCF, 4

The ratios simplify to the same fraction.

METHOD 2   Look for a common multiplier relating the two ratios.

$$\overset{\times 2}{\frac{250}{4}} = \frac{500}{8}$$ The numerator and denominator of the ratios are related by the same multiplier, 2.

The ratios are equivalent.

✓ CHOOSE Your Method

Determine whether the ratios are equivalent.

c. 20 nails for every 5 shingles, 12 nails for every 3 shingles

d. 2 cups flour to 8 cups sugar, 8 cups flour to 14 cups sugar

Real-World EXAMPLE

3 **BASEBALL** Derek Jeter of the New York Yankees had 32 hits out of 93 times at bat. Jorge Posada had 11 hits out of 31 times at bat. Are these ratios equivalent? Justify your answer.

Derek Jeter

$$32 : 93 = \frac{32}{93}$$

Jorge Posada

$$11 : 31 = \frac{11 \times 3}{31 \times 3} \text{ or } \frac{33}{93}$$

Since $\frac{32}{93} \neq \frac{33}{93}$, the ratios are not equivalent.

✓ CHECK Your Progress

e. **SWIMMING** A community pool requires there to be at least 3 lifeguards for every 20 swimmers. There are 60 swimmers and 9 lifeguards at the pool. Is this the correct number of lifeguards based on the above requirement? Justify your answer.

Real-World Link . . . .
In 2006, the New York Yankees had a 0.285 batting average as a team.
Source: Major League Baseball

CHECK Your Understanding

Example 1
(p. 282)

FIELD TRIPS Use the information in the table to write each ratio as a fraction in simplest form.

1. adults : students
2. students : buses
3. buses : people
4. adults : people

Field Trip Statistics	
Students	180
Adults	24
Buses	4

Example 2
(p. 283)

Determine whether the ratios are equivalent. Explain.

5. 12 out of 20 doctors agree
   6 out of 10 doctors agree

6. 2 DVDs to 7 CDs
   10 DVDs to 15 CDs

Example 3
(p. 283)

7. **SHOPPING** A grocery store has a brand-name cereal on sale at 2 boxes for $5. You buy 6 boxes and are charged $20. Based on the price ratio indicated, were you charged the correct amount? Justify your answer.

Practice and Problem Solving

HOMEWORK HELP

For Exercises	See Examples
8–17	1
18–21	2
22–23	3

SOCCER Use the Madison Mavericks team statistics to write each ratio as a fraction in simplest form.

8. wins : losses
9. losses : ties
10. losses : games played
11. wins : games played

Madison Mavericks Team Statistics	
Wins	10
Losses	12
Ties	8

CARNIVALS Use the following information to write each ratio as a fraction in simplest form.

At its annual carnival, Brighton Middle School had 6 food booths and 15 games booths. A total of 66 adults and 165 children attended. The carnival raised a total of $1,600. Of this money, $550 came from ticket sales.

12. children : adults
13. food booths : games booths
14. children : games booths
15. booths : money raised
16. people : children
17. non-ticket sale money : total money

Determine whether the ratios are equivalent. Explain.

18. 20 female lions to 8 male lions,
    34 female lions to 10 male lions

19. $4 for every 16 ounces,
    $10 for every 40 ounces

20. 27 students to 6 microscopes,
    18 students to 4 microscopes

21. 8 roses to 6 babies breath,
    12 roses to 10 babies breath

22. **BAKING** It is recommended that a ham be baked 1 hour for every 2 pounds of meat. Latrell baked a 9-pound ham for 4.5 hours. Did he follow the above recommendation? Justify your answer.

23. **FISHING** Kamala catches two similar looking fish. The larger fish is 12 inches long and 3 inches wide. The smaller fish is 6 inches long and 1 inch wide. Do these fish have an equivalent length to width ratio? Justify your answer.

MEASUREMENT The *aspect ratio* of a television is a ratio comparing the width and height. A wide screen television has an aspect ratio of 16:9. Televisions without the same aspect ratio crop the image to fit the screen. Determine which television sizes have a full 16:9 image. Justify your answers.

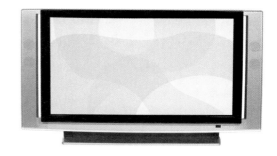

24. 32″ × 18″   **25.** 71″ × 42″   **26.** 48″ × 36″

MAMMALS For Exercises 27 and 28, use the information below.

Mammal	Average Brain Weight (lb)	Average Body Weight (lb)
Adult Human	3	150
Adult Orca Whale	12	5,500

27. How much greater is the average weight of an adult orca whale's brain than the average weight of an adult human's brain?

28. Find the brain-to-body weight ratio for each mammal. Are these ratios equivalent? If not, which mammal has the greater brain-to-body weight ratio? Justify your answer and explain its meaning.

29. MUSIC The pitch of a musical note is measured by the number of sound waves per second, or *hertz*. If the ratio of the frequencies of two notes can be simplified, the two notes are harmonious. Use the information at the right to find if notes E and G are harmonious. Explain.

 **E : 330 Hertz   G : 396 Hertz**

30. FOOD The ratio of the number of cups of chopped onion to the number of cups of chopped cilantro in a salsa recipe is 4:3. If the recipe calls for $\frac{2}{3}$ cup chopped onion, how many cups of chopped cilantro are needed?

ANALYZE TABLES For Exercises 31–33, use the table below that shows the logging statistics for three areas of forest.

Area	Estimated Number of Trees Left to Grow	Estimated Number of Trees Removed for Timber
A	440	1,200
B	1,625	3,750
C	352	960

31. For which two areas was the growth-to-removal ratio the same? Explain.

32. Which area had the greatest growth-to-removal ratio? Justify your answer.

33. Find the additional number of trees that should be planted and left to grow in area A so that its growth-to-removal ratio is the same as area B's. Justify your answer.

Real-World Link . . . .
An orca whale, also called a killer whale, is not really a whale, but a dolphin. Its average birth weight is 300 pounds.

NYSCC • NYSMT
Extra Practice, pp. 681, 709

34. **FIND THE ERROR** Cleveland and Luis are determining whether the ratios $\frac{6}{4}$ and $\frac{18}{16}$ are equivalent. Who is correct? Explain.

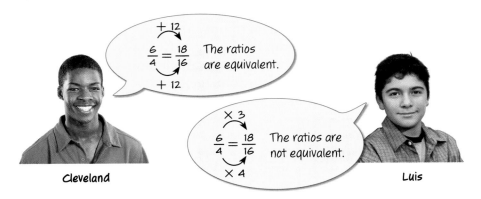

Cleveland

Luis

35. **CHALLENGE** Find the missing number in the following pattern. Explain your reasoning. (*Hint*: Look at the ratios of successive numbers.)

20, 40, 120, 480, ■

36. **WRITING IN MATH** Refer to the application in Exercises 31–33. What would a growth-to-removal ratio greater than 1 indicate?

NYSMT PRACTICE 7.M.5

37. Which of the following ratios does *not* describe a relationship between the marbles in the jar?

   A 8 white : 5 black
   B 2 white : 5 black
   C 5 black : 13 total
   D 8 white : 13 total

38. A class of 24 students has 15 boys. What ratio compares the number of girls to boys in the class?

   F 3 : 5   **H** 3 : 8
   G 5 : 3   **J** 8 : 3

Spiral Review

39. Find $1\frac{4}{7} \div 1\frac{5}{6}$. Write in simplest form. (Lesson 5-7)

ALGEBRA Solve each equation. Check your solution. (Lesson 5-6)

40. $\frac{y}{4} = 7$ 41. $\frac{1}{3}x = \frac{5}{9}$ 42. $4 = \frac{p}{2.7}$ 43. $2\frac{5}{6} = \frac{1}{2}a$

44. **MONEY** Grant and his brother put together their money to buy a present for their mom. If they had a total of $18 and Grant contributed $10, how much did his brother contribute? (Lesson 3-2)

▷ **GET READY for the Next Lesson**

PREREQUISITE SKILL Divide. (p. 676)

45. $9.8 \div 2$ 46. $\$4.30 \div 5$ 47. $\$12.40 \div 40$ 48. $27.36 \div 3.2$

6-2 Rates

MAIN IDEA

Determine unit rates.

NYS Core Curriculum

7.M.5 Calculate unit price using proportions
7.M.6 Compare unit prices *Also addresses 7.PS.10, 7.A.6*

New Vocabulary

rate
unit rate

NY Math Online

glencoe.com

• Extra Examples
• Personal Tutor
• Self-Check Quiz

▷ **MINI Lab**

Choose a partner and take turns taking each other's pulse for 2 minutes.

1. Count the number of beats for each of you.

2. Write the ratio *beats* to *minutes* as a fraction.

A ratio that compares two quantities with different kinds of units is called a **rate**.

$$\frac{160 \text{ beats}}{2 \text{ minutes}}$$ ← The units *beats* and *minutes* are different.

When a rate is simplified so that it has a denominator of 1 unit, it is called a **unit rate**.

$$\frac{80 \text{ beats}}{1 \text{ minute}}$$ ← The denominator is 1 unit.

The table below shows some common unit rates.

Rate	Unit Rate	Abbreviation	Name
$\frac{\text{number of miles}}{1 \text{ hour}}$	miles per hour	mi/h or mph	average speed
$\frac{\text{number of miles}}{1 \text{ gallon}}$	miles per gallon	mi/gal or mpg	gas mileage
$\frac{\text{number of dollars}}{1 \text{ pound}}$	price per pound	dollars/lb	unit price
$\frac{\text{number of dollars}}{1 \text{ hour}}$	dollars per hour	dollars/h	hourly wage

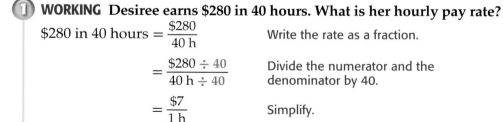 **Real-World EXAMPLE** Find a Unit Rate

① **WORKING** Desiree earns $280 in 40 hours. What is her hourly pay rate?

$280 in 40 hours $= \dfrac{\$280}{40 \text{ h}}$ Write the rate as a fraction.

$= \dfrac{\$280 \div 40}{40 \text{ h} \div 40}$ Divide the numerator and the denominator by 40.

$= \dfrac{\$7}{1 \text{ h}}$ Simplify.

Desiree's hourly pay rate is $7.

 **Your Progress**

Find each unit rate. Round to the nearest hundredth if necessary.

a. $300 for 6 hours

b. 220 miles on 8 gallons

2 **JUICE** Find the unit price if it costs $2 for eight juice boxes. Round to the nearest cent if necessary.

$2 for eight boxes $= \dfrac{\$2}{8 \text{ boxes}}$ Write the rate as a fraction.

$= \dfrac{\$2 \div 8}{8 \text{ boxes} \div 8}$ Divide the numerator and the denominator by 8.

$= \dfrac{\$0.25}{1 \text{ box}}$ Simplify.

The unit price is $0.25 per juice box.

CHECK Your Progress

c. **ESTIMATION** Find the unit price if a 4-pack of mixed fruit sells for $2.12.

Unit rates are useful when you want to make comparisons.

 NYSMT EXAMPLE Compare Using Unit Rates

3 The prices of 3 different bags of dog food are given in the table. Which size bag has the lowest price per pound?

A the 40-lb bag

B the 20-lb bag

C the 8-lb bag

D All three bag sizes have the same price per pound.

Dog Food Prices	
Bag Size (pounds)	**Price**
40	$49.00
20	$23.44
8	$9.88

Test-Taking Tip

Alternative Method
One 40-lb bag is equivalent to two 20-lb bags or five 8-lb bags. The cost for one 40-lb bag is $49, the cost for two 20-lb bags is about 2 × $23 or $46, and the cost for five 8-lb bags is about 5 × $10 or $50. So the 20-lb bag has the lowest price per pound.

Read the Item

To determine the lowest price per pound, find and compare the unit price for each size bag.

Solve the Item

40-pound bag $49.00 ÷ 40 pounds = $1.225 per pound

20-pound bag $23.44 ÷ 20 pounds = $1.172 per pound

8-pound bag $9.88 ÷ 8 pounds = $1.235 per pound

At $1.172 per pound, the 20-pound bag sells for the lowest price per pound.

The answer is B.

d. Tito wants to buy some peanut butter to donate to the local food pantry. If Tito wants to save as much money as possible, which brand should he buy?

Peanut Butter Sales	
Brand	**Sale Price**
Nutty	12 ounces for $2.19
Grandma's	18 ounces for $2.79
Bee's	28 ounces for $4.69
Save-A-Lot	40 ounces for $6.60

    F Nutty, because the quality of the peanut butter is better

    G Grandma's, because the price per ounce is about $0.16

    H Bee's, because the price per ounce is about $0.14

    J Save-A-Lot, because he wants to buy 40 ounces

Real-World Link . . . .
Face paint can be made from 1 teaspoon cornstarch and $\frac{1}{2}$ teaspoon each of water and cold cream.

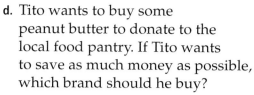 **Real-World EXAMPLE**   Use a Unit Rate

4 **FACE PAINTING** Lexi painted 3 faces in 12 minutes at the Crafts Fair. At this rate, how many faces can she paint in 40 minutes?

Find the unit rate. Then multiply this unit rate by 40 to find the number of faces she can paint in 40 minutes.

$$3 \text{ faces in } 12 \text{ minutes} = \frac{3 \text{ faces} \div 12}{12 \text{ min} \div 12} = \frac{0.25 \text{ faces}}{1 \text{ min}} \quad \text{Find the unit rate.}$$

$$\frac{0.25 \text{ faces}}{1 \text{ min}} \cdot 40 \text{ min} = 10 \text{ faces} \quad \text{Divide out the common units.}$$

Lexi can paint 10 faces in 40 minutes.

 CHECK Your Progress

e. **SCHOOL SUPPLIES** Kimbel bought 4 notebooks for $6.32. At this same unit price, how much would he pay for 5 notebooks?

CHECK Your Understanding

Examples 1, 2
(pp. 287–288)

Find each unit rate. Round to the nearest hundredth if necessary.

  1. 90 miles on 15 gallons   **2.** 1,680 kilobytes in 4 minutes

  3. 5 pounds for $2.49   **4.** 152 feet in 16 seconds

Example 3
(pp. 288–289)

5. **MULTIPLE CHOICE** Four stores offer customers bulk CD rates. Which store offers the best buy?

    A CD Express   **C** Music Place

    B CD Rack   **D** Music Shop

Bulk CD Offers	
Store	**Offer**
CD Express	4 CDs for $60
Music Place	6 CDs for $75
CD Rack	5 CDs for $70
Music Shop	3 CDs for $40

Example 4
(p. 289)

6. **TRAVEL** After 3.5 hours, Pasha had traveled 217 miles. At this same speed, how far will she have traveled after 4 hours?

Find each unit rate. Round to the nearest hundredth if necessary.

7. 360 miles in 6 hours

8. 6,840 customers in 45 days

9. 152 people for 5 classes

10. 815 Calories in 4 servings

11. 45.5 meters in 13 seconds

12. $7.40 for 5 pounds

13. $1.12 for 8.2 ounces

14. 144 miles in 4.5 gallons

15. **ESTIMATION** Estimate the unit rate if 12 pairs of socks sell for $5.79.

16. **ESTIMATION** Estimate the unit rate if a 26-mile marathon was completed in 5 hours.

17. **SPORTS** The results of a swim meet are shown. Who swam the fastest? Explain your reasoning.

Name	Event	Time (s)
Tawni	50-m Freestyle	40.8
Pepita	100-m Butterfly	60.2
Susana	200-m Medley	112.4

18. **MONEY** A grocery store sells three different packages of bottled water. Which package costs the least per bottle? Explain your reasoning.

6-pack for $3.79

9-pack for $4.50

12-pack for $6.89

NUTRITION For Exercises 19 and 20, use the table at the right.

19. Which soft drink has about twice the amount of sodium per ounce than the other two? Explain.

20. Which soft drink has the least amount of sugar per ounce? Explain.

Soft Drink Nutritional Information			
Soft Drink	Serving Size (oz)	Sodium (mg)	Sugar (g)
A	12	40	22
B	8	24	15
C	7	42	30

21. **WORD PROCESSING** Ben can type 153 words in 3 minutes. At this rate, how many words can he type in 10 minutes?

22. **FABRIC** Marcus buys 3 yards of fabric for $7.47. Later he realizes that he needs 2 more yards. How much will he pay for the extra fabric?

23. **ESTIMATION** A player scores 87 points in 6 games. At this rate, about how many points would she score in the next 4 games?

Real-World Link · · · · ·
North Carolina has approximately 8.9 million people living in 48,718 square miles.
Source: U.S. Census Bureau

24. **JOBS** Dalila earns $94.20, for working 15 hours as a holiday helper wrapping gifts. If she works 18 hours the next week, how much money will she earn?

25. **POPULATION** Use the information at the left. What is the *population density* or number of people per square mile in North Carolina?

Rate of Change and Slope

MAIN IDEA

Identify rate of change and slope using tables and graphs.

NYS Core Curriculum

Preparation for 8.G.13 Determine the slope of a line from a graph and explain the meaning of slope as a constant rate of change

New Vocabulary

rate of change
slope

NY Math Online

glencoe.com

• Extra Examples
• Personal Tutor
• Self-Check Quiz

▷ **GET READY** for the Lesson

HEIGHTS The table shows Stephanie's height at ages 9 and 12.

Age (yr)	9	12
Height (in.)	53	59

1. What is the change in Stephanie's height from ages 9 to 12?

2. Over what number of years did this change take place?

3. Write a rate that compares the change in Stephanie's height to the change in age. Express your answer as a unit rate and explain its meaning.

A **rate of change** is a rate that describes how one quantity changes in relation to another. A rate of change is usually expressed as a unit rate.

EXAMPLE Find Rate of Change from a Table

1. **FUNDRAISING** The table shows the amount of money a Booster Club made washing cars for a fundraiser. Use the information to find the rate of change in dollars per car.

Cars Washed	
Number	**Money ($)**
5	40
10	80
15	120
20	160

+5 ⟨ ⟩ +40
+5 ⟨ ⟩ +40
+5 ⟨ ⟩ +40

Find the unit rate to determine the rate of change.

$$\frac{\text{change in money}}{\text{change in cars}} = \frac{40 \text{ dollars}}{5 \text{ cars}} \quad \text{The money earned increases by } \$40 \text{ for every 5 cars.}$$

$$= \frac{8 \text{ dollar}}{1 \text{ car}} \quad \text{Write as a unit rate.}$$

So, the number of dollars earned increases by $8 for every car washed.

✓ **CHECK** Your Progress

a. **PLANES** The table shows the number of miles a plane traveled while in flight. Use the information to find the approximate rate of change in miles per minute.

Time (min)	30	60	90	120
Distance (mi)	290	580	870	1,160

EXAMPLE Find Rate of Change from a Graph

Reading Math

Ordered Pairs The ordered pair (2, 120) represents traveling 120 miles in 2 hours.

② **DRIVING** The graph represents the distance traveled while driving on a highway. Use the graph to find the rate of change in miles per hour.

To find the rate of change, pick any two points on the line, such as (1, 60) and (2, 120).

$$\frac{\text{change in miles}}{\text{change in hours}} = \frac{(120 - 60) \text{ miles}}{(2 - 1) \text{ hours}}$$

$$= \frac{60 \text{ miles}}{1 \text{ hour}}$$

The distance increases by 60 miles in 1 hour. So, the rate of traveling on a highway is 60 miles per hour.

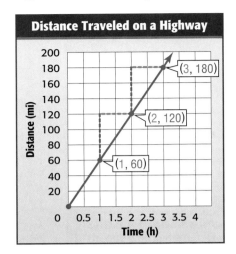

CHECK Your Progress

b. **DRIVING** Use the graph to find the rate of change in miles per hour while driving in the city.

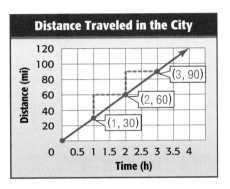

Notice that the graph in Example 2 about driving on a highway represents a rate of change of 60 mph. The graph in Check Your Progress about driving in the city is not as steep. It represents a rate of change of 30 mph.

The constant rate of change in *y* with respect to the constant change *x* is also called the slope of a line. Slope is a number that tells how steep the line is. The slope is the same for any two points on a straight line.

Slope   **Key Concept**

Slope is the rate of change between any two points on a line.

$$\text{slope} = \frac{\text{change in } y}{\text{change in } x} \leftarrow \text{vertical change} \atop \leftarrow \text{horizontal change}$$

$$= \frac{2}{1} \text{ or } 2$$

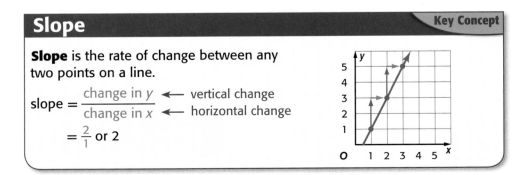

Real-World EXAMPLE   Find Slope

PHYSICAL SCIENCE The table below shows the relationship between the number of seconds y it takes to hear the thunder after a lightning strike and the distance x you are from the lightning.

Distance (x)	0	1	2	3	4	5
Seconds (y)	0	5	10	15	20	25

3 **Graph the data. Then find the slope of the line. Explain what the slope represents.**

$\text{slope} = \dfrac{\text{change in } y}{\text{change in } x}$ Definition of slope

$= \dfrac{25 - 10}{5 - 2}$ Use (2, 10) and (5, 25).

$= \dfrac{15}{3}$ ⟵ seconds ⟵ miles

$= \dfrac{5}{1}$ Simplify.

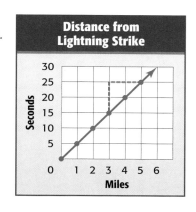

So, for every 5 seconds between a lightning flash and the sound of the thunder, there is 1 mile between you and the lightning strike.

Real-World Link . . . .
Lightning strikes somewhere on the surface of Earth about 100 times every second.
Source: National Geographic

CHECK Your Progress

c. WATER Graph the data. Then find the slope of the line. Explain what the slope represents.

Water Level Loss	
Week	**Water Loss (cm)**
1	1.5
2	3
3	4.5
4	6

CHECK Your Understanding

Example 1
(p. 293)

1. Use the information in the table to find the rate of change in degrees per hour.

Temperature (°F)	54	57	60	63
Time	6 A.M.	8 A.M.	10 A.M.	12 P.M.

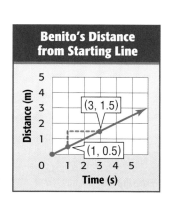

Example 2
(p. 294)

2. DISTANCE The graph shows Benito's distance from the starting line. Use the graph to find the rate of change.

Example 3
(p. 295)

3. **SNACKS** The table below shows the number of small packs of fruit snacks y per box x. Graph the data. Then find the slope of the line. Explain what the slope represents.

Boxes (x)	3	5	7	9
Packs (y)	24	40	56	72

Practice and Problem Solving

HOMEWORK HELP

For Exercises	See Examples
4–6	1
7, 8	2
9, 10	3

For Exercises 4 and 5, find the rate of change for each table.

4.
Time (s)	Distance (m)
0	6
1	12
2	18
3	24

5.
Time (h)	Wage ($)
0	0
1	9
2	18
3	27

6. The number of minutes included in different cell phone plans and the costs are shown in the table. What is the approximate rate of change in cost per minute?

Cost ($)	38	50	62	74	86
Minutes	1,000	1,500	2,000	2,500	3,000

For Exercises 7 and 8, find the rate of change for each graph.

7.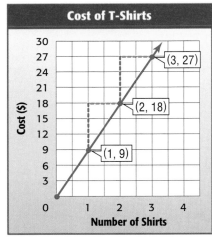

8.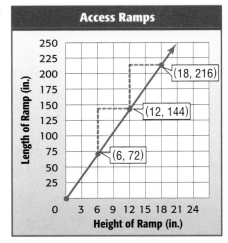

9. **CYCLING** The table shows the distance y Cheryl traveled in x minutes while competing in the cycling portion of a triathlon. Graph the data. Then find the slope of the line. Explain what the slope represents.

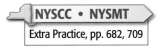
Extra Practice, pp. 682, 709

Time (min)	45	90	135	180
Distance (km)	5	10	15	20

10. **MAPS** The table shows the key for a map. Graph the data. Then find the slope of the line.

Distance on Map (in.)	2	4	6	8
Actual Distance (mi)	40	80	120	160

11. **WATER** At 1:00, the water level in a pool is 13 inches. At 2:30, the water level is 28 inches. What is the rate of change?

12. **MONEY** Dwayne opens a savings account with $75. He makes the same deposit every month and makes no withdrawals. After 3 months, he has $150. After 6 months, he has $300. After 9 months, he has $450 dollars. What is the rate of change?

H.O.T. Problems

13. **OPEN ENDED** Make a table where the rate of change is 6 inches for every foot.

14. **WRITING IN MATH** Write a problem to represent a rate of change of $15 per item.

NYSMT PRACTICE 8.G.13

15. Use the information in the table to find the rate of change.

Number of Apples	Number of Seeds
3	30
7	70
11	110

A $\frac{10}{1}$ C $\frac{40}{4}$

B $\frac{1}{10}$ D $\frac{4}{40}$

16. **SHORT RESPONSE** Find the slope of the line below that shows the distance Jairo traveled while jogging.

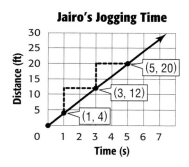

Jairo's Jogging Time

(5, 20)
(3, 12)
(1, 4)

Spiral Review

17. **GROCERIES** Three pounds of pears cost $3.57. At this rate, how much would 10 pounds cost? (Lesson 6-2)

Write each ratio as a fraction in simplest form. (Lesson 6-1)

18. 9 feet in 21 minutes   19. 36 calls in 2 hours   20. 14 SUVs out of 56 vehicles

▷ **GET READY for the Next Lesson**

PREREQUISITE SKILL Solve. (Page 674)

21. 2.5×20 22. 3.5×4 23. $104 \div 16$ 24. $4{,}200 \div 2{,}000$

Measurement: Changing Customary Units

MAIN IDEA

Change units in the customary system.

NYS Core Curriculum

7.M.3 Identify customary and metric **units of mass   7.M.4 Convert mass within a given system** *Also addresses 7.M.2, 7.M.9, 7.M.12, 7.M.13*

New Vocabulary

unit ratio

NY Math Online

glencoe.com

• Extra Examples
• Personal Tutor
• Self-Check Quiz

▷ **GET READY** for the Lesson

ANIMALS The table shows the approximate weights in tons of several large land animals. One ton is equivalent to 2,000 pounds.

Animal	Weight (T)
Grizzly Bear	1
White Rhinoceros	4
Hippopotamus	5
African Elephant	8

You can use a *ratio table*, which have columns filled with ratios that have the same value, to convert each weight from tons to pounds.

1. Copy and complete the ratio table. The first two ratios are done for you.

   ×4

Tons	1	4	5	8
Pounds	2,000	8,000		

   ×4

   To produce equivalent ratios, multiply the quantities in each row by the same number.

2. Then graph the ordered pairs (tons, pounds) from the table. Label the horizontal axis *Weight in Tons* and the vertical axis *Weight in Pounds*. Connect the points. What do you notice about the graph of these data?

The relationships among the most commonly used customary units of length, weight, and capacity are shown in the table below.

Customary Units			Key Concept
Type of Measure	**Larger Unit** →		**Smaller Unit**
Length	1 foot (ft)	=	12 inches (in.)
	1 yard (yd)	=	3 feet
	1 mile (mi)	=	5,280 feet
Weight	1 pound (lb)	=	16 ounces (oz)
	1 ton (T)	=	2,000 pounds
Capacity	1 cup (c)	=	8 fluid ounces (fl oz)
	1 pint (pt)	=	2 cups
	1 quart (qt)	=	2 pints
	1 gallon (gal)	=	4 quarts

Each of the relationships above can be written as a unit ratio. Like a unit rate, a **unit ratio** is one in which the denominator is 1 unit.

$$\frac{3 \text{ ft}}{1 \text{ yd}} \qquad \frac{2{,}000 \text{ lb}}{1 \text{ T}} \qquad \frac{4 \text{ qt}}{1 \text{ gal}}$$

Notice that the numerator and denominator of each fraction above are equivalent, so the value of each ratio is 1. You can multiply by a unit ratio of this type to *convert* or change from larger units to smaller units.

Study Tip

Multiplying by 1
Although the number and units changed in Example 1, because the measure is multiplied by 1, the value of the converted measure is the same as the original.

EXAMPLES Convert Larger Units to Smaller Units

1 Convert 20 feet into inches.

Since 1 foot = 12 inches, the unit ratio is $\frac{12 \text{ in.}}{1 \text{ ft}}$.

$$20 \text{ ft} = 20 \text{ ft} \cdot \frac{12 \text{ in.}}{1 \text{ ft}} \qquad \text{Multiply by } \frac{12 \text{ in.}}{1 \text{ ft}}.$$

$$= 20 \text{ ft} \cdot \frac{12 \text{ in.}}{1 \text{ ft}} \qquad \begin{array}{l}\text{Divide out common units, leaving} \\ \text{the desired unit, inches.}\end{array}$$

$$= 20 \cdot 12 \text{ in. or } 240 \text{ in.} \qquad \text{Multiply.}$$

So, 20 feet = 240 inches.

2 GARDENING Clarence mixes $\frac{1}{4}$ cup of fertilizer with soil before planting each bulb. How many ounces of fertilizer does he use per bulb?

$$\frac{1}{4} \text{ c} = \frac{1}{4} \text{ c} \cdot \frac{8 \text{ fl oz}}{1 \text{ c}} \qquad \begin{array}{l}\text{Since 1 cup} = 8 \text{ fluid ounces, multiply by} \\ \frac{8 \text{ fl oz}}{1 \text{ c}}. \text{ Then, divide out common units.}\end{array}$$

$$= \frac{1}{4} \cdot 8 \text{ fl oz or } 2 \text{ fl oz} \qquad \text{Multiply.}$$

So, 2 fluid ounces of fertilizer are used per bulb.

CHECK Your Progress

Complete.

a. $36 \text{ yd} = \blacksquare \text{ ft}$ b. $\frac{3}{4} \text{ T} = \blacksquare \text{ lb}$ c. $1\frac{1}{2} \text{ qt} = \blacksquare \text{ pt}$

Review Vocabulary

reciprocal The product of a number and its reciprocal is 1; *Example:* The reciprocal of $\frac{3}{5}$ is $\frac{5}{3}$. (Lesson 5-7)

To convert from smaller units to larger units, multiply by the reciprocal of the appropriate unit ratio.

EXAMPLES Convert Smaller Units to Larger Units

3 Convert 15 quarts into gallons.

Since 1 gallon = 4 quarts, the unit ratio is $\frac{4 \text{ qt}}{1 \text{ gal}}$, and its reciprocal is $\frac{1 \text{ gal}}{4 \text{ qt}}$.

$$15 \text{ qt} = 15 \text{ qt} \cdot \frac{1 \text{ gal}}{4 \text{ qt}} \qquad \text{Multiply by } \frac{1 \text{ gal}}{4 \text{ qt}}.$$

$$= 15 \text{ qt} \cdot \frac{1 \text{ gal}}{4 \text{ qt}} \qquad \begin{array}{l}\text{Divide out common units, leaving} \\ \text{the desired unit, gallons.}\end{array}$$

$$= 15 \cdot \frac{1}{4} \text{ gal or } 3.75 \text{ gal} \qquad \begin{array}{l}\text{Multiplying 15 by } \frac{1}{4} \text{ is the same as} \\ \text{dividing 15 by 4.}\end{array}$$

4 **COSTUMES** Umeka needs $4\frac{1}{2}$ feet of fabric to make a costume for a play. How many yards of fabric does she need?

$$4\frac{1}{2}\text{ ft} = 4\frac{1}{2}\text{ ft} \cdot \frac{1\text{ yd}}{3\text{ ft}}$$ Since 1 yard = 3 feet, multiply by $\frac{1\text{ yd}}{3\text{ ft}}$. Then, divide out common units.

$$= \frac{\overset{3}{\cancel{9}}}{2} \cdot \frac{1}{\underset{1}{\cancel{3}}}\text{ yd}$$ Write $4\frac{1}{2}$ as an improper fraction. Then divide out common factors.

$$= \frac{3}{2}\text{ yd or } 1\frac{1}{2}\text{ yd}$$ Multiply.

So Umeka needs $1\frac{1}{2}$ yards of fabric.

CHECK Your Progress

Complete.

d. 2,640 ft = ▓ mi

e. 5 pt = ▓ qt

f. 100 oz = ▓ lb

g. 76c = ▓ gal

h. 3 c = ▓ pt

i. 18 in. = ▓ ft

j. **FOOD** A 3-pound pork loin can be cut into 10 smaller pork chops of equal weight. How many ounces does each pork chop weigh?

You can also convert from one rate to another by multiplying by a unit rate or its reciprocal.

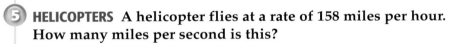
Real-World EXAMPLE   **Convert Rates**

5 **HELICOPTERS** A helicopter flies at a rate of 158 miles per hour. How many miles per second is this?

Since 1 hour = 3,600 seconds, multiply by $\frac{1\text{ h}}{3,600\text{ s}}$.

$$\frac{158\text{ mi}}{\text{h}} = \frac{158\text{ mi}}{1\text{ h}} \times \frac{1\text{ h}}{3,600\text{ s}}$$ Multiply by $\frac{1\text{ h}}{3,600\text{ s}}$.

$$= \frac{158\text{ mi}}{1\text{ }\cancel{h}} \times \frac{1\text{ }\cancel{h}}{3,600\text{ s}}$$ Divide out common units.

$$\approx \frac{0.04\text{ mi}}{1\text{ s}}$$ Simplify.

So, a helicopter flies approximately 0.04 mile per second.

CHECK Your Progress

k. **FISH** A swordfish can swim at a rate of 60 miles per hour. How many feet per hour is this?

l. **WALKING** Ava walks at a speed of 7 feet per second. How many feet per hour is this?

44. Which situation is represented by the graph?

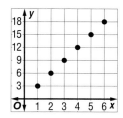

  A conversion of inches to yards

  B conversion of feet to inches

  C conversion of miles to feet

  D conversion of yards to feet

45. How many cups of milk are shown below?

  F $\frac{3}{4}$ c   **H** $2\frac{1}{2}$ c

  G $1\frac{1}{4}$ c   **J** 10 c

46. How many ounces are in $7\frac{3}{4}$ pounds?

  A 124 oz   **C** 120 oz

  B 122 oz   **D** 112 oz

Spiral Review

47. Use the information in the table to find the rate of change in dollars per hour. (Lesson 6-3)

Wage ($)	0	9	18	27
Time (h)	0	1	2	3

48. GROCERIES Find the unit price if Anju spent $2 for six oranges. Round to the nearest cent if necessary. (Lesson 6-2)

49. MEASUREMENT By doubling just the length of the rectangular ice skating rink in Will's backyard from 16 to 32 feet, he increased its area from 128 square feet to 256 square feet. Find the width of both rinks. (Lesson 3-6)

ALGEBRA For Exercises 50–52, use the pay stub at the right. (Lesson 3-3)

50. Write and solve an equation to find the regular hourly wage.

51. Write and solve an equation to find the overtime hourly wage.

52. Write and solve an equation to find how many times greater Grace's overtime hourly wage is than her regular hourly wage.

Martin, Grace		Employee #: 4211
Description:	**Hours:**	**Earnings ($):**
Regular hours:	40	300.00
Overtime hours:	2	22.50

GET READY for the Next Lesson

PREREQUISITE SKILL Multiply. (p. 674)

53. 14.5×8.2 **54.** 7.03×4.6 **55.** 9.29×15.3 **56.** 1.84×16.7

Measurement: Changing Metric Units

MAIN IDEA

Change metric units of length, capacity, and mass.

NYS Core Curriculum

7.M.3 Identify customary and metric units of mass
7.M.4 Convert mass within a given system
Also addresses 7.M.2, 7.M.9, 7.M.12, 7.M.13

New Vocabulary

metric system
meter
liter
gram
kilogram

NY Math Online

glencoe.com

• Extra Examples
• Personal Tutor
• Self-Check Quiz

▷ **MINI Lab**

The lengths of two objects are shown below.

Object	Length (millimeters)	Length (centimeters)
paper clip	45	4.5
CD case	144	14.4

1. Select three other objects. Find and record the width of all five objects to the nearest millimeter and tenth of a centimeter.

2. Compare the measurements of the objects, and write a rule that describes how to convert from millimeters to centimeters.

3. Measure the length of your classroom in meters. Make a conjecture about how to convert this measure to centimeters. Explain.

The **metric system** is a decimal system of measures. The prefixes commonly used in this system are kilo-, centi-, and milli-.

Prefix	Meaning In Words	Meaning In Numbers
kilo-	thousands	1,000
centi-	hundredths	0.01
milli-	thousandths	0.001

In the metric system, the base unit of *length* is the **meter** (m). Using the prefixes, the names of other units of length are formed. Notice that the prefixes tell you how the units relate to the meter.

Unit	Symbol	Relationship to Meter	
kilometer	km	1 km = 1,000 m	1 m = 0.001 km
meter	m	1 m = 1 m	
centimeter	cm	1 cm = 0.01 m	1 m = 100 cm
millimeter	mm	1 mm = 0.001 m	1 m = 1,000 mm

The **liter** (L) is the base unit of *capacity*, the amount of dry or liquid material an object can hold. The **gram** (g) measures *mass*, the amount of matter in an object. The prefixes can also be applied to these units. Whereas the meter and liter are the base units of length and capacity, the base unit of mass is the **kilogram** (kg).

To change a metric measure of length, mass, or capacity from one unit to another, you can use the relationship between the two units and multiplication by a power of 10.

EXAMPLES Convert Units in the Metric System

Study Tip

Metric Conversions
When converting from a larger unit to a smaller unit, the power of ten being multiplied will be greater than 1.

When converting from a smaller unit to a larger unit, the power of ten will be less than 1.

① Convert 4.5 liters to milliliters.

You need to convert liters to milliliters. Use the relationship 1 L = 1,000 mL.

1 L = 1,000 mL	Write the relationship.
4.5 × 1 L = 4.5 × 1,000 mL	Multiply each side by 4.5 since you have 4.5 L.
4.5 L = 4,500 mL	To multiply 4.5 by 1,000, move the decimal point 3 places to the right.

② Convert 500 millimeters to meters.

You need to convert millimeters to meters. Use the relationship 1 mm = 0.001 m.

1 mm = 0.001 m	Write the relationship.
500 × 1 mm = 500 × 0.001 m	Multiply each side by 500 since you have 500 mm.
500 mm = 0.5 m	To multiply 500 by 0.001, move the decimal point 3 places to the left.

 **CHECK Your Progress**

Complete.

a. 25.4 g = ▊ kg

b. 158 mm = ▊ m

Real-World Link . . . . .
The maximum weight of a grizzly bear is 521.64 kilograms.
Source: North American Bear Center

Real-World EXAMPLE

③ BEARS The California Grizzly Bear was designated the official state animal in 1953. Use the information at the left to find the maximum weight of a grizzly bear in grams.

You are converting kilograms to grams. Since the maximum weight of a grizzly bear is 521.64 kilograms, use the relationship 1 kg = 1,000 g.

1 kg = 1,000 g	Write the relationship.
521.64 × 1 kg = 521.64 × 1,000 g	Multiply each side by 521.64 since you have 521.64 kg.
521.64 kg = 521,640 g	To multiply 521.64 by 1,000, move the decimal point 3 places to the right.

So, the maximum weight of a grizzly bear is 521,640 grams.

CHECK Your Progress

c. **FOOD** A bottle contains 1.75 liters of juice. How many milliliters is this?

To convert measures between customary units and metric units, use the relationships below.

Customary and Metric Relationships   Key Concept

Type of Measure	Customary	→	Metric
Length	1 inch (in.)	≈	2.54 centimeters (cm)
	1 foot (ft)	≈	0.30 meter (m)
	1 yard (yd)	≈	0.91 meter (m)
	1 mile (mi)	≈	1.61 kilometers (km)
Weight/Mass	1 pound (lb)	≈	453.6 grams (g)
	1 pound (lb)	≈	0.4536 kilogram (kg)
	1 ton (T)	≈	907.2 kilograms (kg)
Capacity	1 cup (c)	≈	236.59 milliliters (mL)
	1 pint (pt)	≈	473.18 milliliters (mL)
	1 quart (qt)	≈	946.35 milliliters (mL)
	1 gallon (gal)	≈	3.79 liters (L)

EXAMPLES **Convert Between Measurement Systems**

4 **Convert 17.22 inches to centimeters. Round to the nearest hundredth if necessary.**

Use the relationship 1 inch ≈ 2.54 centimeters.

$$1 \text{ inch} \approx 2.54 \text{ cm} \qquad \text{Write the relationship.}$$

$$17.22 \times 1 \text{ in.} \approx 17.22 \times 2.54 \text{ cm} \qquad \text{Multiply each side by 17.22 since you have 17.22 in.}$$

$$17.22 \text{ in.} \approx 43.7388 \text{ cm} \qquad \text{Simplify.}$$

So, 17.22 inches is approximately 43.74 centimeters.

Study Tip

Alternative Method
When converting 17.22 inches to centimeters, you can use the relationship 1 in. ≈ 2.54 cm or the unit ratio $\frac{2.54 \text{ cm}}{1 \text{ in.}}$.

5 **Convert 828.5 milliliters to cups. Round to the nearest hundredth if necessary.**

Since 1 cup ≈ 236.59 milliliters, multiply by $\frac{1 \text{ c}}{236.59 \text{ mL}}$.

$$828.5 \text{ mL} \approx 828.5 \text{ mL} \cdot \frac{1 \text{ c}}{236.59 \text{ mL}} \qquad \text{Multiply by } \frac{1 \text{ c}}{236.59 \text{ mL}}.$$

$$\approx \frac{828.5 \text{ c}}{236.59} \text{ or } 3.5 \text{ c} \qquad \text{Simplify.}$$

So, 828.5 milliliters is approximately 3.5 cups.

✓CHECK Your Progress

Complete. Round to the nearest hundredth if necessary.

d. 7.44 c ≈ ▓ mL　　　**e.** 22.09 lb ≈ ▓ kg　　　**f.** 35.85 L ≈ ▓ gal

 Real-World EXAMPLE Convert with Rates

6 **LIGHT** The speed of light is about 186,000 miles per second. Find the approximate speed of light in kilometers per second.

Since 1 mile ≈ 1.61 kilometers, multiply by $\frac{1.61 \text{ km}}{1 \text{ mi}}$.

$$\frac{186{,}000 \text{ mi}}{\text{s}} \approx \frac{186{,}000 \text{ mi}}{1 \text{ s}} \times \frac{1.61 \text{ km}}{1 \text{ mi}} \quad \text{Multiply by } \frac{1.61 \text{ km}}{1 \text{ mi}}.$$

$$\approx \frac{299{,}460 \text{ km}}{1 \text{ s}} \quad \text{Simplify.}$$

So, the speed of light is approximately 299,460 kilometers per second.

 CHECK Your Progress

g. **RUNNING** Chuck runs at a speed of 3 meters per second. About how many feet per second does Chuck run?

 Real-World Link . . . .
Before the seventeenth century, people thought light was transmitted instantly. Now, we know that light is too fast for the delay to be noticeable.
Source: University of California, Riverside

 Your Understanding

Examples 1, 2, 4, 5
(pp. 305–306)

Complete. Round to the nearest hundredth if necessary.

1. 3.7 m = ▓ cm
2. 550 m = ▓ km
3. 1,460 mg = ▓ g
4. 2.34 kL = ▓ L
5. 9.36 yd ≈ ▓ m
6. 11.07 pt ≈ ▓ mL
7. 58.14 kg ≈ ▓ lb
8. 38.44 cm ≈ ▓ in.

Examples 3, 6
(pp. 305, 307)

9. **SPORTS** About how many feet does a team of athletes run in a 1,600-meter relay race?

Practice and Problem Solving

HOMEWORK HELP

For Exercises	See Examples
10–23	1, 2, 4, 5
24–27	3, 6

Complete. Round to the nearest hundredth if necessary.

10. 720 cm = ▓ m
11. 983 mm = ▓ m
12. 3.2 m = ▓ cm
13. 0.03 g = ▓ mg
14. 997 g = ▓ kg
15. 82.1 g = ▓ kg
16. 9.1 L = ▓ mL
17. 130.5 kL = ▓ L
18. 3.75 c ≈ ▓ mL
19. 41.8 in. ≈ ▓ cm
20. 156.25 lb ≈ ▓ kg
21. 9.5 gal ≈ ▓ L
22. 680.4 g ≈ ▓ lb
23. 4.725 m ≈ ▓ ft

24. **WATERFALLS** At 979 meters tall, Angel Falls in Venezuela is the highest waterfall in the world. How many kilometers tall is the waterfall?

25. **FOOD** An 18-ounce jar contains 510 grams of grape jelly. How many kilograms of grape jelly does the jar contain?

26. CYCLING Ramon rides his bike at a rate of 8 kilometers per hour. About how many miles per hour can Ramon ride his bike?

27. BIRDS A gull can fly at a speed of 22 miles per hour. About how many meters per hour can a gull fly?

Complete. Round to the nearest hundredth if necessary.

28. 8.18 qt $\approx$ ☐ L

29. 15.09 km $\approx$ ☐ yd

30. 72.26 cm $\approx$ ☐ ft

31. 0.445 T $\approx$ ☐ g

Order each set of measures from least to greatest.

32. 0.02 km, 50 m, $3{,}000$ cm

33. 660 mL, 0.06 L, 6.6 kL

34. 0.32 kg, 345 g, $35{,}100$ mg

35. $2{,}650$ mm, 130 cm, 5 m

36. ANALYZE TABLES The table shows the lengths of bridges in the United States. Which bridges are about 1 kilometer in length? Justify your answer.

37. CARPENTRY Jacinta needs a 2.5-meter pole for a birdfeeder that she is building. How many centimeters will she need to cut off a 3-meter pole in order to use it for the birdfeeder?

Bridge	Length (m)
Mackinac, MI	1,158
George Washington, NY	1,067
Tacoma Narrows II, WA	853
Oakland Bay, CA	704
Pennybacker, TX	345
Sunshine Skyway, FL	8,712
Golden Gate, CA	2,780

NYSCC • NYSMT
Extra Practice, pp. 683, 709

38. BAKING A bakery uses 900 grams of peaches per cobbler. How many cobblers can be made using 10.5 pounds of peaches?

H.O.T. Problems

39. FIND THE ERROR Jake and Gerardo are converting 3.25 kilograms to grams. Who is correct? Explain your reasoning.

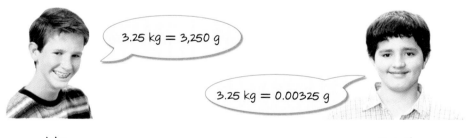

3.25 kg = 3,250 g

3.25 kg = 0.00325 g

Jake

Gerardo

CHALLENGE For Exercises 40–42, use the following information.
The metric prefix *giga* refers to something one billion times larger than the base unit.

40. How many meters are in one gigameter?

41. About how many miles are in one gigameter? Round to the nearest hundredth.

42. The distance from Earth to the Sun is approximately 93 million miles. About how many gigameters is this? Round to the nearest hundredth.

43. **WRITING IN MATH** Explain why it makes sense to multiply by a power of 10 that is greater than 1 when changing from a larger unit to a smaller unit.

NYSMT PRACTICE 7.M.3, 7.M.4

44. The table shows the mass of four wireless telephones. Find the approximate total mass of the telephones in kilograms.

Telephone Owner	Mass (g)
Elena	100.4
Kevin	70.8
Marissa	95.6
Corey	120.4

A 0.39 kilogram

B 3.9 kilograms

C 39.0 kilograms

D 390.0 kilograms

45. Which relationship between the given units of measure is correct?

F One gram is $\frac{1}{100}$ of a centigram.

G One meter is $\frac{1}{100}$ of a centimeter.

H One gram is $\frac{1}{1,000}$ of a kilogram.

J One milliliter is $\frac{1}{100}$ of a liter.

Spiral Review

46. **MEASUREMENT** A certain car weighs 3,200 pounds. What is the weight of the car in tons? (Lesson 6-4)

47. **MEASUREMENT** The table shows the number of inches per foot. Graph the data. Then find the slope of the line. Explain what slope represents. (Lesson 6-3)

Feet (x)	1	2	3	4
Inches (y)	12	24	36	48

Add or subtract. Write in simplest form. (Lesson 5-3)

48. $3\frac{4}{7} + 1\frac{1}{7}$ **49.** $8\frac{3}{5} - 2\frac{2}{5}$ **50.** $9\frac{1}{6} + 4\frac{3}{8}$ **51.** $11\frac{7}{10} - 5\frac{3}{4}$

52. **BASKETBALL** Jared made thirty-seven percent of his free-throw attempts during basketball practice. Write this percent as a decimal. (Lesson 4-7)

▷ **GET READY for the Next Lesson**

PREREQUISITE SKILL **Solve each equation.** (Lesson 3-3)

53. $5 \cdot 4 = x \cdot 2$ **54.** $9 \cdot 24 = 27 \cdot x$ **55.** $x \cdot 15 = 12 \cdot 4$ **56.** $8\frac{1}{2} \cdot x = 11 \cdot 17$

6-6 Algebra: Solving Proportions

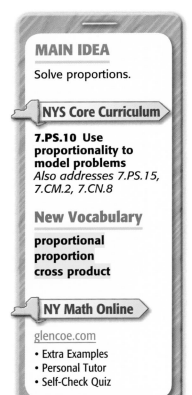

MAIN IDEA

Solve proportions.

NYS Core Curriculum

7.PS.10 Use proportionality to model problems
Also addresses 7.PS.15, 7.CM.2, 7.CN.8

New Vocabulary

proportional
proportion
cross product

NY Math Online

glencoe.com

• Extra Examples
• Personal Tutor
• Self-Check Quiz

▷ **GET READY** for the Lesson

NUTRITION The amount of calcium in different servings of milk is shown.

1. Write the rate $\dfrac{\text{calcium}}{\text{number of servings}}$ for each amount of milk.

2. Find the number of milligrams per cup for each serving size. What do you notice?

Two quantities are **proportional** if they have a constant rate or ratio. In the example above, notice that the number of servings and amount of calcium change or *vary* in the same way.

The unit rates for these different-sized servings are the same, a constant 300 milligrams per serving. So, the amount of calcium is proportional to the serving size.

Proportion		**Key Concept**
Words	A **proportion** is an equation stating that two ratios or rates are equivalent.	
Symbols	**Numbers**	**Algebra**
	$\dfrac{1}{2} = \dfrac{3}{6}, \dfrac{8\text{ ft}}{10\text{ s}} = \dfrac{4\text{ ft}}{5\text{ s}}$	$\dfrac{a}{b} = \dfrac{c}{d}$, where $b, d \neq 0$

Consider the following proportion.

$$\frac{a}{b} = \frac{c}{d}$$

$$\frac{a}{\cancel{b}} \cdot \cancel{b}d = \frac{c}{\cancel{d}} \cdot b\cancel{d} \qquad \text{Multiply each side by } bd.$$

$$ad = bc \qquad \text{Simplify.}$$

The products ad and bc are called the **cross products** of this proportion. The cross products of any proportion are equal. You can compare unit rates or cross products to identify proportional relationships.

Reading Math

Nonproportial
When ratios do not form a
proportion, they are called
nonproportional.

EXAMPLE Identify Proportional Relationships

① **RECREATION** A carousel makes 4 complete turns after 64 seconds and 5 complete turns after 76 seconds. Based on this information, is the number of turns made by this carousel proportional to the time in seconds? Explain.

METHOD 1 Compare unit rates.

$$\frac{\text{seconds}}{\text{complete turns}} \longrightarrow \frac{64 \text{ s}}{4 \text{ turns}} = \frac{16 \text{ s}}{1 \text{ turn}} \qquad \frac{76 \text{ s}}{5 \text{ turns}} = \frac{15.2 \text{ s}}{1 \text{ turn}}$$

Since the unit rates are not equal, the number of turns is not proportional to the time in seconds.

METHOD 2 Compare ratios by comparing cross products.

$$\frac{\text{seconds}}{\text{complete turns}} \longrightarrow \frac{64}{4} \stackrel{?}{=} \frac{76}{5} \longleftarrow \frac{\text{seconds}}{\text{complete turns}}$$

$$64 \cdot 5 \stackrel{?}{=} 4 \cdot 76 \qquad \text{Find the cross products.}$$

$$320 \neq 304 \qquad \text{Multiply.}$$

Since the cross products are not equal, the number of turns is not proportional to the time in seconds.

☑ **CHOOSE** Your Method

Determine if the quantities in each pair of ratios are proportional. Explain.

a. 60 voted out of 100 registered and 84 voted out of 140 registered

b. $12 for 16 yards of fabric and $9 for 24 yards fabric

You can also use cross products to find a missing value in a proportion. This is known as *solving the proportion*.

Study Tip

Mental Math
Some proportions can be
solved using mental math.

$$\frac{2.5}{10} = \frac{x}{30}$$

×3

$$\frac{2.5}{10} = \frac{7.5}{30}$$

×3

EXAMPLES Solve a Proportion

② Solve $\frac{21}{5} = \frac{c}{7}$.

$$\frac{21}{5} = \frac{c}{7} \qquad \text{Write the proportion.}$$

$$21 \cdot 7 = 5 \cdot c \qquad \text{Find the cross products.}$$

$$147 = 5c \qquad \text{Multiply.}$$

$$\frac{147}{5} = \frac{5c}{5} \qquad \text{Divide each side by 5.}$$

$$29.4 = c \qquad \text{Simplify.}$$

Check for Reasonableness Since $\frac{21}{5} \approx \frac{20}{5}$ or $\frac{4}{1}$ and $\frac{29.4}{7} \approx \frac{28}{7}$ or $\frac{4}{1}$, the answer is reasonable. ✔

3 Solve $\frac{2.6}{13} = \frac{8}{n}$.

$\frac{2.6}{13} = \frac{8}{n}$ Write the proportion.

$2.6 \cdot n = 13 \cdot 8$ Find the cross products.

$2.6n = 104$ Multiply.

$\frac{2.6n}{2.6} = \frac{104}{2.6}$ Divide each side by 2.6.

$n = 40$ Simplify.

✓ **CHECK Your Progress**

c. $\frac{16}{k} = \frac{2}{3}$ **d.** $\frac{2}{6} = \frac{5}{h}$ **e.** $\frac{10}{k} = \frac{2.5}{4}$

🌐 Real-World EXAMPLE

4 **MEDICINE** For every 18 people who have a sore throat, there are 2 people who actually have strep throat. If 72 patients have sore throats, how many of these would you expect to have strep throat?

METHOD 1   Write and solve a proportion.

Let s represent strep throat.

$\frac{2 \text{ strep throats}}{18 \text{ sore throats}} = \frac{s}{72 \text{ sore throats}}$ Write a proportion.

$2 \cdot 72 = 18 \cdot s$ Find the cross products.

$144 = 18s$ Multiply.

$8 = s$ Divide each side by 18.

🌐 **Real-World Career** · · · ·
How Does a Physician's Assistant Use Math?
A physician's assistant uses math when calculating safe dosages of medications.

NY Math Online

For more information visit, glencoe.com.

METHOD 2   Find and use a unit rate or ratio.

$\frac{2 \text{ strep throats} \div 2}{18 \text{ sore throats} \div 2} = \frac{1}{9}$ The ratio of strep throats to sore throats is 1 : 9.

Words	For every 9 sore throats, there is 1 strep throat.
Variable	Let s represent the number of strep throats.
Equation	$s = \frac{1}{9} \cdot 72$

$s = \frac{1}{9} \cdot 72$ or 8   Multiply.

So, you would expect 8 people to have strep throat.

✓ **CHOOSE Your Method**

f. **RUNNING** Salvador can run 120 meters in 24 seconds. At this rate, how many seconds will it take him to run a 300-meter race?

Example 1
(p. 311)

Determine if the quantities in each pair of ratios are proportional. Explain.

1. 2 adults for 10 children and 3 adults for 12 children

2. 12 inches by 8 inches and 18 inches by 12 inches

3. 8 feet in 21 seconds and 12 feet in 31.5 seconds

4. $5.60 for 5 pairs of socks and $7.12 for 8 pairs of socks

Examples 2, 3
(pp. 311–312)

Solve each proportion.

5. $\dfrac{5}{6} = \dfrac{t}{18}$

6. $\dfrac{6}{k} = \dfrac{24}{28}$

7. $\dfrac{21}{5} = \dfrac{c}{7}$

8. $\dfrac{15}{w} = \dfrac{2}{5}$

9. $\dfrac{3}{n} = \dfrac{2.7}{18}$

10. $\dfrac{0.2}{3} = \dfrac{3}{d}$

Example 4
(p. 312)

11. **GROCERIES** Orange juice is on sale at 3 half-gallons for $5. At this rate, find the cost of 5 half-gallons of orange juice to the nearest cent.

12. **TRAVEL** Franco drove 203 miles in 3.5 hours. At this rate, how long will it take him to drive another 29 miles to the next town?

Practice and Problem Solving

HOMEWORK HELP	
For Exercises	**See Examples**
13–20	1
21–32	2, 3
33–36	4

Determine if the quantities in each pair of ratios are proportional. Explain.

13. 20 children from 6 families to 16 children from 5 families

14. 5 pounds of dry ice melts in 30 hours and 4 pounds melts in 24 hours

15. 16 winners out of 200 entries and 28 winners out of 350 entries

16. 5 meters in 7 minutes and 25 meters in 49 minutes

17. 1.4 tons produced every 18 days and 10.5 tons every 60 days

18. 3 inches for every 4 miles and 7.5 inches for every 10 miles

19. **READING** Leslie reads 25 pages in 45 minutes. After 60 minutes, she has read a total of 30 pages. Is her time proportional to the number of pages she reads? Explain.

20. **PETS** A store sells 2 hamsters for $11 and 6 hamsters for $33. Is the cost proportional to the number of hamsters sold? Explain.

Solve each proportion.

21. $\dfrac{3}{8} = \dfrac{b}{40}$

22. $\dfrac{x}{12} = \dfrac{12}{4}$

23. $\dfrac{c}{7} = \dfrac{18}{42}$

24. $\dfrac{5}{k} = \dfrac{10}{22}$

25. $\dfrac{3}{8} = \dfrac{n}{4}$

26. $\dfrac{15}{4} = \dfrac{3}{8}$

27. $\dfrac{45}{5} = \dfrac{d}{7}$

28. $\dfrac{30}{a} = \dfrac{8}{20}$

29. $\dfrac{1.6}{m} = \dfrac{2}{3}$

30. $\dfrac{4.5}{5} = \dfrac{t}{7}$

31. $\dfrac{2.5}{4.5} = \dfrac{7.5}{x}$

32. $\dfrac{3.8}{5.2} = \dfrac{7.6}{z}$

33. **SCHOOL** If 4 notebooks weigh 2.8 pounds, how much do 6 of the same notebooks weigh?

34. **COOKING** There are 6 teaspoons in 2 tablespoons. How many teaspoons are in 1.5 tablespoons?

35. **SCIENCE** The ratio of salt to water in a certain solution is 4 to 15. If the solution contains 6 ounces of water, how many ounces of salt does it contain?

36. **CONCERTS** Alethia purchased 7 advanced tickets for herself and her friends to a concert and paid $164.50. If the total cost of tickets to the concert is proportional to the number purchased, how many tickets to the same concert did Serefina purchase if she paid a total of $94?

ANALYZE GRAPHS For Exercises 37–40, use the graph. It shows the cost of several pizzas, with and without a delivery fee.

37. What do the points (3, 15) and (5, 25) represent on the graph? Is this situation proportional? Explain.

38. What do the points (2, 13) and (4, 23) represent on the graph? Is this situation proportional? Explain.

Pizza Cost

with delivery fee (4, 23) (5, 25) (2, 13) (3, 15) no delivery fee

Cost ($): 30 25 20 15 10 5 0

Number of Pizzas: 1 2 3 4 5 6

Real-World Link
Film for an IMAX projection system passes through the projector at the rate of 330 feet per minute or 5.5 feet per second.
Source: IMAX Corporation

39. What is the slope of each line? What does the slope represent?

40. What is the delivery fee? Explain.

41. **SAVINGS** Pao spent $140 of his paycheck and put the remaining $20 in his savings account. If the number of dollars he spends is proportional to the number he saves, how much of a $156-paycheck will he put into savings?

42. **MOVIES** After 30 seconds, 720 frames of film have passed through a movie projector. At this rate, what is the approximate running time in minutes of a movie made up of 57,000 frames of film?

43. **SCHOOL** There are 325 students and 13 teachers at a school. Next school year, the enrollment is expected to increase by 100 students. Write and solve a proportion to find the number of teachers that must be hired so the student-teacher ratio remains the same.

NYSCC • NYSMT
Extra Practice, pp. 683, 709

44. **FIND THE DATA** Refer to the Data File on pages 16–19. Choose some data and write a real-world problem in which you would solve a proportion.

H.O.T. Problems

45. **Which One Doesn't Belong?** Identify the rate that is not proportional to the other three. Explain your reasoning.

| $4.50 for 5 lb | $2.88 for 3.2 lb | $5.70 for 6 lb | $4.86 for 5.4 lb |

46. **CHALLENGE** In a cleaning solution, the ratio of bleach to water is 1 : 5. If there are 36 cups of cleaning solution, how many cups of water are needed? Explain your reasoning.

47. **SELECT A TECHNIQUE** Sweet corn is on sale at $2.50 for a dozen at a farmer's market. Select one or more of the following technique(s) to determine how many ears you can buy for $10. Then use this technique to solve the problem.

| mental math | estimation | number sense |

48. **WRITING IN MATH** Explain why the cross products of a proportion are equal. Use the term *multiplicative inverse* in your explanation.

NYSMT PRACTICE 7.PS.10

49. Mirma gives away 84 flyers over a 3-hour period. If the number of flyers she is able to give away per hour remains the same, which proportion can be used to find x, the number of flyers that she would give away over a 5-hour period?

A $\frac{3}{84} = \frac{x}{5}$ C $\frac{5}{3} = \frac{84}{x}$

B $\frac{84}{3} = \frac{x}{5}$ D $\frac{3}{84} = \frac{x}{8}$

50. A recipe that makes 16 muffins calls for $\frac{1}{2}$ cup of flour. How much flour is needed to make 3 dozen muffins using this recipe?

F $1\frac{1}{8}$ c

G 1 c

H $1\frac{1}{4}$ c

J $1\frac{1}{2}$ c

Spiral Review

51. **MEASUREMENT** Felicia bought 5 pounds of onions. About how many kilograms of onions did she buy? (Lesson 6-5)

MEASUREMENT Complete. (Lesson 6-4)

52. 5 qt = ■ pt

53. $3\frac{1}{2}$ lb = ■ oz

54. 28 c = ■ qt

Multiply. Write in simplest form. (Lesson 5-5)

55. $3\frac{1}{2} \times 5\frac{7}{8}$

56. $1\frac{2}{3} \times 5\frac{4}{5}$

57. $2\frac{1}{4} \times 7\frac{5}{8}$

▷ GET READY for the Next Lesson

58. **PREREQUISITE SKILL** Mr. Andres is filling up his car with gas that costs $2.50 per gallon. His car's gas gauge before filling up is shown at the right. If his car's gas tank holds 16 gallons, about how much will Mr. Andres pay to fill up his tank? Use the *eliminate possibilities* strategy. (Lesson 5-4)

A $15.00   C $27.00

B $25.00   D $35.00

Math Lab
Inverse Proportionality

MAIN IDEA

Graph inverse variations.

NYS Core Curriculum

Preparation for 8.A.3 Describe a situation involving relationships that matches a given graph **Preparation for 8.A.4** Create a graph given a description or an expression for a situation involving a linear or nonlinear relationship *Also addresses 7.PS.10*

New Vocabulary

inverse proportion

Jackie walks at an average rate of 4 miles per hour. This situation can be represented by the equation $d = 4t$, where d is the distance in miles, and t is the time in hours. The number of miles is proportional to the number of hours because the ratios are equal.

Let's vary the situation a little. Suppose Jackie wants to walk 12 miles each day, at varying speeds. How long will it take her to walk 12 miles? The answer depends on how fast she walks.

ACTIVITY

STEP 1 Copy and complete the table for the equation $12 = rt$.

t (hours)	12	8	6	▒	▒	▒	▒
r (miles per hour)	1	1.5	2	2.5	3	3.5	4

STEP 2 Copy and complete the graph of the ordered pairs from Step 1. Connect the line with a smooth curve. The first three points are done for you.

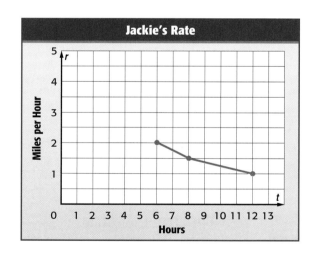

ANALYZE THE RESULTS

1. Is the time proportional to the rate? Explain why or why not.

2. When the product of two variables is a constant, the relationship forms an **inverse proportion.** Which situation is an inverse proportion: Jackie walking at 4 miles per hour or walking 12 miles at varying rates? Identify the constants in each situation.

3. Mr. Anwar agrees to pay Rob a flat fee of $240 to do some yard work. Mr. Sloan agrees to pay Rob's brother $10 per hour to do some yard work. Make a table of ordered pairs and a graph for each situation. Then decide whether each relationship is a proportion or an inverse proportion.

RECIPE For Exercises 1–3, use the information in the table to write each ratio as a fraction in simplest form. (Lesson 6-1)

Cherry Punch Recipe	
Cherry Juice	4 cups
Apple Juice	2 cups
Ginger Ale	16 cups

1. cherry juice : apple juice

2. apple juice : ginger ale

3. cherry juice : ginger ale

Determine whether the ratios are equivalent. Explain. (Lesson 6-1)

4. 6 out of 9 words spelled correctly
   2 out of 3 words spelled correctly

5. 150 athletes to 15 coaches
   3 athletes to 1 coach

6. 24 points in 4 games
   72 points in 8 games

7. **MULTIPLE CHOICE** Which amount of chocolate shown in the table has the best unit price? (Lesson 6-2)

Weight (oz)	Cost ($)
12	2.50
18	3.69
24	4.95
30	6.25

   A 12 oz
   B 18 oz
   C 24 oz
   D 36 oz

8. Use the graph to find the rate of change in cost per magazine. (Lesson 6-3)

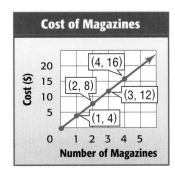

Cost of Magazines
(4, 16)
(2, 8)
(3, 12)
(1, 4)
Cost ($)
Number of Magazines

Complete. (Lesson 6-4)

9. $42 \text{ ft} = \blacksquare \text{ yd}$

10. $9 \text{ pt} = \blacksquare \text{ qt}$

11. $7{,}600 \text{ lb} = \blacksquare \text{ T}$

12. $7\frac{1}{2} \text{ gal} = \blacksquare \text{ qt}$

13. **MULTIPLE CHOICE** Which situation is best represented by the graph? (Lesson 6-4)

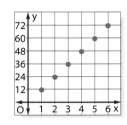

   F conversion of inches to yards
   G conversion of feet to inches
   H conversion of inches to miles
   J conversion of yards to feet

Complete. Round to the nearest hundredth if necessary. (Lesson 6-5)

14. $12.5 \text{ mi} \approx \blacksquare \text{ km}$

15. $4.75 \text{ gal} \approx \blacksquare \text{ L}$

16. $76 \text{ cm} \approx \blacksquare \text{ in.}$

17. $31.8 \text{ kg} \approx \blacksquare \text{ lb}$

Determine if the quantities in each pair of ratios are proportional. Explain. (Lesson 6-6)

18. 8 pages in 5 minutes and 40 pages in 25 minutes

19. 40 blank CDs for $9.60 and 24 blank CDs for $4.80

Solve each proportion. (Lesson 6-6)

20. $\frac{3}{d} = \frac{12}{20}$

21. $\frac{7}{8} = \frac{m}{48}$

22. $\frac{w}{8} = \frac{1}{3}$

23. **MEASUREMENT** It took 45 minutes to fill a circular pool of uniform depth to a level of 18 inches. At this rate, how long will it take to fill the pool to a level of 35 inches? (Lesson 6-6)

6-7 Problem-Solving Investigation

MAIN IDEA: Solve problems by drawing a diagram.

 NYSCC **7.PS.2 Construct appropriate extensions to problem situations** *Also addresses 7.PS.6, 7.PS.11*

P.S.I. TEAM ✚

e-Mail: DRAW A DIAGRAM

CACEY: I dropped a ball from a height of 12 feet. It hits the ground and bounces up half as high as it fell. This is true for each successive bounce.

YOUR MISSION: Draw a diagram to find the height the ball reaches after the fourth bounce.

Understand	You know the ball is dropped from a height of 12 feet. It bounces up half as high as it falls.
Plan	Draw a diagram to show the height of the ball after each bounce.
Solve	 The ball reaches a height of $\frac{3}{4}$ foot after the fourth bounce.
Check	Start at 12 feet. Multiply by $\frac{1}{2}$ for each bounce: $12 \cdot \frac{1}{2} \cdot \frac{1}{2} \cdot \frac{1}{2} \cdot \frac{1}{2} = \frac{12}{16}$ or $\frac{3}{4}$. So, the solution is correct.

In the Solve diagram: 12 ft, 6 ft, 3 ft, $1\frac{1}{2}$ ft, $\frac{3}{4}$ ft, bounces labeled 1, 2, 3, 4.

Analyze The Strategy

1. Determine what height a ball would reach after the fourth bounce if it is dropped from 12 feet and bounces up $\frac{2}{3}$ as high. Draw a new diagram for this situation.

2. **WRITING IN MATH** Write a problem that could be solved by drawing a diagram. Exchange your problem with a classmate and solve.

Solve Exercises 3–5. Use the *draw a diagram* strategy.

3. **TRAVEL** Mr. Garcia has driven 60 miles, which is $\frac{2}{3}$ of the way to his sister's house. How much farther does he have to drive to get to his sister's house?

4. **DISTANCE** Alejandro and Pedro are riding their bikes to school. After 1 mile, they are $\frac{4}{5}$ of the way there. How much farther do they have to go?

5. **VOLUME** A swimming pool is being filled with water. After 25 minutes, $\frac{1}{6}$ of the swimming pool is filled. How much longer will it take to completely fill the pool, assuming the water rate is constant?

Use any strategy to solve Exercises 6–10. Some strategies are shown below.

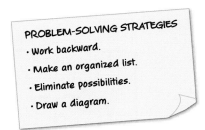

PROBLEM-SOLVING STRATEGIES
- Work backward.
- Make an organized list.
- Eliminate possibilities.
- Draw a diagram.

6. **BASEBALL** Of Lee's baseball cards, $\frac{1}{5}$ show California players. Of these, $\frac{3}{8}$ show San Diego Padres players. Is the fraction of Lee's collection that show Padres players $\frac{23}{40}$, $\frac{4}{13}$, or $\frac{3}{40}$?

7. **GAMES** Eight members of a chess club are having a tournament. In the first round, every player will play a chess game against every other player. How many games will be in the first round of the tournament?

8. **MEASUREMENT** Kiaya is adding a 2-inch border to the length and width of a photograph as shown.

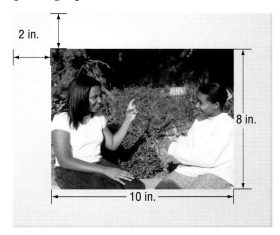

Which expression represents the area of the border to be added to the original photograph?

A $(8 + 4)(10 + 4)$

B $(8 + 4)(10 + 4) - (8)(10)$

C $(8 - 4)(10 - 4)$

D $(8 - 4)(10 - 4) - (8)(10)$

9. **RACES** Anna, Isabela, Mary, and Pilar ran a race. Anna is just ahead of Pilar. Pilar is two places behind Isabela. Isabela is a few seconds behind the leader, Mary. Use the table to place the girls in order from first to last.

10. **FRACTIONS** Marta ate a quarter of a whole pie. Edwin ate $\frac{1}{4}$ of what was left. Cristina then ate $\frac{1}{3}$ of what was left. What fraction of the pie remains?

Scale Drawings

MAIN IDEA

Solve problems involving scale drawings.

NYS Core Curriculum

7.PS.10 Use proportionality to model problems 7.M.1 Calculate distance using a map scale *Also addresses 7.R.3*

New Vocabulary

scale drawing
scale model
scale
scale factor

NY Math Online

glencoe.com
• Concepts In Motion
• Extra Examples
• Personal Tutor
• Self-Check Quiz

▷ **MINI Lab**

• Measure the length of each item in a room, such as a gymnasium.

• Record each length to the nearest $\frac{1}{2}$ foot.

1. Let 1 unit on the grid paper represent 2 feet. So, 4 units = 8 feet. Convert all your measurements to units.

2. On grid paper, make a drawing of your gymnasium like the one shown at the right.

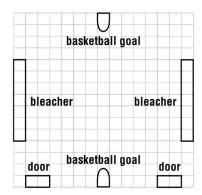

A map is an example of a scale drawing. **Scale drawings** and **scale models** are used to represent objects that are too large or too small to be drawn or built at actual size. The **scale** gives the ratio that compares the measurements of the drawing or model to the measurements of the real object. The measurements on a drawing or model are proportional to measurements of the actual object.

EXAMPLE Use a Map Scale

① **MAPS** What is the actual distance between Hagerstown and Annapolis?

Step 1 Use a centimeter ruler to find the map distance between the two cities. The map distance is about 4 centimeters.

Step 2 Write and solve a proportion using the scale. Let d represent the actual distance between the cities.

$$
\begin{array}{cc}
\textbf{Scale} & \textbf{Length}
\end{array}
$$

$$
\begin{array}{ll}
\text{map} \longrightarrow \\
\text{actual} \longrightarrow
\end{array}
\frac{1 \text{ centimeter}}{24 \text{ miles}} = \frac{4 \text{ centimeters}}{d \text{ miles}}
\begin{array}{l}
\longleftarrow \text{ map} \\
\longleftarrow \text{ actual}
\end{array}
$$

$$1 \times d = 24 \times 4 \qquad \text{Cross products}$$

$$d = 9.6 \qquad \text{Simplify.}$$

The distance between the cities is about 9.6 miles.

Example 1
(pp. 320–321)

GEOGRAPHY Find the actual distance between each pair of cities in New Mexico. Use a ruler to measure.

1. Carlsbad and Artesia

2. Hobbs and Eunice

3. Artesia and Eunice

4. Lovington and Carlsbad

Example 2
(p. 321)

BLUEPRINTS For Exercises 5 and 6, use the blueprint. Each square has a side length of $\frac{1}{4}$ inch.

5. What is the actual length of the pool?

6. What is the actual width of the pool?

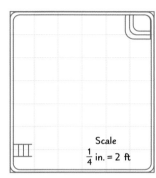

Scale
$\frac{1}{4}$ in. = 2 ft

Example 3
(p. 322)

BRIDGES For Exercises 7 and 8, use the following information.

A engineer makes a model of the bridge using a scale of 1 inch = 3 yards.

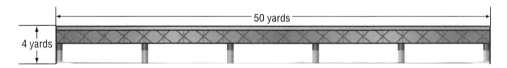

50 yards

4 yards

7. What is the length of the model?

8. What is the height of the model?

Example 4
(p. 322)

Find the scale factor of each scale drawing or model.

9.

1 inch = 4 feet

10.

1 centimeter = 15 millimeters

11. **CITY PLANNING** In the aerial view of a city block at the right, the length of Main Street is 2 inches. If Main Street's actual length is 2 miles, find the scale factor of the drawing.

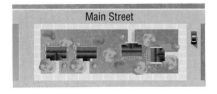

Main Street

GEOGRAPHY Find the actual distance between each pair of locations in South Carolina. Use a ruler to measure.

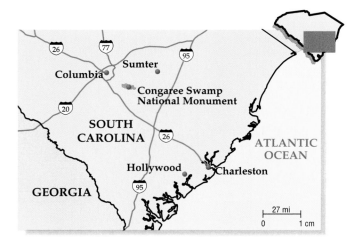

12. Columbia and Charleston

13. Hollywood and Sumter

14. Congaree Swamp and Charleston

15. Sumter and Columbia

For Exercises 16–18, use the blueprint of an apartment at the right. Each square has a side length of $\frac{1}{4}$ inch.

16. What is the actual length of the living room?

17. Find the actual dimensions of the master bedroom.

18. Find the scale factor for this blueprint.

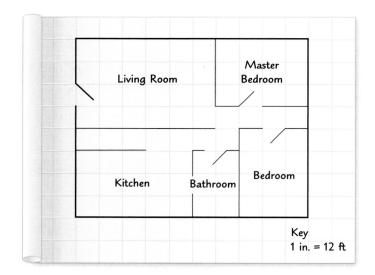

Find the length of each model. Then find the scale factor.

19.

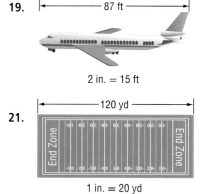

87 ft

2 in. = 15 ft

20.

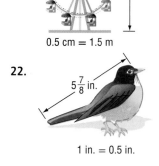

36 m

0.5 cm = 1.5 m

21.

120 yd

End Zone

End Zone

1 in. = 20 yd

22.

$5\frac{7}{8}$ in.

1 in. = 0.5 in.

Real-World Link . . . .
The statue of Thomas Jefferson inside the Jefferson Memorial in Washington, D.C., was made using a scale of 1 foot : 3 feet.

Source: National Register of Historic Places

23. **GEOGRAPHY** A map of Bakersfield, California, has a scale of 1 inch to 5 miles. If the city is $5\frac{1}{5}$ inches across on the map, what is the actual distance across the actual city? Use estimation to check your answer.

24. **TREES** A model of a tree is made using a scale of 1 inch : 25 feet. What is the height of the actual tree if the height of the model is $4\frac{3}{8}$ inches?

25. **STATUES** Refer to the information at the left. Find the scale factor and the actual height of Thomas Jefferson if the height of the statue is 19 feet.

26. **GEOGRAPHY** Lexington and Elizabethtown, Kentucky, are 79 miles apart. If the distance on the map is $2\frac{1}{2}$ inches, find the scale of the map.

27. **RESEARCH** Find the dimensions of any U.S. presidential monument. Give an appropriate scale that can be used to make a scale model of the monument. State the dimensions of the model using your scale.

28. **BUILDINGS** If you are making a model of your bedroom, which would be an appropriate scale: 1 inch = 2 feet, or 1 inch = 12 feet?

29. **LIFE SCIENCE** A scale drawing of a red blood cell is shown below. If the blood cell's actual diameter is 0.008 millimeter, use a ruler to find the scale factor of the drawing.

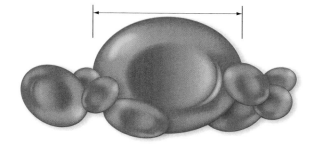

NYSCC • NYSMT
Extra Practice, pp. 684, 709

H.O.T. Problems

30. **OPEN ENDED** On grid paper, create a scale drawing of a room in your home. Include the scale that you used.

31. **CHALLENGE** Montoya constructed three models, A, B, and C, of the same figure, with scales of 0.5 cm = 1 mm, 1.5 mm = 4 cm, and 0.25 cm = 2.5 mm, respectively.

   a. Which model is larger than the actual figure? Justify your answer.

   b. Which model is smaller than the actual figure? Justify your answer.

   c. Which model is the same size as the actual figure? Justify your answer.

32. **REASONING** Compare and contrast the terms *scale* and *scale factor*. Include an example in your comparison.

33. **WRITING IN MATH** Explain how you could use estimation to find the actual distance between San Diego, California, and Seattle, Washington, on a map.

34. A scale drawing of a doctor's office is shown.

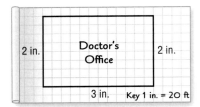

2 in.   Doctor's Office   2 in.

3 in.   Key 1 in. = 20 ft

What are the actual dimensions of the doctor's office?

A 24 feet by 48 feet

B 30 feet by 52 feet

C 40 feet by 60 feet

D 37.5 feet by 65 feet

35. A certain map has a scale of $\frac{1}{4}$ inch = 30 miles. How many miles are represented by 4 inches on this map?

F 480 miles

G 120 miles

H 30 miles

J 16 miles

36. Ernesto drew a map of his school. He used a scale of 1 inch : 50 feet. What distance on Ernesto's map should represent the 625 feet between the cafeteria and the science lab?

A 8 in.

B 10.5 in.

C 12.5 in.

D 15 in.

Spiral Review

37. FAMILY At Nelia's family reunion, $\frac{4}{5}$ of the people are 18 years of age or older. Half of the remaining people are under 12 years old. If 20 children are under 12 years old, how many people are at the reunion? Use the *draw a diagram* strategy. (Lesson 6-7)

Solve each proportion. (Lesson 6-6)

38. $\frac{5}{7} = \frac{a}{35}$

39. $\frac{12}{p} = \frac{36}{45}$

40. $\frac{3}{9} = \frac{21}{k}$

41. JOGGING The table shows the number of miles Tonya jogged each week for the past several weeks. Estimate the total number of miles she jogged. (Lesson 5-1)

Week	Miles
1	$7\frac{1}{6}$
2	$8\frac{3}{4}$
3	10
4	$12\frac{1}{4}$
5	$6\frac{2}{3}$

Find the LCM of each set of numbers. (Lesson 4-8)

42. 2, 4

43. 4, 8, 12

44. 3, 7, 5

45. 5, 10, 15

46. 2, 6, 9

47. 3, 15, 20

▷ GET READY for the Next Lesson

PREREQUISITE SKILL Divide. Write in simplest form. (Lesson 5-7)

48. $2\frac{3}{4} \div 10$

49. $4\frac{1}{3} \div 10$

50. $30\frac{2}{3} \div 100$

51. $87\frac{1}{2} \div 100$

Spreadsheet Lab
Scale Drawings

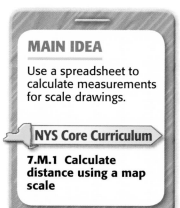

A computer spreadsheet is a useful tool for calculating measures for scale drawings. You can change the scale factors and the dimensions, and the spreadsheet will automatically calculate the new values.

ACTIVITY

Suppose you want to make a scale drawing of your school. Set up a spreadsheet like the one shown below. In this spreadsheet, the actual measures are in feet, and the scale drawing measures are in inches.

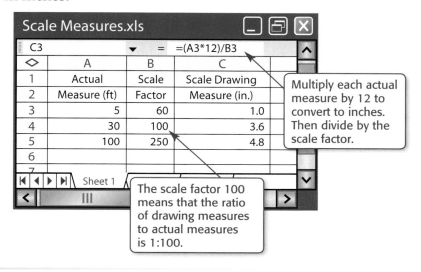

Scale Measures.xls

	A	B	C
1	Actual	Scale	Scale Drawing
2	Measure (ft)	Factor	Measure (in.)
3	5	60	1.0
4	30	100	3.6
5	100	250	4.8
6			
7			

C3 = =(A3*12)/B3

Multiply each actual measure by 12 to convert to inches. Then divide by the scale factor.

The scale factor 100 means that the ratio of drawing measures to actual measures is 1:100.

ANALYZE THE RESULTS

1. The length of one side of the school building is 100 feet. If you use a scale factor of 1:250, what is the length on your scale drawing?

2. The length of a classroom is 30 feet. What is the scale factor if the length of the classroom on a scale drawing is 3.6 inches?

3. Calculate the length of a 30-foot classroom on a scale drawing if the scale factor is 1:10.

4. The width of a hallway is 20 feet. What is the scale factor if the width of the hallway on a scale drawing is 2.5 inches?

5. Suppose the actual measures of your school are given in meters. Describe how you could use a spreadsheet to calculate the scale drawing measures in centimeters using a scale factor of 1:50.

6. Choose three rooms in your home and use a spreadsheet to make scale drawings. First, choose an appropriate scale and calculate the scale factor. Include a sketch of the furniture drawn to scale in each room.

Fractions, Decimals, and Percents

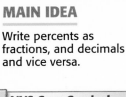

MAIN IDEA

Write percents as fractions, and decimals and vice versa.

NYS Core Curriculum

Preparation for 8.N.3
Read, write, and identify percents less than 1% and greater than 100%

NY Math Online

glencoe.com

• Extra Examples
• Personal Tutor
• Self-Check Quiz

▷ **GET READY** for the Lesson

SURVEYS The graph shows the results of a survey about favorite type of TV show.

1. What percent of the teens chose comedy?

2. Write this percent as a ratio in simplest form.

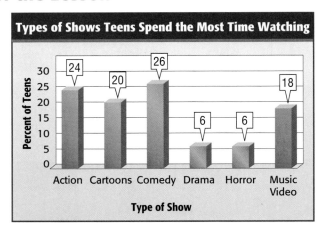

Types of Shows Teens Spend the Most Time Watching

Percent of Teens

Action 24 Cartoons 20 Comedy 26 Drama 6 Horror 6 Music Video 18

Type of Show

Source: Kids USA Survey

In Lesson 4-6, you wrote percents like 26% as fractions by writing fractions with denominators of 100 and then simplifying. You can use the same method to write percents like $8\frac{1}{3}\%$ and 190% as fractions.

EXAMPLES Percents as Fractions

1 Write $8\frac{1}{3}\%$ as a fraction in simplest form.

$$8\frac{1}{3}\% = \frac{8\frac{1}{3}}{100} \qquad \text{Write a fraction.}$$

$$= 8\frac{1}{3} \div 100 \qquad \text{Divide.}$$

$$= \frac{25}{3} \div 100 \qquad \text{Write } 8\frac{1}{3} \text{ as an improper fraction.}$$

$$= \frac{25}{3} \cdot \frac{1}{100} \qquad \text{Multiply by the reciprocal of 100, which is } \frac{1}{100}.$$

$$= \frac{25}{300} \text{ or } \frac{1}{12} \qquad \text{Simplify.}$$

2 **TOYS** A collectible action figure sold for 190% of its original price. Write this percent as a fraction in simplest form.

$$190\% = \frac{190}{100} \qquad \text{Definition of percent}$$

$$= \frac{19}{10} \text{ or } 1\frac{9}{10} \qquad \text{Simplify.}$$

> A percent greater than 100 is equal to a number greater than 1.

So, the toy sold for $1\frac{9}{10}$ of its original price.

 **CHECK Your Progress**

Write each percent as a fraction in simplest form.

a. 150% b. $17\frac{1}{2}\%$ c. $33\frac{1}{3}\%$

To write a fraction like $\frac{8}{25}$ as a percent, multiply the numerator and the denominator by a number so that the denominator is 100. If the denominator is not a factor of 100, you can write fractions as percents by using a proportion.

 **EXAMPLES** **Fractions as Percents**

③ Write $\frac{4}{15}$ as a percent. Round to the nearest hundredth.

Estimate $\frac{4}{15}$ is about $\frac{4}{16}$, which equals $\frac{1}{4}$ or 25%.

$\frac{4}{15} = \frac{n}{100}$ Write a proportion.

$400 = 15n$ Find the cross products.

$\frac{400}{15} = \frac{15n}{15}$ Divide each side by 15.

$26.67 \approx n$ Simplify.

So, $\frac{4}{15}$ is about 26.67%.

Check for Reasonableness   26.67% ≈ 25% ✔

④ **BLOGGING** In Boston, $\frac{89}{100,000}$ residents blog. Write this fraction as a percent.

$\frac{89}{100,000} = \frac{n}{100}$ Write a proportion.

$8,900 = 100,000n$ Find the cross products.

$\frac{8,900}{100,000} = \frac{100,000n}{100,000}$ Divide each side by 100,000.

$0.089 = n$ Simplify.

> A percent less than 1% is equal to a number less than 0.01 or $\frac{1}{100}$.

So, 0.089% of Boston's residents blog.

Study Tip

Choose the Method
To write a fraction as a percent,

• use multiplication when a fraction has a denominator that is a factor of 100,

• use a proportion for any type of fraction.

 CHECK Your Progress

Write each fraction as a percent. Round to the nearest hundredth if necessary.

d. $\frac{2}{15}$ e. $\frac{7}{1,600}$ f. $\frac{17}{25}$

Study Tip

Look Back You can review writing fractions as decimals in lesson 4-3.

In this lesson, you have written percents as fractions and fractions as percents. In Chapter 4, you wrote percents and fractions as decimals. You can also write a fraction as a percent by first writing the fraction as a decimal and then writing the decimal as a percent.

Percents, fractions, and decimals are different names that represent the same number.

EXAMPLES Fractions as Percents

5 Write $\frac{5}{6}$ as a percent. Round to the nearest hundredth.

$\frac{5}{6} = 0.833333333...$ Write $\frac{5}{6}$ as a decimal.

$\approx 83.33\%$ Multiply by 100 and add the %.

6 **BOOKS** Bryce has read $\frac{3}{5}$ of a book. What percent of the book has he read?

$\frac{3}{5} = 0.6$ Write the fraction as a decimal.

$= 60\%$ Multiply by 100 and add the %.

So, Bryce has read 60% of the book.

 CHECK Your Progress

Write each fraction as a percent. Round to the nearest hundredth if necessary.

g. $\frac{5}{16}$ h. $\frac{7}{12}$ i. $\frac{2}{9}$

j. **LAWNS** Mika is mowing lawns to earn extra money. She has mowed 6 out of 13 lawns. What percent of the lawns has she mowed?

Some fractions with denominators that are not factors of 100 are used often in everyday situations. It is helpful to memorize these fractions and their equivalent decimals and percents. These common equivalents are shown below.

Common Equivalents					Key Concept
Fraction	Decimal	Percent	Fraction	Decimal	Percent
$\frac{1}{3}$	$0.\overline{3}$	$33\frac{1}{3}\%$	$\frac{3}{8}$	0.375	$37\frac{1}{2}\%$
$\frac{2}{3}$	$0.\overline{6}$	$66\frac{2}{3}\%$	$\frac{5}{8}$	0.625	$62\frac{1}{2}\%$
$\frac{1}{8}$	0.125	$12\frac{1}{2}\%$	$\frac{7}{8}$	0.875	$87\frac{1}{2}\%$

CHECK Your Understanding

Examples 1, 2
(pp. 328–329)

Write each percent as a fraction in simplest form.

1. 135%
2. 18.75%
3. $7\frac{1}{2}$%
4. $66\frac{2}{3}$%

5. **FOOD** Steven and Rebecca ate 62.5% of a pizza. What fraction of the pizza did they eat?

Examples 3–5
(pp. 329–330)

Write each fraction as a percent. Round to the nearest hundredth if necessary.

6. $\frac{3}{4}$
7. $\frac{4}{2,500}$
8. $\frac{4}{11}$
9. $\frac{1}{9}$

Example 6
(p. 330)

10. **SCHOOL** Moses has finished 11 out of 15 homework questions. To the nearest hundredth, what percent of the homework is complete?

Practice and Problem Solving

HOMEWORK HELP	
For Exercises	See Examples
11–14, 19	1
15–18, 20	2
21–32	3–6
33–34	3

Write each percent as a fraction in simplest form.

11. 62.5%
12. 6.2%
13. 28.75%
14. 56.25%
15. $33\frac{1}{3}$%
16. $16\frac{2}{3}$%
17. $93\frac{3}{4}$%
18. $78\frac{3}{4}$%

19. **ENVIRONMENT** Freshwater from lakes only accounts for 0.1% of the world's water supply. Write this percent as a fraction in simplest form.

20. **ATTENDANCE** At last year's spring dance, $78\frac{1}{3}$% of the student body attended. What fraction of the student body is this?

Write each fraction as a percent. Round to the nearest hundredth if necessary.

21. $\frac{111}{20}$
22. $\frac{180}{25}$
23. $\frac{30}{8}$
24. $\frac{210}{40}$
25. $\frac{29}{30}$
26. $\frac{8}{9}$
27. $\frac{5}{7}$
28. $\frac{1}{16}$
29. $\frac{1}{800}$
30. $\frac{57}{20,000}$
31. $\frac{5}{1,200}$
32. $\frac{7}{1,500}$

33. **FOOD** The size of a large milkshake is $\frac{7}{5}$ times the size of a medium milkshake. Write $\frac{7}{5}$ as a percent.

34. **PETS** In a class, 28 out of 32 students had a pet. What percent is this?

Replace each ● with >, <, or = to make a true statement.

35. 0.86 ● $\frac{7}{8}$
36. $\frac{9}{20}$ ● 45%
37. 5% ● 0.004

Order each set of numbers from least to greatest.

38. $\frac{1}{4}$, 22%, 0.3, 0.02
39. 0.48, $\frac{1}{2}$%, 0.5, $\frac{2}{5}$

Real-World Link . . . . .
Out of the 50 states, 23 states border an ocean or the Gulf of Mexico.
Source: The US50

40. **GEOGRAPHY** Use the information at the left. What percent of the states in the United States do *not* border an ocean or the Gulf of Mexico?

41. **FIND THE DATA** Refer to the Data File on pages 16–19. Choose some data and write a real-world problem in which you would write a fraction as a percent.

CARS For Exercises 42 and 43, use the table, which shows the percent of people in a recent survey who kept the listed items in their car.

Items in Car	Percent of People
Pen/Pencil	73.0%
Cassette Tapes/CDs	66.1%
First-Aid Kit	38.2%
Sports Equipment	28.9%

42. What fraction of people kept a first-aid kit in their car?

43. Approximately 26 out of 125 people surveyed kept a hairbrush in their car. Is this greater or less than the percent who kept sports equipment? Explain.

NYSCC • NYSMT
Extra Practice, pp. 684, 709

H.O.T. Problems

44. **CHALLENGE** For what value of x does $\frac{1}{x} = x\%$?

45. **WRITING IN MATH** Explain why 80%, 0.8, and $\frac{4}{5}$ all represent the same value.

NYSMT PRACTICE 8.N.3

46. Ms. Gallagher made 64 ounces of punch. The punch contained 17 ounces of apple juice. Which equation can be used to find x, the percent of apple juice in the punch?

  A $\frac{x}{100} = \frac{64}{17}$

  B $\frac{x}{17} = \frac{64}{100}$

  C $\frac{x}{64} = \frac{17}{100}$

  D $\frac{x}{100} = \frac{17}{64}$

47. A group of 150 students were asked if they own a pet. The results are shown.

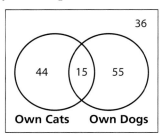

What percent of those surveyed own dogs?

  F 50%   H 36.7%

  G 46.7%   J 30%

Spiral Review

48. **COMPUTERS** Designers are creating a larger model of the computer memory board. They use a scale of 20 inches = 1 inch. If the actual board is $5\frac{1}{4}$ inches long, what is the length of the model? (Lesson 6-8)

49. **RECIPES** Camila is making soup. She has added $\frac{2}{3}$ of the ingredients. If she has added 4 ingredients, how many more does she have to add to be finished? (Lesson 6-7)

▶ **GET READY** for the Next Lesson

PREREQUISITE SKILL Write each fraction in simplest form. (Lesson 4-4)

50. $\frac{8}{10}$ 51. $\frac{45}{100}$ 52. $\frac{450}{100}$ 53. $\frac{175}{100}$

GET READY to Study

FOLDABLES
Study Organizer

Be sure the following Big Ideas are noted in your Foldable.

Ratios
Rates
Rate of Change and Slope
Customary/ Metric Units
Proportions
Scale
Fractions, Decimals, and Percents

BIG Ideas

Ratios and Rates (Lessons 6-1 and 6-2)
• A ratio is a comparison of two quantities by division.

• A rate is a ratio comparing two quantities with different kinds of units.

Slope and Rate of Change (Lesson 6-3)
• Slope is the rate of change between any two points on a line.

Changing Customary Units (Lesson 6-4)
• To convert from larger units to smaller units, multiply by the appropriate unit ratio.

• To convert from smaller units to larger units, multiply by the reciprocal of the appropriate unit ratio.

Changing Metric Units (Lesson 6-5)
• When converting metric units, multiply by the appropriate power of 10.

• When converting between customary and metric units, use the appropriate unit ratio or relationship.

Proportions (Lesson 6-6)
• Proportions are equations that state two ratios or rates are equivalent.

Scale Drawings (Lesson 6-8)
• Scale drawings represent something that is too large or too small to be drawn at actual size.

Key Vocabulary

cross products (p. 310)
equivalent ratios (p. 283)
inverse proportion (p. 316)
proportion (p. 310)
proportional (p. 310)
rate (p. 287)
rate of change (p. 293)
ratio (p. 282)
scale (p. 320)
scale drawing (p. 320)
scale factor (p. 322)
scale model (p. 320)
slope (p. 294)
unit rate (p. 287)
unit ratio (p. 299)

Vocabulary Check

Choose the term from the list above that best matches each phrase.

1. a comparison of two quantities by division

2. two ratios that have the same value

3. a ratio of two measurements with different units

4. an equation that shows that two ratios or rates are equivalent

5. used to represent something that is too large or too small for an actual-size drawing

6. the ratio of the distance on a map to the actual distance

7. a scale written as a ratio in simplest form without units of measurement

8. the constant rate of change in y with respect to the constant change in x

9. a rate that is simplified so that it has a denominator of 1

10. two quantities that have a constant rate or ratio

Lesson-by-Lesson Review

6-1 **Ratios** (pp. 282–286)

7.M.5,
7.M.6

Write each ratio as a fraction in simplest form.

11. 16 dogs : 12 cats   **12.** 5 ft : 25 ft

13. 50 boys : 75 girls   **14.** 36 ft : 6 ft.

15. Determine whether the ratios 18 out of 24 and 5 out of 20 are equivalent.

Example 1 Write the ratio 32 to 18 as a fraction in simplest form.

32 to 18 $= \dfrac{32}{18}$ Write the ratio as a fraction.

$= \dfrac{16}{9}$ Simplify.

Example 2 Determine whether 5 : 6 and 15 : 18 are equivalent.

$5 : 6 = \dfrac{5}{6}$ $15 : 18 = \dfrac{15}{18}$ or $\dfrac{5}{6}$

The ratios in simplest form both equal $\dfrac{5}{6}$.
So, 5 : 6 and 15 : 18 are equivalent.

6-2 **Rates** (pp. 287–292)

7.M.5,
7.M.6

Find each unit rate.

16. $23.75 for 5 pounds

17. 810 miles in 9 days

18. SHAMPOO Which bottle of shampoo shown at the right costs the least per ounce?

Bottle	Price
8 oz	$1.99
12 oz	$2.59
16 oz	$3.19

Example 3 Find the unit price of a 16-ounce box of pasta that is on sale for 96 cents.

16-ounce box for 96 cents $= \dfrac{96 \text{ cents} \div 16}{16 \text{ ounces} \div 16}$

$= \dfrac{6 \text{ cents}}{1 \text{ ounce}}$

The unit price is 6 cents per ounce.

6-3 **Rate of Change and Slope** (pp. 293–297)

8.G.13

Complete.

19. MONEY The table shows the amount of money José saved over a period of time. Graph the data. Then find the slope of the line. Explain what the slope represents.

Amount ($)	30	60	90
Weeks	1	2	3

Example 4 Find the rate of change in degrees per hour.

| | +1 | +1 |
Time (h)	0	1	2
Temperature (C°)	50	52	54
	+2	+2

$\dfrac{\text{change in temperature}}{\text{change in time}} = \dfrac{2° \text{ Celsius}}{1 \text{ hour}}$

So, the temperature increases by 2° Celsius each hour.

LAWN CARE For Exercises 1 and 2, use the following information to write each ratio as a fraction in simplest form.

A bag of fertilizer nutrients contains 18 pounds of nitrogen, 6 pounds of phosphorus, and 12 pounds of potassium.

1. nitrogen : potassium

2. phosphorus : nitrogen

Find each unit rate. Round to the nearest hundredth if necessary.

3. 24 greeting cards for $4.80

4. 330 miles on 15 gallons of gasoline

5. **MULTIPLE CHOICE** The population of bacteria in 4 different-sized lab dishes are given. Which dish has the lowest density of bacteria or bacteria per square inch?

Dish	Bacteria	Dish Area
1	100	205 sq in.
2	50	125 sq in.
3	35	75 sq in.
4	180	300 sq in.

A Dish 1   C Dish 3
B Dish 2   D Dish 4

6. The graph shows the relationship between time and water level of a pool. Find the rate of change.

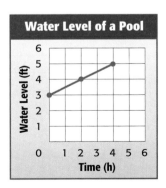

Water Level of a Pool

MEASUREMENT Complete. Round to the nearest tenth if necessary.

7. $7.62 \text{ yd} \approx \blacksquare \text{ m}$

8. $50.8 \text{ lb} \approx \blacksquare \text{ kg}$

9. $3,600 \text{ mL} \approx \blacksquare \text{ qt}$

10. $19.25 \text{ m} \approx \blacksquare \text{ ft}$

ALGEBRA Solve each proportion.

11. $\frac{2}{3} = \frac{x}{42}$

12. $\frac{t}{21} = \frac{15}{14}$

13. **NUTRITION** If an 8-ounce glass of orange juice has 72 milligrams of vitamin C, how much vitamin C is in a 7-ounce glass?

14. **MAPS** The table shows the key for a map. Graph the data. Then find the slope of the line.

Distance on Map (cm)	1	2	3	4
Actual Distance (km)	20	40	60	80

BLUEPRINTS For Exercises 15 and 16, use the following blueprint of a room.

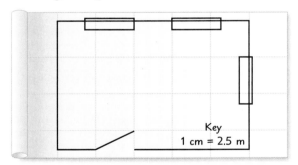

Key
1 cm = 2.5 m

15. Use a centimeter ruler to find the length of the wall with 2 windows.

16. How wide would a 1.4-meter-wide dresser appear on this drawing?

Write each fraction as a percent. Round to the nearest hundredth if necessary.

17. $\frac{5}{8}$

18. $\frac{7}{15}$

19. **GUM** A company test-marketed 7 new flavors of gum last year. Only 2 of these received favorable ratings. Which represents the percent of flavors that did *not* receive favorable ratings?

20. **STOCKS** A company's stock increased 0.83% last month. Write 0.83% as a decimal.

PART 1　Multiple Choice

Read each question. Then fill in the correct answer on the answer sheet provided by your teacher or on a sheet of paper.

1. Francesca typed 496 words in 8 minutes. Which of the following is a correct understanding of this rate?

   A On average, it takes 62 minutes for Francesca to type one word.

   B On average, Francesca can type 62 words in 8 minutes.

   C On average, Francesca can type 62 words in one minute.

   D On average, Francesca can type 8 words in one minute.

2. The table shows the prices of 3 different boxes of cereal. Which box of cereal has the highest price per ounce?

Cereal Box Size (ounces)	Price ($)
48	5.45
32	3.95
20	3.10

   F The 20-ounce box

   G The 32-ounce box

   H The 48-ounce box

   J All three boxes have the same price per ounce.

3. A bakery sells 6 bagels for a total of $2.99 and 4 muffins for a total of $3.29. If you bought 4 dozen bagels and 16 muffins, what is the total cost of the bagels and muffins, not including tax?

   A $64.60　　　　**C** $31.10

   B $37.08　　　　**D** $26.50

4. Mrs. Black is making 2 pasta salads for a picnic. The first pasta salad requires $4\frac{2}{3}$ cups of pasta, and the second pasta salad requires $\frac{1}{3}$ cup more than the first. Which of the following equations can be used to find n, the number of cups of pasta needed for the second recipe?

   F $n = 4\frac{2}{3} \div \frac{1}{3}$

   G $n = 4\frac{2}{3} + \frac{1}{3}$

   H $n = 4\frac{2}{3} - \frac{1}{3}$

   J $n = 4\frac{2}{3} \times \frac{1}{3}$

5. Simplify the expression below.
   $$8 + 3(15 - 5) - 3^2$$

   A 101　　　　**C** 39

   B 44　　　　**D** 29

6. A shoe store had to increase prices. The table shows the regular price r and the new price n of several shoes. Which of the following formulas can be used to calculate the new price?

Shoe	Regular Price (r)	New Price (n)
A	$25.00	$27.80
B	$30.00	$32.80
C	$35.00	$37.80
D	$40.00	$42.80

   F $n = r - 2.80$　　　　**H** $n = r \times 0.1$

   G $n = r + 2.80$　　　　**J** $n = r \div 0.1$

7. Annika can run 2 miles in 15 minutes. At this rate, about how long will it take her to run $3\frac{1}{2}$ miles?

   A 26 minutes　　　　**C** 36 minutes

   B 32 minutes　　　　**D** 45 minutes

8. A building is 55 meters tall. About how tall is the building in feet (ft) and inches (in.)? (1 meter ≈ 39 inches)

   F   179 ft 0 in.

   G   178 ft 9 in.

   H   178 ft 8 in.

   J   178 ft 6 in.

9. You can drive your car 21.75 miles with one gallon of gasoline. How many miles can you drive with 13.2 gallons of gasoline?

   A   13.2

   B   21.75

   C   150.2

   D   287.1

10. The table shows the number of yards of material Leah used each day last week. What was the total number of yards Leah used last week?

Day	Material (yd)
Monday	2.3
Tuesday	$1\frac{3}{4}$
Wednesday	2.8
Thursday	3.1
Friday	$3\frac{1}{4}$
Saturday	1.7
Sunday	$4\frac{1}{2}$

   F   19.4 yd

   G   17 yd

   H   16.5 yd

   J   16 yd

PART 2   Short Response/Grid In

Record your answers on the answer sheet provided by your teacher or on a sheet of paper.

11. Some employees work 40 hours a week. If there are 168 hours in one week, about what part of the week do they work?

12. During a visit to his favorite bookstore, Kevin bought 3 hardback books priced at $14.99 each and 4 paperbacks priced at $7.99 each. Find the total of Kevin's purchase, in dollars, before tax is included.

PART 3   Extended Response

Record your answers on the answer sheet provided by your teacher or on a sheet of paper. Show your work.

13 Pistachios cost $3.99 a pound at the local health food store.

   a. Set up a proportion to find the cost of 3 pounds.

   b. Solve the proportion. How much do 3 pounds of pistachios cost?

   c. If the pistachios are on sale for $3.09 a pound, how much money will you save if you buy 3 pounds of pistachios?

TEST-TAKING TIP

Question 13 When a question involves information from a previous part of a question, make sure to check that information before you move on.

NEED EXTRA HELP?													
If You Missed Question...	1	2	3	4	5	6	7	8	9	10	11	12	13
Go to Lesson...	6-2	6-2	1-1	5-2	1-4	1-10	6-2	6-8	6-6	5-2	6-9	6-7	6-2
NYS Core Curriculum	7.M.5	7.M.5	7.PS.14	6.N.16	7.N.11	7.A.10	7.M.5	7.PS.10	7.PS.10	6.N.16	8.N.3	7.PS.2	7.M.5

CHAPTER 7

Applying Percents

New York State Core Curriculum

Preparation for 8.N.4 Apply percents to: tax, percent increase/decrease, simple interest, sale price, commission, interest rates, gratuities

Key Vocabulary

percent equation (p. 361)

percent of change (p. 369)

percent proportion (p. 350)

Real-World Link

Boogie Boards You can buy a boogie board in Myrtle Beach, South Carolina, for $25. You will also pay a sales tax of 5%.

FOLDABLES
Study Organizer

Applying Percents Make this Foldable to help you organize your notes. Begin with a piece of 11" by 17" paper.

① **Fold** the paper in half lengthwise.

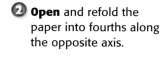

② **Open** and refold the paper into fourths along the opposite axis.

③ **Trace** along the fold lines and label each section with a lesson title or number.

7-1	7-2
7-3	7-4
7-5	7-6
7-7	7-8

GET READY for Chapter 7

Diagnose Readiness You have two options for checking Prerequisite Skills.

Option 2

NY Math Online Take the Online Readiness Quiz at glencoe.com.

Option 1

Take the Quick Quiz below. Refer to the Quick Review for help.

QUICK Quiz

Multiply. (Prior Grade)

1. $300 \times 0.02 \times 8$ 2. $85 \times 0.25 \times 3$

3. $560 \times 0.6 \times 4.5$ 4. $154 \times 0.12 \times 5$

5. **MONEY** If Nicole saves $0.05 every day, how much money will she have in 3 years? (Prior Grade)

Simplify. Write as a decimal. (Prior Grade)

6. $\dfrac{22 - 8}{8}$ 7. $\dfrac{50 - 33}{50}$ 8. $\dfrac{35 - 7}{35}$

9. **BASEBALL CARDS** Tim has 56 baseball cards. He gives 14 of them away. What decimal represents the portion he has left? (Prior Grade)

ALGEBRA Solve. Round to the nearest tenth if necessary. (Lesson 3-3)

10. $0.4m = 52$ 11. $21 = 0.28a$

12. $13 = 0.06s$ 13. $0.95z = 37$

Write each percent as a decimal.
(Lesson 4-7)

14. 40% 15. 17% 16. 110%

17. 157% 18. 3.25% 19. 7.5%

20. **FOOD** Approximately 92% of a watermelon is water. What decimal represents this amount?
(Lesson 4-7)

QUICK Review

Example 1

Evaluate $240 \times 0.03 \times 5$.

$240 \times 0.03 \times 5$
$= 7.2 \times 5$ Multiply 240 by 0.03.
$= 36$ Simplify.

Example 2

Simplify $\dfrac{17 - 8}{8}$. Write as a decimal.

$\dfrac{17 - 8}{8} = \dfrac{9}{8}$ Subtract 8 from 17.
$= 1.125$ Divide 9 by 8.

Example 3

Solve $0.6k = 7.8$

$0.6k = 7.8$ Write the equation.
$k = 13$ Divide each side by 0.6.

Example 4

Write 9.8% as a decimal.

$9.8\% = 0.098$ Move the decimal point two places to the left and remove the percent symbol.

Math Lab
Percent of a Number

Do you enjoy shopping? If so, you may have seen sales or other discounts represented as percents. For example, consider the following situation. A backpack is on sale for 30% off the original price. If the original price of the backpack is $50, how much will you save?

In this situation, you know the percent. You need to find what part of the original price you will save. In this lab, you will use a model to find the percent of a number or *part* of a whole.

ACTIVITY

① **Find 30% of $50 using a model.**

STEP 1 Draw a 1-by-10 rectangle as shown on grid paper. Label the units on the right from 0% to 100% as shown.

Part	Percent
	0%
	10%
	20%
	30%
	40%
	50%
	60%
	70%
	80%
	90%
	100%

STEP 2 Since $50 represents the original price, mark equal units from $0 to $50 on the left side of the model as shown.

STEP 3 Draw a line from 30% on the right side to the left side of the model as shown and shade the portion of the rectangle above this line.

Part	Percent
$0	0%
$5	10%
$10	20%
$15	30%
$20	40%
$25	50%
$30	60%
$35	70%
$40	80%
$45	90%
$50	100%

The model shows that 30% of $50 is $15. So, you will save $15.

✔ CHECK Your Progress

Draw a model to find the percent of each number.

    a. 20% of 120   **b.** 60% of 70   **c.** 90% of 400

Suppose a bicycle is on sale for 35% off the original price. How much will you save if the original price of the bicycle is $180?

ACTIVITY

2 Find 35% of $180 using a model.

STEP 1 Draw a 1-by-10 rectangle as shown on grid paper. Label the units on the right from 0% to 100% as shown.

Part	Percent
	0%
	10%
	20%
	30%
	40%
	50%
	60%
	70%
	80%
	90%
	100%

Study Tip

Equal Units
For the model at the right, use an interval of $18 since $180 ÷ 10 = $18.

STEP 2 The original price is $180. So, mark equal units from $0 to $180 on the left side of the model as shown.

STEP 3 Draw a line from 35% on the right side to the left side of the model.

Part	Percent
$0	0%
$18	10%
$36	20%
$54	30%
$72	40%
$90	50%
$108	60%
$126	70%
$144	80%
$162	90%
$180	100%

The model shows that 35% of $180 is halfway between $54 and $72, or $63.

So, you will save $63.

CHECK Your Progress

Draw a model to find the percent of each number. If it is not possible to find an exact answer from the model, estimate.

d. 25% of 140   **e.** 7% of 50   **f.** 0.5% of 20

ANALYZE THE RESULTS

1. Tell how to determine the units that get labeled on the left side of a percent model.

2. Explain how to find 40% of 30 using a model.

3. **REASONING** How does knowing 10% of a number help you find the percent of the number when the percent is a multiple of 10%?

7-1 Percent of a Number

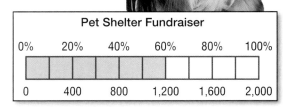

MAIN IDEA

Find the percent of a number.

NYS Core Curriculum

Preparation for 8.N.3
Read, write, and identify percents less than 1% and greater than 100% *Also addresses 7.PS.15, 7.CM.2, 7.CM.5*

NY Math Online

glencoe.com

• Extra Examples
• Personal Tutor
• Self-Check Quiz

▷ **GET READY** for the Lesson

PETS Some students are collecting money for a local pet shelter. The model shows that they have raised 60% of their $2,000 goal or $1,200.

Pet Shelter Fundraiser					
0%	20%	40%	60%	80%	100%
0	400	800	1,200	1,600	2,000

1. Sketch the model and label using decimals instead of percents.

2. Sketch the model using fractions instead of percents.

3. Use these models to write two multiplication sentences that are equivalent to 60% of 2,000 = 1,200.

To find the percent of a number such as 60% of 2,000, you can use one of the following methods.

• Write the percent as a fraction and then multiply, or

• Write the percent as a decimal and then multiply.

EXAMPLE Find the Percent of a Number

1 **Find 5% of 300.**

To find 5% of 300, you can use either method.

METHOD 1 Write the percent as a fraction.

$5\% = \dfrac{5}{100}$ or $\dfrac{1}{20}$

$\dfrac{1}{20}$ of $300 = \dfrac{1}{20} \times 300$ or 15

METHOD 2 Write the percent as a decimal.

$5\% = \dfrac{5}{100}$ or 0.05

0.05 of $300 = 0.05 \times 300$ or 15

So, 5% of 300 is 15.

 **CHOOSE** Your Method

Find the percent of each number.

a. 40% of 70   b. 15% of 100   c. 55% of 160

EXAMPLE   Use Percents Greater Than 100%

2 **Find 120% of 75.**

Study Tip

Check for Reasonableness
120% is a little more than
100%. So, the answer
should be a little more
than 100% of 75 or a little
more than 75.

METHOD 1   **Write the percent as a fraction.**

$120\% = \dfrac{120}{100}$ or $\dfrac{6}{5}$

$\dfrac{6}{5}$ of $75 = \dfrac{6}{5} \times 75$

$\quad\quad\quad = \dfrac{6}{5} \times \dfrac{75}{1}$ or 90

METHOD 2   **Write the percent as a decimal.**

$120\% = \dfrac{120}{100}$ or 1.2

1.2 of $75 = 1.2 \times 75$ or 90

So, 120% of 75 is 90. Use a model to check the answer.

✓ CHOOSE Your Method

Find each number.

d. 150% of 20   **e.** 160% of 35

Real-World EXAMPLE

3 **ANALYZE GRAPHS** Refer to
the graph. If 275 students
took the survey, how many
can be expected to have 3
televisions each in their houses?

To find 23% of 275, write the
percent as a decimal. Then
multiply.

23% of $275 = 23\% \times 275$

$\quad\quad\quad\quad\quad = 0.23 \times 275$

$\quad\quad\quad\quad\quad = 63.25$

So, about 63 students can be expected to have 3 televisions each
in their houses.

Survey Results of Number of Televisions in House	
0	2%
1	9%
2	17%
3	23%
4	20%
More than 4	25%

 = 5%

✓ CHECK Your Progress

f. **ANALYZE GRAPHS** Refer to the graph above. Suppose 455 students
took the survey. How many can be expected to have more than
4 televisions each in their houses?

Find each number. Round to the nearest tenth if necessary.

Examples 1–2
(pp. 344–345)

1. 8% of 50
2. 95% of 40
3. 42% of 263
4. 110% of 70
5. 115% of 20
6. 130% of 78

Example 3
(p. 345)

7. **TAXES** Mackenzie wants to buy a new backpack that costs $50. If the tax rate is 6.5%, how much tax will she pay when she buys the backpack?

Practice and Problem Solving

For Exercises	See Examples
8–13, 20–25	1
14–19	2
26–27	3

HOMEWORK HELP

Find each number. Round to the nearest tenth if necessary.

8. 65% of 186
9. 45% of $432
10. 23% of $640
11. 54% of 85
12. 12% of $230
13. 98% of 15
14. 130% of 20
15. 175% of 10
16. 150% of 128
17. 250% of 25
18. 108% of $50
19. 116% of $250
20. 3.2% of 40
21. 5.4% of 65
22. 23.5% of 128
23. 75.2% of 130
24. 67.5% of 76
25. 18.5% of 500

26. **BASEBALL** Tomás got on base 60% of the times he was up to bat. If he was up to bat 5 times, how many times did he get on base?

27. **TELEVISION** In a recent year, 17.7% of households watched the finals of a popular reality series. There are 110.2 million television households in the United States. How many households watched the finals?

Find each number. Round to the nearest hundredth if necessary.

28. $\frac{4}{5}$% of 500
29. $5\frac{1}{2}$% of 60
30. $20\frac{1}{4}$% of 3
31. 1,000% of 99
32. 100% of 79
33. 520% of 100
34. 0.15% of 250
35. 0.3% of 80
36. 0.28% of 50

37. **TIPPING** A customer wants to tip 15% of the restaurant bill. How much change should there be after the tip if the customer pays with a $50 bill?

Sal's Bistro	
Herbed Salmon	$16.25
Chicken Pasta	15.25
Iced Tea	1.75
Iced Tea	1.75
Total	$35.00

38. **INTERNET** A family pays $19 each month for Internet access. Next month, the cost will increase 5%. After this increase, what will be the cost for the Internet access?

39. **BUSINESS** A store sells a certain brand of a lawn mower for $275. Next year, the cost of the lawn mower will increase by 8%. What will be the cost of the lawn mower next year?

ANALYZE GRAPHS For Exercises 40–42, use the graph below that shows the results of a poll of 2,632 listeners. Round to the nearest whole number.

Radio Listeners

Favorite Places to Listen

40. How many people listen to the radio during work?

41. How many people like to listen to the radio while they are at the gym?

42. Determine how many more people listen to the radio in the car than at home.

Use mental math to find each percent. Justify your answer.

43. 53% of 60

44. 24% of 48

45. 75% of 19

ANALYZE GRAPHS For Exercises 46–49, use the graph that shows the results of a favorite fruit survey.

46. How many people were surveyed?

47. Of those surveyed, how many people prefer peaches?

48. Which type of fruit did more than 100 people prefer?

49. Of those surveyed, how many people did *not* prefer cherries? Explain how you arrived at the answer.

250 people were asked which type of fruit they preferred.

44%   32%   24%

Berries   Peaches   Cherries

Real-World Link . . . .
California, the largest producer of peaches, produces about 60% of all the peaches, grown in the U.S.
Source: California Tree Fruit Agreement

50. SHOPPING The Leather Depot sells a certain leather coat for $179.99. If sales tax is 6.25%, what will be the approximate total cost of the coat?

51. SCHOOL Suppose there are 20 questions on a multiple-choice test. If 25% of the answers are choice B, how many of the answers are *not* choice B?

52. COMMISSION In addition to her salary, Ms. Lopez earns a 3% *commission*, or fee paid based on a percent of her sales, on every vacation package that she sells. One day, she sold the three vacation packages shown. What was her total commission?

NYSCC • NYSMT
Extra Practice, pp. 684, 710

Package #1	Package #2	Package #3
$2,375	$3,950	$1,725

H.O.T. Problems

53. **OPEN ENDED** Give two examples of real-world situations in your life in which you would find the percent of a number.

54. **SELECT A TECHNIQUE** Maggie uses a $50 gift card to buy a pair of shoes that costs $24.99 and a purse that costs $19.99. If the tax rate is 7%, will the gift card cover the entire purchase? Select and use one or more of the following techniques to solve the problem. Justify your selection(s).

| mental math | number sense | estimation |

55. **CHALLENGE** Suppose you add 10% of a number to the number, and then you subtract 10% of the total. Is the result *greater than*, *less than*, or *equal to* the original number? Explain your reasoning.

56. **WRITING IN MATH** Explain which method you prefer to use to find the percent of a number: write the percent as a fraction or write the percent as a decimal. Explain your reasoning.

NYSMT PRACTICE 8.N.3

57. Reggie has memorized 60% of the 50 state capitals for a social studies test. How many more capitals does Reggie need to memorize before the test?

   A 35   **C** 20

   B 30   **D** 18

58. **SHORT RESPONSE** Tanner has 200 baseball cards. Of those, 42% are in mint condition. How many of the cards are *not* in mint condition?

Spiral Review

59. **PETS** In Rebecca's class, 17 out of 24 students have pets. What percent of the students have pets? Round to the nearest percent. (Lesson 6-9)

60. **MODELS** On a scale model of a building, 3 in. = 12 ft. If the model is 8 inches tall, how tall is the actual building? (Lesson 6-8)

Add or subtract. Write in simplest form. (Lesson 6-2)

61. $\dfrac{7}{10} - \dfrac{1}{10}$

62. $\dfrac{20}{21} - \dfrac{3}{7}$

63. $\dfrac{5}{6} - \dfrac{1}{8}$

64. **ALGEBRA** What are the next three numbers in the pattern 3, 10, 17, 24, …? (Lesson 1-9)

▷ **GET READY for the Next Lesson**

PREREQUISITE SKILL Solve each equation. (Lesson 3-3)

65. $12b = 144$

66. $9x = 630$

67. $8,100 = 100k$

READING to SOLVE PROBLEMS

Meaning of Percent

When you solve percent problems, look for three parts: the *part*, the *whole*, and the *percent*. Consider this example.

The table at the right shows the results of a survey about favorite flavor of sugarless gum.

Favorite Flavor of Sugarless Gum	
Flavor	**Number**
Cinnamon	10
Peppermint	18
Watermelon	12
Total	40

- ⊙ **Part**

    Ten students chose cinnamon as their favorite.

- ⊙ **Whole**

    Forty students were surveyed.

- ⊙ **Percent**

    25% of the students who were surveyed (10 out of 40) chose cinnamon as their favorite.

Using all three parts, 25% of 40 is 10.

NYSCC **7.CM.11** Draw conclusions about mathematical ideas through decoding, comprehension, and interpretation of mathematical visuals, symbols, and technical writing

PRACTICE

Identify each statement as the *part*, the *whole*, or the *percent*. Then write a sentence using all three parts.

1. The table at the right shows the results of a survey about which "bugs" people dislike most.
    a. Fifty people were surveyed.
    b. 60% disliked spiders the most.
    c. Thirty people disliked spiders.

Least Favorite "Bug"	
Kind	**Number**
Centipede	2
Cockroach	18
Spider	30
Total	50

2. Suppose you find a sale at the mall.
    a. Everything was 20% off.
    b. The original price of a jacket was $30.
    c. You saved $6.

3. You and your family are eating at a restaurant.
    a. The meal cost $34.
    b. You want to leave a tip of 15%.
    c. The tip is $5.10.

4. Your sister plays basketball.
    a. She usually makes 75% of her free throws.
    b. In the last game, she made 6 free throws.
    c. She had 8 free throws.

The Percent Proportion

MAIN IDEA

Solve problems using the percent proportion.

NYS Core Curriculum

Preparation for 8.N.4
Apply percents to: tax, percent increase/decrease, simple interest, sale price, commission, interest rates, gratuities

New Vocabulary

percent proportion

NY Math Online

glencoe.com

• Extra Examples
• Personal Tutor
• Self-Check Quiz
• Reading in the Content Area

▷ **GET READY** for the Lesson

MONSTER TRUCKS The tires on a monster truck weigh approximately 3,600 pounds. The entire truck weighs about 11,000 pounds.

1. Write the ratio of tire weight to total weight as a fraction.

2. Use a calculator to write the fraction as a decimal to the nearest hundredth.

3. What percent of the monster truck's weight is the tires?

In a **percent proportion**, one ratio or fraction compares part of a quantity to the whole quantity, also called the *base*. The other ratio is the equivalent percent written as a fraction with a denominator of 100.

4 out of 5 is 80%.

$$\frac{\text{part}}{\text{whole}} \rightarrow \frac{4}{5} = \frac{80}{100} \Big\} \text{ percent}$$

When given two of these pieces of information—part, whole, or percent—you can use the proportion to find the missing information.

EXAMPLE Find the Percent

① **What percent of $15 is $9?**

The number 15 comes after the word *of*, so the whole is 15. You are asked to find the percent, so the part is the remaining number, 9.

Words	What percent of $15 is $9?
Variable	Let *n*% represent the percent.
Proportion	$\dfrac{\text{part}}{\text{whole}} \rightarrow \dfrac{9}{15} = \dfrac{n}{100} \Big\}$ percent

$\dfrac{9}{15} = \dfrac{n}{100}$ Write the proportion.

$9 \cdot 100 = 15 \cdot n$ Find the cross products.

$900 = 15n$ Simplify.

$\dfrac{900}{15} = \dfrac{15n}{15}$ Divide each side by 15.

$60 = n$

So, $9 is 60% of $15.

CHECK Your Progress

Find each number. Round to the nearest tenth if necessary.

a. What percent of 25 is 20?   **b.** $12.75 is what percent of $50?

EXAMPLE   Find the Part

② **What number is 40% of 120?**

The percent is 40%. Since the number 120 comes after the word *of*, the whole is 120. You are asked to find the part.

Words	What number is 40% of 120?
Variable	Let p represent the part.
Proportion	$\dfrac{\text{part} \rightarrow}{\text{whole} \rightarrow}\ \dfrac{p}{120} = \dfrac{40}{100}\Big\}$ percent

$$\frac{p}{120} = \frac{40}{100} \qquad \text{Write the proportion.}$$

$$p \cdot 100 = 120 \cdot 40 \qquad \text{Find the cross products.}$$

$$100p = 4{,}800 \qquad \text{Simplify.}$$

$$\frac{100p}{100} = \frac{4{,}800}{100} \qquad \text{Divide each side by 100.}$$

$$p = 48$$

So, 48 is 40% of 120.

 CHECK Your Progress

Find each number. Round to the nearest tenth if necessary.

c. What number is 5% of 60?   **d.** 12% of 85 is what number?

Study Tip

The Percent Proportion
The part usually comes before or after the word *is* and the whole usually comes before or after the word *of*.

EXAMPLE   Find the Whole

③ **18 is 25% of what number?**

The percent is 25%. The words *what number* come after the word *of*. So, you are asked to find the whole. Thus, 18 is the part.

Words	18 is 25% of what number?
Variable	Let w represent the whole.
Proportion	$\dfrac{\text{part} \rightarrow}{\text{whole} \rightarrow}\ \dfrac{18}{w} = \dfrac{25}{100}\Big\}$ percent

(continued on the next page)

$$\frac{18}{w} = \frac{25}{100}$$ Write the proportion.

$18 \cdot 100 = w \cdot 25$ Find the cross products.

$1{,}800 = 25w$ Simplify.

$$\frac{1{,}800}{25} = \frac{25w}{25}$$ Divide each side by 25.

$72 = w$

So, 18 is 25% of 72.

✔ CHECK Your Progress

Find each number. Round to the nearest tenth if necessary.

e. 40% of what number is 26? f. 80 is 75% of what number?

Real-World EXAMPLE

4 **ANIMALS** The average adult male Western Lowland gorilla eats about 33.5 pounds of fruit each day. How much food does the average adult male gorilla eat each day?

You know that 33.5 pounds of fruit is 67% of the total amount eaten daily. So, the problem asks 33.5 is 67% of what number. Thus, you need to find the whole.

Western Lowland Gorilla's Diet	
Food	**Percent**
Fruit	67%
Seeds, Leaves, Stems, and Pith	17%
Insects/Insect Larvae	16%

$$\frac{33.5}{w} = \frac{67}{100}$$ Write the proportion.

$33.5 \cdot 100 = w \cdot 67$ Find the cross products.

$3{,}350 = 67w$ Simplify.

$$\frac{3{,}350}{67} = \frac{67w}{67}$$ Divide each side by 67.

$50 = w$

So, the average adult male gorilla eats 50 pounds of food each day.

✔ CHECK Your Progress

g. **ZOO** If 200 of the 550 reptiles in a zoo are on display, what percent of the reptiles are on display? Round to the nearest whole number.

Real-World Link . . . .
Male Western Lowland gorillas weigh about 350–400 pounds. Females weigh about 160–200 pounds.
Source: Columbus Zoo and Aquarium

Types of Percent Problems		Key Concept
Type	**Example**	**Proportion**
Find the Percent	What percent of 6 is 3?	$\frac{3}{6} = \frac{n}{100}$
Find the Part	What number is 50% of 6?	$\frac{p}{6} = \frac{50}{100}$
Find the Whole	3 is 50% of what number?	$\frac{3}{w} = \frac{50}{100}$

Examples 1–3
(pp. 350–352)

Find each number. Round to the nearest tenth if necessary.

1. What percent of 50 is 18?

2. What percent of $90 is $9?

3. What number is 2% of 35?

4. What number is 25% of 180?

5. 9 is 12% of which number?

6. 62 is 90.5% of what number?

Example 4
(p. 352)

7. **MEASUREMENT** If a box of Brand A cereal contains 10 cups of cereal, how many more cups of cereal are in a box of Brand B cereal?

Practice and Problem Solving

For Exercises	See Examples
8–11	1, 2
12–17	3
18–21	4
22, 23	5

Find each number. Round to the nearest tenth if necessary.

8. What percent of 60 is 15?

9. $3 is what percent of $40?

10. What number is 15% of 60?

11. 12% of 72 is what number?

12. 9 is 45% of what number?

13. 75 is 20% of what number?

14. **SCHOOL** Roman has 2 red pencils in his backpack. If this is 25% of the total number of pencils, how many pencils are in his backpack?

15. **BASKETBALL** Lisa and Michelle scored 48% of their team's points. If their team had a total of 50 points, how many points did they score?

16. **SHOES** A pair of sneakers are on sale as shown. This is 75% of the original price. What was the original price of the shoes?

17. **BOOKS** Of the 60 books on a bookshelf, 24 are nonfiction. What percent of the books are nonfiction?

Sale Price
$51

Find each number. Round to the nearest hundredth if necessary.

18. What percent of 25 is 30?

19. What number is 8.2% of 50?

20. 40 is 50% of what number?

21. 12.5% of what number is 24?

22. What number is 0.5% of 8?

23. What percent of 300 is 0.6?

24. **BUSINESS** The first week of June, there were 404 customers at an ice cream parlor. Eight weeks later, the number of customers was 175% of this amount. How many customers were there eight weeks later?

25. **MONEY** Ajamu saves 40% of his allowance each week. If he saves $16 in 5 weeks, how much allowance does Ajamu receive each week?

26. **SCHOOL** A class picture includes 95% of the students. Seven students were absent. How many students are in the class?

ASTRONOMY For Exercises 27–29, use the table shown.

27. Mercury's radius is what percent of Jupiter's radius?

28. If the radius of Mars is about 13.7% of Neptune's radius, what is the radius of Neptune?

Planet	Radius (km)
Mercury	2,440
Mars	3,397
Jupiter	71,492

NYSCC • NYSMT
Extra Practice, pp. 685, 710

29. Earth's radius is about 261.4% of Mercury's radius. What is the radius of Earth?

H.O.T. Problems

30. **OPEN ENDED** Write a proportion that can be used to find the percent scored on a science quiz that has 10 questions.

31. **CHALLENGE** Without calculating, arrange the following from greatest to least value. Justify your reasoning.

$$20\% \text{ of } 100, \quad 20\% \text{ of } 500, \quad 5\% \text{ of } 100$$

32. **WRITING IN MATH** Create a problem involving a percent that can be solved by using the proportion $\frac{3}{b} = \frac{15}{100}$.

NYSMT PRACTICE 8.N.4

33. Of the 273 students in a school, 95 volunteered to work the book sale. About what percent of the students did *not* volunteer?

 A 55%

 B 65%

 C 70%

 D 75%

34. A customer at a restaurant leaves a tip of $2.70. This amount is 15% of the total bill. Which equation can be used to find x, the total amount of the food bill?

 F $\frac{2.70}{15} = \frac{15}{100}$ H $\frac{15}{2.70} = \frac{x}{100}$

 G $\frac{2.70}{x} = \frac{15}{100}$ J $\frac{x}{2.70} = \frac{15}{100}$

Spiral Review

Find each number. Round to the nearest tenth if necessary. (Lesson 7-1)

35. What is 25% of 120?

36. Find 45% of 70.

37. **PLANTS** A plant was 275% taller than the month before. What decimal represents this percent? (Lesson 6-9)

MEASUREMENT Complete. (Lesson 6-4)

38. 3,000 lb = ▦ T

39. 36 in. = ▦ ft

40. $4\frac{1}{2}$ lb = ▦ oz

▷ **GET READY** for the Next Lesson

PREREQUISITE SKILL Multiply. (Lesson 5-5)

41. $\frac{1}{2} \cdot 60$

42. $\frac{3}{4} \cdot 28$

43. $\frac{2}{5} \cdot 45$

7-3 Percent and Estimation

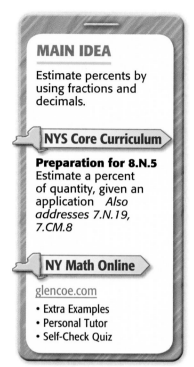

MAIN IDEA

Estimate percents by using fractions and decimals.

NYS Core Curriculum

Preparation for 8.N.5
Estimate a percent of quantity, given an application *Also addresses 7.N.19, 7.CM.8*

NY Math Online

glencoe.com

• Extra Examples
• Personal Tutor
• Self-Check Quiz

▷ GET READY for the Lesson

MUSIC Refer to the graph below.

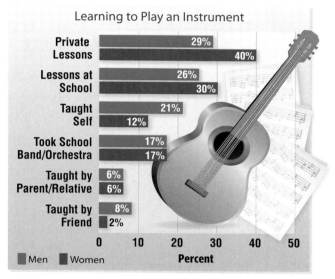

Learning to Play an Instrument

	Men	Women
Private Lessons	29%	40%
Lessons at School	26%	30%
Taught Self	21%	12%
Took School Band/Orchestra	17%	17%
Taught by Parent/Relative	6%	6%
Taught by Friend	8%	2%

1. What fraction of women took lessons at school? If 200 women were surveyed, how many of them took lessons at school?

2. Use a fraction to estimate the number of men who took lessons at school. Assume 200 men were surveyed.

Sometimes an exact answer is not needed when using percents. One way to estimate the percent of a number is to use a fraction.

Real-World EXAMPLE

1 **SPORTS** In a recent year, quarterback Carson Palmer completed 62% of his passes. He threw 520 passes. About how many did he complete?

$$62\% \text{ of } 520 \approx 60\% \text{ of } 520 \qquad 62\% \approx 60\%$$

$$\approx \frac{3}{5} \cdot 520 \qquad 60\% = \frac{6}{10} \text{ or } \frac{3}{5}$$

$$\approx 312 \qquad \text{Multiply.}$$

So, Carson Palmer completed about 312 out of 520 passes.

✓ CHECK Your Progress

a. **REPTILES** Box turtles have been known to live for 120 years. American alligators have been known to live 42% as long as box turtles. About how long can an American alligator live?

Another method for estimating the percent of a number is to first find 10% of the number and then multiply. For example, 70% = 7 • 10%. So, 70% of a number equals 7 times 10% of the number.

Real-World EXAMPLE

② **MONEY** Marita decides to leave a 20% tip on a restaurant bill of **$14.72. About how much money should she tip the restaurant server?**

You need to estimate 20% of $14.72.

METHOD 1   **Use a fraction to estimate.**

20% is $\frac{2}{10}$ or $\frac{1}{5}$.

20% of $14.72 $\approx \frac{1}{5}$ • $15.00   20% = $\frac{1}{5}$ and round $14.72 to $15.00.

$\approx$ $3.00   Multiply.

METHOD 2   **Use 10% of a number to estimate.**

Step 1   Find 10% of the number.
$14.72 is about $15.00.
10% of $15.00 = 0.1 • $15.00   To multiply by 10%, move the
= $1.50   decimal point one place to the left.

Step 2   Multiply.
20% of $15.00 is 2 times 10% of $15.00.
2 • $1.50 = $3.00

So, Marita should tip the restaurant server about $3.00.

 **Your Method**

b. **MONEY** Dante plans to put 80% of his paycheck into a savings account. His paycheck this week was $295. About how much money will he put into his savings account?

You can also estimate percents of numbers when the percent is greater than 100 or the percent is less than 1.

EXAMPLES   Percents Greater Than 100 or Less Than 1

③ **Estimate 122% of 50.**

122% is about 120%.

120% of 50 = (100% of 50) + (20% of 50)   120% = 100% + 20%

$= (1 \cdot 50) + \left(\frac{1}{5} \cdot 50\right)$ 100% = 1 and 20% = $\frac{1}{5}$

= 50 + 10 or 60   Simplify.

So, 122% of 50 is about 60.

Real-World Link . . . . .
In a recent year, the Internal Revenue Service estimated that Americans paid $15.37 billion in tips.

4 Estimate $\frac{1}{4}$% of 589.

$\frac{1}{4}$% is one fourth of 1%. 589 is about 600.

1% of 600 = 0.01 · 600   Write 1% as 0.01.

         = 6   To multiply by 1%, move the decimal point two places to the left.

One fourth of 6 is $\frac{1}{4}$ · 6 or 1.5. So, $\frac{1}{4}$% of 589 is about 1.5.

 **CHECK Your Progress**

Estimate.

c. 174% of 200   **d.** 298% of 45   **e.** 0.25% of 789

 **Real-World EXAMPLE**

5 **CELL PHONES** In a recent year, there were about 200 million people in the U.S. with cell phones. Of those, about 0.5% used their phone as an MP3 player. Estimate the number of people who used their phone as an MP3 player.

0.5% is half of 1%.

1% of 200 million = 0.01 · 200,000,000

                = 2,000,000

So, 0.5% of 200,000,000 is about $\frac{1}{2}$ of 2,000,000 or 1,000,000.

So, about 1,000,000 people used their phone as an MP3 player.

 **CHECK Your Progress**

f. **ATTENDANCE** Last year, 639 students attended a summer camp. Of those who attended this year, 0.9% also attended last year. About how many people attended the camp two years in a row?

 **Your Understanding**

Examples 1–4 (pp. 355–357) **Estimate.**

1. 52% of 10   **2.** 7% of 20   **3.** 38% of 62

4. 79% of 489   **5.** 151% of 70   **6.** $\frac{1}{2}$% of 82

Example 1 (p. 355) **7.** **BUSINESS** A bicycle store increases its prices by 23%. About how much more will a customer pay for a bicycle that originally costs $200?

Example 2 (p. 356) **8.** **BIRTHDAYS** Of the 78 teenagers at a youth camp, 63% have birthdays in the spring. About how many have birthdays in the spring?

Example 5 (p. 357) **9.** **GEOGRAPHY** About 0.8% of the land in Maine is federally owned. If Maine is 19,847,680 acres, about how many acres are federally owned?

Estimate.

HOMEWORK HELP	
For Exercises	See Examples
10–21	1, 3
22–23	2
24–25	3
26–27, 30	4
28–29, 31	5

10. 47% of 70

11. 21% of 90

12. 39% of 120

13. 76% of 180

14. 57% of 29

15. 92% of 104

16. 24% of 48

17. 28% of 121

18. 88% of 207

19. 62% of 152

20. 65% of 152

21. 72% of 238

22. **MONEY** Jessica spent $42 at the hair salon. About how much money should she tip the hair stylist if she wants to leave a 15% tip?

23. **HEALTH** You use 43 muscles to frown. When you smile, you use 32% of these same muscles. About how many muscles do you use when you smile?

Estimate.

24. 132% of 54

25. 224% of 320

26. $\frac{1}{2}$% of 412

27. $\frac{3}{4}$% of 168

28. 0.4% of 510

29. 0.9% of 74

30. **GEOGRAPHY** The United States has 12,383 miles of coastline. If $\frac{4}{5}$% of the U.S. coastline is located in Georgia, about how many miles of coastline are in Georgia?

31. **BIRDS** During migration, 450,000 sandhill cranes stop to rest in Nebraska. About 0.6% of these cranes stop to rest in Oregon. About how many sandhill cranes stop in Oregon during migration?

Estimate.

32. 67% of 8.7

33. 54% of 76.8

34. 32% of 89.9

35. 10.5% of 238

36. 22.2% of 114

37. 98.5% of 45

ANALYZE GRAPHS For Exercises 38–40, use the graph shown.

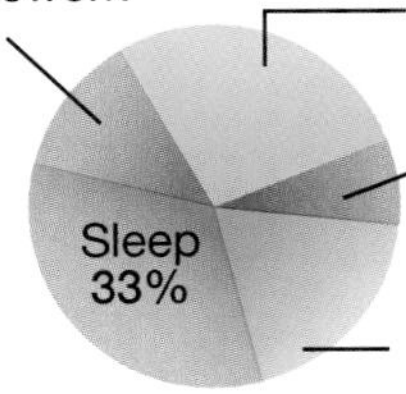

Amanda's Day

Homework 13%
School 27%
Extracurricular Activities 8%
Sleep 33%
Other 19%

Real-World Link . . . . .
About 75% of the entire world sandhill crane population stops to rest in Nebraska during migration.
Source: *World Book of Records*

38. About how many hours does Amanda spend doing her homework each day?

39. About how many more hours does Amanda spend sleeping than doing the activities in the "other" category? Justify your answer.

40. What is the approximate number of minutes Amanda spends each day on extracurricular activities?

41. ANALYZE GRAPHS 2,075 tennis fans were asked to name the greatest all time female tennis player. The top five responses are shown. About how many more people chose Martina Navratilova than Steffi Graf?

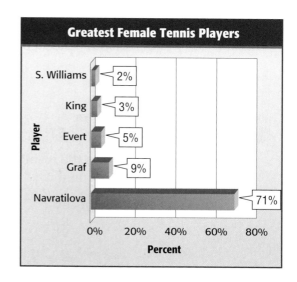

Greatest Female Tennis Players

42. ANIMALS The average white rhinoceros gives birth to a single calf that weighs about 3.8% as much as its mother. If the mother rhinoceros weighs 3.75 tons, about how many pounds does its calf weigh?

43. POPULATION According to the 2000 U.S. Census, about 7.8% of the people in Minnesota live in Minneapolis. If the population of Minnesota is about 4,920,000, estimate the population of Minnesota.

CLEANING For Exercises 44 and 45, use the following information.

A cleaning solution is made up of 0.9% chlorine bleach.

44. About how many ounces of bleach are in 189 ounces of cleaning solution?

NYSCC • NYSMT
Extra Practice, pp. 685, 710

45. About how many ounces of bleach would be found in 412 ounces of cleaning solution?

H.O.T. Problems

46. OPEN ENDED Write a real-world problem in which the answer can be found by estimating 12% of 50.

47. CHALLENGE Explain how you could find $\frac{3}{8}\%$ of $800.

48. FIND THE ERROR Tom and Elsa are estimating 1.5% of 210. Who is correct? Explain.

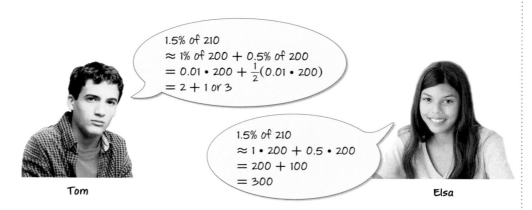

Tom:
1.5% of 210
≈ 1% of 200 + 0.5% of 200
= 0.01 · 200 + $\frac{1}{2}$(0.01 · 200)
= 2 + 1 or 3

Elsa:
1.5% of 210
≈ 1 · 200 + 0.5 · 200
= 200 + 100
= 300

49. NUMBER SENSE Is an estimate for the percent of a number *always*, *sometimes*, or *never* greater than the actual percent of the number? Give an example or a counterexample to support your answer.

50. **WRITING IN MATH** Estimate 22% of 136 using two different methods. Justify the steps used in each method.

51. The graph shows the results of a survey of 510 students.

Pet Preferences

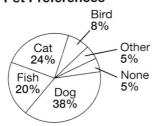

Bird 8%
Cat 24%
Other 5%
Fish 20%
None 5%
Dog 38%

Which is the best estimate for the number of students who prefer cats?

A 75   C 225

B 125   D 450

52. Megan is buying an entertainment system for $1,789.43. The speakers are 39.7% of the total cost. Which is the best estimate for the cost of the speakers?

F $540   H $720

G $630   J $810

53. Daniel decides to leave a 20% tip for a restaurant bill of $28.92. About how much money should he tip the restaurant server?

A $2.00   C $4.00

B $3.00   D $6.00

Spiral Review

Find each number. Round to the nearest tenth if necessary. (Lesson 7-2)

54. 6 is what percent of 15?

55. Find 72% of 90.

56. What number is 120% of 60?

57. 35% of what number is 55?

58. **HEALTH** Adults have 32 teeth. Children have 62.5% as many teeth as adults. How many teeth do children have? (Lesson 7-1)

Estimate. (Lesson 5-1)

59. $\dfrac{8}{9} + \dfrac{1}{12}$

60. $\dfrac{4}{7} + \dfrac{7}{16}$

61. $\dfrac{7}{8} - \dfrac{7}{16}$

62. $\dfrac{4}{5} - \dfrac{9}{10}$

Solve. Round to the nearest tenth if necessary. (Lesson 3-3)

63. $40 = 0.8x$

64. $10r = 61$

65. $0.07t = 25$

66. $56 = 0.32n$

▶ GET READY for the Next Lesson

Solve each equation. Check your solution. (Lesson 3-3)

67. $14 = n \cdot 20$

68. $25 = n \cdot 40$

69. $28.5 = n \cdot 38$

70. $36 = n \cdot 80$

7-4 Algebra: The Percent Equation

MAIN IDEA

Solve problems by using the percent equation.

NYS Core Curriculum

Preparation for 8.N.4
Apply percents to: tax, percent increase/decrease, simple interest, sale price, commission, interest rates, gratuities

New Vocabulary

percent equation

NY Math Online

glencoe.com

• Extra Examples
• Personal Tutor
• Self-Check Quiz

▷ **GET READY** for the Lesson

ARTHROPODS There are about 854,000 different species of spiders, insects, crustaceans, millipedes, and centipedes on Earth. The graph shows that 88% of the total numbers of species of arthropods are insects.

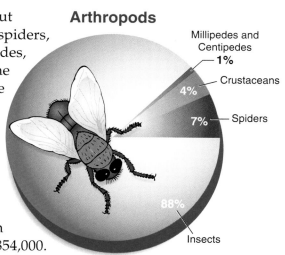

Arthropods

Millipedes and Centipedes — 1%
Crustaceans 4%
Spiders 7%
Insects 88%

1. Use the percent proportion to find how many species are insects.

2. Express the percent of insects as a decimal. Then multiply the decimal by 854,000.

In Lesson 7-2, you used a percent proportion to find the missing part, percent, or whole. You can also use an equation. The percent equation is another form of the percent proportion.

$$\frac{part}{whole} = percent$$

The percent must be written as a decimal or fraction.

$$\frac{part}{whole} \cdot whole = percent \cdot whole$$

Multiply each side by the whole.

$$part = percent \cdot whole$$

This form is called the **percent equation**.

EXAMPLE Find the Part

1 **What number is 12% of 150?** **Estimate** 12% of 150 ≈ 0.1 · 150 or 15

Write 12% as a decimal, 0.12. The whole is 150. You need to find the part. Let p represent the part.

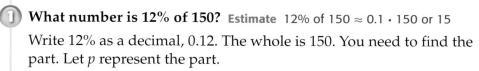

part = percent · whole

| p | = | 0.12 | · | 150 | Write the percent equation. |
| p | = | 18 | | | Multiply. The part is 18. |

So, 18 is 12% of 150.   **Check for Reasonableness** 18 is close to 15. ✔

Study Tip

Percent Equation
A percent must always be converted to a decimal or a fraction when it is used in an equation.

 CHECK Your Progress

Write an equation for each problem. Then solve. Round to the nearest tenth if necessary.

a. What is 6% of 19?   b. Find 72% of 90.

2 **21 is what percent of 40?** Estimate $\frac{21}{40} \approx \frac{1}{2}$ or 50%

The part is 21. The whole is 40. You need to find the percent. Let n represent the percent.

$$\underbrace{\text{part}}_{} = \underbrace{\text{percent}}_{} \cdot \underbrace{\text{whole}}_{}$$

21	=	n $\cdot$ 40	Write the percent equation.
$\frac{21}{40}$	=	$\frac{40n}{40}$	Divide each side by 40.
0.525	=	n	Since n represents the decimal form, the percent is 52.5%.

So, 21 is 52.5% of 40.

Check for Reasonableness 52.5% $\approx$ 50% ✔

Study Tip

Percent
Remember to write the decimal as a percent in your final answer.

 CHECK Your Progress

Write an equation for each problem. Then solve. Round to the nearest tenth if necessary.

c. 35 is what percent of 70? **d.** What percent of 125 is 75?

e. What percent of 40 is 9? **f.** 27 is what percent of 150?

---

3 **13 is 26% of what number?** Estimate $\frac{1}{4}$ of 48 = 12

The part is 13. The percent is 26, which when written as a decimal, is 0.26. You need to find the whole. Let w represent the whole.

$$\underbrace{\text{part}}_{} = \underbrace{\text{percent}}_{} \cdot \underbrace{\text{whole}}_{}$$

13	=	0.26 $\cdot$ w	Write the percent equation. 26% = 0.26
$\frac{13}{0.26}$	=	$\frac{0.26w}{0.26}$	Divide each side by 0.26.
50	=	w	The number is 50.

So, 13 is 26% of 50.

Check for Reasonableness 50 is close to 48. ✔

 CHECK Your Progress

Write an equation for each problem. Then solve. Round to the nearest tenth if necessary.

g. 39 is 84% of what number? **h.** 26% of what number is 45?

i. 14% of what number is 7? **j.** 24 is 32% of what number?

Real-World EXAMPLE

4 **CELL PHONES** A survey found that 25% of people age 18–24 gave up their home phone and only use a cell phone. If 3,264 people only use a cell phone, how many people were surveyed?

Words	3,264 people is 25% of what number of people?
Variable	Let *n* represent the number of people.
Equation	3,264 = 0.25 • *n*

$3{,}264 = 0.25 \cdot n$ Write the percent equation. 25% = 0.25

$\dfrac{3{,}264}{0.25} = \dfrac{0.25n}{0.25}$ Divide each side by 0.25. Use a calculator.

$13{,}056 \approx n$ Simplify.

The number of people surveyed is about 13,056.

CHECK Your Progress

k. **POPULATION** The Louisville-Jefferson County metropolitan area contains 17.2% of the population of Kentucky. If the population of Kentucky is about 4,040,000 people, what is the population of the Louisville-Jefferson County metropolitan area?

Types of Percent Problems — Concept Summary

Type	Example	Proportion
Find the Part	What number is 50% of 6?	$p = 0.5 \cdot 6$
Find the Percent	3 is what percent of 6?	$3 = n \cdot 6$
Find the Whole	3 is 50% of what number?	$3 = 0.5 \cdot w$

CHECK Your Understanding

Examples 1–3
(pp. 361–362)

Write an equation for each problem. Then solve. Round to the nearest tenth if necessary.

1. What number is 88% of 300?

2. What number is 12% of 250?

3. 75 is what percent of 150?

4. 24 is what percent 120?

5. 3 is 12% of what number?

6. 84 is 60% of what number?

Example 4
(p. 363)

7. **BUSINESS** A local bakery sold 60 loaves of bread in one day. If 65% of these were sold in the afternoon, how many loaves were sold in the afternoon?

HOMEWORK HELP	
For Exercises	See Examples
8–11	1
12–15	2
16–19	3
20–23	4

Write an equation for each problem. Then solve. Round to the nearest tenth if necessary.

8. What number is 65% of 98?

9. Find 39% of 65.

10. Find 24% of 25.

11. What number is 53% of 470?

12. 9 is what percent of 45?

13. What percent of 96 is 26?

14. What percent of 392 is 98?

15. 30 is what percent of 64?

16. 33% of what number is 1.45?

17. 84 is 75% of what number?

18. 17 is 40% of what number?

19. 80% of what number is 64?

20. **BOOKS** Emma bought 6 new books for her collection. This increased her collection by 12%. How many books did she have before her purchases?

21. **VIDEO GAMES** A store sold 550 video games during the month of December. If this made up 12.5% of their yearly video game sales, about how many video games did the store sell all year?

22. **MEASUREMENT** The length of Giselle's arm is 27 inches. The length of her lower arm is 17 inches. About what percent of Giselle's arm is her lower arm?

23. **LOBSTERS** Approximately 0.02% of North Atlantic lobsters are born bright blue in color. Out of 5,000 North Atlantic lobsters, how many would you expect to be blue in color?

Write an equation for each problem. Then solve. Round to the nearest tenth if necessary.

24. Find 135% of 64.

25. What number is 0.4% of 82.1?

26. 450 is 75.2% of what number?

27. What percent of 200 is 230?

28. **SALARY** Suppose you earn $6 per hour at your part-time job. What will your new hourly rate be after a 2.5% raise?

ANALYZE GRAPHS About 672 million metric tons of corn was produced worldwide in 2006. For Exercises 29–31, use the graph at the right.

29. About what percent of corn was produced in the United States?

30. About what percent of corn was produced in Mexico?

31. What percent of the world's corn production do Brazil and China make together?

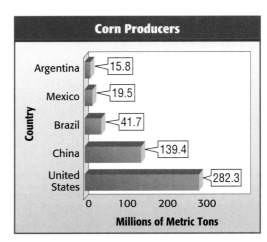

Corn Producers

Country	Millions of Metric Tons
Argentina	15.8
Mexico	19.5
Brazil	41.7
China	139.4
United States	282.3

32. MEASUREMENT A pool that holds 320 cubic feet of water is 84% filled to capacity. Another pool that holds 400 cubic feet of water is 82.5% filled to capacity. Which pool contains more water? How much more?

H.O.T. Problems

33. OPEN ENDED Write a percent problem for which the percent is greater than 100% and the part is known. Use the percent equation to solve your problem to find the base.

34. CHALLENGE If you need to find the percent of a number, explain how you can predict whether the part will be less than, greater than, or equal to the number.

35. WRITING IN MATH Compare the percent equation and the percent proportion. Then explain when it might be easier to use the percent equation rather than the percent proportion.

NYSMT PRACTICE 8.N.4

36. In a survey, 100 students were asked to choose their favorite take-out food. The table shows the results.

Favorite Take-Out Food	
Type of Food	**Percent**
Pizza	40
Sandwiches	32
Fried chicken	28

Based on this data, predict how many out of 1,800 students would choose sandwiches.

A 504   C 680

B 576   D 720

37. If 60% of a number is 18, what is 90% of the number?

F 3   H 27

G 16   J 30

38. Taryn's grandmother took her out to dinner. If the dinner was $34 and she left a 20% tip, how much money did Taryn's grandmother spend?

A $6.80   C $39.50

B $27.20   D $40.80

Spiral Review

39. RESTAURANT Mitchell spent $13 on dinner. About how much money should he tip the server if he wants to leave a 15% tip? (Lesson 7-3)

Find each number. Round to the nearest hundredth if necessary. (Lesson 7-2)

40. What percent of 15 is 20?

41. 20.5% of what number is 35?

42. What number is 0.5% of 10?

▷ **GET READY for the Next Lesson**

43. PREREQUISITE SKILL To estimate the age of a dog in human years, count the first year as 15 human years, the second year as 10 human years, and all of the following years as 3 human years. How old in human years is a 6-year-old dog? (Lesson 1-1)

Problem-Solving Investigation

MAIN IDEA: Solve problems by determining reasonable answers.

 7.N.19 Justify the reasonableness of answers using estimation **7.PS.12** Interpret solutions within the given constraints of a problem

P.S.I. TEAM +

e-Mail: DETERMINE REASONABLE ANSWERS

Doug: My dad painted 25% of my bedroom in 28 minutes. I think the whole project will take about 3 hours.

YOUR MISSION: Determine whether it is reasonable for Doug's dad to paint the bedroom in 3 hours.

Understand	Twenty-five percent of the room has been painted in 28 minutes. Doug thinks it will take a total of 3 hours to paint the whole room.
Plan	Since 25% or $\frac{1}{4}$ of the room was painted in about 30 minutes, use a model of 25%.
Solve	Round 28 minutes to 30 minutes 25% [ ] [ ] [ ] ⟶ 25% 25% 25% 25% 30 min 30 min 30 min 30 min 30 min 30 minutes × 4 = 120 minutes 120 min = 2 h So, 2 hours would be a better estimate than 3 hours.
Check	Thirty minutes is $\frac{1}{2}$ hour. Since $\frac{1}{2} \times 4 = 2$, 2 hours is reasonable answer. ✔

Analyze The Strategy

1. Describe other problem-solving strategies that you could use to determine whether answers are reasonable.

2. **WRITING IN MATH** Write two word problems. One should have a reasonable answer and the other should not.

NYSCC • NYSMT

Extra Practice, pp. 686, 710.

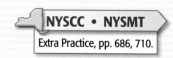

Determine reasonable answers for Exercises 3–6.

3. **SAVING** Aliayah saves $11 each month for her class trip. What is a reasonable estimate for the amount of money she will have saved after a year: about $100, $120, or $160? Explain.

4. **SCHOOL** Of 423 students, 57.6% live within 5 miles of the school. What is a reasonable estimate for the number of students living within 5 miles of the school? Explain.

5. **EXERCISE** A survey showed that 61% of middle school students do some kind of physical activity every day. If there are 828 middle school students in your school, would the number of students who exercise be about 300, 400, or 500? Explain.

6. **ANALYZE GRAPHS** A travel agency surveyed 140 families about their favorite vacation spots. Is 60, 70, or 80 families a reasonable estimate for the number of families that did *not* choose Hawaii?

Favorite Vacation Spots

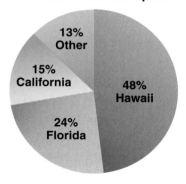

Use any strategy to solve Exercises 7–13. Some strategies are shown below.

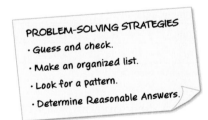

PROBLEM-SOLVING STRATEGIES
· Guess and check.
· Make an organized list.
· Look for a pattern.
· Determine Reasonable Answers.

7. **COINS** John has 10 coins that total $0.83. What are the coins?

8. **ANALYZE GRAPHS** Refer to the graph. A pie is set out to cool. Is it reasonable to estimate that the pie will be 90°F after ten minutes of cooling?

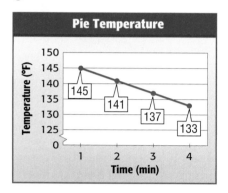

9. **SHOPPING** Deshawn wants to buy a shirt that has a regular price of $41, but is now on sale for 25% off. Is $25, $30, or $35 the best estimate for the cost of the shirt?

10. **BOWLING** In bowling, you get a spare when you knock down the ten pins in two throws. How many possible ways are there to get a spare?

11. **TIPS** Shawnda decides to leave a 20% tip on a restaurant bill of $17.50. How much should she tip the restaurant server?

12. **FUNDRAISER** During a popcorn sale for a fundraiser, the soccer team gets to keep 25% of the sales. One box of popcorn sells for $1.50, and the team has sold 510 boxes so far. Has the team raised a total of $175?

13. **MEASUREMENT** How many square yards of carpet are needed to carpet the two rooms described below? Explain.

Room	Dimensions
living room	15 ft by 18 ft
TV room	18 ft by 20 ft

Find each number. Round to the nearest tenth if necessary. (Lesson 7-1)

1. Find 17% of 655.

2. What is 235% of 82?

3. Find 75% of 160.

4. What number is 162.2% of 55?

5. **MULTIPLE CHOICE** Ayana has 220 coins in her piggy bank. Of those, 45% are pennies. How many coins are not pennies? (Lesson 7-1)

   A 121   C 109

   B 116   D 85

Find each number. Round to the nearest tenth if necessary. (Lesson 7-2)

6. What percent of 84 is 12?

7. 15 is 25% of what number?

8. 85% of 252 is what number?

ANALYZE GRAPHS For Exercises 9 and 10, refer to the graph that shows the results of a survey of 200 students' favorite DVDs.

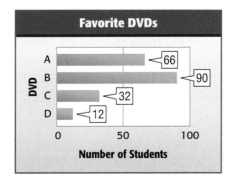

9. What percent of students preferred DVD A?

10. Which DVD did about 15% of students prefer?

Estimate. (Lesson 7-3)

11. 20% of 392   12. 78% of 112

13. 52% of 295   14. 30% of 42

15. 79% of 88   16. 41.5% of 212

17. **MULTIPLE CHOICE** A football player has made about 75% of the field goals he has attempted in his career. If he attempts 41 field goals in one season, about how many would he be expected to make?

   A 35   C 25

   B 30   D 20

Write an equation for each problem. Then solve. Round to the nearest tenth if necessary. (Lesson 7-4)

18. What number is 35% of 72?

19. 16.1 is what percent of 70?

20. 27.2 is 68% of what number?

21. 16% of 32 is what number?

22. 55% of what number is 1.265?

23. 17 is 40% of what number?

24. **ANALYZE TABLES** The table shows the costs of owning a dog over an average 11-year lifespan. What percent of the total cost is veterinary bills? (Lesson 7-4)

Dog Ownership Costs	
Item	**Cost ($)**
Food	4,020
Veterinary Bills	3,930
Grooming, Equipment	2,960
Training	1,220
Other	2,470

Source: American Kennel Club

25. **SHOPPING** A desktop computer costs $849.75 and the hard drive is 61.3% of the total cost. What is a reasonable estimate for the cost of the hard drive? (Lesson 7-5)

7-6 Percent of Change

MAIN IDEA

Find the percent of increase or decrease.

NYS Core Curriculum

Preparation for 8.N.4
Apply percents to: tax, percent increase/ decrease, simple interest, sale price, commission, interest rates, gratuities

New Vocabulary

percent of change
percent of increase
percent of decrease

NY Math Online

glencoe.com

• Concepts In Motion
• Extra Examples
• Personal Tutor
• Self-Check Quiz

▷ **MINI Lab**

You can model a 50% increase using straws.

100%

50%

150%

• Begin with two straws. The first straw will represent 100%.
• Cut the second straw in half. One part represents 50%.
• Tape the two straws together. This new straw represents a 50% increase or 150% of the original straw.

Model each percent of change.

1. 25% increase   2. 75% increase   3. 30% increase

4. Describe a model that represents a 100% increase, a 200% increase, and a 300% increase.

5. Describe how this process would change to show percent of decrease.

One way to describe a change in quantities is to use percent of change.

Percent of Change   Key Concept

Words	A **percent of change** is a ratio that compares the change in quantity to the original amount.
Equation	$\text{percent of change} = \dfrac{\text{amount of change}}{\text{original amount}}$

The percent of change is based on the original amount. If the original quantity is increased, then it is called a **percent of increase**. If the original quantity is decreased, then it is called a **percent of decrease**.

$$\textbf{percent of increase} = \frac{\textbf{amount of increase}}{\textbf{original amount}} \longleftarrow \text{new} - \text{original}$$

$$\textbf{percent of decrease} = \frac{\textbf{amount of decrease}}{\textbf{original amount}} \longleftarrow \text{original} - \text{new}$$

EXAMPLE Find Percent of Increase

1 GASOLINE Find the percent of change in the cost of gasoline from 1970 to 2007. Round to the nearest whole percent if necessary.

1970

2007

Since the 2007 price is greater than the 1970 price, this is a percent of increase. The amount of increase is $2.85 − $1.30 or $1.55.

$$\text{percent of increase} = \frac{\text{amount of increase}}{\text{original amount}}$$

$$= \frac{\$1.55}{\$1.30} \qquad \text{Substitution}$$

$$\approx 1.19 \qquad \text{Simplify.}$$

$$\approx 119\% \qquad \text{Write 1.19 as a percent.}$$

The cost of gasoline increased 119% from 1970 to 2007.

CHECK Your Progress

a. **MEASUREMENT** Find the percent of change from 10 yards to 13 yards.

 Study Tip

Percents
In the percent of change formula, the decimal representing the percent of change must be written as a percent.

EXAMPLE Find Percent of Decrease

2 DVD RECORDER Yusuf bought a DVD recorder for $280. Now, it is on sale for $220. Find the percent of change in the price. Round to the nearest whole percent if necessary.

Since the new price is less than the original price, this is a percent of decrease. The amount of decrease is $280 − $220 or $60.

$$\text{percent of decrease} = \frac{\text{amount of decrease}}{\text{original amount}}$$

$$= \frac{\$60}{\$280} \qquad \text{Substitution}$$

$$\approx 0.21 \qquad \text{Simplify.}$$

$$\approx 21\% \qquad \text{Write 0.21 as a percent.}$$

The price of the DVD recorder decreased by about 21 percent.

CHECK Your Progress

b. **MONEY** Find the percent of change from $20 to $15.

3 The table shows about how many people attended the home games of a high school football team for five consecutive years. Which statement is supported by the information in the table?

Attendance of Home Games	
Year	Total Attendance (thousands)
2003	16.6
2004	16.4
2005	15.9
2006	17.4
2007	17.6

A The attendance in 2006 was 15% greater than the attendance in 2005.

B The greatest decrease in attendance occurred from 2003 to 2004.

C The attendance in 2005 was 3% less than the attendance in 2004.

D The greatest increase in attendance occurred from 2006 to 2007.

Read the Item

You need to determine which statement is best supported by the information given in the table.

Solve the Item

- Check **A**. The percent of change from 2005 to 2006 was $\frac{17.4 - 15.9}{15.9}$ or about 10%, not 15%.

- Check **B**.
  From 2003 to 2004, the decrease was $16.6 - 16.4$ or 0.2.
  From 2004 to 2005, the decrease was $16.4 - 15.9$ or 0.5.
  This statement is not supported by the information.

- Check **C**. The percent of change from 2004 to 2005 was $\frac{16.4 - 15.9}{16.4}$ or about 3%. This statement is supported by the information.

- Check **D**.
  From 2005 to 2006, the increase was $17.4 - 15.9$ or 1.5.
  From 2006 to 2007, the increase was $17.6 - 17.4$ or 0.2.
  This statement is not supported by the information.

The solution is **C**.

Test-Taking Tip

Check the Results If you have time, check all of the choices given. By doing so, you will verify that your choice is correct.

 CHECK **Your Progress**

c. Which of the following represents the greatest percent of change?

F A savings account that had $500 now has $470.

G An MP3 player that stored 15 GB now stores 30 GB.

H A plant grew from 3 inches to 8 inches in one month.

J An airplane ticket that was originally priced at $345 is now $247.

Find each percent of change. Round to the nearest whole percent if necessary. State whether the percent of change is an *increase* or a *decrease*.

Examples 1, 2
(p. 370)

1. 30 inches to 24 inches

2. 20.5 meters to 35.5 meters

3. $126 to $150

4. $75.80 to $94.75

Example 3
(p. 371)

5. **MULTIPLE CHOICE** The table shows the number of youth 7 years and older who played soccer from 1998 to 2006. Which statement is supported by the information in the table?

Playing Soccer	
Year	**Number (millions)**
1998	13.2
2000	12.9
2002	13.7
2004	13.3
2006	14.0

Source: National Sporting Goods Association

A The greatest decrease in the number of players occurred from 1998 to 2000.

B There were 7% fewer youth playing soccer in 2004 than in 2002.

C The number of players in 2002 was 6% greater than the number of players in 2000.

D There were 10% more youth playing soccer in 2000 than in 1998.

Practice and Problem Solving

HOMEWORK HELP

For Exercises	See Examples
6–7, 14–15 18–19	1
8–13 16–17	2
39, 40	3

For Exercises 6–19, find each percent of change. Round to the nearest whole percent if necessary. State whether the percent of change is an *increase* or a *decrease*.

6. 15 yards to 18 yards

7. 100 acres to 140 acres

8. $12 to $6

9. 48 notebooks to 14 notebooks

10. 125 centimeters to 87.5 centimeters

11. $15.60 to $11.70

12. 1.6 hours to 0.95 hour

13. 132 days to 125.4 days

14. $240 to $320

15. 624 feet to 702 feet

16. **BOOKS** On Monday, Kenya spent 60 minutes reading her favorite book. Today, she spent 45 minutes reading this book.

17. **EXERCISE** Three months ago, Ernesto could walk 2 miles in 40 minutes. Today he can walk 2 miles in 25 minutes.

18. **SCHOOL** Last school year the enrollment of Gilboa Middle School was 465 students. This year the enrollment is 525.

19. **MONEY** Jake had $782 in his checking account. He now has $798.

Find each percent of change. Round to the nearest whole percent if necessary. State whether the percent of change is an *increase* or a *decrease*.

20. $\frac{1}{2}$ to $\frac{1}{4}$

21. $\frac{4}{6}$ to $\frac{1}{6}$

22. $\frac{1}{5}$ to $\frac{4}{5}$

23. $\frac{2}{3}$ to $\frac{5}{3}$

MEASUREMENT For Exercises 24 and 25, refer to the rectangle at the right. Suppose the side lengths are doubled.

24. Find the percent of change in the perimeter.

25. Find the percent of change in the area.

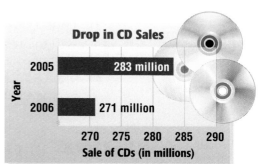

8 cm

3 cm

26. **MUSIC PHONES** Between 2006 and 2007, music phone owners increased from 6.8 million to 33 million. Find the percent of increase. Round to the nearest whole percent.

27. **FIND THE DATA** Refer to the Data File on pages 16–19. Choose some data and write a real-world problem in which you would find the percent of change.

28. **ANALYZE GRAPHS** Use the graphic shown to find the percent of change in CD sales from 2005 to 2006.

29. **SHOES** In 2009, shoe sales for a certain company were $25.9 billion. Sales are expected to increase by about 20% from 2009 to 2010. Find the projected amount of shoe sales in 2010.

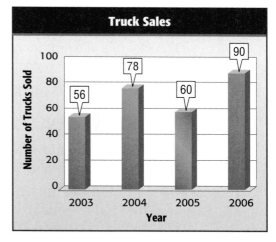

Drop in CD Sales

2005 — 283 million
2006 — 271 million

Sale of CDs (in millions)

Source: Fox News

30. **BABYSITTING** The table shows how many hours Catalina spent babysitting during the months of April and May. If Catalina charges $6.50 per hour, what is the percent of change in the amount of money earned from April to May?

Month	Hours Worked
April	40
May	32

ANALYZE GRAPHS For Exercises 31–33, refer to the graph.

31. Find the percent of decrease of truck sales from 2004 to 2005. Round to the nearest whole percent.

32. Find the percent of increase of truck sales from 2003 to 2004. Round to the nearest whole percent.

33. Between which two consecutive years is the percent of increase the greatest? What is the percent of increase? Round to the nearest whole percent.

Truck Sales

2003: 56
2004: 78
2005: 60
2006: 90

Number of Trucks Sold / Year

NYSCC • NYSMT
Extra Practice, pp. 686, 710.

H.O.T. Problems

34. **OPEN ENDED** Write a percent of change problem using the quantities 14 and 25, and state whether there is a percent of increase or decrease. Find the percent of change.

35. **NUMBER SENSE** The costs of two different sound systems were decreased by $10. The original costs of the systems were $90 and $60, respectively. Without calculating, which had greater percent of decrease? Explain.

36. **FIND THE ERROR** Sade and Trish are finding the percent of change from $52 to $125. Who is correct? Explain.

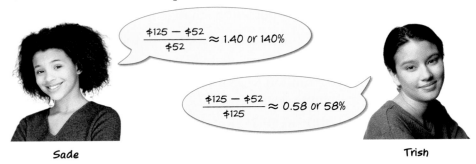

$$\frac{\$125 - \$52}{\$52} \approx 1.40 \text{ or } 140\%$$

$$\frac{\$125 - \$52}{\$125} \approx 0.58 \text{ or } 58\%$$

Sade   Trish

37. **CHALLENGE** If a quantity increases by 10% and then decreases by 10%, will the result be the original quantity? Explain.

38. **WRITING IN MATH** Explain how you know whether a percent of change is a percent of increase or a percent of decrease.

NYSMT PRACTICE 8.N.4

39. Which of the following represents the least percent of change?

   A A coat that was originally priced at $90 is now $72.

   B A puppy who weighed 6 ounces at birth now weighs 96 ounces.

   C A child grew from 54 inches to 60 inches in 1 year.

   D A savings account increased from $500 to $550 in 6 months.

40. If each dimension of the rectangle is doubled, what is the percent of increase in the area?

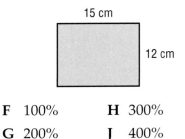

   F 100%   **H** 300%

   G 200%   **J** 400%

Spiral Review

ALGEBRA Write an equation for each problem. Then solve. Round to the nearest tenth if necessary. (Lesson 7-4)

41. **FOOD** Of 823 students 47.2% of the students chose pizza as their favorite food. What is a reasonable estimate for the number of students who chose pizza as their favorite food? Explain. (Lesson 7-5)

42. 30% of what number is 17?

43. What is 21% of 62?

44. **SHOPPING** Four pounds of pecans cost $12.75. How much is this per pound? (Lesson 6-2)

▷ **GET READY for the Next Lesson**

PREREQUISITE SKILL Write each percent as a decimal. (Lesson 6-8)

45. 6.5%

46. $5\frac{1}{2}\%$

47. $8\frac{1}{4}\%$

48. $6\frac{3}{4}\%$

Sales Tax and Discount

MAIN IDEA

Solve problems involving sales tax and discount.

NYS Core Curriculum

Preparation for 8.N.4
Apply percents to: tax, percent increase/ decrease, simple interest, sale price, commission, interest rates, gratuities

New Vocabulary

sales tax
discount

NY Math Online

glencoe.com

• Extra Examples
• Personal Tutor
• Self-Check Quiz

▷ **GET READY** for the Lesson

KAYAKS Horatio plans to buy a new kayak that costs $1,849. He lives in North Carolina where there is a 4.25% sales tax.

1. Calculate the sales tax by finding 4.25% of $1,849. Round to the nearest cent.

2. What will be the total cost including the sales tax?

3. Multiply 1.0425 and 1,849. How does the result compare to your answer in Exercise 2?

Sales tax is an additional amount of money charged on items that people buy. The total cost of an item is the regular price plus the sales tax.

EXAMPLE Find the Total Cost

1 **ELECTRONICS** A DVD player costs $140, and the sales tax is 5.75%. What is the total cost of the DVD player?

METHOD 1 Add sales tax to the regular price.

First, find the sales tax.

5.75% of $\$140 = 0.0575 \times 140$ Write 5.75% as a decimal.
$= 8.05$ The sales tax is $8.05.

Next, add the sales tax to the regular price.
$8.05 + $140 = $148.05

METHOD 2 Add the percent of tax to 100%.

$100\% + 5.75\% = 105.75\%$ Add the percent of tax to 100%.

The total cost is 105.75% of the regular price.

105.75% of $\$140 = 1.0575 \times \$140$ Write 105.75% as a decimal.
$= \$148.05$ Multiply.

So, the total cost of the DVD player is $148.05.

☑ **CHOOSE** Your Method

a. **CLOTHES** What is the total cost of a sweatshirt if the regular price is $42 and the sales tax is $5\frac{1}{2}\%$?

Study Tip

Sales Tax and Discount
If both are represented as percents, sales tax is a percent of increase, and discount is a percent of decrease.

Discount is the amount by which the regular price of an item is reduced. The sale price is the regular price minus the discount.

Find the Sale Price

2 **BOOGIE BOARDS** A boogie board that has a regular price of $69 is on sale at a 35% discount. What is the sale price of the boogie board?

METHOD 1 **Subtract the discount from the regular price.**

First, find the amount of the discount.

35% of $69 = 0.35 · $69   Write 35% as a decimal.
         = $24.15   The discount is $24.15.

Next, subtract the discount from the regular price.
$69 − $24.15 = $44.85

METHOD 2 **Subtract the percent of discount from 100%.**

100% − 35% = 65%   Subtract the discount from 100%.

The sale price is 65% of the regular price.

65% of $69 = 0.65 · $69   Write 65% as a decimal.
         = 44.85   Multiply.

So, the sale price of the boogie board is $44.85.

 CHOOSE Your Method

b. MUSIC A CD that has a regular price of $15.50 is on sale at a 25% discount. What is the sale price of the CD?

Study Tip

Percent Equation
Remember that, in the percent equation, the percent must be written as a decimal. Since the sale price is 70% of the original price, use 0.7 to represent 70% in the percent equation.

EXAMPLE **Find the Original Price**

3 **CELL PHONES** A cell phone is on sale for 30% off. If the sale price is $239.89, what is the original price?

The sale price is 100% − 30% or 70% of the original price.

Words	$239.89 is 70% of what price?
Variable	Let p represent the original price.
Equation	$239.89 = 0.7 × p$

$239.89 = 0.7p$ Write the equation.

$\dfrac{239.89}{0.7} = \dfrac{0.7p}{0.7}$ Divide each side by 0.7.

$342.70 = p$ Simplify.

The original price is $342.70.

 CHECK Your Progress

c. Find the original price if the sale price of the cell phone is $205.50.

CHECK Your Understanding

Find the total cost or sale price to the nearest cent.

Example 1
(p. 375)

1. $2.95 notebook; 5% tax

2. $46 shoes; 2.9% tax

Example 2
(p. 376)

3. $1,575 computer; 15% discount

4. $119.50 skateboard; 20% off

Example 3
(p. 376)

5. **IN-LINE SKATES** A pair of in-line skates is on sale for $90. If this price represents a 9% discount from the original price, what is the original price to the nearest cent?

Practice and Problem Solving

HOMEWORK HELP

For Exercises	See Examples
6–13	1–2
14–17	3

Find the total cost or sale price to the nearest cent.

6. $58 ski lift ticket; 20% discount

7. $1,500 computer; 7% tax

8. $99 CD player; 5% tax

9. $12.25 pen set; 60% discount

10. $4.30 makeup; 40% discount

11. $7.50 meal; 6.5% tax

12. $39.60 sweater; 33% discount

13. $89.75 scooter; $7\frac{1}{4}$% tax

14. **COSMETICS** A bottle of hand lotion is on sale for $2.25. If this price represents a 50% discount from the original price, what is the original price to the nearest cent?

15. **TICKETS** At a movie theater, the cost of admission to a matinee is $5.25. If this price represents a 30% discount from the evening price, find the evening price to the nearest cent.

Find the original price to the nearest cent.

16. calendar: discount, 75%
    sale price, $2.25

17. telescope: discount, 30%
    sale price, $126

18. **VIDEO GAMES** What is the sales tax of a $178.90 video game system if the tax rate is 5.75%?

19. **RESTAURANTS** A restaurant bill comes to $28.35. Find the total cost if the tax is 6.25% and a 20% tip is left on the amount before tax.

SKATEBOARDS For Exercises 20–22, use the information in the table at the right.

A skateboard costs $320, not including the sales tax.

State	2007 Sales Tax Rate
Washington	6.5%
Kansas	5.3%
North Carolina	4.25%

Source: Federation of Tax Administrators

20. What is the total cost of the skateboard, including tax in Washington?

21. What is the total cost of the skateboard, including tax, in North Carolina?

NYSCC • NYSMT
Extra Practice, pp. 686, 710

22. A store in Kansas has the skateboard on sale for 20% off. If the sales tax is calculated after the discount, what is the cost of the skateboard?

H.O.T. Problems

23. **CHALLENGE** A gift store is having a sale in which all items are discounted 20%. Including tax, Colin paid $21 for a picture frame. If the sales tax rate is 5%, what was the original price of the picture frame?

24. **OPEN ENDED** Give an example of the regular price of an item and the total cost including sales tax if the tax rate is 5.75%.

25. **Which One Doesn't Belong?** In each pair, the first value is the regular price of an item and the second value is the sale price. Identify the pair that does not have the same percent of discount as the other three. Explain.

| $24, $18 | $50, $25 | $12, $9 | $80, $60 |

26. **WRITING IN MATH** Describe two methods for finding the sale price of an item that is discounted 30%. Which method do you prefer? Explain.

NYSMT PRACTICE 8.N.4

27. A computer software store is having a sale. The table shows the regular price, r, and the sales price, s, of various items.

Item	Regular Price (r)	Sale Price (s)
A	$5.00	$4.00
B	$8.00	$6.40
C	$10.00	$8.00
D	$15.00	$12.00

Which formula can be used to calculate the sale price?

A $s = r \times 0.2$ C $s = r \times 0.8$

B $s = r - 0.2$ D $s = r - 0.8$

28. A chair that costs $210 was reduced by 40% for a one day sale. After the sale, the sale price was increased by 40%. What is the price of the chair?

F $176.40 H $205.50

G $185.30 J $210.00

29. Juanita paid $10.50 for a T-shirt at the mall. It was on sale for 30% off. What was the original price before the discount?

A $3.15 C $15.00

B $7.35 D $35.00

Spiral Review

Find each percent of change. Round to the nearest whole percent if necessary. State whether the percent of change is an *increase* or *decrease*. (Lesson 7-6)

30. 4 hours to 6 hours

31. $500 to $456

32. 20.5 meters to 35.5 meters

33. **TRAVEL** Out of a 511-mile trip, Mya drove about 68% on Monday. Determine a reasonable estimate for the number of miles she drove on Monday. (Lesson 7-5)

▷ **GET READY for the Next Lesson**

PREREQUISITE SKILL Multiply. Write in simplest form. (Lesson 5-5)

34. $\frac{2}{7} \cdot \frac{4}{5}$

35. $\frac{1}{8} \cdot \frac{4}{9}$

36. $\frac{6}{11} \cdot \frac{9}{24}$

7-8 Simple Interest

MAIN IDEA

Solve problems involving simple interest.

NYS Core Curriculum

Preparation for 8.N.4
Apply percents to: tax, percent increase/ decrease, simple interest, sale price, commission, interest rates, gratuities *Also addresses 7.CN.8*

New Vocabulary

principal
simple interest

NY Math Online

glencoe.com

• Extra Examples
• Personal Tutor
• Self-Check Quiz

▷ **GET READY** for the Lesson

INVESTING Suni plans to save the $200 she received for her birthday. The graphs shows the average yearly rates at three different banks.

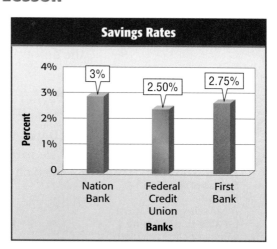

1. Calculate 2.50% of $200 to find the amount of money Suni can earn in one year at Federal Credit Bank.

2. Calculate 2.75% of $200 to find the amount of money Suni can earn in one year at First Bank.

Principal is the amount of money deposited or borrowed.

Simple interest is the amount paid or earned for the use of money. To find simple interest I, use the following formula.

EXAMPLES Find Interest Earned

CHECKING Arnold has $580 in a savings account that pays 3% simple interest. How much interest will he earn in each amount of time?

① **5 years**

$I = prt$	Formula for simple interest
$I = 580 \cdot 0.03 \cdot 5$	Replace p with $580, r$ with 0.03, and t with 5.
$I = 87$	Simplify.

Arnold will earn $87 in interest in 5 years.

② **6 months**

6 months $= \frac{6}{12}$ or 0.5 year	Write the time as years.
$I = prt$	Formula for simple interest
$I = 580 \cdot 0.03 \cdot 0.5$	$p = \$580, r = 0.03, t = 0.5$
$I \approx 8.7$	Simplify.

Arnold will earn $8.70 in interest in 6 months.

Real-World Career....

How does a Car Salesperson Use Math?
A car salesperson must be able to determine values of cars, calculate interest rates, and determine monthly payments.

NY Math Online

For more information go to glencoe.com.

Study Tip

Fractions of Years Remember to express 1 month as $\frac{1}{12}$ year in the formula.

✅ **CHECK Your Progress**

SAVINGS Jenny has $1,560 in a savings account that pays 2.5% simple interest. How much interest will she earn in each amount of time?

a. 3 years

b. 6 months

The formula $I = prt$ can also be used to find the interest owed when you borrow money. In this case, p is the amount of money borrowed, and t is the amount of time the money is borrowed.

EXAMPLE Find Interest Paid on a Loan

③ **LOANS** Rondell's parents borrow $6,300 from the bank for a new car. The interest rate is 6% per year. How much simple interest will they pay if they take 2 years to repay the loan?

$I = prt$	Formula for simple interest
$I = 6{,}300 \cdot 0.06 \cdot 2$	Replace p with $6,300, r with 0.06, and t with 2.
$I = 756$	Simplify.

Rondell's parents will pay $756 in interest in 2 years.

✅ **CHECK Your Progress**

c. **LOANS** Mrs. Hanover borrows $1,400 at a rate of 5.5% per year. How much simple interest will she pay if it takes 8 months to repay the loan?

EXAMPLE Find Total Paid on a Credit Card

④ **CREDIT CARDS** Derrick's dad bought new tires for $900 using a credit card. His card has an interest rate of 19%. If he has no other charges on his card and does not pay off his balance at the end of the month, how much money will he owe after one month?

$I = prt$	Formula for simple interest
$I = 900 \cdot 0.19 \cdot \frac{1}{12}$	Replace p with $900, r with 0.19, and t with $\frac{1}{12}$.
$I = 14.25$	Simplify.

The interest owed after one month is $14.25. So, the total amount owed would be $900 + $14.25 or $914.25.

✅ **CHECK Your Progress**

d. **CREDIT CARDS** An office manager charged $425 worth of office supplies on a charge card with an interest rate of 9.9%. How much money will he owe if he makes no other charges on the card and does not pay off the balance at the end of the month?

CHECK Your Understanding

Examples 1, 2
(pp. 379–380)

Find the simple interest earned to the nearest cent for each principal, interest rate, and time.

1. $640, 3%, 2 years
2. $1,500, 4.25%, 4 years
3. $580, 2%, 6 months
4. $1,200, 3.9%, 8 months

Example 3
(p. 380)

Find the simple interest paid to the nearest cent for each loan, interest rate, and time.

5. $4,500, 9%, 3.5 years
6. $290, 12.5%, 6 months

Example 4
(p. 380)

7. **FINANCES** The Masters family financed a computer that costs $1,200. If the interest rate is 19%, how much will the family owe after one month if no payments are made?

Practice and Problem Solving

For Exercises	See Examples
8–9	1
10–11	2
12–15	3
16–17	4

HOMEWORK HELP

Find the simple interest earned to the nearest cent for each principal, interest rate, and time.

8. $1,050, 4.6%, 2 years
9. $250, 2.85%, 3 years
10. $500, 3.75%, 4 months
11. $3,000, 5.5%, 9 months

Find the simple interest paid to the nearest cent for each loan, interest rate, and time.

12. $1,000, 7%, 2 years
13. $725, 6.25%, 1 year
14. $2,700, 8.2%, 3 months
15. $175.80, 12%, 8 months

16. **CREDIT CARDS** Leon charged $75 at an interest rate of 12.5%. How much will Leon have to pay after one month if he makes no payments?

17. **TRAVEL** A family charged $1,345 in travel expenses. If no payments are made, how much will they owe after one month if the interest rate is 7.25%?

BANKING For Exercises 18 and 19, use the table.

18. What is the simple interest earned on $900 for 9 months?

19. Find the simple interest earned on $2,500 for 18 months.

Home Savings and Loan	
Time	**Rate**
6 months	2.4%
9 months	2.9%
12 months	3.0%
18 months	3.1%

INVESTING For Exercises 20 and 21, use the following information.

Ramon has $4,200 to invest for college.

20. If Ramon invests $4,200 for 3 years and earns $630, what was the simple interest rate?

21. Ramon's goal is to have $5,000 after 4 years. Is this possible if he invests with a rate of return of 6%? Explain.

NYSCC • NYSMT
Extra Practice, pp. 687, 711.

22. OPEN ENDED Suppose you earn 3% on a $1,200 deposit for 5 years. Explain how the simple interest is affected if the rate is increased by 1%. What happens if the time is increased by 1 year?

23. CHALLENGE Mrs. Antil deposits $800 in a savings account that earns 3.2% interest annually. At the end of the year, the interest is added to the principal or original amount. She keeps her money in this account for three years without withdrawing any money. Find the total in her account after each year for three years.

24. **WRITING IN MATH** List the steps you would use to find the simple interest on a $500 loan at 6% interest rate for 18 months. Then find the simple interest.

NYSMT PRACTICE 8.N.4

25. Jada invests $590 in a money market account. Her account pays 7.2% simple interest. If she does not add or withdraw any money again, how much interest will Jada's account earn after 4 years of simple interest?

A $75.80

B $158.67

C $169.92

D $220.67

26. Mr. Sprockett borrows $3,500 from his bank to buy a used car. The loan has a 7.4% annual simple interest rate. If it takes Mr. Sprockett two years to pay back the loan, what is the total amount he will be paying?

F $3,012

G $4,018

H $4,550

J $3,598

Spiral Review

27. Find the total cost of a $19.99 DVD if the tax rate is 7%. (Lesson 7-7)

Find each percent of change. Round to the nearest whole percent if necessary. State whether the percent of change is an increase or decrease. (Lesson 7-6)

28. 35 birds to 45 birds

29. 60 inches to 38 inches

30. $2.75 to $1.80

Divide. Write in simplest form. (Lesson 5-7)

31. $\frac{3}{5} \div \frac{1}{2}$

32. $\frac{4}{7} \div \frac{5}{8}$

33. $2\frac{2}{3} \div 1\frac{1}{4}$

Problem Solving in Art   Real-World Unit Project

It's Golden! It's time to complete your project. Use the information and data you have gathered about the Golden Ratio to prepare a Power Point presentation. Be sure to include your reports and calculations in your presentation.

NY Math Online Unit Project at glencoe.com

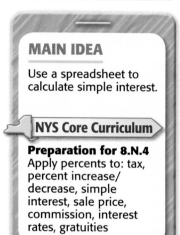
Spreadsheet Lab
Simple Interest

A computer spreadsheet is a useful tool for quickly calculating simple interest for different values of principal, rate, and time.

ACTIVITY

Max plans on opening a "Young Savers" account at his bank. The current rate on the account is 4%. He wants to see how different starting balances, rates, and times will affect his account balance. To find the balance at the end of 2 years for different principal amounts, he enters the values B2 = 4 and C2 = 2 into the spreadsheet below.

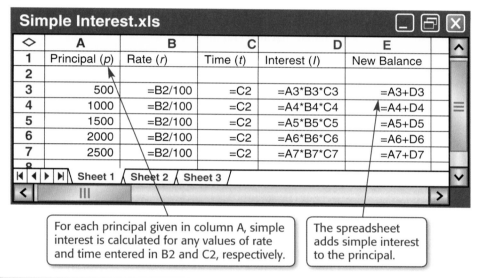

Simple Interest.xls				
A	**B**	**C**	**D**	**E**
1 Principal (*p*)	Rate (*r*)	Time (*t*)	Interest (*I*)	New Balance
2				
3 500	=B2/100	=C2	=A3*B3*C3	=A3+D3
4 1000	=B2/100	=C2	=A4*B4*C4	=A4+D4
5 1500	=B2/100	=C2	=A5*B5*C5	=A5+D5
6 2000	=B2/100	=C2	=A6*B6*C6	=A6+D6
7 2500	=B2/100	=C2	=A7*B7*C7	=A7+D7

Sheet 1 / Sheet 2 / Sheet 3

For each principal given in column A, simple interest is calculated for any values of rate and time entered in B2 and C2, respectively.

The spreadsheet adds simple interest to the principal.

ANALYZE THE RESULTS

1. Why is the rate in column B divided by 100?

2. What is the balance in Max's account after 2 years if the principal is $1,500 and the simple interest rate is 4%?

3. How much interest does Max earn in 2 years if his account has a principal of $2,000 and a simple interest rate of 4%?

4. Is the amount of principal proportional to the interest Max earns if his account earns 4% simple interest over 2 years? Explain.

5. Is the amount of principal proportional to the balance in Max's account if it earns 4% simple interest over 2 years? Explain.

6. What entries for cells B2 and C2 would you use to calculate the simple interest on a principal of $1,500 at a rate of 7% for a 9-month period?

7. What is the balance of this account at the end of the 9 months?

Study Organizer

GET READY to Study

Be sure the following Big Ideas are noted in your Foldable.

7-1	7-2
7-3	7-4
7-5	7-6
7-7	7-8

BIG Ideas

Percent of a Number (Lesson 7-1)
• To find the percent of a number, first write the percent as either a fraction or decimal and then multiply.

Percent Proportion (Lesson 7-2)

$$\frac{part}{whole} = \frac{n}{100} \Big\} percent$$

Percent and Estimation (Lesson 7-3)
• One way to estimate the percent of a number is to use a fraction. The other way is to first find 10% of the number and then multiply.

Percent Equation (Lesson 7-4)

$$part = percent \cdot whole$$

Percent of Change (Lesson 7-6)
• A percent of change is a ratio that compares the change in quantity to the original amount.

$$percent\ of\ change = \frac{amount\ of\ change}{original\ amount}$$

Sales Tax and Discount (Lesson 7-7)
• Sales tax is an additional amount of money charged on items. The total cost of an item is the regular price plus the sales tax.

• Discount is the amount by which the regular price of an item is reduced. The sale price is the regular price minus the discount.

Simple Interest (Lesson 7-8)
• Simple interest is the amount paid or earned for the use of money.

$$I = prt$$

Key Vocabulary

discount (p. 375)
percent equation (p. 361)
percent of change (p. 369)
percent of decrease (p. 369)
percent of increase (p. 369)

percent proportion (p. 350)
principal (p. 379)
sales tax (p. 375)
simple interest (p. 379)

Vocabulary Check

State whether each sentence is *true* or *false*. If *false*, replace the underlined word or number to make a true sentence.

1. The sale price of a discounted item is the regular price <u>minus</u> the discount.

2. A ratio that compares the change in quantity to the original amount is called the <u>percent of change</u>.

3. A <u>percent proportion</u> compares part of a quantity to the whole quantity using a percent.

4. The formula for simple interest is <u>$I = prt$</u>.

5. A method for estimating the percent of a number is to find <u>21%</u> of the number and then multiply.

6. The equation part = percent • whole is known as the <u>principal</u> equation.

7. The <u>principal</u> is the amount of money deposited or borrowed.

8. A <u>tax</u> is the amount by which the regular price of an item is reduced.

9. To find a percent of increase, compare the amount of the increase to the <u>new</u> amount.

10. If the new amount is greater than the original amount, then the percent of change is percent of <u>decrease</u>.

Lesson-by-Lesson Review

7-1   **Percent of a Number** (pp. 344–348)

8.N.3

Find each number. Round to the nearest tenth if necessary.

11. Find 78% of 50.

12. 45.5% of 75 is what number?

13. What is 225% of 60?

14. 0.75% of 80 is what number?

Example 1   Find 24% of 200.

24% of 200

$= 24\% \times 200$ Write the expression.

$= 0.24 \times 200$ Write 24% as a decimal.

$= 48$ Multiply.

So, 24% of 200 is 48.

7-2   **The Percent Proportion** (pp. 350–354)

8.N.4

Find each number. Round to the nearest tenth if necessary.

15. **SOCCER** A soccer team lost 30% of their games. If they played 20 games, how many did they win?

16. 6 is what percent of 120?

17. Find 0.8% of 35.

18. What percent of 375 is 40?

19. **PHONE SERVICE** A family pays $21.99 each month for their long distance phone service. This is 80% of the original price of the phone service. What is the original price of the phone service? Round to the nearest cent if necessary.

Example 2   What percent of 90 is 18?

$\dfrac{18}{90} = \dfrac{n}{100}$ Write the proportion.

$18 \cdot 100 = 90 \cdot n$ Find the cross products.

$1{,}800 = 90n$ Simplify.

$\dfrac{1{,}800}{90} = \dfrac{90n}{90}$ Divide each side by 90.

$20 = n$ So, 18 is 20% of 90.

Example 3   52 is 65% of what number?

$\dfrac{52}{w} = \dfrac{65}{100}$ Write the proportion.

$52 \cdot 100 = w \cdot 65$ Find the cross products.

$5{,}200 = 65w$ Simplify.

$\dfrac{5{,}200}{65} = \dfrac{65w}{65}$ Divide each side by 65.

$80 = w$ So, 52 is 65% of 80.

7-3   **Percent and Estimation** (pp. 355–360)

8.N.5

Estimate.

20. 25% of 81   21. 33% of 122

22. 77% of 38   23. 19.5% of 96

Estimate by using 10%.

24. 12% of 77   25. 88% of 400

26. **BOOKS** About 26% of the 208 books in Deja's collection are nonfiction. Estimate how many of Deja's books are nonfiction.

Example 4   Estimate 52% of 495.

$52\% \approx 50\%$ or $\dfrac{1}{2}$, and $495 \approx 500$.

52% of $495 \approx \dfrac{1}{2} \cdot 500$ or 250

So, 52% of 495 is about 250.

Example 5   Estimate 68% of 80.

10% of $80 = 0.1 \cdot 80$ or 8   Find 10% of 80.

68% is about 70%.

$7 \cdot 8 = 56$ 70% of 80 ≈ 7 · (10% of 80)

So, 68% of 80 is about 56.

7-4 | Algebra: The Percent Equation (pp. 361–365)

8.N.4

Write an equation for each problem. Then solve. Round to the nearest tenth if necessary.

27. 32 is what percent of 50?

28. 65% of what number is 39?

29. Find 42% of 300.

30. 7% of 92 is what number?

31. 12% of what number is 108?

32. **SALONS** A local hair salon increased their sales of hair products by about 12.5% this week. If they sold 48 hair products, how many hair products did they sell last week?

Example 6 27 is what percent of 90?

27 is the part and 90 is the base.

Let n represent the percent.

$$part = percent \cdot base$$
$$27 = n \cdot 90 \quad \text{Write an equation.}$$
$$\frac{27}{90} = \frac{90n}{90} \quad \text{Divide each side by 90.}$$
$$0.3 = n \quad \text{The percent is 30\%.}$$

So, 27 is 30% of 90.

7-5 | PSI: Determine Reasonable Answers (pp. 366–367)

7.N.19, 7.PS.12

Determine a reasonable answer for each problem.

33. **CABLE TV** In a survey of 1,813 consumers, 18% said that they would be willing to pay more for cable if they got more channels. Is 3.3, 33, or 333 a reasonable estimate for the number of consumers willing to pay more for cable?

34. **SCHOOL** There are 880 students at Medina Middle School. If 68% of the students are involved in sports, would the number of students involved in sports be about 510, 630, or 720?

35. **VACATION** Suppose you are going on vacation for $689 and the airfare accounts for 43.5% of the total cost. What is a reasonable cost of the airfare?

Example 7 Mr. Swanson harvested 1,860 pounds of apples from one orchard, 1,149 pounds from another, and 905 pounds from a third. The apples will be placed in crates that hold 42 pounds of apples. Will Mr. Swanson need 100, 200, or 400 crates?

Since an exact answer is not needed, we can estimate the total of pounds.

1,860	→	1,900
1,149	→	1,100
+ 905	→	+ 900
		3,900

Since 3,900 ÷ 40 is about 100, it is reasonable that 100 crates need to be ordered.

7-6 Percent of Change (pp. 369–374)

8.N.4

Find each percent of change. Round to the nearest whole percent if necessary. State whether the percent of change is an *increase* or *decrease*.

36. original: 172
    new: 254

37. original: $200
    new: $386

38. original: 75
    new: 60

39. original: $49.95
    new: $54.95

40. Tyree bought a collectible comic book for $49.62 last year. This year, he sold it for $52.10. Find the percent of change of the price of the comic book. Round to the nearest percent.

Example 8 A magazine that originally cost $2.75 is now $3.55. Find the percent of change. Round to the nearest whole percent.

The new price is greater than the original price, so this is a percent of increase.

amount of increase $= 3.55 - 2.75$ or 0.80

$$\text{percent of increase} = \frac{\text{amount of increase}}{\text{original amount}}$$

$$= \frac{0.80}{2.75} \quad \text{Substitution}$$

$$\approx 0.29 \quad \text{Simplify.}$$

The percent of increase is about 29%.

7-7 Sales Tax and Discount (pp. 375–378)

8.N.4

Find the total cost or sale price to the nearest cent.

41. $25 backpack; 7% tax

42. $210 bicycle; 15% discount

43. $8,000 car; $5\frac{1}{2}$% tax

44. $40 sweater; 33% discount

Find the percent of discount to the nearest percent.

45. shirt: regular price: $42
    sale price: $36

46. boots: regular price: $78
    sale price: $70

47. **MONEY** At the media store a certain DVD normally costs $21.99. This week the DVD is on sale for 25% off. Tara buys the DVD and pays using a $20 bill. Not including tax, how much change will she receive to the nearest cent?

Example 9 A new computer system is priced at $2,499. Find the total cost if the sales tax is 6.5%.

First, find the sales tax.

6.5% of $2,499 $= 0.065 \cdot 2,499$

$$\approx 162.44$$

Next, add the sales tax. The total cost is $162.44 + 2,499$ or $2,661.44.

Example 10 A pass at a water park is $58. At the end of the season, the same pass costs $46.40. What is the percent of discount?

$58 - 46.40 = 11.60$ Find the amount of discount.

Next, find what percent of 58 is 11.60.

$11.60 = n \cdot 58$ Write an equation.

$0.2 = n$ Divide each side by 58.

The percent of discount is 20%.

7-8 Simple Interest (pp. 379–382)

8.N.4

Find the interest earned to the nearest cent for each principal, interest rate, and time.

48. $475, 5%, 2 years

49. $5,000, 10%, 3 years

50. $2,500, 11%, $1\frac{1}{2}$ years

51. **SAVINGS** Tonya deposited $450 into a savings account earning 3.75% annual simple interest. How much interest will she earn in 6 years?

Find the interest paid to the nearest cent for each loan balance, interest rate, and time.

52. $3,200, 8%, 4 years

53. $1,980, 21%, 9 months

54. **CREDIT CARDS** David bought a computer for $600 using his credit card. The interest rate on his credit card is 19%. How much will he pay in all for the computer, if he pays off the balance at the end of 2 years?

Example 11 Find the interest earned on $400 at 9% for 3 years.

$I = prt$	Simple interest formula
$I = 400 \cdot 0.09 \cdot 3$	$p = \$400, r = 0.09, t = 3$
$I = 108$	Simplify.

The interest earned is $108.

Example 12 Elisa has a loan for $1,300. The interest rate is 7%. If she pays it off in 6 months, how much interest will she pay?

$I = prt$	Simple interest formula
$I = 1,300 \cdot 0.07 \cdot 0.5$	$p = \$1,300, r = 0.07,$ $t = 0.5$
$I = 45.5$	Simplify.

The interest she will pay after 6 months is $45.50.

Find each number. Round to the nearest tenth if necessary.

1. Find 55% of 164.

2. What is 355% of 15?

3. Find 25% of 80.

4. **MULTIPLE CHOICE** Of 365 students, 210 bought a hot lunch. About what percent of the students did *not* buy a hot lunch?

   A 35%   **C** 56%
   B 42%   **D** 78%

Estimate.

5. 18% of 246   6. 145% of 81

7. 71% of 324   8. 56% of 65.4

9. **COMMUNICATION** Theresa makes a long distance phone call and talks for 50 minutes. Of these minutes, 25% were spent talking to her brother. Would the time spent talking with her brother be about 8, 12, or 15 minutes? Explain your reasoning.

Write an equation for each problem. Then solve. Round to the nearest tenth if necessary.

10. Find 14% of 65.

11. What number is 36% of 294?

12. 82% of what number is 73.8?

13. 75 is what percent of 50?

Find each percent of change. Round to the nearest whole percent if necessary. State whether the percent of change is an *increase* or a *decrease*.

14. $60 to $75

15. 145 meters to 216 meters

16. 48 minutes to 40 minutes

FOOD For Exercises 17 and 18, use the table below. It shows the results of a survey in which 175 students were asked what type of food they wanted for their class party.

Type of Food	Percent
Subs	32%
Tex-Mex	56%
Italian	12%

17. How many of the 175 students chose Italian food for their class party?

18. How many students chose Tex-Mex food for the party?

Find the total cost or sale price to the nearest cent.

19. $2,200 computer, $6\frac{1}{2}$% sales tax

20. $16 hat, 55% discount

21. $35.49 jeans, 33% discount

Find the simple interest earned to the nearest cent for each principal, interest rate, and time.

22. $750, 3%, 4 years

23. $1,050, 4.6%, 2 years

24. $2,600, 4%, 3 months

25. **MULTIPLE CHOICE** Mr. Jackson borrows $3,500 to renovate his home. His loan has an annual simple interest rate of 15%. If he pays off the loan after 6 months, about how much will he pay in all?

   F $3,763
   G $3,500
   H $3,720
   J $4,025

PART 1 Multiple Choice

Read each question. Then fill in the correct answer on the answer document provided by your teacher or on a sheet of paper.

1. Sarah wants to buy pillows for her living room. Which store offers the best buy on pillows?

Store	Sale Price
A	3 pillows for $40
B	4 pillows for $50
C	2 pillows for $19
D	1 pillow for $11

A Store A
B Store B
C Store C
D Store D

2. The graph below shows the attendance at a summer art festival from 2002 to 2007. If the trend in attendance continues, which is the best prediction of the attendance at the art festival in 2010?

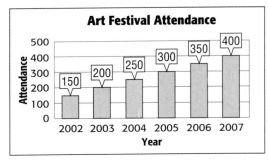

F Fewer than 200
G Between 500 and 600
H Between 700 and 800
J More than 800

3. At their annual car wash, the science club washes 30 cars in 45 minutes. At this rate, how many cars will they wash in 1 hour?

A 40
B 45
C 50
D 60

4. The cost of Ken's haircut was $23.95. If he wants to give his hair stylist a 15% tip, about how much of a tip should he leave?

F $2.40
G $3.60
H $4.60
J $4.80

5. At a pet store, 38% of the animals are dogs. If there are a total of 88 animals at the pet store, which equation can be used to find x, the number of dogs at the pet store?

A $\frac{x}{88} = \frac{100}{38}$

B $\frac{38}{88} = \frac{100}{x}$

C $\frac{x}{88} = \frac{38}{100}$

D $\frac{100}{88} = \frac{x}{38}$

6. An architect made a model of an office building using a scale of 1 inch equals 3 meters. If the height of the model is 12.5 inches, which of the following represents the actual height of the building?

F 40.0 m
G 37.5 m
H 36.0 m
J 28.4 m

7. Mrs. Stewart painted the door to her deck. The door is a rectangle with length x feet and width y feet. In the middle of the door, there is a rectangular panel of glass that measures 5 feet by 2 feet. Which expression gives the painted area of the door in square feet?

A $x + y - 10$
B $xy + 10$
C $xy - 10$
D $x + y + 10$

8. At a grocery store, half-gallons of milk are on sale 5 for $4. Find the cost of 7 half-gallons of milk to the nearest cent.

   F $2.86   H $5.40

   G $4.75   J $5.60

9. If point B is translated 3 units to the left and 2 units up, what will be point B's new coordinates?

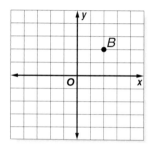

   A (−3, 2)   C (4, −1)

   B (5, 0)   D (−1, 4)

10. In Nadia's DVD collection, she has 8 action DVDs, 12 comedy DVDs, 7 romance DVDs, and 3 science fiction DVDs. What percent of Nadia's DVD collection are comedies?

    F 25%   H 35%

    G 30%   J 40%

11. Cassandra bought 2 dozen juice boxes priced at 6 juice boxes for $2.29 and 24 snack packages priced at 8 snack packages for $6.32. What is the total amount, not including tax, she spent on juice boxes and snack packages?

    A $34.44   C $28.12

    B $32.15   D $25.83

PART 2   Short Response/Grid In

Record your answers on the answer sheet provided by your teacher or on a sheet of paper.

12. The average cost of a 2-bedroom apartment in Grayson was $625 last year. This year, the average cost is $650. What is the percent of increase from last year to this year?

13. A necklace regularly sells for $18.00. The store advertises a 15% discount. What is the sale price of the necklace in dollars?

PART 3   Extended Response

Record your answers on the answer sheet provided by your teacher or on a sheet of paper. Show your work.

TEST-TAKING TIP

Question 14 Remember to show all of your work. You may be able to get partial credit for your answers, even if they are not entirely correct.

14. Cable Company A increases their rates from $98 a month to $101.92 a month.

    a. What is the percent of increase?

    b. Cable Company B offers their cable for $110 dollars a month, but gives a 10% discount for new customers. Describe two ways to find the cost for new customers.

    c. If you currently use Cable Company A, would it make sense to change to Cable Company B?

NEED EXTRA HELP?														
If You Missed Question...	1	2	3	4	5	6	7	8	9	10	11	12	13	14
Go to Lesson...	6-2	2-6	6-2	7-1	7-2	6-8	3-6	6-6	2-3	7-5	6-1	7-6	7-7	7-6
NYS Core Curriculum	7.M.5	7.N.3	7.M.5	8.N.3	8.N.4	7.PS.10	7.A.6	7.PS.10	6.G.10	7.PS.12	7.M.5	8.N.4	8.N.4	8.N.4

Unit 4

Statistics, Data Analysis, and Probability

Focus
Use statistical measures and probability to describe data.

CHAPTER 8
Statistics: Analyzing Data

BIG Idea Use measures of central tendency and range to describe a set of data.

BIG Idea Create and read graphs that depict data.

CHAPTER 9
Probability

BIG Idea Students use probability and proportions to make predictions.

Problem Solving in Science

Real-World Unit Project

Math Genes What's math have to do with genetics? Well, you're about to find out. You'll research basic genetics and learn how to use a Punnett square. Then you'll create sample genes for pet traits. You'll make predictions based on the pets' traits to determine the traits of their offspring. So, put on your lab coat and grab your math tool kit to begin this adventure.

NY Math Online Log on to glencoe.com to begin.

CHAPTER 8

Statistics: Analyzing Data

New York State Core Curriculum

7.S.6 Read and interpret data represented graphically (pictograph, bar graph, histogram, line graph, double line/bar graphs or circle graph)

Key Vocabulary

histogram (p. 416)

measures of central tendency (p. 402)

range (p. 397)

scatter plot (p. 427)

Real-World Link

Amusement Parks Hershey Park in Pennsylvania has over 60 rides and attractions. You can use a bar graph to display and then compare the speeds of these rides.

Statistics: Analyzing Data Make this Foldable to help you organize your notes. Begin with nine sheets of notebook paper.

① **Fold** 9 sheets of paper in half along the width.

② **Cut** a 1" tab along the left edge through one thickness.

③ **Glue** the 1" tab down. Write the lesson number and title on the front tab.

④ **Repeat** Steps 2 and 3 for the remaining sheets. Staple them together on the glued tabs to form a booklet.

8-1
Line Plots

GET READY for Chapter 8

Diagnose Readiness You have two options for checking Prerequisite Skills.

Option 2

NY Math Online Take the Online Readiness Quiz at glencoe.com.

Option 1

Take the Quick Quiz below. Refer to the Quick Review for help.

QUICK Quiz

Order from least to greatest. (Lesson 4-9)

1. 96.2, 96.02, 95.89

2. 5.61, 5.062, 5.16

3. 22.02, 22, 22.012

4. **JEANS** A store sells boot-cut jeans for $49.97, classic for $49.79, and flared for $47.99. Write these prices in order from least to greatest. (Lesson 4-9)

Order from greatest to least. (Lesson 4-9)

5. 74.65, 74.67, 74.7

6. 1.26, 1.026, 10.26

7. 3.304, 3.04, 3.340

Evaluate each expression. (Lesson 1-4)

8. $\dfrac{23 + 44 + 37 + 45}{4}$

9. $\dfrac{1.7 + 2.6 + 2.4 + 3.1 + 1.8}{5}$

10. **PIZZA** Four friends ordered a large pizza for $14.95, a salad for $3.75, and two bottles of soda for $2.25 each. If they split the cost evenly, how much does each person owe? (Lesson 1-4)

QUICK Review

Example 1

Order 47.7, 47.07, and 40.07 from least to greatest.

47.7
47.07   Line up the decimal points
40.07   and compare place value.
↑

The numbers in order from least to greatest are 40.07, 47.07, and 47.7.

Example 2

Order 2.08, 20.8, 0.28 from greatest to least.

 2.08
20.8   Line up the decimal points
 0.28   and compare place value.
↑

The numbers in order from greatest to least are 20.8, 2.08, and 0.28.

Example 3

Evaluate $\dfrac{3.4 + 4.5 + 3.8}{3}$.

$\dfrac{3.4 + 4.5 + 3.8}{3} = \dfrac{11.7}{3}$ Add 3.4, 4.5, and 3.8.

$= 3.9$ Divide 11.7 by 3.

Line Plots

MAIN IDEA

Display and analyze data using a line plot.

NYS Core Curriculum

7.S.6 Read and interpret data represented graphically (pictograph, bar graph, histogram, line graph, double line/bar graphs or circle graph) **7.S.4 Calculate the range for a given set of data** *Also addresses 7.A.7, 7.S.1, 7.CM.5*

New Vocabulary

statistics
data
line plot
outlier
cluster
range
analyze

NY Math Online

glencoe.com

• Concepts In Motion
• Extra Examples
• Personal Tutor
• Self-Check Quiz
• Reading in the Content Area

▷ **GET READY for the Lesson**

BUILDINGS The table shows the number of stories in 20 of the tallest buildings in Boston, Massachusetts.

1. Do any of the values seem much greater or much less than the other data values?

2. Do some of the buildings have the same height? Is this easy to see? Explain.

Boston's Tallest Buildings (Number of Stories)				
60	38	40	35	38
46	26	41	36	52
33	33	32	37	37
46	40	36	40	32

Source: Emporis Buildings

Statistics deals with collecting, organizing, and interpreting data. **Data** are pieces of information, which are often numerical. One way to show how data are spread out is to use a line plot. A **line plot** is a diagram that shows the data on a number line.

EXAMPLE Display Data Using a Line Plot

① **BUILDINGS** Make a line plot of the data shown above.

Step 1 Draw a number line. The shortest building in the table has 26 stories, and the tallest has 60. You can use a scale of 25 to 65 and an interval of 5. Other scales and intervals could also be used.

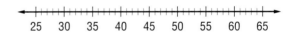

Step 2 Put an × above the number that represents the number of stories in each building. Include a title.

**Boston's Tallest Buildings
Number of Stories**

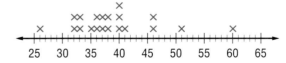

✔ **CHECK Your Progress**

a. **BUILDINGS** The number of stories in the 15 tallest buildings in the world are listed at the right. Display the data in a line plot.

World's Tallest Buildings (Number of Stories)				
101	88	88	110	88
88	80	69	102	78
72	54	73	85	80

Source: *The World Almanac*

You can make some observations about the *distribution* of data, or how data are grouped together or spread out. Consider the line plot below.

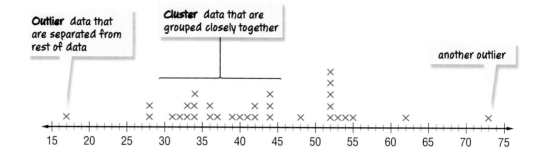

Outlier data that are separated from rest of data

Cluster data that are grouped closely together

another outlier

Vocabulary Link
Range

Everyday Meaning in music, all the notes between the high and low notes, as in a singer with a wide range.

Math Use the difference between the greatest number and least number in a set of data.

·In a line plot, you can easily find the **range**, or spread, of the data, which is the difference between the greatest and least numbers. When you **analyze** data, you use these observations to describe and compare data.

EXAMPLES Use a Plot to Analyze Data

2 **ANIMALS** The line plot below shows the life spans for different animals. Identify any clusters, gaps, and outliers and find the range.

Average Life Spans

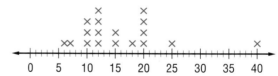

Many of the data cluster between 10 and 12 years.

There is a gap between 25 and 40 years.

Since 40 is apart from the rest of the data, it is an outlier.

The greatest age is 40 years, and the least age is 6 years. So, the range of the ages is 40 − 6 or 34.

Study Tip

Clusters You can describe a cluster by stating a range of values or by giving a single value around which the data appear to be grouped.

3 **Describe how the range would change if the data value 54 was added to the data set in Example 2.**

The greatest age would change to 54, and the least age would remain the same at 6. So, the range of the ages would change from 34 to 54 − 6 or 48.

CHECK Your Progress

Refer to Example 1.

b. Identify any clusters, gaps, and outliers and find the range.

c. Describe how the range would change if the data value 50 was added to the data set.

CHECK Your Understanding

Example 1
(p. 396)

Display each set of data in a line plot.

1.

Costs of Video Games ($)			
20	29	40	50
45	20	50	50
20	25	50	40

2.

Sizes of Tennis Shoes					
8	10	9	8	7	6
9	10	9	6	5	7
7	8	11	6	8	7

MUSIC For Exercises 3 and 4, analyze the line plot below.

Number of Music CDs Owned

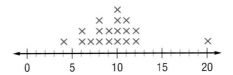

Example 2
(p. 397)

3. Identify any clusters, gaps, and outliers and find the range of the data.

Example 3
(p. 397)

4. Describe how the range would change if the data value 3 was added to the data set.

SURVEYS For Exercises 5–8, analyze the line plot at the right and use the information below.

Jamie asked her classmates how many glasses of water they drink on a typical day. The results are shown.

Glasses of Water Consumed

```
        ×  ×
     ×  ×  ×           ×
     ×  ×  ×           ×
     ×  ×  ×  ×  ×  ×
   ┼──┼──┼──┼──┼──┼──
   0  1  2  3  4  5
```

Example 2
(p. 397)

5. What was the most frequent response?

6. What was the least frequent response?

7. What is the range?

Example 3
(p. 397)

8. Describe how the range would change if an additional data value of 4 was added to the data set.

Practice and Problem Solving

Display each set of data in a line plot.

HOMEWORK HELP

For Exercises	See Examples
9–12	1
13–20	2–3

9.

Snowfall (in.)				
2	10	1	5	2
4	1	2	3	4
6	3	12	2	1

10.

Drink Size (oz)				
12	16	8	24	32
20	12	12	16	24
8	20	48	16	12

11.

Basketball Scores (points)				
101	105	99	130	120
100	108	126	135	98
120	122	115	129	97

12.

Ages of Students (y)					
12	13	13	13	12	14
13	12	13	13	12	12
13	14	12	13	12	12

··WEATHER For Exercises 13–16, analyze the line plot that shows the record high temperatures recorded by weather stations in each of the fifty states.

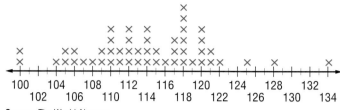

Record High Temperatures (°F)

Source: *The World Almanac*

13. What is the range of the data?

14. What temperature occurred most often?

15. Identify any clusters, gaps, or outliers.

16. Describe how the range of the data would change if 134°F were not part of the data set.

MOVIES For Exercises 17–20, analyze the line plot below that shows the number of digital video discs various students have in their DVD collection.

Digital Video Disc (DVD) Collection

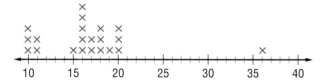

17. Find the range of the data.

18. What number of DVDs occurred most often?

19. How many students have more than 15 DVDs in their collection?

20. Describe how the range would change if the data value 38 was added to the data set.

Determine whether each statement is *sometimes*, *always*, or *never* true. Explain your reasoning.

21. If a new piece of data is added to a data set, the range will change.

22. If there is a cluster, it will appear in the center of the line plot.

BOOKS For Exercises 23–25, analyze the line plot at the right.

23. How many students read 4 or more books?

24. How many more students read 1–2 books than 5–6 books?

25. About what percent of the students read less than 5 books?

Number of Books Read

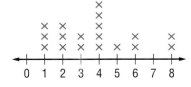

GOVERNMENT For Exercises 26–28, refer to the table showing the number of representatives for each of the Northeast states.

The House of Representatives Northeast States					
State	Number	State	Number	State	Number
CT	5	MA	10	PA	19
DE	1	NH	2	RI	2
ME	2	NJ	13	VT	1
MD	8	NY	29		

26. Display the data in a line plot.

27. Find the range and determine any clusters, gaps, or outliers.

28. Use the line plot to summarize the data.

29. The number of House of Representatives for the ten Southwestern states is quite different. The Southwestern states have 4, 8, 53, 7, 7, 3, 3, 5, 32, and 3 representatives. Display this data in a line plot. Compare this line plot to the line plot you made in Exercise 26. Include a discussion about clusters, outliers, range, and gaps in data.

NYSCC • NYSMT
Extra Practice, pp. 687, 711.

30. **COLLECT THE DATA** Conduct a survey of your classmates to determine how many hours of television they watch on a typical school night. Then display and analyze the data in a line plot. Use your analysis of the data to write a convincing argument about television viewing on a school night.

H.O.T. Problems

31. **REASONING** Explain how the inclusion or exclusion of outliers affects the computation of the range of a data set.

32. **FIND THE ERROR** Elena and Rashaun are analyzing the data shown in the line plot at the right. Who is correct? Explain.

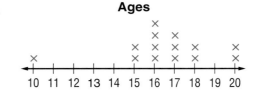

Ages

greatest data value: 20
least data value: 10

greatest data value: 16
least data value: 10

Elena

Rashaun

33. **CHALLENGE** Compare and contrast line plots and frequency tables. Include a discussion about what they show and when it is better to use each one.

34. **WRITING IN MATH** The number of fundraising items sold by two grades is shown. Describe which grade is more consistent and explain how you know.

Fundraising Items

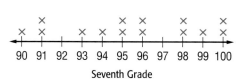

Seventh Grade

Fundraising Items

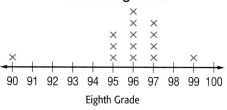

Eighth Grade

NYSMT PRACTICE 7.S.6, 7.S.4

35. The graph shows the weight of the emperor penguins at a zoo.

Emperor Penguins Weight (kg)

```
                        ×
          ×           × × ×        ×
  ×       ×   × × × ×  × × ×        ×
  +---+---+---+---+---+---+---+---+---+---+---+---+---+
 23  25  27  29  31  33  35  37  39  41  43  45  47  49
```

Which statement is *not* valid?

A More than half of these penguins weigh at least 41 kilograms.

B There are 16 emperor penguins at the zoo.

C Of these penguins, 30% weigh between 30 and 38 kilograms.

D The range of the emperor penguins' weight is 49 − 23 or 26 kilograms.

36. The table shows the math scores for 24 students in Mr. Baker's class.

Math Test Scores							
90	86	96	89	85	91	82	89
100	65	73	85	85	93	77	93
71	70	75	80	82	99	84	75

How would the range of the test scores change if a score of 83 was added?

F The range would remain unchanged at 45.

G The range would remain unchanged at 35.

H The range would change from 45 to 83.

J The range would change from 35 to 17.

Spiral Review

Find the interest earned to the nearest cent for each principal, interest rate, and time. (Lesson 7-8)

37. $300, 10%, 2 years

38. $900, 5.5%, 4.5 years

39. **BASEBALL CARDS** What is the total cost of a package of baseball cards if the regular price is $4.19 and the sales tax is 6.5%? (Lesson 7-7)

40. Solve $m + 18 = 33$ mentally. (Lesson 1-7)

▷ **GET READY for the Next Lesson**

PREREQUISITE SKILL Add or divide. Round to the nearest tenth if necessary.

41. $16 + 14 + 17$

42. $4.6 + 2.5 + 9$

43. $\dfrac{202}{16}$

44. $\dfrac{255}{7}$

Measures of Central Tendency and Range

MAIN IDEA

Describe a set of data using mean, median, mode, and range.

NYS Core Curriculum

7.S.5 Select the appropriate measure of central tendency 7.S.1 Identify and collect data using a variety of methods *Also addresses 7.S.6, 7.R.7*

New Vocabulary

measures of central tendency
mean
median
mode

NY Math Online

glencoe.com
• Extra Examples
• Personal Tutor
• Self-Check Quiz

▷ **MINI Lab**

The number of pennies in each cup represents Jack's scores for five math quizzes.

    8   7   9   6   10

Move the pennies among the cups so that each cup has the same number of pennies.

1. What was Jack's average score for the five quizzes?

2. If Jack scores 14 points on the next quiz, how many pennies would be in each cup?

A number used to describe the *center* of a set of data is a **measure of central tendency**. The most common of these measures is the mean.

Mean	Key Concept
Words	The **mean** of a set of data is the sum of the data divided by the number of items in the data set. The mean is also referred to as *average*.
Examples	data set: 1 cm, 1 cm, 5 cm, 2 cm, 2 cm, 4 cm, 2 cm, 5 cm mean: $\dfrac{1 + 1 + 5 + 2 + 2 + 4 + 2 + 5}{8}$ or 2.75 cm

EXAMPLE Find the Mean

① **QUIZ SCORES** The table shows the quiz scores for 16 students. Find the mean.

$$\text{mean} = \frac{47 + 40 + \ldots + 44}{16} \quad \substack{\leftarrow \text{ sum of data} \\ \leftarrow \text{ number of data items}}$$

$$= \frac{714}{16} \text{ or } 44.625$$

Quiz Scores			
47	40	43	45
49	41	49	44
43	41	44	49
44	50	41	44

 CHECK Your Progress

a. **MONEY** Adam earned $14, $10, $12, $15, and $13 by doing chores around the house. What is the mean amount Adam earned doing these chores?

Two other common measures of central tendency are median and mode.

Median   Key Concept

Words	In a data set that has been ordered from least to greatest, the **median** is the middle number if there is an odd number of data items. If there is an even number of data items, the median is the mean of the two numbers closest to the middle.	
Example	data set: 7 yd, 11 yd, 15 yd, 17 yd, 20 yd, 20 yd	
	median: $\dfrac{15 + 17}{2}$ or 16 yd	The median divides the data in half.

Mode

Words	The **mode** of a set of data is the number that occurs most often. If there are two or more numbers that occur most often, all of them are modes.
Example	data set: 50 mi, 45 mi, 45 mi, 52 mi, 49 mi, 56 mi, 56 mi
	modes: 45 mi and 56 mi

Median
Everyday Use the middle paved or planted section of a highway, as in median strip.

Math Use the middle number of the ordered data.

EXAMPLE Find the Mean, Median, and Mode

② MOVIE RENTALS The number of DVDs rented during one week at Star Struck Movie Rental is shown in the table. What are the mean, median, and mode of the data?

Star Struck Movie Rental Daily DVD Rentals						
S	M	T	W	TH	F	S
55	34	35	34	57	78	106

mean: $\dfrac{55 + 34 + 35 + 34 + 57 + 78 + 106}{7} = \dfrac{399}{7}$ or 57

median: 34, 34, 35, 55, 57, 78, 106   First, write the data in order.
                    ↓
                  median

mode: 34   It is the only value that occurs more than once.

The mean is 57 DVDs, the median is 55 DVDs, and the mode is 34 DVDs.

CHECK Your Progress

b. **BICYCLES** The sizes of the bicycles owned by the students in Ms. Garcia's class are listed in the table. What are the mean, median, and mode of the data?

Students' Bicycle Sizes (in.)			
20	24	20	26
24	24	24	26
24	29	26	24

c. **FOOTBALL** The points scored in each game by Darby Middle School's football team for 9 games are 21, 35, 14, 17, 28, 14, 7, 21, and 14. Find the mean, median, and mode.

Lesson 8-2 Measures of Central Tendency and Range **403**

3 The maximum length in feet of several whales is listed below.

46, 53, 33, 53, 79

If the maximum length of the Blue Whale, 98 feet, is added to this list, which of the following statements would be true?

A The mode would decrease.   **C** The mean would increase.

B The median would decrease.   **D** The mean would decrease.

Read the Item

You are asked to identify which statement would be true if the data value 98 was added to the data set.

Solve the Item

Use number sense to eliminate possibilities.
The mode, 53, will remain unchanged since the new data value occurs only once. So, eliminate answer choice A.

Since the new data value is greater than each value in the data set, the median will not decrease. So, eliminate answer choice B.

The remaining two answer choices refer to the mean. Since 98 is greater than each value in the data set, the mean will increase, not decrease. So, the answer is C.

✓ CHECK Your Progress

d. If the maximum length of the Orca Whale, 30 feet, is added to the list in Example 3, which of the following statements would be true?

F The mode would decrease.   **H** The mean would increase.

G The median would increase.   **J** The mean would decrease.

In addition to the mean, median, and mode, you can also use the range to describe a set of data. Below are some guidelines for using these measures.

Mean, Median, Mode, and Range	Concept Summary
Measure	**Most Useful When...**
Mean	• data set has no outliers
Median	• data set has outliers • there are no big gaps in the middle of the data
Mode	• data set has many identical numbers
Range	• describing the spread of the data

Study Tip

Median
When there is an odd number of data, the median is the middle number of the ordered set. When there is an even number of data, the median is the mean of the two middle numbers.

EXAMPLE Choose Mean, Median, Mode, or Range

④ **PLANTS** The line plot shows the height of desert cacti. Would the mean, median, mode, or range best represent the heights?

Heights of Desert Cacti (ft)

```
          ×
          ×
          ×
      ×   ×               ×
   ×××  ×       ×       ×           ×                    ×
   +--+--+--+--+--+--+--+--+--+--+--+--+--+--+--+--+--
   0  2  4  6  8 10 12 14 16 18 20 22 24 26 28 30
```

mean: $\dfrac{1 + 2 + 2 + ... + 30}{14}$ or 8.8

median: $\dfrac{\text{7th term} + \text{8th term}}{2} = \dfrac{5 + 5}{2}$ or 5

mode: 5

range: 30 − 1 or 29

The mean of 8.8 misrepresents the score. The median or mode represents the height of the cacti well.

✓ CHECK Your Progress

e. **GAMES** The table shows the cost of various board games. Would the mean, median, mode, or range best represent the costs? Explain.

Board Game Costs ($)

12	15	40	22
14	40	15	17
20	18	40	19
16	21	19	16

✓ CHECK Your Understanding

Examples 1, 2
(pp. 402–403)

Find the mean, median, and mode for each set of data. Round to the nearest tenth if necessary.

1. Miles traveled on the weekend: 29, 14, 80, 59, 78, 30, 59, 69, 55, 50

2.
Team	Number of Wins
Eagles	10
Hawks	8
Zipps	9
Falcons	11

3. **Minutes Spent Walking**

```
   ×       ×
   ×  ×  ×     ×       ×          ×
   ×  ×  ×  ×  ×     ×         ×  ×
   +--+--+--+--+--+--+--+--+--+--+--
  22 23 24 25 26 27 28 29 30 31 32
```

Example 3
(p. 404)

4. **MULTIPLE CHOICE** During the week, the daily low temperatures were 52°F, 45°F, 51°F, 45°F, and 48°F. If Saturday's low temperature of 51°F is added, which statement about the data set would be true?

  A The mean would decrease.   **C** The mode would increase.

  B The median would decrease.   **D** The mode would decrease.

Example 4
(p. 405)

5. **SHOES** The line plot shows the price of athletic shoes. Which measure best describes the data: mean, median, mode, or range? Explain.

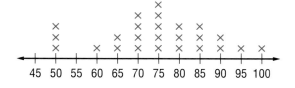

```
                           ×
                        ×  ×
                        ×  ×  ×  ×
         ×              ×  ×  ×  ×  ×
         ×           ×  ×  ×  ×  ×  ×
         ×        ×  ×  ×  ×  ×  ×  ×  ×  ×
   +--+--+--+--+--+--+--+--+--+--+--+--
  45 50 55 60 65 70 75 80 85 90 95 100
```

Practice and Problem Solving

HOMEWORK HELP

For Exercises	See Examples
6–11	1, 2
30–32	3
12–13	4

Find the mean, median, and mode for each set of data. Round to the nearest tenth if necessary.

6. Number of dogs groomed each week: 65, 56, 57, 75, 76, 66, 64

7. Daily number of boats in a harbor: 93, 84, 80, 91, 94, 90, 78, 93, 80

8. Scores earned on a math test: 95, 90, 92, 94, 91, 90, 98, 88, 89, 100

9. Prices of books: $10, $18, $11, $6, $6, $5, $10, $11, $46, $7, $6, $8

10.

Cost	Number of Coats
$75	8
$80	3
$85	6

11. **Springdale Middle School Basketball Scores**

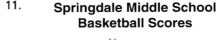

12. **MUSIC** The line plot shows the number of weeks that songs have been on the Top 20 Country Songs list. Would the mean, median, mode, or range best represent the data? Explain.

**Country Songs
Number of Weeks in Top 20**

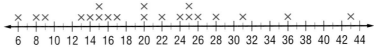

13. **SPACE** Twenty-seven countries have sent people into space. The table shows the number of individuals from each country. Which measure best describes the data: mean, median, mode, or range? Explain.

People in Space								
267	1	9	8	1	1	1	1	1
97	1	1	1	3	1	1	2	1
11	2	1	1	5	1	1	1	1

Source: *The World Almanac*

Real-World Link
The International Space Station measures 356 feet by 290 feet, and contains almost an acre of solar panels.
Source: *The World Almanac*

Find the mean, median, and mode for each set of data. Round to the nearest tenth if necessary.

14. Weight in ounces of various insects: 6.1, 5.2, 7.2, 7.2, 3.6, 9.0, 6.5, 7.4, 5.4

15. Prices of magazines: $3.50, $3.75, $3.50, $4.00, $3.00, $3.50, $3.25

16. Daily low temperatures: −2°F, −8°F, −2°F, 0°F, −1°F, 1°F, −2°F, −1°F

REASONING Determine whether each statement is *always*, *sometimes*, or *never* true about the data set {8, 12, 15, 23}. Explain your reasoning.

17. If a value greater than 23 is added, the mean will increase.

18. If a value less than or equal to 8 is added, the mean will decrease.

19. If a value between 8 and 23 is added, the mean will remain unchanged.

DINOSAURS For Exercises 20–22, use the lengths of the dinosaurs shown below.

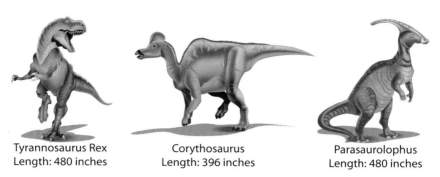

Tyrannosaurus Rex
Length: 480 inches

Corythosaurus
Length: 396 inches

Parasaurolophus
Length: 480 inches

20. What is the mean length of the dinosaurs?

21. One of the largest dinosaurs ever is the Brachiosaurus. Its length was 960 inches. If this data value is added to the lengths of the dinosaurs above, how will it affect the mean? Explain your reasoning.

22. Which measure best describes the data if the length of the Brachiosaurus is included: mean, median, mode, or range? If the length of the Brachiosaurus is *not* included? Explain any similarities or differences.

23. **SPORTS** The table shows the points scored by a lacrosse team so far this season. The team will play 14 games this season. How many points need to be scored during the last game so that the average number of points scored this season is 12? Explain.

Hawks Lacrosse Team Points Scored						
11	15	12	10	10	10	13
14	13	13	10	15	12	

NYSCC • NYSMT
Extra Practice, pp. 687, 711.

24. **FIND THE DATA** Refer to the Data File on pages 16–19. Choose some data and then describe it using the mean, median, mode, and range.

H.O.T. Problems

25. **OPEN ENDED** Give an example of a set of data in which the mean is not the best representation of the data set. Explain why not.

26. **Which One Doesn't Belong?** Identify the term that does not have the same characteristic as the other three. Explain your reasoning.

mean   median   range   mode

27. **REASONING** Determine whether the median is *sometimes, always,* or *never* part of the data set. Explain your reasoning.

28. **CHALLENGE** Without calculating, would the mean, median, or mode be most affected by eliminating 1,000 from the data shown? Which would be the least affected? Explain your reasoning.

50, 100, 75, 60, 75, 1,000, 90, 100

29. **WRITING IN MATH** According to the U.S. Census Bureau, the typical number of family members per household is 2.59. State whether this measure is a mean or mode. Explain how you know.

30. The table below shows the number of soup labels collected in one week by each homeroom in grade 7.

Classroom	Number of Soup Labels
Mr. Martin	138
Ms. Davis	125
Mr. Cardona	89
Mrs. Turner	110
Mr. Wilhelm	130
Mrs. LaBash	?

Which number could be added to the set of data in order for the mode and median of the set to be equal?

A 89   C 125

B 110   D 130

31. An antique dealer purchased 5 antiques for a total of $850.00. He later bought another antique for $758.00. What is the mean cost of all the antiques?

F $151.60   H $268.00

G $170.00   J $321.60

32. Gina found the mean and median of the following list of numbers.

5, 7, 7

If the number 11 was added to this list, which of the following statements would be true?

A The mean would increase.

B The mean would decrease.

C The median would increase.

D The median would decrease.

Spiral Review

33. TEMPERATURE The table shows record high temperatures for Kentucky in July. Make a line plot of the data. (Lesson 8-1)

July Temperatures				
99	98	96	98	98
97	100	103	103	103
100	95	100	103	105

Find the simple interest earned to the nearest cent for each principal, interest rate, and time. (Lesson 7-8)

34. $1,250, 3.5%, 2 years

35. $569, 5.5%, 4 months

36. FOOD The United States produced almost 11 billion pounds of apples in a recent year. Use the information in the graph to find how many pounds of apples were used to make juice and cider. (Lesson 7-1)

37. Name the property shown by the statement $4 \times 6 = 6 \times 4$. (Lesson 1-8)

Uses of Apples in the United States

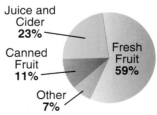

Juice and Cider **23%**

Canned Fruit **11%**

Fresh Fruit **59%**

Other **7%**

Source: U.S. Apple Association

▷ **GET READY for the Next Lesson**

PREREQUISITE SKILL Name the place value of the underlined digit. (p. 669)

38. 5̲81   **39.** 6,29̲5   **40.** 4̲,369   **41.** 2.8̲4

Graphing Calculator Lab
Mean and Median

MAIN IDEA

Use technology to calculate the mean and median of a set of data.

NYS Core Curriculum

7.S.5 Select the appropriate measure of central tendency
7.S.1 Identify and collect data using a variety of methods *Also addresses 7.S.6*

You can more efficiently calculate the mean and median of a large set of data using a graphing calculator.

ACTIVITY

COMPUTERS Kendrick surveys thirty seventh graders and asks them how many times they had to wait longer than 5 minutes during the previous week to use a computer in the school library. The results are shown below.

Number of Times a Student Had to Wait to Use the Library Computer									
5	2	9	1	1	2	1	2	5	2
3	4	2	1	4	0	4	2	2	5
4	2	2	3	2	1	3	9	5	2

Find the mean and median of the data.

STEP 1 Clear list L1 by pressing [STAT] [ENTER] [▲] [CLEAR] [ENTER]

STEP 2 Enter the number of times students had to wait in L1. Press 5 [ENTER] 2 [ENTER] . . . 2 [ENTER].

STEP 3 Display a list of statistics for the data by pressing [STAT] [▶] [ENTER] 1 [ENTER].

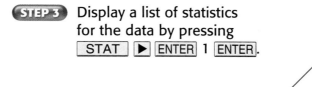

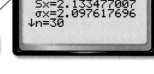

The first value, *x*, is the mean.

Use the down arrow key to locate **Med**. The mean number of times a student waited was 3 and the median number of times was 2.

ANALYZE THE RESULTS

1. **WRITING IN MATH** Kendrick claims that, on average, students had to wait more than 5 minutes about 3 times last week. Based on your own analysis of the data, write a convincing argument to dispute his claim. (*Hint:* Create and use a line plot of the data to support your argument.)

2. **COLLECT THE DATA** Collect some numerical data from your classmates. Then use a graphing calculator to calculate the mean and median of the data. After analyzing the data, write a convincing argument to support a claim you can make about your data.

Stem-and-Leaf Plots

MAIN IDEA

Display and analyze data in a stem-and-leaf plot.

NYS Core Curriculum

7.S.6 Read and interpret data represented graphically (pictograph, bar graph, histogram, line graph, double line/bar graphs or circle graph) **7.A.7 Draw the graphic representation of a pattern from an equation or from a table of data** *Also addresses 7.S.1, 7.CN.8*

New Vocabulary

stem-and-leaf plot
leaf
stem

NY Math Online

glencoe.com

• Extra Examples
• Personal Tutor
• Self-Check Quiz

▶ GET READY for the Lesson

BIRDS The table shows the average chick weight in grams of sixteen different species of birds.

1. Which chick weight is the lightest?

2. How many of the weights are less than 10 grams?

Chick Weight (g)			
19	6	7	10
11	13	18	25
21	12	5	12
20	21	11	12

In a **stem-and-leaf plot**, the data are organized from least to greatest. The digits of the least place value usually form the **leaves**, and the next place-value digits form the **stems**.

EXAMPLE   Display Data in a Stem-and-Leaf Plot

1 **BIRDS** Display the data in the table above in a stem-and-leaf plot.

Step 1 Choose the stems using digits in the tens place, 0, 1, and 2. The least value, 5, has 0 in the tens place. The greatest value, 25, has 2 in the tens place.

Step 2 List the stems from least to greatest in the *Stem* column. Write the leaves, the ones digits to the right of the corresponding stems.

Stem	Leaf
0	6 7 5
1	9 0 1 3 8 2 2 1 2
2	5 1 0 1

Step 3 Order the leaves and write a *key* that explains how to read the stems and leaves. Include a title.

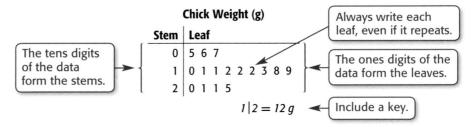

Chick Weight (g)

Stem	Leaf
0	5 6 7
1	0 1 1 2 2 2 3 8 9
2	0 1 1 5

The tens digits of the data form the stems.

Always write each leaf, even if it repeats.

The ones digits of the data form the leaves.

1|2 = 12 g   Include a key.

✓ CHECK Your Progress

a. **HOMEWORK** The number of minutes the students in Mr. Blackwell's class spent doing their homework one night is shown. Display the data in a stem-and-leaf plot.

Homework Time (min)				
42	5	75	30	45
47	0	24	45	51
56	23	39	30	49
58	55	75	45	35

Stem-and-leaf plots are useful in analyzing data because you can see all the data values, including the greatest, least, mode, and median value.

EXAMPLE Describe Data

(2) **CHESS** The stem-and-leaf plot shows the number of chess matches won by members of the Avery Middle School Chess Team. Find the range, median, and mode of the data.

Chess Matches Won	
Stem	Leaf
0	8 8 9
1	9
2	0 0 2 4 4 8 9
3	1 1 2 4 5 5 6 6 7 7 8
4	0 0 0 3 8 9
5	2 4
6	1

$3|2 = 32$ *wins*

range: greatest wins − least wins
= 61 − 8 or 53

median: middle value, or 35 wins

mode: most frequent value, 40

CHECK Your Progress

b. **BIRDS** Find the range, median, and mode of the data in Example 1.

Real-World Career . . . .
How Does a Sports Scout Use Math?
Sports scouts review game records and statistics and evaluate athletes' skills.

NY Math Online

For more information, go to glencoe.com.

EXAMPLE Effect of Outliers

(3) **SPORTS** The stem-and-leaf plot shows the number of points scored by a college basketball player. Which measure of central tendency is most affected by the outlier?

Basketball Points	
Stem	Leaf
0	2
1	2 2 3 5 8
2	0 0 1 1 3 4 6 6 6 8 9
3	0 0 1

$1|2 = 12$ *points*

The mode, 26, is not affected by the inclusion of the outlier, 2.

Calculate the mean and median each without the outlier, 2. Then calculate them including the outlier and compare.

	without the outlier	including the outlier
mean:	$\frac{12 + 12 + ... + 31}{19} \approx 22.37$	$\frac{2 + 12 + 12 + ... + 31}{20} = 21.35$
median:	23	22

The mean decreased by 22.37 − 21.35, or 1.02, while the median decreased by 23 − 22, or 1. Since 1.02 > 1, the mean is most affected.

CHECK Your Progress

c. **CHESS** Refer to Example 2. If an additional student had 84 wins, which measure of central tendency would be most affected?

Example 1
(p. 410)

Display each set of data in a stem-and-leaf plot.

1.

Height of Trees (ft)				
15	25	8	12	20
10	16	15	8	18

2.

Cost of Shoes ($)				
42	47	19	16	21
23	25	25	29	31
33	34	35	39	48

Examples 2, 3
(p. 411)

CAMP The stem-and-leaf plot at the right shows the ages of students in a pottery class.

Ages of Students

Stem	Leaf
0	9 9 9
1	0 1 1 1 1 2 2 3 3 4

$1|0 = 10$ years

3. What is the range of the ages of the students?

4. Find the median and mode of the data.

5. If an additional student was 6 years old, which measure of central tendency would be most affected?

Practice and Problem Solving

HOMEWORK HELP

For Exercises	See Examples
6–9	1
10, 11, 13, 14, 16–18	2
12, 15, 19	3

Display each set of data in a stem-and-leaf plot.

6.

Quiz Scores (%)			
70	96	72	91
80	80	79	93
76	95	73	93
90	93	77	91

7.

Low Temperatures (°F)				
15	13	28	32	38
30	31	13	36	35
38	32	38	24	20

8.

Floats at Annual Parade			
151	158	139	103
111	134	133	154
157	142	149	159

9.

School Play Attendance			
225	227	230	229
246	243	269	269
267	278	278	278

CYCLING The number of Tour de France titles won by eleven countries is shown.

Tour de France Titles Won by Eleven Countries

Stem	Leaf
0	1 1 1 2 2 4 8 9
1	0 8
2	
3	6

$0|4 = 4$ titles

10. Find the range of titles won.

11. Find the median and mode of the data.

12. Which measure of central tendency is most affected by the outlier?

ELECTRONICS For Exercises 13–15, use the stem-and-leaf plot that shows the costs of various DVD players at an electronics store.

Costs of DVD Players

Stem	Leaf
8	2 5 5
9	9 9
10	0 0 2 5 6 8
11	0 0 5 5 5 9 9
12	5 7 7

$11|5 = 115

13. What is the range of the prices?

14. Find the median and mode of the data.

15. If an additional DVD player cost $153, which measure of central tendency would be most affected?

Bar Graphs and Histograms

MAIN IDEA

Display and analyze data using bar graphs and histograms.

NYS Core Curriculum

7.S.6 Read and interpret data represented graphically (pictograph, **bar graph, histogram,** line graph, double line/ bar graphs or circle graph) **7.S.1 Identify and collect data using a variety of methods** *Also addresses 7.A.7*

New Vocabulary

bar graph
histogram

NY Math Online

glencoe.com

• Extra Examples
• Personal Tutor
• Self-Check Quiz

▷ **GET READY for the Lesson**

FOOTBALL The table shows 5 teams that scored the greatest number of points in the Super Bowl.

Team	Score
San Francisco	55
Dallas	52
Tampa Bay	48
Chicago	46
Washington	42

Source: National Football League

1. What is the greatest and least scores in the table?

2. How can you summarize the data with a visual representation?

3. Do any of these representations show both the team and its score?

A **bar graph** is a method of comparing data by using solid bars to represent quantities.

EXAMPLE Display Data Using a Bar Graph

① **Display the data in the table above in a bar graph.**

Step 1 Draw a horizontal axis and a vertical axis. Label the axes as shown. In this case, the scale on the vertical axis is chosen so that it includes all the scores. Add a title.

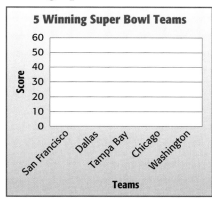

Step 2 Draw a bar to represent each category. In this case, a bar is used to represent the score of each team.

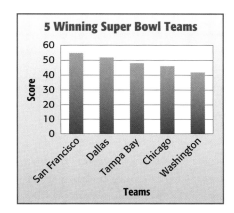

 Your Progress

a. FLOWERS The table shows the diameters of the world's largest flowers. Display the data in a bar graph.

Flower	Maximum Size (in.)
Rafflesia	36
Sunflower	19
Giant Water Lily	18
Brazilian Dutchman	14
Magnolia	10

Source: *Book of World Records*

Reading Math

Frequency
Frequency refers to the number of data items in a particular interval. In Example 2, the frequency of 7 in the third row means that there are 7 games with a score of 31–40.

A special kind of bar graph, called a **histogram**, uses bars to represent the frequency of numerical data that have been organized in intervals.

EXAMPLE Display Data Using a Histogram

2 **FOOTBALL** The winning scores of twenty recent Super Bowl games have been organized into a frequency table. Display the data in a histogram.

Score	Frequency
11–20	3
21–30	4
31–40	7
41–50	4
51–60	2

Source: National Football League

Study Tip

Histograms
Because the intervals are equal, all of the bars have the same width, with no space between them.

Step 1 Draw and label horizontal and vertical axes. Add a title.

Step 2 Draw a bar to represent the frequency of each interval.

The three highest bars represent a majority of the data. From the graph, you can easily see that most of the scores were between 21 and 50 points.

b. **EARTHQUAKES** The magnitudes of the largest U.S. earthquakes are organized into the frequency table shown. Display the data in a histogram.

Magnitude	Frequency
7.0–7.4	4
7.5–7.9	14
8.0–8.4	5
8.5–8.9	2
9.0–9.4	1

Source: National Earthquake Information Center

Study Tip

Alternate Method
You can also use a proportion to find the percent in Example 4.

$$\frac{6}{30} = \frac{x}{100}$$
$$6 \cdot 100 = 30 \cdot x$$
$$600 = 30x$$
$$20 = x$$

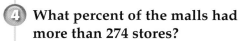
EXAMPLES Interpret Histograms and Bar Graphs

MALLS The histogram shows the number of stores in the largest malls in the U.S.

3 How many malls are represented in the histogram? Justify your answer.

Find the sum of the heights of the bars in the histogram.
So, $10 + 14 + 4 + 1 + 1 = 30$.

Source: Directory of Major Malls

4 What percent of the malls had more than 274 stores?

$\dfrac{6}{30}$ ⟶ number of malls with more than 274 stores
 ⟶ total number of malls

$\dfrac{6}{30} = 0.2$ Write the fraction as a decimal.

$0.2 = 20\%$ Write the decimal as a percent.

So, 20% of the malls had more than 274 stores.

✓ CHECK Your Progress

NASCAR The histogram shows the average winning times for the Daytona 500.

c. How many winning times are represented in the histogram? Explain your reasoning.

d. What percent of the winning speeds were faster than 150 miles per hour?

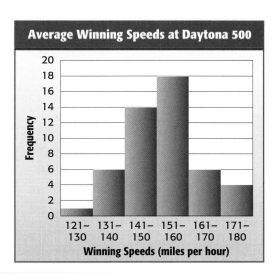

Examples 1, 2
(pp. 415–417)

Select the appropriate graph to display each set of data: bar graph or histogram. Then display the data in the appropriate graph.

1.

State Sales Tax Rates	
Percent	**States**
2.0–2.9	1
3.0–3.9	0
4.0–4.9	12
5.0–5.9	12
6.0–6.9	16
7.0–7.9	4

Source: Federation of Tax Administrators

2.

Men's Grand Slam Titles	
Player	**Titles**
Pete Sampras	14
Roy Emerson	12
Bjorn Borg	11
Rod Laver	11
Andre Agassi	8

Source: Book of World Records

Examples 3, 4
(p. 417)

TEXTBOOKS For Exercises 3 and 4, use the bar graph that shows the average number of pages in various textbooks.

3. On average, which textbook has the least number of pages?

4. Is it reasonable to say that on average, a health textbook has half as many pages as a science textbook? Explain.

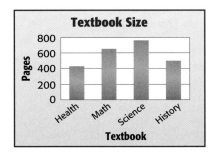

Practice and Problem Solving

HOMEWORK HELP

For Exercises	See Examples
5–8	1, 2
9–15	3, 4

Select the appropriate graph to display each set of data: bar graph or histogram. Then display the data in the appropriate graph.

5.

Most Threatened Reptiles	
Country	**Number of Species**
Australia	38
China	31
Indonesia	28
U.S.	27
India	25

Source: Top 10 of Everything

6.

Home Run Leaders	
Home Runs	**Frequency**
31–36	1
37–42	4
43–48	7
49–54	5
55–60	3

7.

Major U.S. Rivers	
Length (mi)	**Frequency**
600–999	15
1,000–1,399	5
1,400–1,799	3
1,800–2,199	3
2,200–2,599	2

Source: The World Almanac

8.

City Skyscrapers	
City	**Skyscrapers**
New York	176
Hong Kong	163
Chicago	81
Shanghai	49
Tokyo	44

Source: Book of World Records

HEIGHT For Exercises 9-12, use the histogram that shows the heights of students in a classroom.

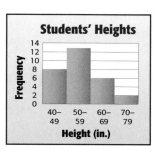

9. How many students are from 50 to 59 inches tall?

10. What percent of students are at least 60 inches tall?

11. Write a statement comparing the 70–79 interval to the 60–69 interval.

12. **COLLECT THE DATA** Conduct a survey of your classmates to determine their height. Then display and analyze the data in a histogram. How does your data compare to the histogram above?

Real-World Link · · · ·
Giraffes live for 10–15 years in the wild, but average 25 years at zoos.

Source: Woodland Park Zoo

ZOOS For Exercises 13–15, use the histogram that shows the attendance at the major U.S. zoos in a recent year.

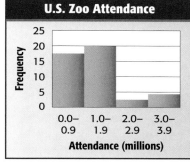

13. About how many zoos does the graph represent?

14. What is the range of attendance for most of the zoos?

15. Compare the number of zoos with 0.0–0.9 million visitors to the number of zoos with 3.0–3.9 million visitors.

Source: The World Almanac

Match each characteristic to the appropriate graph(s).

16. data display based upon place value

17. shows the frequency of data on a number line

18. compares data using solid bars

19. data are organized using intervals

a. line plot

b. histogram

c. stem-and-leaf plot

d. bar graph

DISTANCE For Exercises 20 and 21, use the histogram shown below.

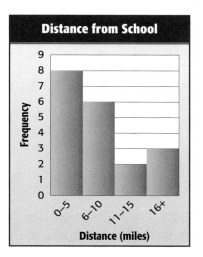

20. How many students live 6–10 miles from school?

21. What percent of students live 16 or more miles from school?

FOOD For Exercises 22 and 23, use the multiple-bar graph that compares boys' and girls' favorite pizza toppings

22. For which topping is the difference in girls' and boys' favorites the greatest? Explain.

23. Describe an advantage of using a multiple-bar graph rather than two separate graphs to compare data.

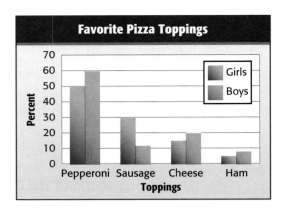

EXERCISE For Exercises 24–27, refer to the graph below.

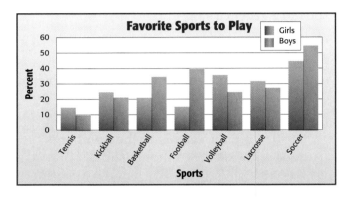

24. Which sport did the girls surveyed prefer the most?

25. Which sport is the least favorite for the boys?

26. Based on this survey, boys prefer football 4 times more than what sport?

27. Is it reasonable to say that almost twice as many boys prefer basketball than girls? Explain.

NYSCC • NYSMT
Extra Practice, pp. 688, 711

H.O.T. Problems

28. **CHALLENGE** The histograms show players' salaries for two major league baseball teams. Compare the salary distributions of the two teams.

29. **DATA SENSE** Describe how to determine the number of values in a data set that is represented by a histogram.

30. **WRITING IN MATH** Can any data set be displayed using a histogram? If yes, explain why. If no, give a counterexample and explain why not.

31. The results of a survey are displayed in the graph.

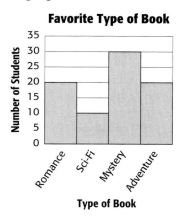

Favorite Type of Book

Which statement is valid about the survey?

A Twice as many students enjoy reading mysteries than romance.

B Most students enjoy reading adventure books.

C Twice as many students enjoy reading romance books than science fiction.

D Half as many students enjoy reading mysteries than romance.

32. **SHORT RESPONSE** The graph shows the average car sales per month at a car dealership.

Average Monthly Sales

What is the best prediction for the number of station wagons the dealer sells in a year?

Spiral Review

SPORTS For Exercises 33 and 34, refer to the table that lists the number of games won by each team in a baseball league.

33. Make a stem-and-leaf plot of the data. (Lesson 8-3)

34. What is the mean, median, and mode of the data? (Lesson 8-2)

Number of Wins					
25	36	46	15	30	53
40	32	17	45	41	31
56	50	52	47	26	40
43	56	51	50	55	50
44	47	53	23	19	

35. **SELECT A TECHNIQUE** The video game that Neil wants to buy costs $50. He has saved $\frac{1}{5}$ of the amount he needs. Which of the following techniques might Neil use to find how much more money he will need to buy the game? Justify your selection(s). Then use the technique(s) to solve the problem. (Lesson 5-5)

| mental math | number sense | estimation |

▷ **GET READY** for the Next Lesson

36. **WEATHER** At 5:00 P.M., the outside temperature was 81°F. At 6:00 P.M., it was 80°F. At 7:00 P.M., it was 79°F. Use the *look for a pattern* strategy to predict the outside temperature at 8:00 P.M. (Lesson 2-7)

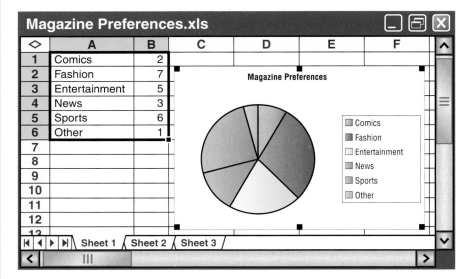

Extend 8-4

Spreadsheet Lab
Circle Graphs

MAIN IDEA

Use technology to create circle graphs.

NYS Core Curriculum

7.S.2 Display data in a circle graph   7.S.6 Read and interpret data represented graphically (pictograph, bar graph, histogram, line graph, double line/bar graphs or **circle graph**) *Also addresses 7.A.7, 7.S.1*

Another type of display used to compare categories of data is a *circle graph*. Circle graphs are useful when comparing parts of a whole.

ACTIVITY

MAGAZINES The table shows the results of a survey in which students were asked to indicate their favorite type of magazine. Use a spreadsheet to make a circle graph of these data.

STEP 1 Enter the data in a spreadsheet as shown.

Magazine Preferences	
Type	**Frequency**
Comics	2
Fashion	7
Entertainment	5
News	3
Sports	6
Other	1

STEP 2 Select the information in cells A1 to B6. Click on the Chart Wizard icon. Choose the Pie chart type. Click Next twice. Enter the title Magazine Preferences. Then click Next and Finish.

ANALYZE THE RESULTS

1. **MAKE A CONJECTURE** Use the graph to determine which types of magazines were preferred by about $\frac{1}{3}$ and 25% of the students surveyed. Explain your reasoning. Then check your answers.

2. **COLLECT THE DATA** Collect some data that can be displayed in either a circle or bar graph. Record the data in a spreadsheet. Then use the spreadsheet to make both types of displays. Which display is more appropriate? Justify your selection.

1. **MULTIPLE CHOICE** The table shows quiz scores of a math class. What is the range of test scores? (Lesson 8-1)

Math Scores						
89	92	67	75	95	89	82
92	88	89	80	91	79	90

A 89   **C** 67
B 82   **D** 28

For Exercises 2–4, use the data below. (Lesson 8-1)

Age Upon Receiving Driver's License									
16	17	16	16	18	21	16	16	18	18
17	25	16	17	17	17	17	16	20	16

2. Make a line plot of the data.

3. Identify any clusters, gaps, or outliers.

4. Describe how the range of data would change if 25 was not part of the data set.

5. **MULTIPLE CHOICE** The table shows the average April rainfall for 12 cities. If the value 4.2 is added to this list, which of the following would be true? (Lesson 8-2)

Average Rain (in.)					
0.5	0.6	1.0	1.0	2.5	3.7
2.6	3.3	2.0	1.4	0.7	0.4

F The mode would increase.

G The mean would increase.

H The mean would decrease.

J The median would decrease.

6. **TREES** The heights, in meters, of several trees are 7.6, 6.8, 6.5, 7.0, 7.9, and 6.8. Find the mean, median, and mode. Round to the nearest tenth if necessary. (Lesson 8-2)

7. **SPEED** Display the data shown in a stem-and-leaf plot and write one conclusion based on the data. (Lesson 8-3)

Car Highway Speeds				
65	72	76	68	65
59	70	69	71	74
68	65	71	74	69

MAMMALS For Exercises 8–10, refer to the stem-and-leaf plot that shows the maximum weight in kilograms of several rabbits.

Maximum Weight of Rabbits (kg)

Stem	Leaf
0	8 9
1	0 2 4 6 8
2	7
3	
4	
5	4

0|8 = 0.8 kg

8. Find the range of weights.

9. Find the median and mode of the data.

10. Which measure of central tendency is most affected by the inclusion of the outlier? Explain.

ATTENDANCE For Exercises 11 and 12, refer to the graph. (Lesson 8-4)

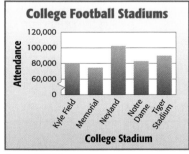

College Football Stadiums

Source: MSNBC

11. About how many people does the graph represent?

12. Which two stadiums house about the same number of people?

Problem-Solving Investigation

MAIN IDEA: Solve problems by using a graph.

NYSCC **7.PS.6** Represent problem situations verbally, numerically, algebraically, and graphically **7.PS.17** Evaluate the efficiency of different representations of a problem

P.S.I. TEAM +

e-Mail: USE A GRAPH

RICK: The table shows the study times and test scores of 13 students in Mrs. Collins's English class.

YOUR MISSION: Use a graph to predict the test score of a student who studied for 80 minutes.

Study Time and Test Scores											
Study Time (min)	120	30	60	95	70	55	90	45	75	60	10
Test Score (%)	98	77	91	93	77	78	95	74	87	83	65

Understand	You know the number of minutes studied. You need to predict the test score.
Plan	Organize the data in a graph so you can easily see any trends.
Solve	The graph shows that as the study times progress, the test scores increase. You can predict that the test score of a student who studied for 80 minutes is about 88%.
Check	Draw a line that is close to as many of the points as possible, as shown. The estimate is close to the line so the prediction is reasonable.

Study Time and Test Scores

(graph: y-axis Test Score (%) with values 65, 75, 85, 95; x-axis Study Time (min) with values 0, 20, 40, 60, 80, 100)

Analyze The Strategy

1. Explain why analyzing a graph is a useful way to quickly make conclusions about a set of data.

2. **WRITING IN MATH** Write a problem in which using a graph would be a useful way to check a solution.

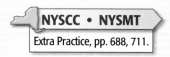
NYSCC • NYSMT
Extra Practice, pp. 688, 711.

For Exercises 3 and 4, use the table. It shows the relationship between Celsius and Fahrenheit temperatures.

Temperature	
Celsius	**Fahrenheit**
0	32
10	50
20	68
30	86
40	104

3. Make a graph of the data.

4. Suppose the temperature is 25° Celsius. Estimate the temperature in Fahrenheit.

5. **FUNDRAISING** The graph shows how many boxes of popcorn were sold by four students for a fundraiser. Which student sold about half as many boxes as Alyssa?

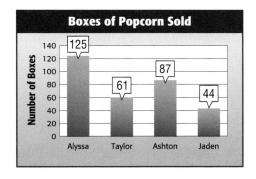

Use any strategy to solve Exercises 6–12. Some strategies are shown below.

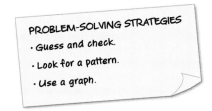

PROBLEM-SOLVING STRATEGIES
- Guess and check.
- Look for a pattern.
- Use a graph.

6. **ALGEBRA** What are the next two numbers in the pattern 8, 18, 38, 78, ...?

7. **EXERCISE** Emma walked 8 minutes on Sunday and plans to walk twice as long each day than she did the previous day. On what day will she walk over 1 hour?

8. **EXERCISE** The graph shows the number of minutes Jacob exercised during one week. According to the graph, which two days did he exercise about the same amount of time?

9. **ALGEBRA** Find two numbers with a sum of 56 and with a product of 783.

10. **HELICOPTERS** A helicopter has a maximum freight capacity of 2,400 pounds. How many crates, each weighing about 75 pounds, can the helicopter hold?

11. **SKATING** Moses and some of his friends are going to the movies. Suppose they each buy nachos and a beverage. They spend $36. How many friends are going to the movies with Moses?

Movie Costs	
Item	**Price**
ticket	$6.00
beverage	$2.25
nachos	$3.75

12. **NUMBER THEORY** A number is squared and the result is 324. Find the number.

Using Graphs to Predict

MAIN IDEA

Display and analyze data using a line plot.

NYS Core Curriculum

7.S.6 Read and interpret data represented graphically (pictograph, bar graph, histogram, **line graph, double line**/**bar graphs** or circle graph)

New Vocabulary

line graph
scatter plot

NY Math Online

glencoe.com

• Extra Examples
• Personal Tutor
• Self-Check Quiz

▷ **MINI Lab**

• Pour 1 cup of water into the drinking glass.

• Measure the height of the water, and record it in a table like the one shown.

• Place 5 marbles in the glass. Measure the height of the water. Record.

Number of Marbles	Height of Water (cm)
0	
5	
10	
15	
20	

• Continue adding marbles, 5 at a time, until there are 20 marbles in the glass. After each time, measure and record the height of the water.

1. By how much did the water's height change after each addition of marbles?

2. Predict the height of the water when 30 marbles are in the drinking glass. Explain how you made your prediction.

3. Test your prediction by placing 10 more marbles in the glass.

4. Draw a graph of the data that you recorded in the table.

You created a line graph in the Mini Lab. **Line graphs** can be useful in predicting future events because they show relationships or trends over time.

EXAMPLES **Use a Line Graph to Predict**

① **TEMPERATURE** The relationship between temperature readings in °C and °F is shown below. Use the line graph to predict the temperature reading 35°C in °F.

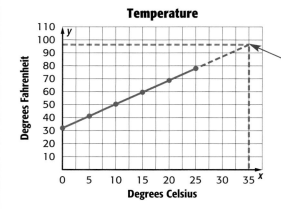

Temperature

Continue the graph with a dotted line in the same direction until you align vertically with 35°C. Graph a point. Find what value in °F corresponds with the point.

The temperature reading 35°C is equivalent to 95°F.

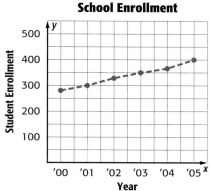

② **SCHOOL** The graph shows the student enrollment at McDaniel Middle School for the past several years. If the trend continues, what will be the enrollment in 2010?

Study Tip

Broken Lines
In Example 2, there are no data points between the points that represent enrollment. So, a broken line was used to help you easily see trends in the data.

School Enrollment

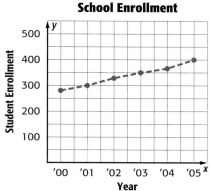

If the trend continues, the enrollment in 2010 will be about 525 students.

✓ CHECK Your Progress

a. **READING** Kerry is reading *The Game of Sunken Places* over summer break. The graph shows the time it has taken her to read the book so far. Predict the time it will take her to read 150 pages.

The Game of Sunken Places

b. **JUICE BOXES** The table shows the number of juice boxes a cafeteria sold in a five-week period. Display the data in a line graph. If the trend continues, how many juice boxes will be sold in week 8?

Juice Box Sales	
Week	**Number Sold**
1	50
2	52
3	56
4	60
5	62

Study Tip

Scatter Plots
In a positive relationship, as the value of *x* increases, so does the value of *y*. In a negative relationship, as the value of *x* increases, the value of *y* decreases.

A **scatter plot** displays two sets of data on the same graph. Like line graphs, scatter plots are useful for making predictions because they show trends in data. If the points on a scatter plot come close to lying on a straight line, the two sets of data are related.

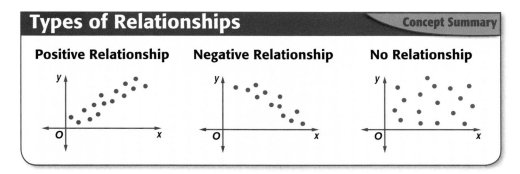

EXAMPLE Use a Scatter Plot to Predict

3 **NASCAR** The scatter plot shows the earnings for the winning driver for the Nextel Cup Series from 1986 to 2006. Predict the winning earnings for the next Nextel Cup Series.

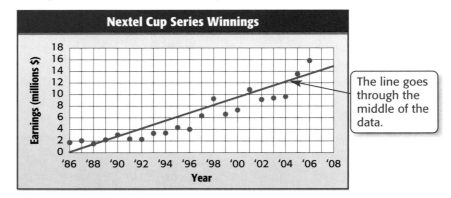

By looking at the pattern, we can predict the winning earnings for 2008 will be about $16,500,000.

CHECK Your Progress

c. **NASCAR** Use the scatter plot above to predict winning earnings for 2010.

CHECK Your Understanding

Examples 1, 2
(pp. 426–427)

POPULATION Delaware is a fast growing city in Ohio. The graph shows its increase in population.

1. Describe the relationship between the two sets of data.

2. If the trend continues, what will be the population in 2010?

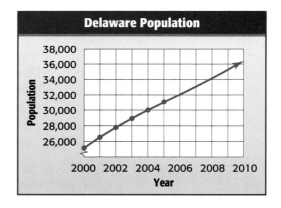

Example 3
(p. 428)

3. **PICNICS** The scatter plot shows the number of people who attended a neighborhood picnic each year. How many people should be expected to attend the picnic in 2007?

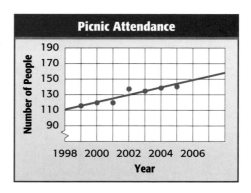

HOMEWORK HELP

For Exercises	See Examples
4–5	1, 2
6–7	3

MONUMENTS For Exercises 4 and 5, use the graph that shows the time it takes Ciro to climb the Statue of Liberty.

4. Predict the time it will take Ciro to climb 354 steps to reach the top.

5. How many steps will he have climbed after 14 minutes?

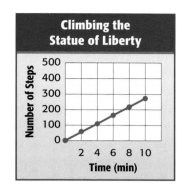

SCHOOL For Exercises 6 and 7, use the graph that shows the time students spent studying for a test and their test score.

6. What score should a student who studies for 1 hour be expected to earn?

7. If a student scored 90 on the test, about how much time can you assume the student spent studying?

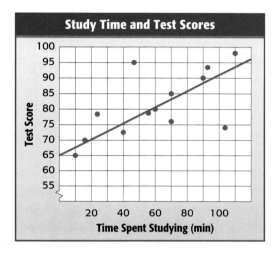

SLEEP For Exercises 8–10, use the table that shows the relationship between hours of sleep before and scores on a math test.

8. Display the data in a scatter plot.

9. Describe the relationship, if any, between the two sets of data.

10. Predict the test score for someone that sleeps 5 hours.

Hours of Sleep	Math Test Score
9	96
8	88
7	76
6	71

BASKETBALL For Exercises 11–13, use the table at the right.

11. Make a scatter plot of the data to show the relationship between free throws made and free throw attempts.

12. Predict the number of free throws made if 500 free throws were attempted.

13. Describe the trend in the data.

Player	Free Throws Made	Free Throw Attempts
T. Duncan	362	568
S. Nash	222	247
Y. Ming	356	413
B. Davis	275	369
A. Jamison	226	307
D. Williams	227	296
D. Howard	390	666
J. Howard	243	294

Source: National Basketball Association

SCHOOLS For Exercises 14 and 15, use the graphic that shows public school teachers' average salaries for the past few years.

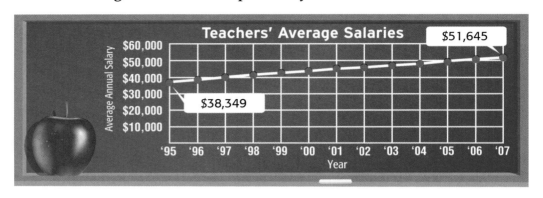

Teachers' Average Salaries

$51,645

$38,349

14. Describe the relationship, if any, between the two sets of data.

15. If the trend continues, what will be the average annual salary in 2011?

16. **POPULATION** The multiple line graph at the right shows the population of San Diego, California, and Phoenix, Arizona, from 1980 to 2005. What can you conclude from the graph?

17. **RESEARCH** Use the Internet or another source to find a real-world example of a scatter plot. Write a description of what the graph displays and extend the graph to show where the data will be in the future.

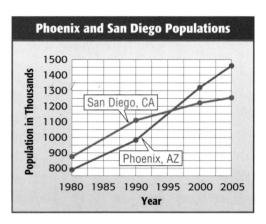

Phoenix and San Diego Populations

NYSCC • NYSMT
Extra Practice, pp. 689, 711

H.O.T. Problems

18. **OPEN ENDED** Name two sets of data that can be graphed on a scatter plot.

19. **Which One Doesn't Belong?** Identify the term that does not have the same characteristic as the other three. Explain your reasoning.

| line plot | mode | bar graph | scatter plot |

20. **CHALLENGE** What can you conclude about the relationship between pet owner age and number of pets shown in the scatter plot at the right?

21. **WRITING IN MATH** Explain how a graph can be used to make predictions.

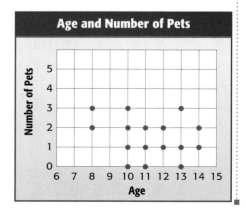

Age and Number of Pets

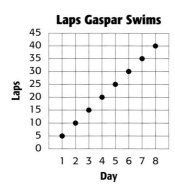
22. The number of laps Gaspar has been swimming each day is shown.

Laps Gaspar Swims

If the trend shown in the graph continues, what is the best prediction for the number of laps he will swim on day 10?

A 50

B 65

C 75

D 100

23. The number of people at the pool at different times during the day is shown.

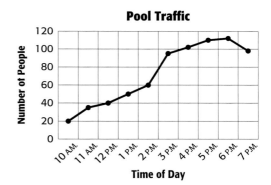

If an extra lifeguard is needed when the number of people at the pool exceeds 100, between which hours is an extra lifeguard needed?

F 10:00 A.M.–12:00 P.M.

G 12:00 P.M.–3:00 P.M.

H 2:00 P.M.–5:00 P.M.

J 4:00 P.M.–6:00 P.M.

Spiral Review

24. **SKATING** Use the *use a graph* strategy to compare the number of people who skate in California to the number of people who skate in Texas. (Lesson 8-5)

25. **COLORS** Of 57 students, 13 prefer the color red, 16 prefer blue, 20 prefer green, and 8 prefer yellow. Display this data in a bar graph. (Lesson 8-4)

Source: National Sporting Goods Association

▷ GET READY for the Next Lesson

PREREQUISITE SKILL Find the mean and median for each set of data. (Lesson 8-2)

26. 89 ft, 90 ft, 74 ft, 81 ft, 68 ft

27. 76°, 90°, 88°, 84°, 82°, 78°

Spreadsheet Lab
Multiple-Line and -Bar Graphs

MAIN IDEA

Use a spreadsheet to make a multiple-line graph and a multiple-bar graph.

NYS Core Curriculum

7.S.3 Convert raw data into double bar graphs and double line graphs

In Lessons 8-4 and 8-6, you interpreted data in a multiple-bar graph and in a multiple-line graph, respectively. You can use a spreadsheet to make these two types of graphs.

ACTIVITY

1 The stopping distances for a car on dry pavement and on wet pavement are shown in the table at the right.

Speed (mph)	Stopping Distance (ft)	
	Dry Pavement	Wet Pavement
50	200	250
60	271	333
70	342	430
80	422	532

Source: Continental Teves

Set up a spreadsheet like the one shown below.

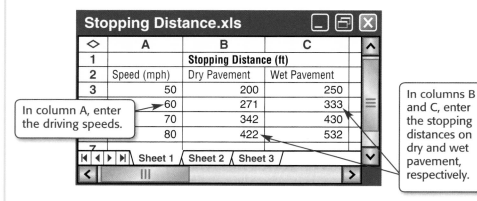

In column A, enter the driving speeds.

In columns B and C, enter the stopping distances on dry and wet pavement, respectively.

The next step is to "tell" the spreadsheet to make a double-line graph for the data.

1. Highlight the data in columns B and C, from B2 through C6.

2. Click on the Chart Wizard icon.

3. Choose the line graph and click Next.

4. To set the *x*-axis, choose the Series tab and press the icon next to the Category (X) axis labels.

This tells the spreadsheet to read the data in columns B and C.

5. On the spreadsheet, highlight the data in column A, from A3 through A6.

6. Press the icon on the bottom of the Chart Wizard box to automatically paste the information.

7. Click Next and enter the chart title and labels for the *x*- and *y*-axes.

8. Click Next and then Finish.

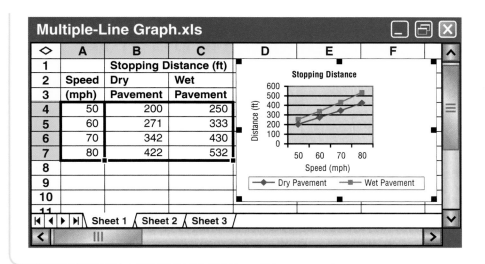

Multiple-Line Graph.xls

	A	B	C
1		Stopping Distance (ft)	
2	Speed	Dry	Wet
3	(mph)	Pavement	Pavement
4	50	200	250
5	60	271	333
6	70	342	430
7	80	422	532
8			
9			
10			

Sheet 1 / Sheet 2 / Sheet 3 /

ACTIVITY

(2) Use the same data to make a multiple-bar graph.

- Highlight the data in columns B and C, from B2 through C6.
- Click on the Chart Wizard icon.
- Click on Column and Next to choose the vertical bar graph.
- Complete steps 4–8 from Activity 1.

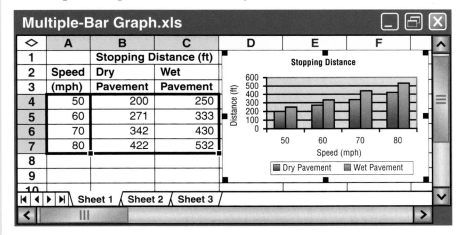

Multiple-Bar Graph.xls

	A	B	C
1		Stopping Distance (ft)	
2	Speed	Dry	Wet
3	(mph)	Pavement	Pavement
4	50	200	250
5	60	271	333
6	70	342	430
7	80	422	532
8			
9			
10			

Sheet 1 / Sheet 2 / Sheet 3 /

ANALYZE THE RESULTS

1. Explain the steps you would take to make a multiple-line graph of the stopping distances that include the speeds 55, 65, and 75.

2. **COLLECT THE DATA** Collect two sets of data that represent the number of boys and the number of girls in your class born in the spring, summer, fall, and winter. Use a spreadsheet to make a multiple-line or -bar graph of the data. Justify your selection.

Using Data to Predict

▷ GET READY for the Lesson

TELEPHONE The circle graph shows the results of a survey in which children ages 8 to 12 were asked whether they have a television in their bedroom.

1. Can you tell how many were surveyed? Explain.

2. Describe how you could use the graph to predict how many students in your school have a television in their bedroom.

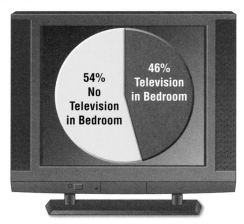

Source: National Institute on Media and the Family

A **survey** is designed to collect data about a specific group of people, called the **population**. If a survey is conducted at random, or without preference, you can assume that the survey represents the population. In this lesson, you will use the results of randomly conducted surveys to make predictions about the population.

🌐 Real-World EXAMPLE

1 **TELEVISION** Refer to the graphic above. **Predict how many out of 1,725 students would not have a television in their bedrooms.**

You can use the percent proportion and the survey results to predict what part p of the 1,725 students have no TV in their bedrooms.

part of the population →

whole population →

$$\frac{p}{1,725} = \frac{54}{100} \Big\} \text{ Survey results: 54\%}$$

$p \cdot 100 = 1,725 \cdot 54$ Find the cross products.

$100p = 93,150$ Simplify.

$\dfrac{100p}{100} = \dfrac{93,150}{100}$ Divide each side by 100.

$p = 931.5$ Simplify.

About 932 students do not have a television in their bedrooms.

✓ CHECK Your Progress

a. **TELEVISION** Refer to the same graphic. Predict how many out of 1,370 students have a television in their bedrooms.

Real-World Link . . . .
A survey found that 85% of people use emoticons on their instant messengers.

2 **INSTANT MESSAGING** Use the information at the left to predict how many of the 2,450 students at Washington Middle School use emoticons on their instant messengers.

You need to predict how many of the 2,450 students use emoticons.

Words	What number of students is 85% of 2,450 students?
Variable	Let n represent the number of students.
Equation	$n = 0.85 \cdot 2,450$

$n = 0.85 \cdot 2,450$ Write the percent equation.

$n = 2,082.5$ Multiply.

About 2,083 of the students use emoticons.

CHECK Your Progress

b. **INSTANT MESSAGING** This same survey found that 59% of people use sound on their instant messengers. Predict how many of the 2,450 students use sound on their instant messengers.

CHECK Your Understanding

Example 1
(p. 434)

SPENDING For Exercises 1 and 2, use the circle graph that shows the results of a poll to which 60,000 teens responded.

1. How many of the teens surveyed said that they would save their money?

2. Predict how many of the approximately 28 million teens in the United States would buy a music CD if they were given $20.

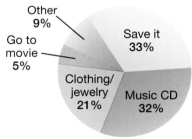

How Would You Spend a Gift of $20?

Other 9%
Go to movie 5%
Save it 33%
Clothing/ jewelry 21%
Music CD 32%

Source: *USA WEEKEND*

Example 2
(p. 435)

FOOD For Exercises 3 and 4, use the bar graph that shows the results of a survey in which students at Vail Middle School were asked their favorite ice cream flavor.

3. Out of 538 students at Vail Middle School, predict how many prefer strawberry ice cream.

4. Predict how many students prefer chocolate ice cream.

Favorite Ice Cream Flavor

Chocolate 19%
Vanilla 61%
Strawberry 11%
Butter Pecan 9%

Flavor

0 10 20 30 40 50 60 70
Percent

RECREATION In a survey, 250 people from a town were asked if they thought the town needed a recreation center. The results are shown in the table.

Recreation Center Needed	
Response	**Percent**
yes	44%
no	38%
undecided	18%

5. Predict how many of the 3,225 people in the town think a recreation center is needed.

6. About how many of the people would be undecided?

7. **MP3 PLAYERS** In a survey, 82% of teens said they own an MP3 player. Predict how many of 346,000 teens do *not* own an MP3 player.

8. **VOLUNTEERING** A survey showed that 90% of teens donate money to a charity during the holidays. Based on that survey, how many teens in a class of 400 will donate money the next holiday season?

Match each situation with the appropriate equation or proportion.

9. 27% of MP3 owners download music weekly. Predict how many MP3 owners out of 238 owners download music weekly.

10. 27 MP3s is what percent of 238 MP3s?

11. 238% of 27 is what number?

a. $n = 27 \cdot 2.38$

b. $\dfrac{27}{100} = \dfrac{p}{238}$

c. $\dfrac{27}{238} = \dfrac{n}{100}$

CATS For Exercises 12 and 13, use the graph that shows the percent of cat owners who train their cats in each category.

12. Out of 255 cat owners, predict how many owners trained their cat not to climb on furniture.

13. Out of 316 cat owners, predict how many more cat owners have trained their cat not to claw on furniture than have trained their cat not to fight with other animals.

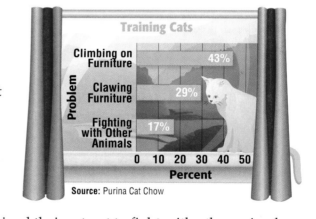

14. **FIND THE DATA** Refer to the Data File on pages 16–19. Choose some data and write a real-world problem in which you could use the percent proportion or percent equation to make a prediction.

H.O.T. Problems

15. **CHALLENGE** A survey found that 80% of teens enjoy going to the movies in their free time. Out of 5,200 teens, predict how many said that they do not enjoy going to the movies in their free time.

16. **OPEN ENDED** Select a newspaper or magazine article that contains a table or graph. Identify the population and explain how you think the results were found.

17. **SELECT A TOOL** A survey showed that 15% of the people in Tennessee over the age of 16 belong to a fitness center. Predict how many of the 5,900,962 people in Tennessee over the age of 16 belong to a fitness center. Select one or more of the following tools to solve the problem. Justify your selection(s).

| make a model | calculator | paper/pencil | real objects |

18. **WRITING IN MATH** Explain how to use a sample to predict what a group of people prefer. Then give an example of a situation in which it makes sense to use a sample.

19. The table shows how students spend time with their family.

How Students Spend Time with Family	
Dinner	34%
TV	20%
Talking	14%
Sports	14%
Taking Walks	4%
Other	14%

Source: Kids USA Survey

Of the 515 students surveyed, predict about how many spend time with their family at dinner.

A 17   C 119

B 34   D 175

20. Yesterday, a bakery baked 54 loaves of bread in 20 minutes. Today, the bakery needs to bake 375 loaves of bread. At this rate, predict how long it will take to bake the bread.

F 1.5 hours   H 3.0 hours

G 2.3 hours   J 3.75 hours

21. Of the 357 students in a freshman class, about 82% plan to go to college. How many students plan on going to college?

A 224   C 314

B 293   D 325

Spiral Review

RUNNING For Exercises 22–24, refer to the table that shows the time it took Dale to run each mile of a 5-mile run.

Mile	Time
1	4 min 19 s
2	4 min 28 s
3	4 min 39 s
4	4 min 54 s
5	5 min 1 s

22. Make a scatter plot of the data. (Lesson 8-6)

23. Describe the relationship, if any, between the two sets of data. (Lesson 8-6)

24. Suppose the trend continues. Predict the time it would take Dale to run a sixth mile. (Lesson 8-5)

Multiply. (Lesson 2-6)

25. -4×6

26. $5 \times (-8)$

27. $-6 \times (-9)$

28. 8×3

GET READY for the Next Lesson

PREREQUISITE SKILL Simplify. (Lesson 1-4)

29. $\dfrac{10 - 7}{10}$

30. $\dfrac{50 - 18}{50}$

31. $\dfrac{22 - 4}{4}$

32. $\dfrac{39 - 15}{15}$

MAIN IDEA

Predict the actions of a larger group by using a sample.

NYS Core Curriculum

7.S.9 Determine the validity of sampling methods to predict outcomes   7.S.10 Predict the outcome of an experiment *Also addresses 7.S.1, 7.S.6, 7.S.8*

New Vocabulary

sample
unbiased sample
simple random sample
biased sample
convenience sample
voluntary response sample

NY Math Online

glencoe.com

• Extra Examples
• Personal Tutor
• Self-Check Quiz

▶ **GET READY** for the Lesson

CELL PHONES The manager of a local cell phone company wants to conduct a survey to determine what kind of musical ring tones people typically use.

What Kind of Musical Ring Tone Do You Use?
Classical
Rock
Rap/Hip-Hop
Dance
Other

1. Suppose she decides to survey the listeners of a rock radio station. Do you think the results would represent the entire population? Explain.

2. Suppose she decides to survey a group of people standing in line for a symphony. Do you think the results would represent the entire population? Explain.

3. Suppose she decides to mail a survey to every 100th household in the area. Do you think the results would represent the entire population? Explain.

The manager of the cell phone company cannot survey everyone. A smaller group called a **sample** is chosen. A sample should be representative of the population.

Population	Sample
United States citizens	registered voters
California residents	homeowners
Six Flags Marine World visitors	teenagers

For valid results, a sample must be chosen very carefully. An **unbiased sample** is selected so that it is representative of the entire population. A simple random sample is the most common type of unbiased sample.

Unbiased Samples		Concept Summary
Type	**Description**	**Example**
Simple Random Sample	Each item or person in the population is as likely to be chosen as any other.	Each student's name is written on a piece of paper. The names are placed in a bowl, and names are picked without looking.

 **Vocabulary Link** · · · · ·

Bias

Everyday Use a tendency or prejudice.

Math Use error introduced by selecting or encouraging a specific outcome.

· · · ·In a **biased sample**, one or more parts of the population are favored over others. Two ways to pick a biased sample are listed below.

Biased Samples		Concept Summary
Type	**Description**	**Example**
Convenience Sample	A convenience sample includes members of a population who are easily accessed.	To represent all the students attending a school, the principal surveys the students in one math class.
Voluntary Response Sample	A voluntary response sample involves only those who want to participate in the sampling.	Students at a school who wish to express their opinion are asked to complete an online survey.

EXAMPLES Determine Validity of Conclusions

Determine whether each conclusion is valid. Justify your answer.

1 **To determine what kind of movies people like to watch, every tenth person who walks into a video rental store is surveyed. The store carries all kinds of movies. Out of 180 customers surveyed, 62 stated that they prefer action movies. The store manager concludes that about a third of all customers prefer action movies.**

The conclusion is valid. Since the population is every tenth customer of a video rental store, the sample is an unbiased random sample.

2 **A television program asks its viewers to visit a Web site to indicate their preference for two presidential candidates. 76% of the viewers who responded preferred candidate A, so the television program announced that most people prefer candidate A.**

The conclusion is not valid. The population is restricted to viewers who have Internet access, it is a voluntary response sample, and is biased. The results of a voluntary response sample do not necessarily represent the entire population.

✓CHECK Your Progress

Determine whether each conclusion is valid. Justify your answer.

a. To determine what people like to do in their leisure time, people at a local mall are surveyed. Of these, 82% said they like to shop. The mall manager concludes that most people like to shop during their leisure time.

b. To determine what kind of sport junior high school students like to watch, 100 students are randomly selected from each of four junior high schools in a city. Of these, 47% like to watch football. The superintendent concludes that about half of all junior high students like to watch football.

In Lesson 8-7, you used the results of a random sampling method to make predictions. In this lesson, you will first determine if a sampling method is valid and if so, use the results to make predictions.

Study Tip

Misleading Predictions
Predictions based on biased samples can be misleading. If the students surveyed were all boys, the predictions generated by the survey would not be valid, since both girls and boys attend the junior high school.

Real-World EXAMPLE

③ MASCOTS The Student Council at a new junior high school surveyed 5 students from each of the 10 homerooms to determine what mascot students would prefer. The results of the survey are shown at the right. If there are 375 students at the school, predict how many students prefer a tiger as the school mascot.

Mascot	Number
Tornadoes	15
Tigers	28
Twins	7

The sample is an unbiased random sample since students were randomly selected. Thus, the sample is valid.

$\frac{28}{50}$ or 56% of the students prefer a tiger. So, find 56% of 375.

$0.56 \times 375 = 210$ 56% of 375 = 0.56 × 375

So, about 210 students would prefer a tiger as the school mascot.

✓ CHECK Your Progress

c. **AIRLINES** During flight, a pilot determined that 20% of the passengers were traveling for business and 80% were traveling for pleasure. If there are 120 passengers on the next flight, how many can be expected to be traveling for pleasure?

✓ CHECK Your Understanding

Examples 1, 2
(p. 439)

Determine whether each conclusion is valid. Justify your answer.

1. To determine the number of umbrellas the average household in the United States owns, a survey of 100 randomly selected households in Arizona is conducted. Of the households, 24 said that they own 3 or more umbrellas. The researcher concluded that 24% of the households in the United States own 3 or more umbrellas.

2. A researcher randomly surveys ten employees from each floor of a large company to determine the number of employees who carpool to work. Of these, 31% said that they carpool. The researcher concludes that most employees do not carpool.

Example 3
(p. 440)

3. **LUNCH** Jared randomly surveyed some students to determine their lunch habits. The results are shown in the table. If there are 268 students in the school, predict how many bring their lunch from home.

Lunch Habit	Number
Bring Lunch from Home	19
Buy Lunch in the Cafeteria	27
Other	4

HOMEWORK HELP	
For Exercises	**See Examples**
4–9	1, 2
10, 11	3

Determine whether each conclusion is valid. Justify your answer.

4. The principal of a high school randomly selects 50 students to participate in a school improvement survey. Of these, 38 said that more world language courses should be offered. As a result, the principal decides to offer both Japanese and Italian language classes.

5. To evaluate their product, the manufacturer of light bulbs inspects the first 50 light bulbs produced on one day. Of these, 2 are defective. The manufacturer concludes that about 4% of light bulbs produced are defective.

6. To evaluate its service, a restaurant asks its customers to call a number and complete a telephone survey. The majority of those who replied said that they prefer broccoli instead of carrots as the vegetable side dish. As a result, the restaurant decides to offer broccoli instead of carrots.

7. To determine which type of pet is preferred by most customers, the manager of a pet store surveys every 15th customer that enters the store.

8. To determine which school dance theme most students favor, 20 students from each grade level at Lakewood Middle School are surveyed. The results are shown in the table. Based on these results, the student council decides that the dance theme should be *Unforgettable*.

Theme	Number
Starry Night	23
Unforgettable	26
At the Hop	11

9. To determine whether 15 boxes of porcelain tea sets have not been cracked during shipping, the owner of an antique store examines the first two boxes. None of the tea sets have been cracked, so the owner concludes that none of the tea sets in the remaining boxes are cracked.

10. **LAWNS** A researcher randomly surveyed 100 households in a small community to determine the number of households that use a professional lawn service. Of these, 27% of households use a professional lawn service. If there are 786 households in the community, how many can be expected to use a professional lawn service?

11. **PASTA** A grocery store asked every 20th person entering the store what kind of pasta they preferred. The results are shown in the table. If the store decides to restock their shelves with 450 boxes of pasta, how many boxes of lasagna should they order?

Pasta	Number
Macaroni	38
Spaghetti	56
Rigatoni	12
Lasagna	44

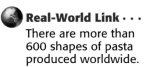

Real-World Link · · · ·
There are more than 600 shapes of pasta produced worldwide.

12. **FURNITURE** The manager of a furniture store asks the first 25 customers who enter the store if they prefer dining room tables made of oak, cherry, or mahogany wood. Of these, 17 said they prefer cherry. If the store manager orders 80 dining room tables in the next shipment, how many should be made of cherry wood?

13. **RADIO** A radio station asks its listeners to dial one of two numbers to indicate their preference for one of two candidates in an upcoming election. Of the responses received, 76% favored candidate A. If there are 1,500 registered voters, how many will vote for candidate A?

14. **HOBBIES** Pedro wants to conduct a survey about the kinds of hobbies that sixth graders enjoy. Describe a valid sampling method he could use.

AMUSEMENT PARKS For Exercises 15 and 16, use the following information.

The manager of an amusement park mailed 2,000 survey forms to households near the park. The results of the survey are shown in the graph at the right.

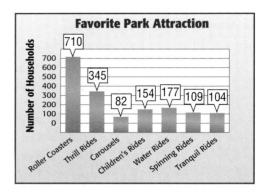

15. Assume the survey is valid. If there are 5,000 park visitors, about how many would prefer water rides?

16. Based on the survey results, the manager concludes that about 36% of park visitors prefer roller coasters. Is this a valid conclusion? Explain.

INTERNET For Exercises 17–19, use the following information.

A survey is to be conducted to find out how many hours students at a school spend on the Internet each weeknight. Describe the sample and explain why each sampling method might not be valid.

17. Ten randomly selected students from each grade level are asked to keep a log during their winter vacation.

18. Randomly selected parents are mailed a questionnaire and asked to return it.

19. A questionnaire is handed out to all students on the softball team.

COMPARE SAMPLES For Exercises 20–23, use the following information.

Suppose you were asked to determine the approximate percent of students in your school who are left-handed without surveying every student in the school.

20. Describe three different samples of the population that you could use to approximate the percent of students who are left-handed.

21. Would you expect the percent of left-handed students to be the same in each of these three samples? Explain your reasoning.

22. Describe any additional similarities and differences in your three samples.

23. You could have surveyed every student in your school to determine the percent of students who are left-handed. Describe a situation in which it makes sense to use a sample to describe aspects of a population instead of using the entire population.

24. **FIND THE DATA** Refer to the Data File on pages 16–19. Choose some data and write a real-world problem in which you would make a prediction based on samples.

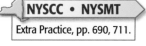
NYSCC • NYSMT
Extra Practice, pp. 690, 711.

25. CHALLENGE Is it possible to create an unbiased random sample that is also a convenience sample? Explain and cite an example, if possible.

26. WRITING IN MATH Explain why the way in which a survey question is asked might influence the results that are obtained. Cite at least two examples in your explanation.

NYSMT PRACTICE 7.S.9, 7.S.10

27. Yolanda wants to conduct a survey to determine what type of salad dressing is preferred by most students at her school. Which of the following methods is the best way for her to choose a random sample of the students at her school?

A Select students in her math class.

B Select members of the Spanish Club.

C Select ten students from each homeroom.

D Select members of the girls basketball team.

28. The manager of a zoo wanted to know which animals are most popular among visitors. She surveyed every 10th visitor to the reptile exhibit. Of these, she found that 75% like snakes. If there are 860 visitors to the zoo, which of the following claims is valid?

F About 645 zoo visitors like snakes.

G The reptile exhibit is the most popular exhibit.

H 25% of zoo visitors prefer mammals.

J No valid prediction can be made since the sample is a convenience sample.

Spiral Review

29. SCHOOL In a survey of 120 randomly selected students at Jefferson Middle School, 34% stated that science was their favorite class. Predict how many of the 858 students in the school would choose science as their favorite class. (Lesson 8-7)

30. HEALTH Use the scatter plot at the right to predict the height of a 16-year-old. (Lesson 8-6)

31. SHOPPING Nora bought a pair of running shoes that was discounted 35%. If the original price of the shoes was $89.90, find the discounted price to the nearest cent. (Lesson 7-7)

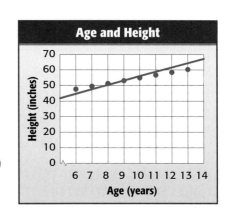

Write each percent as a fraction in simplest form. (Lesson 6-9)

32. 17%

33. 62.5%

34. 12.8%

▷ **GET READY for the Next Lesson**

PREREQUISITE SKILL Determine whether each statement is *true* or *false*. (Lesson 8-6)

35. The vertical scale on a line graph must have equal intervals.

36. You do not need to label the axes of a line graph.

Misleading Statistics

MAIN IDEA

Recognize when statistics and graphs are misleading.

NYS Core Curriculum

7.S.7 Identify and explain misleading statistics and graphs
7.R.7 Investigate relationships between different representations and their impact on a given problem

NY Math Online

glencoe.com

• Extra Examples
• Personal Tutor
• Self-Check Quiz

▶ **GET READY for the Lesson**

HOCKEY The graph shows the all-time Stanley Cup playoff leaders.

1. According to the size of the hockey players, how many times more points does Mark Messier appear to have than Jari Kurri? Explain.

2. Do you think this graph is representative of the players' number of points?

Source: *ESPN Sports Almanac*

Graphs let readers analyze and interpret data easily, but are sometimes drawn to influence conclusions by misrepresenting the data. The use of different scales can influence conclusions drawn from graphs.

EXAMPLE Changing the Interval of Graphs

1 **SCHOOL DANCES** The graphs show how the price of spring dance tickets increased.

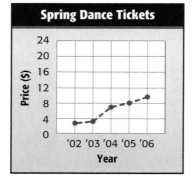

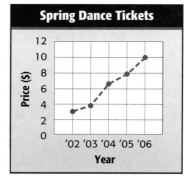

Do the graphs show the same data? If so, explain how they differ.
The graphs show the same data. However, the graphs differ in that Graph A uses an interval of 4, and Graph B uses an interval of 2.

Which graph makes it appear that the prices increased more rapidly?
Graph B makes it appear that the prices increased more rapidly even though the price increase is the same.

Which graph might Student Council use to show that while ticket prices have risen, the increase is not significant? Explain. They might use Graph A. The scale used on the vertical axis of this graph makes the increase appear less significant.

✓ CHECK Your Progress

a. **BUSINESS** The line graphs show monthly profits of a company from October to March. Which graph suggests that the business is extremely profitable? Is this a valid conclusion? Explain.

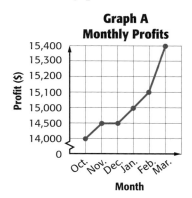

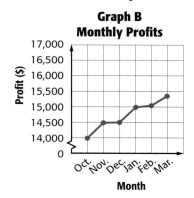

Sometimes the data used to create the display comes from a biased sample. In these cases, the data and the display are both biased and should be considered invalid.

EXAMPLE   Identify Biased Displays

② **FITNESS** The president of a large company mailed a survey to 500 of his employees in order to determine if they use the fitness room at work. The results are shown in the graph. Identify any sampling errors and explain why the sample and the display might be biased.

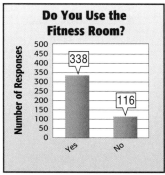

Not all of the surveys were returned since 338 + 116 < 500. This is a biased, voluntary response sample. The sample is not representative of the entire population since only those who wanted to participate in the survey are involved in the sampling.

The display is biased because the data used to create the display came from a biased sample.

b. **MOVIES** The manager of a movie theater asked 100 of his customers what they like to do on a Saturday night. The results are shown in the graph. Identify any sampling errors and explain why the sample and the display might be biased.

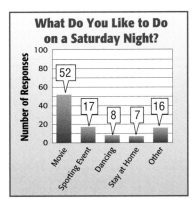

Statistics can also be used to influence conclusions.

EXAMPLE Misleading Statistics

3 **MARKETING** Refer to the table that gives the height of roller coasters at an amusement park. The park boasts that the average height of their roller coasters is 170 feet. Explain how this is misleading.

Park Rollercoaster Heights	
Coaster	**Height (ft)**
Viper	109
Monster	135
Red Zip	115
Tornado	365
Riptide	126

Real-World Link . . . .
The tallest roller coaster in the world is the Kingda Ka in Jackson, New Jersey, with a height of 456 feet.

Source: Ultimate Roller Coaster

mean: $\dfrac{109 + 135 + 115 + 365 + 126}{5} = \dfrac{850}{5}$
$= 170$

median: 109, 115, (126), 135, 365

mode: none

The average used by the park was the mean. This measure is much greater than most of the heights listed because of the outlier, 365 feet. So, it is misleading to use this measure to attract visitors.

A more appropriate measure to describe the data would be the median, 126 feet, which is closer to the height of most of the coasters.

 CHECK Your Progress

c. **FOOD** A restaurant claims its average menu price is $3.50. Use the table to explain how this is misleading.

Menu	
Burger	$4.00
Fish Sandwich	$4.45
Chicken Sandwich	$4.35
Garden Salad	$3.90
Coffee	$0.80

Example 1
(pp. 444–445)

1. **BASEBALL** Refer to the graphs below. Which graph suggests that Cy Young had three times as many wins as Jim Galvin? Is this a valid conclusion? Explain.

Graph A

Most Career Wins by a Pitcher

Pitchers: Cy Young, Walter Johnson, Grover Alexander, Christy Mathewson, Jim Gavin

Wins: 0, 200, 400, 600

Graph B

Most Career Wins by a Pitcher

Pitchers: Cy Young, Walter Johnson, Grover Alexander, Christy Mathewson, Jim Gavin

Wins: 350, 400, 450, 500, 550

Example 2
(pp. 445–446)

2. **PHONES** The manager of a telephone company mailed a survey to 400 households asking each household how they prefer to pay their monthly bill. The results are shown in the graph at the right. Identify any sampling errors and explain why the sample and the display might be biased.

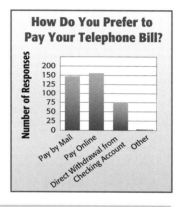

How Do You Prefer to Pay Your Telephone Bill?

Number of Responses: 200, 175, 150, 125, 100, 75, 50, 25, 0

Pay by Mail, Pay Online, Direct Withdrawal from Checking Account, Other

Example 3
(p. 446)

3. **TUNNELS** The table lists the five largest land vehicle tunnels in the U.S. Write a convincing argument for which measure of central tendency you would use to emphasize the average length of the tunnels.

U.S. Vehicle Tunnels on Land	
Name	**Length (ft)**
Anton Anderson Memorial	13,300
E. Johnson Memorial	8,959
Eisenhower Memorial	8,941
Allegheny	6,072
Liberty Tubes	5,920

Practice and Problem Solving

HOMEWORK HELP

For Exercises	See Examples
4, 8	1
5, 9	2
6, 7	3

4. **GAS** The bar graph shows monthly gas prices for 2006–2007. Why is the graph misleading?

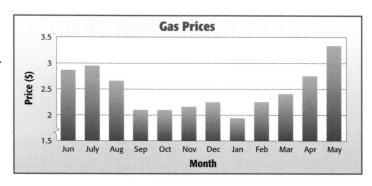

Gas Prices

Price ($): 3.5, 3, 2.5, 2, 1.5

Month: Jun, July, Aug, Sep, Oct, Nov, Dec, Jan, Feb, Mar, Apr, May

5. **SCHOOL** To determine how often his students are tardy, Mr. Kessler considered his first period class. The results are shown in the graph at the right. Identify any sampling errors and explain why the sample and the display might be biased.

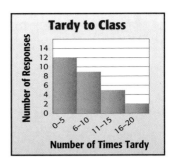

Tardy to Class

Number of Responses / *Number of Times Tardy* (0–5, 6–10, 11–15, 16–20)

TRAVEL For Exercises 6 and 7, use the table.

6. Find the mean, median, and mode of the data. Which measure might be misleading in describing the average annual number of visitors who visit these sights? Explain.

7. Which measure would be best if you wanted a value close to the most number of visitors? Explain.

Annual Sight-Seeing Visitors	
Sight	**Visitors***
Cape Cod	4,600,000
Grand Canyon	4,500,000
Lincoln Memorial	4,000,000
Castle Clinton	4,600,000
Smoky Mountains	10,200,000

Source: *The World Almamac*
*Approximation

8. **STOCK** The graphs below show the increases and decreases in the monthly closing prices of Skateboard Depot's stock.

Graph A

Monthly Stock Prices

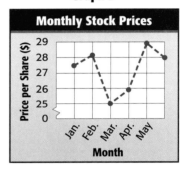

Graph B

Monthly Stock Prices

Suppose you are a stockbroker and want to show a customer that the price of the stock has been fairly stable since January. Write a convincing argument as to which graph you should show the customer.

9. **MANUFACTURING** To evaluate their product, the manager of an assembly line inspects the first 100 batteries that are produced out of 30,000 total batteries produced that day. He displays the results in the graph at the right and then releases it to the local newspaper. Identify any sampling errors and explain why the sample and the display might be biased.

Battery Production

Percent of Batteries / *Product*
95 — Excellent Condition
3 — Does Not Hold a Charge
2 — Broken or Cracked Casing

APARTMENTS For Exercises 10 and 11, create a display that would support each argument given the monthly costs to rent an apartment for the last five years are $500, $525, $560, $585, and $605.

NYSCC • NYSMT

Extra Practice, pp. 690, 711.

10. Rent has remained fairly stable.

11. Rent has increased dramatically.

Lesson-by-Lesson Review

8-1 **Line Plots** (pp. 396–401)

7.S.6,
7.S.4

Display each set of data in a line plot. Identify any clusters, gaps, or outliers.

9. 10°, 12°, 10°, 8°, 13°, 10°, 8°, 12°.

10. 7 ft, 8 ft, 8 ft , 9 ft, 14 ft, 9 ft, 8 ft, 7 ft

11. Number of Calories: 43, 41, 42, 45, 43, 42, 43, 46, 44, 44

Example 1 Display the test scores 72, 75, 72, 74, 73, 68, 73, 74, 74, 75, and 73 in a line plot. Identify any clusters, gaps, or outliers.

Test Scores

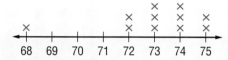

There is a cluster from 72 to 75, a gap between 68 and 72, and an outlier at 68.

8-2 **Measures of Central Tendency and Range** (pp. 402–408)

7.S.6,
7.R.7

Find the mean, median, and mode. Round to the nearest tenth if necessary.

12. Number of siblings: 2, 3, 4, 3, 4, 3, 8, 7, 2

13. 89°, 76°, 93°, 100°, 72°, 86°, 74°

14. **MONEY** Which measure, mean, median, mode, or range best represents the amount of money students spent on clothing?

Money Spent ($)			
125	108	172	136
121	112	218	172

Example 2 Find the mean, median, and mode for the following college students' ages: 23, 22, 19, 19, and 20.

mean: $\dfrac{23 + 22 + 19 + 19 + 20}{5}$ or 20.6 years

median: 20, the middle value of the ordered set

mode: 19, the data value that occurs most often

8-3 **Stem-and-Leaf Plots** (pp. 410–414)

7.S.6,
7.A.7

Display each set of data using a stem-and-leaf plot.

15. Hours worked: 29, 54, 31, 26, 38, 46, 23, 21, 32, 37

16. Number of points: 75, 83, 78, 85, 87, 92, 78, 53, 87, 89, 91

17. Birth dates: 9, 5, 12, 21, 18, 7, 16, 24, 11, 10, 3, 14

Example 3 Display the number of pages read 12, 15, 17, 20, 22, 22, 23, 25, 27, and 35 in a stem-and-leaf plot.

The tens digits form the stems, and the ones digits form the leaves.

Pages Read

Stem	Leaf
1	2 5 7
2	0 2 2 3 5 7
3	5

2|3 = 23 pages

8-4 Bar Graphs and Histograms (pp. 415–421)

7.S.6,
7.S.1

ATTENDANCE For Exercises 18–20, refer to the graph.

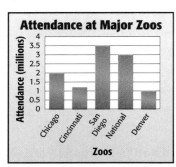

18. Which zoo did the most people attend?

19. About what was the total attendance for all five zoos?

20. Write a statement comparing the attendance at the National Zoo to the attendance at the Denver Zoo.

Example 4 The graph shows the number of bordering states for each of the fifty states.

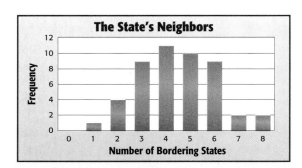

What is the greatest number of states that border any state?

8 states

How many states are bordered by 6 states?

9 states

8-5 PSI: Use a Graph (pp. 424–425)

7.PS.6,
7.PS.17

STATUES For Exercises 21 and 22, use the graph that shows the heights of free-standing statues in the world.

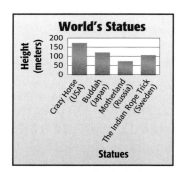

21. Which statue is the tallest?

22. Compare the height of the Motherland statue to the height of the Crazy Horse statue.

Example 5 The graph shows the results of a survey about favorite vacation places.

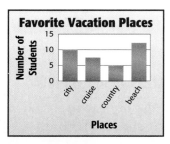

Which place was favored by most students?

The beach was favored by 12 students, which was the greatest number.

Mixed Problem Solving
For mixed problem-solving practice,
see page 711.

8-6

7.S.6

Using Graphs to Predict (pp. 426–431)

PHONE CALLS For Exercises 23 and 24, use the graph showing the number of people in a family and the number of weekly calls.

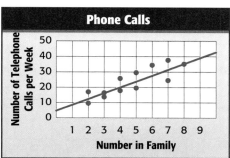

Phone Calls

23. Describe the relationship between the two sets of data.

24. Predict the number of weekly phone calls for a family of 10.

Example 6 The scatter plot below shows the keyboarding speeds in words per minute of 12 students.

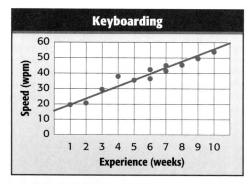

Keyboarding

Describe the relationship between the two sets of data.

The graph shows a positive relationship. That is, as the weeks pass, speed increases.

8-7

7.S.6

Using Data to Predict (pp. 434–437)

CAREERS For Exercises 25 and 26, use the table that shows the results of a university survey of incoming freshmen.

Career Goal	Percent
Elementary teacher	5.5%
Engineer	6.4%

25. Predict how many of the 3,775 freshmen would choose a career as an elementary teacher.

26. How many of the 3,775 freshmen would you expect to choose a career as an engineer?

27. **SHOES** A survey showed that 72% of teens bought new athletic shoes for the new school year. Based on that survey, how many teens in group of 225 bought new athletic shoes for the new school year?

Example 7 The circle graph shows the results of a survey to which 150 students at McAuliffe Middle School responded. Predict how many of the 644 students at the school have after-school jobs.

**Do You Have an
After-School Job?**

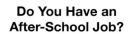

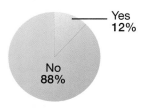

Yes
12%

No
88%

Find 12% of 644.

$n = 0.12 \cdot 644$ Write an equation.

$\quad = 77.28$ Multiply.

So, you could predict that about 77 students at McAuliffe Middle School have after-school jobs.

8-8 **Using Sampling to Predict** (pp. 438–443)

7.S.9,
7.S.10

Determine whether each conclusion is valid. Justify your answer.

28. To determine the number of vegetarians in a city, a restaurant owner surveys the first 50 customers who enter the restaurant. Of these, 6 said they are vegetarians, so the owner concludes that about 12% of the city's population are vegetarians.

29. The principal of a junior high school randomly surveys 40 students from each grade level to determine how many students are interested in after-school tutoring. Of these, 88% are interested, so the principal decides to offer after-school tutoring.

30. **BOOKS** The owner of a bookstore surveyed every 10th person that entered the store to determine her customers' preferred type of book. Of these, 32% preferred mysteries. If the owner will order 500 new books, about how many should be mysteries?

Example 8 To determine the preference of her customers, a florist mails surveys to 100 of her customers. The results are shown in the table. Based on these results, the florist decides to stock more roses.

Type of Flower	Number
Roses	45
Tulips	26
Lilacs	17

This conclusion is invalid. This is a biased voluntary response sample. Not all surveys were returned.

8-9 **Misleading Statistics** (pp. 444–449)

7.S.7,
7.R.7

31. **UTILITIES** The line graph shows the monthly electric bill for the condominium that Toshiko is interested in renting. Why is the graph misleading?

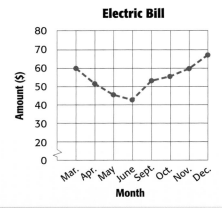

Example 9 The graph shows the number of motorcycles produced in each decade. Why might this graph be misleading?

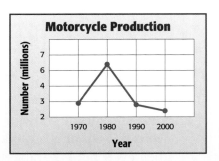

The scale does not start at zero and does not increase by the same amount each time.

For Exercises 1 and 2, use the line plot that shows the number of hours students spend listening to the radio per week.

Number of Radio Hours

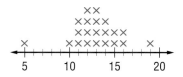

1. Identify any clusters, gaps, or outliers.

2. Describe how the range of data would change if 5 was not part of the data set.

3. **INSECTS** The lengths in inches of several insects are given below. Find the mean, median, and mode of the data set. Round to the nearest tenth if necessary.

    0.75, 1.24, 0.95, 2.6, 1.18, 1.3

Display each data set in a stem-and-leaf plot.

4. 37°, 59°, 26°, 42°, 57°, 53°, 31°, 58°

5. $461, $422, $430, $425, $425, $467, $429

6. **MULTIPLE CHOICE** Refer to the data below. Which of the following statements is true concerning the measures of central tendency?

    41, 45, 42, 38, 77, 44, 36, 43

    A The mode is most affected by the inclusion of the outlier.
    B The median is not affected by the inclusion of the outlier.
    C The mean is most affected by the inclusion of the outlier.
    D None of the measures of central tendency are affected by the inclusion of the outlier.

7. **GRADES** Make a histogram for the following French test grades: 95, 76, 82, 90, 83, 76, 79, 82, 95, 85, 93, 81, and 63.

8. **EMPLOYMENT** The line graph shows the percent of women who had jobs outside the home from 1975 to 2000. Use the graph to predict the number of women who will have jobs outside the home in 2010.

9. **AMUSEMENT PARKS** A researcher asked 250 students at Lake Valley Middle School to dial one of four telephone numbers to indicate their preference for the type of amusement park rides that they enjoy. Of these, 19% said they prefer the Ferris wheel. The researcher concludes that about $\frac{1}{5}$ of the students at Lake Valley Middle School prefer the Ferris wheel.

10. **MULTIPLE CHOICE** The line graph shows ship sales at Marvin's Marina in thousands of dollars. Which of the following statements best tells why the graph is misleading?

    F The graph's title is misleading.
    G The intervals on the horizontal scale are inconsistent.
    H The graph does not show any data.
    J The vertical axis is not labeled.

PART 1 Multiple Choice

Read each question. Then fill in the correct answer on the answer sheet provided by your teacher or on a sheet of paper.

1. Ed's Used Car Lot bought 5 used cars for $32,000. The business later bought another used car for $4,600. What was the mean cost of all of the used cars?

   A $3,200.00   **C** $6,100.00

   B $4,600.00   **D** $8,500.00

2. A fitness club charges a membership fee of $50 plus $25 each month you belong to the club. Which expression could be used to find the total cost of belonging to the club for 10 months?

   F $50(10) + 25$

   G $50 - 25(10)$

   H $50 + 25(10)$

   J $50(10) + 25(10)$

3. Sierra has 11.5 yards of fabric. She will use 20% of the fabric to make a flag. How many yards of fabric will she use?

   A 9.2 yd   **C** 4.5 yd

   B 8.6 yd   **D** 2.3 yd

4. Ms. Thompson made 17 liters of punch for a party. The punch contained 5 liters of orange juice. Which equation could be used to find y, the percent of orange juice in the punch?

   F $\dfrac{17}{5} = \dfrac{y}{100}$

   G $\dfrac{5}{17} = \dfrac{y}{100}$

   H $\dfrac{5}{17} = \dfrac{100}{y}$

   J $\dfrac{17}{y} = \dfrac{100}{5}$

5. The number of students in each grade level at Hampton Middle School is shown in the graph below.

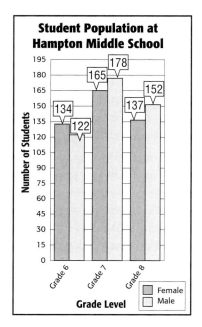

Which statement is true based on this information?

A There are more total students in the 6th grade than there are in the 8th grade.

B The female 7th graders outnumber the male 8th graders.

C The student population decreases by the same amount as each grade level increases.

D There are more female students in each grade level than male students.

6. A football team scored 20, 32, 28, 21, and 24 points in their first five games. How many points should they score in the next game so that the median and mode scores are equal?

   F 32   **H** 21

   G 24   **J** 20

7. Regina priced six MP3 players. The prices are shown below.

$120.00, $90.00, $75.00, $105.00, $85.00, $150.00

What is the median price?

A $90.00

B $97.50

C $104.17

D $105.00

8. Which of the following is the prime factorization of the lowest common denominator of $\frac{3}{8} + \frac{5}{6}$?

F 2×3

G $2 \times 2 \times 2$

H $2 \times 2 \times 2 \times 3$

J $2 \times 2 \times 2 \times 2 \times 3$

9. If a is a negative number and b is a negative number, which of the following expressions is always negative?

A $a + b$

B $a - b$

C $a \times b$

D $a \div b$

10. The numbers of monthly minutes Gary used on his cell phone for the last eight months are shown below.

400, 550, 450, 620, 550, 600, 475, 425

What is the mode of this data?

F 550 H 450

G 475 J 400

PART 2 Short Response/Grid In

Record your answers on the answer sheet provided by your teacher or on a sheet of paper.

11. Taylor spends between $125 and $200 per month on food. Which is the best estimate of how much she spends on food in 6 months?

12. A store has 2,545 CDs. On Saturday, the store sold $12\frac{1}{2}\%$ of the CDs. What fraction of the CDs were sold on Saturday?

PART 3 Extended Response

Record your answers on the answer sheet provided by your teacher or a sheet of paper. Show your work.

13. The table shows how values of a painting increased over ten years.

Year	Value	Year	Value
1997	$350	2002	$1,851
1998	$650	2003	$2,151
1999	$950	2004	$2,451
2000	$1,200	2005	$2,752
2001	$1,551	2006	$3,052

a. Make a line graph of the data.

b. Use the graph to predict what the value of the painting will be in 2010.

TEST-TAKING TIP

Question 13 If you find that you cannot answer every part of an open-ended question, do as much as you can. You may earn partial credit.

NEED EXTRA HELP?													
If You Missed Question...	1	2	3	4	5	6	7	8	9	10	11	12	13
Go to Lesson...	8-2	1-6	7-1	7-2	8-6	8-2	8-2	4-1	2-4	8-2	1-1	6-8	8-6
NY Core Curriculum	7.S.6	7.N.11	8.N.3	8.N.4	7.S.6	7.S.6	7.S.6	7N.8	7.N.13	7.S.6	7.PS.14	7.PS.10	7.S.6

CHAPTER 9

Probability

New York State Core Curriculum

7.S.8 Interpret data to provide the basis for predictions and to establish experimental probabilities

Key Vocabulary

compound event (p. 492)

independent event (p. 492)

probability (p. 460)

sample space (p. 465)

 Real-World Link

Sports Before a football game, there is a coin toss to determine which team gets the ball first. Probability tells you that each team has a 50% chance of winning the coin toss.

Study Organizer

Probability Make this Foldable to help you organize your notes. Begin with five sheets of $8\frac{1}{2}''$ by 11" paper.

① **Stack** five sheets of paper $\frac{3}{4}$ inch apart.

② **Roll** up bottom edges so that all tabs are the same size.

③ **Crease** and staple along the fold.

④ **Write** the chapter title on the front. Label each tab with a lesson number and title. Label the last tab *Vocabulary*.

GET READY for Chapter 9

Diagnose Readiness You have two options for checking Prerequisite Skills.

Option 2

NY Math Online Take the Online Readiness Quiz at glencoe.com.

Option 1

Take the Quick Quiz below. Refer to the Quick Review for help.

QUICK Quiz

Multiply. (Prior Grade)

1. 7×15
2. 24×6
3. 13×4
4. 8×21
5. 5×32
6. 30×8
7. $7 \times 6 \times 5$
8. $8 \times 7 \times 6$
9. $6 \times 5 \times 4 \times 3$
10. $4 \times 3 \times 2 \times 1$
11. $10 \times 9 \times 8 \times 7$
12. $11 \times 10 \times 9$

13. **JOBS** If you earn $9 an hour and work 5 hours each day for 7 days, how much have you earned? (Prior Grade)

Write each fraction in simplest form. Write *simplified* if the fraction is already in simplest form. (Lesson 4-4)

14. $\dfrac{8}{12}$
15. $\dfrac{3}{18}$
16. $\dfrac{4}{9}$
17. $\dfrac{5}{15}$

18. **SLEEP** If the average adult gets 8 hours of sleep, what fraction of the day, in simplest form, is spent asleep? (Lesson 4-4)

Find each value. (Prior Grade)

19. $\dfrac{6 \times 5}{3 \times 2}$
20. $\dfrac{9 \times 8 \times 7}{5 \times 4 \times 3}$
21. $\dfrac{4 \times 3 \times 2}{3 \times 2 \times 1}$
22. $\dfrac{7 \times 6 \times 5 \times 4}{4 \times 3 \times 2 \times 1}$

QUICK Review

Example 1 Multiply $7 \times 6 \times 5 \times 4$.

$$7 \times 6 \times 5 \times 4 = 42 \times 5 \times 4 \quad \text{Multiply from left to right.}$$
$$= 210 \times 4$$
$$= 840$$

Example 2 Write $\dfrac{21}{28}$ in simplest form.

$$\overset{\div 7}{\dfrac{21}{28}} = \dfrac{3}{4} \quad \begin{array}{l}\text{Divide the numerator and} \\ \text{denominator by the GCF, 7.}\end{array}$$
$$\div 7$$

Example 3 Find the value of $\dfrac{6 \times 5 \times 4}{3 \times 2 \times 1}$.

$$\dfrac{6 \times 5 \times 4}{3 \times 2 \times 1} = \dfrac{120}{6} \quad \begin{array}{l}\leftarrow \text{Multiply the numerator.} \\ \leftarrow \text{Multiply the denominator.}\end{array}$$
$$= 20 \quad \text{Simplify.}$$

Simple Events

MAIN IDEA

Find the probability of a simple event.

NYS Core Curriculum

Preparation for 7.S.9
Determine the validity of sampling methods to predict outcomes
Preparation for 7.S.10
Predict the outcome of an experiment

New Vocabulary

outcome
simple event
probability
random
complementary event

NY Math Online

glencoe.com

• Concepts In Motion
• Extra Examples
• Personal Tutor
• Self-Check Quiz

▶ **GET READY** for the Lesson

FOOD A cheesecake has four slices of each type as shown.

Cheesecake	
original	raspberry
chocolate	turtle

1. What fraction of the cheesecake is chocolate? Write in simplest form.

2. Suppose your friend gives you the first piece of cheesecake without asking which type you prefer. Are your chances of getting original the same as getting raspberry?

An **outcome** is any one of the possible results of an action. A **simple event** is one outcome or a collection of outcomes. For example, getting a piece of chocolate cheesecake is a simple event. The chance of that event happening is called its **probability**.

Probability   Key Concept

Words   If all outcomes are equally likely, the probability of a simple event is a ratio that compares the number of favorable outcomes to the number of possible outcomes.

Symbols $P(\text{event}) = \dfrac{\text{number of favorable outcomes}}{\text{number of possible outcomes}}$

EXAMPLE Find Probability

1 What is the probability of rolling an even number on a number cube marked with 1, 2, 3, 4, 5, and 6 on its faces?

$$P(\text{even number}) = \frac{\text{even numbers possible}}{\text{total numbers possible}}$$
$$= \frac{3}{6} \text{ or } \frac{1}{2}$$

The probability of rolling an even number is $\frac{1}{2}$, 0.5, or 50%.

✓ **CHECK Your Progress**

Use the number cube above to find each probability. Write as a fraction in simplest form.

a. $P(\text{odd number})$ **b.** $P(5 \text{ or } 6)$ **c.** $P(\text{prime number})$

Outcomes occur at **random** if each outcome occurs by chance. For example, rolling a number on a number cube occurs at random.

 Real-World EXAMPLE

2 **TALENT COMPETITION** Simone and her three friends were deciding how to pick the song they will sing for their school's talent show. They decide to roll a number cube. The person with the lowest number chooses the song. If her friends rolled a 6, 5, and 2, what is the probability that Simone will get to choose the song?

The possible outcomes of rolling a number cube are 1, 2, 3, 4, 5, and 6.

In order for Simone to be able to choose the song, she will need to roll a 1.

Let $P(A)$ be the probability that Simone chooses the song.

$$P(A) = \frac{\text{number of favorable outcomes}}{\text{number of possible outcomes}}$$

$$= \frac{1}{6} \qquad \text{There are 6 possible outcomes, and 1 of them is favorable.}$$

The probability that Simone will choose the song is $\frac{1}{6}$, or about 17%.

Real World Link . . . . .

Founded in 1842, the New York Philharmonic is the oldest symphony orchestra in the United States. They have performed over 14,000 concerts.

Source: New York Philharmonic

✓ **CHECK Your Progress**

MUSIC The table shows the numbers of brass instrument players in the New York Philharmonic. Suppose one brass instrument player is randomly selected to be a featured performer. Find the probability of each event. Write as a fraction in simplest form.

New York Philharmonic Brass Instrument Players	
Horn	6
Trombone	4
Trumpet	3
Tuba	1

Source: New York Philharmonic

d. $P(\text{trumpet})$ **e.** $P(\text{brass})$

f. $P(\text{flute})$ **g.** $P(\text{horn or tuba})$

The probability that an event will happen can be any number from 0 to 1, including 0 and 1, as shown on the number line below. Notice that probabilities can be written as fractions, decimals, or percents.

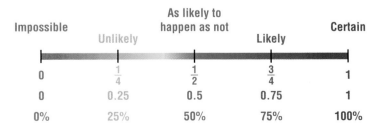

Either Simone will go first or she will *not* go first. These two events are **complementary events**. The sum of the probabilities of an event and its complement is 1 or 100%. In symbols, $P(A) + P(not\ A) = 1$.

Study Tip

Complement of an Event
The probability that event A will not occur is noted as P(not A). Since P(A) + P(not A) = 1, P(not A) = 1 − P(A). P(not A) is read as the probability of the complement of A.

EXAMPLE Complementary Events

 TALENT COMPETITION Refer to Example 2. Find the probability that Simone will *not* choose the song.

The probability that Simone will *not* choose the song is the complement of the probability that Simone will choose the song.

$$P(A) + P(not\ A) = 1 \qquad \text{Definition of complementary events}$$
$$\frac{1}{6} + P(not\ A) = 1 \qquad \text{Replace } P(A) \text{ with } \frac{1}{6}.$$
$$\underline{-\frac{1}{6} \qquad\qquad -\frac{1}{6}} \qquad \text{Subtract } \frac{1}{6} \text{ from each side.}$$
$$P(not\ A) = \frac{5}{6} \qquad 1 - \frac{1}{6} \text{ is } \frac{6}{6} - \frac{1}{6} \text{ or } \frac{5}{6}$$

The probability that Simone will *not* choose the song is $\frac{5}{6}$, or about 83%.

✓ **CHECK Your Progress**

SCHOOL Ramón's teacher uses a spinner similar to the one shown at the right to determine the order in which each group will make their presentation. Use the spinner to find each probability. Write as a fraction in simplest form.

h. $P(not\ \text{group } 4)$ **i.** $P(not\ \text{group } 1 \text{ or group } 3)$

✓ **CHECK Your Understanding**

Example 1
(p. 460)
Use the spinner to find each probability. Write as a fraction in simplest form.

1. $P(M)$ **2.** $P(Q \text{ or } R)$ **3.** $P(\text{vowel})$

Examples 2, 3
(pp. 461–462)
MARBLES Robert has a bag that contains 7 blue, 5 purple, 12 red, and 6 orange marbles. Find each probability if he draws one marble at random from the bag. Write as a fraction in simplest form.

4. $P(\text{purple})$ **5.** $P(\text{red or orange})$ **6.** $P(\text{green})$

7. $P(not\ \text{blue})$ **8.** $P(not\ \text{red or orange})$ **9.** $P(not\ \text{yellow})$

Example 3
(p. 462)
10. SURVEYS Shanté asked her classmates how many pets they own. The responses are in the table. If a student in her class is selected at random, what is the probability that the student does *not* own 3 or more pets?

Number of Pets	Response
None	6
1–2	15
3 or more	4

HOMEWORK HELP

For Exercises	See Examples
11–14	1
17–22	2
15–16, 23–26	3

A set of 20 cards is numbered 1, 2, 3, . . ., 20. Suppose you pick a card at random without looking. Find the probability of each event. Write as a fraction in simplest form.

11. $P(1)$

12. $P(3 \text{ or } 13)$

13. $P(\text{multiple of 3})$

14. $P(\text{even number})$

15. $P(not\ 20)$

16. $P(not\ \text{a factor of 10})$

RAFFLE The table shows those students in seventh grade who entered in the school drawing to win lunch with the principal. Suppose that only one student is randomly selected to win. Find the probability of each event. Write as a fraction in simplest form.

Lunch Raffle	
girls	25
boys	15
Room 10	10
Room 11	16
Room 12	14

17. $P(\text{girl})$

18. $P(\text{boy})$

19. $P(\text{Room 12})$

20. $P(\text{Room 10})$

21. $P(\text{girl or boy})$

22. $P(\text{Room 11})$

23. $P(not\ \text{Room 10})$

24. $P(\text{Room 10 or 11})$

25. SOUP A cupboard contains 20 soup cans. Seven are tomato, 4 are cream of mushroom, 5 are chicken, and 4 are vegetable. If one can is chosen at random from the cupboard, what is the probability that it is *neither* cream of mushroom *nor* vegetable soup? Write as a percent.

26. VIDEOS In a drawing, one name is randomly chosen from a jar of 75 names to receive free video rentals for a month. If Enola entered her name 8 times, what is the probability that she is *not* chosen to receive the free rentals? Write as a fraction in simplest form.

27. PETS The graph shows the last 33 types of pets that were purchased at a local pet store. Based on this, what is the probability that the next pet purchased will be a dog? Write as a fraction in simplest form.

28. GAMES For a certain game, the probability of choosing a card with the number 13 is 0.008. What is the probability of *not* choosing card 13?

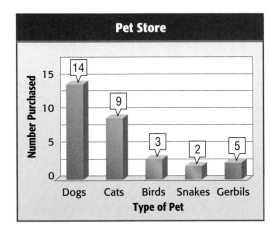

29. WEATHER The forecast for tomorrow says that there is a 37% chance of rain. Describe the complementary event and predict its probability.

30. ANNOUNCEMENTS There are 90 students in the seventh grade. Fifty-two of those students are girls. If one student will be chosen at random to read the morning announcements for the week, what is the probability that the student is a boy?

NYSCC • NYSMT
Extra Practice, pp. 690, 712

31. **REASONING** A *leap year* has 366 days and occurs in non-century years that are evenly divisible by 4. The extra day is added as February 29th. Determine whether each probability is 0 or 1. Explain your reasoning.

    a. P(there will be 29 days in February in 2032)

    b. P(there will be 29 days in February in 2058)

32. **CHALLENGE** A bag contains 6 red marbles, 4 blue marbles, and 8 green marbles. How many marbles of each color should be added so that the total number of marbles is 27, but the probability of randomly selecting one marble of each color remains unchanged? Explain your reasoning.

33. **Which One Doesn't Belong?** Identify the pair of probabilities that does not represent probabilities of complementary events. Explain your reasoning.

    | $\frac{3}{5}, \frac{2}{5}$ | 0.625, $\frac{3}{8}$ | $\frac{6}{8}, \frac{1}{4}$ | 0.33, 0.44 |

34. **WRITING IN MATH** Marissa has 5 black T-shirts, 2 purple T-shirts, and 1 orange T-shirt. Without calculating, determine whether each of the following probabilities is reasonable if she randomly selects one T-shirt. Explain your reasoning.

    a. P(black T-shirt) $= \frac{1}{3}$ b. P(orange T-shirt) $= \frac{4}{5}$ c. P(purple T-shirt) $= \frac{1}{4}$

NYSMT PRACTICE 7.S.9, 7.S.10

35. A bag contains 8 blue marbles, 15 red marbles, 10 yellow marbles, and 3 brown marbles. If a marble is randomly selected, what is the probability that it will be brown?

    A 0.27   **C** 0.08$\overline{3}$

    B 11%   **D** $\frac{3}{8}$

36. What is the probability of the spinner landing on a number less than 3?

    F 25%   **H** 50%

    G 37.5%   **J** 75%

Spiral Review

37. **RAIN** The scatter plot shows the relationship between rainfall and lawn growth. Why might the graph be misleading? (Lesson 8-9)

38. **PARKS** A researcher randomly selected 100 households near a city park. Of these, 26% said they visit the park daily. If there are 500 households near the park, about how many visit it daily? (Lesson 8-8)

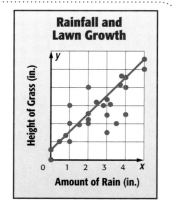

Rainfall and Lawn Growth

▷ **GET READY** for the Next Lesson

PREREQUISITE SKILL Write each fraction in simplest form. (Lesson 4-4)

39. $\frac{2}{6}$ 40. $\frac{6}{8}$ 41. $\frac{15}{30}$ 42. $\frac{6}{16}$ 43. $\frac{18}{32}$

9-2 Sample Spaces

MAIN IDEA

Find sample spaces and probabilities.

NYS Core Curriculum

7.S.8 Interpret data to provide the basis for predictions and to establish experimental probabilities

sample space tree diagram

NY Math Online

glencoe.com

- Extra Examples
- Personal Tutor
- Self-Check Quiz

▷ **MINI Lab**

Here is a probability game for two players.

- Place two green marbles into Bag A. Place one green and one red marble into Bag B.

- Without looking, player 1 chooses a marble from each bag. If both marbles are the same color, player 1 wins a point. If the marbles are different colors, player 2 wins a point. Record your results and place the marbles back in the bag.

- Player 2 then pulls a marble from each bag and records the results. Continue alternating turns until each player has pulled from the bag 10 times. The player with the most points wins.

1. Make a conjecture. Do you think this is a fair game? Explain.

2. Now, play the game. Who won? What was the final score?

The set of all of the possible outcomes in a probability experiment is called the **sample space**. A table or grid can be used to list the outcomes in the sample space.

EXAMPLE Find the Sample Space

① **ICE CREAM** A vendor sells vanilla and chocolate ice cream. Customers can choose from a waffle or sugar cone and either hot fudge or caramel topping. Find the sample space for all possible orders of one scoop of ice cream in a cone with one topping.

Make a table that shows all of the possible outcomes.

Outcomes		
Vanilla	Waffle	Hot Fudge
Vanilla	Waffle	Caramel
Vanilla	Sugar	Hot Fudge
Vanilla	Sugar	Caramel
Chocolate	Waffle	Hot Fudge
Chocolate	Waffle	Caramel
Chocolate	Sugar	Hot Fudge
Chocolate	Sugar	Caramel

✓ **CHECK Your Progress**

a. **PETS** The animal shelter has both male and female Labradors in yellow, brown, or black. Find the sample space for all possible Labradors available at the shelter.

A **tree diagram** can also be used to display the sample space.

 NYSMT EXAMPLE

② A car comes in silver, red, or purple as a convertible or hardtop. Which list shows all possible color-top outcomes?

A

Outcomes	
silver	convertible
silver	hardtop
red	convertible
red	hardtop
purple	convertible
purple	hardtop

C

Outcomes	
silver	convertible
red	hardtop
purple	convertible
silver	hardtop
purple	convertible

B

Outcomes	
silver	convertible
red	hardtop
purple	convertible

D

Outcomes	
silver	convertible
red	hardtop
purple	convertible
silver	hardtop

Test-Taking Tip

Educated Guess Find out if there is a penalty for incorrect answers. If there is no penalty, making an educated guess can only increase your score, or at worst, leave your score the same.

Read the Item

The car comes in 3 colors: silver, red, or purple, and 2 tops: convertible or hardtop. Find all of the color-top combinations.

Solve the Item

Make a tree diagram to show the sample space.

There are 6 different color-top combinations.
The answer is A.

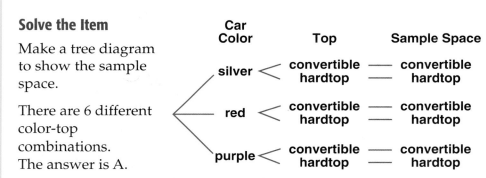

 CHECK Your Progress

b. For dessert, you can have apple or cherry pie with or without ice cream. Which list shows all the possible outcomes of choosing pie and ice cream?

F

Outcomes	
apple	with ice cream
cherry	without ice cream

H

Outcomes	
apple	without ice cream
cherry	with ice cream

G

Outcomes	
apple	with ice cream
apple	without ice cream
cherry	with ice cream
cherry	without ice cream

J

Outcomes	
apple	with ice cream
cherry	with ice cream
apple	without ice cream

You can use a table or a tree diagram to find the probability of an event.

EXAMPLE Find Probability

Study Tip

Fair Game A fair game is one in which each player has an equal chance of winning. This game is a fair game.

3 **GAMES** Refer to the Mini Lab at the start of this lesson. Find the sample space. Then find the probability that player 2 wins.

There are 4 equally likely outcomes with 2 favoring each player. So, the probability that player 2 wins is $\frac{2}{4}$, or $\frac{1}{2}$.

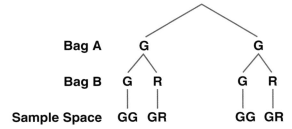

CHECK Your Progress

c. **GAMES** Marcos tosses three coins. If all three coins show up heads, Marcos wins. Otherwise, Kara wins. Find the sample space. Then find the probability that Marcos wins.

CHECK Your Understanding

Examples 1, 2
(pp. 465–466)

For each situation, find the sample space using a table or tree diagram.

1. A number cube is rolled twice.

2. A pair of brown or black sandals are available in sizes 7, 8, or 9.

Example 2
(p. 466)

3. **MULTIPLE CHOICE** Sandwiches can be made with ham or turkey on rye, white, or sourdough breads. Which list shows all the possible outcomes?

A

Outcomes	
ham	rye
turkey	white
ham	sourdough
ham	rye
turkey	white
turkey	sourdough

C

Outcomes	
ham	rye
turkey	rye
ham	white
turkey	white
ham	sourdough
turkey	sourdough

B

Outcomes	
ham	rye
turkey	white
turkey	sourdough

D

Outcomes	
ham	rye
turkey	white
ham	sourdough

Example 3
(p. 467)

4. **GAMES** Brianna spins a spinner with four sections of equal size twice, labeled A, B, C, and D. If letter A is spun at least once, Brianna wins. Otherwise, Odell wins. Find the probability that Odell wins.

HOMEWORK HELP

For Exercises	See Examples
5–12	1, 2
28	2
13–14	3

For each situation, find the sample space using a table or tree diagram.

5. tossing a coin, and spinning the spinner from the choices at the right

6. choosing left or right and either boots, gym shoes, or dress shoes

7. tossing a coin and rolling a number cube

8. picking a number from 1 to 5 and choosing the color red, white, or blue

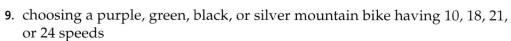

9. choosing a purple, green, black, or silver mountain bike having 10, 18, 21, or 24 speeds

10. choosing a letter from the word SPACE and choosing a consonant from the word MATH

11. **CLOTHES** Jerry can buy a school T-shirt with either short sleeves or long sleeves in either gray or white and in small, medium, or large. Find the sample space for all possible T-shirts he can buy.

12. **FOOD** Three-course dinners can be made from the menu shown. Find the sample space for a dinner consisting of an appetizer, entrée, and dessert.

Appetizers	Entrees	Desserts
Soup	Steak	Carrot Cake
Salad	Chicken	Cheesecake
	Fish	

For each game, find the sample space. Then find the indicated probability.

13. Elba tosses a quarter, a dime, and a nickel. If tails comes up at least twice, Steve wins. Otherwise Elba wins. Find P(Elba wins).

14. Ming rolls a number cube, tosses a coin, and chooses a card from two cards marked A and B. If an even number and heads appears, Ming wins, no matter which card is chosen. Otherwise Lashonda wins. Find P(Ming wins).

Real-World Link
The average family size in the United States is 2.59 people.
Source: U.S. Census Bureau

FAMILIES Mr. and Mrs. Romero are expecting triplets. Suppose the chance of each child being a boy is 50% and of being a girl is 50%. Find each probability.

15. P(all three children will be boys)

16. P(at least one boy and one girl)

17. P(two boys and one girl)

18. P(at least two girls)

19. P(the first two born are boys and the last born is a girl)

20. **GAMES** The following is a game for two players. Find the probability that each player wins.

- Three counters are labeled according to the table at the right.

- Toss the three counters.

- If exactly 2 counters match, Player 1 scores a point. Otherwise, Player 2 scores a point.

	Side 1	Side 2
Counter 1	red	blue
Counter 2	red	yellow
Counter 3	blue	yellow

UNIFORMS For Exercises 21 and 22, use the information below.

The University of Oregon's football team has many different uniforms. The coach can choose from four colors of jerseys and pants: green, yellow, white, and black. There are three helmet options: green, white, and yellow. Also, there are four colors of socks and two colors of shoes.

21. How many jersey/pant combinations are there?

22. If the coach picks a jersey/pant combination at random, what is the probability he will pick a yellow jersey with green pants?

23. **RESEARCH** Use the Internet or another resource to find the number of jerseys and pants your favorite college or professional sports team has as part of its uniform. How many jersey/pants combinations are there for the team you chose?

NYSCC • NYSMT

Extra Practice, pp. 691, 712

H.O.T. Problems

24. **CHALLENGE** Refer to Exercise 20. A *fair game* is one in which each player has an equal chance of winning. Adjust the scoring of the game so that it is fair.

25. **SELECT A TOOL** Mei wants to determine the probability of guessing correctly on two true-false questions on her history test. Which of the following tools might Mei use to determine the probability of answering both questions correctly by guessing? Justify your selection(s). Then use the tool(s) to solve the problem.

| draw a model | calculator | real objects |

26. **FIND THE ERROR** Rhonda and Elise are finding all the possible unique outcomes of rolling an even number on a number cube and landing on A or B on the spinner shown. Who is correct? Explain your reasoning.

Rhonda

Outcomes	
Number Cube	Spinner
2	A
2	B
4	C
4	A
6	B
6	C

Elise

Outcomes	
Number Cube	Spinner
2	A
2	B
4	A
4	B
6	A
6	B

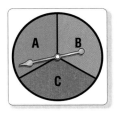

27. **WRITING IN MATH** Describe a game between two players using one coin in which each player has an equal chance of winning.

28. Mr. Zajac will choose one student from each of the two groups below to present their history reports to the class.

Group 1	Group 2
Julia	Keith
Antoine	Isabel
Greg	

Which set shows all the possible choices?

A {(Julia, Keith), (Antoine, Keith), (Greg, Keith)}

B {(Julia, Antoine), (Antoine, Greg), (Isabel, Keith)}

C {(Julia, Keith), (Antoine, Keith), (Greg, Keith), (Julia, Isabel), (Antoine, Isabel), (Greg, Isabel)}

D {(Isabel, Antoine), (Keith, Greg), (Julia, Isabel), (Keith, Antoine)}

Spiral Review

PROBABILITY A spinner is equally likely to stop on each of its regions numbered 1 to 20. Find each probability as a fraction in simplest form. (Lesson 9-1)

29. a prime number **30.** GCF(12, 18) **31.** multiple of 2 or 3

32. *not* a multiple of 4 **33.** factor of 10 or 6 **34.** *not* an even number

UTILITIES For Exercises 35 and 36, use the graph that shows the prices for natural gas charged by an Illinois natural gas supplier. (Lesson 8-6)

35. What trend seems to be revealed in the line graph?

36. What problems might there be in using this information to predict future prices of natural gas?

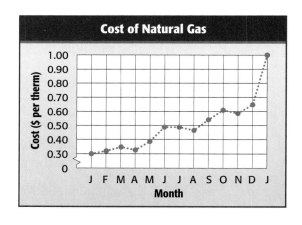

37. SAVINGS Amare invests $3,400 in a savings account that pays 3.5 % simple interest. After four years, how much money will be in his account?

Find each number. Round to the nearest tenth if necessary. (Lesson 7-1)

38. 43% of 266 **39.** 17% of 92 **40.** 2.5% of 44

▷ GET READY for the Next Lesson

PREREQUISITE SKILL Multiply.

41. 7 · 22 **42.** 11 · 16 **43.** 23 · 20 **44.** 131 · 4

9-3 The Fundamental Counting Principle

MAIN IDEA

Use multiplication to count outcomes and find probabilities.

NYS Core Curriculum

Reinforcement of 6.S.11 Determine the number of possible outcomes for a compound event by using the fundamental counting principle and use this to determine the probabilities of events when the outcomes have equal probability

New Vocabulary

Fundamental Counting Principle

NY Math Online

glencoe.com

• Extra Examples
• Personal Tutor
• Self-Check Quiz
• Reading in the Content Area

▷ **GET READY** for the Lesson

SALES The Shoe Warehouse sells sandals in different colors and styles.

Color	Style
black	platform
brown	slides
tan	wedges
white	
red	

1. According to the table, how many colors of sandals are available?

2. How many styles are available?

3. Find the product of the two numbers you found in Exercises 1 and 2.

4. Draw a tree diagram to find the number of different color and style combinations. How does the number of outcomes compare to the product you found above?

In the activity above, you discovered that multiplication, instead of a tree diagram, can be used to find the number of possible outcomes in a sample space. This is called the **Fundamental Counting Principle**.

Fundamental Counting Principle   Key Concept

Words   If event M has m possible outcomes and event N has n possible outcomes, then event M followed by event N has $m \times n$ possible outcomes.

EXAMPLE   Find the Number of Outcomes

① **Find the total number of outcomes when a coin is tossed and a number cube is rolled.**

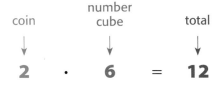

coin   number cube   total

$$2 \cdot 6 = 12 \qquad \text{Fundamental Counting Principle}$$

There are 12 different outcomes.

Check Draw a tree diagram to show the sample space.

✓ **CHECK Your Progress**

a. Find the total number of outcomes when choosing from bike helmets that come in three colors and two styles.

The Fundamental Counting Principle can be used to find the number of possible outcomes and solve probability problems in more complex problems, when there are more than two events.

Real-World EXAMPLE

2 **JEANS** The Jean Shop sells young men's jeans in different sizes, styles, and lengths, as shown in the table. Find the number of jeans available. Then find the probability of selecting a size 32 × 34 slim fit.

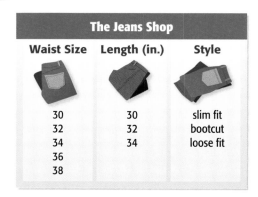

The Jeans Shop		
Waist Size	**Length (in.)**	**Style**
30	30	slim fit
32	32	bootcut
34	34	loose fit
36		
38		

$$\underset{\text{sizes}}{5} \cdot \underset{\text{length}}{3} \cdot \underset{\text{style}}{3} = \underset{\text{total}}{45}$$

Fundamental Counting Principle

There are 45 different types of jeans to choose from. Out of the 45 possible outcomes, only one is favorable. So, the probability of randomly selecting a 32 × 34 slim fit is $\frac{1}{45}$.

Study Tip

Jean size
In mens' jeans, the size is labeled waist × length. So, a 32 × 34 is a 32-inch waist with a 34-inch length.

CHECK Your Progress

b. JEANS If the Jean Shop adds relaxed fit jeans to its selection, find the number of available jeans. Then find the probability of randomly selecting a 36 × 30 relaxed fit pair of jeans.

CHECK Your Understanding

Example 1
(p. 471)

Use the Fundamental Counting Principle to find the total number of outcomes in each situation.

1. tossing a quarter, a dime, and a nickel

2. choosing scrambled, sunny-side up, or poached eggs with bacon or sausage and milk or orange juice

3. choosing a number on a number cube and picking a marble from the bag at the right

Example 2
(p. 472)

4. **CLOTHES** Beth has 3 sweaters, 4 blouses, and 6 skirts that coordinate. Find the number of different outfits consisting of a sweater, blouse, and skirt that are possible. Then find the probability of randomly selecting a particular sweater-blouse-skirt outfit.

Use the Fundamental Counting Principle to find the total number of outcomes in each situation.

5. choosing a bagel with one type of cream cheese from the list shown in the table

Bagels	Cream Cheese
plain	plain
blueberry	chive
cinnamon raisin	sun-dried tomato
garlic	

6. choosing a number from 1 to 20 and a color from 7 colors

7. picking a month of the year and a day of the week

8. choosing from a comedy, horror, or action movie each shown in four theaters

9. rolling a number cube and tossing two coins

10. choosing iced tea in regular, raspberry, lemon, or peach flavors; sweetened or unsweetened; and in a glass or a plastic container

11. **ROADS** Two roads, Broadway and State, connect the towns of Eastland and Harping. Three roads, Park, Fairview, and Main, connect the towns of Harping and Johnstown. Find the number of possible routes from Eastland to Johnstown that pass through Harping. Then find the probability that State and Fairview will be used if a route is selected at random.

12. **APPLES** An orchard makes apple nut bread, apple pumpkin nut bread, and apple buttermilk bread using 6 different varieties of apples, including Fuji. Find the number of possible bread choices. Then find the probability of selecting a Fuji apple buttermilk bread if a customer buys a loaf of bread at random.

Real-World Link . . . .
A popular orchard in Wapato, Washington, grows 7 varieties of apples including Granny Smith and Fuji.
Source: All About Apples

13. **GAMES** Find the number of possible outcomes if five number cubes are rolled at one time during a board game.

14. **PASSWORDS** Find the number of possible choices for a 2-digit password that is greater than 19. Then find the number of possible choices for a 4-digit Personal Identification Number (PIN) if the digits *cannot* be repeated.

15. **T-SHIRTS** A store advertises that they have a different T-shirt for each day of the year. The store offers 32 different T-shirt designs and 11 choices of color. Is the advertisement true? Explain.

16. **ANALYZE TABLES** The table shows cell phone options offered by a wireless phone company. If a phone with one payment plan and one accessory is given away at random, predict the probability that it will be Brand B and have a headset. Explain your reasoning.

Phone Brands	Payment Plans	Accessories
Brand A	Individual	Leather Case
Brand B	Family	Car Mount
Brand C	Business	Headset
	Government	Travel Charger

H.O.T. Problems

17. **CHALLENGE** Determine the number of possible outcomes for tossing one coin, two coins, and three coins. Then determine the number of possible outcomes for tossing n coins. Describe the strategy you used.

18. **Which One Doesn't Belong?** Identify the choices for event M and N that do *not* result in the same number of outcomes as the other two. Explain your reasoning.

> 9 drinks, 8 desserts   18 shirts, 4 pants   10 groups, 8 activities

19. **WRITING IN MATH** Explain when you might choose to use the Fundamental Counting Principle to find the number of possible outcomes and when you might choose to use a tree diagram.

NYSMT PRACTICE 6.S.11

20. A bakery offers white, chocolate, or yellow cakes with white or chocolate icing. There are also 24 designs that can be applied to a cake. If all orders are equally likely, what is the probability that a customer will order a white cake with white icing in a specific design?

   A $\frac{1}{30}$ **C** $\frac{1}{120}$

   B $\frac{1}{64}$ **D** $\frac{1}{144}$

21. **SHORT RESPONSE** Hat World sells 9 different styles of hats in several different colors for 2 different sports teams. If the company makes 108 kinds of hats, how many different colors do they make?

Spiral Review

22. **SCHOOL** Horacio can choose from 2 geography, 3 history, and 2 statistics classes. Find the sample space for all possible schedules. (Lesson 9-2)

PROBABILITY Find the probability that the spinner shown at the right will stop on each of the following. Write as a fraction in simplest form. (Lesson 9-1)

23. a vowel

24. red

Order each set of numbers from least to greatest. (Lesson 6-8)

25. 27%, $\frac{1}{5}$, 0.22, 20.1

26. $\frac{19}{20}$, 88%, 0.85, $\frac{3}{4}$

▷ **GET READY** for the Next Lesson

PREREQUISITE SKILL Multiply.

27. $3 \cdot 2 \cdot 1$ 28. $9 \cdot 8 \cdot 7$ 29. $5 \cdot 4 \cdot 3 \cdot 2$ 30. $7 \cdot 6 \cdot 5 \cdot 4$

9-4 Permutations

MAIN IDEA

Find the number of permutations of a set of objects and find probabilities.

NYS Core Curriculum

Reinforcement of 6.S.11 Determine the number of possible outcomes for a compound event by using the fundamental counting principle and use this to determine the probabilities of events when the outcomes have equal probability

New Vocabulary

permutation

NY Math Online

glencoe.com

• Extra Examples
• Personal Tutor
• Self-Check Quiz

▷ MINI Lab

How many different ways are there to arrange your first 3 classes if they are math, science, and language arts?

STEP 1 Write math, science, and language arts on the index cards.

MATH	SCIENCE	LANGUAGE ARTS

STEP 2 Find and record all arrangements of classes by changing the order of the index cards.

1. When you first started to make your list, how many choices did you have for your first class?

2. Once your first class was selected, how many choices did you have for the second class? then, the third class?

A **permutation** is an arrangement, or listing, of objects in which order is important. In the example above, the arrangement science, math, language arts is a permutation of math, science, language arts because the order of the classes is different. You can use the Fundamental Counting Principle to find the number of possible permutations.

EXAMPLE Find a Permutation

1. **SCHEDULES** Find the number of possible arrangements of classes in the Mini Lab above using the Fundamental Counting Principle.

There are **3** choices for the first class.

There are **2** choices that remain for the second class.

There is **1** choice that remains for the third class.

$$3 \cdot 2 \cdot 1 = 6 \longleftarrow \text{The number of permutations of 3 classes}$$

There are 6 possible arrangements, or permutations, of the 3 classes.

✓ CHECK Your Progress

a. **VOLLEYBALL** In how many ways can the starting six players of a volleyball team stand in a row for a picture?

You can use a permutation to find the probability of an event.

 **Find Probability**

 SWIMMING The finals of the Northwest Swimming League features 8 swimmers. If each swimmer has an equally likely chance of finishing in the top two, what is the probability that Yumii will be in first place and Paquita in second place?

Northwest League Finalists	
Octavia	Eden
Natasha	Paquita
Calista	Samantha
Yumii	Lorena

There are **8** choices for first place.

There are **7** choices that remain for second place.

8 • 7 = 56 ◄—— The number of permutations of the 2 places

There are 56 possible arrangements, or permutations, of the 2 places.

Since there is only one way of having Yumii come in first and Paquita second, the probability of this event is $\frac{1}{56}$.

CHECK Your Progress

b. **LETTERS** Two different letters are randomly selected from the letters in the word *math*. What is the probability that the first letter selected is *m* and the second letter is *h*?

CHECK Your Understanding

Example 1
(p. 475)

1. **AMUSEMENT PARKS** Seven friends are waiting to ride the new roller coaster. In how many ways can they board the ride, once it is their turn?

2. **COMMITTEES** In how many ways can a president, vice president, and secretary be randomly selected from a class of 25 students?

Example 2
(p. 476)

3. **DVDs** You have five seasons of your favorite TV show on DVD. If you randomly select two of them from a shelf, what is the probability that you will select season one first and season two second?

4. **PASSWORDS** A password consists of four letters, of which none are repeated. What is the probability that a person could guess the entire password by randomly selecting the four letters?

HOMEWORK HELP

For Exercises	See Examples
5–8	1
9–12	2

5. CONTESTS In the Battle of the Bands contest, in how many ways can the four participating bands be ordered?

6. CODES A garage door code has 5 digits. If no digit is repeated, how many codes are possible?

7. LETTERS How many permutations are possible of the letters in the word *friend*?

8. NUMBERS How many different 3-digit numbers can be formed using the digits 9, 3, 4, 7, and 6? Assume no number can be used more than once.

9. CAPTAINS The members of the Evergreen Junior High Quiz Bowl team are listed at the right. If a captain and an assistant captain are chosen at random, what is the probability that Walter is selected as captain and Mi-Ling as co-captain?

Evergreen Junior High Quiz Bowl Team	
Jamil	Luanda
Savannah	Mi-Ling
Tucker	Booker
Ferdinand	Nina
Walter	Meghan

10. BASEBALL Adriano, Julián, and three of their friends will sit in a row of five seats at a baseball game. If each friend is equally likely to sit in any seat, what is the probability that Adriano will sit in the first seat and Julián will sit in the second seat?

11. GAMES Alex, Aiden, Dexter, and Dion are playing a video game. If they each have an equally likely chance of getting the highest score, what is the probability that Dion will get the highest score and Alex the second highest?

12. BLOCKS A child has wooden blocks with the letters *G*, *R*, *T*, *I*, and *E*. Find the probability that the child randomly arranges the letters in the order *TIGER*.

Real-World Link . . . . .
In 2006, Stephen Clarke set the record for the fastest time to carve a face into a pumpkin. It took him 24.03 seconds.

Source: Guinness World Records

13. SELECT A TOOL Refer to the information in the table. There are four ribbons passed out to participants in the Perfect Pumpkin competition. Which of the following tools might you use to find the number of ways a ribbon can be awarded to a pumpkin in the Perfect Pumpkin competition? Justify your selection(s) to solve the problem.

Pumpkin Festival	
Competition	**Number of Participants**
Perfect Pumpkin	24
Biggest Pumpkin	17
Heaviest Pumpkin	15
Pumpkin Carving	5

NYSCC • NYSMT
Extra Practice, pp. 691, 712

calculator   mental math   estimation

14. **PHOTOGRAPHY** A family discovered they can stand in a row for their portrait in 720 different ways. How many members are in the family?

15. **STUDENT ID** Hamilton Middle School assigns a four-digit identification number to each student. The number is made from the digits 1, 2, 3, and 4, and no digit is repeated. If assigned randomly, what is the probability that an ID number will end with a 3?

H.O.T. Problems

16. **CHALLENGE** There are 1,320 ways for three students to win first, second, and third place during a debate match. How many students are there on the debate team? Explain your reasoning.

17. **WRITING IN MATH** Describe a real-world situation that has 6 permutations. Justify your answer.

NYSMT PRACTICE 6.S.11

18. The five finalists in a random drawing are shown. Find the probability that Sean is awarded first prize and Teresa is awarded second prize.

   A $\frac{1}{5}$

   B $\frac{1}{10}$

   C $\frac{2}{5}$

   D $\frac{1}{20}$

Finalists
Cesar
Teresa
Sean
Nikita
Alfonso

19. A baseball coach is deciding on the batting order for his nine starting players with the pitcher batting last. How many batting orders are possible?

   F 8

   G 72

   H 40,320

   J 362,880

Spiral Review

20. **BREAKFAST** Find the total number of outcomes if you can choose from 8 kinds of bagels, 3 toppings, and 4 beverages. (Lesson 9-3)

21. **LUNCH** Make a tree diagram showing different ways to make a sandwich with turkey, ham, or salami and either cheddar or Swiss cheese. (Lesson 9-2)

22. **PROBABILITY** What is the probability of rolling a number greater than four on a number cube? (Lesson 9-1)

Find each product. Write in simplest form. (Lesson 5-5)

23. $\frac{4}{5} \times 2\frac{1}{3}$

24. $11\frac{1}{8} \times \frac{1}{2}$

25. $4\frac{5}{6} \times \frac{7}{8}$

▷ **GET READY for the Next Lesson**

PREREQUISITE SKILL Find each value. (Lesson 1-4)

26. $\frac{5 \cdot 4}{2 \cdot 1}$

27. $\frac{8 \cdot 7 \cdot 6}{3 \cdot 2 \cdot 1}$

28. $\frac{5 \cdot 4 \cdot 3}{4 \cdot 3 \cdot 2}$

29. $\frac{10 \cdot 9 \cdot 8 \cdot 7}{8 \cdot 7 \cdot 6 \cdot 5}$

A number cube is rolled. Find each probability. Write as a fraction in simplest form. (Lesson 9-1)

1. P(an odd number)

2. P(a number not greater than 4)

3. P(a number less than 6)

4. P(a multiple of 2)

5. **BOOKS** Brett owns 5 science-fiction books, 3 biographies, and 12 mysteries. If he randomly picks a book to read, what is the probability that he will *not* pick a science-fiction book? Write as a percent.

For each situation, find the sample space using a table or tree diagram. (Lesson 9-2)

6. Two coins are tossed.

7. The spinner shown is spun, and a digit is randomly selected from the number 803.

8. **MULTIPLE CHOICE** At a diner, a customer can choose from eggs or pancakes as an entrée and from ham or sausage as a side. Which set shows all the possible choices of one entrée and one side? (Lesson 9-2)

   A {(eggs, pancakes), (ham, sausage)}

   B {(eggs, ham), (eggs, sausage), (pancakes, ham), (pancakes, sausage)}

   C {(eggs, ham), (eggs, pancakes), (sausage, pancakes)}

   D {(eggs, ham), (pancakes, sausage)}

9. **GAMES** Abbey rolls a number cube and chooses a card from among cards marked *A*, *B*, and *C*. If an odd number and a vowel turn up, Abbey wins. Otherwise, Benny wins. Find the sample space. Then find the probability that Benny wins. (Lesson 9-2)

For Exercises 10 and 11, use the Fundamental Counting Principle to find the total number of possible outcomes in each situation. (Lesson 9-3)

10. A customer chooses a paper color, size, and binding style for some copies.

Color	Size	Binding
white	8.5″ x 11″	paper clip
yellow	8.5″ x 14″	binder clip
green	8.5″ x 17″	staple

11. A number cube is rolled and three coins are tossed.

12. **CARS** A certain car model comes in the colors in the table and either automatic or manual transmission. Find the probability that a randomly selected car will have a black exterior, tan interior, and manual transmission if all combinations are equally likely. (Lesson 9-3)

Exterior	Interior
Black	Gray
White	Tan
Red	
Silver	

13. **COMPETITION** How many ways can 6 swimmers come in first, second, or third place?

14. **MULTIPLE CHOICE** Noriko packed five different sweaters for her weekend vacation. If she randomly selects one sweater to wear each day, what is the probability that she will select the brown sweater on Friday, the orange sweater on Saturday, and the pink sweater on Sunday? She will not wear each sweater more than once. (Lesson 9-4)

   F $\dfrac{1}{60}$

   G $\dfrac{1}{120}$

   H $\dfrac{3}{5}$

   J $\dfrac{1}{5}$

9-5 Combinations

MAIN IDEA

Find the number of combinations of a set of objects and find probabilities.

NYS Core Curriculum

7.S.8 Interpret data to provide the basis for predictions and to establish experimental probabilities

New Vocabulary

combination

NY Math Online

glencoe.com
• Extra Examples
• Personal Tutor
• Self-Check Quiz

▷ GET READY for the Lesson

FOOD Mr. Rius is making a salad for a party. He has tomatoes, green peppers, cucumbers, and radishes. He starts with lettuce and then decides to add three of the above ingredients to his salad.

1. Use the first letter of each vegetable to list all of the permutations of the ingredients added to the lettuce. How many are there?

2. Cross out any arrangement that contains the same letters as another one in the list. How many are there now?

3. Explain the difference between the two lists.

An arrangement, or listing, of objects in which order is not important is called a **combination**. In the activity above, choosing cucumbers and radishes is the same as choosing radishes and cucumbers.

Permutations and combinations are related. You can find the number of combinations of objects by dividing the number of permutations of the entire set by the number of ways the smaller set can be arranged.

EXAMPLE   Find the Number of Combinations

1 **FOOD** Terrence's Pizza Parlor is offering the special shown in the table. How many different two-topping pizzas are possible?

This is a combination problem because the order of the toppings on the pizza is not important.

Today's Special:
Large two-topping pizza for $14.99

Toppings
Pepperoni
Sausage
Green Peppers
Onions
Mushrooms

METHOD 1   Make a list.

Use the first letter of each topping to list all of the permutations of the toppings taken two at a time. Then cross out the pizzas that are the same as another one.

p, s	s̶,̶ p̶	g̶,̶ p̶	o̶,̶ p̶	m̶,̶ p̶
p, o	s, o	g̶,̶ s̶	o̶,̶ s̶	m̶,̶ s̶
p, m	s, m	g̶,̶ o̶	o, m	m̶,̶ o̶
p, g	s, g	g̶,̶ m̶	o, g	m, g

So, there are 10 different two-topping pizzas.

METHOD 2	Use a permutation.

Step 1   Find the number of permutations of the entire set.

$5 \cdot 4 = 20$ A permutation of 5 toppings, taken 2 at time

Step 2   Find the number of permutations of the smaller set.

$2 \cdot 1 = 2$ Number of ways to arrange 2 toppings

Step 3   Find the number of combinations.

$\frac{20}{2}$ or 10 Divide the number of permutations of the entire set by the number of permutations of each smaller set.

So, there are 10 different two-topping pizzas.

✓ CHOOSE Your Method

a. **FOOD** How many different three-topping pizzas are possible if Terrence adds ham and anchovies to the topping choices?

Real-World EXAMPLES

 BASKETBALL A one-on-one basketball tournament has six players from the regional finals. Each player will play every opponent once. The 2 players with the best records will then play in the championship. How many one-on-one games will be played?

Find the number of ways 2 players can be chosen from a group of 6.

There are $6 \cdot 5$ ways to choose 2 people. → $\frac{6 \cdot 5}{2 \cdot 1} = \frac{30}{2} = 15$
There are $2 \cdot 1$ ways to arrange 2 people.

There are 15 games plus 1 final game to determine the overall winner. So, there will be 16 games played.

Check Make a diagram in which each person is represented by a point. Draw a line segment between each pair of points to represent all the games. This produces 15 line segments, or 15 games. Then add the final game to make a total of 16 games.

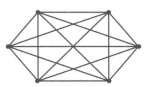

Real-World Link . . . .
The Last Man Standing One-on-One Basketball Tournament is held in New York. The top 16 players from preliminary competition play in a championship tournament in Madison Square Garden.

 If the players in Example 2 are selected at random, what is the probability that player 3 will play player 6 in the first game?

There are 15 possible first games and only one favorable outcome. So, the probability is $\frac{1}{15}$.

✓ CHECK Your Progress

b. **BASKETBALL** What is the probability that player 3 will play against player 12 in the first game if 12 people play in the tournament?

Example 1
(pp. 480–481)

1. **STICKERS** In how many ways can you pick 2 stickers from a package of 7?

2. **CRAFT FAIR** In how many ways can 3 out of 10 students be chosen to present their projects at a craft fair?

Examples 2, 3
(p. 481)

PAINTING For Exercises 3 and 4, use the information below.

Jade is going to paint her room two different colors from among white, gray, sage, or yellow.

3. How many combinations of two paint colors are there?

4. Find the probability that two colors chosen randomly will be white and sage.

Practice and Problem Solving

HOMEWORK HELP	
For Exercises	**See Examples**
5–8	1, 2
9–10	3

5. **VOLUNTEERS** In how many ways can you select 4 volunteers out of 10?

6. **ART** In how many ways can four drawings out of 15 be chosen for display?

7. **INTERNET** Of 12 Web sites, in how many ways can you choose to visit 6?

8. **SPORTS** On an 8-member volleyball team, how many different 6-player starting teams are possible?

9. **STUDENT COUNCIL** The students listed are members of Student Council. Three will be chosen at random to form a committee. Find the probability that the three students chosen will be Placido, Maddie, and Akira.

Roster
Leon
Placido
Maddie
Adrahan
Matt
Akira

10. **FOOD** At a hot dog stand, customers can select three toppings from among chili, onions, cheese, mustard, or relish. What is the probability that three toppings selected at random will include onions, mustard, and relish?

11. **MUSIC** Marissa practiced the five pieces listed at the right for a recital. Find the number of different ways that three pieces will be randomly chosen for her to play. Then find the probability that all three were composed by Beethoven.

Recital Piece	Composer
Fur Elise	Ludwig van Beethoven
First Piano Sonata	Sergei Rachmaninoff
The Four Seasons	Antonio Vivaldi
Cappricio	Ludwig van Beethoven
Moonlight Sonata	Ludwig van Beethoven

Tell whether each problem represents a *permutation* or a *combination*. Then solve the problem.

12. Six children remain in a game of musical chairs. If two chairs are removed, how many different groups of four children can remain?

13. How many ways can first and second chair positions be awarded in a band that has 10 flute players?

 **NYSCC • NYSMT**
Extra Practice, pp. 692, 712

14. How many ways can 12 books be stacked in a single pile?

15. **CHALLENGE** How many people were at a party if each person shook hands exactly once with every other person and there were 105 handshakes?

16. **FIND THE ERROR** Ling and Daniela are calculating the number of ways that a 4-member committee can be chosen from an 8-member club. Who is correct? Explain your reasoning.

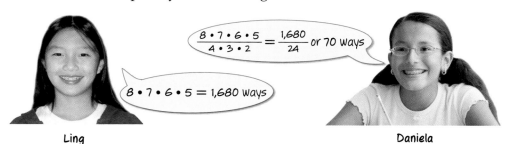

Ling: $8 \cdot 7 \cdot 6 \cdot 5 = 1,680$ ways

Daniela: $\dfrac{8 \cdot 7 \cdot 6 \cdot 5}{4 \cdot 3 \cdot 2} = \dfrac{1,680}{24}$ or 70 ways

17. **WRITING IN MATH** Write about a real-world situation that can be solved using a combination. Then solve the problem.

NYSMT PRACTICE 7.S.8

18. Three cheerleaders will be randomly selected to represent the squad at a game. If there are 12 cheerleaders, find the probability that the three members chosen are Kameko, Lynn, and Tory.

    A $\dfrac{1}{220}$ C $\dfrac{1}{4}$

    B $\dfrac{3}{110}$ D $\dfrac{1}{3}$

19. Four students are to be chosen from a roster of 9 students to attend a science camp. In how many ways can these 4 students be chosen?

    F 5   H 126

    G 36   J 3,024

Spiral Review

20. **TRACK** Six sprinters are entered in a 100-meter dash. In how many different ways can the race be completed? Assume there are no ties. (Lesson 9-4)

21. **T-SHIRTS** A clothing company makes T-shirts with the choices as shown in the table. How many different T-shirts are possible? (Lesson 9-3)

Size	Color	Style	Material
S	Orange	Crew	Cotton
M	Pink	V-Neck	Polyester
L	Lime Green	Tank-Top	
XL			

Estimate. (Lesson 5-1)

22. $\dfrac{1}{10} + \dfrac{7}{8}$

23. $\dfrac{5}{12} - \dfrac{1}{9}$

▷ GET READY for the Next Lesson

PREREQUISITE SKILL For Exercises 24 and 25, use the graph. (Lesson 8-5)

24. How many students were surveyed?

25. Find the probability that a student's favorite picnic game is sack racing.

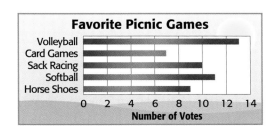

Favorite Picnic Games

9-6 Problem-Solving Investigation

MAIN IDEA: Solve problems by acting it out.

 7.PS.8 Understand how to break a complex problem into simpler parts or use a similar problem type to solve a problem

P.S.I. TEAM +

e-Mail: ACT IT OUT

EDDIE: I've been practicing free throws every day after school. Now I can make an average of 3 out of every 4 free throws I try. I wonder how many times I usually make two free throws in a row.

YOUR MISSION: Act it out to determine the probability that Eddie makes two free throws in a row.

Understand	You know that Eddie makes an average of 3 out of every 4 free throws. You could have Eddie actually make free throws, but that requires a basketball hoop. You could also act it out with a spinner.
Plan	Spin a spinner, numbered 1 to 4, two times. If the spinner lands on 1, 2, or 3, he makes the free throw. If the spinner lands on 4, he doesn't make it. Repeat the experiment 10 times.
Solve	Spin the spinner and make a table of the results.

Trials	1	2	3	4	5	6	7	8	9	10
First Spin	4	1	4	3	1	2	2	1	3	2
Second Spin	2	3	3	2	1	4	1	4	3	3

	The highlighted columns show that six out of the 10 trials resulted in two free throws in a row. So, the probability is 60%.
Check	Repeat the experiment several times to see whether the results agree.

Analyze The Strategy

1. Explain whether the results of the experiment would be the same if it were repeated.

2. **WRITING IN MATH** Write a problem that can be solved by acting it out. Then solve the problem by acting it out.

Mixed Problem Solving

For Exercises 3–6, use the *act it out* strategy.

3. **TESTS** Determine whether using a spinner with four equal sections is a good way to answer a 5-question multiple-choice quiz if each question has choices A, B, C, and D. Justify your answer.

4. **BOOKS** There are 6 students in a book club. Two of them order books, and the delivery comes to the classroom teacher. However, the teacher cannot remember which 2 students ordered the books. Is it a good idea for the teacher to randomly pass out the books to any two students? Explain. What is the probability that the teacher will give the correct books to the students?

5. **RUNNING** Six runners are entered in a race. Assuming no ties, in how many different ways can first and second places be awarded?

6. **MOVIES** In how many different ways can four friends sit in a row of four seats at the movies if two of the friends insist on sitting next to each other?

Use any strategy to solve Exercises 7–14. Some strategies are shown below.

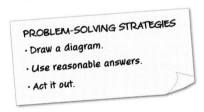

PROBLEM-SOLVING STRATEGIES
· Draw a diagram.
· Use reasonable answers.
· Act it out.

7. **FESTIVALS** The Student Council will have a booth set up at the town festival. They surveyed 160 students to find out their preference for the booth. The results are shown below. Is 35, 65, or 95 a reasonable answer for the number of students who would prefer a dunking booth? Explain.

Town Festival Survey Results

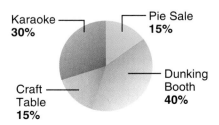

Karaoke 30%
Pie Sale 15%
Dunking Booth 40%
Craft Table 15%

8. **TRANSPORTATION** There are 4 students waiting at a subway stop. How many different ways can they board the subway?

9. **ALGEBRA** The pattern below is known as Pascal's Triangle. Would 1, 6, 10, 10, 6, 1 be a reasonable conjecture for the numbers in the 6th row? Justify your answer.

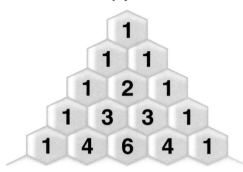

10. **CHESS** A chess tournament is held, and 32 students participate. The tournament will be single-elimination, which means if a player loses one match, he or she will be eliminated. How many total games will be played in the tournament?

11. **SCHOOL** Suppose rolling an even number on a number cube corresponds to an answer of *true* and rolling an odd number corresponds to an answer of *false*. Determine whether rolling this number cube is a good way to answer a 5-question true-false quiz. Justify your answer.

12. **CABLE** A cable company is running a special for new customers. For the first 3 months they will get a discount of 20% off their regular bill. Would $50, $60, or $70 be a reasonable estimate for the first bill if their regular bill is $95? Explain.

13. **MONEY** If you save $1.25 every week in a piggy bank, how much money will you have after 3 years?

14. **MUSIC** Liseta has an equal number of jazz, country, rap, pop, and R&B songs on her MP3 player. She listens to her MP3 player on random mode on both Wednesday and Thursday nights. What is the probability Liseta will hear a rap song first on Wednesday night?

9-7 Theoretical and Experimental Probability

MAIN IDEA

Find and compare experimental and theoretical probabilities.

NYS Core Curriculum

7.S.8 Interpret data to provide the basis for predictions and to establish experimental probabilities   7.S.10 Predict the outcome of an experiment   *Also addresses 7.S.12, 7.CN.8*

New Vocabulary

theoretical probability
experimental probability

NY Math Online

glencoe.com

• Extra Examples
• Personal Tutor
• Self-Check Quiz

▷ MINI Lab

Follow the steps to determine how many times doubles are expected to turn up when two number cubes are rolled.

Step 1   Use the table to help you find the expected number of times doubles should turn up when rolling two number cubes 36 times. The top row represents one number cube, and the left column represents the other number cube.

	1	2	3	4	5	6
1	1, 1	1, 2	1, 3	1, 4	1, 5	1, 6
2	2, 1	2, 2	2, 3	2, 4	2, 5	2, 6
3	3, 1	3, 2	3, 3	3, 4	3, 5	3, 6
4	4, 1	4, 2	4, 3	4, 4	4, 5	4, 6
5	5, 1	5, 2	5, 3	5, 4	5, 5	5, 6
6	6, 1	6, 2	6, 3	6, 4	6, 5	6, 6

Step 2   Roll two number cubes 36 times. Record the number of times doubles turn up.

1. Compare the number of times you *expected* to roll doubles with the number of times you *actually* rolled doubles.

2. Write the probability of rolling doubles out of 36 rolls using the number of times you *expected* to roll doubles from Step 1. Then write the probability of rolling doubles out of 36 rolls using the number of times you *actually* rolled doubles from Step 2.

In the Mini Lab above, you found both the theoretical probability and the experimental probability of rolling doubles using two number cubes. **Theoretical probability** is based on what *should* happen when conducting a probability experiment. This is the probability you have been using since Lesson 9-1. **Experimental probability** is based on what *actually* occurred during such an experiment.

Theoretical Probability	Experimental Probability
$\dfrac{6}{36}$ ← 6 rolls *should* occur	$\dfrac{n}{36}$ ← *n* rolls *actually* occurred

The theoretical probability and the experimental probability of an event may or may not be the same. As the number of times an experiment is conducted increases, the theoretical probability and the experimental probability should become closer in value.

EXAMPLE Experimental Probability

1 Two number cubes are rolled 75 times, and a sum of 9 is rolled 10 times. What is the experimental probability of rolling a sum of 9?

$$P(9) = \frac{\text{number of times a sum of 9 occurs}}{\text{total number of rolls}}$$

$$= \frac{10}{75} \text{ or } \frac{2}{15}$$

The experimental probability of rolling a sum of 9 is $\frac{2}{15}$.

 CHECK Your Progress

a. In the above experiment, what is the experimental probability of rolling a sum that is *not* 9?

b. In the above experiment, what is the experimental probability of rolling a sum that is *not* six if a sum of six was rolled 18 times?

 EXAMPLES Experimental and Theoretical Probability

Study Tip

Trials
A trial is one experiment in a series of successive experiments.

2 The graph shows the results of an experiment in which a spinner with 3 equal sections is spun sixty times. Find the experimental probability of spinning red for this experiment.

The graph indicates that the spinner landed on red 24 times, blue 15 times, and green 21 times.

$$P(\text{red}) = \frac{\text{number of times red occurs}}{\text{total number of spins}}$$

$$= \frac{24}{60} \text{ or } \frac{2}{5}$$

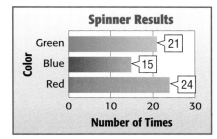

The experimental probability of spinning red was $\frac{2}{5}$.

3 Compare the experimental probability you found in Example 2 to its theoretical probability.

The spinner has three possible outcomes, red, blue, or green.

So, the theoretical probability of spinning red is $\frac{1}{3}$. Since $\frac{2}{5} \approx \frac{1}{3}$, the experimental probability is close to the theoretical probability.

 CHECK Your Progress

c. Refer to Example 2. If the spinner was spun 3 more times and landed on green each time, find the experimental probability of spinning green for this experiment.

d. Compare the experimental probability you found in Exercise c to its theoretical probability.

Theoretical and experimental probability can be used to make predictions about future events.

Real-World EXAMPLES Predict Future Events

(4) MEDIA A media buyer examines last year's DVD sales to decide how many DVDs of each type to buy this year. Last year's sales are shown in the table. What is the probability that a person buys a comedy DVD?

DVDs Sold	
Type	**Number**
action	670
comedy	580
drama	450
horror	300

There were 2,000 DVDs sold and 580 chose comedy. So, the probability is $\frac{580}{2,000}$ or $\frac{29}{100}$.

(5) Suppose the media buyer expects to sell 5,000 DVDs this year. How many drama DVDs should she buy?

$\frac{450}{2,000} = \frac{x}{5,000}$ Write a proportion.

$450 \cdot 5,000 = 2,000 \cdot x$ Find the cross products.

$2,250,000 = 2,000x$ Multiply.

$1,125 = x$ Divide each side by 2,000.

She should buy about 1,125 drama DVDs.

 CHECK Your Progress

e. What is the probability that a person buys a horror DVD?

f. If the media buyer expects to sell 3,000 DVDs this year, about how many action movies should she buy?

CHECK Your Understanding

Examples 1–3
(p. 487)

For Exercises 1–3, a coin is tossed 50 times, and it lands on heads 28 times.

1. Find the experimental probability of the coin landing heads.

2. Find the theoretical probability of the coin landing heads.

3. Compare the probabilities in Exercises 1 and 2.

FOOD For Exercises 4 and 5, use the table showing the types of muffins that customers bought one morning from their local bakery.

Muffin	Number of People
blueberry	22
poppyseed	17
banana	11

Example 4
(p. 488)

4. What is the probability that a customer buys a blueberry muffin?

5. If 100 customers buy muffins tomorrow, about how many would you expect to buy a banana muffin?

HOMEWORK HELP

For Exercises	See Examples
6–7	1–3
8–9	4
10–11	5

For Exercises 6 and 7, a number cube is rolled 20 times and lands on 1 two times and on 5 four times.

6. Find the experimental probability of landing on 5. Compare the experimental probability to the theoretical probability.

7. Find the experimental probability of *not* landing on 1. Compare the experimental probability to the theoretical probability.

ZOO For Exercises 8–11, use the graph of a survey of 70 zoo visitors who were asked to name their favorite animal exhibit.

8. What is the probability that the elephant exhibit is someone's favorite?

9. What is the probability that the bear exhibit is someone's favorite?

10. Suppose 540 people visit the zoo. Predict how many people will choose the monkey exhibit as their favorite.

11. Suppose 720 people visit the zoo. Predict how many people will choose the penguin exhibit as their favorite.

What is your Favorite Animal Exhibit?		
Exhibit	**Tally**	**Frequency**
bears	卌 I	6
elephants	卌 卌 卌 II	17
monkeys	卌 卌 卌 卌 I	21
penguins	卌 卌 III	13
snakes	卌 卌 III	13

For Exercises 12–14, a spinner with three equal-sized sections marked A, B, and C is spun 100 times.

Section	Frequency
A	24
B	50
C	26

12. What is the theoretical probability of landing on A?

13. The results of the experiment are shown in the table. What is the experimental probability of landing on A? of landing on C?

14. Make a drawing of what the spinner might look like based on its experimental probabilities. Explain your reasoning.

GIFTS For Exercises 15–17, use the graph at the right.

15. What is the probability that a mother will receive a gift of flowers or plants? Write the probability as a fraction.

16. Out of 400 mothers that receive gifts, predict how many will receive flowers or plants.

17. Out of 750 mothers that receive gifts, is it reasonable to expect 250 mothers to receive jewelry? Why or why not?

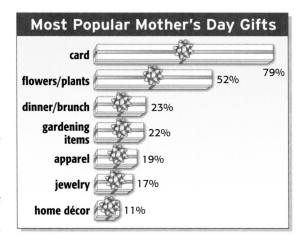

Most Popular Mother's Day Gifts

card 79%
flowers/plants 52%
dinner/brunch 23%
gardening items 22%
apparel 19%
jewelry 17%
home décor 11%

Source: Carlton Cards

NYSCC • NYSMT

Extra Practice, pp. 692, 712

H.O.T. Problems

18. **CHALLENGE** The experimental probability of a coin landing on heads is $\frac{7}{12}$. If the coin landed on tails 30 times, find the number of tosses.

19. **REASONING** Twenty sharpened pencils are placed in a box containing an unknown number of unsharpened pencils. Suppose 15 pencils are taken out at random, of which five are sharpened. Based on this, is it reasonable to assume that the number of unsharpened pencils was 40? Explain your reasoning.

20. **WRITING IN MATH** Compare and contrast experimental probability and theoretical probability.

NYSMT PRACTICE 7.S.8, 7.S.10

21. The frequency table shows Mitch's record for the last thirty par-3 holes he has played.

Mitch's Golf Results	
Score	Number of Holes
2	4
3	14
4	9
5	3

Based on this record, what is the probability that Mitch will score a 2 or 3 on the next par-3 hole?

A $\frac{7}{9}$ C $\frac{3}{10}$

B $\frac{3}{5}$ D $\frac{9}{50}$

22. J.R. tossed a coin 100 times and graphed the results.

Tossing a Coin

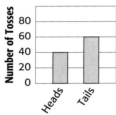

Based on this information, what is the experimental probability of tossing tails on the next toss?

F $\frac{1}{5}$ H $\frac{3}{5}$

G $\frac{2}{3}$ J $\frac{4}{5}$

Spiral Review

23. **SPINNERS** In how many ways could the colors in the spinner shown be arranged so that red and blue remain in the same places? (Lesson 9-6)

24. **BASEBALL** How many ways can a baseball coach select four starting pitchers from a pitching staff of eight? (Lesson 9-5)

25. **CLOTHES** A pair of jeans comes in 4 different styles, 3 different colors, and 5 different sizes. How many unique outcomes are possible? (Lesson 9-3)

▷ **GET READY for the Next Lesson**

PREREQUISITE SKILL Write each fraction as a percent. (Lesson 6-9)

26. $\frac{4}{5}$ 27. $\frac{11}{20}$ 28. $\frac{7}{8}$ 29. $\frac{39}{50}$

490 Chapter 9 Probability

Extend
9-7

Probability Lab
Simulations

MAIN IDEA

Investigate experimental
probability by conducting
a simulation.

▶ **NYS Core Curriculum**

**7.S.8 Interpret data to
provide the basis for
predictions and to
establish experimental
probabilities 7.S.10
Predict the outcome
of an experiment**
*Also addresses 7.S.11,
7.S.12, 7.R.3*

A *simulation* is a way of acting out a problem situation. Simulations often use models to act out events that would be difficult or impractical to perform. In this lab, you will simulate purchasing a box of cereal and getting one of four possible prizes inside.

ACTIVITY

STEP 1 Place four different colored cubes into a paper bag.

STEP 2 Without looking, draw a cube from the bag, record its color, and then place the cube back in the bag.

STEP 3 Repeat steps 1 and 2 until you have drawn a cube from the bag a total of four times.

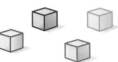

ANALYZE THE RESULTS

1. Based on your results, predict the probability of getting each prize.

2. What is the theoretical probability of getting each prize?

3. How do your probabilities in Exercises 1 and 2 compare?

4. **MAKE A PREDICTION** Predict the probability of selecting all four prizes in four boxes of cereal.

5. Repeat the simulation above 20 times. Use this data to predict the probability of selecting all four prizes in four boxes of cereal.

6. Calculate the experimental probability described in Exercise 5 using the combined data of five different groups. How does this probability compare with your prediction?

7. Describe a simulation that could be used to predict the probability of taking a five question true/false test and getting all five questions correct by guessing. Choose from among two-sided counters, number cubes, coins, or spinners as your model.

8. **COLLECT THE DATA** Conduct 50 trials of the experiment you described in Exercise 7. Then calculate the experimental probability of getting all five questions correct by guessing.

Extend 9-7 Probability Lab: Simulations   **491**

9-8 Compound Events

GET READY for the Lesson

MONEY Reginald has 3 state quarters, one from Colorado, one from Montana, and one from Washington.

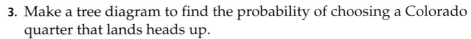

1. If Reginald picks one quarter without looking, what is the probability it is from Colorado?

2. Suppose he tosses the coin. What is the probability it lands heads up?

3. Make a tree diagram to find the probability of choosing a Colorado quarter that lands heads up.

4. How are the answers to Exercises 1, 2, and 3 related?

In the example above, choosing a quarter and tossing heads is a compound event. A **compound event** consists of two or more simple events. Since choosing a quarter does not affect tossing heads, the two events are called **independent events**. The outcome of one event does not affect the outcome of the other event.

EXAMPLE Independent Events

① **A coin is tossed, and the spinner shown is spun. Find the probability of tossing heads and spinning a consonant.**

List the sample space. Use H for heads and T for tails.

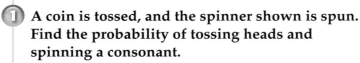

| H, A | H, B | H, C |
| T, A | T, B | T, C |

$P(\text{H and a consonant}) = \dfrac{\text{number of times heads and a consonant occurs}}{\text{number of possible outcomes}}$

$P(\text{H and a consonant}) = \dfrac{2}{6} \text{ or } \dfrac{1}{3}$

So, the probability is $\dfrac{1}{3}$ or about 33%.

CHECK Your Progress

A number cube is rolled, and the spinner in Example 1 is spun. Find each probability.

a. $P(4 \text{ and a consonant})$ b. $P(\text{odd and a B})$

The probability in Example 1 can also be found by multiplying the probabilities of each event. $P(\text{H}) = \frac{1}{2}$ and $P(\text{consonant}) = \frac{1}{3}$ and so $P(\text{H and C}) = \frac{1}{2} \cdot \frac{1}{3}$ or $\frac{1}{6}$. This leads to the following.

Probability of Independent Events Key Concept

Words	The probability of two independent events can be found by multiplying the probability of the first event by the probability of the second event.
Symbols	$P(A \text{ and } B) = P(A) \cdot P(B)$

Real-World EXAMPLE

② **SHOPPING** Brianna is buying a new outfit. She is choosing among 2 red, 1 purple, 3 pink, or 4 yellow tops. For pants, she is choosing between jeans or capris. If Brianna chooses a top and capris at random, find the probability that she chooses a yellow top and jeans.

$P(\text{yellow top and jeans})$

$$= P(\text{yellow top}) \cdot P(\text{jeans})$$

$$= \frac{4}{10} \cdot \frac{1}{2} \qquad \begin{array}{l}\text{4 out of 10 tops are yellow.}\\ \text{1 out of 2 pants are jeans.}\end{array}$$

$$= \frac{^2 4}{10} \cdot \frac{1}{2_1} = \frac{2}{10} \text{ or } \frac{1}{5} \quad \text{Simplify.}$$

So, the probability is $\frac{1}{5}$ or about 20%.

Study Tip

Reasonable Answer
You can check your answer in Example 2 by listing the sample space or by making a tree diagram.

CHECK Your Progress

c. **SHOPPING** If khakis and shorts are added to Brianna's pant choices, find the probability that she chooses a red top and capris.

If the outcome of one event affects the outcome of a second event, the events are called **dependent events**. Just as in independent events, the probabilities of dependent events can be found by multiplying the probabilities of each event. However, now the probability of the second event depends on the fact that the first event has already occurred.

Probability of Dependent Events Key Concept

Words	If two events, A and B, are dependent, then the probability of both events occurring is the product of the probability of A and the probability of B after A occurs.
Symbols	$P(A \text{ and } B) = P(A) \cdot P(B \text{ following } A)$

EXAMPLE Dependent Events

3 There are 2 red, 5 green, and 8 yellow marbles in a jar. Martina randomly selects two marbles without replacing the first marble. What is the probability that she selects two green marbles?

Since the first marble is not replaced, the first event affects the second event. These are dependent events.

$$P(\text{first marble is green}) = \frac{5}{15} \quad \longleftarrow \text{ number of green marbles} \\ \longleftarrow \text{ total number of marbles}$$

$$P(\text{second marble is green}) = \frac{4}{14} \quad \left\{\begin{array}{l}\text{number of green marbles after}\\\text{one green marble is removed}\end{array}\right. \\ \left\{\begin{array}{l}\text{total number of marbles after}\\\text{one green marble is removed}\end{array}\right.$$

$$P(\text{two green marbles}) = \frac{\overset{1}{\cancel{5}}}{\underset{3}{\cancel{15}}} \cdot \frac{\overset{2}{\cancel{4}}}{\underset{7}{\cancel{14}}} \text{ or } \frac{2}{21}$$

So, the probability of selecting two green marbles is $\frac{2}{21}$, or about 9.5%.

 CHECK Your Progress

d. There are 4 blueberry, 6 raisin, and 2 plain bagels in a bag. Javier randomly selects two bagels without replacing the first bagel. Find the probability that he selects a raisin bagel and then a plain bagel.

Sometimes two events cannot happen at the same time. For example, when a coin is tossed, the outcome of heads cannot happen at the same time as tails. Either heads *or* tails will turn up. Tossing heads and tossing tails are examples of **disjoint events**, or events that cannot happen at the same time. Disjoint events are also called *mutually exclusive events*.

Study Tip

Disjoint Events
When finding the probabilities of disjoint events, the word *or* is usually used.

 EXAMPLE Disjoint Events

4 A number cube is rolled. What is the probability of rolling an odd number or a 6?

These are disjoint events since it is impossible to roll an odd number and a 6 at the same time.

$$P(\text{odd number or 6}) = \frac{4}{6} \quad \longleftarrow \text{ There are four favorable outcomes: 1, 3, 5, or 6.} \\ \longleftarrow \text{ There are 6 total possible outcomes.}$$

So, the probability of rolling an odd number or a 6 is $\frac{4}{6}$, or $\frac{2}{3}$.

 CHECK Your Progress

e. Twenty-six cards are labeled, each with a letter of the alphabet, and placed in a box. A single card is randomly selected. What is the probability that the card selected will be labeled with the letter M or the letter T?

Study Tip

Probability
The probability of two disjoint events is the sum of the two individual probabilities. The probability of two independent events is the product of the two individual probabilities.

Notice that the probability in Example 4 can also be found by adding the probabilities of each event.

Probability of Disjoint Events | Key Concept

Words   If two events, A and B, are disjoint, then the probability that either A or B occurs is the sum of their probabilities.

Symbols $P(A \text{ or } B) = P(A) + P(B)$

CHECK Your Understanding

Example 1
(p. 492)

A number cube is rolled, and the spinner is spun. Find each probability.

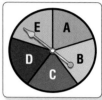

1. $P(5 \text{ and } E)$
2. $P(2 \text{ and vowel})$
3. $P(3 \text{ and a consonant})$
4. $P(\text{factor of 6 and D})$

Example 2
(p. 493)

5. **CLOTHES** Loretta has 2 pairs of black pants, 3 pairs of blue pants, and 1 pair of tan pants. She also has 4 white and 2 red shirts. If Loretta chooses a pair of pants and a shirt at random, what is the probability that she will choose a pair of black pants and a white shirt?

6. **MARBLES** A jar contains 12 marbles. Four are red, 3 are white, and 5 are blue. A marble is randomly selected, its color recorded, and then the marble is returned to the jar. A second marble is randomly selected. Find the probability that both marbles selected are blue.

Example 3
(p. 494)

7. The digits 0–9 are each written on a slip of paper and placed in a hat. Two slips of paper are randomly selected, without replacing the first. What is the probability that the number 0 is drawn first and then a 7 is drawn?

Example 4
(p. 494)

A number cube is rolled. Find each probability.

8. $P(4 \text{ or } 5)$
9. $P(3 \text{ or even number})$
10. $P(1 \text{ or multiple of 2})$
11. $P(6 \text{ or number less than 3})$

Practice and Problem Solving

HOMEWORK HELP	
For Exercises	**See Examples**
12–17	1
18–19	2
20–21	3
22–25	4

A coin is tossed, and a number cube is rolled. Find each probability.

12. $P(\text{heads and 1})$
13. $P(\text{tails and multiple of 3})$

A set of five cards is labeled 1–5. A second set of ten cards contains the following colors: 2 red, 3 purple, and 5 green. One card from each set is selected. Find each probability.

14. $P(5 \text{ and green})$
15. $P(\text{odd and red})$
16. $P(\text{prime and purple})$
17. $P(\text{even and yellow})$

18. **MUSIC** Denzel is listening to a CD that contains 12 songs. If he presses the random button on his CD player, what is the probability that the first two songs played will be the first two songs listed on the album?

19. **JUICE POPS** Lakita has two boxes of juice pops with an equal number of pops in each flavor. Find the probability of randomly selecting a grape juice pop from the first box and randomly selecting a juice pop from the second box that is *not* grape.

20. **FRUIT** Francesca randomly selects two pieces of fruit from a basket containing 8 oranges and 4 apples without replacing the first fruit. Find the probability that she selects two oranges.

21. **SCHOOL** The names of 24 students, of which 14 are girls and 10 are boys, in Mr. Santiago's science class are written on cards and placed in a jar. Mr. Santiago randomly selects two cards without replacing the first to determine which students will present their lab reports today. Find the probability that two boys are selected.

A day of the week is randomly selected. Find each probability.

22. P(Monday or Tuesday)

23. P(a day beginning with T or Friday)

24. P(a weekday or Saturday)

25. P(Wednesday or a day with 6 letters)

A coin is tossed twice, and a letter is randomly picked from the word *event*. Find each probability.

26. P(two heads and T)

27. P(tails, *not* tails, consonant)

28. P(heads, tails, *not* V)

29. P(two tails and vowel)

FAMILY For Exercises 30–32, use the fact that the probability for a boy or a girl is each $\frac{1}{2}$.

30. Copy and complete the table that gives the probability that all the children in a family are boys given the number of children in the family.

31. Predict the probability that, in a family of ten children, all ten are boys.

32. Predict the probability that, in a family of n children, all n are boys.

Number of Children	P(all boys)
1	$\frac{1}{2}$
2	$\frac{1}{2} \cdot \frac{1}{2}$ or $\frac{1}{4}$
3	▨
4	▨
5	▨

33. **LIGHTING** Gene has set two of his lights on timers. He always has at least one light on between the hours of 8:00 P.M. and 7:00 A.M. Light A is on 30% of the time, and Light B is on 70% of the time. What is the probability both lights are on at the same time?

34. **RESEARCH** The *contiguous* United States consists of all states excluding Alaska and Hawaii. If one of these contiguous states is chosen at random, what is the probability that it will end with the letter A or O? Write as a percent.

NYSCC • NYSMT
Extra Practice, pp. 693, 712

CHALLENGE For Exercises 35 and 36, use the spinner.

35. Use a tree diagram to construct the sample space of all the possible outcomes of three successive spins.

36. Suppose the spinner is designed so that for each spin there is a 40% probability of spinning red and a 20% chance of spinning blue. What is the probability of spinning two reds and then one blue?

37. **WRITING IN MATH** A shelf has books A, B, and C on it. You pick a book at random, place it on a table, and then pick a second book. Explain why the probability that you picked books A and B is *not* $\frac{1}{9}$.

NYSMT PRACTICE 6.S.9

38. A jar contains 8 white marbles, 4 green marbles, and 2 purple marbles. If Darla picks one marble from the jar without looking, what is the probability that it will be either white or purple?

 A $\frac{5}{7}$ **C** $\frac{2}{7}$

 B $\frac{4}{7}$ **D** $\frac{1}{7}$

39. What is the probability of spinning a red, the number 1, and the letter A on the three spinners below?

 F $\frac{1}{3}$ **G** $\frac{1}{32}$ **H** $\frac{1}{12}$ **J** $\frac{1}{64}$

Spiral Review

40. **PROBABILITY** Ella is going to roll a number cube 30 times. How many times should she expect to roll a number greater than 2? (Lesson 9-7)

41. **CHORES** This weekend, Brennen needs to do laundry, mow the lawn, and clean his room. How many different ways can he do these three chores? (Lesson 9-6)

ALGEBRA Evaluate each expression if $a = 6$, $b = -4$, and $c = -3$. (Lesson 1-6)

42. $9c$ 43. $-8a$ 44. $2bc$ 45. $5b^2$

Problem Solving in Science   🌐 **Real-World Unit Project**

Math Genes It's time to complete your project. Use the information and data you have gathered about genetics and pet traits to prepare a poster. Be sure to include a chart displaying your data with your project.

 **NY Math Online** **Unit Project at** glencoe.com

Study Guide and Review

Study Organizer

GET READY to Study

Be sure the following Big Ideas are noted in your Foldable.

Probability
9-1 Simple Events
9-2 Sample Spaces
9-3 The Fundamental Counting Principle
9-4 Permutations
9-5 Combinations
9-6 Act it Out
9-7 Theoretical & Experimental Probability
9-8 Compound Events
Vocabulary

BIG Ideas

Probability (Lesson 9-1)
• The probability of a simple event is a ratio that compares the number of favorable outcomes to the number of possible outcomes.

Fundamental Counting Principle (Lesson 9-3)
• If event *M* has *m* possible outcomes and is followed by event *N* that has *n* possible outcomes, then the event *M* followed by *N* has $m \times n$ possible outcomes.

Theoretical and Experimental Probability (Lesson 9-7)
• Theoretical probability is based on what *should* happen when conducting a probability experiment.
• Experimental probability is based on what *actually occurred* during a probability experiment.

Independent Events (Lesson 9-8)
• The probability of two independent events can be found by multiplying the probability of the first event by the probability of the second event.

Dependent Events (Lesson 9-8)
• If two events, *A* and *B*, are dependent, then the probability of both events occurring is the product of the probability of *A* and the probability of *B* after *A* occurs.

Disjoint Events (Lesson 9-8)
• If two events are disjoint, then the probability that either event will occur is the sum of their individual probabilities.

Key Vocabulary

combination (p. 480)
complementary events (p. 462)
compound event (p. 492)
dependent events (p. 493)
disjoint events (p. 494)
experimental probability (p. 486)
Fundamental Counting Principle (p. 471)
independent events (p. 492)
outcome (p. 460)
permutation (p. 475)
probability (p. 460)
random (p. 461)
sample space (p. 465)
simple event (p. 460)
theoretical probability (p. 486)
tree diagram (p. 466)

Vocabulary Check

State whether each sentence is *true* or *false*. If *false*, replace the underlined word or number to make a true sentence.

1. <u>Compound events</u> consist of two or more simple events.

2. A <u>random</u> outcome is an outcome that occurs by chance.

3. *P*(not *A*) is read the <u>permutation</u> of the complement of *A*.

4. The Fundamental Counting Principle counts the number of possible outcomes using the operation of <u>addition</u>.

5. Events in which the outcome of the first event does not affect the outcome of the other event(s) are <u>simple events</u>.

6. The <u>sample space</u> of an event is the set of outcomes not included in the event.

7. Events that cannot occur at the same time are called <u>dependent</u> events.

Lesson-by-Lesson Review

9-1 Simple Events (pp. 460–464)

7.S.9,
7.S.10

A bag of animal crackers contains 5 monkeys, 4 giraffes, 6 elephants, and 3 tigers. Suppose you draw a cracker at random. Find the probability of each event. Write as a fraction in simplest form.

8. P(monkey)

9. P(tiger)

10. P(giraffe or elephant)

11. P(*not* monkey)

12. P(monkey, giraffe, or elephant)

13. **ARRIVALS** The probability that a plane will arrive at the airport on time is $\frac{23}{25}$. Find the probability that the plane will *not* arrive on time. Write as a percent.

Example 1 What is the probability of rolling a number less than 3 on a number cube?

$$P(1 \text{ or } 2) = \frac{\text{numbers less than 3}}{\text{total number of possible outcomes}}$$

$$= \frac{2}{6} \quad \text{Two numbers are less than 3.}$$

$$= \frac{1}{3} \quad \text{Simplify.}$$

Therefore, $P(1 \text{ or } 2) = \frac{1}{3}$.

9-2 Sample Spaces (pp. 465–470)

7.S.8

For each situation, find the sample space using a table or tree diagram.

14. rolling a number cube and tossing a coin

15. choosing from pepperoni, mushroom, or cheese pizza and water, juice, or milk

16. **GAMES** Eliza and Zeke are playing a game in which Zeke spins the spinner shown and rolls a number cube. If the sum of the numbers is less than six, Eliza wins. Otherwise Zeke wins. Find the sample space. Then find the probability that Zeke wins.

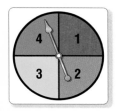

Example 2 Ginger and Micah are playing a game in which a coin is tossed twice. If heads comes up exactly once, Ginger wins. Otherwise, Micah wins. Find the sample space. Then find the probability that Ginger wins.

Make a tree diagram.

First Toss	Second Toss	Sample Space	
H	H	HH	Micah wins.
	T	HT	Ginger wins.
T	H	TH	Ginger wins.
	T	TT	Micah wins.

There are four equally likely outcomes with 2 favoring each player. The probability that Ginger wins is $\frac{2}{4}$ or $\frac{1}{2}$.

9-3 **The Fundamental Counting Principle** (pp. 471–474)

6.S.11

Use the Fundamental Counting Principle to find the total number of outcomes in each situation.

17. rolling two number cubes

18. creating an outfit from 6 different shirts and 4 different pants

19. **TUXEDOS** A tuxedo shop offers a tuxedo in three colors, black, gray, and white. The tie can be a bow tie or a regular tie. The tuxedo can come with tails or no tails. If a tuxedo is selected at random, what is the probability that it will be black, with a bow tie, and no tails?

Example 3 Use the Fundamental Counting Principle to find the total number of outcomes for a coin that is tossed four times.

There are 2 possible outcomes, heads or tails, each time a coin is tossed. For a coin that is tossed four times, there are $2 \cdot 2 \cdot 2 \cdot 2$, or 16 outcomes.

Example 4 Find the probability that, in a family of four children, all four children are girls.

There are 16 outcomes. There is one possible outcome resulting in four girls. So, the probability that all four children are girls is $\frac{1}{16}$.

9-4 **Permutations** (pp. 475–478)

6.S.11

20. **PORTRAITS** In how many ways can a family of five arrange themselves in a line for a family portrait?

21. **LETTERS** How many permutations are there of the letters in the word *computer*?

22. **RUNNING** Jacinda and Raul are entered in a race with 5 other runners. If each runner is equally likely to win, what is the probability that Jacinda will finish first and Raul will finish second?

Example 5 Nathaniel needs to choose two of the chores shown to do after school. If he is equally likely to choose the chores, what is the probability that he will walk the dog first and rake the leaves second?

Chores
Walk the Dog
Do Homework
Clean the Kitchen
Rake the Leaves

There are $4 \cdot 3$, or 12, arrangements in which Nathaniel can complete the chores. There is one way in which he will walk the dog first and rake the leaves second. So, the probability that he will walk the dog first and rake the leaves second is $\frac{1}{12}$.

Mixed Problem Solving
For mixed problem-solving practice, see page 712.

9-5 **Combinations** (pp. 480–483)

7.S.8

23. **DVD** In how many ways can Toni select 2 DVDs from the 15 in her collection?

24. **SPORTS** How many ways can a coach select 3 players from a roster of 9?

25. **GAMES** Marcus has enough money on his gift card to purchase 3 new games from the 14 displayed in the New Release section. In how many ways can Marcus select 3 different games?

26. **QUIZ** Frances must answer 3 of the 5 questions on a quiz, numbered 1–5. What is the probability that Frances will answer questions 2, 3, and 4?

Example 6 Caitlin and Román are playing a game in which Román chooses four different numbers from 1–15. What is the probability that Caitlin will guess all four numbers correctly?

There are $15 \cdot 14 \cdot 13 \cdot 12$ permutations of four numbers chosen from 15 numbers. There are $4 \cdot 3 \cdot 2 \cdot 1$ ways to arrange the 4 numbers.

$$\frac{15 \cdot 14 \cdot 13 \cdot 12}{4 \cdot 3 \cdot 2 \cdot 1} = \frac{32,760}{24} \text{ or } 1,365$$

There are 1,365 ways to choose four numbers from 15 numbers. There is one way to guess all four numbers correctly, so the probability that Caitlin will guess all four numbers correctly is $\frac{1}{1,365}$.

9-6 **PSI: Act It Out** (pp. 484–485)

7.PS.8

Solve each problem. Use the *act it out* strategy.

27. **QUIZ** Determine whether tossing a coin is a good way to answer a 6-question true-false quiz. Justify your answer.

28. **FAMILY PORTRAIT** In how many ways can the Maxwell family pose for a portrait if Mr. and Mrs. Maxwell are sitting in the middle and their three children are standing behind them?

29. **AMUSEMENT PARK** In how many ways can 4 friends be seated in 2 rows of 2 seats each on a roller coaster if Judy and Harold must ride together?

Example 7 In how many ways can three females and two males sit in a row of five seats at a concert if the females must sit in the first three seats?

Place five desks or chairs in a row. Have three females and two males sit in any of the seats as long as the females sit in the first three seats. Continue rearranging until you find all the possibilities. Record the results.

F_1	F_2	F_3	M_1	M_2	F_2	F_3	F_1	M_1	M_2
F_1	F_2	F_3	M_2	M_1	F_2	F_3	F_1	M_2	M_1
F_1	F_3	F_2	M_1	M_2	F_3	F_2	F_1	M_1	M_2
F_1	F_3	F_2	M_2	M_1	F_3	F_2	F_1	M_2	M_1
F_2	F_1	F_3	M_1	M_2	F_3	F_1	F_2	M_1	M_2
F_2	F_1	F_3	M_2	M_1	F_3	F_1	F_2	M_2	M_1

There are 12 possible arrangements.

CHAPTER 9 Study Guide and Review

7.S.8, 7.S.10

9-7 Theoretical and Experimental Probability (pp. 486–490)

The student council surveyed their classmates to find what they want to eat for their end-of-year celebration. The results are shown in the table. Find the experimental probability of each event.

Meal	Number of Students
pizza	26
chicken	20
hamburger	13
hot dog	5

30. P(hot dog)   31. P(chicken)

32. P(hamburger)   33. P(hot dog or hamburger)

34. P(*not* pizza)   35. P(pizza or chicken)

36. **PROBABILITY** If a spinner has four equal sections labeled 1–4, what is the theoretical probability of landing on 2?

Example 8 A coin is tossed 65 times, and it lands on tails 40 times. What is the experimental probability of the coin landing on heads?

The coin landed on heads 25 times.

$$P(\text{heads}) = \frac{\text{number of times heads occurs}}{\text{total number of possible outcomes}}$$
$$= \frac{25}{65} \text{ or } \frac{5}{13}$$

So, the experimental probability of the coin landing on heads is $\frac{4}{15}$ or about 27%.

9-8 Compound Events (pp. 492–497)

6.S.9

A bag contains 6 green, 8 white, and 2 blue counters. Two are randomly drawn. Find each probability if the first counter is replaced before the second counter is drawn. Then find each probability if the first counter is not replaced.

37. P(green, blue)

38. P(2 white)

39. P(blue, *not* white)

40. P(white, *not* green)

41. **PROBABILITY** A coin is tossed and a number cube is rolled. Find the probability that tails and a number less than 5 comes up.

42. **COMPUTERS** A computer randomly generates a digit from 0–9. Find the probability that an odd number or the number 8 is generated.

Example 9 A number cube is rolled and a spinner with 8 equal sections labeled A through H is spun. Find the probability of rolling an even number and spinning a vowel.

$$P(\text{even}) = \frac{3}{6}$$

$$P(\text{vowel}) = \frac{2}{8}$$

$$\frac{3}{6} \cdot \frac{2}{8} = \frac{6}{48} \text{ or } \frac{1}{8}$$

So, the probability of rolling an even number and spinning a vowel is $\frac{1}{8}$, or about 12.5%.

Preparing for NYSMT
For test-taking strategies and
practice, see pages 716–733.

7. Hanako spins each spinner shown below once. Find the total possible letter/number combinations that could have resulted from Hanako's spins. **C**

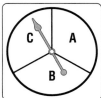

First Spinner   Second Spinner

A 3   **C** 9

B 6   **D** 12

8. Juan rolled a number cube four times. Each time, the number 3 appeared. If Juan rolls the number cube one more time, what is the probability that the number 3 will appear?

F $\frac{2}{3}$ **H** $\frac{1}{6}$

G $\frac{1}{2}$ **J** $\frac{5}{6}$

TEST-TAKING TIP

Question 8 You may want to find your own answer before looking at the answer choices. This keeps you from choosing an answer that looks correct, but is still wrong.

9. The owner of a fruit stand has x pounds of apples on display. She sells 30 pounds and then adds $4y$ pounds of apples to the display. Which of the following expressions represents the weight in pounds of the apples that are now on the display?

A $x + 30 + 4y$

B $x - 30 + 4y$

C $x + 30 - 4y$

D $x - 30 - 4y$

PART 2   Short Response/Grid In

Record your answers on the answer sheet provided by your teacher or on a sheet of paper.

10. A building is 182 meters tall. About how tall is the building in feet and inches? (1 meter ≈ 39 inches)

11. Wilson has 7 different pieces of fruit in his refrigerator. If he randomly selects 3 pieces of fruit, how many possible unique combinations are there?

12. In how many ways can you select 3 flowers out of 9?

PART 3   Extended Response

Record your answers on the answer sheet provided by your teacher or on a sheet of paper. Show your work.

13. Audrey is going on vacation. She packs 4 shirts and 3 pairs of pants. The shirts are red, green, yellow, and pink. The pants are black, white, and brown.

   a. Make a tree diagram that shows all of the outfits that Audrey can make.

   b. If Audrey chooses a shirt and pair of pants at random, what is the probability that she will wear a red shirt with white pants on the first day of her vacation?

   c. Audrey does not repeat outfits. What is the probability that she will wear a yellow shirt with black pants on the first day and a green shirt with brown pants on the second day?

NEED EXTRA HELP?													
If You Missed Question...	1	2	3	4	5	6	7	8	9	10	11	12	13
Go to Lesson...	9-2	1-1	7-1	8-2	8-4	5-7	9-2	9-8	3-1	6-8	9-2	9-5	9-8
NYS Core Curriculum	7.S.8	7.PS.14	8.N.3	7.S.6	7.S.6	6.N.18	7.S.8	6.S.9	7.A.1	7.PS.10	7.S.8	7.S.8	6.S.9

Unit 5
Geometry and Measurement

Focus
Use formulas to determine surface areas and volumes of three-dimensional shapes.

CHAPTER 10
Geometry: Polygons

BIG Idea Identify and describe properties of two-dimensional figures.

CHAPTER 11
Measurement: Two- and Three-Dimensional Figures

BIG Idea Use formulas to determine area and volume of two- and three-dimensional figures.

BIG Idea Derive and determine the area of a circle.

CHAPTER 12
Looking Ahead to Next Year: Geometry and Measurement

BIG Idea Find length using the Pythagorean Theorem.

BIG Idea Find surface areas of rectangular prisms and cylinders.

506

Problem Solving in Life Skills

Real-World Unit Project

Design That House Are you ready to become an architect? You've been selected to design and decorate a dream house. Along the way, you'll determine actual angle measures and wall lengths of the blueprint you'll be creating of your dream house. Your journey will begin soon, so pack your geometry tool kit.

NY Math Online Log on to glencoe.com to begin.

Geometry: Polygons

New York State Core Curriculum

7.G.7 Find a missing angle when given angles of a quadrilateral

7.S.2 Display data in a circle graph

Key Vocabulary

complementary angles (p. 514)

line of symmetry (p. 558)

similar figures (p. 540)

supplementary angles (p. 514)

🌐 Real-World Link

Tulips Holland, Michigan, hosts a Tulip Time Festival each May. Geometry is used to explain how a tulip shows rotational symmetry.

Geometry: Polygons Make this Foldable to help you organize your notes. Begin with a piece of 11" by 17" paper.

① **Fold** a 2" tab along the long side of the paper.

② **Unfold** the paper and fold in thirds widthwise.

③ **Open** and draw lines along the folds. Label the head of each column as shown. Label the front of the folded table with the chapter title.

What I Know About Polygons	What I Need to Know	What I've Learned

GET READY for Chapter 10

Diagnose Readiness You have two options for checking Prerequisite Skills.

Option 2

NY Math Online Take the Online Readiness Quiz at glencoe.com.

Option 1

Take the Quick Quiz below. Refer to the Quick Review for help.

QUICK Quiz

Multiply or divide. Round to the nearest hundredth if necessary.
(Prior Grade)

1. 360×0.85 **2.** $48 \div 191$

3. $24 \div 156$ **4.** 0.37×360

5. $33 \div 307$ **6.** 0.69×360

Solve each equation. (Lesson 3-2)

7. $122 + x + 14 = 180$

8. $45 + 139 + k + 17 = 360$

9. SCHOOL There are 180 school days at Lee Middle School. If school has been in session for 62 days and there are 13 days until winter break, how many school days are after the break? (Lesson 3-2)

Solve each proportion. (Lesson 6-5)

10. $\dfrac{4}{a} = \dfrac{3}{9}$ **11.** $\dfrac{7}{16} = \dfrac{h}{32}$

12. $\dfrac{5}{8} = \dfrac{15}{y}$ **13.** $\dfrac{t}{42} = \dfrac{6}{7}$

14. READING Sandra can read 28 pages of a novel in 45 minutes. At this rate, how many pages can she read in 135 minutes?
(Lesson 6-6)

QUICK Review

Example 1
Find 0.92×360.

$$
\begin{array}{r}
360 \\
\times\ 0.92 \quad \leftarrow \text{two decimal places} \\
\hline
720 \\
+\ 32400 \\
\hline
331.20 \quad \leftarrow \text{two decimal places}
\end{array}
$$

So, $0.92 \times 360 = 331.2$.

Example 2
Solve the equation.
$46 + 90 + p = 180$.

$46 + 90 + p =$	180	Write the equation.
$136 + p =$	180	Add 46 and 90.
$-\,136$	$-\,136$	Subtract 136 from
$p =$	44	each side.

The solution to the equation $46 + 90 + p = 180$ is $p = 44$.

Example 3
Solve the proportion $\dfrac{3}{8} = \dfrac{g}{48}$.

$\dfrac{3}{8} = \dfrac{g}{48}$ Write a proportion.

$\overset{\times\,6}{\dfrac{3}{8}} = \dfrac{18}{48}$ Since $8 \times 6 = 48$, multiply 3 by 6 to find g.

So, $g = 18$.

MAIN IDEA

Classify angles and identify vertical and adjacent angles.

NYS Core Curriculum

Preparation for 8.G.1
Identify pairs of vertical angles as congruent
Preparation for 8.G.4
Determine angle pair relationships when given two parallel lines cut by a transversal

New Vocabulary

angle
degrees
vertex
congruent angles
right angle
acute angle
obtuse angle
straight angle
vertical angles
adjacent angles

NY Math Online

glencoe.com

• Extra Examples
• Personal Tutor
• Self-Check Quiz
• Reading in the Content Area

▷ **GET READY** for the Lesson

ROLLER COASTERS The angles of descent of a roller coaster are shown.

1. The roller coaster at the right shows two angles of descent. Draw an angle between 44° and 70°.

2. Some roller coasters have an angle of descent that is 90°, known as a vertical angle of descent. Draw a vertical angle of descent.

An **angle** has two sides that share a common endpoint and is measured in units called **degrees**. If a circle were divided into 360 equal-sized parts, each part would have an angle measure of 1 degree (1°).

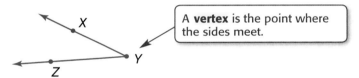

A **vertex** is the point where the sides meet.

An angle can be named in several ways. The symbol for angle is ∠.

EXAMPLE Naming Angles

① **Name the angle at the right.**

- Use the vertex as the middle letter and a point from each side.
  ∠ABC or ∠CBA

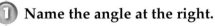

- Use the vertex only.
  ∠B

- Use a number.
  ∠1

The angle can be named in four ways: ∠ABC, ∠CBA, ∠B, or ∠1.

✓ **CHECK** Your Progress

a. Name the angle shown in four ways.

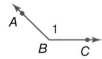

Angles are classified according to their measure. Two angles that have the same measure are said to be **congruent**.

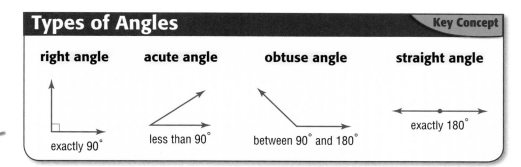

Types of Angles   **Key Concept**

right angle	acute angle	obtuse angle	straight angle
exactly 90°	less than 90°	between 90° and 180°	exactly 180°

EXAMPLES Classify Angles

Classify each angle as *acute, obtuse, right,* or *straight.*

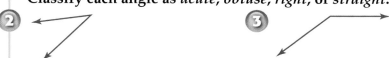

2 The angle is less than 90°, so it is an acute angle.

3 The angle is between 90° and 180°, so it is an obtuse angle.

✓**CHECK Your Progress**

Classify each angle as *acute, obtuse, right,* or *straight.*

b.   c.   d.

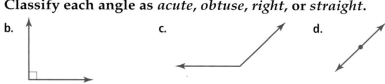

Vertical Angles   **Key Concept**

Words   Two angles are **vertical** if they are opposite angles formed by the intersection of two lines.

Examples

∠1 and ∠3 are vertical angles.
∠2 and ∠4 are vertical angles.

Adjacent Angles

Words   Two angles are **adjacent** if they share a common vertex, a common side, and do not overlap.

Examples

Adjacent angle pairs are ∠1 and ∠2, ∠2 and ∠3, ∠3 and ∠4, and ∠4 and ∠1.

∠5 and ∠6 are adjacent angles.

4 **INTERSECTIONS** Identify a pair of vertical angles in the diagram at the right. Justify your response.

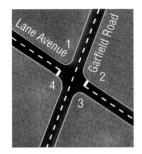

Since ∠2 and ∠4 are opposite angles formed by the intersection of two lines, they are vertical angles. Similarly, ∠1 and ∠3 are also vertical angles.

CHECK Your Progress

Refer to the diagram at the right. Identify each of the following. Justify your response.

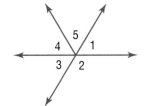

e. a pair of vertical angles

f. a pair of adjacent angles

CHECK Your Understanding

Examples 1–3
(pp. 510–511)

Name each angle in four ways. Then classify the angle as *acute*, *right*, *obtuse*, **or** *straight*.

1.

2.
(diagram: S, R, T, angle 3)

Example 4
(p. 512)

3. **RAILROADS** Identify a pair of vertical angles on the railroad crossing sign. Justify your response.

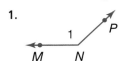

Practice and Problem Solving

HOMEWORK HELP

For Exercises	See Examples
4–9	1–3
10–17	4

Name each angle in four ways. Then classify the angle as *acute*, *right*, *obtuse*, **or** *straight*.

4.
(diagram: A, B, C, angle 4)

5.
(diagram: D, E, F, angle 5)

6.
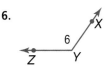

7.

8.
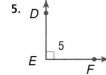

9.
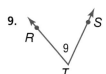

For Exercises 10–15, refer to the diagram at the right.
Identify each angle pair as *adjacent,* *vertical,* **or** *neither.*

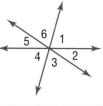

10. ∠2 and ∠5   **11.** ∠4 and ∠6   **12.** ∠3 and ∠4

13. ∠5 and ∠6   **14.** ∠1 and ∠3   **15.** ∠1 and ∠4

GEOGRAPHY For Exercises 16 and 17, use the diagram at the right and the following information.

The corner where the states of Utah, Arizona, New Mexico, and Colorado meet is called the Four Corners.

NYSCC • NYSMT
Extra Practice, pp. 693, 713

16. Identify a pair of vertical angles. Justify your response.

17. Identify a pair of adjacent angles. Justify your response.

H.O.T. Problems

CHALLENGE For Exercises 18 and 19, determine whether each statement is *true* or *false*. If the statement is true, provide a diagram to support it. If the statement is false, explain why.

18. A pair of obtuse angles can also be vertical angles.

19. A pair of straight angles can also be adjacent angles.

20. **WRITING IN MATH** Describe the differences between vertical and adjacent angles.

NYSMT PRACTICE   8.G.1, 8.G.4

21. Which word best describes the angle marked in the figure?

angle
YIELD

A acute

B obtuse

C right

D straight

22. Which statement is true?

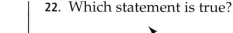
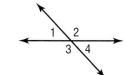

F ∠1 and ∠4 are adjacent angles.

G ∠2 and ∠3 are vertical angles.

H ∠3 and ∠4 are vertical angles.

J ∠2 and ∠3 are adjacent angles.

Spiral Review

A coin is tossed twice and a number cube is rolled. Find each probability. (Lesson 9-8)

23. P(2 heads and 6)   **24.** P(1 head, 1 tail, and a 3)   **25.** P(2 tails and *not* 4)

26. **PROBABILITY** A spinner is spun 20 times, and it lands on the color red 5 times. What is the experimental probability of *not* landing on red? (Lesson 9-7)

▷ **GET READY for the Next Lesson**

ALGEBRA Solve each equation. Check your solution. (Lesson 3-2)

27. $44 + x = 90$ **28.** $117 + x = 180$ **29.** $90 = 36 + x$ **30.** $180 = 75 + x$

Complementary and Supplementary Angles

▷ **MINI Lab**

GEOMETRY Refer to ∠A shown at the right.

1. Classify it as *acute, right, obtuse,* or *straight.*

2. Copy the angle onto a piece of paper. Then draw a ray that separates the angle into two congruent angles. Label these angles ∠1 and ∠2.

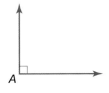

3. What is $m\angle 1$ and $m\angle 2$?

4. What is the sum of $m\angle 1$ and $m\angle 2$?

5. Copy the original angle onto a piece of paper. Then draw a ray that separates the angle into two non-congruent angles. Label these angles ∠3 and ∠4.

6. What is true about the sum of $m\angle 3$ and $m\angle 4$?

7. Complete Exercises 1–6 for ∠B shown at the right.

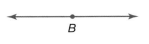

A special relationship exists between two angles whose sum is 90°. A special relationship also exists between two angles whose sum is 180°.

Complementary Angles		**Key Concept**
Words	Two angles are **complementary** if the sum of their measures is 90°.	
Examples		

$m\angle 1 + m\angle 2 = 90°$ $55° + 35° = 90°$

Supplementary Angles	
Words	Two angles are **supplementary** if the sum of their measures is 180°.
Examples	

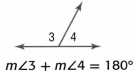 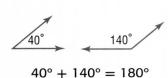

$m\angle 3 + m\angle 4 = 180°$ $40° + 140° = 180°$

Reading Math

Angle Measure The notation $m\angle 1$ is read *the measure of angle 1.*

You can use these relationships to identify complementary and supplementary angles.

Identify each pair of angles as *complementary,* *supplementary,* **or** *neither.*

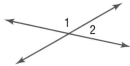

∠1 and ∠2 form a straight angle. So, the angles are supplementary.

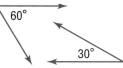

60° + 30° = 90°
The angles are complementary.

CHECK Your Progress

Identify each pair of angles as *complementary,* *supplementary,* **or** *neither.*

a.

b.

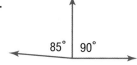

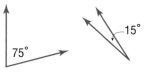

You can use angle relationships to find missing measures.

EXAMPLE Find a Missing Angle Measure

Reading Math

Perpendicular Lines or sides that meet to form right angles are perpendicular.

 ALGEBRA Find the value of *x*.

Since the two angles form a right angle, they are complementary.

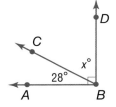

Words	The sum of the measures of ∠ABC and ∠CBD	is	90°.
Variable	Let *x* represent the measure of ∠CBD.		
Equation	28 + *x*	=	90

$$28 + x = 90 \qquad \text{Write the equation.}$$
$$\underline{-\,28 \qquad\quad -\,28} \qquad \text{Subtract 28 from each side.}$$
$$x = 62$$

So, the value of *x* is 62.

CHECK Your Progress

c. **ALGEBRA** Find the value of *x*.

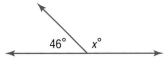

d. **ALGEBRA** If ∠*J* and ∠*K* are complementary and the measure of ∠*K* is 65°, what is the measure of ∠*J*?

Examples 1, 2
(p. 515)

Identify each pair of angles as *complementary*, *supplementary*, or *neither*.

1.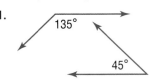

2.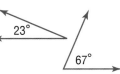

Example 3
(p. 515)

3. **ALGEBRA** Find the value of *x*.

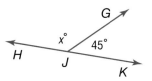

Practice and Problem Solving

HOMEWORK HELP	
For Exercises	**See Examples**
4–9	1, 2
10–11	3

Identify each pair of angles as *complementary*, *supplementary*, or *neither*.

4.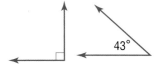

5.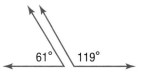

6.

7.

8.

9.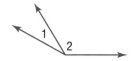

10. **ALGEBRA** If ∠A and ∠B are complementary and the measure of ∠B is 67°, what is the measure of ∠A?

11. **ALGEBRA** What is the measure of ∠J if ∠J and ∠K are supplementary and the measure of ∠K is 115°?

12. **SCHOOL SUPPLIES** What is the measure of the angle given by the opening of the scissors, *x*?

13. **SKATEBOARDING** A skateboard ramp forms a 43° angle as shown. Find the unknown angle.

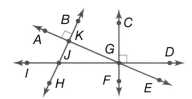

Use the figure at the right to name the following.

14. a pair of supplementary angles

15. a pair of complementary angles

16. a pair of vertical angles

NYSCC • NYSMT
Extra Practice, pp. 693, 713

GEOMETRY For Exercises 17–20, use the figure at the right.

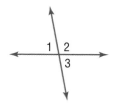

17. Are ∠1 and ∠2 vertical angles, adjacent angles, or neither? ∠2 and ∠3? ∠1 and ∠3?

18. Write an equation representing the sum of $m\angle1$ and $m\angle2$. Then write an equation representing the sum of $m\angle2$ and $m\angle3$.

19. Solve the equations you wrote in Exercise 18 for $m\angle1$ and $m\angle3$, respectively. What do you notice?

20. **MAKE A CONJECTURE** Use your answer to Exercise 19 to make a conjecture as to the relationship between vertical angles.

H.O.T. Problems

21. **CHALLENGE** Angles E and F are complementary. If $m\angle E = x - 10$ and $m\angle F = x + 2$, find the measure of each angle.

22. **WRITING IN MATH** Describe a strategy for determining whether two angles are *complementary*, *supplementary*, or *neither* without knowing or measuring each angle using a protractor.

NYSMT PRACTICE 8.G.2, 8.G.3

23. In the figure below, $m\angle YXZ = 35°$ and $m\angle WXV = 40°$. What is $m\angle ZXW$?

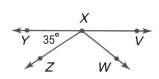

A 180°   **C** 75°

B 105°   **D** 15°

24. Which is a true statement about angles 1 and 2 shown below?

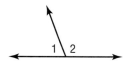

F ∠1 is complementary to ∠2.

G ∠1 and ∠2 are vertical angles.

H ∠1 is supplementary to ∠2.

J Both angles are obtuse.

Spiral Review

25. Name the angle at the right in four ways. Then classify it as *acute*, *right*, *obtuse*, or *straight*. (Lesson 10-1)

26. **MEASUREMENT** A house for sale has a rectangular lot with a length of 250 feet and a width of 120 feet. What is the area of the lot? (Lesson 3-6)

GET READY for the Next Lesson

PREREQUISITE SKILL Multiply or divide. Round to the nearest hundredth if necessary. (pp. 674 and 676)

27. $0.62 \cdot 360$ **28.** $360 \cdot 0.25$ **29.** $17 \div 146$ **30.** $63 \div 199$

10-3

Statistics: Display Data in a Circle Graph

▷ **GET READY** for the Lesson

VEGETABLES The students at Pine Ridge Middle School were asked to identify their favorite vegetable. The table shows the results of the survey.

Favorite Vegetable	
Vegetable	**Percent**
Carrots	45%
Green Beans	23%
Peas	17%
Other	15%

1. Explain how you know that each student only selected one favorite vegetable.

2. If 400 students participated in the survey, how many students preferred carrots?

A graph that shows data as parts of a whole is called a **circle graph**. In a circle graph, the percents add up to 100.

EXAMPLE Display Data in a Circle Graph

1 **VEGETABLES** Display the data above in a circle graph.

• There are 360° in a circle. Find the degrees for each part.

45% of 360° = 0.45 · 360° or 162°

23% of 360° = 0.23 · 360° or 83°   Round to the nearest whole degree.

17% of 360° = 0.17 · 360° or about 61°

15% of 360° = 0.15 · 360° or about 54°

• Draw a circle with a radius as shown. Then use a protractor to draw the first angle, in this case 162°. Repeat this step for each section or *sector*.

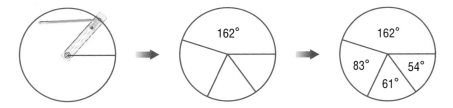

• Label each section of the graph with the category and percent it represents. Give the graph a title.

Check   The sum of the angle measures should equal to 360°.
162° + 83° + 61° + 54° = 360°

Favorite Vegetable

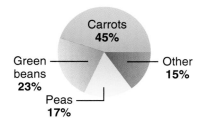

a. **SCIENCE** The table shows the present composition of Earth's atmosphere. Display the data in a circle graph.

Composition of Earth's Atmosphere	
Element	**Percent**
Nitrogen	78%
Oxygen	21%
Other gases	1%

When constructing a circle graph, you may need to first convert the data to ratios and decimals and then to degrees and percents.

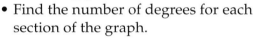 **EXAMPLE** Construct a Circle Graph

② **ANIMALS** The table shows endangered species in the United States. Make a circle graph of the data.

Species	Number of Species
Mammals	68
Birds	77
Reptiles	14
Amphibians	11

Source: U.S. Fish & Wildlife Service

- Find the total number of species:
  $68 + 77 + 14 + 11 = 170$.

- Find the ratio that compares each number with the total. Write the ratio as a decimal rounded to the nearest hundredth.

  mammals: $\frac{68}{170} \approx 0.40$ birds: $\frac{77}{170} \approx 0.45$

  reptiles: $\frac{14}{170} \approx 0.08$ amphibians: $\frac{11}{170} \approx 0.06$

- Find the number of degrees for each section of the graph.

  mammals: $0.40 \cdot 360° = 144°$

  birds: $0.45 \cdot 360° \approx 162°$

  reptiles: $0.08 \cdot 360° \approx 29°$

  amphibians: $0.06 \cdot 360° \approx 22°$

  Because of rounding, the sum of the degrees is 357°.

- Draw the circle graph.

  $0.40 = 40\%$, $0.45 = 45\%$,
  $0.08 = 8\%$, $0.06 = 6\%$

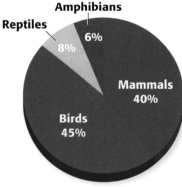

Check   After drawing the first two sections, you can measure the last section of a circle graph to verify that the angles have the correct measures.

Real-World Link . . . .
The Carolina Northern and Virginian Northern Flying Squirrel are both endangered. The northern flying squirrel is a small nocturnal gliding mammal that is about 10 to 12 inches in total length and weighs about 3–5 ounces.

Source: U.S. Fish and Wildlife Service

 CHECK Your Progress

b. **OLYMPICS** The number of Winter Olympic medals won by the U.S. from 1924 to 2006 is shown in the table. Display the data in a circle graph.

U.S. Winter Olympic Medals	
Type	**Number**
Gold	78
Silver	81
Bronze	59

AUTOMOBILES The graph shows the percent of automobiles registered in the western United States in a recent year.

U.S. Registered Automobiles in West

Washington 13%
Oregon 6%
Nevada 3%
California 78%

Source: Bureau of Transportation Statistics

3 Which state had the most registered automobiles?

The largest section of the circle is the one representing California. So, California has the most registered automobiles.

4 If 24.0 million automobiles were registered in these states, how many more automobiles were registered in California than Oregon?

California: 78% of 24.0 million → 0.78 × 24.0, or 18.72 million

Oregon: 6% of 24.0 million → 0.06 × 24.0, or 1.44 million

There were 18.72 million − 1.44 million, or 17.28 million more registered automobiles in California than in Oregon.

Study Tip

Check for Reasonableness
To check Example 4, you can estimate and solve the problem another way.

78% − 6% ≈ 70%
70% of 24 is 17

Since 17.28 is about 17, the answer is reasonable.

 CHECK Your Progress

c. Which state had the least number of registered automobiles? Explain.
d. What was the total number of registered automobiles in Washington and Oregon?

CHECK Your Understanding

Examples 1, 2
(pp. 518–519)

Display each set of data in a circle graph.

1.

Blood Types in the U.S.	
Blood Type	**Percent**
O	44%
A	42%
B	10%
AB	4%

Source: Stanford School of Medicine

2.

Favorite Musical Instrument	
Type	**Number of Students**
Piano	54
Guitar	27
Drum	15
Flute	24

Examples 3, 4
(p. 520)

COLORS For Exercises 3 and 4, use the graph that shows the results of a survey.

3. What color is most favored?

4. If 400 people were surveyed, how many more people favored purple than red?

Favorite Color

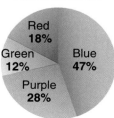

Red 18%
Green 12%
Blue 47%
Purple 28%

Display each set of data in a circle graph.

5.

U.S. Steel Roller Coasters	
Type	**Percent**
Sit down	86%
Inverted	8%
Other	6%

6.

U.S. Orange Production	
State	**Orange Production**
California	18%
Florida	81%
Texas	1%

Source: National Agriculture Statistics Service

7.

Animals in Pet Store	
Animal	**Number of Pets**
Birds	13
Cats	11
Dogs	9
Fish	56
Other	22

8.

Favorite Games	
Type of Game	**Number of Students**
Card	7
Board	9
Video	39
Sports	17
Drama	8

LANDFILLS For Exercises 9-11, use the circle graph that shows what is in United States landfills.

9. What takes up the most space in landfills?

10. About how many times more paper is there than food and yard waste?

11. If a landfill contains 200 million tons of trash, how much of it is plastic?

What is in U.S. Landfills?

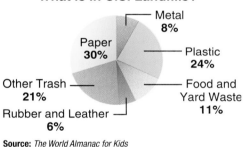

- Metal 8%
- Paper 30%
- Plastic 24%
- Other Trash 21%
- Rubber and Leather 6%
- Food and Yard Waste 11%

Source: *The World Almanac for Kids*

MONEY For Exercises 12–14, use the graph that shows the results of a survey about a common currency for North America.

Do Americans Favor Common North American Currency?

No 53%

Yes 43%

Don't Know 4%

Source: Coinstar

12. What percent of Americans favor a common North American currency?

13. Based on these results, about how many of the approximately 298 million Americans would say "Don't Know" in response to this survey?

14. About how many more Americans oppose a common currency than favor it?

DATA SENSE For each graph, find the missing values.

15. **Dog Expenses**

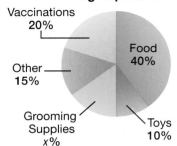

Vaccinations 20%
Food 40%
Other 15%
Grooming Supplies x%
Toys 10%

16. **Family Budget**

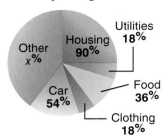

Utilities 18%
Other x%
Housing 90%
Car 54%
Food 36%
Clothing 18%

Select an appropriate type of graph to display each set of data: line graph, bar graph, or circle graph. Then display the data using the graph.

17.

Top 5 Presidential Birth States	
Place	**Presidents**
Virginia	8
Ohio	7
Massachusetts	4
New York	4
Texas	3

Source: *The World Book of Facts*

18.

Tanya's Day	
Activity	**Percent**
School	25%
Sleep	33%
Homework	12%
Sports	8%
Other	22%

GEOGRAPHY For Exercises 19–21, use the table.

19. Display the data in a circle graph.

20. Use your graph to find which two lakes equal the size of Lake Superior.

21. Compare the size of Lake Ontario to the size of Lake Michigan.

Sizes of U.S. Great Lakes	
Lake	**Size (sq mi)**
Erie	9,930
Huron	23,010
Michigan	22,400
Ontario	7,520
Superior	31,820

POLITICS For Exercises 22 and 23, use the graph and information below.

A group of students were asked whether people their age could make a difference in the political decisions of elected officials.

22. How many students participated in the survey?

23. Write a convincing argument explaining whether or not it is reasonable to say that 50% more students said they could make a difference than those who said they could not make a difference.

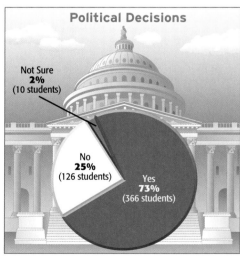

Political Decisions

Not Sure 2% (10 students)
No 25% (126 students)
Yes 73% (366 students)

Source: *Mom's Life* and *Mothering Magazine*

24. **FIND THE DATA** Refer to the Data File on pages 16–19. Choose some data that can be displayed in a circle graph. Then display the data in a circle graph and write one statement analyzing the data.

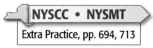

NYSCC • NYSMT
Extra Practice, pp. 694, 713

H.O.T. Problems

25. CHALLENGE The graph shows the results of a survey about students' favorite school subject. About what percent of those surveyed said that math was their favorite subject? Explain your reasoning.

26. COLLECT THE DATA Collect some data from your classmates that can be represented in a circle graph. Then create the circle graph and write one statement analyzing the data.

27. ❨ **WRITING IN** ❩ **MATH** The table shows the percent of people that like each type of fruit juice. Can the data be represented in a circle graph? Justify your answer.

Favorite Subject

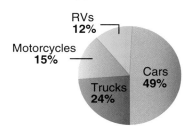

Fruit Juice	Percent
Apple	54%
Grape	48%
Orange	37%
Cranberry	15%

NYSMT PRACTICE 7.M.8, 7.S.2

28. The graph shows the type of vehicles that used Highway 82 during one month.

Types of Vehicles

RVs
12%

Motorcycles
15%

Trucks
24%

Cars
49%

Which statement is true according to the circle graph shown?

A More cars used the highway than RVs and trucks combined.

B More than half the vehicles that used the highway were cars.

C More RVs used the highway than trucks.

D More trucks used the highway than cars.

Spiral Review

29. GEOMETRY Refer to the diagram at the right. Identify a pair of vertical angles. (Lesson 10-1)

30. ALGEBRA $\angle A$ and $\angle B$ are complementary. If $m\angle A = 15°$, find $m\angle B$. (Lesson 10-2)

▷ **GET READY** for the Next Lesson

PREREQUISITE SKILL Solve each equation. (Lesson 3-2)

31. $x + 112 = 180$ **32.** $50 + t = 180$ **33.** $180 = 79 + y$ **34.** $180 = h + 125$

10-4 Triangles

MAIN IDEA

Identify and classify triangles.

NYS Core Curriculum

Preparation for 7.G.7
Find a missing angle when given angles of a quadrilateral

New Vocabulary

triangle
congruent segments
acute triangle
right triangle
obtuse triangle
scalene triangle
isosceles triangle
equilateral triangle

NY Math Online

glencoe.com

• Extra Examples
• Personal Tutor
• Self-Check Quiz

▷ **MINI Lab**

STEP 1 Use a straightedge to draw a triangle with three acute angles. Label the angles A, B, and C. Cut out the triangle.

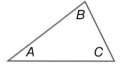

STEP 2 Fold ∠A, ∠B, and ∠C so the vertices meet on the line between angles A and C.

1. What kind of angle is formed where the three vertices meet?

2. Repeat the activity with another triangle. Make a conjecture about the sum of the measures of the angles of any triangle.

A **triangle** is a figure with three sides and three angles. The symbol for triangle is △. There is a relationship among the three angles in a triangle.

Angles of a Triangle		Key Concept
Words	The sum of the measures of the angles of a triangle is 180°.	**Model** 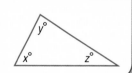
Algebra	$x + y + z = 180$	

EXAMPLE Find a Missing Measure

1 **ALGEBRA** Find $m\angle Z$ in the triangle.

Since the sum of the angle measures in a triangle is 180°, $m\angle Z + 43° + 119° = 180°$.

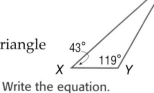

$$m\angle Z + 43° + 119° = 180°$$ Write the equation.

$$m\angle Z + 162° = 180°$$ Simplify.

$$\underline{-162° = -162°}$$ Subtract 162° from each side.

$$m\angle Z = 18°$$

So, $m\angle Z$ is 18°.

✓ **CHECK Your Progress**

a. **ALGEBRA** In △ABC, if $m\angle A = 25°$ and $m\angle B = 108°$, what is $m\angle C$?

NYSMT EXAMPLE

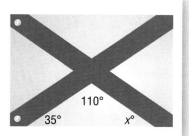

2 The Alabama state flag is constructed with the triangles shown. What is the missing measure in the design?

A 145°	**C** 35°
B 75°	**D** 25°

Read the Item

To find the missing measure, write and solve an equation.

Solve the Item

$$x + 110 + 35 = 180 \quad \text{The sum of the measures is 180.}$$
$$x + 145 = 180 \quad \text{Simplify.}$$
$$\underline{-145 = -145} \quad \text{Subtract 145 from each side.}$$
$$x = 75$$

The answer is C.

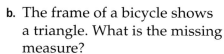 **CHECK Your Progress**

b. The frame of a bicycle shows a triangle. What is the missing measure?

F 31°	**H** 45°
G 40°	**J** 50°

Every triangle has at least two acute angles. One way you can classify a triangle is by using the third angle. Another way to classify triangles is by their sides. Sides with the same length are **congruent segments**.

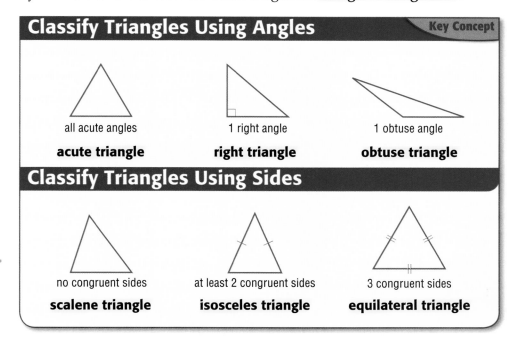

Classify Triangles Using Angles — Key Concept

all acute angles — **acute triangle**

1 right angle — **right triangle**

1 obtuse angle — **obtuse triangle**

Classify Triangles Using Sides

no congruent sides — **scalene triangle**

at least 2 congruent sides — **isosceles triangle**

3 congruent sides — **equilateral triangle**

Study Tip

Congruent Segments
The marks on the sides of the triangle indicate that those sides are congruent.

③ Classify the marked triangle at the right by its angles and by its sides.

The triangle on the side of a house has one obtuse angle and two congruent sides. So, it is an obtuse, isosceles triangle.

✓ **CHECK Your Progress**

c.
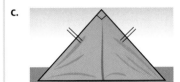

d.

Real-World Link · · · ·
There are two main types of roofs—flat and pitched. Most houses have pitched, or sloped, roofs. A pitched roof generally lasts 15 to 20 years.

Source: National Association of Certified Home Inspectors

EXAMPLES Draw Triangles

④ Draw a triangle with one right angle and two congruent sides. Then classify the triangle.

Draw a right angle. The two segments should be congruent.

Connect the two segments to form a triangle.

The triangle is a right isosceles triangle.

⑤ Draw a triangle with one obtuse angle and no congruent sides. Then classify the triangle.

Draw an obtuse angle. The two segments of the angle should have different lengths.

Connect the two segments to form a triangle.

The triangle is an obtuse scalene triangle.

✓ **CHECK Your Progress**

Draw a triangle that satisfies each set of conditions below. Then classify each triangle.

e. a triangle with three acute angles and three congruent sides

f. a triangle with one right angle and no congruent sides

Example 1
(p. 524)

Find the value of x.

1.

2.
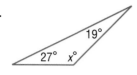

3.

4. **ALGEBRA** Find $m\angle T$ in $\triangle RST$ if $m\angle R = 37°$ and $m\angle S = 55°$.

Example 2
(p. 525)

5. **MULTIPLE CHOICE** A triangle is used in the game of pool to rack the pool balls. Find the missing measure of the triangle.

A 30°   C 60°

B 40°   D 75°

Example 3
(p. 526)

NATURE Classify the marked triangle in each object by its angles and by its sides.

6.

7.

8.

Examples 4, 5
(p. 526)

DRAWING TRIANGLES For Exercises 9 and 10, draw a triangle that satisfies each set of conditions. Then classify each triangle.

9. a triangle with three acute angles and two congruent sides

10. a triangle with one obtuse angle and two congruent sides

Practice and Problem Solving

HOMEWORK HELP	
For Exercises	**See Examples**
11–18, 47, 48	1–2
19–26	3
27–30	4, 5

Find the value of x.

11.
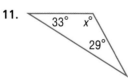

12.
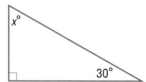

13.
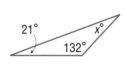

14.
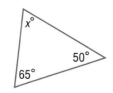

15.
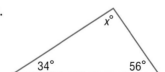

16.
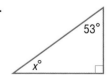

17. **ALGEBRA** Find $m\angle Q$ in $\triangle QRS$ if $m\angle R = 25°$ and $m\angle S = 102°$.

18. **ALGEBRA** In $\triangle EFG$, $m\angle F = 46°$ and $m\angle G = 34°$. What is $m\angle E$?

Classify the marked triangle in each object by its angles and by its sides.

19.

20.

21.

22.

23.

24.

25. **ART** The sculpture at the right is entitled *Texas Triangles*. It is located in Lincoln, Massachusetts. What type of triangle is shown: *acute*, *right*, or *obtuse*?

Source: DeCordova Museum and Sculpture Park

26. **ARCHITECTURE** Use the photo at the left to classify the side view of the Transamerica building by its angles and by its sides.

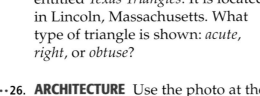

DRAWING TRIANGLES For Exercises 27–30, draw a triangle that satisfies each set of conditions. Then classify each triangle.

27. a triangle with three acute angles and no congruent sides

28. a triangle with one obtuse angle and two congruent sides

29. a triangle with three acute angles and three congruent sides

30. a triangle with one right angle and no congruent sides

Find the missing measure in each triangle with the given angle measures.

31. $80°$, $20.5°$, $x°$

32. $75°$, $x°$, $50.2°$

33. $x°$, $10.8°$, $90°$

34. $45.5°$, $x°$, $105.6°$

35. $x°$, $140.1°$, $18.6°$

36. $110.2°$, $x°$, $35.6°$

37. **ALGEBRA** Find the third angle measure of a right triangle if one of the angles measures is $10°$.

38. **ALGEBRA** What is the third angle measure of a right triangle if one of the angle measures is $45.8°$?

ALGEBRA Find the value of x in each triangle.

39.

40.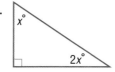

41.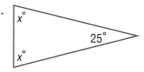

H.O.T. Problems

42. **CHALLENGE** Apply what you know about triangles to find the missing angle measures in the figure.

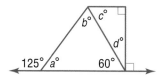

43. **OPEN ENDED** Draw an acute scalene triangle. Describe the angles and sides of the triangle.

REASONING Determine whether each statement is *sometimes*, *always*, or *never* true. Justify your answer.

44. It is possible for a triangle to have two right angles.

45. It is possible for a triangle to have two obtuse angles.

46. **WRITING IN MATH** An equilateral triangle not only has three congruent sides, but also has three congruent angles. Based on this, explain why it is impossible to draw an equilateral triangle that is either right or obtuse.

NYSMT PRACTICE 7.G.7

47. How would you find $m\angle R$?

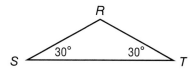

A Add 30° to 180°.

B Subtract 60° from 180°.

C Subtract 30° from 90°.

D Subtract 180° from 60°.

48. Which of the following is an acute triangle?

Spiral Review

49. **SURVEYS** A circle graph shows that 28% of people chose tea as their favorite drink. What is the measure of the angle of the tea section of the graph? (Lesson 10-3)

Classify each pair of angles as *complementary*, *supplementary*, or *neither*. (Lesson 10-2)

50.

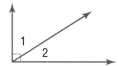

51.

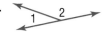

52.

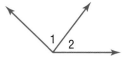

▷ **GET READY** for the Next Lesson

53. **PREREQUISITE SKILL** If Jade buys 5 notebooks at $1.75 each, what will be the total cost of the notebooks; about $6, $7, or $9? (Lesson 7-5)

Problem-Solving Investigation

MAIN IDEA: Solve problems by using logical reasoning.

 7.PS.16 Justify solution methods through logical argument *Also addresses 7.RP.6, 7.RP.8.*

P.S.I. TEAM +

e-Mail: USE LOGICAL REASONING

SANTOS: I know that at least two sides of an isosceles triangle are congruent. It also looks like two of the angles in an isosceles triangle are congruent.

YOUR MISSION: Use logical reasoning to find if the angles in an isosceles triangles are congruent.

Understand	Isosceles triangles have at least two congruent sides. We need to find if there is a relationship between the angles.
Plan	Draw several isosceles triangles and measure their angles.
Solve	In each triangle, two angles are congruent. So, it seems like an isosceles triangle has two congruent angles.
Check	Try drawing several more isosceles triangles and measuring their angles. Although this is not a proof, it is likely your conclusion is valid.

Analyze The Strategy

1. When you use *inductive reasoning*, you make a rule after seeing several examples. When you use *deductive reasoning*, you use a rule to make a decision. What type of reasoning did Santos use to solve the problem? Explain your reasoning.

2. Explain how the *look for a pattern* strategy is similar to inductive reasoning.

Mixed Problem Solving

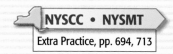

For Exercises 3–5, use logical reasoning to solve the problem. Justify your response.

3. **GEOMETRY** Draw several scalene triangles and measure their angles. What do you notice about the measures of the angles of a scalene triangle?

4. **HOUSE NUMBERS** Rico's house number contains four digits. The digits are 5, 8, 3, and 2. If his house number is odd, divisible by 3, and the middle two numbers are a perfect square, what is his house number?

5. **FRUIT** Julio, Rashanda, and Perry each brought a fruit with their lunch. The fruits brought were a mango, a banana, and an orange. If Perry did not bring a banana and Julio brought a mango, what type of fruit did each student bring?

Use any strategy to solve Exercises 6–14. Some strategies are shown below.

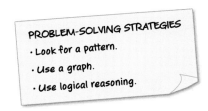

PROBLEM-SOLVING STRATEGIES
· Look for a pattern.
· Use a graph.
· Use logical reasoning.

6. **GEOMETRY** Draw several rectangles and measure their diagonals. Find a relationship between the diagonals of a rectangle.

7. **MONEY** Corelia has twice as many quarters in her purse as dimes and half as many nickels as dimes. If she has 12 quarters in her purse, what is the total amount of money she has?

8. **ALGEBRA** Find the next three numbers in the pattern below.

71, 64, 57, 50, ▥, ▥, ▥

9. **MEASUREMENT** The large square has been divided into 9 squares. The lengths of the squares are given. Find the area of the entire square.

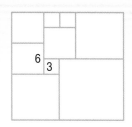

10. **READING** Sonia read 10 pages of a 150-page book on Monday. She plans to read twice as many pages each day than she did the previous day. On what day will she finish the book?

11. **SUPPLIES** Bianca has $55 to buy school supplies. She bought a backpack for $23.50, a combination lock for $6.25, and 4 binders that are $3.99 each. If the costs include tax and mechanical pencils are $2.50 per pack, how many packs can she buy?

12. **BOWLING** Colin and three of his friends are going bowling, and they have a total of $70 to spend. Suppose they buy a large pizza, four beverages, and each rent bowling shoes. How many games can they bowl if they all bowl the same number of games?

Bowling Costs	
Item	**Price**
large pizza	$15.75
beverage	$1.50
shoe rental	$3.50
game	$4.00

13. **STATISTICS** David has earned scores of 73, 85, 91, and 82 on the first four out of five math tests for the grading period. He would like to finish the grading period with a test average of at least 82. What is the minimum score David needs to earn on the fifth test in order to achieve his goal?

14. **ALLOWANCE** Samantha earns $520 in allowance yearly. Her parents promise to give her a $60 raise each year. At this rate, what will her yearly allowance be in 4 years?

Geometry Lab
Investigating Quadrilaterals

Four-sided figures are called *quadrilaterals*. In this lab, you will explore
the properties of different types of quadrilaterals.

ACTIVITY

STEP 1 Draw the quadrilaterals shown on grid paper.

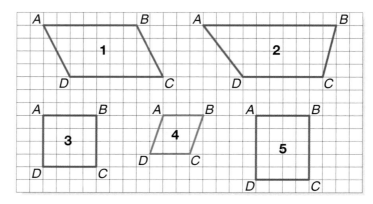

STEP 2 Use a ruler and a protractor to measure the sides and
angles of each quadrilateral. Record your results in a table.

Quadrilateral	m∠A	m∠B	m∠C	m∠D	AB	BC	CD	DA
1								
2								

ANALYZE THE RESULTS

1. Describe any similarities or patterns in the angle measurements.

2. Describe any similarities or patterns in the side measurements.

3. **MAKE A VENN DIAGRAM** Cut out the quadrilaterals you drew in the
   activity. Then sort them into categories according to their similarities
   and differences. Arrange and record your categories in a two-circled
   Venn diagram. Be sure to label each circle with its category.

4. Create two other Venn diagrams illustrating two different ways of
   categorizing these quadrilaterals.

5. **WRITING IN MATH** Did you find shapes that did not fit a category?
   Where did you place these shapes? Did any shapes have properties
   allowing them to belong to more than one category? Could you
   arrange these quadrilaterals into a three-circled Venn diagram?
   If so, how?

10-6 Quadrilaterals

MAIN IDEA

Identify and classify quadrilaterals.

NYS Core Curriculum

7.G.7 Find a missing angle when given angles of a quadrilateral

New Vocabulary

quadrilateral
parallelogram
trapezoid
rhombus

NY Math Online

glencoe.com

• Extra Examples
• Personal Tutor
• Self-Check Quiz

GET READY for the Lesson

VIDEO GAMES The general shape of a video game controller is shown.

1. Describe the angles inside the four-sided figure.

2. Which sides of the figure appear to be parallel?

3. Which sides appear to be congruent?

A **quadrilateral** is a closed figure with four sides and four angles. Quadrilaterals are named based on their sides and angles. The diagram shows how quadrilaterals are related. Notice how it goes from the most general to the most specific.

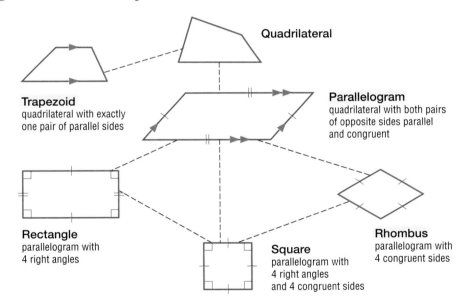

Trapezoid
quadrilateral with exactly one pair of parallel sides

Quadrilateral

Parallelogram
quadrilateral with both pairs of opposite sides parallel and congruent

Rectangle
parallelogram with 4 right angles

Square
parallelogram with 4 right angles and 4 congruent sides

Rhombus
parallelogram with 4 congruent sides

The name that *best* describes a quadrilateral is the one that is most specific.

• If a quadrilateral has all the properties of a parallelogram and a rhombus, then the *best* description of the quadrilateral is a rhombus.

• If a quadrilateral has all the properties of a parallelogram, rhombus, rectangle, and square, then the *best* description of the quadrilateral is a square.

Study Tip

Parallel Lines The sides with matching arrows are parallel.

EXAMPLES **Draw and Classify Quadrilaterals**

Draw a quadrilateral that satisfies each set of conditions. Then classify each quadrilateral with the name that best describes it.

1 **a parallellogram with four right angles and four congruent sides**

Draw one right angle. The two segments should be congruent.

Draw a second right angle that shares one of the congruent segments. The third segment drawn should be congruent to the first two segments drawn.

Study Tip

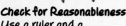

Check for Reasonableness
Use a ruler and a protractor to measure the sides and angles to verify that your drawing satisfies the given conditions.

Connect the fourth side of the quadrilateral. All four angles should be right angles, and all four sides should be congruent.

The figure is a square.

2 **a quadrilateral with opposite sides parallel**

Draw two parallel sides of equal length.

Connect the endpoints of these two sides so that two new parallel sides are drawn.

The figure is a parallelogram.

 CHECK Your Progress

Draw a quadrilateral that satisfies each set of conditions. Then classify each quadrilateral with the name that best describes it.

a. a quadrilateral with exactly one pair of parallel sides

b. a parallellogram with four congruent sides

A quadrilateral can be separated into two triangles, *A* and *B*. Since the sum of the angle measures of each triangle is 180°, the sum of the angle measures of the quadrilateral is 2 · 180, or 360°.

Angles of a Quadrilateral		**Key Concept**
Words	The sum of the measures of the angles of a quadrilateral is 360°.	**Model**
Algebra	$w + x + y + z = 360$	

EXAMPLE Find a Missing Measure

3 **ALGEBRA** Find the value of x in the quadrilateral shown.

Write and solve an equation.

Words	The sum of the measures is 360°.
Variable	Let x represent the missing measure.
Equation	$85 + 73 + 59 + x = 360$

$$85 + 73 + 59 + x = 360 \qquad \text{Write the equation.}$$
$$217 + x = 360 \qquad \text{Simplify.}$$
$$\underline{-217 = -217} \qquad \text{Subtract 217 from each side.}$$
$$x = 143$$

So, the missing angle measure is 143°.

Study Tip
Check for Reasonableness
Since $\angle x$ is an obtuse angle, $m\angle x$ should be between 90° and 180°. Since $90° < 143° < 180°$, the answer is reasonable.

CHECK Your Progress

c. ALGEBRA Find the value of x in the quadrilateral shown.

CHECK Your Understanding

Examples 1, 2
(p. 534)

Classify each quadrilateral with the name that best describes it.

1.
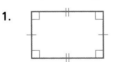

2.

3.

4. **BOATS** The photo shows a sailboat called a schooner. What type of quadrilateral does the indicated sail best represent?

Example 3
(p. 535)

5. **ALGEBRA** In quadrilateral $DEFG$, $m\angle D = 57°$, $m\angle E = 78°$, $m\angle G = 105°$. What is $m\angle F$?

ALGEBRA Find the missing angle measure in each quadrilateral.

6.
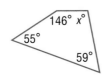

7.

8.
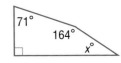

Practice and Problem Solving

Classify each quadrilateral with the name that best describes it.

HOMEWORK HELP	

For Exercises	See Examples
9–14, 23–24	1, 2
15–22	3

9.

10.

11.

12.

13.

14.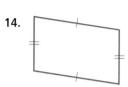

ALGEBRA Find the missing angle measure in each quadrilateral.

15.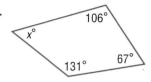

16.

17.

18.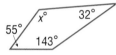

19.

20.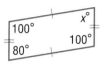

21. **ALGEBRA** Find $m\angle B$ in quadrilateral $ABCD$ if $m\angle A = 87°$, $m\angle C = 135°$, and $m\angle D = 22°$.

22. **ALGEBRA** What is $m\angle X$ in quadrilateral $WXYZ$ if $m\angle W = 45°$, $m\angle Y = 128°$, and $\angle Z$ is a right angle?

23. **LANDSCAPE** Identify the shapes of the bricks used in the design at the right. Use the name that *best* describes each brick.

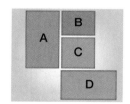

24. **MEASUREMENT** Find each of the missing angle measures a, b, c, and d in the figure at the right. Justify your answers.

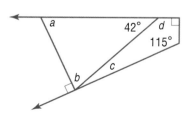

Find the missing measure in each quadrilateral with the given angle measures.

25. $37.5°$, $78°$, $115.4°$, $x°$

26. $x°$, $108.3°$, $49.8°$, $100°$

27. $25.5°$, $x°$, $165.9°$, $36.8°$

28. $79.1°$, $120.8°$, $x°$, $65.7°$

ART For Exercises 29–31, identify the types of triangles and quadrilaterals used in each quilt block pattern. Use the names that *best* describe the figures.

29.

30.

31.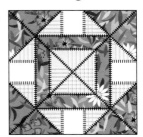

DRAWING QUADRILATERALS Determine whether each figure described below can be drawn. If the figure can be drawn, draw it. If not, explain why not.

32. a quadrilateral that is both a rhombus and a rectangle

33. a trapezoid with three right angles

34. a trapezoid with two congruent sides

ALGEBRA Find the value of x in each quadrilateral.

NYSCC • NYSMT
Extra Practice, pp. 694, 713

35.
| $2x°$ | | $2x°$ |
| $2x°$ | | $2x°$ |

36. $135°$ $135°$ $x°$ $x°$

37. $100°$ $x°$ / $x°$ $100°$

H.O.T. Problems

CHALLENGE For Exercises 38 and 39, refer to the table that gives the properties of several parallelograms. Property A states that both pairs of opposite sides are parallel and congruent.

Parallelogram	Properties
1	A, C
2	A, B, C
3	A, B

38. If property C states that all four sides are congruent, classify parallelograms 1–3. Justify your response.

39. If parallelogram 3 is a rectangle, describe Property B. Justify your response.

REASONING Determine whether each statement is *sometimes*, *always*, or *never* true. Explain your reasoning.

40. A quadrilateral is a trapezoid.

41. A trapezoid is a parallelogram.

42. A square is a rectangle.

43. A rhombus is a square.

44. **FIND THE ERROR** Isabelle and John are describing a rectangle. Who is more accurate? Explain.

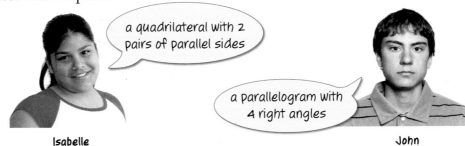

a quadrilateral with 2 pairs of parallel sides

Isabelle

a parallelogram with 4 right angles

John

45. **WRITING IN MATH** The diagonals of a rectangle are congruent, and the diagonals of a rhombus are perpendicular. Based on this information, what can you conclude about the diagonals of a square? of a parallelogram? Explain your reasoning.

46. Identify the name that does *not* describe the quadrilateral shown.

 A square

 B rectangle

 C rhombus

 D trapezoid

47. Which statement is always true about a rhombus?

 F It has 4 right angles.

 G The sum of the measures of the angles is 180°.

 H It has exactly one pair of parallel sides.

 J It has 4 congruent sides.

48. Which of the following is a correct drawing of a quadrilateral with all sides congruent and with four right angles?

 A

 B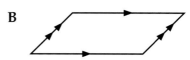

 C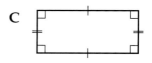

 D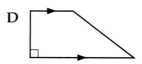

Spiral Review

49. **REASONING** Neva, Sophie, and Seth have a turtle, a dog, and a hamster for a pet, but not in that order. Sophie's pet lives in a glass aquarium and does not have fur. Neva never has to give her pet a bath. Who has what pet? Use the *logical reasoning* strategy. (Lesson 10-5)

Classify each triangle by its angles and by its sides. (Lesson 10-4)

50.    51.   52.

53. **LETTERS** How many permutations are possible of the letters in the word *Fresno?* (Lesson 9-4)

Find the sales tax or discount to the nearest cent. (Lesson 7-7)

54. $54 jacket; 7% sales tax   55. $23 hat; 15% discount

▷ **GET READY** for the Next Lesson

PREREQUISITE SKILL Solve each proportion. (Lesson 6-6)

56. $\frac{3}{5} = \frac{x}{75}$ 57. $\frac{a}{7} = \frac{18}{42}$ 58. $\frac{7}{9} = \frac{28}{m}$ 59. $\frac{3.5}{t} = \frac{16}{32}$ 60. $\frac{3}{6} = \frac{c}{5}$

Name each angle in four ways. Then classify each angle as *acute*, *obtuse*, *right*, or *straight*. (Lesson 10-1)

1.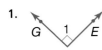

2.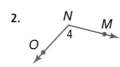

3. **MULTIPLE CHOICE** Which angle is complementary to ∠CBD? (Lesson 10-2)

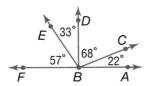

A ∠ABC   C ∠DBE

B ∠FBC   D ∠EBF

4. **SOCCER** Display the data in a circle graph. (Lesson 10-3)

Injuries of High School Girls' Soccer Players	
Position	**Percent**
Halfbacks	37%
Fullbacks	23%
Forward Line	28%
Goalkeepers	12%

EDUCATION For Exercises 5 and 6, use the circle graph that shows the percent of students by grade level in U.S. schools. (Lesson 10-3)

Grade Level of U.S. Students

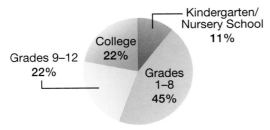

5. In which grades are most students?

6. How many times as many students are there in grades 1–8 than in grades 9–12?

ALGEBRA Find the value of x. (Lesson 10-4)

7.

8.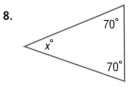

9. **MULTIPLE CHOICE** In triangle ABC, $m\angle A = 62°$ and $m\angle C = 44°$. What is $m\angle B$? (Lesson 10-4)

F 90°   H 64°

G 74°   J 42°

10. **RACES** Norberto, Isabel, Fiona, Brock, and Elizabeth were the first five finishers of a race. From the given clues, find the order in which they finished. Use the *logical reasoning* strategy. (Lesson 10-5)

- Norberto passed Fiona just before the finish line.
- Elizabeth finished 5 seconds ahead of Norberto.
- Isabel crossed the finish line after Fiona.
- Brock was fifth at the finish line.

Classify the quadrilateral with the name that best describes it. (Lesson 10-6)

11.

12.

ALGEBRA Find the value of x in each quadrilateral. (Lesson 10-6)

13.

14.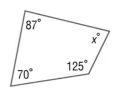

15. **ALGEBRA** What is $m\angle A$ in quadrilateral $ABCD$ if $m\angle B = 36°$, $m\angle C = 74°$, and $\angle D$ is a right angle? (Lesson 10-6)

Similar Figures

MAIN IDEA

Determine whether figures are similar and find a missing length in a pair of similar figures.

NYS Core Curriculum

7.PS.10 Use proportionality to model problems

New Vocabulary

similar figures
corresponding sides
corresponding angles
indirect measurement

NY Math Online

glencoe.com

• Extra Examples
• Personal Tutor
• Self-Check Quiz

▷ MINI Lab

The figures in each pair below have the same shape but different sizes. Copy each pair onto dot paper. Then find the measure of each angle using a protractor and the measure of each side using a centimeter ruler.

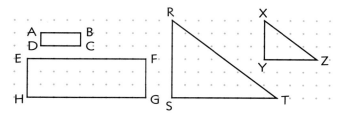

1. $\overline{AB}$ on the smaller rectangle matches $\overline{EF}$ on the larger rectangle. Name all pairs of matching sides in each pair of figures.
   The notation $\overline{AB}$ means the segment with endpoints at A and B.

2. Write each ratio in simplest form.
   The notation AB means the *measure* of segment AB.

   a. $\dfrac{AB}{EF}$, $\dfrac{BC}{FG}$, $\dfrac{DC}{HG}$, $\dfrac{AD}{EH}$ b. $\dfrac{RS}{XY}$; $\dfrac{ST}{YZ}$; $\dfrac{RT}{XZ}$

3. What do you notice about the ratios of matching sides?

4. Name all pairs of matching angles in the figures above. What do you notice about the measure of these angles?

5. **MAKE A CONJECTURE** about figures that have the same shape but not necessarily the same size.

Figures that have the same shape but not necessarily the same size are **similar figures**. In the figures below, triangle RST is similar to triangle XYZ. We write this as $\triangle RST \sim \triangle XYZ$.

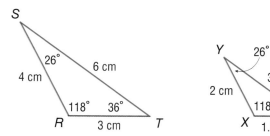

The matching sides are $\overline{ST}$ and $\overline{YZ}$, $\overline{SR}$ and $\overline{YX}$, and $\overline{RT}$ and $\overline{XZ}$. The sides of similar figures that "match" are called **corresponding sides**.

The matching angles are $\angle S$ and $\angle Y$, $\angle R$ and $\angle X$, and $\angle T$ and $\angle Z$. The angles of similar figures that "match" are called **corresponding angles**.

The Mini Lab illustrates the following statements.

Similar Figures Key Concept

Words If two figures are similar, then
- the corresponding sides are proportional, and
- the corresponding angles are congruent.

Models
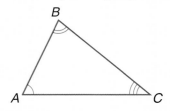 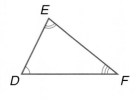

Symbols $\triangle ABC \sim \triangle DEF$ The symbol $\sim$ means *is similar to*.

corresponding sides: $\dfrac{AB}{DE} = \dfrac{BC}{EF} = \dfrac{AC}{DF}$

corresponding angles: $\angle A \cong \angle D;\ \angle B \cong \angle E;\ \angle C \cong \angle F$

Reading Math

Geometry Symbols

$\sim$ is similar to
$\cong$ is congruent to

EXAMPLE **Identify Similar Figures**

1. Which trapezoid below is similar to trapezoid *DEFG*?

 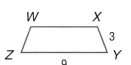 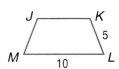

Find the ratios of the corresponding sides to see if they form a constant ratio.

Trapezoid *PQRS*   **Trapezoid *WXYZ***   **Trapezoid *JKLM***

$\dfrac{EF}{QR} = \dfrac{4}{6}$ or $\dfrac{2}{3}$ $\dfrac{EF}{XY} = \dfrac{4}{3}$ $\dfrac{EF}{KL} = \dfrac{4}{5}$

$\dfrac{FG}{RS} = \dfrac{12}{14}$ or $\dfrac{6}{7}$ $\dfrac{FG}{YZ} = \dfrac{12}{9}$ or $\dfrac{4}{3}$ $\dfrac{FG}{LM} = \dfrac{12}{10}$ or $\dfrac{6}{5}$

Not similar   Similar   Not similar

So, trapezoid *WXYZ* is similar to trapezoid *DEFG*.

CHECK Your Progress

a. Which triangle below is similar to triangle *DEF*?

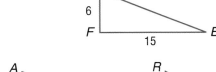

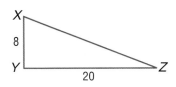 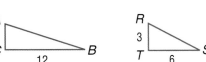

EXAMPLE **Find Side Measures of Similar Triangles**

2 If $\triangle RST \sim \triangle XYZ$, find the length of $\overline{XY}$.

Since the two triangles are similar, the ratios of their corresponding sides are equal. Write and solve a proportion to find XY.

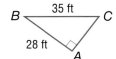

$\dfrac{RT}{XZ} = \dfrac{RS}{XY}$ Write a proportion.

$\dfrac{6}{18} = \dfrac{4}{n}$ Let n represent the length of $\overline{XY}$. Then substitute.

$6n = 18(4)$ Find the cross products.

$6n = 72$ Simplify.

$n = 12$ Divide each side by 6.   The length of $\overline{XY}$ is 12 meters.

Reading Math

Geometry Symbols

Just as the measure of angle A can be written as $m\angle A$, there is a special way to indicate the measure of a segment. The measure of $\overline{AB}$ is written as AB, without the bar over it.

 CHECK Your Progress

b. If $\triangle ABC \sim \triangle EFD$, find the length of $\overline{AC}$.

Indirect measurement uses similar figures to find the length, width, or height of objects that are too difficult to measure directly.

 Real-World EXAMPLE

3 **GEYSERS** Old Faithful in Yellowstone National Park shoots water 60 feet into the air that casts a shadow of 42 feet. What is the height of a nearby tree that casts a shadow 63 feet long? Assume the triangles are similar.

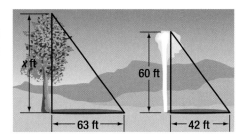

Tree   **Old Faithful**

$\dfrac{x}{63} = \dfrac{60}{42}$ ← height   ← shadow

$42x = 60(63)$ Find the cross products.

$42x = 3{,}780$ Simplify.

$x = 90$ Divide each side by 42.

The tree is 90 feet tall.

 CHECK Your Progress

c. **PHOTOGRAPHY** Destiny wants to resize a 4-inch wide by 5-inch long photograph for the school newspaper. It is to fit in a space that is 2 inches wide. What is the length of the resized photograph?

5 in. x

2 in.

4 in.

Example 1
(p. 541)

1. Which rectangle below is similar to rectangle *ABCD*?

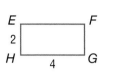

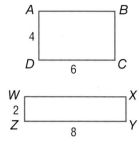

Example 2
(p. 542)

ALGEBRA Find the value of *x* in each pair of similar figures.

2.
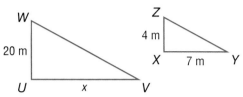

3.
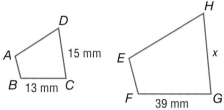

Example 3
(p. 542)

4. **SHADOWS** A flagpole casts a 20-foot shadow. At the same time, Humberto, who is 6 feet tall, casts a 5-foot shadow. What is the height of the flagpole? Assume the triangles are similar.

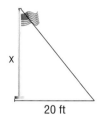

Practice and Problem Solving

HOMEWORK HELP	
For Exercises	**See Examples**
5–6	1
7–10	2
11–12	3

5. Which triangle below is similar to triangle *FGH*?

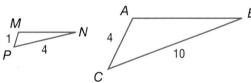

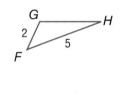

6. Which parallelogram below is similar to parallelogram *HJKM*?

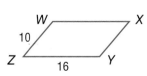

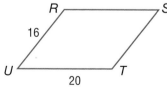

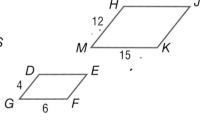

ALGEBRA Find the value of *x* in each pair of similar figures.

7.
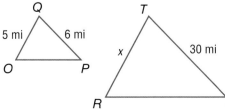

8.
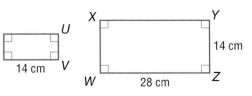

ALGEBRA Find the value of x in each pair of similar figures.

9.

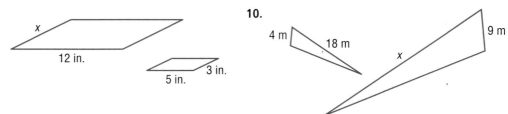

10.

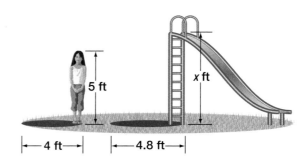

11. **PARKS** Ruth is at the park standing next to a slide. Ruth is 5 feet tall, and her shadow is 4 feet long. If the shadow of the slide is 4.8 feet long, what is the height of the slide? Assume the triangles are similar.

12. **FURNITURE** A child's desk is made so that it is a replica of a full-size adult desk. Suppose the top of the full-size desk measures 54 inches long by 36 inches wide. If the top of a child's desk is 24 inches wide and is similar to the full-size desk, what is the length?

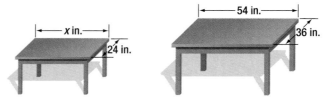

ALGEBRA Find the value of x in each pair of similar figures.

13.

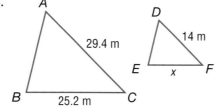

14.

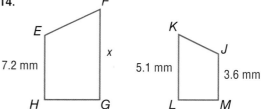

SKYCRAPERS For Exercises 15 and 16, use the information below and at the left.

Tricia has a miniature replica of the Empire State Building. The replica is 10 inches tall, and the height of the observation deck is 8.3 inches.

15. About how tall is the actual observation deck?

16. Tricia's sister has a larger replica in which the height of the observation deck is 12 inches. How tall is the larger replica?

NYSCC • NYSMT
Extra Practice, pp. 695, 713

17. **MEASUREMENT** The ratio of the length of square A to the length of square B is 3:5. If the length of square A is 18 meters, what is the perimeter of square B?

CHALLENGE For Exercises 18 and 19, use the following information.

Two rectangles are similar. The ratio of their corresponding sides is 1:4.

18. Find the ratio of their perimeters.

19. What is the ratio of their areas?

20. **WRITING IN MATH** Write a problem about a real-world situation that could be solved using proportions and the concept of similarity. Then use what you have learned in this lesson to solve the problem.

NYSMT PRACTICE 7.PS.10

21. Which rectangle is similar to the rectangle shown?

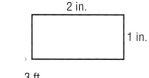

A

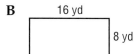
B

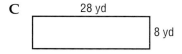
C

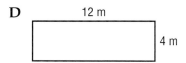
D

22. Which of the following equations is a correct use of cross-multiplication in solving the proportion $\frac{12}{15} = \frac{m}{6}$?

F $12 \cdot m = 15 \cdot 6$

G $12 \cdot 6 = m \cdot 15$

H $12 \cdot 15 = m \cdot 6$

J $12 \div 6 = m \div 15$

23. Horatio is 6 feet tall and casts a shadow 3 feet long. What is the height of a nearby tower if it casts a shadow 25 feet long at the same time?

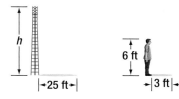

A 25 feet   **C** 50 feet

B 45 feet   **D** 75 feet

Spiral Review

GEOMETRY Classify the quadrilateral using the name that *best* describes it. (Lesson 10-6)

24.

25.

26.

27. MEASUREMENT A triangular-shaped sail has angle measures of 44° and 67°. Find the measure of the third angle. (Lesson 10-4)

▷ **GET READY for the Next Lesson**

PREREQUISITE SKILL Solve each equation. (Lesson 3-3)

28. $5a = 120$ **29.** $360 = 4x$ **30.** $940 = 8n$ **31.** $6t = 720$

10-8 Polygons and Tessellations

MAIN IDEA

Classify polygons and determine which polygons can form a tessellation.

NYS Core Curriculum

7.A.9 Build a pattern to develop a rule for determining the sum of the interior angles of polygons *Also addresses 7.CN.8*

New Vocabulary

polygon
pentagon
hexagon
heptagon
octagon
nonagon
decagon
regular polygon
tessellation

NY Math Online

glencoe.com
• Extra Examples
• Personal Tutor
• Self-Check Quiz

▷ GET READY for the Lesson

POOLS Prairie Pools designs and builds swimming pools in various shapes and sizes. The shapes of five swimming pool styles are shown in their catalog.

  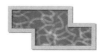

Aquarius　　**Kidney**　　**Roman**　　**Oval**　　**Rustic**

1. In the pool catalog, the Aquarius and the Roman styles are listed under Group A. The remaining three pools are listed under Group B. Describe one difference between the shapes of the pools in the two groups.

2. Create your own drawing of the shape of a pool that would fit into Group A. Group B.

A **polygon** is a simple, closed figure formed by three or more straight line segments. A *simple figure* does not have lines that cross each other. You have drawn a *closed figure* when your pencil ends up where it started.

Polygons	Not Polygons
• Line segments are called sides. • Sides meet only at their endpoints. • Points of intersection are called vertices.	• Figures with sides that cross each other. • Figures that are open. • Figures that have curved sides.

A polygon can be classified by the number of sides it has.

Words	pentagon	hexagon	heptagon	octagon	nonagon	decagon
Number of Sides	5	6	7	8	9	10
Models						

A **regular polygon** has all sides congruent and all angles congruent. Equilateral triangles and squares are examples of regular polygons.

EXAMPLES Classify Polygons

Determine whether each figure is a polygon. If it is, classify the polygon and state whether it is regular. If it is not a polygon, explain why.

The figure has 6 congruent sides and 6 congruent angles. It is a regular hexagon.

The figure is not a polygon since it has a curved side.

 CHECK Your Progress

Determine whether each figure is a polygon. If it is, classify the polygon and state whether it is regular. If it is *not* a polygon, explain why.

a.

b.

The sum of the measures of the angles of a triangle is 180°. You can use this relationship to find the measures of the angles of regular polygons.

EXAMPLE Angle Measures of a Polygon

③ **ALGEBRA** Find the measure of each angle of a regular pentagon.

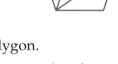

- Draw all of the diagonals from one vertex as shown and count the number of triangles formed.

- Find the sum of the angle measures in the polygon.

  number of triangles formed × 180° = sum of angle measures in polygon

  $$3 \times 180° = 540°$$

- Find the measure of each angle of the polygon. Let n represent the measure of one angle in the pentagon.

  $5n = 540$ There are five congruent angles.

  $n = 108$ Divide each side by 5.

The measure of each angle in a regular pentagon is 108°.

 CHECK Your Progress

Find the measure of an angle in each polygon.

c. regular octagon   d. equilateral triangle

A repetitive pattern of polygons that fit together with no overlaps or holes is called a **tessellation**. The surface of these bricks is an example of a tessellation of squares.

The sum of the measures of the angles where the vertices meet in a tessellation is 360°.

$$4 \times 90° = 360°$$

Real-World EXAMPLE

4 DESIGN Ms. Evans wants to design a floor tile using pentagons. Can Ms. Evans make a tessellation using pentagons?

The measure of each angle in a regular pentagon is 108°.

The sum of the measures of the angles where the vertices meet must be 360°. So, solve $108°n = 360$.

$108n = 360$ Write the equation.

$\dfrac{108n}{108} = \dfrac{360}{108}$ Divide each side by 108.

$n = 3.3$ Use a calculator.

Since 108° does not divide evenly into 360°, the sum of the measures of the angles where the vertices meets is not 360°. So, Ms. Evans cannot make a tessellation using the pentagons.

Check

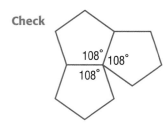

CHECK Your Progress

e. DESIGN Can Ms. Evans use tiles that are equilateral triangles to cover the floor? Explain.

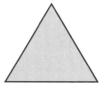

Examples 1, 2
(p. 547)

Determine whether each figure is a polygon. If it is, classify the polygon and state whether it is regular. If it is *not* a polygon, explain why.

1.

2.

3.

Example 3
(p. 547)

Find the measure of an angle in each polygon if the polygon is regular. Round to the nearest tenth of a degree if necessary.

4. hexagon

5. heptagon

Example 4
(p. 548)

6. **ART** In art class, Trisha traced and then cut several regular octagons out of tissue paper. Can she use the figures to create a tessellation? Explain.

Practice and Problem Solving

HOMEWORK HELP	
For Exercises	**See Examples**
7–12	1, 2
13–16	3
17–18	4

Determine whether each figure is a polygon. If it is, classify the polygon and state whether it is regular. If it is *not* a polygon, explain why.

7.

8.

9.

10.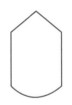

11.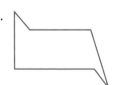

12.

Find the measure of an angle in each polygon if the polygon is regular. Round to the nearest tenth of a degree if necessary.

13. decagon

14. nonagon

15. quadrilateral

16. 11-gon

17. **TOYS** Marty used his magnetic building set to build the regular decagon at the right. Assume he has enough building parts to create several of these shapes. Can the figures be arranged in a tessellation? Explain.

18. **COASTERS** Paper coasters are placed under a beverage glass to protect the table surface. The coasters are shaped like regular heptagons. Can the coasters be arranged in a tessellation? Explain your reasoning.

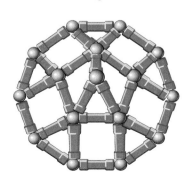

Classify the polygons that are used to create each tessellation.

19.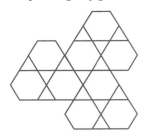

20.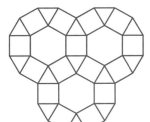

21.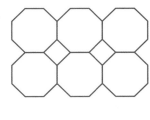

22. What is the perimeter of a regular nonagon with sides 4.8 centimeters?

23. Find the perimeter of a regular pentagon having sides $7\frac{1}{4}$ yards long.

24. **ART** The mosaic shown at the right is from a marble floor in Venice, Italy. Name the polygons used in the floor.

25. **SIGNS** Refer to the photo at the left. Stop signs are made from large sheets of steel. Suppose one sheet of steel is large enough to cut nine signs. Can all nine signs be arranged on the sheet so that none of the steel goes to waste? Explain.

26. **RESEARCH** Use the Internet or another source to find the shape of other road signs. Name the type of sign, its shape, and state whether or not it is regular.

SCHOOL For Exercises 27–29, use the information below and the graphic of the cafeteria tray shown.

A company designs cafeteria trays so that four students can place their trays around a square table without bumping corners. The top and bottom sides of the tray are parallel.

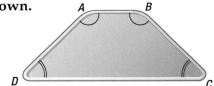

27. Classify the shape of the tray.

28. If $\angle A \cong \angle B$, $\angle C \cong \angle D$, and $m\angle A = 135°$, find $m\angle B$, $m\angle C$, and $m\angle D$.

NYSCC • NYSMT
Extra Practice, pp. 695, 713

29. Name the polygon formed by the outside of four trays when they are placed around the table with their sides touching. Justify your answer.

H.O.T. Problems

30. **REASONING** *True* or *False*? Only a regular polygon can tessellate a plane.

31. **OPEN ENDED** Draw examples of a pentagon and a hexagon that represent real-world objects.

32. **CHALLENGE** You can make a tessellation with equilateral triangles. Can you make a tessellation with any isosceles or scalene triangles? If so, explain your reasoning and make a drawing.

33. **WRITING IN MATH** Analyze the parallelogram at the right and then explain how you know the parallelogram can be used by itself to make a tessellation.

34. **SHORT RESPONSE** What is the measure of ∠1?

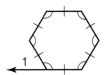

35. Which statement is *not* true about polygons?

A A polygon is classified by the number of sides it has.

B The sides of a polygon overlap.

C A polygon is formed by 3 or more line segments.

D The sides of a polygon meet only at its endpoints.

Spiral Review

For Exercises 36 and 37, use the figures at the right.

36. **ALGEBRA** The quadrilaterals are similar. Find the value of x. (Lesson 10-7)

37. **GEOMETRY** Classify figure $ABCD$. (Lesson 10-6)

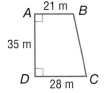

38. **PROBABILITY** Two students will be randomly selected from a group of seven to present their reports. If Carla and Pedro are in the group of 7, what is the probability that Carla will be selected first and Pedro selected second? (Lesson 9-4)

39. **RUNNING** Use the information in the table to find the rate of change. (Lesson 6-3)

Distance (m)	15	30	45	60
Time (s)	10	20	30	40

Add or subtract. Write each sum or difference in simplest form. (Lesson 5-3)

40. $3\frac{2}{9} + 5\frac{4}{9}$ **41.** $5\frac{1}{3} - 2\frac{1}{6}$ **42.** $1\frac{3}{7} + 6\frac{1}{4}$ **43.** $9\frac{4}{5} - 4\frac{7}{8}$

▷ **GET READY for the Next Lesson**

PREREQUISITE SKILL Graph and label each point on the same coordinate plane. (Lesson 2-3)

44. $A(-2, 3)$ **45.** $B(4, 3)$ **46.** $C(2, -1)$ **47.** $D(-4, -1)$

Geometry Lab
Tessellations

In this lab, you will create tessellations.

ACTIVITY

STEP 1 Draw a square on the back of an index card. Then draw a triangle on the inside of the square and a trapezoid on the bottom of the square as shown.

STEP 2 Cut out the square. Then cut out the triangle and slide it from the right side of the square to the left side of the square. Cut out the trapezoid and slide it from the bottom to the top of the square.

STEP 3 Tape the figures together to form a pattern.

STEP 4 Trace this pattern onto a sheet of paper as shown to create a tessellation.

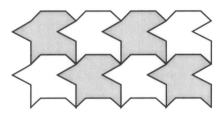

✓ **CHECK Your Progress**

Create a tessellation using each pattern.

a. 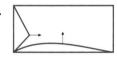   b.   c.

ANALYZE THE RESULTS

1. Design and describe your own tessellation pattern.

2. **MAKE A CONJECTURE** *Congruent figures* have corresponding sides of equal length and corresponding angles of equal measure. Explain how congruent figures are used in your tessellation.

10-9 Translations

MAIN IDEA

Graph translations of polygons on a coordinate plane.

NYS Core Curriculum

Preparation for 8.G.10
Draw the image of a figure under a translation

New Vocabulary

transformation
translation
congruent figures

NY Math Online

glencoe.com

• Concepts In Motion
• Extra Examples
• Personal Tutor
• Self-Check Quiz

▷ MINI Lab

STEP 1 Trace a parallelogram-shaped pattern block onto a coordinate grid. Label the vertices *ABCD*.

STEP 2 Slide the pattern block over 5 units to the right and 2 units down.

STEP 3 Trace the figure in its new position. Label the vertices *A′*, *B′*, *C′*, and *D′*.

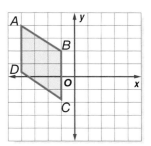

1. Trace the horizontal and vertical path between corresponding vertices. What do you notice?

2. Add 5 to each *x*-coordinate of the vertices of the original figure. Then subtract 2 from each *y*-coordinate of the vertices of the original figure. What do you notice?

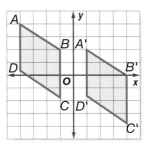

A **transformation** maps one figure onto another. When you move the figure without turning it, the motion is called a **translation**.

When translating a figure, every point of the original figure is moved the same distance and in the same direction.

EXAMPLE   Graph a Translation

① Translate quadrilateral *HIJK* 2 units left and 5 units up. Graph quadrilateral *H′ I′ J′ K′*.

• Move each vertex of the figure 2 units left and 5 units up. Label the new vertices *H′*, *I′*, *J′*, and *K′*.

• Connect the vertices to draw the trapezoid. The coordinates of the vertices of the new figure are *H′*(−4, 1), *I′*(−3, 4), *J′*(1, 3), and *K′*(−1, 1).

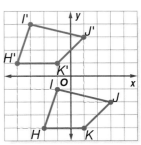

✓ CHECK Your Progress

a. Translate quadrilateral *HIJK* 4 units up and 2 units right. Graph quadrilateral *H′ I′ J′ K′*.

Prime Symbols Use prime symbols for vertices in a transformed image.

$A \rightarrow A'$
$B \rightarrow B'$
$C \rightarrow C'$

A' is read *A prime*.

When a figure has been translated, the original figure and the translated figure, or *image*, are congruent. **Congruent figures** have the same size and same shape, and the corresponding sides and angles have equal measures.

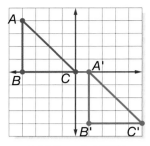

$\triangle ABC \cong \triangle A'B'C'$

You can increase or decrease the coordinates of the vertices of a figure by a fixed amount to find the coordinates of the translated vertices.

A *positive* integer describes a translation right or up on a coordinate plane. A *negative* integer describes a translation left or down.

EXAMPLE   Find Coordinates of a Translation

2 Triangle *LMN* has vertices $L(-1, -2)$, $M(6, -3)$, and $N(2, -5)$. Find the vertices of $\triangle L'M'N'$ after a translation of 6 units left and 4 units up. Then graph the figure and its translated image.

The vertices can be found by adding -6 to the *x*-coordinates and 4 to the *y*-coordinates.

Study Tip

Transformations
A transformation of the form $(x', y') = (x + a, y + b)$ is a translation which moves the point (x, y) *a* units horizontally and *b* units vertically.

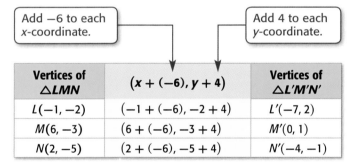

Vertices of △LMN	$(x + (-6), y + 4)$	Vertices of △L'M'N'
$L(-1, -2)$	$(-1 + (-6), -2 + 4)$	$L'(-7, 2)$
$M(6, -3)$	$(6 + (-6), -3 + 4)$	$M'(0, 1)$
$N(2, -5)$	$(2 + (-6), -5 + 4)$	$N'(-4, -1)$

Use the vertices of △LMN and of △L'M'N' to graph each triangle.

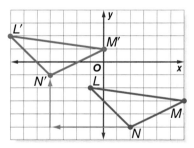

✓ CHECK Your Progress

b. Triangle *TUV* has vertices $T(6, -3)$, $U(-2, 0)$, and $V(-1, 2)$. Find the vertices of $\triangle T'U'V'$ after a translation of 3 units right and 4 units down. Then graph the figure and its translated image.

In Example 2, △LMN was translated 6 units left and 4 units up. This translation can be described using the ordered pair $(-6, 4)$. In Check Your Progress b., △TUV was translated 3 units right and 4 units down. This translation can be described using the ordered pair $(3, -4)$.

Example 1
(p. 553)

1. Translate △ABC 3 units left and 3 units down. Graph △A'B'C'.

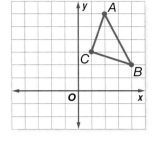

Example 2
(p. 554)

Quadrilateral *DEFG* has vertices *D*(1, 0), *E*(−2, −2), *F*(2, 4), and *G*(6, −3). Find the vertices of *D'E'F'G'* after each translation. Then graph the figure and its translated image.

2. 4 units right, 5 units down

3. 6 units right

4. **MAPS** Nakos explores part of the Denver Zoo in Colorado as shown. He starts at the felines and then visits the hoofed animals. Describe this translation in words and as an ordered pair.

Practice and Problem Solving

HOMEWORK HELP

For Exercises	See Examples
5–6	1
7–12	2

5. Translate △*HIJ* 2 units right and 6 units down. Graph △*H'I'J'*.

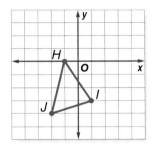

6. Translate rectangle *KLMN* 1 unit left and 3 units up. Graph rectangle *K'L'M'N'*.

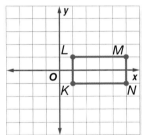

Triangle *PQR* has vertices *P*(0, 0), *Q*(5, −2), and *R*(−3, 6). Find the vertices of *P'Q'R'* after each translation. Then graph the figure and its translated image.

7. 6 units right, 5 units up

8. 8 units left, 1 unit down

9. 3 units left

10. 9 units down

GAMES When playing the game shown at the right, the player can move horizontally or vertically across the board. Describe each of the following as a translation in words and as an ordered pair.

11. Green player

12. Orange player

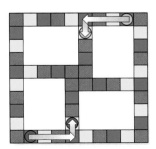

13. Triangle *ABC* is translated 2 units left and 3 units down. Then the translated figure is translated 3 units right. Graph the resulting triangle.

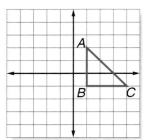

14. Parallelogram *RSTU* is translated 3 units right and 5 units up. Then the translated figure is translated 2 units left. Graph the resulting parallelogram.

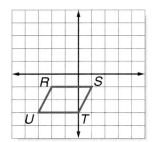

15. **ART** Marjorie Rice creates art using tessellations. At the right is her artwork of fish. Explain how translations and tessellations were used in the figure.

16. **RESEARCH** Use the Internet or another source to find other pieces of art that contain tessellations of translations. Describe how the artists incorporated both ideas into their work.

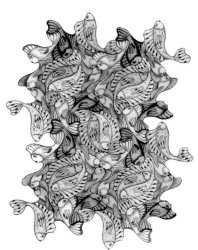

17. Triangle *FGH* has vertices $F(7, 6)$, $G(3, 4)$, $H(1, 5)$. Find the coordinates of $\triangle F'G'H'$ after a translation $1\frac{1}{2}$ units right and $3\frac{1}{2}$ units down. Then graph the figure and its translated image.

REASONING The coordinates of a point and its image after a translation are given. Describe the translation in words and as an ordered pair.

NYSCC • NYSMT
Extra Practice, pp. 696, 713

18. $N(0, -3) \rightarrow N'(2, 2)$

19. $M(2, 4) \rightarrow M'(-3, 1)$

20. $P(-2, -1) \rightarrow P'(3, -2)$

21. $Q(-4, 0) \rightarrow Q'(1, 4)$

H.O.T. Problems

22. **CHALLENGE** Is it possible to make a tessellation with translations of an equilateral triangle? Explain your reasoning.

23. **Which One Doesn't Belong?** Identify the transformation that is not the same as the other three. Explain your reasoning.

A B C D

24. **WRITING IN MATH** Triangle *ABC* is translated 4 units right and 2 units down. Then the translated image is translated again 7 units left and 5 units up. Describe the final translated image in words.

25. Which graph shows a translation of the letter Z?

A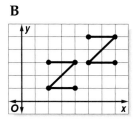

C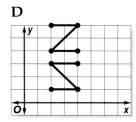

B

D

26. If point A is translated 4 units left and 3 units up, what will be the coordinates of point A in its new position?

F $(4, 4)$

G $(-5, 5)$

H $(-5, -1)$

J $(-4, 3)$

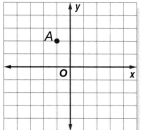

Spiral Review

27. **GEOMETRY** What is the name of a polygon with eight sides? (Lesson 10-8)

28. **GEOMETRY** The triangles at the right are similar. What is the measure of $\angle F$? (Lesson 10-7)

29. **FOOD** For dinner, you can choose one of two appetizers, one of four entrées, and one of three desserts. How many possible unique dinners can you choose? (Lesson 9-3)

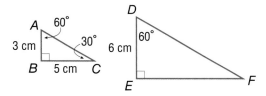

For each set of data, describe how the range would change if the value 15 was added to the data set. (Lesson 8-2)

30. $\{8, 17, 32\}$ 31. $\{22, 38, 41, 77\}$ 32. $\{10, 10, 19\}$ 33. $\{7, 11, 13\}$

Write each percent as a decimal. (Lesson 4-7)

34. 83.8%   35. 56.7%   36. 3.8%   37. 102.6%

▷ **GET READY for the Next Lesson**

PREREQUISITE SKILL Determine whether each figure can be folded in half so that one side matches the other. Write *yes* or *no*.

38.   39.   40.   41.

10-10 Reflections

MAIN IDEA

Identify figures with line symmetry and graph reflections on a coordinate plane.

NYS Core Curriculum

Preparation for 8.G.9
Draw the image of a figure under a reflection over a given line

New Vocabulary

line symmetry
line of symmetry
reflection
line of reflection

NY Math Online

glencoe.com

• Extra Examples
• Personal Tutor
• Self-Check Quiz

▷ MINI Lab

VISION Scientists have determined that the human eye uses symmetry to see. It is possible to understand what you are looking at, even if you do not see all of it.

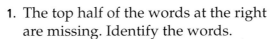
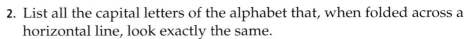

1. The top half of the words at the right are missing. Identify the words.

2. List all the capital letters of the alphabet that, when folded across a horizontal line, look exactly the same.

3. On a piece of paper, write the bottom half of other words that, when reflected across a horizontal line, look exactly the same.

Figures that match exactly when folded in half have **line symmetry**. The figures at the right have line symmetry.

Each fold line is called a **line of symmetry**.

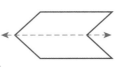

Real-World EXAMPLES

GRAPHIC DESIGN Determine whether each figure has line symmetry. If so, copy the figure and draw all lines of symmetry.

1

2 no symmetry

3

✓ CHECK Your Progress

a.

b.

c.

Study Tip

Congruent Figures As with translations, the original figure and the reflected image are congruent.

A **reflection** is a mirror image of the original figure. It is the result of a transformation of a figure over a line called a **line of reflection**.

EXAMPLE Reflect a Figure Over the *x*-axis

④ Triangle *ABC* has vertices *A*(5, 2), *B*(1, 3), and *C*(−1, 1). Graph the figure and its reflected image over the *x*-axis. Then find the coordinates of the vertices of the reflected image.

The *x*-axis is the line of reflection. So, plot each vertex of *A′B′C′* the same distance from the *x*-axis as its corresponding vertex on *ABC*.

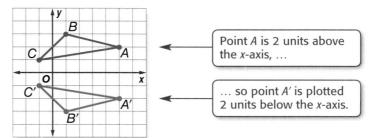

Point *A* is 2 units above the *x*-axis, ...

... so point *A′* is plotted 2 units below the *x*-axis.

The coordinates are *A′*(5, −2), *B′*(1, −3), and *C′*(−1, −1).

CHECK Your Progress

d. Rectangle *GHIJ* has vertices *G*(3, −4), *H*(3, −1), *I*(−2, −1), and *J*(−2, −4). Graph the figure and its image after a reflection over the *x*-axis. Then find the coordinates of the reflected image.

EXAMPLE Reflect a Figure Over the *y*-axis

⑤ Quadrilateral *KLMN* has vertices *K*(2, 3), *L*(5, 1), *M*(4, −2), and *N*(1, −1). Graph the figure and its reflected image over the *y*-axis. Then find the coordinates of the vertices of the reflected image.

The *y*-axis is the line of reflection. So, plot each vertex of *K′L′M′N′* the same distance from the *y*-axis as its corresponding vertex on *KLMN*.

Point *K′* is 2 units to the left of the *y*-axis.

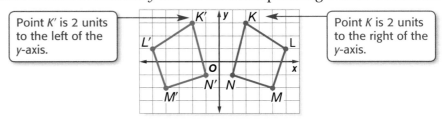

Point *K* is 2 units to the right of the *y*-axis.

The coordinates are *K′*(−2, 3), *L′*(−5, 1), *M′*(−4, −2), and *N′*(−1, −1).

CHECK Your Progress

e. Triangle *PQR* has vertices *P*(1, 5), *Q*(3, 7), and *R*(5, −1). Graph the figure and its reflection over the *y*-axis. Then find the coordinates of the reflected image.

Examples 1–3
(p. 558)

Determine whether each figure has line symmetry. If so, copy the figure and draw all lines of symmetry.

1.
2.
3.

4. **INSECTS** Identify the number of lines of symmetry in the photo of the butterfly at the right.

Example 4
(p. 559)

Graph each figure and its reflection over the *x*-axis. Then find the coordinates of the reflected image.

5. $\triangle ABC$ with vertices $A(5, 8)$, $B(1, 2)$, and $C(6, 4)$

6. quadrilateral $DEFG$ with vertices $D(-4, 6)$, $E(-2, -3)$, $F(2, 2)$, and $G(4, 9)$

Example 5
(p. 559)

Graph each figure and its reflection over the *y*-axis. Then find the coordinates of the reflected image.

7. $\triangle QRS$ with vertices $Q(2, -5)$, $R(5, -5)$, and $S(2, 3)$

8. parallelogram $WXYZ$ with vertices $W(-4, -2)$, $X(-4, 3)$, $Y(-2, 4)$, and $Z(-2, -1)$

Practice and Problem Solving

HOMEWORK HELP

For Exercises	See Examples
9–14 23–24	1, 3
15–18	4
19–22	5

Determine whether each figure has line symmetry. If so, copy the figure and draw all lines of symmetry.

9.
10.
11.

12.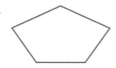
13.
14.

Graph each figure and its reflection over the *x*-axis. Then find the coordinates of the reflected image.

15. TUV with vertices $T(-6, -1)$, $U(-2, -3)$, and $V(5, -4)$

16. MNP with vertices $M(2, 1)$, $N(-3, 1)$, and $P(-1, 4)$

17. square $ABCD$ with vertices $A(2, 4)$, $B(-2, 4)$, $C(-2, 8)$, and $D(2, 8)$

18. $WXYZ$ with vertices $W(-1, -1)$, $X(4, 1)$, $Y(4, 5)$, and $Z(1, 7)$

Graph each figure and its reflection over the *y*-axis. Then find the coordinates of the reflected image.

19. △*RST* with vertices *R*(−5, 3), *S*(−4, −2), and *T*(−2, 3)

20. △*GHJ* with vertices *G*(4, 2), *H*(3, −4), and *J*(1, 1)

21. parallelogram *HIJK* with vertices *H*(−1, 3), *I*(−1, −1), *J*(2, −2), and *K*(2, 2)

22. quadrilateral *DEFG* with vertices *D*(1, 0), *E*(1, −5), *F*(4, −1), and *G*(3, 2)

23. GATES Describe the location of the line(s) of symmetry in the photograph of Brandenburg Gate in Berlin, Germany.

24. FLAGS Flags of some countries have line symmetry. Of the flags shown below, which flags have line symmetry? Copy and draw all lines of symmetry.

| Nigeria | Ghana | Japan | Mexico |

Real-World Link . . . .
A violin is usually around 14 inches long.

25. MUSIC Use the photo at the left to determine how many lines of symmetry the body of a violin has.

For Exercises 26–29, use the graph shown at the right.

26. Identify the pair(s) of figures for which the *x*-axis is the line of reflection.

27. For which pair(s) of figures is the line of reflection the *y*-axis?

28. What type of transformation do figures *B* and *C* represent?

29. Describe the possible transformation(s) required to move figure *A* onto figure *D*.

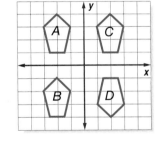

30. △*RST* is reflected over the *x*-axis and then translated 3 units to the left and 2 units down. Graph the resulting triangle.

31. △*MNP* is translated 2 units right and 3 units up. Then the translated figure is reflected over the *y*-axis. Graph the resulting triangle.

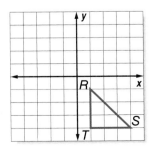
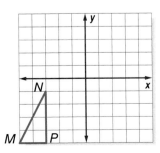

NYSCC • NYSMT
Extra Practice, pp. 696, 713

The coordinates of a point and its image after a reflection are given. Describe the reflection as over the x-axis or y-axis.

32. $A(-3, 5) \rightarrow A'(3, 5)$

33. $M(3, 3) \rightarrow M'(3, -3)$

34. $X(-1, -4) \rightarrow X'(-1, 4)$

35. $W(-4, 0) \rightarrow W'(4, 0)$

H.O.T. Problems

36. OPEN ENDED Make a tessellation using a combination of translations and reflections of polygons. Explain your method.

37. CHALLENGE Triangle JKL has vertices $J(-7, 4)$, $K(7, 1)$, and $L(2, -2)$. Without graphing, find the new coordinates of the vertices of the triangle after a reflection first over the x-axis and then over the y-axis.

38. WRITING IN MATH Draw a figure on a coordinate plane and its reflection over the y-axis. Explain how the x- and y-coordinates of the reflected figure relate to the x- and y-coordinates of the original figure. Then repeat, this time reflecting the figure over the x-axis.

NYSMT PRACTICE ▷ 8.G.9

39. The figure shown was transformed from quadrant II to quadrant III.

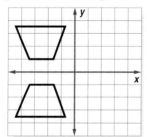

This transformation best represents which of the following?

A translation   **C** reflection

B tessellation   **D** rotation

40. If $ABCD$ is reflected over the x-axis and translated 5 units to the right, which is the resulting image of point B?

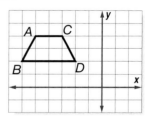

F $(-1, -2)$ **H** $(-1, 2)$

G $(-11, 2)$ **J** $(11, 2)$

Spiral Review

41. GEOMETRY Triangle FGH has vertices $F(-3, 7)$, $G(-1, 5)$, and $H(-2, 2)$. Graph the figure and its image after a translation 4 units right and 1 unit down. Write the ordered pairs for the vertices of the image. (Lesson 10-9)

42. GEOMETRY Melissa wishes to construct a tessellation for a wall hanging made only from regular decagons. Is this possible? Explain. (Lesson 10-8)

Estimate. (Lesson 5-1)

43. $\frac{4}{9} + 8\frac{1}{9}$

44. $\frac{1}{9} \times \frac{2}{5}$

45. $12\frac{1}{4} \div 5\frac{6}{7}$

CHAPTER 10 Study Guide and Review

FOLDABLES Study Organizer

GET READY to Study

Be sure the following Big Ideas are noted in your Foldable.

What I Know About Polygons	What I Need to Know	What I've Learned

BIG Ideas

Angles (Lessons 10-1 and 10-2)
• Two angles are adjacent if they have the same vertex, share a common side, and do not overlap.
• Two angles are vertical if they are opposite angles formed by the intersection of two lines.
• Two angles are complementary if the sum of their measures is 90°.
• Two angles are supplementary if the sum of their measures is 180°.

Triangles (Lesson 10-4)
• The sum of the measures of the angles of a triangle is 180°.

Quadrilaterals (Lesson 10-6)
• The sum of the measures of the angles of a quadrilateral is 360°.

Similar Figures (Lesson 10-7)
• If two figures are similar then the corresponding sides are proportional and the corresponding angles are congruent.

Transformations (Lessons 10-9 and 10-10)
• When translating a figure, every point in the original figure is moved the same distance in the same direction.
• When reflecting a figure, every point in the original figure is the same distance from the line of reflection as its corresponding point on the original figure.

Key Vocabulary

acute angle (p. 511)
angle (p. 510)
circle graph (p. 518)
complementary angles (p. 514)
congruent angles (p. 511)
congruent figures (p. 554)
degrees (p. 510)
hexagon (p. 546)
indirect measurement (p. 542)
line of symmetry (p. 558)
line symmetry (p. 558)
obtuse angle (p. 511)
octagon (p. 546)
parallelogram (p. 533)
pentagon (p. 546)
polygon (p. 546)
quadrilateral (p. 533)
reflection (p. 559)
regular polygon (p. 546)
rhombus (p. 533)
right angle (p. 511)
similar figures (p. 540)
straight angle (p. 511)
supplementary angles (p. 514)
tessellation (p. 548)
transformation (p. 553)
translation (p. 553)
trapezoid (p. 533)
triangle (p. 524)
vertex (p. 510)

Vocabulary Check

State whether each sentence is *true* or *false*. If *false*, replace the underlined word or number to make a true sentence.

1. Two angles with measures adding to 180° are called <u>complementary angles</u>.
2. A <u>hexagon</u> is a polygon with 6 sides.
3. An angle with a measure of less than 90° is called a <u>right angle</u>.
4. The <u>vertex</u> is where the sides of an angle meet.
5. The point (3, −2) when translated up 3 units and to the left 5 units becomes <u>(6, −7)</u>.
6. A <u>trapezoid</u> has both pairs of opposite sides parallel.

Lesson-by-Lesson Review

10-1 Angle Relationships (pp. 510–513)

8.G.1, 8.G.4

For Exercises 7 and 8, refer to the figure at the right to identify each pair of angles. Justify your response.

7. a pair of vertical angles

8. a pair of adjacent angles

Example 1 Refer to the figure below. Identify a pair of vertical angles.

∠1 and ∠4 are opposite angles formed by the intersection of two lines.

∠1 and ∠4 are vertical angles.

10-2 Complementary and Supplementary Angles (pp. 514–517)

8.G.2, 8.G.3

Classify each pair of angles as *complementary*, *supplementary*, or *neither*.

9.

10.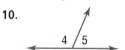

Example 2 Find the value of x.

$$x + 27 = 90$$
$$\underline{-27 = -27}$$
$$x \quad = \quad 63$$

10-3 Statistics: Display Data in a Circle Graph (pp. 518–523)

7.M.8, 7.S.2

11. **COLORS** The table shows favorite shades of blue. Display the set of data in a circle graph.

Shade	Percent
Navy	35%
Sky/Light Blue	30%
Aquamarine	17%
Other	18%

Example 3 Which pizza was chosen by about twice as many people as supreme?

Pepperoni was chosen by about twice as many people as supreme.

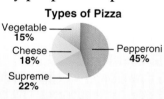

Types of Pizza

Vegetable 15%
Cheese 18%
Supreme 22%
Pepperoni 45%

10-4 Triangles (pp. 524–529)

7.G.7

ALGEBRA Find the value of x.

12.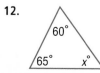

13.

Example 4 Find the value of x.

$$x + 64 + 67 = 180$$
$$x + 131 = 180$$
$$\underline{-131 = -131}$$
$$x \quad = \quad 49$$

10-5 PSI: Logical Reasoning (pp. 530–531)

7.PS.16

14. **SPORTS** Donnie, Jenna, Milo, and Barbara play volleyball, field hockey, golf, and soccer but not in that order. Use the clues given below to find the sport each person plays.

- Donnie does not like golf, volleyball, or soccer.
- Neither Milo nor Jenna likes golf.
- Milo does not like soccer.

15. **FOOD** Angelo's Pizza Parlor makes square pizzas. After baking, the pizzas are cut along one diagonal into two triangles. Classify the triangles made.

Example 5 Todd, Virginia, Elaine, and Peter are siblings. Todd was born after Peter, but before Virginia. Elaine is the oldest. Who is the youngest in the family?

Use logical reasoning to determine the youngest of the family.

You know that Elaine is the oldest, so she is first on the list. Todd was born after Peter, but before Virginia. So, Peter was second and then Todd was born. Virginia is the youngest of the family.

10-6 Quadrilaterals (pp. 533–538)

7.G.7

Classify the quadrilateral with the name that best describes it.

16.    17.

18. **GEOMETRY** What quadrilateral does not have opposite sides congruent?

Example 6 Classify the quadrilateral using the name that *best* describes it.

The quadrilateral is a parallelogram with 4 right angles and 4 congruent sides. It is a square.

10-7 Similar Figures (pp. 540–545)

7.PS.10

Find the value of x in each pair of similar figures.

19. 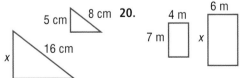   20.

21. **FLAGPOLES** Hiroshi is 1.6 meters tall and casts a shadow 0.53 meter in length. How tall is a flagpole if it casts a shadow 2.65 meters in length?

Example 7

Find the value of x in the pair of similar figures.

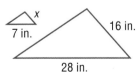

$\dfrac{7}{28} = \dfrac{x}{16}$ Write a proportion.

$28 \cdot x = 7 \cdot 16$ Find the cross products.

$28x = 112$ Simplify.

$\dfrac{28x}{28} = \dfrac{112}{28}$ Divide each side by 28.

$x = 4$ Simplify.

So, the value of x is 4.

Mixed Problem Solving
For mixed problem-solving practice, see page 713.

10-8 Polygons and Tessellations (pp. 546–551)

7.A.9

Determine whether each figure is a polygon. If it is, classify the polygon and state whether it is regular. If it is *not* a polygon, explain why.

22.

23.

24. **ALGEBRA** Find the measure of each angle of a regular 12-gon.

Example 8 Determine whether the figure is a polygon. If it is, classify the polygon and state whether it is regular. If it is *not* a polygon, explain why.

Since the polygon has 5 congruent sides and 5 congruent angles, it is a regular pentagon.

10-9 Translations (pp. 553–557)

7.A.9

Triangle *PQR* has coordinates $P(4, -2)$, $Q(-2, -3)$, and $R(-1, 6)$. Find the coordinates of $P'Q'R'$ after each translation. Then graph each translation.

25. 6 units left, 3 units up

26. 4 units right, 1 unit down

27. 3 units left

28. 7 units down

Example 9 Find the coordinates of $\triangle G'H'I'$ after a translation of 2 units left and 4 units up.

The vertices of $\triangle G'H'I'$ are $G'(-2, 7)$, $H'(2, 3)$, and $I'(-4, 1)$.

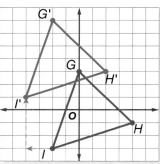

10-10 Reflections (pp. 558–562)

8.G.9

Find the coordinates of each figure after a reflection over the given axis. Then graph the figure and its reflected image.

29. $\triangle RST$ with coordinates $R(-1, 3)$, $S(2, 6)$, and $T(6, 1)$; *x*-axis

30. parallelogram *ABCD* with coordinates $A(1, 3)$, $B(2, -1)$, $C(5, -1)$, and $D(4, 3)$; *y*-axis

31. rectangle *EFGH* with coordinates $E(4, 2)$, $F(-2, 2)$, $G(-2, 5)$, and $H(4, 5)$; *x*-axis

Example 10 Find the coordinates of $\triangle C'D'E'$ after a reflection over the *y*-axis. Then graph its reflected image.

The vertices of $\triangle C'D'E'$ are $C'(-3, 4)$, $D'(-2, 1)$, and $E'(-5, 3)$.

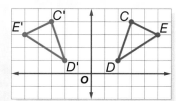

Name each angle in four ways. Then classify each angle as *acute, obtuse, right,* or *straight.*

1.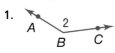

2.

Classify each pair of angles as *complementary, supplementary,* or *neither.*

3.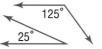

4.

5. **GEOMETRY** Classify the angle pair at the right as *vertical, adjacent,* or *neither*.

6. **MULTIPLE CHOICE** The table shows the results of a survey. The results are to be displayed in a circle graph. Which statement about the graph is *not* true?

Favorite Type of Bagels	
Type	Students
blueberry	8
cinnamon raisin	9
everything	18
plain	32

A About 12% of students chose blueberry as their favorite bagel.
B The blueberry section on the graph will have an angle measure of about 43°.
C The everything and plain sections on the circle graph form supplementary angles.
D Plain bagels were preferred more than any other type of bagel.

ALGEBRA Find the missing measure in each triangle with the given angle measures.

7. 75°, 25.5°, $x°$

8. 23.5°, $x°$, 109.5°

9. **ALGEBRA** Numbers ending in zero or five are divisible by five. Are the numbers 25, 893, and 690 divisible by 5? Use the *logical reasoning* strategy.

ALGEBRA Find the value of x in each quadrilateral.

10.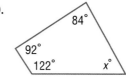

11.

12. **ART** A drawing is enlarged so that it is 14 inches long and 11 inches wide. If the original length of the drawing is 8 inches, what is its width?

13. **GEOMETRY** Can a regular heptagon, with angle measures that total 900°, be used by itself to make a tessellation? Explain.

14. **MULTIPLE CHOICE** Which quadrilateral does *not* have opposite sides congruent?

F parallelogram   H trapezoid
G square   J rectangle

15. **ALGEBRA** Square *ABCD* is shown. What are the vertices of *A'B'C'D'* after a translation 2 units right and 2 units down? Graph the translated image.

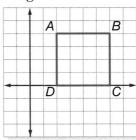

16. **GEOMETRY** Draw a figure with one line of symmetry. Then draw a figure with no lines of symmetry.

PART 1 **Multiple Choice**

Read each question. Then fill in the correct answer on the answer document provided by your teacher or on a sheet of paper.

1. Which of the following two angles are complementary?

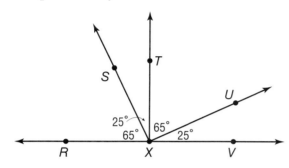

A ∠RXS and ∠TXU

B ∠SXT and ∠TXU

C ∠RXS and ∠SXV

D ∠SXR and ∠SXV

2. A square is divided into 9 congruent squares. Which of the following methods can be used to find the area of the larger square, given the area of one of the smaller squares?

F Multiply the area of the larger square by 9.

G Add 9 to the area of one of the smaller squares.

H Multiply the area of one of the smaller squares by 9.

J Add the area of the larger square to the sum of the areas of each of the 9 smaller squares.

3. Which of the following groups does *not* contain equivalent fractions, decimals, and percents?

A $\frac{9}{20}$, 0.45, 45%

B $\frac{3}{10}$, 0.3, 30%

C $\frac{7}{8}$, 0.875, 87.5%

D $\frac{1}{100}$, 0.1, 1%

4. The table below shows all the possible outcomes when tossing two fair coins at the same time.

1st Coin	2nd Coin
H	H
H	T
T	H
T	T

Which of the following must be true?

F The probability that both coins have the same outcome is $\frac{1}{4}$.

G The probability of getting at least one tail is higher than the probability of getting two heads.

H The probability that exactly one coin will turn up heads is $\frac{3}{4}$.

J The probability of getting at least one tail is lower than the probability of getting two tails.

5. Seth has $858.60 in his savings account. He plans to spend 15% of his savings on a bicycle. Which of the following represents the amount Seth plans to spend on the bicycle?

A $182.79   **C** $128.79

B $171.72   **D** $122.79

6. A manager took an employee to lunch. If the lunch was $48 and she left a 20% tip, how much money did she spend on lunch?

   F $68.00   **H** $55.80

   G $57.60   **J** $38.40

7. What is the measure of ∠1 in the figure below?

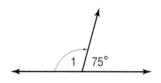

   A 15°   **C** 100°

   B 25°   **D** 105°

8. Josiah found the mean and median of the following list of numbers.

   $$11, 17, 17$$

   If the number 25 is added to this list, then which of the following statements would be true?

   F The mean would increase.

   G The mean would decrease.

   H The median would increase.

   J The median would decrease.

TEST-TAKING TIP

Question 8 Sometimes, it is not necessary to perform any calculations in order to answer the question correctly. In Question 8, you can use number sense to eliminate certain answer choices. Not having to perform calculations can help save time during a test.

PART 2   Short Response/Grid In

Record your answers on the answer sheet provided by your teacher or on a sheet of paper.

9. Three students are to be chosen from 8 auditions to star in the school play. In how many ways can these 3 students be chosen?

PART 3   Extended Response

Record your answers on the answer sheet provided by your teacher or on a sheet of paper. Show your work.

10. Use triangle XYZ to answer the following questions.

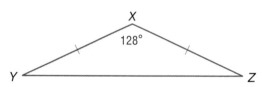

   a. Classify angle X.

   b. Classify angle Y.

   c. Classify the triangle by its sides and angles.

   d. If $\angle Y$ is congruent to $\angle Z$, find the measure of $\angle Z$. Explain.

   e. Can triangle XYZ be used by itself to make a tesselation? If so, include a drawing of the tesselation. If not, explain why not.

NEED EXTRA HELP?

If You Missed Question...	1	2	3	4	5	6	7	8	9	10
Go to Lesson...	10-1	10-5	6-8	9-1	7-1	7-4	10-2	8-2	9-5	10-5
NYS Core Curriculum	8.G.1	7.PS.16	7.PS.10	7.S.9	8.N.3	8.N.4	8.G.2	7.S.6	7.S.8	7.PS.16

CHAPTER 11

Measurement: Two- and Three-Dimensional Figures

New York State Core Curriculum

7.G.3 Identify the two-dimensional shapes that make up the faces and bases of three-dimensional shapes (prisms, cylinders, cones, and pyramids)

7.G.2 Calculate the volume of prisms and cylinders, using a given formula and a calculator

Key Vocabulary

circumference (p. 584)

cylinder (p. 604)

pyramid (p. 603)

volume (p. 613)

🌐 Real-World Link

Architecture If you visit the Jay Pritzker Pavilion in Chicago, Illinois, you will see examples of three-dimensional figures used in architecture.

Measurement: Two- and Three-Dimensional Figures Make this Foldable to help you organize your notes. Begin with a sheet of $8\frac{1}{2}$" by 11" construction paper and two sheets of notebook paper.

① **Fold** the construction paper in half lengthwise. Label the chapter title on the outside.

② **Fold** the sheets of notebook paper in half lengthwise. Then fold top to bottom twice.

③ **Open** the notebook paper. Cut along the second folds to make four tabs.

④ **Glue** the uncut notebook paper side by side onto the construction paper. Label each tab as shown.

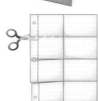

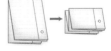

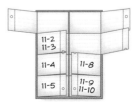

GET READY for Chapter 11

Diagnose Readiness You have two options for checking Prerequisite Skills.

Option 2

NY Math Online Take the Online Readiness Quiz at glencoe.com.

Option 1

Take the Quick Quiz below. Refer to the Quick Review for help.

QUICK Quiz

Evaluate each expression. (Prior Grade)

1. 8×17

2. 5.6×9.8

3. $12 \times 4 \times 26$

4. $4.5 \times 3.2 \times 1.7$

5. $\left(\frac{1}{2}\right)(11)(14)$

6. $\left(\frac{1}{2}\right)(8.8)(2.3)$

7. **SHOPPING** Margo bought 3 sweaters that originally cost $27.78 each. If they were on sale for half the price, what was the total cost of the sweaters, not including tax? (Prior Grade)

Evaluate each expression. (Lesson 1-2)

8. 3^2

9. 11 squared

10. 5 to the third power

11. 6 to the second power

12. **CHESS** If a chessboard has 8^2 squares, how many squares is this? (Lesson 1-2)

Use the π button on your calculator to evaluate each expression. Round to the nearest tenth.
(Prior Grade)

13. $\pi \times 4$

14. $\pi \times 13.8$

15. $(2)(\pi)(5)$

16. $(2)(\pi)(1.7)$

17. $\pi \times 9^2$

18. $\pi \times 6^2$

QUICK Review

Example 1 Find $1.2 \cdot 3.4$.

$$
\begin{array}{r}
1.2 \quad \leftarrow \quad \text{1 decimal place} \\
\times 3.4 \quad \leftarrow \quad \underline{+1 \text{ decimal place}} \\
\hline
48 \\
36 \\
\hline
4.08 \quad \leftarrow \quad \text{2 decimal places}
\end{array}
$$

Example 2 Find $\left(\frac{1}{2}\right)(26)(19)$.

$\left(\frac{1}{2}\right)(26)(19) = (13)(19)$ Multiply $\frac{1}{2}$ by 26.

$\qquad\qquad\quad = 247$ Multiply 13 by 19.

Example 3 Evaluate 7^3.

$7^3 = 7 \cdot 7 \cdot 7$ or 343

Example 4 Evaluate 2 to the fourth power.

2 to the fourth power is written 2^4.

$2^4 = 2 \cdot 2 \cdot 2 \cdot 2$ or 16

Example 5 Use the π button on your calculator to evaluate $\pi \times 5^2$. Round to the nearest tenth.

$\pi \cdot 5^2 = \pi \cdot 25$ $5^2 = 25$

$\qquad\quad = 78.5$ Multiply π by 25.

MAIN IDEA

Find the areas of parallelograms.

NYS Core Curriculum

Reinforcement of 6.G.2 Determine the area of triangles and quadrilaterals (squares, rectangles, rhombi, and trapezoids) and develop formulas

New Vocabulary

base
height

NY Math Online

glencoe.com

• Extra Examples
• Personal Tutor
• Self-Check Quiz

▷ **MINI Lab**

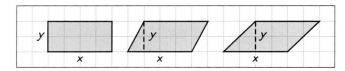

1. What is the value of x and y for each parallelogram?
2. Count the grid squares to find the area of each parallelogram.
3. On grid paper, draw three different parallelograms in which $x = 5$ units and $y = 4$ units. Find the area of each.
4. **MAKE A CONJECTURE** Explain how to find the area of a parallelogram if you know the values of x and y.

You can find the area of a parallelogram by using the values for the base and height, as described below.

The **base** is any side of a parallelogram.

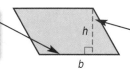

The **height** is the length of the segment perpendicular to the base with endpoints on opposite sides.

Area of a Parallelogram   Key Concept

Words   The area A of a parallelogram equals the product of its base b and height h.

Model

Symbols $A = bh$

EXAMPLES   Find the Area of a Parallelogram

① **Find the area of the parallelogram.**

Estimate $A = 13 \cdot 6$ or 78 cm^2

$A = bh$ Area of a parallelogram

$A = 13 \cdot 5.8$ Replace b with 13 and h with 5.8.

$A = 75.4$ Multiply.

The area is 75.4 square centimeters.

Check for Reasonableness $75.4 \approx 78$ ✔

5.8 cm

13 cm

② **Find the area of the parallelogram.**

The base is 11 inches, and the height is 9 inches.

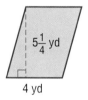

11 in.
9 in. $9\frac{1}{2}$ in.

Estimate $A = 10 \cdot 10$ or 100 in²

$A = bh$ Area of a parallelogram

$A = 11 \cdot 9$ Replace b with 11 and h with 9.

$A = 99$ Multiply.

The area of the parallelogram is 99 square inches.

Check for Reasonableness $99 \approx 100$ ✔

✓ CHECK Your Progress

Find the area of each parallelogram.

a.
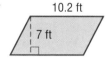
10.2 ft
7 ft

b.
15.5 ft
$5\frac{1}{4}$ yd

4 yd

Real-World EXAMPLE

③ **WEATHER** The map shows the region of a state that is under a tornado warning. What is the area of this region?

KNOX   ADAMS
27.5 mi
LAKE
30.6 mi
FOX   LUCAS

Estimate $A = 30 \cdot 30$ or 900 mi²

$A = bh$ Area of a parallelogram

$A = 27.5 \cdot 30.6$ Replace b with 27.5 and h with 30.6.

$A = 841.5$ Multiply.

The area of the region is 841.5 square miles.

Check for Reasonableness $841.5 \approx 900$ ✔

✓ CHECK Your Progress

c. **SAFETY** A street department painted the pavement markings shown at the right on the surface of a highway to reduce traffic speeds and crashes. What is the area inside one of the markings?

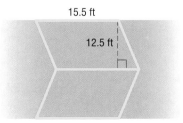
15.5 ft
12.5 ft

Example 1
(p. 572)

Find the area of each parallelogram. Round to the nearest tenth if necessary.

1.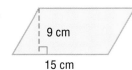
9 cm
15 cm

2.
0.75 m
1.5 m

Example 2
(p. 573)

3.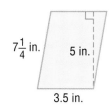
$7\frac{1}{4}$ in.   5 in.
3.5 in.

4.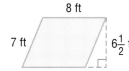
8 ft
7 ft $6\frac{1}{2}$ ft

Example 3
(p. 573)

5. **ART** Martina designs uniquely-shaped ceramic wall tiles. What is the area of the tile shown?

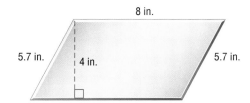
8 in.
5.7 in.   4 in.   5.7 in.

Practice and Problem Solving

Find the area of each parallelogram. Round to the nearest tenth if necessary.

For Exercises	See Examples
6–9, 12–13	1
10–11	2
14–15	3

6.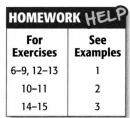
16 ft
16 ft

7.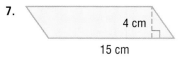
4 cm
15 cm

8.
21 mm
20.4 mm

9.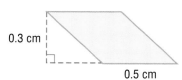
0.3 cm
0.5 cm

10.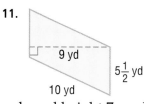
12 in.   14 in.
$17\frac{1}{4}$ in.

11.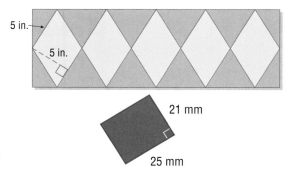
9 yd $5\frac{1}{2}$ yd
10 yd

12. What is the area of a parallelogram with base $10\frac{3}{4}$ yards and height 7 yards?

13. Find the area of a parallelogram with base 12.5 meters and height 15.25 meters.

14. **WALLPAPER** The design of the wallpaper border at the right contains parallelograms. How much space on the border is covered by the parallelograms?

5 in.
5 in.

15. **PATTERN BLOCKS** What is the area of the parallelogram-shaped pattern block shown at the right?

21 mm
25 mm

16. PARKING SPACES A city ordinance requires that each parking space has a minimum area of 162 square feet. Do the measurements of the parking spaces shown below meet the requirements? Explain.

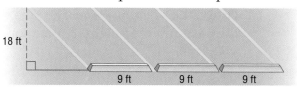

18 ft

9 ft   9 ft   9 ft

Find the area of each parallelogram. Round to the nearest tenth if necessary.

17.

18 in.
1 ft

18.

4 yd

15 ft

19.

18 in.
1.5 yd

GEOGRAPHY Estimate the area of each state.

20. 225 mi
300 mi
★
MISSOURI

21. 350 mi
TENNESSEE ★
120 mi

22. ALGEBRA A parallelogram has an area of 75 square feet. Find the base of the parallelogram if the height is 3 feet.

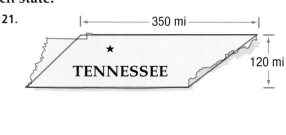

$A = 75 \text{ ft}^2$ 3 ft
b ft

23. ALGEBRA What is the height of a parallelogram if the base is 24 inches and the area is 360 square inches?

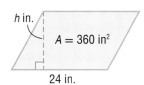

h in.
$A = 360 \text{ in}^2$
24 in.

QUILTING For Exercises 24 and 25, use the four quilt blocks shown and following information.

Each quilt block uses eight parallelogram-shaped pieces of fabric that have a height of $1\frac{1}{2}$ inches and a base of $4\frac{2}{3}$ inches.

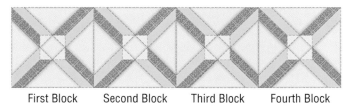

First Block   Second Block   Third Block   Fourth Block

24. Find the amount of fabric in square inches needed to make the parallelogram-shaped pieces for one quilt block.

25. How much fabric is needed to make the parallelogram pieces for a quilt that is made using 30 blocks? Write in square feet. (*Hint*: 144 in^2 = 1 ft^2)

NYSCC • NYSMT
Extra Practice, pp. 697, 714

H.O.T. Problems

26. **OPEN ENDED** Draw three different parallelograms, each with an area of 24 square units.

27. **CHALLENGE** *True* or *False*? The area of a parallelogram doubles if you double the base and the height. Explain or give a counterexample to support your answer.

28. **WRITING IN MATH** Compare and contrast the formula for the area of a rectangle and a parallelogram.

NYSMT PRACTICE 6.G.2

29. A scale drawing of a piece of land is shown below.

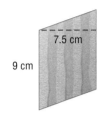

7.5 cm

9 cm

What is the actual area of the land if the scale on the drawing is 10 km = 1 cm?

A 67.5 cm²   C 675 km²

B 67.5 km²   D 6,750 km²

30. Ms. Cruz painted one wall in her living room. She did not paint the window.

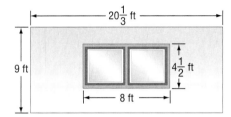

$20\frac{1}{3}$ ft

9 ft

$4\frac{1}{2}$ ft

8 ft

Which of the following is closest to the painted area of the wall in square feet?

F 180 ft²   H 120 ft²

G 150 ft²   J 34 ft²

Spiral Review

For Exercises 31 and 32, refer to △ABC at the right. Find the vertices of △A′B′C′ after each transformation. Then graph the triangle and its reflected or translated image.

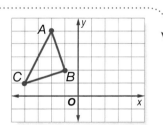

31. △ABC reflected over the y-axis (Lesson 10-10)

32. △ABC translated 5 units right and 1 unit up (Lesson 10-9)

33. **GEOMETRY** Explain how to determine whether a regular octagon tessellates the plane. (Lesson 10-8)

34. **FRUIT** If 2 out of every 3 pieces of fruit in a fruit basket are oranges, how many oranges are there if the fruit basket has 18 pieces of fruit? (Lesson 6-5)

▷ **GET READY for the Next Lesson**

PREREQUISITE SKILL Find each value. (Lesson 1-4)

35. $6(4 + 10)$ **36.** $\frac{1}{2}(8)(8)$ **37.** $\frac{1}{2}(24 + 15)$ **38.** $\frac{1}{2}(5)(13 + 22)$

Measurement Lab
Triangles and Trapezoids

MAIN IDEA

Find the areas of parallelograms.

NYS Core Curriculum

Reinforcement of 6.G.2 Determine the area of triangles and quadrilaterals (squares, rectangles, rhombi, and trapezoids) and develop formulas
Reinforcement of 6.G.3 Use a variety of strategies to find the area of regular and irregular polygons

ACTIVITY

STEP 1 On grid paper, draw a triangle with a base of 6 units and a height of 3 units. Label the base b and the height h as shown.

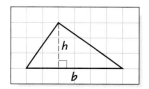

STEP 2 Fold the grid paper in half and cut out the triangle through both sheets so that you have two congruent triangles.

STEP 3 Turn the second triangle upside down and tape it to the first triangle.

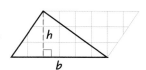

ANALYZE THE RESULTS

1. What figure is formed by the two triangles?

2. Write the formula for the area of the figure. Then find the area.

3. What is the area of each of the triangles? How do you know?

4. Repeat the activity above, drawing a different triangle in Step 1. Then find the area of each triangle.

5. Compare the area of a triangle to the area of a parallelogram with the same base and height.

6. **MAKE A CONJECTURE** Write a formula for the area of a triangle with base b and height h.

For Exercises 7–10, refer to the information below.

On grid paper, cut out two identical trapezoids. Label the bases b_1 and b_2, respectively, and label the heights h. Then turn one trapezoid upside down and tape it to the other trapezoid as shown.

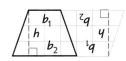

7. Write an expression to represent the base of the parallelogram.

8. Write a formula for the area A of the parallelogram using b_1, b_2, and h.

9. How does the area of each trapezoid compare to the area of the parallelogram?

10. **MAKE A CONJECTURE** Write a formula for the area A of a trapezoid with bases b_1 and b_2, and height h.

Area of Triangles and Trapezoids

11-2

MAIN IDEA

Find the areas of triangles and trapezoids.

NYS Core Curriculum

Reinforcement of 6.G.2 Determine the area of triangles and quadrilaterals (squares, rectangles, rhombi, and trapezoids) and develop formulas

NY Math Online

glencoe.com

• Concepts In Motion
• Extra Examples
• Personal Tutor
• Self-Check Quiz

▷ **MINI Lab**

• Draw a parallelogram with a base of 6 units and a height of 4 units.

• Draw a diagonal as shown.

• Cut out the parallelogram.

1. What is the area of the parallelogram?

2. Cut along the diagonal. What is true about the triangles formed?

3. What is the area of each triangle?

4. If the area of a parallelogram is bh, then write an expression for the area A of each of the two congruent triangles that form the parallelogram.

You can find the area of a triangle by using the base and height.

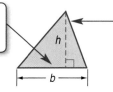

The base of a triangle can be any of its sides.

The height is the perpendicular distance from the vertex opposite the base to the line containing the base.

Area of a Triangle		**Key Concept**
Words	The area A of a triangle equals half the product of its base b and height h.	**Model**
Symbols	$A = \frac{1}{2}bh$	

EXAMPLE Find the Area of a Triangle

1 **Find the area of the triangle shown.**

Estimate $A = \frac{1}{2}(10)(7)$ or 35

$A = \frac{1}{2}bh$ Area of a triangle

$A = \frac{1}{2}(10)(6.5)$ Replace b with 10 and h with 6.5.

$A = 32.5$ Multiply.

The area of the triangle is 32.5 square meters.

6.5 m

10 m

Check for Reasonableness $32.5 \approx 35$ ✔

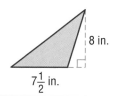

CHECK Your Progress

Find the area of each triangle. Round to the nearest tenth if necessary.

a.

11 ft

14 ft

b.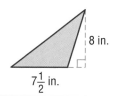

8 in.

$7\frac{1}{2}$ in.

A trapezoid has two bases, b_1 and b_2. The bases are always the two sides that are parallel. The height of a trapezoid is the perpendicular distance between the bases.

Reading Math

Subscripts Read b_1 as *b sub 1*. Read b_2 as *b sub 2*. The subscripts mean that b_1 and b_2 represent different variables.

Area of a Trapezoid Key Concept

Words The area A of a trapezoid equals half the product of the height h and the sum of the bases $b_1 + b_2$.

Model

b_1

h

b_2

Symbols $A = \frac{1}{2}h(b_1 + b_2)$

EXAMPLE Find the Area of a Trapezoid

2 Find the area of the trapezoid.

The bases are 5 inches and 12 inches.
The height is 7 inches.

5 in.

7 in.

12 in.

$A = \frac{1}{2}h(b_1 + b_2)$ Area of a trapezoid

$A = \frac{1}{2}(7)(5 + 12)$ Replace h with 7, b_1 with 5, and b_2 with 12.

$A = \frac{1}{2}(7)(17)$ Add 5 and 12.

$A = 59.5$ Multiply.

The area of the trapezoid is 59.5 square inches.

CHECK Your Progress

Find the area of each trapezoid. Round to the nearest tenth if necessary.

c.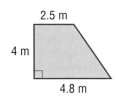

2.5 m

4 m

4.8 m

d.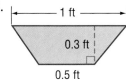

1 ft

0.3 ft

0.5 ft

Real-World EXAMPLE

3 **GEOGRAPHY** The shape of Iron County, Michigan, resembles a trapezoid. Find the approximate area of this county.

$A = \frac{1}{2}h(b_1 + b_2)$ Area of a trapezoid

$A = \frac{1}{2}(42)(22 + 34)$ Replace h with 42, b_1 with 22, and b_2 with 34.

$A = \frac{1}{2}(42)(56)$ Add 22 and 34.

$A = 1{,}176$ Multiply.

The area of Iron County is approximately 1,176 square miles.

CHECK Your Progress

e. GEOGRAPHY The shape of the state of Arkansas resembles a trapezoid. Find the approximate area of Arkansas.

CHECK Your Understanding

Examples 1, 2
(pp. 578–579)

Find the area of each figure. Round to the nearest tenth if necessary.

1.

3 in.

4 in.

2.

16.5 m

←—12.8 m—→

3.

7 ft

8 ft

15.6 ft

Example 3
(p. 580)

4. HOCKEY In the National Hockey League, goaltenders can play the puck behind the goal line only in a trapezoid-shaped area, as shown at the right. Find the area of the trapezoid.

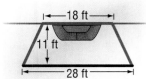

←—18 ft—→

11 ft

←———— 28 ft ————→

Find the area of each figure. Round to the nearest tenth if necessary.

5.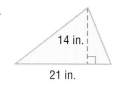
14 in.
21 in.

6.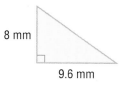
8 mm
9.6 mm

7.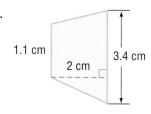
1.1 cm
2 cm
3.4 cm

8.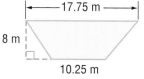
17.75 m
8 m
10.25 m

9.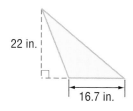
22 in.
16.7 in.

10.
15 ft
$8\frac{1}{2}$ ft
10 ft
23 ft

11. **GEOGRAPHY** The shape of Idaho is roughly triangular with a base of 380 miles and a height of 500 miles. Find the approximate area of Idaho.

12. **ALGEBRA** Find the area of a trapezoid with bases 13 inches and 15 inches and a height of 7 inches.

500 mi
IDAHO
380 mi

ALGEBRA **Find the height of each figure.**

13.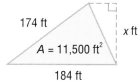
174 ft
x ft
A = 11,500 ft²
184 ft

14.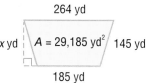
264 yd
x yd
A = 29,185 yd²
145 yd
185 yd

Draw and label each figure. Then find the area.

15. a triangle with no right angles and an area less than 12 square centimeters

16. a trapezoid with a right angle and an area greater than 40 square meters

17. a trapezoid with no right angles and an area less than 25 square feet

18. **TENTS** A play tent is shown at the right. How much fabric was used to make the front and back of the play tent?

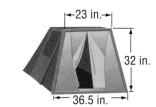
23 in.
32 in.
36.5 in.

LANDSCAPING **For Exercises 19 and 20, use the diagram that shows the lawn that surrounds an office building.**

19. What is the area of the lawn?

20. If one bag of grass seed covers 2,000 square feet, how many bags are needed to seed the lawn?

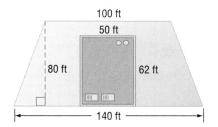
100 ft
50 ft
80 ft
62 ft
140 ft

21. CHALLENGE Triangle *ABC* has a base of 4 units and a height of 8 units. Triangle *DEF* has twice the base and height. Describe how the ratio of the bases is related to the ratio of the areas.

22. REASONING Apply what you know about rounding to explain how to estimate the height *h* of the trapezoid shown if the area is 235.5 in^2.

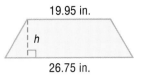

23. WRITING IN MATH Describe the relationship between the area of a parallelogram and the area of a triangle with the same height and base.

NYSMT PRACTICE ▷ **6.G.2**

24. △*FGH* and △*JKM* are similar.

 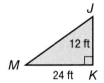

Which choice shows the equations that can be used to find the area of △*FGH*?

A $\dfrac{18}{24} = \dfrac{12}{x}$ and then $\dfrac{1}{2}(18x)$

B $\dfrac{18}{24} = \dfrac{x}{12}$ and then $18x$

C $\dfrac{18}{24} = \dfrac{x}{12}$ and then $\dfrac{1}{2}(18x)$

D $\dfrac{18}{24} = \dfrac{12}{x}$ and then $18x$

25. SHORT RESPONSE Randy was hired to put in sod on a piece of land shaped like a trapezoid as shown. How many square feet of sod are needed?

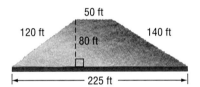

Spiral Review

26. MEASUREMENT Find the area of a parallelogram having a base of 2.3 inches and a height of 1.6 inches. Round to the nearest tenth. (Lesson 11-1)

27. GEOMETRY Graph △*JLK* with vertices *J*(−1, −4), *L*(3, −2) and *K*(1, 1), and its reflection over the *x*-axis. Write the ordered pairs for the vertices of the new figure. (Lesson 10-10)

Find each number. Round to the nearest tenth if necessary. (Lesson 7-2)

28. What number is 56% of 600?

29. 24.5 is what percent of 98?

30. 72 is 45% of what number?

31. 62.5% of 250 is what number?

▷ **GET READY for the Next Lesson**

PREREQUISITE SKILL Use a calculator to find each product to the nearest tenth. (Lesson 1-4)

32. $\pi \cdot 13$ **33.** $\pi \cdot 29$ **34.** $\pi \cdot 16^2$ **35.** $\pi \cdot 4.8^2$

Measurement Lab
Circumference of Circles

MAIN IDEA

Find a relationship between circumference and diameter.

NYS Core Curriculum

7.G.1 Calculate the radius or diameter, given the circumference or area of a circle

In this lab, you will investigate how the *circumference*, or the distance around a circle, is related to its *diameter*, or the distance across a circle through its center.

ACTIVITY

STEP 1 Use a ruler to measure the diameter of a circular object. Record the length in a table like the one shown below.

Object	Diameter (cm)	Circumference (cm)

STEP 2 Make a small mark at the edge of the circular object. Place a measuring tape on a flat surface. Place the mark you made on the circular object at the beginning of the measuring tape. Roll the object along the tape for one revolution, until you reach the mark again.

STEP 3 Record the length in the table. This is the circumference.

STEP 4 Repeat this activity with circular objects of various sizes.

ANALYZE THE RESULTS

1. For each object, divide the circumference by the diameter. Add another column to your table and record the results. Round to the nearest tenth if necessary.

2. What do you notice about the ratio of each circumference to each diameter?

3. Graph the ordered pair (diameter, circumference) on a coordinate plane for each object. What do you notice?

4. Use the graph to predict the circumference of a circular object that has a diameter of 18 centimeters.

5. **MAKE A CONJECTURE** Write a rule describing how you would find the circumference C of a circle if you know the diameter d.

6. Use your rule to approximate the circumference of a circular object that has a diameter of 45 centimeters.

11-3 Circles and Circumference

MAIN IDEA

Find the circumference of circles.

NYS Core Curriculum

7.G.1 Calculate the radius or diameter, given the circumference or area of a circle

New Vocabulary

circle
center
diameter
circumference
radius
π (pi)

NY Math Online

glencoe.com

• Extra Examples
• Personal Tutor
• Self-Check Quiz

▷ **GET READY** for the Lesson

CLOCKS Big Ben is a famous clock tower in London, England. The diameter of the clock face is 23 feet.

1. Which point appears at the middle of Big Ben?

2. How does the distance from A to C compare to the distance from B to D?

3. Find the distance from D to C.

A **circle** is the set of all points in a plane that are the same distance from a given point, called the **center**.

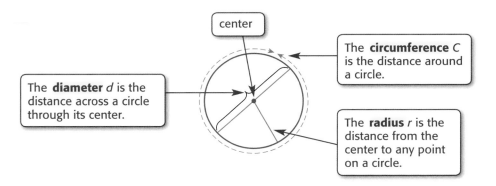

The **diameter** d is the distance across a circle through its center.

center

The **circumference** C is the distance around a circle.

The **radius** r is the distance from the center to any point on a circle.

The diameter of a circle is 2 times the radius, or $d = 2r$. Another relationship that is true of all circles is $\frac{C}{d} = 3.1415926\ldots$. This nonterminating and nonrepeating number is represented by the Greek letter **π (pi)**. An approximation often used for π is 3.14.

<table>
<tr><td colspan="2">**Circumference of a Circle**   Key Concept</td></tr>
<tr><td>**Words**</td><td>The circumference C of a circle is equal to its diameter d times π, or 2 times its radius r times π.</td></tr>
<tr><td>**Symbols**</td><td>$C = \pi d$ or $C = 2\pi r$</td></tr>
</table>

When finding the circumference of a circle, it is necessary to use an approximation of π since its exact value cannot be determined.

584   Chapter 11 Measurement: Two- and Three-Dimensional Figures

Three-Dimensional Figures

MAIN IDEA

Build three-dimensional figures given the top, side, and front views.

NYS Core Curriculum

7.G.3 Identify the two-dimensional shapes that make up the faces and bases of three-dimensional shapes (prisms, cylinders, cones, and pyramids)

New Vocabulary

three-dimensional figure
face
edge
lateral face
vertex (vertices)
prism
base
pyramid
cone
cylinder
sphere
center

NY Math Online

glencoe.com
• Extra Examples
• Personal Tutor
• Self-Check Quiz

▷ **GET READY** for the Lesson

Study the shape of each common object below.
Then compare and contrast the properties of each object.

Many common shapes are **three-dimensional figures**. That is, they have length, width, and depth (or height). Some terms associated with three-dimensional figures are shown below.

A **face** is a flat surface.

The **edges** are the segments formed by intersecting faces.

The sides are called **lateral faces**.

The edges intersect at the **vertices**.

Two types of three-dimensional figures are prisms and pyramids.

Prisms and Pyramids
Key Concept

Figure	Properties
Prism	• Has at least three lateral faces that are parallelograms. • The top and bottom faces, called the **bases**, are congruent parallel polygons. • The shape of the base tells the name of the prism. Rectangular prism Triangular prism Square prism or cube
Pyramid	• Has at least three lateral faces that are triangles. • Has only one base, which is a polygon. • The shape of the base tells the name of the pyramid. Triangular pyramid Square pyramid

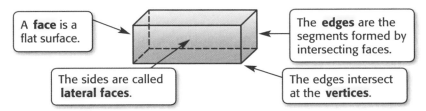

Some three-dimensional figures have curved surfaces.

Figure	Properties	
Cone	• Has only one base. • The base is a circle. • Has one vertex.	
Cylinder	• Has only two bases. • The bases are congruent circles. • Has no vertices and no edges.	
Sphere	• All of the points on a sphere are the same distance from the **center**. • No faces, bases, edges, or vertices.	center

Study Tip

Three-Dimensional Figures
In three-dimensional figures, dashed lines are used to indicate edges that are hidden from view.

Study Tip

Prisms and Pyramids
Prisms and pyramids are examples of polyhedra, or solids with flat surfaces that are polygonal regions. Cones, cylinders, and spheres are not examples of polyhedra.

EXAMPLES Classify Three-Dimensional Figures

For each figure, identify the shape of the base(s). Then classify the figure.

  The figure has one circular base, no edge, and one vertex.

The figure is a cone.

  The base and all other faces are squares.

The figure is a square prism or cube.

CHECK Your Progress

a.

b.

Real-World EXAMPLE

 CAMERAS Classify the shape of the body of the digital camera, not including the lens, as a three-dimensional figure.

The body of the camera is a rectangular prism.

CHECK Your Progress

c. Classify the shape of the zoom lens as a three-dimensional figure.

Examples 1, 2
(p. 604)

For each figure, identify the shape of the base(s). Then classify the figure.

1.

2.

3.

Example 3
(p. 604)

4. **SPORTS** An official major league baseball has 108 stitches. Classify the shape of a baseball as a three-dimensional figure.

Practice and Problem Solving

For each figure, identify the shape of the base(s). Then classify the figure.

HOMEWORK HELP	
For Exercises	**See Examples**
5–8	1–2
9–10	3

5.

6.

7.

8.

9. **FOOD** What three-dimensional figure describes the item at the right?

10. **SCHOOL SUPPLIES** Classify the shape of your math textbook as a three-dimensional figure.

For each figure, identify the shape of the base(s). Then classify the figure.

11.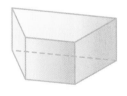

12.

13.

14. **SCHOOL SUPPLIES** The model of the pencil shown is made of two geometric figures. Classify these figures.

15. **HOUSES** The model of the house shown is made of two geometric figures. Classify these figures.

NYSCC • NYSMT
Extra Practice, pp. 699, 714.

16. **REASONING** Two sets of figures were sorted according to a certain rule. The figures in Set A follow the rule and the figures in Set B do not follow the rule. Describe the rule.

Set A	Prism	Pyramid	Cube
Set B	Cylinder	Cone	Sphere

17. **CHALLENGE** What figure is formed if only the height of a cube is increased? Draw a figure to justify your answer.

18. **OPEN ENDED** Select one three-dimensional figure in which you could use the term *congruent* to describe the bases of the figure. Then write a sentence using *congruent* to describe the figure.

19. **WRITING IN MATH** Apply what you know about the properties of geometric figures to compare and contrast cones and pyramids.

NYSMT PRACTICE 7.G.3

20. Which statement is true about all triangular prisms?

   A All of the edges are congruent line segments.

   B There are exactly 6 faces.

   C The bases are congruent triangles.

   D All of the faces are triangles.

21. Which figure is shown?

   F triangular pyramid

   G square pyramid

   H rectangular pyramid

   J triangular prism

Spiral Review

22. **MEASUREMENT** Find the area of the figure shown at the right if each triangle has a height of 3.5 inches and the square has side lengths of 4 inches. (Lesson 11-6)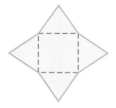

23. **MEASUREMENT** Find the area of a circle with a radius of 5.7 meters. Round to the nearest tenth. (Lesson 11-4)

ALGEBRA Find the missing angle measure in each quadrilateral. (Lesson 10-6)

24.

25.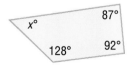

26.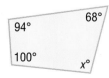

▷ **GET READY for the Next Lesson**

PREREQUISITE SKILL Describe the shape seen when each object is viewed from the top.

27. number cube

28. cereal box

29. soup can

Geometry Lab
Three-Dimensional Figures

Cubes are examples of three-dimensional figures because they have length, width, and depth. In this lab, you will use centimeter cubes to build other three-dimensional figures.

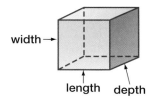

width→ length depth

ACTIVITY

The top view, side view, and front view of a three-dimensional figure are shown below. Use centimeter cubes to build the figure. Then make a sketch of the figure.

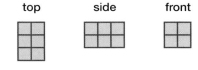

top   side   front

STEP 1 Use the top view to build the figure's base.

STEP 2 Use the side view to complete the figure.

STEP 3 Use the front view to check the figure.

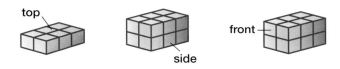

top   side   front

✔ CHECK Your Progress

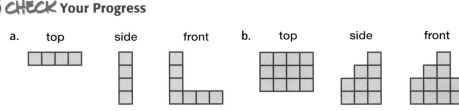

a.   top   side   front   b.   top   side   front

ANALYZE THE RESULTS

1. Explain how you began building the figures in Exercises a and b.

2. Determine whether there is more than one way to build each model. Explain your reasoning.

3. Build two different models that would look the same from two views, but not the third view. Draw a top view, side view, and front view of each model.

4. Describe a real-world situation where it might be necessary to draw a top, side, and front view of a three-dimensional figure.

Drawing Three-Dimensional Figures

11-8

MAIN IDEA

Draw a three-dimensional figure given the top, side, and front views.

NYS Core Curriculum

7.G.3 Identify the two-dimensional shapes that make up the faces and bases of three-dimensional shapes (prisms, cylinders, cones, and pyramids)

NY Math Online

glencoe.com

- Concepts In Motion
- Extra Examples
- Personal Tutor
- Self-Check Quiz

▷ **GET READY** for the Lesson

MONUMENTS The front view of the Wright Brothers Memorial in Kittyhawk, North Carolina, is shown.

1. Describe the two-dimensional figure(s) that make up the front view.

2. The monument is a three-sided building. Sketch what you think the top view might look like.

You can draw different views of three-dimensional figures. The most common views drawn are the top, side, and front views.

EXAMPLE Draw a Three-Dimensional Figure

① Draw a top, a side, and a front view of the figure at the right.

The top view is a triangle.

The side and front view are rectangles.

top	side	front
	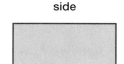	

✔ **CHECK Your Progress**

Draw a top, a side, and a front view of each solid.

a.    b.

Real-World EXAMPLE

2 **VIDEO GAMES** Draw a top, a side, and a front view of the video console shown.

The top view is a rectangle.

The side and front views are two rectangles.

✔ CHECK Your Progress

c. **TENTS** Draw a top, a side, and a front view of the tent shown.

The top, side, and front views of a three-dimensional figure can be used to draw a corner view of the figure.

EXAMPLE Draw a Three-Dimensional Figure

3 Draw a corner view of the three-dimensional figure whose top, side, and front views are shown.

top   side   front

Step 1 Use the top view to draw the base of the figure, a 1-by-3 rectangle.

Step 2 Add edges to make the base a solid figure.

Step 3 Use the side and front views to complete the figure.

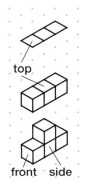

✔ CHECK Your Progress

d. Draw a corner view of the three-dimensional figure whose top, side, and front views are shown.

top   side   front

Example 1
(p. 608)

Draw a top, a side, and a front view of each solid.

1.

2.

Example 2
(p. 609)

3. **SCIENCE** A transparent prism can be used to refract or disperse a beam of light. Draw a top, a side, and a front view of the prism shown at the right.

Example 3
(p. 609)

4. Draw a corner view of the three-dimensional figure whose top, side, and front views are shown.

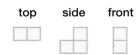

top   side   front

Practice and Problem Solving

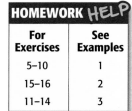

HOMEWORK HELP	
For Exercises	**See Examples**
5–10	1
15–16	2
11–14	3

Draw a top, a side, and a front view of each solid.

5.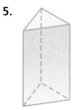

6.

7.

8.

9.

10.

Draw a corner view of each three-dimensional figure whose top, side, and front views are shown. Use isometric dot paper.

11. top   side   front

12. top   side   front

13. top   side   front

14. top   side   front

11-9 Volume of Prisms

MAIN IDEA

Find the volumes of rectangular and triangular prisms.

NYS Core Curriculum

7.G.2 Calculate the volume of prisms and cylinders, **using a given formula and a calculator** *Also addresses 7.CN.8*

New Vocabulary

volume
rectangular prism
triangular prism

NY Math Online

glencoe.com
• Extra Examples
• Personal Tutor
• Self-Check Quiz
• Reading in the Content Area

▷ MINI Lab

• On a piece of grid paper, cut out a square that is 10 centimeters on each side.

• Cut a 1-centimeter square from each corner. Fold the paper and tape the corners together to make a box.

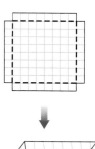

1. What is the area of the base, or bottom, of the box? What is the height of the box?

2. How many centimeter cubes fit in the box?

3. What do you notice about the product of the base area and the height of the box?

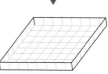

The **volume** of a three-dimensional figure is the measure of space occupied by it. It is measured in cubic units such as cubic centimeters (cm^3) or cubic inches (in^3). The volume of the figure at the right can be shown using cubes.

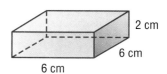

The bottom layer, or base, has 6 · 6 or 36 cubes.

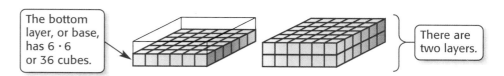

There are two layers.

It takes 36 · 2 or 72 cubes to fill the box. So, the volume of the box is 72 cubic centimeters.

The figure above is a rectangular prism. A **rectangular prism** is a prism that has rectangular bases.

Volume of a Rectangular Prism		**Key Concept**
Words	The volume V a rectangular prism is the area of the base B times the height h. It is also the product of the length ℓ, the width w, and the height h.	Model 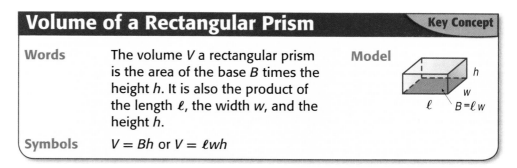
Symbols	$V = Bh$ or $V = \ell wh$	

You can use the formula $V = Bh$ or $V = \ell wh$ to find the volume of a rectangular prism.

Volume of a Rectangular Prism

1 **Find the volume of the rectangular prism.**

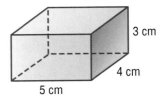

3 cm
4 cm
5 cm

$V = \ell w h$ Volume of a prism

$V = 5 \cdot 4 \cdot 3$ $\ell = 5$, $w = 4$, and $h = 3$

$V = 60$ Multiply.

The volume is 60 cubic centimeters or 60 cm^3.

CHECK Your Progress

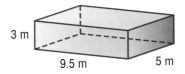

3 m
9.5 m
5 m

a. Find the volume of the rectangular prism at the right.

Real-World EXAMPLE

2 **MARKETING** A company needs to decide which size lunch box to manufacture. Which lunch box shown will hold more food?

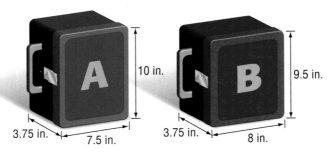

A   10 in.
3.75 in.   7.5 in.

B   9.5 in.
3.75 in.   8 in.

Find the volume of each lunch box. Then compare.

Lunch Box A

$V = \ell w h$ Volume of a rectangular prism

$V = 7.5 \cdot 3.75 \cdot 10$ $\ell = 7.5$, $w = 3.75$, and $h = 10$

$V = 281.25 \text{ in}^3$ Multiply.

Lunch Box B

$V = \ell w h$ Volume of a rectangular prism

$V = 8 \cdot 3.75 \cdot 9.5$ $\ell = 8$, $w = 3.75$, and $h = 9.5$

$V = 285 \text{ in}^3$ Multiply.

Since 285 in^3 > 281.25 in^3, Lunch Box B will hold more food.

CHECK Your Progress

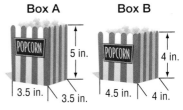

Box A   Box B
POPCORN   5 in.   POPCORN   4 in.
3.5 in.   3.5 in.   4.5 in.   4 in.

b. **PACKAGING** A movie theater serves popcorn in two different container sizes. Which container holds more popcorn? Justify your answer.

Study Tip

Height Do not confuse the height of the triangular base with the height of the prism.

A **triangular prism** is a prism that has triangular bases. The diagram below shows that the volume of a triangular prism is also the product of the area of the base B and the height h of the prism.

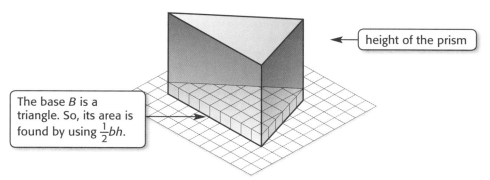

height of the prism

The base B is a triangle. So, its area is found by using $\frac{1}{2}bh$.

Volume of a Triangular Prism
Key Concept

Words	The volume V of a triangular prism is the area of the base B times the height h.	**Model**
Symbols	$V = Bh$	

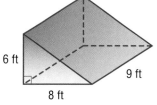

EXAMPLE Volume of a Triangular Prism

Study Tip

Base Before finding the volume of a prism, identify the base. In Example 3, the base is a triangle so you replace B with $\frac{1}{2}bh$.

3 **Find the volume of the triangular prism shown.**

The area of the triangle is $\frac{1}{2} \cdot 6 \cdot 8$ so replace B with $\frac{1}{2} \cdot 6 \cdot 8$.

$V = Bh$ Volume of a prism

$V = \left(\frac{1}{2} \cdot 6 \cdot 8\right)h$ Replace B with $\frac{1}{2} \cdot 6 \cdot 8$.

$V = \left(\frac{1}{2} \cdot 6 \cdot 8\right)9$ The height of the prism is 9.

$V = 216$ Multiply.

The volume is 216 cubic feet or 216 ft^3.

6 ft

9 ft

8 ft

CHECK Your Progress

Find the volume of each triangular prism.

c. 7 in.

5 in.

4 in.

d.

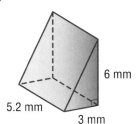

6 mm

5.2 mm

3 mm

CHECK Your Understanding

Example 1
(p. 614)

Find the volume of each prism. Round to the nearest tenth if necessary.

1.
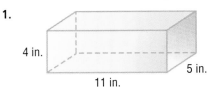
4 in.
11 in.
5 in.

2.
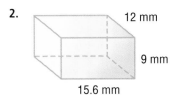
12 mm
9 mm
15.6 mm

Example 3
(p. 615)

3.
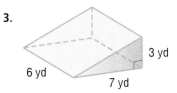
3 yd
6 yd
7 yd

4.
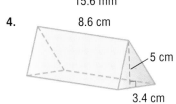
8.6 cm
5 cm
3.4 cm

Example 2
(p. 614)

5. **STORAGE** One cabinet measures 3 feet by 2.5 feet by 5 feet. A second measures 4 feet by 3.5 feet by 4.5 feet. Which cabinet has the greater volume?

Practice and Problem Solving

HOMEWORK HELP	
For Exercises	**See Examples**
6–9	1
14–15	2
10–13	3

Find the volume of each prism. Round to the nearest tenth if necessary.

6.
6 in.
20 in.
8 in.

7.
10 ft
3 ft
3 ft

8.
12 mm
9 mm
15.6 mm

9.
12.5 cm
4.2 cm
4.5 cm

10.
9 ft
8 ft
11 ft

11.
9 m
6 m
4 m

12.
2.8 yd
4.5 yd
6 yd

13.
3.4 mm
4.8 mm
2.5 mm

14. **PACKAGING** A soap company sells laundry detergent in two different containers. Which container holds more detergent? Justify your answer.

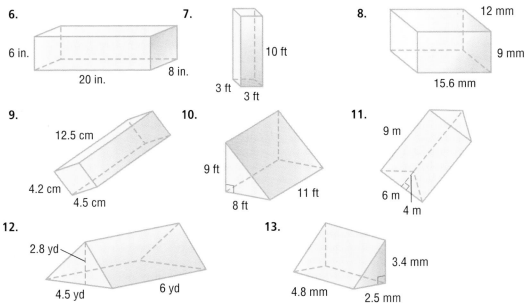
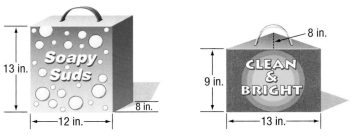

Soapy Suds
13 in.
8 in.
12 in.

CLEAN & BRIGHT
8 in.
9 in.
13 in.

Real-World Math . . . .
The Flatiron Building in New York City resembles a triangular prism.

15. **TOYS** A toy company makes rectangular sandboxes that measure 6 feet by 5 feet by 1.2 feet. A customer buys a sandbox and 40 cubic feet of sand. Did the customer buy too much or too little sand? Justify your answer.

Find the volume of each prism.

16.
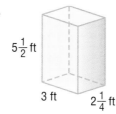
$5\frac{1}{2}$ ft
3 ft
$2\frac{1}{4}$ ft

17.
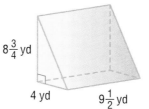
$8\frac{3}{4}$ yd
4 yd
$9\frac{1}{2}$ yd

18.
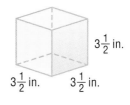
$3\frac{1}{2}$ in.
$3\frac{1}{2}$ in.
$3\frac{1}{2}$ in.

ARCHITECTURE For Exercises 19 and 20, use the diagram at the right that shows the approximate dimensions of the Flatiron Building in New York City.

19. What is the approximate volume of the Flatiron Building?

20. The building is a 22-story building. Estimate the volume of each story.

21. **ALGEBRA** The base of a rectangular prism has an area of 19.4 square meters and a volume of 306.52 cubic meters. Write an equation that can be used to find the height h of the prism. Then find the height of the prism.

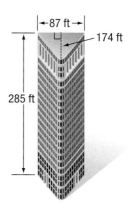
|← 87 ft →|
174 ft
285 ft

ESTIMATION Estimate to find the approximate volume of each prism.

22.
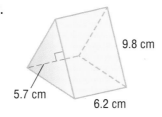
9.8 cm
5.7 cm
6.2 cm

23.
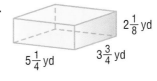
$2\frac{1}{8}$ yd
$5\frac{1}{4}$ yd
$3\frac{3}{4}$ yd

24. **MONEY** The diagram shows the dimensions of an office. It costs about 11¢ per year to air condition one cubic foot of space. On average, how much does it cost to air condition the office for one month?

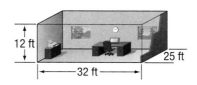
12 ft
25 ft
32 ft

25. **MEASUREMENT** The Garrett family is building a pool in the shape of a rectangular prism in their backyard. The pool will cover an area 18 feet by 25 feet and will hold 2,700 cubic feet of water. If the pool is equal depth throughout, find that depth.

NYSCC • NYSMT
Extra Practice, pp. 700, 714.

26. CHALLENGE How many cubic inches are in a cubic foot?

27. REASONING Two rectangular prisms are shown at the right. When the dimensions of Prism A are doubled, does the volume also double? Explain your reasoning.

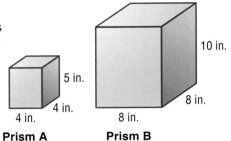

Prism A   Prism B

28. **WRITING IN MATH** Explain the similarities and differences in finding the volume of a rectangular prism and a triangular prism.

NYSMT PRACTICE 7.G.2

29. A fish aquarium is shown below.

What is the volume of the aquarium?

A 168 ft³   C 2,016 ft³

B 342 ft³   D 4,032 ft³

30. Use a ruler to measure the dimensions of the paper clip box in centimeters.

Which is closest to the volume of the box?

F 1.5 cm³   H 4.5 cm³

G 2.5 cm³   J 5.5 cm³

Spiral Review

31. GEOMETRY The top, side, and front view of a three-dimensional figure are shown at the right. Draw a corner view of the figure. (Lesson 11-8)

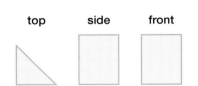

top   side   front

For each figure, identify the shape of the base(s). Then classify the figure. (Lesson 11-7)

32.

33.

34.

35. RATES A car travels 180 miles in 3.6 hours. What is the average rate of speed in miles per hour? (Lesson 6-2)

▷ **GET READY** for the Next Lesson

PREREQUISITE SKILL Estimate. (page 674)

36. $3.14 \cdot 6$ **37.** $5 \cdot 2.7^2$ **38.** $9.1 \cdot 8.3$ **39.** $3.1 \cdot 1.75^2 \cdot 2$

11-10 Volume of Cylinders

MAIN IDEA

Find the volumes of cylinders.

NYS Core Curriculum

7.G.2 Calculate the volume of prisms and cylinders, using a given formula and a calculator

NY Math Online

glencoe.com

• Extra Examples
• Personal Tutor
• Self-Check Quiz

▷ **MINI Lab**

Set a vegetable can on a piece of grid paper and trace around the base, as shown at the right.

1. Estimate the number of centimeter cubes that would fit at the bottom of the can. Include parts of cubes.

2. If each layer is 1 centimeter high, how many layers would it take to fill the cylinder?

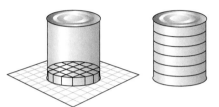

3. **MAKE A CONJECTURE** How can you find the volume of the can?

As with prisms, the area of the base of a cylinder tells the number of cubic units in one layer. The height tells how many layers there are in the cylinder.

Volume of a Cylinder		**Key Concept**
Words	The volume *V* of a cylinder with radius *r* is the area of the base *B* times the height *h*.	**Model**
Symbols	$V = Bh$ where $B = \pi r^2$ or $V = \pi r^2 h$	

EXAMPLE Find the Volume of a Cylinder

① **Find the volume of the cylinder. Round to the nearest tenth.**

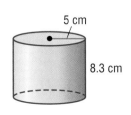

$V = \pi r^2 h$ Volume of a cylinder

$V = \pi (5)^2 (8.3)$ Replace *r* with 5 and *h* with 8.3.

Use a calculator.

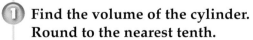 651.8804756

The volume is about 651.9 cubic centimeters.

✓ CHECK Your Progress

Find the volume of each cylinder. Round to the nearest tenth.

a.
3 in.

1.8 in.

b. 2.4 m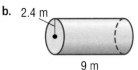

9 m

Study Tip

Circles Recall that the radius is half the diameter.

Real-World EXAMPLE

② **WEATHER** The decorative rain gauge shown has a height of 13 centimeters and a diameter of 3 centimeters. How much water can the rain gauge hold?

$V = \pi r^2 h$ Volume of a cylinder

$V = \pi(1.5)^2 13$ Replace r with 1.5 and h with 13.

$V \approx 91.8$ Simplify.

The rain gauge can hold about 91.8 cubic centimeters.

3 cm

13 cm

✓ **CHECK Your Progress**

c. **PAINT** Find the volume of a cylinder-shaped paint can that has a diameter of 4 inches and a height of 5 inches.

✓ **CHECK Your Understanding**

Find the volume of each cylinder. Round to the nearest tenth.

Example 1
(pp. 619–620)

1.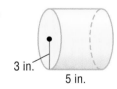

3 in.
5 in.

2. 1.5 cm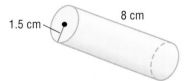
8 cm

3.
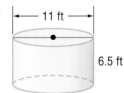
← 11 ft →

6.5 ft

Example 2
(p. 620)

4. **CONTAINERS** A can of concentrated orange juice has the dimensions shown at the right. Find the volume of the can of orange juice to the nearest tenth.

5. **CANDLES** A scented candle is in the shape of a cylinder. The radius is 4 centimeters and the height is 12 centimeters. Find the volume of the candle.

15 cm

7 cm

HOMEWORK HELP

For Exercises	See Examples
6–10	1
16–17	2

Find the volume of each cylinder. Round to the nearest tenth.

6.
4 in.
8 in.

7.
9 ft
16 ft

8.
24 mm
5 mm

9.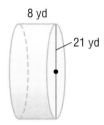
8 yd
21 yd

10.
13.3 cm
2 cm

11.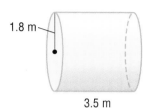
1.8 m
3.5 m

12. diameter = 15 mm
    height = 4.8 mm

13. diameter = 4.5 m
    height = 6.5 m

14. radius = 6 ft
    height = $5\frac{1}{3}$ ft

15. radius = $3\frac{1}{2}$ in.
    height = $7\frac{1}{2}$ in.

16. **WATER BOTTLE** What is the volume of a cylinder-shaped water bottle that has a radius of $1\frac{1}{4}$ inches and a height of 7 inches?

17. **BIRDS** A cylinder-shaped bird feeder has a diameter of 4 inches and a height of 18 inches. How much bird seed can the feeder hold?

Find the volume of each cylinder. Round to the nearest tenth.

18.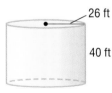
26 ft
40 ft

19.
75 m
46 m

20.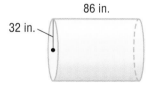
86 in.
32 in.

Real-World Link . . . . .
Bird feeders can attract many species of birds. There are over 800 species of birds in North America.

ESTIMATION Match each cylinder with its approximate volume.

21. radius = 4.1 ft, height = 5 ft

22. diameter = 8 ft, height = 2.2 ft

23. diameter = 6.2 ft, height = 3 ft

24. radius = 2 ft, height = 3.8 ft

a. 91 ft^3

b. 48 ft^3

c. 111 ft^3

d. 264 ft^3

25. **POTTERY** A vase in the shape of a cylinder has a diameter of 11 centimeters and a height of 250 millimeters. Find the volume of the vase to the nearest cubic centimeter. Use 3.14 for π.

26. **BAKING** Which will hold more cake batter, the rectangular pan or the two round pans? Explain.

27. **ALGEBRA** Cylinder A has a radius of 4 inches and a height of 2 inches. Cylinder B has a radius of 2 inches. What is the height of Cylinder B if both cylinders have the same volume?

ANALYZE TABLES For Exercises 28 and 29, use the table at the right and the following information.

The volume, using 3.14 for π, of four cylinders is shown at the right.

Radius (cm)	Height (cm)	Volume (cm)³
2	4	50.24
4	8	401.92
8	16	3,215.36
16	32	25,722.88

28. Describe how the radius and the height increase for each successive cylinder.

NYSCC • NYSMT
Extra Practice, pp. 701, 714.

29. As the radius and the height increase, how does the volume of each cylinder increase?

H.O.T. Problems

30. **CHALLENGE** Two equal-size sheets of construction paper are rolled along the length and along the width, as shown. Which cylinder do you think has the greater volume? Explain.

31. **OPEN ENDED** Draw and label a cylinder that has a larger radius, but less volume than the cylinder shown at the right.

8 cm

16 cm

32. **NUMBER SENSE** What is the ratio of the volume of a cylinder to the volume of a cylinder having twice the height but the same radius?

33. **NUMBER SENSE** Suppose cylinder A has the same height but twice the radius of cylinder B. What is the ratio of the volume of cylinder B to cylinder A?

34. **WRITING IN MATH** Explain how the formula for the volume of a cylinder is similar to the formula for the volume of a rectangular prism.

35. The oatmeal container shown has a diameter of $3\frac{1}{2}$ inches and a height of 9 inches.

Which is closest to the number of cubic inches it will hold when filled?

A 32   C 75.92

B 42.78   D 86.55

36. Which statement is true about the volumes of the cylinders shown?

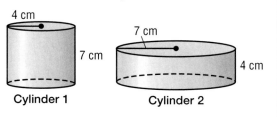

F The volume of cylinder 1 is greater than the volume of cylinder 2.

G The volume of cylinder 2 is greater than the volume of cylinder 1.

H The volumes are equal.

J The volume of cylinder 1 is twice the volume of cylinder 2.

Spiral Review

37. MEASUREMENT Find the volume of a rectangular prism with a length of 6 meters, a width of 4.9 meters, and a height of 5.2 meters. (Lesson 11-9)

Draw a corner view of each three-dimensional figure using the top, side, and front views shown. (Lesson 11-8)

38.

39.

PROBABILITY A coin is tossed and a number cube is rolled. Find the probability of each of the following. (Lesson 9-8)

40. P(heads and 4)

41. P(tails and an odd number)

42. P(heads and *not* 5)

43. P(*not* tails and not 2)

44. TEST SCORES The list gives the scores on a recent history test. Find each measure of central tendency and range. Round to the nearest tenth if necessary. Then state which measure best represents the data. Explain your reasoning. (Lesson 8-2)

History Test Scores									
78	92	83	88	89	91	96	72	74	99
81	88	86	95	73	97	78	78	60	
84	85	90	92	98	74	76	80	83	

Graphing Calculator Lab
Graphing Geometric Relationships

In this lab, you will use a TI-83/84 Plus graphing calculator to analyze geometric relationships among the base, height, and area of several parallelograms.

ACTIVITY

1 **STEP 1** Draw five parallelograms that each have a height of 4 centimeters on centimeter grid paper.

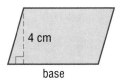

4 cm

base

STEP 2 Copy and complete the table shown for each parallelogram.

Base (cm)	Height (cm)	Area (cm^2)
	4	
	4	
	4	
	4	
	4	

STEP 3 Next enter the data into your graphing calculator. Press `STAT` 1 and enter the length of each base in L1. Then enter the area of each parallelogram in L2.

STEP 4 Turn on the statistical plot by pressing `2nd` [STAT PLOT] `ENTER` `ENTER`. Select the scatter plot and enter or confirm L1 as the Xlist and L2 as the Ylist.

STEP 5 Graph the data by pressing `ZOOM` 9. Use the Trace feature and the left and right arrow keys to move from one point to another.

ANALYZE THE RESULTS

1. What does an ordered pair on your graph represent?

2. Sketch and describe the shape of the graph.

3. **MAKE A CONJECTURE** Write an equation for your graph. Check your equation by pressing `Y=`, entering your equation into Y1, and then pressing `Graph`. What does this equation mean?

4. As the length of the base of the parallelogram increases, what happens to its area? Does this happen at a constant rate? How can you tell this from the table? from the graph?

ACTIVITY

2 **STEP 1** Draw five rectangles that each have an area of 36 square centimeters on centimeter grid paper. The length should be greater than or equal to its width.

$A = 36$ cm² width

length

STEP 2 Copy and complete the table shown for each rectangle.

Base (cm)	Height (cm)	Area (cm²)
		36
		36
		36
		36
		36

STEP 3 Clear list L1 and L2 by pressing [STAT] 4 [2nd] [L1], [2nd] [L2] [ENTER]. Then press [STAT] 1 and enter the length of each rectangle in L1 and the width of each rectangle in L2.

STEP 4 Follow Steps 4 and 5 of Activity 1 to graph the data.

ANALYZE THE RESULTS

5. What does an ordered pair on your graph represent?

6. Sketch and describe the shape of the graph.

7. **MAKE A CONJECTURE** Write an equation for your graph. Use the calculator to graph and check your equation. What does this equation mean?

8. As the length of the rectangle increases, what happens to its width? Does this happen at a constant rate? How can you tell this from the table? from the graph?

9. **MAKE A PREDICTION** Draw five cubes with different edge lengths. Predict the shape of the graph of the relationship between the edge length and volume of the cube.

10. Create a table to record the edge length and volume of each cube. Then graph the data to show the relationship between the edge length and volume of the cube. Sketch and describe the shape of the graph.

11. **MAKE A CONJECTURE** Write an equation for your graph. Use the calculator to graph and check your equation. What does this equation mean?

12. If the length of the cube's edge doubles, what happens to the volume? Explain.

FOLDABLES® Study Organizer

GET READY to Study

Be sure the following Big Ideas are noted in your Foldable.

Chapter 11 Measurement: Two- & Three-Dimensional Figures

BIG Ideas

Area (Lessons 11-1, 11-2, and 11-6)

  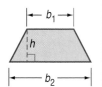

parallelogram	triangle	trapezoid
$A = bh$	$A = \frac{1}{2}bh$	$A = \frac{1}{2}h(b_1 + b_2)$

Circles (Lessons 11-3 and 11-4)

• circumference
$C = \pi d$ or $C = 2\pi r$

• area
$A = \pi r^2$

Volume (Lessons 11-9 and 11-10)

rectangular prism	triangular prism	cylinder
$B = \ell w$		$B = \pi r^2$
$V = Bh$		$V = Bh$ or $\pi r^2 h$

Key Vocabulary

base (p. 572)	prism (p. 603)
center (p. 584)	pyramid (p. 603)
circle (p. 584)	radius (p. 584)
circumference (p. 584)	rectangular prism (p. 613)
composite figure (p. 596)	sector (p. 590)
cone (p. 604)	semicircle (p. 596)
cylinder (p. 604)	sphere (p. 604)
diameter (p. 584)	three-dimensional figure (p. 603)
edge (p. 603)	
face (p. 603)	triangular prism (p. 615)
height (p. 572)	vertex (p. 603)
lateral face (p. 603)	volume (p. 613)
π (pi) (p. 584)	

Vocabulary Check

Choose the correct term or number to complete each sentence.

1. A (rectangular prism, rectangle) is a three-dimensional figure that has three sets of parallel congruent sides.

2. The (volume, surface area) of a three-dimensional figure is the measure of the space occupied by it.

3. $A = \frac{1}{2}h(b_1 + b_2)$ is the formula for the area of a (triangle, trapezoid).

4. Volume is measured in (square, cubic) units.

5. A (cylinder, prism) is a three-dimensional figure that has two congruent, parallel circles as its bases.

6. The volume of a rectangular prism is found by (adding, multiplying) the length, the width, and the height.

7. The formula for the area of a (square, circle) is $A = \pi r^2$.

Lesson-by-Lesson Review

11-1 **Area of Parallelograms** (pp. 572–576)

6.G.2

Find the area of each parallelogram. Round to the nearest tenth if necessary.

8.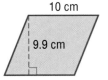
10 cm
9.9 cm

9.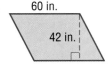
60 in.
42 in.

10. **ALGEBRA** A parallelogram has an area of 57 square inches. Find the base of the parallelogram if the height is 6 inches.

Example 1 Find the area of a parallelogram if the base is 15 inches and the height is 8 inches.

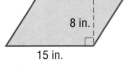
8 in.
15 in.

$A = bh$ — Area of a parallelogram
$A = 15 \cdot 8$ — Replace b with 15 and h with 8.
$A = 120 \text{ in}^2$. — Multiply.

11-2 **Area of Triangles and Trapezoids** (pp. 578–582)

6.G2

Find the area of each figure. Round to the nearest tenth if necessary.

11.
12 ft
6 ft

12.
5 in.
5 in.
10 in.

13. trapezoid: bases 22 yd and 35 yd
    height 18.5 yd

14. **ALGEBRA** The area of a triangle is 26.9 square inches. If the base is 9.6 inches, what is the height of the triangle?

Example 2 Find the area of a triangle with a base of 8 meters and a height of 11.2 meters.

$A = \frac{1}{2}bh$ — Area of a triangle
$A = \frac{1}{2}(8)(11.2)$ or 44.8 m^2 — $b = 8, h = 11.2$

Example 3 Find the area of the trapezoid.

10 in.
3 in.
2 in.

$A = \frac{1}{2}h(b_1 + b_2)$
$A = \frac{1}{2}(3)(2 + 10)$ — $h = 3, b_1 = 2, b_2 = 10$
$A = \frac{1}{2}(3)(12)$ or 18 in^2 — Simplify.

11-3 **Circles and Circumference** (pp. 584–588)

7.G.1

Find the circumference of each circle. Use 3.14 or $\frac{22}{7}$ for π. Round to the nearest tenth if necessary.

15. radius = 12 in.   16. diameter = 28 m
17. diameter = $8\frac{2}{5}$ ft   18. radius = 4.4 cm

19. **LIFE SCIENCE** A circular nest built by bald eagles has a diameter of $9\frac{1}{2}$ feet. Find the nest's circumference.

Example 4 Find the circumference of a circle with a diameter of 12.2 meters.

12.2 m

$C = \pi d$ — Circumference of a circle
$C \approx 3.14(12.2)$ — $\pi \approx 3.14$ and $d = 12.2$
$C \approx 38.3$ — Multiply.

The circumference is about 38.3 meters.

11-4 Area of Circles (pp. 589–593)

6.G.7

Find the area of each circle. Round to the nearest tenth.

20. radius = 11.4 in.

21. diameter = 44 cm

22. **CONCERTS** During an outdoor concert, a band can be heard within a 2-mile radius. What is the area of the region that can hear the concert? Round to the nearest tenth.

Example 5 Find the area of a circle with a radius of 5 inches.

$A = \pi r^2$ Area of a circle

$A = \pi(5)^2$ Replace r with 5.

$A \approx 78.5$ Multiply.

The area of the circle is about 78.5 square inches.

11-5 PSI: Solve a Simpler Problem (pp. 594–595)

7.PS.8

23. **BAKING** The local baker can make 10 cakes in 2 days. How many cakes can 8 bakers make working at the same rate in 20 days?

24. **TRAVEL** Mrs. Whitmore left Chicago at 6:45 A.M. and arrived in St. Louis at 11:15 A.M., driving a distance of approximately 292 miles. Find her approximate average speed.

25. **SHOPPING** Mercedes spent $175.89 over the weekend. Of the money she spent, 40% was spent on shoes. About how much money was *not* spent on shoes?

Example 6 A total of 950 residents voted on whether to build a neighborhood playground. Of those that voted, 70% voted for the playground. How many residents voted for the playground?

Find 10% of 950 and then use the result to find 70% of 950.

10% of 950 = 95

Since there are seven 10s in 70%, multiply 95 by 7.

So, 95 × 7 or 665 residents voted for the playground.

11-6 Area of Composite Figures (pp. 596–599)

6.G.3

Find the area of each figure. Round to the nearest tenth if necessary.

26. 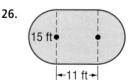   27.

28. **PATIOS** The McAllister family is installing a brick patio like the one shown. What is the total area of the patio?

Example 7 Find the area of the figure. Round to the nearest tenth.

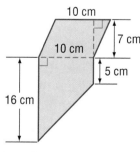

The figure can be separated into a parallelogram and a trapezoid.

parallelogram: $A = 10 \cdot 7$ or 70

trapezoid: $A = \frac{1}{2}(10)(16 + 5)$ or 105

The area is 70 + 105 or 175 cm².

11-7 **Three-Dimensional Figures** (pp. 603–606)

7.G.3

For each figure, identify the shape of the base(s). Then classify the figure.

29. 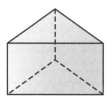   30.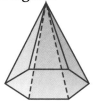

31. **VEGETABLES** Classify the shape of a can of green beans.

DOGHOUSES For Exercises 32 and 33, use the figure of the doghouse shown.

32. What geometric figure is represented by the main part of the doghouse?

33. Identify the top figure of the doghouse.

Example 8 For the figure, identify the shape of the base(s). Then classify the figure.

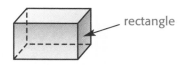

Since the figure has 6 rectangular faces, parallel bases, and a rectangular base, the figure is a rectangular prism.

Example 9 Classify the shape of a basketball.

The figure has no faces, bases, edges, or vertices. The figure is a sphere.

11-8 **Drawing Three-Dimensional Figures** (pp. 608–612)

7.G.3

Draw a top, a side, and a front view of each solid.

34.   35.

36. Draw a corner view of the three-dimensional figure whose top, side, and front views shown. Use isometric dot paper.

37. **CRAFTS** Alejandra put all of her craft supplies in the box shown. Draw a top, a side, and a front view of the box.

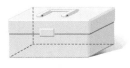

Example 10 Draw a corner view of the three-dimensional figure using the top, side, and front views shown.

top   side   front

The side view is a square. The top and front views are rectangles.

The figure drawn is a rectangular prism.

11-9 **Volume of Prisms** (pp. 613–618)

7.G.2

Find the volume of each prism. Round to the nearest tenth if necessary.

38.

3.6 m
1.4 m
2.9 m

39.

$8\frac{1}{2}$ in.
7 in.
$10\frac{3}{4}$ in.

40. CEREAL A box of cereal is 8.5 inches long, 12.5 inches tall, and 3.5 inches wide. What is the maximum amount of cereal the box can contain?

41. TRUCKS The dimensions of the bed of a dump truck are length 20 feet, width 7 feet, and height $9\frac{1}{2}$ feet. What is the volume of the bed of the dump truck?

Example 11 A local city provides residents with a rectangular container for recycling products. Find the volume of the rectangular container.

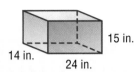

15 in.
14 in.
24 in.

$V = \ell wh$	Volume of a rectangular prism
$V = (24)(14)(15)$	Replace ℓ with 24, w with 14 and h with 15.
$V = 5,040$	Multiply.

The volume of the rectangular container is 5,040 cubic inches.

11-10 **Volume of Cylinders** (pp. 619–623)

7.G.2

Find the volume of each cylinder. Round to the nearest tenth.

42.

8.7 km
17 km

43.
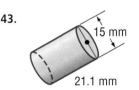
15 mm
21.1 mm

44. CONTAINERS A can of soup has a diameter of 3.5 inches, and a height of 5 inches. Find the volume of the soup can. Use 3.14 for π.

45. COOKIES Mrs. Delagado stores cookies in a cylinder-shaped jar that has a height of 12 inches and a diameter of 10 inches. Find the volume to the nearest cubic inch. Use 3.14 for π.

Example 12 Marquez stores his toys in a cylinder-shaped can like the one shown below. Find the volume of the cylinder-shaped can. Round to the nearest tenth.

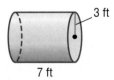

3 ft
7 ft

$V = \pi r^2 h$	Volume of a cylinder
$V = \pi 3^2 (7)$	Replace r with 3 and h with 7.
$V \approx 197.8$	Multiply.

The volume of the cylinder-shaped can is 197.8 cubic feet.

Find the area of each figure. Round to the nearest tenth if necessary.

1.
9.6 ft
8 ft

2.
$7\frac{1}{3}$ ft
15 ft

3.
8 yd
6 yd
5 yd

4.
16 ft
12 ft

5. **MEASUREMENT** Mrs. Torres has a circular rug underneath her dining room table. What is the approximate circumference of the rug if it has a radius of $3\frac{1}{2}$ yards?

Find the area of each circle. Round to the nearest tenth.

6. radius = 9 ft

7. diameter = 5.2 m

8. **MULTIPLE CHOICE** A fountain is in the shape of a circle. If the fountain has a diameter of 8.8 meters, which equation could be used to find the area of the base of the fountain?

A $A = \pi \times 8.8^2$ **C** $A = 2 \times \pi \times 4.4$

B $A = \pi \times 4.4^2$ **D** $A = \pi \times 8.8$

9. **MEASUREMENT** The Gruseser family wants to build a swimming pool like the one shown below. If they have 85 square feet of land available for the pool, do they have enough space for the swimming pool? Justify your response.

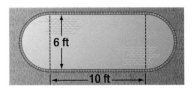
6 ft
10 ft

For each figure, identify the shape of the base(s). Then classify the figure.

10.

11.

12. **GEOMETRY** Classify the shape of a roll of paper towels.

13. **GEOMETRY** Which geometric shape has at least three lateral faces that are triangles and only one base?

Draw a top, a side, and a front view of each solid.

14.

15.

Find the volume of each prism and cylinder. Round to the nearest tenth.

16.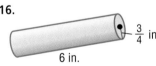
$\frac{3}{4}$ in.
6 in.

17.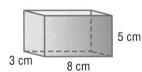
5 cm
3 cm
8 cm

18.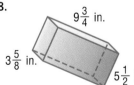
$9\frac{3}{4}$ in.
$3\frac{5}{8}$ in.
$5\frac{1}{2}$ in.

19.
6 ft
12 ft

20. **MULTIPLE CHOICE** A cylinder-shaped coffee mug has a radius of 4 centimeters and a height of 10 centimeters. How much coffee is contained in the mug if it is only half full?

F 0.2 in³   **H** 3.4 in³

G 1.5 in³   **J** 3.8 in³

PART 1   Multiple Choice

Read each question. Then fill in the correct answer on the answer document provided by your teacher or on a sheet of paper.

1. Stephanie shaded part of a circle like the one shown below. What is the approximate area of the sector?

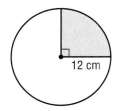

12 cm

A 113 cm²

B 364 cm²

C 452 cm²

D 728 cm²

TEST-TAKING TIP

Question 1 Many tests include a Mathematics Chart in the test booklet. Refer to the chart for area and volume formulas.

2. The circular floor rug shown has a diameter of 6 feet. Which expression can be used to find its circumference, C, in feet?

6 ft

F $C = 3 \times \pi$

G $C = 3^2 \times \pi$

H $C = 6 \times \pi$

J $C = 2 \times 6 \times \pi$

3. Angle D and angle E are complementary angles. If $\angle D$ is 35°, what is the measure of $\angle E$?

A 35°   **C** 90°

B 55°   **D** 145°

4. If the corresponding angles of 2 trapezoids are congruent and the lengths of the corresponding sides of the trapezoids are proportional, the trapezoids are —

F regular   **H** symmetric

G congruent   **J** similar

5. Which description shows the relationship between a term and n, its position in the sequence?

Position	1	2	3	4	5	n
Value of Term	2	5	8	11	14	

A Add 1 to n.

B Multiply n by 2 and add 3.

C Multiply n by 3 and subtract 1.

D Add 9 to n.

6. A metal toolbox has a length of 11 inches, a width of 5 inches, and a height of 6 inches. What is the volume of the toolbox?

F 22 in³   **H** 210 in³

G 121 in³   **J** 330 in³

7. A bag contains 5 red, 2 yellow, and 8 blue marbles. Xavier removed one blue marble from the bag and did not put it back. He then randomly removed another marble. What is the probability that the second marble removed was blue?

A $\frac{8}{14}$ **C** $\frac{1}{2}$

B $\frac{8}{15}$ **D** $\frac{7}{15}$

8. In the figure shown, all the corners form right angles. What is the area of the figure in square feet?

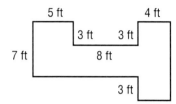

F 91 ft²

G 107 ft²

H 115 ft²

J 122 ft²

9. The circles shown have radii of 4 feet and 8 feet.

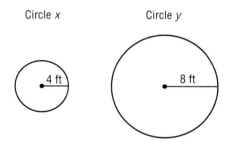

What is $\dfrac{\text{circumference of Circle } x}{\text{circumference of Circle } y}$?

A $\dfrac{\pi}{4}$ **C** $\dfrac{1}{4}$

B $\dfrac{\pi}{2}$ **D** $\dfrac{1}{2}$

10. Katie has 3 apples to serve to her friends. If Katie serves each friend $\frac{1}{3}$ of a whole apple, how many friends can she serve?

F 1   **H** 9

G 3   **J** 12

PART 2   **Short Response/Grid In**

Record your answers on the answer sheet provided by your teacher or on a sheet of paper.

11. Ms. Williams recorded the time it took four of her top students to complete a math quiz. What is the median time in minutes for these four students?

Math Quiz Times	
Student	**Time (minutes)**
1	12.8
2	23.1
3	19.6
4	15.7

PART 3   **Extended Response**

Record your answers on the answer sheet provided by your teacher or on a sheet of paper. Show your work.

12. A plastic cylinder-shaped pipe has the dimensions shown below.

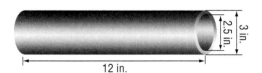

a. How much liquid can the pipe contain?

b. Describe how you could find the amount of plastic needed to make the pipe.

c. Use the method you described in part b to find the amount of plastic in the pipe.

NEED EXTRA HELP?												
If You Missed Question...	1	2	3	4	5	6	7	8	9	10	11	12
Go to Lesson...	11-4	11-6	10-1	10-6	1-9	11-9	9-1	11-6	11-3	5-7	8-2	11-2
NYS Core Curriculum	6.G.7	6.G.3	8.G.1	7.G.7	7.PS.4	7.G.2	7.S.9	6.G.3	7.G.1	6.N.18	7.S.6	6.G.2

CHAPTER 12

Geometry and Measurement

New York State Core Curriculum

7.G.8 Use the Pythagorean Theorem to determine the unknown length of a side of a right triangle

7.G.4 Determine the surface area of prisms and cylinders, using a calculator and a variety of methods

Key Vocabulary

hypotenuse (p. 640)

irrational number (p. 637)

Pythagorean Theorem (p. 640)

surface area (p. 649)

🌐 Real-World Link

Spaghetti The shape of many spaghetti boxes are rectangular prisms, and the shape of many cans are cylinders. You can use the formula $S = 2lw + 2lh + 2wh$ to find the surface area of a box of spaghetti given the length l, the width w, and the height h of the box.

FOLDABLES®
Study Organizer

Geometry and Measurement Make this Foldable to help you organize your notes. Begin with a piece of 11" by 17" paper.

① Fold the paper in fourths lengthwise.

② Open and fold a 2" tab along the short side. Then fold the rest in half.

③ Draw lines along the folds and label as shown.

Ch. 12	Rectangular Prisms	Cylinders
Draw Examples		
Find Volume		
Find Surface Area		

GET READY for Chapter 12

Diagnose Readiness You have two options for checking Prerequisite Skills.

Option 2

NY Math Online Take the Online Readiness Quiz at glencoe.com.

Option 1

Take the Quick Quiz below. Refer to the Quick Review for help.

QUICK Quiz

Evaluate each expression. (Lesson 1-2)

1. 4^2 2. 7^2

3. 13^2 4. 24^2

5. $5^2 + 8^2$ 6. $10^2 + 6^2$

7. $9^2 + 12^2$ 8. $15^2 + 17^2$

9. **AGES** Samuel's mother is 7^2 years old, and his grandmother is 9^2 years old. Find the sum of their ages. (Lesson 1-2)

Evaluate the expression $2ab + 2bc + 2ac$ for each value of the variables indicated. (Lesson 1-6)

10. $a = 4, b = 5, c = 8$

11. $a = 2, b = 7, c = 11$

12. $a = 3.1, b = 2.4, c = 9.9$

13. $a = 2.1, b = 1.7, c = 4.6$

Use the π button on your calculator to evaluate each expression below. Round to the nearest tenth. (Lesson 11-3)

14. $(2)(\pi)(3^2) + (2)(\pi)(3)(8)$

15. $(2)(\pi)(7^2) + (2)(\pi)(7)(5)$

QUICK Review

Example 1
Evaluate $3^2 + 5^2$.

$3^2 + 5^2 = 9 + 25$ Evaluate 3^2 and 5^2.
$\qquad = 34$ Add 9 and 25.

Example 2
Evaluate the expression $2ab + 2bc + 2ac$ for $a = 3, b = 5$, and $c = 6$.

$2ab + 2bc + 2ac$
$= 2(3)(5) + 2(5)(6) + 2(3)(6)$ Replace a with 3, b with 5, and c with 6.

$= 30 + 60 + 36$ Multiply.
$= 126$ Add.

Example 3
Use the π button on your calculator to evaluate $(2)(\pi)(4^2) + (2)(\pi)(4)(6)$. Round to the nearest tenth.

$(2)(\pi)(4^2) + (2)(\pi)(4)(6)$
$= (2)(\pi)(16) + (2)(\pi)(4)(6)$ Evaluate 4^2.
$= (32)(\pi) + (48)(\pi)$ Multiply.
≈ 251.3 Multiply and add.

12-1 Estimating Square Roots

▷ MINI Lab

Estimate the square root of 27.

- Arrange 27 tiles into the largest square possible. You will use 25 tiles and 2 will remain.

- Add tiles to make the next larger square. So, add 9 tiles to make a square with 36 tiles.

- The square root of 27 is between 5 and 6. Since 27 is much closer to 25 than 36, we can expect that the square root of 27 is closer to 5 than 6.

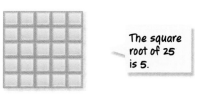

The square root of 25 is 5.

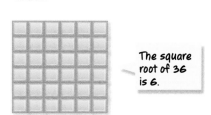

The square root of 36 is 6.

Use algebra tiles to estimate the square root of each number to the nearest whole number.

1. 40   2. 28   3. 85   4. 62

5. Describe another method that you could use to estimate the square root of a number.

The square root of a perfect square is an integer. You can estimate the square root of a number that is *not* a perfect square.

EXAMPLE Estimate a Square Root

① **Estimate $\sqrt{78}$ to the nearest whole number.**

List some perfect squares.

1, 4, 9, 16, 25, 36, 49, 64, 81, ...

$$64 < 78 < 81$$ 78 is between the perfect squares 64 and 81.
$$\sqrt{64} < \sqrt{78} < \sqrt{81}$$ Find the square root of each number.
$$8 < \sqrt{78} < 9$$ $\sqrt{64} = 8$ and $\sqrt{81} = 9$

So, $\sqrt{78}$ is between 8 and 9. Since 78 is much closer to 81 than to 64, the best whole number estimate is 9.   Verify with a calculator.

✔ CHECK Your Progress

a. Estimate $\sqrt{50}$ to the nearest whole number.

A number that cannot be expressed as the quotient of two integers is an **irrational number**.

Vocabulary Link . . . .
Irrational

Everyday Use lacking usual or normal clarity, as in irrational thinking

Math Use a number that cannot be expressed as the quotient of two integers

Irrational Numbers $\sqrt{2}$, π, 0.636336333...

The square root of any number that is not a perfect square is an irrational number. You can use a calculator to estimate square roots that are irrational numbers.

EXAMPLE Graph Square Roots on a Number Line

② Graph $\sqrt{42}$ on a number line.

[2nd] [√] 42 [ENTER]   6.480740698
$\sqrt{42} \approx 6.5$

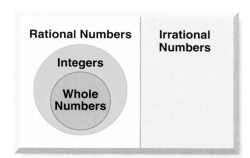

Check for Reasonableness
$6^2 = 36$ and $7^2 = 49$. Since 42 is between 36 and 49, the answer, 6.5, is reasonable.

CHECK Your Progress

Graph each square root on a number line.

b. $\sqrt{6}$ c. $\sqrt{23}$ d. $\sqrt{309}$

The Venn diagram shows the relationship among sets of numbers.

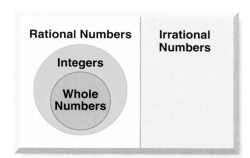

Whole Numbers: 0, 1, 2, 3, ...
Integers: ..., −2, −1, 0, 1, 2, ...
Rational Numbers: $\frac{1}{2}$, 0.25, −0.2, 0.333...
Irrational Numbers: π, $\sqrt{2}$, 0.124543...

CHECK Your Understanding

Example 1
(p. 636)

Estimate each square root to the nearest whole number.

1. $\sqrt{39}$ 2. $\sqrt{106}$ 3. $\sqrt{90}$ 4. $\sqrt{140}$

Example 2
(p. 637)

Graph each square root on a number line.

5. $\sqrt{7}$ 6. $\sqrt{51}$ 7. $\sqrt{135}$ 8. $\sqrt{462}$

9. **MEASUREMENT** The diagram at the right shows the floor plan of a square kitchen. What is the approximate length of one side of the kitchen floor to the nearest tenth?

Area = 105 ft² | x
x

Estimate each square root to the nearest whole number.

10. $\sqrt{11}$ 11. $\sqrt{20}$ 12. $\sqrt{35}$ 13. $\sqrt{65}$

14. $\sqrt{89}$ 15. $\sqrt{116}$ 16. $\sqrt{137}$ 17. $\sqrt{409}$

Graph each square root on a number line.

18. $\sqrt{15}$ 19. $\sqrt{8}$ 20. $\sqrt{44}$ 21. $\sqrt{89}$

22. $\sqrt{160}$ 23. $\sqrt{573}$ 24. $\sqrt{645}$ 25. $\sqrt{2,798}$

26. **MEASUREMENT** The bottom of the square baking pan has an area of 67 square inches. What is the approximate length of one side of the pan?

27. **ALGEBRA** What whole number is closest to $\sqrt{m - n}$ if $m = 45$ and $n = 8$?

Estimate each square root to the nearest whole number.

28. $\sqrt{925}$ 29. $\sqrt{2,480}$ 30. $\sqrt{1,610}$ 31. $\sqrt{6,500}$

Find each square root to the nearest tenth.

32. $\sqrt{0.25}$ 33. $\sqrt{0.49}$ 34. $\sqrt{1.96}$ 35. $\sqrt{2.89}$

ALGEBRA For Exercises 36 and 37, estimate each expression to the nearest tenth if $a = 8$ and $b = 3.7$.

36. $\sqrt{a + b}$ 37. $\sqrt{6b - a}$

STAMPS For Exercises 38 and 39, use the information below.

The Special Olympics' commemorative stamp is square in shape with an area of 1,008 square millimeters.

38. Find the length of one side of the postage stamp to the nearest tenth.

39. What is the length of one side in centimeters?

40. **ALGEBRA** The formula $D = 1.22 \times \sqrt{h}$ can be used to estimate the distance D in miles you can see from a point h feet above Earth's surface. Use the formula to find the distance D in miles you can see from the top of a 120-foot hill. Round to the nearest tenth.

41. **FIND THE DATA** Refer to the Data File on pages 16–19. Choose some data and write a real-world problem in which you would estimate a square root.

H.O.T. Problems

42. **Which One Doesn't Belong?** Identify the number that does not have the same characteristic as the other three. Explain your reasoning.

$\sqrt{5}$	π	$\sqrt{81}$	0.535335333...

43. **OPEN ENDED** Select three numbers with square roots between 4 and 5.

44. **NUMBER SENSE** Explain why 8 is the best whole number estimate for $\sqrt{71}$.

CHALLENGE A cube root of a number is one of three equal factors of that number. Estimate the cube root of each number to the nearest whole number.

45. $\sqrt[3]{9}$ 46. $\sqrt[3]{26}$ 47. $\sqrt[3]{120}$ 48. $\sqrt[3]{500}$

49. **WRITING IN MATH** Apply what you know about numbers to explain why $\sqrt{30}$ is an irrational number.

NYSMT PRACTICE   7.N.16, 7.N.18

50. Reina wrote four numbers on a piece of paper. She then asked her friend Tyron to select the number closest to 5. Which number should he select?

| $\sqrt{56}$ $\sqrt{48}$ $\sqrt{37}$ $\sqrt{28}$ |

A $\sqrt{56}$

B $\sqrt{48}$

C $\sqrt{37}$

D $\sqrt{28}$

51. Which of the following is an irrational number?

F $\sqrt{25}$ H -13

G $\sqrt{7}$ J $\frac{4}{5}$

52. **SHORT RESPONSE** If the area of a square is 169 square inches, what is the length of the side of the square?

Spiral Review

53. **MEASUREMENT** Find the volume of a can of vegetables with a diameter of 3 inches and a height of 4 inches. Round to the nearest tenth. (Lesson 11-10)

54. **MEASUREMENT** A rectangular prism is 14 inches long, 4.5 inches wide, and 1 inch high. What is the volume of the prism? (Lesson 11-9)

GEOMETRY For Exercises 55–58, use the graph at the right. Classify the angle that represents each category as *acute, obtuse, right,* or *straight.* (Lesson 10-1)

55. 30–39 hours   56. 1–29 hours

57. 40 hours   58. 41–50 hours

▷ **GET READY** for the Next Lesson

PREREQUISITE SKILL Solve each equation.
(Lesson 1-7)

59. $7^2 + 5^2 = c$ 60. $4^2 + b = 36$

61. $3^2 + a = 25$ 62. $9^2 + 2^2 = c$

Hours Worked in a Typical Week

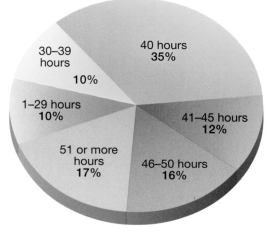

30–39 hours 10%
40 hours 35%
1–29 hours 10%
41–45 hours 12%
51 or more hours 17%
46–50 hours 16%

Source: Heldrich Work Trends Survey

The Pythagorean Theorem

MAIN IDEA

Find length using the Pythagorean Theorem.

NYS Core Curriculum

7.G.8 Use the Pythagorean Theorem to determine the unknown length of a side of a right triangle **7.G.5** Identify the right angle, hypotenuse, and legs of a right triangle *Also addresses 7.G.6, 7.CN.8*

New Vocabulary

leg
hypotenuse
Pythagorean Theorem

NY Math Online

glencoe.com
• Extra Examples
• Personal Tutor
• Self-Check Quiz

▷ MINI Lab

Three squares with sides 3, 4, and 5 units are used to form the right triangle shown.

1. Find the area of each square.

2. How are the squares of the sides related to the areas of the squares?

3. Find the sum of the areas of the two smaller squares. How does the sum compare to the area of the larger square?

4. Use grid paper to cut out three squares with sides 5, 12, and 13 units. Form a right triangle with these squares. Compare the sum of the areas of the two smaller squares with the area of the larger square.

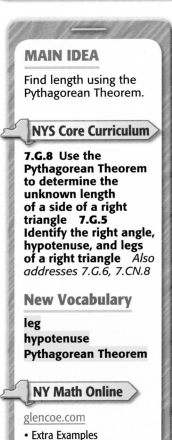

3 units

5 units

4 units

In a right triangle, the sides have special names.

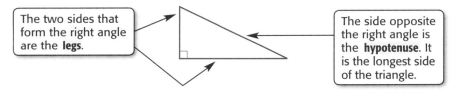

The two sides that form the right angle are the **legs**.

The side opposite the right angle is the **hypotenuse**. It is the longest side of the triangle.

The **Pythagorean Theorem** describes the relationship between the length of the hypotenuse and the lengths of the legs.

Pythagorean Theorem		Key Concept
Words	In a right triangle, the square of the length of the hypotenuse equals the sum of the squares of the lengths of the legs.	**Model** 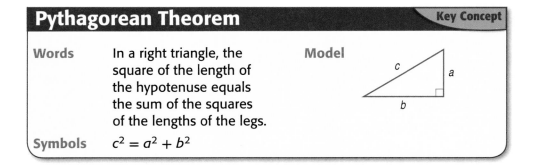
Symbols	$c^2 = a^2 + b^2$	

When using the Pythagorean Theorem, you will encounter equations that involve square roots. Every positive number has both a positive and a negative square root. By the definition of square roots, if $n^2 = a$, then $n = \pm\sqrt{a}$. The notation $\pm\sqrt{}$ indicates both the positive and negative square root of a number. You can use this relationship to solve equations that involve squares.

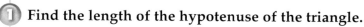

EXAMPLE   Find the Length of the Hypotenuse

 Find the length of the hypotenuse of the triangle.

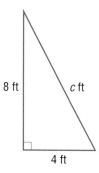

$c^2 = a^2 + b^2$ Pythagorean Theorem

$c^2 = 8^2 + 4^2$ Replace a with 8 and b with 4.

$c^2 = 64 + 16$ Evaluate 8^2 and 4^2.

$c^2 = 80$ Add.

$c = \pm\sqrt{80}$ Definition of square root

$c \approx \pm 8.9$ Simplify.

The length of the hypotenuse is about 8.9 feet.

Study Tip

Check for Reasonableness
You can eliminate -8.9 as a solution because the length of a side of a triangle cannot be a negative number.

8 ft c ft

4 ft

✅ CHECK Your Progress

a. Find the length of the hypotenuse of a right triangle with legs 5 yards and 7 yards. Round to the nearest tenth.

Real-World EXAMPLE

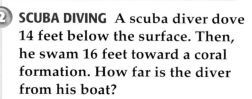

2 **SCUBA DIVING** A scuba diver dove 14 feet below the surface. Then, he swam 16 feet toward a coral formation. How far is the diver from his boat?

The diver's distance from the boat is the hypotenuse of a right triangle. Write and solve an equation for x.

$c^2 = a^2 + b^2$ Pythagorean Theorem

$x^2 = 14^2 + 16^2$ Replace c with x, a with 14, and b with 16.

$x^2 = 196 + 256$ Evaluate 14^2 and 16^2.

$x^2 = 452$ Add.

$x = \pm\sqrt{452}$ Definition of square root

$x \approx \pm 21.3$ Simplify.

The diver's distance from the boat is about 21.3 feet.

Real-World Career . . . .
How Does a Professional Diver Use Math? Professional divers must use formulas to compute pressure and air supply in order to determine safe diving depths and dive times.

NY Math Online

For more information, go to glencoe.com.

✅ CHECK Your Progress

b. **SOFTBALL** A softball diamond is a square measuring 60 feet on each side. How far does a player on second base throw when she throws from second base to home? Round to the nearest tenth.

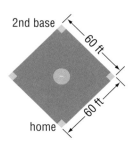

2nd base

60 ft

60 ft

home

You can also use the Pythagorean Theorem to find the measure of a leg if the measure of the other leg and the hypotenuse are known.

EXAMPLE Find the Length of a Leg

③ Find the missing measure of the triangle. Round to the nearest tenth if necessary.

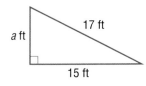

The missing measure is the length of a leg.

$$c^2 = a^2 + b^2 \qquad \text{Pythagorean Theorem}$$
$$13^2 = 5^2 + b^2 \qquad \text{Replace } a \text{ with 5 and } c \text{ with 13.}$$
$$169 = 25 + b^2 \qquad \text{Evaluate } 13^2 \text{ and } 5^2.$$
$$\underline{-25 = -25} \qquad \text{Subtract 25 from each side.}$$
$$144 = b^2 \qquad \text{Simplify.}$$
$$\pm\sqrt{144} = b \qquad \text{Definition of square root}$$
$$12 = b \qquad \text{Simplify.}$$

The length of the leg is 12 centimeters.

✔CHECK Your Progress

c.

d.

e. $b = 7$ in., $c = 25$ in.

NYSMT EXAMPLE

Test-Taking Tip

Formulas Some formulas will be given to you during the test. It is a good idea to familiarize yourself with the formulas before the test.

④ Mr. Thomson created a mosaic tile in the shape of a square to place in his kitchen.

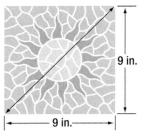

Which is closest to the length of the diagonal of the tile?

A 10 in. **C** 15 in.

B 13 in. **D** 17 in.

Read the Item

You need to use the Pythagorean Theorem to find the length of the diagonal.

Solve the Item

$c^2 = a^2 + b^2$ Pythagorean Theorem

$c^2 = 9^2 + 9^2$ Replace a with 9 and b with 9.

$c^2 = 81 + 81$ Evaluate 9^2 and 9^2.

$c^2 = 162$ Add.

$c = \pm\sqrt{162}$ Definition of square root

$c \approx \pm 12.7$ Simplify.

The length is about 12.7 inches.

The answer choice closest to 12.7 inches is 13 inches. So, the answer is B.

✔ CHECK Your Progress

f. A painter leans a ladder against the side of a building. How far from the bottom of the building is the top of the ladder?

    F 38.2 ft   **H** 21.8 ft

    G 28.0 ft   **J** 20.0 ft

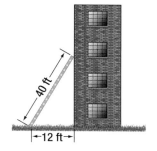

✔ CHECK Your Understanding

Examples 1, 3
(pp. 641–642)

Find the missing measure of each triangle. Round to the nearest tenth if necessary.

1.

c mm, 10 mm, 24 mm

2.

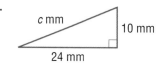

19 in., a in., 31 in.

3. $b = 21$ cm, $c = 28$ cm

4. $a = 11$ yd, $b = 12$ yd

Example 2
(p. 641)

5. **ARCHITECTURE** What is the width of the the fence gate shown at the right? Round to the nearest tenth.

2.5 ft   4.7 ft

Example 4
(pp. 642–643)

6. **MULTIPLE CHOICE** A company designed a public play area in the shape of a square. The play area will include a pathway, as shown. Which is closest to the length of the pathway?

    A 100 yd   **C** 140 yd

    B 125 yd   **D** 175 yd

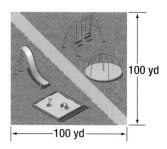

100 yd   100 yd

HOMEWORK HELP

For Exercises	See Examples
7–8, 11–12, 15–16	1
17–20	2
9–10, 13–14	3
26–27	4

Find the missing measure of each triangle. Round to the nearest tenth if necessary.

7.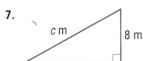

8.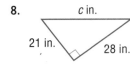

9.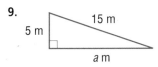

10.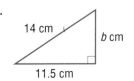

11.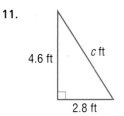

12.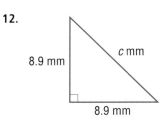

13. $a = 2.4$ yd, $c = 3.7$ yd

14. $b = 8.5$ m, $c = 10.4$ m

15. $a = 7$ in., $b = 24$ in.

16. $a = 13.5$ mm, $b = 18$ mm

MEASUREMENT For Exercises 17 and 18, find each distance to the nearest tenth.

17.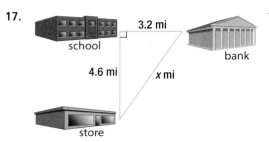

18.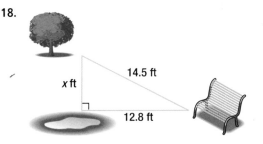

SPORTS For Exercises 19 and 20, find the length or width of each piece of sports equipment. Round to the nearest tenth.

19.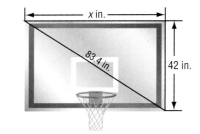

20.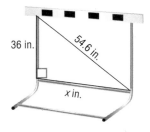

21. **MEASUREMENT** A barn door is 10 feet wide and 15 feet tall. A square plank 16 feet on each side must be taken through the doorway. Can the plank fit through the doorway? Justify your answer.

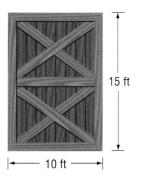

22. **MEASUREMENT** On a weekend trip around California, Sydney left her home in Modesto and drove 75 miles east to Yosemite National Park, then 70 miles south to Fresno, and finally 110 miles west to Monterey Bay. About how far is she from her starting point? Justify your answer with a drawing.

NYSCC • NYSMT

Extra Practice, pp. 702, 715.

23. **CHALLENGE** What is the length of the diagonal shown in the cube at the right?

6 in.

x in.

24. **FIND THE ERROR** Marcus and Aisha are writing an equation to find the missing measure of the triangle at the right. Who is correct? Explain.

8 cm

21 cm

x cm

$21^2 = 8^2 + x^2$

$x^2 = 21^2 + 8^2$

Marcus

Aisha

25. **WRITING IN MATH** Write a problem about a real-world situation in which you would use the Pythagorean Theorem.

NYSMT PRACTICE 7.G.8, 7.G.5

26. Which triangle has sides a, b, and c so that the relationship $a^2 + b^2 = c^2$ is true?

A
 a c b

B a c b

C a b c

D b a c

27. An isosceles right triangle has legs that are each 8 inches long. About how long is the hypotenuse?

F 12.8 inches

G 11.3 inches

H 8 inches

J 4 inches

Spiral Review

28. **ESTIMATION** Which is closer to $\sqrt{55}$: 7 or 8? (Lesson 12-1)

29. **MEASUREMENT** A cylinder-shaped popcorn tin has a height of 1.5 feet and a diameter of 10 inches. Find the volume to the nearest cubic inch. (Lesson 11-10)

Write each percent as a decimal. (Lesson 4-7)

30. 45%   31. 8%   32. 124%   33. 265%

▷ **GET READY for the Next Lesson**

34. **PREREQUISITE SKILL** The average person takes about 15 breaths per minute. At this rate, how many breaths does the average person take in one week? Use the *solve a simpler problem* strategy. (Lesson 11-5)

Problem-Solving Investigation

MAIN IDEA: Solve problems by making a model.

 7.PS.7 Understand that there is no one right way to solve mathematical problems but that different methods have advantages and disadvantages

P.S.I. TEAM +

e-Mail: MAKE A MODEL

AYITA: I am decorating the school's gymnasium for the spring dance with cubes that will hang from the ceiling.

YOUR MISSION: Make a model to find how much cardboard will be needed for each cube if the edge of one cube measures 12 inches.

Understand	You know that each cube is 12 inches long.
Plan	Make a cardboard model of a cube with sides 12 inches long. You will also need to determine where to put tabs so that all of the edges are glued together.
Solve	Start with a cube, then unfold it, to show the pattern. You know that 5 of the edges don't need tabs because they are the fold lines. The remaining 7 edges need a tab. Use $\frac{1}{2}$-inch tabs. 7×12 in. $\times \frac{1}{2}$ in. $= 42$ in^2 area of 7 tabs 6×12 in. $\times 12$ in. $= \underline{864 \text{ in}^2}$ area of 6 faces $\phantom{6 \times 12 \text{ in.} \times 12 \text{ in.} = }906$ in^2 total area So, 906 square inches of cardboard is needed to make one cube.
Check	Make another cube to determine whether all the edges can be glued together using your model.

Analyze The Strategy

1. How can making a model be useful when solving a word problem?

2. **WRITING IN MATH** Write a problem that can be solved by making a model. Then solve the problem.

For Exercises 3–5, make a model to solve the problem.

3. **CARS** Fiona counted the number of vehicles in the parking lot at a store. She counted a total of 12 cars and motorcycles. If there was a total of 40 wheels, how many cars and motorcycles were there?

4. **ART** Miguel is making a drawing of his family room for a school project. The room measures 18 feet by 21 feet. If he uses a scale of 1 foot = $\frac{1}{2}$ inch, what are the dimensions of the family room on the drawing?

5. **MEASUREMENT** Francis has a photo that measures 10 inches by $8\frac{1}{2}$ inches. If the frame he uses is $1\frac{1}{4}$ inches wide, what is the perimeter of the framed picture?

Use any strategy to solve Exercises 6–13. Some strategies are shown below.

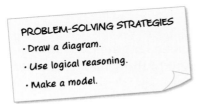

PROBLEM-SOLVING STRATEGIES
· Draw a diagram.
· Use logical reasoning.
· Make a model.

6. **DONATIONS** Hickory Point Middle School collected money for a local shelter. The table shows the total amount collected by each grade level. Suppose the school newspaper reported that $5,000 was collected. Is this estimate reasonable? Explain.

Grade	Dollars Collected
sixth	1,872
seventh	2,146
eighth	1,629

7. **TRACK** Wei can jog one 400-meter lap in $1\frac{1}{3}$ minutes. How long will it take her to run 1,600 meters at the same rate?

8. **BIRD HOUSES** About how many square inches of the bird house will be painted if only the outside of the wood is painted?

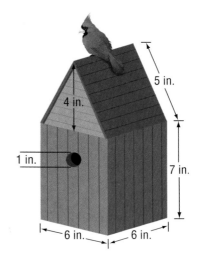

5 in.

4 in.

1 in.

7 in.

6 in.   6 in.

9. **BOXES** Juliet is placing 20 cereal boxes that measure 8 inches by 2 inches by 12 inches on a shelf that is 3 feet long and 11 inches deep. What is a possible arrangement for the boxes on the shelf?

10. **MONEY** At the beginning of the week, Marissa had $45.50. She spent $2.75 each of five days on lunch, bought a sweater for $14.95, and Jacob repaid her $10 that he owed her. How much money does she have at the end of the week?

11. **MEASUREMENT** How many square feet of wallpaper are needed to cover a wall that measures $15\frac{1}{4}$ feet by $8\frac{3}{4}$ feet and has a window that measures 2 feet by 4 feet?

12. **BASEBALL** A regulation baseball diamond is a square with an area of 8,100 square feet. If it is laid out on a field that is 172 feet wide and 301 feet long, how much greater is the distance around the whole field than the distance around the diamond?

13. **DVDs** Marc currently has 68 DVDs in his collection. By the end of the next four months, he wants to have 92 DVDs in his collection. How many DVDs must he buy each month to obtain his goal?

Estimate each square root to the nearest whole number. (Lesson 12-1)

1. $\sqrt{32}$

2. $\sqrt{80}$

3. $\sqrt{105}$

4. $\sqrt{230}$

MEASUREMENT **Estimate the side length of each square to the nearest whole number.** (Lesson 12-1)

5.
Area = 14 m²

6.
Area = 110 ft²

Graph each square root on a number line.
(Lesson 12-1)

7. $\sqrt{18}$

8. $\sqrt{230}$

9. **MULTIPLE CHOICE** Imani is playing a review game in math class. She needs to pick the card that is labeled with a number closest to 8. Which should she pick? (Lesson 12-1)

$\sqrt{37}$ $\sqrt{52}$ $\sqrt{70}$ $\sqrt{83}$

A $\sqrt{37}$

B $\sqrt{52}$

C $\sqrt{70}$

D $\sqrt{83}$

Find the length of the hypotenuse of each triangle. Round to the nearest tenth if necessary. (Lesson 12-2)

10.
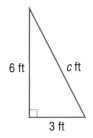
6 ft   c ft
3 ft

11. 7 cm
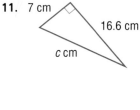
16.6 cm
c cm

Find the missing measure of each triangle. Round to the nearest tenth if necessary.
(Lesson 12-2)

12.
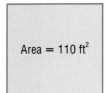
10 in.
25 in.
a in.

13.
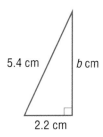
5.4 cm   b cm
2.2 cm

14. **MEASUREMENT** On a computer monitor, the diagonal measure of the screen is 17 inches.

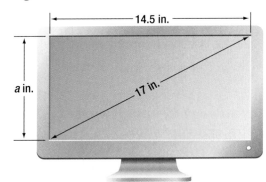
14.5 in.
a in.
17 in.

If the screen length is 14.5 inches, what is the height of the screen to the nearest tenth?
(Lesson 12-2)

15. **MULTIPLE CHOICE** Eduardo jogs 5 kilometers north and 5 kilometers west. To the nearest kilometer, how far is he from his starting point? (Lesson 12-2)

F 25 km   **H** 7 km

G 10 km   **J** 5 km

16. **SCIENCE** A certain type of bacteria doubles every hour. If there are two bacteria initially in a sample, how many will be present after five hours? Use the *make a model* strategy. (Lesson 12-3)

17. **SCALE MODELS** A scale model is made of a building measuring 120 feet long, 75 feet wide, and 45 feet high. If the scale is 1 inch = 15 feet, what are the dimensions of the model? Use the *make a model* strategy. (Lesson 12-3)

Surface Area of Rectangular Prisms

▷ **MINI Lab**

• Use the cubes to build a rectangular prism with a length of 8 centimeters.

• Count the number of squares on the outside of the prism. The sum is the *surface area*.

1. Record the dimensions, volume, and surface area in a table.

2. Build two more prisms using all of the cubes. For each, record the dimensions, volume, and surface area.

3. Describe the prisms with the greatest and least surface areas.

The sum of the areas of all of the surfaces, or faces, of a three-dimensional figure is the **surface area**.

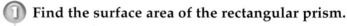

Surface Area of a Rectangular Prism		**Key Concept**
Words	The surface area S of a rectangular prism with length ℓ, width w, and height h is the sum of the areas of its faces.	**Model**
Symbols	$S = 2\ell w + 2\ell h + 2wh$	

EXAMPLES Find Surface Area

① **Find the surface area of the rectangular prism.**

There are three pairs of congruent faces.

• top and bottom
• front and back
• two sides

Faces	Area
top and bottom	$2(5 \cdot 4) = 40$
front and back	$2(5 \cdot 3) = 30$
two sides	$2(3 \cdot 4) = 24$
sum of the areas	$40 + 30 + 24 = 94$

The surface area is 94 square centimeters.

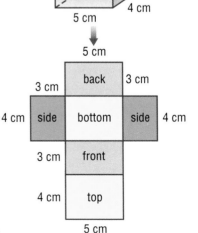

2 Find the surface area of the rectangular prism.

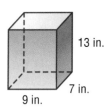

13 in.

7 in.

9 in.

Replace ℓ, with 9, w with 7, and h with 13.

surface area $= 2\ell w + 2\ell h + 2wh$

$\qquad = 2 \cdot 9 \cdot 7 + 2 \cdot 9 \cdot 13 + 2 \cdot 7 \cdot 13$

$\qquad = 126 + 234 + 182$ Multiply first. Then add.

$\qquad = 542$

The surface area of the prism is 542 square inches.

Study Tip

Surface Area
When you find the surface area of a three-dimensional figure, the units are square units, not cubic units.

✓ **CHECK** Your Progress

Find the surface area of each rectangular prism.

a.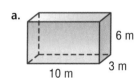

6 m

10 m

3 m

b.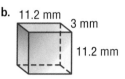

11.2 mm

3 mm

11.2 mm

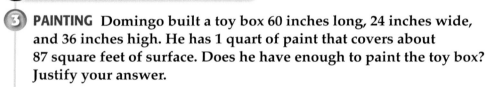

Real-World EXAMPLE

3 **PAINTING** Domingo built a toy box 60 inches long, 24 inches wide, and 36 inches high. He has 1 quart of paint that covers about 87 square feet of surface. Does he have enough to paint the toy box? Justify your answer.

STEP 1   Find the surface area of the toy box.

Replace ℓ with 60, w with 24, and h with 36.

surface area $= 2\ell w + 2\ell h + 2wh$

$\qquad\qquad = 2 \cdot 60 \cdot 24 + 2 \cdot 60 \cdot 36 + 2 \cdot 24 \cdot 36$

$\qquad\qquad = 8{,}928 \text{ in}^2$

STEP 2   Find the number of square inches the paint will cover.

$1 \text{ ft}^2 = 1 \text{ ft} \times 1 \text{ ft}$ Replace 1 ft with 12 in.

$\qquad = 12 \text{ in.} \times 12 \text{ in.}$ Multiply.

$\qquad = 144 \text{ in}^2$

So, 87 square feet is equal to 87×144 or 12,528 square inches.

Since $12{,}528 > 8{,}928$, Domingo has enough paint.

Study Tip

Consistent Units
Since the surface area of the toy box is expressed in inches, convert 87 ft² to square inches so that all measurements are expressed using the same units.

✓ **CHECK** Your Progress

c. **BOXES** The largest corrugated cardboard box ever constructed measured about 23 feet long, 9 feet high, and 8 feet wide. Would 950 square feet of paper be enough to cover the box? Justify your answer.

d. **BOXES** If 1 foot was added to each dimension of the largest corrugated cardboard box ever constructed, would 950 square feet of paper still be enough to cover the box? Justify your answer.

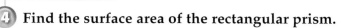

EXAMPLE Use the Pythagorean Theorem

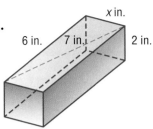

4 **Find the surface area of the rectangular prism.**

The width and height of the prism are given. To find the surface area, you need to find the length of the prism. Notice that the diagonal, length, and width of the top face of the prism form a right triangle.

$c^2 = a^2 + b^2$	Pythagorean Theorem
$7^2 = 6^2 + x^2$	Replace c with 7, a with 6, and b with x.
$49 = 36 + x^2$	Evaluate 7^2 and 6^2.
$49 - 36 = 36 + x^2 - 36$	Subtract 36 from each side.
$13 = x^2$	Simplify.
$\pm\sqrt{13} = x$	Definition of square root
$\pm 3.6 \approx x$	Simplify.

Study Tip

Square Roots
The equation $13 = x^2$ has two solutions, 3.6 and −3.6. However, the length of the prism must be positive, so choose the positive solution.

The length of the prism is about 3.6 inches. Find the surface area.

surface area $= 2\ell w + 2\ell h + 2wh$

$= 2(3.6)(6) + 2(3.6)(2) + 2(6)(2)$ or 81.6

The surface area of the prism is about 81.6 square inches.

✔ **CHECK Your Progress**

e. Find the surface area of the rectangular prism to the nearest tenth.

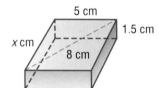

✔ **CHECK Your Understanding**

Examples 1, 2
(pp. 649–650)

Find the surface area of each rectangular prism. Round to the nearest tenth if necessary.

1.
   4 ft
   3 ft
   6 ft

2. 8.2 cm
   5.5 cm
   3.4 cm

Example 3
(p. 650)

3. **GIFTS** Megan is wrapping a gift. She places it in a box 8 inches long, 2 inches wide, and 11 inches high. If Megan bought a roll of wrapping paper that is 1 foot wide and 2 feet long, did she buy enough paper to wrap the gift? Justify your answer.

Example 4
(p. 651)

4. **MEASUREMENT** Find the surface area of the rectangular prism at the right. Round to the nearest tenth if necessary.

4 m   10 m
3 m
x m

Lesson 12-4 Surface Area of Rectangular Prisms **651**

Find the surface area of each rectangular prism. Round to the nearest tenth if necessary.

HOMEWORK HELP	
For Exercises	**See Examples**
5–10	1, 2
11–12	3
13–14	4

5.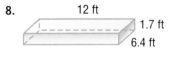

8 cm
9 cm
5 cm

6.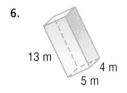

13 m
4 m
5 m

7.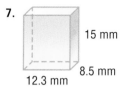

15 mm
8.5 mm
12.3 mm

8.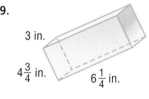

12 ft
1.7 ft
6.4 ft

9.

3 in.
$4\frac{3}{4}$ in.
$6\frac{1}{4}$ in.

10.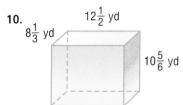

$12\frac{1}{2}$ yd
$8\frac{1}{3}$ yd
$10\frac{5}{6}$ yd

11. **BOOKS** When making a book cover, Anwar adds an additional 20 square inches to the surface area to allow for overlap. How many square inches of paper will Anwar use to make a book cover for a book 11 inches long, 8 inches wide, and 1 inch high?

12. **FENCES** If one gallon of paint covers 350 square feet, will 8 gallons of paint be enough to paint the inside and outside of the fence shown once? Explain.

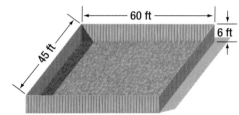

60 ft
6 ft
45 ft

Find the surface area of each rectangular prism. Round to the nearest tenth if necessary.

13.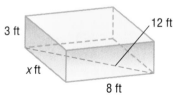

3 ft
12 ft
x ft
8 ft

14.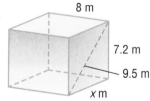

8 m
7.2 m
9.5 m
x m

15. **MUSIC** To the nearest tenth, find the approximate amount of plastic covering the outside of the CD case.

16. **MEASUREMENT** What is the surface area of a rectangular prism that has a length of 6.5 centimeters, a width of 2.8 centimeters, and a height of 9.7 centimeters?

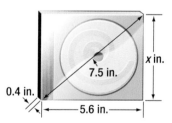

x in.
7.5 in.
0.4 in.
5.6 in.

17. **ALGEBRA** Write a formula for the surface area s of a cube in which each side measures x units.

18. **PACKAGING** A company will make a cereal box with whole number dimensions and a volume of 100 cubic centimeters. If cardboard costs $0.05 per 100 square centimeters, what is the least cost to make 100 boxes?

19. **REASONING** The bottom and sides of a pool in the shape of a rectangular prism will be painted blue. The length, width, and height of the pool are 18 feet, 12 feet, and 6 feet, respectively. Explain why the number of square feet to be painted is *not* equivalent to the expression $2(18)(12) + 2(18)(6) + 2(12)(6)$.

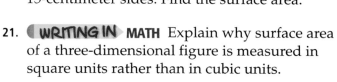

20. **CHALLENGE** The figure at the right is made by placing a cube with 12-centimeter sides on top of another cube with 15-centimeter sides. Find the surface area.

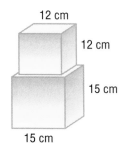

NYSCC • NYSMT
Extra Practice, pp. 703, 715

21. **WRITING IN MATH** Explain why surface area of a three-dimensional figure is measured in square units rather than in cubic units.

NYSMT PRACTICE 7.G.4

22. Which of the following expressions represents the surface area of a cube with side length w?

　A w^3

　B $6w^2$

　C $6w^3$

　D $2w + 4w^2$

23. How much cardboard is needed to make a box with a length of 2.5 feet, a width of 1.6 feet, and a height of 2 feet?

　F 37.5 square feet

　G 24.4 square feet

　H 8 square feet

　J 6.1 square feet

Spiral Review

24. **MEASUREMENT** A rectangular-shaped yard that measures 50 feet by 70 feet is bordered by a flowerbed that is 2 feet wide. What is the perimeter of the entire yard? Use the *make a model* strategy. (Lesson 12-3)

25. **MEASUREMENT** What is the missing measure of a right triangle in which $a = 13$ feet and $c = 18$ feet? Round to the nearest tenth. (Lesson 12-2)

26. **MEASUREMENT** What is the volume of the cylinder shown at the right? Round to the nearest tenth. (Lesson 11-10)

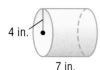

▷ **GET READY** for the Next Lesson

PREREQUISITE SKILL Find the area of each circle. Round to the nearest tenth. (Lesson 11-4)

27.

28.

29. diameter = 13.6 yd

30. radius = 23 km

Measurement Lab
Changes in Scale

Suppose you have a model of a rectangular prism and you are asked to create a similar model with dimensions that are twice as large. In this lab, you will investigate how the scale factor that relates the lengths in two similar objects affects how the surface areas and volumes are related.

ACTIVITY

 STEP 1 Draw a cube on dot paper that measures 1 unit on each side. Calculate the volume and the surface area of the cube. Then record the data in a table like the one shown below.

1 unit

STEP 2 Double the side lengths of the cube. Calculate the volume and the surface area of this cube. Record the data in your table.

2 units

STEP 3 Triple the side lengths of the original cube. Now each side measures 3 units long. Calculate the volume and the surface area of the cube and record the data.

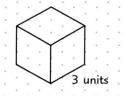
3 units

STEP 4 For each cube, write a ratio comparing the side length and the volume. Then write a ratio comparing the side length and the surface area. The first one is done for you.

Side Length (units)	Volume (units³)	Surface Area (units²)	Ratio of Side Length to Volume	Ratio of Side Length to Surface Area
1	$1^3 = 1$	$6(1^2) = 6$	1 : 1	1 : 6
2				
3				
4				
5				
s				

 **CHECK Your Progress**

a. Complete the table above.

EXAMPLE Find the Surface Area of a Cylinder

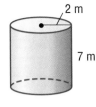

1 Find the surface area of the cylinder.
Round to the nearest tenth.

$S = 2\pi r^2 + 2\pi rh$ Surface area of a cylinder

    $= 2\pi(2)^2 + 2\pi(2)(7)$ Replace *r* with 2 and *h* with 7.

    ≈ 113.1 Simplify.

The surface area is about 113.1 square meters.

 CHECK Your Progress

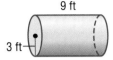

  a. Find the surface area of the cylinder.
     Round to the nearest tenth.

 Real-World EXAMPLE

2 **CAROUSELS** A circular fence that is 2 feet high is to be built around the outside of a carousel. The distance from the center of the carousel to the edge of the fence will be 35 feet. How much fencing material is needed to make the fence around the carousel?

The radius of the circular fence is 35 feet. The height is 2 feet.

$S = 2\pi rh$ Curved surface of a cylinder

   $= 2\pi(35)(2)$ Replace *r* with 35 and *h* with 2.

   ≈ 439.8 Simplify.

So, about 439.8 square feet of material is needed to make the fence.

Real-World Link . . . .
Of the 3,000 to 4,000 wooden carousels carved in America between 1885 and 1930, fewer than 150 operate today.
Source: National Carousel Association

 CHECK Your Progress

  b. **DESIGN** Find the area of the label of a can of tuna with a radius of 5.1 centimeters and a height of 2.9 centimeters.

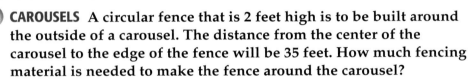 **CHECK Your Understanding**

Example 1
(p. 657)

Find the surface area of each cylinder. Round to the nearest tenth.

1.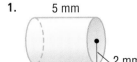
5 mm
2 mm

2.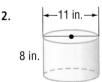
←11 in.→
8 in.

Example 2
(p. 657)

3. **STORAGE** The height of a water tank is 10 meters, and it has a diameter of 10 meters. What is the surface area of the tank?

HOMEWORK HELP

For Exercises	See Examples
4–9	1
10–11	2

Find the surface area of each cylinder. Round to the nearest tenth.

4.
6 yd
10 yd

5.
12.5 m
9 m

6.
3 ft
18 ft

7.
8.7 mm
5.6 mm

8.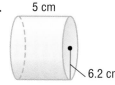
5 cm
6.2 cm

9.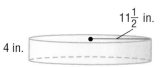
$11\frac{1}{2}$ in.
4 in.

10. **CANDLES** A cylindrical candle has a diameter of 4 inches and a height of 7 inches. What is the surface area of the candle?

11. **PENCILS** Find the surface area of an unsharpened cylindrical pencil that has a radius of 0.5 centimeter and a height of 19 centimeters.

ESTIMATION **Estimate the surface area of each cylinder**

12.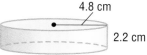
4.8 cm
2.2 cm

13.
8.2 m
3.7 m

14.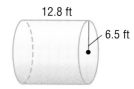
12.8 ft
6.5 ft

15. **BAKING** Mrs. Jones baked a cake 5 inches high and 9 inches in diameter. If Mrs. Jones covers the top and sides of the cake with frosting, find the area that the frosting covers to the nearest tenth.

16. **PACKAGING** The mail tube shown is made of cardboard and has plastic end caps. Approximately what percent of the surface area of the mail tube is cardboard?

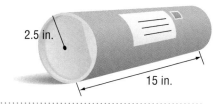
2.5 in.
15 in.

NYSCC • NYSMT

Extra Practice, pp. 703, 715

H.O.T. Problems

17. **CHALLENGE** If the height of a cylinder is doubled, will its surface area also double? Explain your reasoning.

18. **WRITING IN MATH** Write a problem about a real-world situation in which you would find the surface area of a cylinder. Be sure to include the answer to your problem.

19. **REASONING** Which has more surface area, a cylinder with radius 6 centimeters and height 3 centimeters or a cylinder with radius 3 centimeters and height 6 centimeters? Explain your reasoning.

20. Stacey has a cylindrical paper clip holder with the net shown. Use a centimeter ruler to measure the dimensions of the net in centimeters.

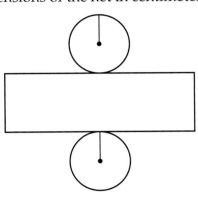

Which is closest to the surface area of the cylindrical paper clip holder?

A 6.0 cm^2

B 6.5 cm^2

C 7.5 cm^2

D 15.5 cm^2

21. The three containers below each hold about 1 liter of liquid. Which container has the greatest surface area?

12 cm
5.2 cm
Container I

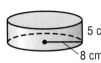
5 cm
8 cm
Container II

10 cm
5.7 cm
Container III

11 cm
5.4 cm
Container IV

F Container I

G Container II

H Container III

J Container IV

Spiral Review

MEASUREMENT Find the surface area of each rectangular prism. (Lesson 12-4)

22.
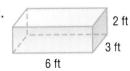
2 ft
3 ft
6 ft

23.
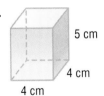
5 cm
4 cm
4 cm

24.
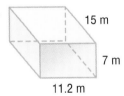
15 m
7 m
11.2 m

MEASUREMENT Find the missing measure of each right triangle. Round to the nearest tenth if necessary. (Lesson 12-2)

25. $a = 8$ in., $b = 10$ in.

26. $a = 12$ ft, $c = 20$ ft

27. $b = 12$ cm, $c = 14$ cm

Problem Solving in Life Skills
Real-World Unit Project

Design That House It's time to complete your project. Use the measurements you have gathered to create a blueprint of your dream house. Be sure to include the scale and actual measurements of your dream house.

NY Math Online > Unit Project at glencoe.com

NY Math Online glencoe.com
- STUDY *TO GO*
- Vocabulary Review

FOLDABLES ▸ **GET READY** to Study
Study Organizer

Be sure the following Big Ideas are noted in your Foldable.

Ch. 12	Rectangular Prisms	Cylinders
Draw Examples		
Find Volume		
Find Surface Area		

BIG Ideas

Irrational Numbers (Lesson 12-1)
- An irrational number is a number that cannot be written as a fraction.

Pythagorean Theorem (Lesson 12-2)
- In a right triangle, the square of the length of the hypotenuse equals the sum of the squares of the lengths of the legs.

Surface Area (Lessons 12-4, 12-5)
- The surface area S of a rectangular prism with length ℓ, width w, and height h is the sum of the areas of the faces. $S = 2\ell w + 2\ell h + 2wh$

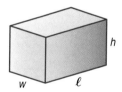

- The surface area of S of a cylinder with height h and a radius r is the sum of the area of the circular bases and the area of the curved surface. $S = 2\pi r^2 + 2\pi rh$

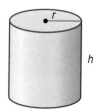

Key Vocabulary

hypotenuse (p. 640)

irrational number (p. 637)

leg (p. 640)

Pythagorean Theorem (p. 640)

surface area (p. 649)

Vocabulary Check

State whether each sentence is *true* or *false*. If *false*, replace the underlined word or number to make a true sentence.

1. The side opposite the right angle in a <u>scalene triangle</u> is called a hypotenuse.

2. Either of the two sides that form the right angle of a right triangle is called a <u>hypotenuse</u>.

3. An <u>irrational number</u> is a number that cannot be expressed as the quotient of two integers.

4. In a right triangle, the square of the length of the hypotenuse equals the <u>difference</u> of the squares of the lengths of the legs.

5. The sum of the areas of all the surfaces of a three-dimensional figure is called the <u>surface area</u>.

6. The formula for finding the surface area of a <u>cylinder</u> is $S = 2\ell w + 2\ell h + 2wh$.

7. Rational numbers include <u>only positive</u> numbers.

8. The <u>Pythagorean Theorem</u> can be used to find the length of the hypotenuse of a right triangle if the measures of both legs are known.

9. To find the surface area of a <u>rectangular prism</u>, you must know the measurements of the height and the radius.

10. The square root of a perfect square is a <u>rational number</u>.

Estimate each square root to the nearest whole number.

1. $\sqrt{500}$
2. $\sqrt{95}$
3. $\sqrt{265}$

Graph each square root on the number line.

4. $\sqrt{570}$
5. $\sqrt{7}$
6. $\sqrt{84}$

7. **MULTIPLE CHOICE** The length of one side of a square sandbox is 7 feet. Which number is closest to the length of the diagonal of the sandbox?

   A $\sqrt{100}$
   B $\sqrt{50}$
   C $\sqrt{14}$
   D $\sqrt{7}$

Find the missing measure of each right triangle. Round to the nearest tenth if necessary.

8. $a = 5$ m, $b = 4$ m
9. $b = 12$ in., $c = 14$ in.
10. $a = 7$ in., $c = 13$ in.

11. **MEASUREMENT** Use the diagram below to find the distance from the library to the post office. Round to the nearest tenth.

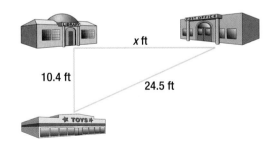

12. **CHAIRS** Chris is responsible for arranging the chairs at the meeting. There are 72 chairs, and he wants to have twice as many chairs in each row as he has in each column. How many chairs should he put in each row? How many rows does he need?

Find the surface area of each rectangular prism and cylinder. Round to the nearest tenth if necessary.

13.
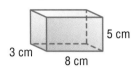

14.

15.
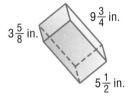

16.

17.

18.

19. **PACKAGING** Mrs. Rodriguez is wrapping a gift. What is the least amount of wrapping paper she will need to wrap the box below?

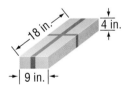

20. **MULTIPLE CHOICE** The dimensions of four containers are given below. Which container has the greatest surface area?

F   9.3 in.   H   9.3 in.
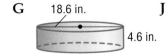   6.6 in.   4.2 in.

G   18.6 in.   J   19.8 in.
   4.6 in.   5.1 in.

CHAPTER 12 NYSMT Practice
Cumulative, Chapters 1–12

NY Math Online glencoe.com
• Test Practice

PART 1 Multiple Choice

Read each question. Then fill in the correct answer on the answer document provided by your teacher or on a sheet of paper.

1. Which of the following three-dimensional figures could be formed from this net?

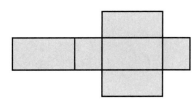

A cube

B rectangular pyramid

C triangular prism

D rectangular prism

2. Which of the following nets could be used to make a cylinder?

F

G

H

J

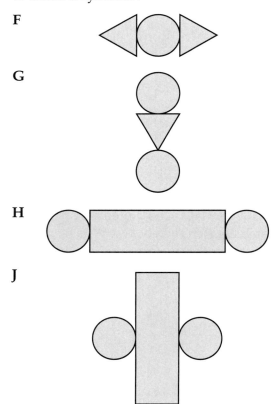

3. Carla has an above-ground swimming pool with a circumference of 20 feet. Which of the following equations could be used to find r, the radius of the pool?

A $r = \frac{10}{\pi}$ C $r = \frac{10}{2\pi}$

B $r = \frac{40}{\pi}$ D $r = \frac{\pi}{20}$

4. Of the following figures that Ryan drew, which 2 figures have the same area?

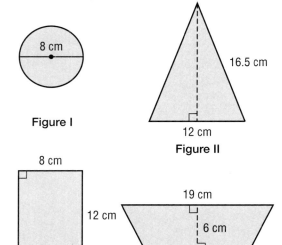

8 cm
16.5 cm
12 cm
8 cm
12 cm
19 cm
6 cm
13 cm
Figure I
Figure II
Figure III
Figure IV

F Figure I and II

G Figure II and III

H Figure II and IV

J Figure III and IV

5. Cassandra drew a circle with a radius of 12 inches and another circle with a radius of 8 inches. What is the approximate difference between the areas of the 2 circles? Use $\pi = 3.14$.

A 452.16 in^2 C 50.24 in^2

B 251.2 in^2 D 25.12 in^2

6. Which equation could be used to find
the area of a circle with a radius of
10 centimeters?

   F $A = 5 \times \pi$

   G $A = \pi \times 5^2$

   H $A = 10 \times \pi$

   J $A = \pi \times 10^2$

7. Dave can run 30 yards in 8.2 seconds.
During a race, he ran 120 yards. If Dave's
rate of speed remained the same, how long
did it take him to run the race?

   A 43 seconds   C 24.6 seconds

   B 32.8 seconds   D 18.4 seconds

8. Which of the following equations gives the
surface area S of a cube with side length m?

   F $S = m^3$

   G $S = 6m^2$

   H $S = 6m$

   J $S = 2m + 4m^2$

TEST-TAKING TIP

Question 9 Be sure to read each
question carefully. In question 9, you are
asked to find which statement is *not* true.

9. Which statement is *not* true about an
equilateral triangle?

   A The sum of the angles is 180°.

   B It has three congruent angles.

   C It has one right angle.

   D It has exactly three congruent sides.

PART 2   Short Response/Grid In

Record your answers on the answer sheet
provided by your teacher or on a sheet of paper.

10. Bill's Electronics bought 5 computers for
a total of $3,000. The business later bought
another computer for $600. What was the
mean price of all the computers?

11. A jar contains 9 yellow marbles and 1 red
marble. Ten students will each randomly
select one marble to determine who goes first
in a game. Whoever picks the red marble goes
first. Lily will pick first and keep the marble
that she picks. Heath will pick second. What
is the probability that Lily will pick a yellow
marble and Heath will pick the red marble?

PART 3   Extended Response

Record your answers on the answer sheet
provided by your teacher or on a sheet of
paper. Show your work.

12. A square with a side of y inches is inside a
square with a side of 6 inches, as shown below.

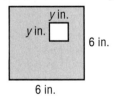

a. Write an expression that can be used
to find the area of the shaded region in
terms of y.

b. If the dimensions of both squares are
doubled, write an expression that could
be used to find the area of the new
shaded region.

NEED EXTRA HELP?

If You Missed Question...	1	2	3	4	5	6	7	8	9	10	11	12
Go to Lesson...	11-8	11-8	11-3	11-2	11-4	11-4	6-6	12-4	10-3	8-2	9-8	11-1
NYS Core Curriculum	7.G.3	7.G.3	7.G.1	6.G.2	6.G.7	6.G.7	7.PS.10	7.G.4	7.M.8	7.S.6	6.S.9	6.G.2

Looking Ahead

to Next Year

Let's Look Ahead

Scientific Notation

Looking Ahead 1

MAIN IDEA

Express numbers in scientific notation and in standard form.

NYS Core Curriculum

7.N.6 Translate numbers from scientific notation into standard form 7.N.7 Compare numbers written in scientific notation *Also addresses 7.N.5.*

New Vocabulary

scientific notation

NY Math Online

glencoe.com

• Personal Tutor
• Self-Check Quiz
• Extra Examples

▷ **GET READY** for the Lesson

More than 425 million pounds of gold has been discovered in the world. If all this gold were in one place, it would form a cube seven stories on each side.

1. Write 425 million in standard form.

2. Complete: $4.25 \times \underline{\quad?\quad} = 425$ million.

When you deal with very large numbers like 425,000,000, it can be difficult to keep track of the zeros. You can express numbers such as this in **scientific notation** by writing the number as the product of a factor and a power of 10.

Scientific Notation		Key Concept
Words	A number is expressed in scientific notation when it is written as the product of a factor and a power of 10. The factor must be greater than or equal to 1 and less than 10.	
Symbols	$a \times 10^n$, where $1 \le a < 10$ and n is an integer	
Examples	$425,000,000 = 4.25 \times 10^8$	

EXAMPLE **Express Large Numbers in Standard Form**

① Express 2.16×10^5 in standard form.

$2.16 \times 10^5 = 2.16 \times 100,000$ $10^5 = 100,000$

$ = \underline{216,000}$ Move the decimal point 5 places to the right.

 CHECK Your Progress

Express each number in standard form.

a. 7.6×10^6 b. 3.201×10^4

Scientific notation is also used to express very small numbers. Study the pattern of products at the right. Notice that multiplying by a negative power of 10 moves the decimal point to the left the same number of places as the absolute value of the exponent.

$1.25 \times 10^2 = 125$
$1.25 \times 10^1 = 12.5$
$1.25 \times 10^0 = 1.25$
$1.25 \times 10^{-1} = 0.125$
$1.25 \times 10^{-2} = 0.0125$
$1.25 \times 10^{-3} = 0.00125$

EXAMPLE Express Small Numbers in Standard Form

2 Express 5.8×10^{-3} in standard form.

$5.8 \times 10^{-3} = 5.8 \times 0.001$ $10^{-3} = 0.001$

$\qquad\quad = 0.0058$ Move the decimal point 3 places to the left.

CHECK Your Progress

c. 4.7×10^{-5} d. 9.0×10^{-4}

To write a number in scientific notation, place the decimal point after the first nonzero digit. Then find the power of 10.

EXAMPLES Express Numbers in Scientific Notation

Express each number in scientific notation.

3 1,457,000

$1{,}457{,}000 = 1.457 \times 1{,}000{,}000$ The decimal point moves 6 places to the left.

$\qquad\qquad = 1.457 \times 10^{6}$ The exponent is positive.

4 0.00063

$0.00063 = 6.3 \times 0.0001$ The decimal point moves 4 places to the right.

$\qquad\quad = 6.3 \times 10^{-4}$ The exponent is negative.

CHECK Your Progress

e. 35,000   f. 0.00722

To compare numbers in scientific notation, first compare the exponents. With positive numbers, any number with a greater exponent is greater. If the exponents are the same, compare the factors.

 Real-World EXAMPLE   Compare Numbers in Scientific Notation

5 **OCEANS** The Atlantic Ocean has an area of 3.18×10^{7} square miles. The Pacific Ocean has an area of 6.4×10^{7} square miles. Which ocean has the greater area?

Since both exponents are the same, compare the factors.

$3.18 < 6.4 \quad \rightarrow \quad 3.18 \times 10^{7} < 6.4 \times 10^{7}$

So, the Pacific Ocean has the greater area.

CHECK Your Progress

g. Replace ● with $<$, $>$, or $=$ to make 4.13×10^{-2} ● 5.0×10^{-3} a true sentence.

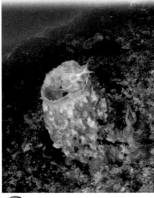

Real-World Link . . . .
At the deepest point in the ocean, the pressure is greater than 8 tons per square inch and the temperature is only a few degrees above freezing.

Source: Ocean Planet Smithsonian

Examples 1, 2
(pp. LA2–LA3)

Express each number in standard form.

1. 3.754×10^5

2. 8.34×10^6

3. 1.5×10^{-4}

4. 2.68×10^{-3}

Examples 3, 4
(p. LA3)

Express each number in scientific notation.

5. 4,510,000

6. 0.00673

7. 0.000092

8. 11,620,000

9. **PHYSICAL SCIENCE** Light travels 300,000 kilometers per second. Write this number in scientific notation.

Example 5
(p. LA3)

10. **TECHNOLOGY** The distance between tracks on a CD and DVD are shown in the table. Which disc has the greater distance between tracks?

Disc	Distance (mm)
CD	1.6×10^{-3}
DVD	7.4×10^{-4}

Replace each ● with <, >, or = to make a true sentence.

11. 2.3×10^5 ● 1.7×10^5

12. 0.012 ● 1.4×10^{-1}

Practice and Problem Solving

HOMEWORK HELP	
For Exercises	See Examples
13–22	1, 2
23–32	3, 4
33–38	5

Express each number in standard form.

13. 6.1×10^4

14. 5.72×10^6

15. 3.3×10^{-1}

16. 5.68×10^{-3}

17. 9.014×10^{-2}

18. 1.399×10^5

19. 2.505×10^3

20. 7.4×10^{-5}

21. **SPIDERS** The diameter of a spider's thread is 1.0×10^{-3} inch. Write this number in standard form.

22. **DINOSAURS** The *Gigantosaurus* dinosaur weighed about 1.4×10^4 pounds. Write this number in standard form.

Express each number in scientific notation.

23. 499,000

24. 2,000,000

25. 0.006

26. 0.0125

27. 50,000,000

28. 39,560

29. 0.000078

30. 0.000425

31. **CHESS** The number of possible ways that a player can play the first four moves in a chess game is 3 billion. Write this number in scientific notation.

32. **SCIENCE** A particular parasite is approximately 0.025 inch long. Write this number in scientific notation.

SPORTS For Exercises 33 and 34, use the table. Determine which category in each pair had a greater amount of sales.

Category	Sales ($)
Camping	1.547×10^9
Golf	3.243×10^9
Tennis	3.73×10^8

Source: National Sporting Goods Assoc.

33. golf or tennis

34. camping or golf

Replace each ● with <, >, or = to make a true sentence.

35. 1.8×10^3 ● 1.9×10^{-1}

36. 5.2×10^2 ● 5000

37. 0.00701 ● 7.1×10^{-3}

38. 6.49×10^4 ● 649×10^2

39. MEASUREMENT The table at the right shows the values of different prefixes that are used in the metric system. Write the units attometer, gigameter, kilometer, nanometer, petameter, and picometer in order from greatest to least measure.

Metric Measures	
Prefix	**Meaning**
atto	10^{-18}
giga	10^9
kilo	10^3
nano	10^{-9}
peta	10^{15}
pico	10^{-12}

40. NUMBER SENSE Write the product of 0.00004 and 0.0008 in scientific notation.

41. NUMBER SENSE Order 6.1×10^4, 6100, 6.1×10^{-5}, 0.0061, and 6.1×10^{-2} from least to greatest.

PHYSICAL SCIENCE For Exercises 42 and 43, use the table.

The table shows the maximum amounts of lava in cubic meters per second that erupted from four volcanoes.

42. How many times greater was the Mount St. Helens eruption than the Ngauruhoe eruption?

43. How many times greater was the Hekla eruption than the Ngauruhoe eruption?

Volcanic Eruptions	
Volcano, Year	**Eruption Rate (m^3/s)**
Mount St. Helens, 1980	2.0×10^4
Ngauruhoe, 1975	2.0×10^3
Hekla, 1970	4.0×10^3
Agung, 1963	3.0×10^4

Source: University of Alaska

Write each number in standard form.

44. $(8 \times 10^0) + (4 \times 10^{-3}) + (3 \times 10^{-5})$

45. $(4 \times 10^4) + (8 \times 10^3) + (3 \times 10^2) + (9 \times 10^1) + (6 \times 10^0)$

H.O.T. Problems

46. CHALLENGE Convert the numbers in each expression to scientific notation. Then evaluate the expression. Express in scientific notation and in decimal notation.

a. $\dfrac{(420{,}000)(0.015)}{0.025}$

b. $\dfrac{(0.078)(8.5)}{0.16(250{,}000)}$

47. REASONING Which is a better estimate for the number of times per year that a person blinks: 6.25×10^{-2} times or 6.25×10^6 times? Explain your reasoning.

48. OPEN ENDED Describe a real-life value or measure using numbers in scientific notation and in standard form.

49. WRITING IN MATH Explain the relationship between a number in standard form and the sign of the exponent when the number is written in scientific notation.

Solving Multi-Step Equations

MAIN IDEA

Solve equations with the variable on each side.

NYS Core Curriculum

7.A.4 Solve multi-step equations by combining like terms, using the distributive property, or moving variables to one side of the equation

NY Math Online

glencoe.com
• Personal Tutor
• Self-Check Quiz
• Extra Examples

▷ **GET READY** for the Lesson

Jeremy bought two used video games for d dollars each. Alyssa bought one used video game for d dollars and a used CD for $5. They both spent the same amount.

amount Jeremy spent		amount Alyssa spent
$2d$	$=$	$d + 5$

1. Use the expressions above to write an equation that represents this situation.

2. Use the *guess-and-check* strategy to solve the equation. What does the solution mean?

To solve equations with variables on each side, use the Addition or Subtraction Property of equality to write an equivalent equation with the variables on one side. Then solve.

EXAMPLES Equations with Variables on Each Side

Solve each equation. Check the solution.

1 $3x + 8 = 4x$

$3x + 8 = 4x$	Write the equation.
$3x - 3x + 8 = 4x - 3x$	Subtract $3x$ from each side.
$8 = x$	Simplify.

The solution is 8. **Check** $3(8) + 8 = 32$ and $4(8) = 32$ ✓

2 $4 + t = 3.5t - 2$

$4 + t = 3.5t - 2$	Write the equation.
$4 + t - t = 3.5t - t - 2$	Subtract t from each side.
$4 = 2.5t - 2$	Simplify.
$4 + 2 = 2.5t - 2 + 2$	Add 2 to each side.
$6 = 2.5t$	Simplify.
$\dfrac{6}{2.5} = \dfrac{2.5t}{2.5}$	Divide each side by 2.5.
$2.4 = t$	Simplify.

The solution is 2.4. **Check** $4 + 2.4 = 6.4$ and $3.5(2.4) - 2 = 6.4$ ✓

 **CHECK** Your Progress

a. $7d - 13 = 3d + 7$ b. $12.6 - x = 2x$

You can use the Distributive Property to remove grouping symbols in equations.

EXAMPLE **Use the Distributive Property**

3 Solve $2n = 10(n - 1)$.

$2n = 10(n - 1)$	Write the equation.
$2n = 10(n) - 10(1)$	Use the Distributive Property.
$2n = 10n - 10$	Simplify.
$2n - 10n = 10n - 10n - 10$	Subtract $10n$ from each side.
$-8n = -10$	Simplify.
$\dfrac{-8n}{-8} = \dfrac{-10}{-8}$	Divide each side by -8.
$n = 1.25$	Simplify.

The solution is 1.25.

Study Tip

Check
Check your solution by replacing *n* with 1.25 in the original equation. Since 2(1.25) = 2.5 and 10(1.25 – 1) = 2.5, the solution is correct.

 CHECK Your Progress

c. $3(k + 2) = 6k$

d. $3(a - 1) = 4(a - 1.5)$

Real-World EXAMPLE **Use an Equation to Solve a Problem**

4 **CAR RENTAL** Suppose you can rent a car for either $35 a day plus $0.45 per mile or for $50 a day plus $0.25 per mile. What number of miles *m* results in the same cost for one day?

$35 a day plus $0.45 per mile		$50 a day plus $0.25 per mile
$35 + 0.45m$	$=$	$50 + 0.25m$

$35 + 0.45m = 50 + 0.25m$	Write the equation.
$35 + 0.45m - 0.25m = 50 + 0.25m - 0.25m$	Subtract $0.25m$ from each side.
$35 + 0.2m = 50$	Simplify.
$35 - 35 + 0.2m = 50 - 35$	Subtract 35 from each side.
$0.2m = 15$	Simplify.
$\dfrac{0.2m}{0.2} = \dfrac{15}{0.2}$	Divide each side by 0.2.
$m = 75$	Simplify.

The cost is the same for 75 miles.

Study Tip

Alternative Method
In Example 4, you can also solve the equation by subtracting 35 from each side first, then subtracting 0.25m from each side.

 CHECK Your Progress

e. **GEOMETRY** A rectangle's length is 30 centimeters longer than its width. The perimeter is 4.6 times the width. Find the dimensions.

Some equations have *no* solution. That is, no value of the variable results in a true sentence. This is shown by the symbol $\varnothing$ or { }. Other equations may have *every* number as the solution.

EXAMPLES No Solution or a Solution Set of All Numbers

Solve each equation.

⑤ $4x + \frac{1}{2} = 4x - 9$

$$4x + \frac{1}{2} = 4x - 9 \qquad \text{Write the equation.}$$

$$4x - 4x + \frac{1}{2} = 4x - 4x - 9 \qquad \text{Subtract } 4x \text{ from each side.}$$

$$\frac{1}{2} = -9 \qquad \text{Simplify.}$$

The sentence $\frac{1}{2} = -9$ is *never* true. So, the solution set is $\varnothing$.

⑥ $8(r + 3) = 2(12 + 4r)$

$$8(r + 3) = 2(12 + 4r) \qquad \text{Write the equation.}$$

$$8r + 24 = 24 + 8r \qquad \text{Use the Distributive Property.}$$

$$8r + 24 - 24 = 24 - 24 + 8r \qquad \text{Subtract 24 from each side.}$$

$$8r = 8r \qquad \text{Simplify.}$$

$$\frac{8r}{8} = \frac{8r}{8} \qquad \text{Divide each side by 8.}$$

$$r = r \qquad \text{Simplify.}$$

The sentence $r = r$ is *always* true. So, the solution set is all numbers.

 CHECK Your Progress

f. $3(x + 1) - 5 = 3x - 2$ \qquad g. $8y - 3 = 5(y - 1) + 3y$

CHECK Your Understanding

Examples 1, 2, 5, 6
(pp. LA6, LA8)

Solve each equation. Check the solution.

1. $4x - 7 = 5x$ \qquad\qquad 2. $4x - 1 = 3x + 2$

3. $3.1w + 5 = 0.8 + w$ \qquad 4. $12 - h = -h + 3$

Example 4
(p. LA7)

5. **BASKETBALL** Abigail has 6 more points than Victoria. Together, they have twice as many points as Hannah. If Hannah has 18 points, how many points does each of the other girls have?

6. **NUMBER THEORY** Twice a number is 220 less than six times the number. What is the number?

Examples 3, 5, 6
(pp. LA7, LA8)

Solve each equation. Check the solution.

7. $2(d + 6) = 3d - 1$ \qquad 8. $g + 1 = 3(g - 3)$

9. $3(2a + 4) = 6(a + 2)$ \qquad 10. $2(t - 3) + 5 = 3(t - 1)$

Solve each equation. Check the solution.

11. $12x = 2x + 40$

12. $4x + 9 = 7x$

13. $4k + 24 = 6k - 10$

14. $2f - 6 = 7f + 24$

15. $12n - 24 = -14n + 28$

16. $n + 0.4 = -n + 1$

17. $-8b + 5 = 7 - 8b - 2$

18. $3y + \frac{1}{3} = 3y - \frac{1}{2}$

19. GEOMETRY The perimeter of a rectangle is 74 inches. Find the dimensions if the length is 7 inches longer than twice the width.

20. NUMBER THEORY Three times the sum of three consecutive integers is 72. What are the integers?

Solve each equation. Check the solution.

21. $3(x + 1) = 21$

22. $5(2c + 7) = 55$

23. $6(3d + 5) = 75$

24. $4(x - 2) = 3(1.5 + x)$

25. $3(s + 22) = 4(s + 12)$

26. $4.2x - 9 = 3(1.2x + 4)$

27. $4(f + 3) + 5 = 17 + 4f$

28. $3n + 4 = 5(n + 2) - 2n$

29. GEOMETRY The triangle and the rectangle have the same perimeter. Find the dimensions of the rectangle.

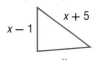

30. RECREATION Jennette begins in-line skating around a park path at a rate of 3 miles per hour. One hour later, one of her friends starts on the same path on her bike, riding at 12 miles per hour. Solve $3t = 12(t - 1)$ to find the time t Jennette's friend catches up to her.

Solve each equation. Check the solution.

31. $-3(4b - 10) = \frac{1}{2}(-24b + 60)$

32. $\frac{3}{4}a + 4 = \frac{1}{4}(3a + 16)$

33. $\frac{d}{0.4} = 2d + 1.24$

34. $\frac{a - 6}{12} = \frac{a - 2}{4}$

H.O.T. Problems

35. CHALLENGE An apple costs the same as 2 oranges. Together, an orange and a banana cost 10¢ more than an apple. What is the cost for one of each fruit?

36. REASONING Describe the steps used to solve $15x = -3x - 9$.

37. OPEN ENDED Write an equation that has parentheses and contains the variable b on both sides. Solve the equation.

38. WRITING IN MATH Describe how the Distributive Property is used to solve equations.

Parallel Lines and Transversals

Looking Ahead 3

MAIN IDEA

Analyze the relationships of angles formed by two parallel lines and a transversal.

NYS Core Curriculum

Preparation for 8.G.4 Determine angle pair relationships when given two parallel lines cut by a transversal

New Vocabulary

parallel lines
transversal
alternate interior angles
alternate exterior angles
corresponding angles
vertical angles
adjacent angles

NY Math Online

glencoe.com
• Personal Tutor
• Self-Check Quiz
• Extra Examples

▶ **GET READY for the Lesson**

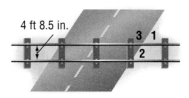

In the United States, the standard distance between rails on a railroad track is 4 feet 8.5 inches. The diagram at the right shows a road crossing over railroad tracks.

1. Measure angles 1 and 2. Record the measures. Then make a conjecture about the measure of angle 3. Then measure the angle to verify your conjecture.

Lines in a plane that never intersect are **parallel lines**. When two parallel lines are intersected by a third line, this line is called a **transversal**.

Names of Special Angles   **Key Concept**

If a pair of parallel lines is intersected by a transversal, this pair of angles are congruent.

Alternate interior angles are on opposite sides of the transversal and inside the parallel lines.

$\angle 3 \cong \angle 5$, $\angle 4 \cong \angle 6$

Alternate exterior angles are on opposite sides of the transversal and outside the parallel lines.

$\angle 1 \cong \angle 7$, $\angle 2 \cong \angle 8$

Corresponding angles are in the same position on the parallel lines in relation to the transversal.

$\angle 1 \cong \angle 5$, $\angle 2 \cong \angle 6$, $\angle 3 \cong \angle 7$, $\angle 4 \cong \angle 8$

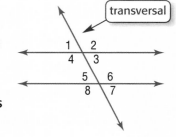

EXAMPLE   Find Measures of Angles

In the figure at the right, $p \parallel q$ and $m\angle 1 = 95°$. Find each measure.

1 $m\angle 7$

$\angle 3$ and $\angle 7$ are alternate interior angles.

$m\angle 7 = m\angle 3 = 95°$

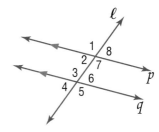

✓ **CHECK Your Progress**

In the figure, $m\angle 4 = 85°$. Find each measure.

a. $m\angle 8$ b. $m\angle 2$

Vertical Angles and Adjacent Angles **Key Concept**

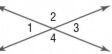

Vertical angles are opposite angles formed by the intersection of two lines. Vertical angles are congruent.

$\angle 1 \cong \angle 3$, $\angle 2 \cong \angle 4$

Two angles that have the same vertex, share a common side, and do not overlap are **adjacent angles**. Adjacent angles are supplementary.

$\angle 1$ and $\angle 2$, $\angle 2$ and $\angle 3$, $\angle 3$ and $\angle 4$, $\angle 4$ and $\angle 1$

EXAMPLES Use Vertical Angles and Adjacent Angles

Reading Math

Congruent Angles
- Angle 1 *is congruent to* angle 2: $\angle 1 \cong \angle 2$.
- The measure of $\angle 1$ *is equal to* the measure of $\angle 2$: $m\angle 1 = m\angle 2$.

In the figure at the right, $m\angle 1 = 70°$. Find each measure.

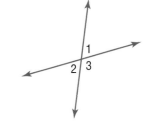

② $m\angle 2$

$\angle 1$ and $\angle 2$ are vertical angles, so they are congruent. $m\angle 2 = m\angle 1 = 70°$

③ $m\angle 3$

$\angle 1$ and $\angle 3$ are adjacent angles, so they are supplementary.

$m\angle 1 + m\angle 3 = 180°$ Definition of supplementary angles

$70 + m\angle 3 = 180$ Replace $m\angle 1$ with 70.

$m\angle 3 = 110$ Subtract 70 from each side.

✓ CHECK Your Progress

In the figure at the right, $m\angle 1 = 52°$. Find each measure.

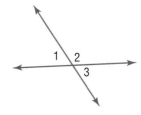

c. $m\angle 2$

d. $m\angle 3$

EXAMPLE Use Angle Relationships

④ **ALGEBRA** In the figure at the right, $r \parallel s$. Find the value of x.

The angles with measures $2x°$ and $110°$ are vertical angles, so they are congruent.

$2x = 110$ Congruent angles have equal measures.

$\dfrac{2x}{2} = \dfrac{110}{2}$ Divide each side by 2.

$x = 55$ Simplify.

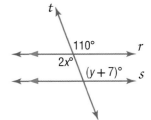

✓ CHECK Your Progress

e. Find the value of y.

CHECK Your Understanding

Examples 1–3
(pp. LA10–LA11)

In the figure at the right, $\ell \parallel m$ and k is a transversal. If $m\angle 1 = 56°$, find each measure.

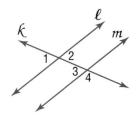

1. $m\angle 2$ 2. $m\angle 3$ 3. $m\angle 4$

In the figure at the right, $p \parallel q$ and ℓ is a transversal. If $m\angle 8 = 60°$, find each measure.

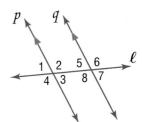

4. $m\angle 1$ 5. $m\angle 3$ 6. $m\angle 5$

Example 4
(p. LA11)

7. **SWIMMING** A swimmer crosses the lanes in a pool and swims from point A to point B, as shown in the figure. What is the value of x?

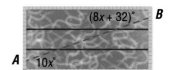

8. **ALGEBRA** Find the value of x in the figure at the right.

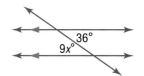

Practice and Problem Solving

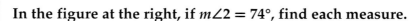

HOMEWORK HELP	
For Exercises	**See Examples**
9–18	1–3
19–26	4

In the figure at the right, if $m\angle 5 = 108°$, find each measure.

9. $m\angle 1$ 10. $m\angle 3$

11. $m\angle 6$ 12. $m\angle 7$

In the figure at the right, if $m\angle 2 = 74°$, find each measure.

13. $m\angle 8$ 14. $m\angle 6$

15. $m\angle 4$ 16. $m\angle 1$

17. **FURNITURE** In the chair at the right, $m\angle 4 = 106°$. Find $m\angle 6$ and $m\angle 3$.

18. **DRIVING** Ambulances can't safely make turns of less than 70°. The angle at the southeast corner of Delavan and Elmwood is 108°. Can an ambulance safely turn the northeast corner of Bidwell and Elmwood? Explain your reasoning.

In the figure at the right, $m\angle 7 = 96°$.
Find each measure.

19. $m\angle 2$

20. $m\angle 5$

21. $m\angle 4$

22. $m\angle 8$

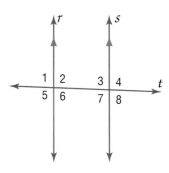

ALGEBRA Find the value of x in each figure.

23.

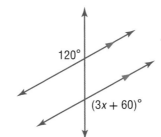

120°

$(3x + 60)°$

24.

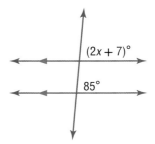

$(2x + 7)°$

85°

ALGEBRA In the figure at the right, $m \parallel \ell$ and
t is a transversal. Find the value of x for each
of the following.

25. $m\angle 2 = 2x + 3$ and $m\angle 4 = 4x - 7$

26. $m\angle 8 = 4x - 32$ and $m\angle 5 = 5x + 50$

27. **FLAGS** The flag at the right is the national
flag of Bosnia. If $m\angle 1 = 135°$, what is $m\angle 2$?
Explain how you found your answer.

28. **ALGEBRA** Find the value of x in the figure at
the right.

29. **ALGEBRA** A transversal intersects two parallel
lines and forms adjacent angles 5 and 6.
If $m\angle 5 = (7x - 11)°$ and $m\angle 6 = (3x + 1)°$, find
the measures of the angles.

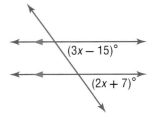

$(3x - 15)°$

$(2x + 7)°$

H.O.T. Problems

30. **CHALLENGE** Suppose two parallel lines are cut by a transversal. How are the
interior angles on the same side of the transversal related?

31. **REASONING** Determine whether the following statement is *sometimes*,
always, or *never* true. Explain your reasoning.

Vertical angles are supplementary.

32. **OPEN ENDED** Draw a pair of adjacent, supplementary angles. Label the
angle measures.

33. **WRITING IN MATH** Summarize the angle relationships that are formed by
parallel lines and a transversal. Describe which angles are congruent.

Congruent Polygons

Looking 4 Ahead

MAIN IDEA

Identify similar and congruent polygons and use congruent polygons to solve problems.

NYS Core Curriculum

7.PS.10 Use proportionality to model problems

New Vocabulary

congruent polygons
similar polygons
corresponding sides
corresponding angles

NY Math Online

glencoe.com

• Personal Tutor
• Self-Check Quiz
• Extra Examples

▷ **GET READY** for the Lesson

Photo 1

Photo 2

Photo 3

1. Name two photos that are the same shape, but not the same size.

2. Name two photos that are not the same shape.

Polygons that are the same size and shape are **congruent polygons**. When polygons have the same shape but may be different in size, they are called **similar polygons**.

EXAMPLES **Identify Congruent and Similar Polygons**

Determine whether each pair of figures appears to be *congruent*, *similar*, or *neither*.

①

The figures are the same size and the same shape. So, they are congruent.

②

The figures are not the same shape. So, they are neither similar nor congruent.

✓ **CHECK Your Progress**

a.

b.

The sides of similar or congruent figures that match are called **corresponding sides**. The angles of similar or congruent figures that match are called **corresponding angles**. Refer to the similar triangles.

• Side *AB* corresponds to side *DE*.
• Side *BC* corresponds to side *EF*.
• Side *CA* corresponds to side *FD*.
• Angles *A*, *B*, and *C* correspond to angles *D*, *E*, and *F*, respectively.

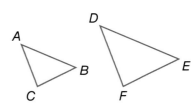

Similar and Congruent Polygons

Key Concept

- Two polygons are *similar* if the corresponding angles are congruent and the ratios of corresponding sides are proportional.
- Two polygons are *congruent* if the corresponding angles are congruent and the corresponding sides are congruent.

EXAMPLES Apply Similarity and Congruence

Determine whether each pair of figures is *congruent*, *similar*, or *neither*.

③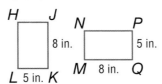

The corresponding sides are congruent. So, the figures are congruent.

④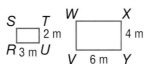

$\dfrac{TU}{XY} = \dfrac{2}{4}$ or $\dfrac{1}{2}$

$\dfrac{RU}{VY} = \dfrac{3}{6}$ or $\dfrac{1}{2}$

Since the ratios of the corresponding sides are equal, the figures are similar.

✓ **CHECK Your Progress**

c. Determine whether the pair of figures at the right is *congruent*, *similar*, or *neither*.

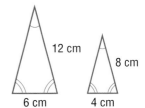

Real-World EXAMPLE **Use Congruent Polygons to Solve Problems**

⑤ **GAMES** The table tennis tables at the right are congruent rectangles. What is the area of table A?

Since the rectangles are congruent, the corresponding sides are congruent. So, $x = 9$. The area of table A is $5 \cdot 9$ or 45 square feet.

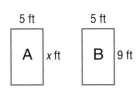

✓ **CHECK Your Progress**

d. Triangle *A* has sides 2 centimeters, 3 centimeters, and 6 centimeters. Triangle *B* has sides 3 centimeters, *x* centimeters, and *y* centimeters. If the triangles are congruent, what is the perimeter of Triangle *B*?

Examples 1, 2
(p. LA14)

Determine whether each pair of figures appears to be *congruent*, *similar*, or *neither*.

1.

2.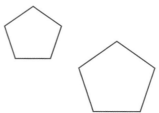

Examples 3, 4
(p. LA15)

3. SCREENS Determine whether the television screens shown below are *congruent*, *similar*, or *neither*.

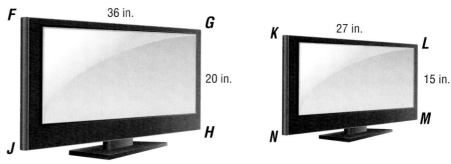

Determine whether each pair of figures is *congruent*, *similar*, or *neither*.

4.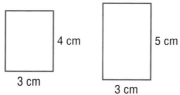

5.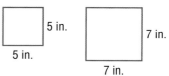

Example 5
(p. LA15)

6. Triangle *ABC* is congruent to triangle *RST*. What is the perimeter of triangle *ABC*?

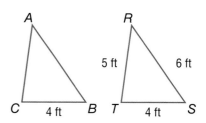

HOMEWORK HELP	
For Exercises	**See Examples**
7–10	1, 2
11–14	3, 4
15, 16	5

Determine whether each pair of figures is *congruent*, *similar*, or *neither*.

7.

8.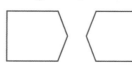

9.

10.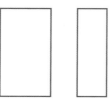

Determine whether each pair of figures is *congruent, similar,* or *neither.*

11.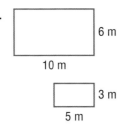

6 m

10 m

3 m

5 m

12.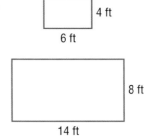

4 ft

6 ft

8 ft

14 ft

13.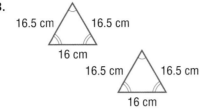

16.5 cm   16.5 cm

16 cm

16.5 cm   16.5 cm

16 cm

14.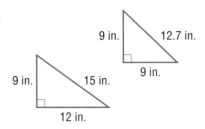

9 in.   12.7 in.

9 in.

9 in.   15 in.

12 in.

15. **MEASUREMENT** Two squares are congruent. If one square has an area of 100 square inches, what is the perimeter of the other square?

16. **MEASUREMENT** Two equilateral triangles are congruent. If one triangle has a perimeter of 24 centimeters, what are the side lengths of the other triangle?

17. **SIGNS** One stop sign has a perimeter of 48 inches. Another stop sign has a perimeter of 72 inches. Determine whether the stop signs are *congruent, similar,* or *neither*. Explain.

Each pair of figures is similar. Find the value of *x*.

18.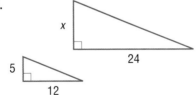

x

5

24

12

19.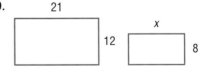

21

12

x

8

20. **CHALLENGE** Analyze the figure to determine the number of congruent triangles and the number of similar triangles there are in the figure. (It may be helpful to copy the figure and outline the triangles.)

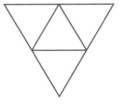

REASONING Tell whether each statement is *sometimes, always,* or *never* true. Explain your reasoning.

21. Two rectangles are similar.

22. Two squares are similar.

23. **OPEN ENDED** Draw two similar triangles and label the vertices. Then write a proportion that compares the corresponding sides.

24. **WRITING IN MATH** Compare and contrast similarity and congruence.

Converse of the Pythagorean Theorem

Looking Ahead 5

MAIN IDEA

Use the converse of the Pythagorean Theorem to determine whether a triangle is a right triangle.

NYS Core Curriculum

7.G.9 Determine whether a given triangle is a right triangle by applying the Pythagorean Theorem and using a calculator

New Vocabulary

converse

NY Math Online

glencoe.com
• Personal Tutor
• Self-Check Quiz
• Extra Examples

▶ **GET READY** for the Lesson

A triangular flower bed is 4 feet by 6 feet by 10 feet.

1. If a right triangle has legs 4 feet and 6 feet, what is the measure of the hypotenuse?

2. Is the flower bed described above a right triangle? Explain.

According to the Pythagorean Theorem, if a triangle is a right triangle, the square of the length of the hypotenuse is equal to the sum of the squares of the lengths of the legs. If you reverse the statements after *if* and *then*, you have formed the **converse** of the Pythagorean Theorem.

Converse of Pythagorean Theorem	Key Concept

Words	If the square of the length of the hypotenuse is equal to the sum of the squares of the lengths of the legs, then a triangle is a right triangle.
Symbols	$c^2 = a^2 + b^2$ **Model**
Example	$5^2 = 3^2 + 4^2$
	$25 = 9 + 16$

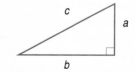

EXAMPLES Identify a Right Triangle

The measures of three sides of a triangle are given. Determine whether each triangle is a right triangle.

1 9 cm, 12 cm, 15 cm

$c^2 = a^2 + b^2$	Pythagorean Theorem
$15^2 \stackrel{?}{=} 9^2 + 12^2$	$c = 15, a = 9, b = 12$
$225 \stackrel{?}{=} 81 + 144$	Evaluate.
$225 = 225$ ✓	Simplify.

The triangle is a right triangle.

2 3 in., 4 in., 6 in.

$c^2 = a^2 + b^2$	Pythagorean Theorem
$6^2 \stackrel{?}{=} 3^2 + 4^2$	$c = 6, a = 3, b = 4$
$36 \stackrel{?}{=} 9 + 16$	Evaluate.
$36 \neq 25$	Simplify.

The triangle is *not* right.

 CHECK Your Progress

a. 4 cm, 9 cm, 10 cm   **b.** 15 mm, 20 mm, 25 mm

Study Tip

Figures
Do not assume that if a triangle looks like a right triangle that it actually is a right triangle.

Real-World Link · · · ·
An *archaeological park* is an archaeological site that has been preserved and is open to the public. Elden Pueblo in Flagstaff, Arizona, is the site of an ancient village that was inhabited from about 1070 to 1275 A.D.

EXAMPLE Identify a Right Triangle

③ **Determine whether the triangle at the right is a right triangle.**

$$c^2 = a^2 + b^2 \qquad \text{Pythagorean Theorem}$$
$$12^2 \stackrel{?}{=} 8^2 + 9^2 \qquad c = 12, a = 8, b = 9$$
$$144 \stackrel{?}{=} 64 + 81 \qquad \text{Evaluate.}$$
$$144 \neq 145 \qquad \text{The triangle is } \textit{not} \text{ a right triangle.}$$

8 ft   12 ft
9 ft

✓ CHECK Your Progress

c. Determine whether the triangle at the right is a right triangle.

17 m
8 m
15 m

Real-World EXAMPLE Use the Converse

④ **ARCHAEOLOGY** An archaeological dig site is shown at the right. Is the site a rectangle?

A rectangle has four right angles. Determine whether the triangle formed is a right triangle.

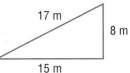

35 ft   21 ft   28 ft

$$c^2 = a^2 + b^2 \qquad \text{Pythagorean Theorem}$$
$$35^2 \stackrel{?}{=} 21^2 + 28^2 \qquad c = 35, a = 21, b = 28$$
$$1225 \stackrel{?}{=} 441 + 784 \qquad \text{Evaluate. Use a calculator.}$$

$1225 = 1225$ The triangle *is* a right triangle. So, the area is a rectangle.

✓ CHECK Your Progress

d. Another dig site is a quadrilateral that measures 12 yards on each side. If the diagonal is 15 yards, is this a square?

✓ CHECK Your Understanding

Examples 1, 2
(p. LA18)

The lengths of three sides of a triangle are given. Determine whether each triangle is a right triangle.

1. $a = 12$ cm, $b = 16$ cm, $c = 20$ cm   **2.** $a = 2$ in., $b = 3$ in., $c = 4$ in.

Example 3
(p. LA19)

Determine whether each triangle is a right triangle.

3.

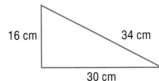

16 cm   34 cm
30 cm

4.

6 mm
4 mm
5 mm

Example 4
(p. LA19)

5. ART A mosaic is made from triangles that have side lengths of 14 millimeters, 18 millimeters, and 20 millimeters. Are the triangles right?

Practice and Problem Solving

HOMEWORK HELP

For Exercises	See Examples
6–13	1, 2
14–17	3
18–19	4

The lengths of three sides of a triangle are given. Determine whether each triangle is a right triangle.

6. $a = 9$ m, $b = 40$ m, $c = 41$ m

7. $a = 12$ km, $b = 35$ km, $c = 37$ km

8. $a = 6$ yd, $b = 11$ yd, $c = 13$ yd

9. $a = 18$ mm, $b = 24$ mm, $c = 30$ mm

10. $a = 5$ cm, $b = 9$ cm, $c = 12$ cm

11. $a = 14$ in., $b = 18$ in., $c = 22$ in.

12. $a = 16$ ft, $b = 30$ ft, $c = 34$ ft

13. $a = 10$ m, $b = 16$ m, $c = 24$ m

Determine whether each triangle is a right triangle.

14.
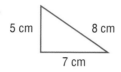
5 cm, 8 cm, 7 cm

15.
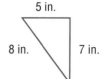
5 in., 8 in., 7 in.

16.
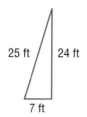
25 ft, 24 ft, 7 ft

17.
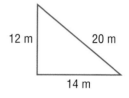
12 m, 20 m, 14 m

18. **GEOMETRY** The circle at the right has a diameter of 51 units. Is the triangle with vertices on the circle a right triangle?

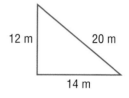
45, 24

19. **TRAVEL** Smithville is 28 miles from Littlefield and 26 miles from Alpine. The distance between Littlefield and Alpine is 32 miles. Do the segments connecting the three cities on the map form a right triangle?

20. **SEWING** A quilt is made from triangular pieces of cloth that have side lengths of 4 inches, 4 inches, and 4 inches. Are the pieces of cloth right triangles?

The lengths of three sides of a triangle are given. Determine whether each triangle is a right triangle.

21. $a = 5$ m, $b = 8$ m, $c = \sqrt{89}$ m

22. $a = 6$ in., $b = \sqrt{76}$ in., $c = \sqrt{112}$ in.

H.O.T. Problems

23. **CHALLENGE** A triangle has side lengths of x units. Is it possible for the triangle to be a right triangle? Explain.

24. **REASONING** Can you determine whether a triangle is a right triangle if you are given two side lengths? Explain.

25. **OPEN ENDED** Choose three numbers. Use the converse of the Pythagorean Theorem to determine whether the numbers could be measures of the sides of a right triangle.

26. **WRITING IN MATH** Explain how to use the converse of the Pythagorean Theorem to determine whether a triangle with three given measures is a right triangle.

Box-and-Whisker Plots

Looking 6 Ahead

MAIN IDEA

Display and interpret data in box-and-whisker plots.

NYS Core Curriculum

7.S.6 Read and interpret data represented graphically (pictograph, bar graph, histogram, line graph, double line/bar graphs or circle graph)

New Vocabulary

quartiles
lower quartile
upper quartile
interquartile range
box-and-whisker plot

NY Math Online

glencoe.com
• Personal Tutor
• Self-Check Quiz
• Extra Examples

▷ **GET READY** for the Lesson

The lengths in millimeters of 10 ghost-faced bats are shown below.

62, 58, 65, 61, 61, 62, 62, 59, 63, 65

1. Order the numbers from least to greatest.

2. Divide the ordered list of numbers in half. What is the median of the lower half of the data?

In a set of data, the **quartiles** are the values that divide the data into four equal parts. The median of the lower half of a set of data is called the **lower quartile** or LQ. The median of the upper half of the data is called the **upper quartile** or UQ.

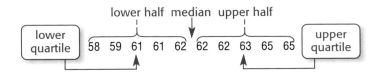

The **interquartile range** is the difference between the upper and lower quartiles. It is the range of the middle half of the data.

EXAMPLE Interquartile Range

① **Find the median, upper and lower quartiles, and the interquartile range for the set of data {8, 11, 6, 23, 7, 2, 20, 4, 16}.**

STEP 1 List the data from least to greatest. Then find the median.

$$2 \quad 4 \quad 6 \quad 7 \quad 8 \quad 11 \quad 16 \quad 20 \quad 23$$
$$\text{median} = 8$$

STEP 2 Find the upper and lower quartiles.

lower half   upper half

$$2 \quad 4 \quad 6 \quad 7 \quad 8 \quad 11 \quad 16 \quad 20 \quad 23$$

$LQ = \frac{4+6}{2}$ or 5   median = 8 $UQ = \frac{16+20}{2}$ or 18

STEP 3 Find the interquartile range. UQ − LQ = 18 − 5 or 13

Study Tip

Common Misconception
You may think that the median always divides the box in half. However, the median may not divide the box in half because the data may be clustered toward one quartile.

☑ **CHECK** Your Progress

a. Find the median, upper and lower quartiles, and the interquartile range for the set of data {35, 27, 21, 37, 54, 47}.

A **box-and-whisker plot** divides a set of data into four parts using the median and quartiles. A *box* is drawn around the quartile values, and *whiskers* extend from each quartile to the extreme data points.

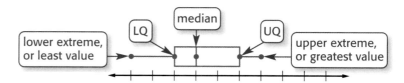

Real-World EXAMPLE Draw a Box-and-Whisker Plot

2 **DRIVING** The list below shows the speeds of eleven cars. Draw a box-and-whisker plot of the data.

25 35 27 22 34 40 20 19 23 25 30

STEP 1 Order the numbers from least to greatest. Then draw a number line that covers the range of the data.

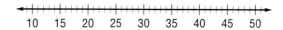

STEP 2 Find the median, the extremes, and the upper and lower quartiles. Mark these points above the number line.

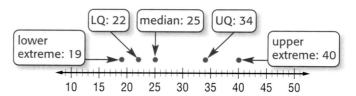

STEP 3 Draw the box so that it includes the quartile values. Draw a vertical line through the median value. Extend the whiskers from each quartile to the extreme data points.

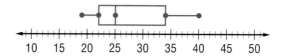

☑ **CHECK** Your Progress

b. Draw a box-and-whisker plot of the data set below.
{$25, $30, $27, $35, $19, $23, $25, $22, $40, $34, $20}

24. **WORD PROCESSING** Find the median, upper and lower quartiles, and the interquartile range for the set of data in the stem-and-leaf plot shown at the right.

Words Typed per Minute

4	0 2
5	1 5 9
6	3 5 7 8
7	2 3 7 8
8	0 1

$4|0 = 40$

TEMPERATURE For Exercises 25–27, use the table. It shows the average monthly temperatures for two cities.

Average Monthly Temperatures (°F)

	J	F	M	A	M	J	J	A	S	O	N	D
Tampa, FL	60	62	67	71	77	81	82	82	81	75	68	62
Caribou, ME	9	12	25	38	51	61	66	63	54	43	31	15

25. Find the low, high, and the median temperatures, and the upper and lower quartiles for each city.

26. On the same number line, draw a box-and-whisker plot for each set of data. Place the Caribou data above the Tampa data.

27. Write a few sentences comparing the average monthly temperatures displayed in the box-and-whisker plots.

ANIMALS For Exercises 28 and 29, use the box-and-whisker plots shown below. They summarizes the weights of Asiatic black bears and pandas.

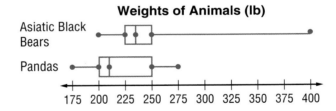

Weights of Animals (lb)

28. How do the weights of the two types of animals compare? Discuss how much most of the animals of each type weigh and the greatest weight of each.

29. Write a sentence or two describing what the length of the box-and-whisker plot tells about the weights of the animals.

H.O.T. Problems

30. **CHALLENGE** Write a set of data that contains 12 values for which the box-and-whisker plot has no whiskers. State the median, lower and upper quartiles, and lower and upper extremes.

31. **REASONING** Determine whether the following statement is *true* or *false*. Explain.

    The median divides the box of a box-and-whisker plot in half.

32. **OPEN ENDED** Write a set of data that, when displayed in a box-and-whisker plot, will result in a long box and short whiskers. Draw the box-and-whisker plot.

33. **WRITING IN MATH** Explain how the information you can learn from a set of data shown in a box-and-whisker plot is different from what you can learn from the same set of data shown in a stem-and-leaf plot.

Student Handbook

Built-In Workbooks

Reference

666

How to Use the Student Handbook

The Student Handbook is the additional skill and reference material found at the end of books. The Student Handbook can help answer these questions.

What If I Need More Practice?
You, or your teacher, may decide that working through some additional problems would be helpful. The **Extra Practice** section provides these problems for each lesson so you have ample opportunity to practice new skills.

What If I Have Trouble with Word Problems?
The **Mixed Problem Solving** portion of the book provides additional word problems that use the skills presented in each chapter. These problems give you real-world situations where the math can be applied.

What If I Need to Prepare for a Standardized Test?
The **Preparing for Standardized Tests** section provides worked-out examples and practice problems for multiple-choice, gridded response, short-response, and extended-response questions.

What If I Forget What I Learned Last Year?
Use the **Concepts and Skills** section to refresh your memory about topics you have learned in other math classes or to prepare for next year.

What If I Forget a Vocabulary Word?
The **English-Spanish Glossary** provides a list of important, or difficult, words used throughout the textbook. It provides a definition in English and Spanish as well as the page number(s) where the word can be found.

What If I Need to Check a Homework Answer?
The answers to the odd-numbered problems are included in **Selected Answers**. Check your answers to make sure you understand how to solve all of the assigned problems.

What If I Need to Find Something Quickly?
The **Index** alphabetically lists the subjects covered throughout the entire textbook and the pages on which each subject can be found.

What If I Forget a Formula?
Inside the back cover of your math book is a list of **Formulas and Symbols** that are used in the book.

Extra Practice

Lesson 1-1

Pages 25–29

Use the four-step plan to solve each problem.

1. The Reyes family rode their bicycles for 9 miles to the park. The ride back was along a different route for 14 miles. How many miles did they ride in all?

2. Four hundred sixty people are scheduled to attend a banquet. If each table seats 8 people, how many tables are needed?

3. A group of 251 people is eating dinner at a school fund-raiser. If each person pays $8.00 for their meal, how much money is raised?

4. Sherita's service charges a monthly fee of $20.00 plus $0.15 per minute. One monthly bill is $31.25. How many minutes did Sherita use during the month?

5. ABC Car Rental charges $25 per day to rent a mid-sized car plus $0.20 per mile driven. Mr. Ruiz rents a mid-sized car for 3 days and drives a total of 72 miles. Find the amount of Mr. Ruiz's bill.

Lesson 1-2

Pages 30–33

Write each power as a product of the same factor.

1. 13^4
2. 9^6
3. 1^7
4. 12^2
5. 5^8
6. 15^4

Evaluate each expression.

7. 5^6
8. 17^3
9. 2^{12}
10. 3^5
11. 1^4
12. 5^3
13. 10^2
14. 2^8
15. 8^2
16. 7^4
17. 20^3
18. 42^3

Write each product in exponential form.

19. $2 \cdot 2 \cdot 2 \cdot 2 \cdot 2$
20. $3 \cdot 3$
21. $1 \cdot 1 \cdot 1 \cdot 1 \cdot 1 \cdot 1$
22. $18 \cdot 18 \cdot 18 \cdot 18$
23. $9 \cdot 9 \cdot 9 \cdot 9 \cdot 9 \cdot 9 \cdot 9 \cdot 9$
24. $10 \cdot 10 \cdot 10 \cdot 10 \cdot 10 \cdot 10$

Lesson 1-3

Pages 34–37

Find the square of each number.

1. 4
2. 19
3. 13
4. 25
5. 9
6. 2
7. 14
8. 24
9. 40
10. 50
11. 100
12. 250

Find each square root.

13. $\sqrt{324}$
14. $\sqrt{900}$
15. $\sqrt{2,500}$
16. $\sqrt{576}$
17. $\sqrt{8,100}$
18. $\sqrt{676}$
19. $\sqrt{100}$
20. $\sqrt{784}$
21. $\sqrt{1,024}$
22. $\sqrt{841}$
23. $\sqrt{2,304}$
24. $\sqrt{3,025}$

Evaluate each expression.

1. $14 - (5 + 7)$
2. $(32 + 10) - 5 \times 6$
3. $(50 - 6) + (12 + 4)$
4. $12 - 2 \cdot 3$
5. $16 + 4 \times 5$
6. $(5 + 3) \times 4 - 7$
7. $2 \times 3 + 9 \times 2$
8. $6 \cdot (8 + 4) \div 2$
9. $7 \times 6 - 14$
10. $8 + (12 \times 4) \div 8$
11. $13 - 6 \cdot 2 + 1$
12. $(80 \div 10) \times 8$
13. $14 - 2 \cdot 7 + 0$
14. $156 - 6 \times 0$
15. $30 - 14 \cdot 2 + 8$
16. $3 \times 4 - 3^2$
17. $10^2 - 5$
18. $3 + (10 - 5 + 1)^2$
19. $(4 + 3)^2 \div 7$
20. 8×10^3
21. $10^4 \times 6$
22. 4.5×10^3
23. 1.8×10^2
24. $3 + 5(1.7 + 2.3)$
25. $4(3.6 + 5.4) - 9$
26. $10 + 3(6.1 + 3.7)$
27. $6(7.5 + 2.1) - 2.3$

Use the *guess and check* strategy to solve each problem.

1. **NUMBERS** A number is divided by 3. Then 8 is added to the quotient. The result is 15. What is the number?

2. **NUMBERS** Benny is thinking of two numbers. Their product is 32 and their difference is 4. Find the numbers.

3. **MONEY** A theater is charging $5 for children under 12 and $8 for everyone else. If the total for a group of people was $36, how many people under the age of 12 were in the group?

4. **PLACE VALUE** Mindy wrote down a decimal number. The digit in the tenth's place is half the digit in the hundredth's place. If the product of the two digits is 18, what is the number?

5. **MONEY** Penny has 14 coins totaling $1.55. She has one more nickel than she has dimes, and three less quarters than nickels. How many quarters, dimes, and nickels does she have if these are the only coin types she has?

6. **SOUVENIRS** A souvenir shop sells standard-sized postcards in packages of 5 and large-sized postcards in packages of 3. If Juan bought 16 postcards, how many packages of each did he buy?

Evaluate each expression if $a = 3$, $b = 4$, $c = 12$, and $d = 1$.

1. $a + b$
2. $c - d$
3. $a + b + c$
4. $b - a$
5. $c - ab$
6. $a + 2d$
7. $b + 2c$
8. ab
9. $a + 3b$
10. $6a + c$
11. $\dfrac{c}{d}$
12. abc
13. $2(a + b)$
14. $\dfrac{2c}{b}$
15. $144 - abc$
16. $2ab$
17. $\dfrac{b}{2}$
18. a^2
19. $c^2 - 100$
20. $a^3 + 3$
21. $2b^2$
22. $b^3 + c$
23. $\dfrac{a^2}{d}$
24. $5a^2 + 2d^2$
25. $\dfrac{4d^2}{b}$
26. $\dfrac{15}{a}$
27. $3a^2$
28. $10d^3$
29. $\dfrac{ab}{c}$
30. $\dfrac{(a + b)}{d}$
31. $2.5b + c$
32. $\dfrac{10}{d}$
33. $\dfrac{(2c + b)}{b}$
34. $\dfrac{(b^2 + 2d)}{a}$
35. $\dfrac{(2c + ab)}{c}$
36. $\dfrac{(3.5c + 2)}{11}$

Lesson 1-7

Pages 49–52

Solve each equation mentally.

1. $b + 7 = 12$
2. $a + 3 = 15$
3. $s + 10 = 23$
4. $9 + n = 13$
5. $20 = 24 - n$
6. $4x = 36$
7. $2y = 10$
8. $15 = 5h$
9. $j \div 3 = 2$
10. $14 = w - 4$
11. $24 \div k = 6$
12. $b - 3 = 12$
13. $c \div 10 = 8$
14. $6 = t \div 5$
15. $14 + m = 24$
16. $3y = 39$
17. $\frac{f}{2} = 12$
18. $16 = 4v$
19. $81 = 80 + a$
20. $9 = \frac{72}{x}$
21. $66 = 22m$
22. $77 - 12 = a$
23. $9k = 81$
24. $95 + d = 100$
25. $b = \frac{72}{6}$
26. $z = 15 + 22$
27. $15b = 225$
28. $43 + s = 57$
29. $4w = 52$
30. $e - 10 = 0$
31. $62 - d = 12$
32. $14f = 14$
33. $48 \div n = 8$
34. $a - 82 = 95$
35. $\frac{x}{2} = 36$
36. $99 = c \div 2$

Lesson 1-8

Pages 53–56

Use the Distributive Property to evaluate each expression.

1. $3(4 + 5)$
2. $(2 + 8)6$
3. $4(9 - 6)$
4. $8(6 - 3)$
5. $5(200 - 50)$
6. $20(3 + 6)$
7. $(20 - 5)8$
8. $50(8 + 2)$
9. $15(1,000 - 200)$
10. $3(2,000 + 400)$
11. $12(1,000 + 10)$
12. $7(1,000 - 50)$

Find each expression mentally. Justify each step.

13. $(5 + 17) + 25$
14. $13 + (22 + 17)$
15. $(8 + 18) + 92$
16. $(11 + 32) + 9$
17. $4 + (15 + 76)$
18. $(25 + 56) + 75$
19. $(4 \cdot 21) \cdot 25$
20. $5 \cdot (40 \cdot 8)$
21. $(2 \cdot 38) \cdot 50$
22. $(12 \cdot 7) \cdot 5$
23. $25 \cdot (12 \cdot 4)$
24. $(15 \cdot 9) \cdot 2$

Lesson 1-9

Pages 57–61

Describe the relationship between the terms in each arithmetic sequence. Then write the next three terms in each sequence.

1. $5, 9, 13, 17, \ldots$
2. $3, 5, 7, 9, \ldots$
3. $10, 15, 20, 25, \ldots$
4. $90, 93, 96, 99, \ldots$
5. $8, 14, 20, 26, \ldots$
6. $4.5, 5.4, 6.3, 7.2, \ldots$
7. $0.3, 0.4, 0.5, \ldots$
8. $2.3, 3.4, 4.5, 5.6, \ldots$
9. $8.9, 9.1, 9.3, 9.5, \ldots$
10. $3, 11, 19, 27, \ldots$
11. $350, 375, 400, 425, \ldots$
12. $620, 635, 650, 665, \ldots$
13. $2, 7, 12, 17, \ldots$
14. $10, 17, 24, 31, \ldots$
15. $0, 7, 14, 21, \ldots$
16. $1, 7, 13, 19, \ldots$
17. $95, 101, 107, 113, \ldots$
18. $9, 90, 171, 252, \ldots$
19. $2.6, 2.8, 3.0, 3.2, \ldots$
20. $4.1, 4.6, 5.1, 5.6, \ldots$
21. $6.6, 7.7, 8.8, 9.9, \ldots$
22. $19.5, 21, 22.5, 24, \ldots$
23. $14.5, 14.8, 15.1, 15.4, \ldots$
24. $0.1, 0.4, 0.7, 1.0, \ldots$

Copy and complete each function table. Then identify the domain and range.

1.

x	2x	y
0		
1		
2		
3		

2.

x	3x + 1	y
1		
2		
3		
4		

3.

x	x − 2	y
3		
4		
5		
6		

4.

x	x + 0.1	y
2		
3		
4		
5		

Write an integer for each situation.

1. seven degrees below zero 2. a loss of 3 pounds 3. a loss of 20 yards
4. a profit of $25 5. 112°F above 0 6. 2,830 feet above sea level

Graph each set of integers on a number line.

7. $\{-2, 0, 2\}$ 8. $\{1, 3, 5\}$ 9. $\{-2, -5, 3\}$ 10. $\{7, -1, 4\}$

Evaluate each expression.

11. $|1|$ 12. $|-8|$ 13. $|0|$ 14. $|-82|$
15. $|64|$ 16. $|-128|$ 17. $|-22| + 5$ 18. $|-40| - 8$
19. $|-18| + |10|$ 20. $|-7| + |-1|$ 21. $|98| - |-5|$ 22. $|-49| - |-10|$

Replace each ● with < or > to make a true sentence.

1. $7 ● -7$ 2. $-8 ● 4$ 3. $-4 ● -9$ 4. $-3 ● 0$
5. $8 ● 10$ 6. $-5 ● -4$ 7. $6 ● -7$ 8. $-12 ● -13$
9. $3 ● 1$ 10. $-2 ● 2$ 11. $7 ● -1$ 12. $-15 ● -20$
13. $-40 ● 30$ 14. $0 ● -3$ 15. $-5 ● 0$ 16. $85 ● -17$

Order the integers from least to greatest.

17. $-2, -8, 4, 10, -6, -12$ 18. $19, -19, -21, 32, -14, 18$
19. $18, 23, 95, -95, -18, -23, 2$ 20. $46, -48, -47, -52, -18, 12$
21. $0, -10, -6, -8, 12$ 22. $-15, 18, -1, 0, 14, -20$

Lesson 2-3

Pages 88–92

Write the ordered pair for each point graphed at the right.
Then name the quadrant or axis on which each point
is located.

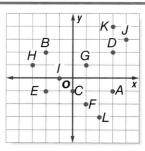

1. A
2. B
3. C
4. D
5. E
6. F
7. G
8. H
9. I
10. J
11. K
12. L

On graph paper, draw a coordinate plane. Then graph and label each point.

13. $N(-4, 3)$
14. $K(2, 5)$
15. $W(-6, -2)$
16. $X(5, 0)$
17. $Y(4, -4)$
18. $M(0, -3)$
19. $Z(-2, 0.5)$
20. $S(-1, -3)$
21. $A(0, 2)$
22. $C(-2, -2)$
23. $E(0, 1)$
24. $G(1, -1)$

Lesson 2-4

Pages 95–99

Add.

1. $-4 + 8$
2. $14 + 16$
3. $-7 + (-7)$
4. $-9 + (-6)$
5. $-18 + 11$
6. $-36 + 40$
7. $42 + (-18)$
8. $-42 + 29$
9. $18 + (-32)$
10. $12 + (-9)$
11. $-24 + 9$
12. $-7 + (-1)$

Evaluate each expression if $a = 6$, $b = -2$, $c = -6$, and $d = 3$.

13. $-96 + a$
14. $b + (-5)$
15. $c + (-32)$
16. $d + 98$
17. $-120 + b$
18. $-120 + c$
19. $5 + b$
20. $a + d$
21. $c + a$
22. $d + (-9)$
23. $b + c$
24. $d + c$

Lesson 2-5

Pages 103–106

Subtract.

1. $3 - 7$
2. $-5 - 4$
3. $-6 - 2$
4. $8 - 13$
5. $6 - (-4)$
6. $12 - 9$
7. $-2 - 23$
8. $63 - 78$
9. $0 - (-14)$
10. $15 - 6$
11. $18 - 20$
12. $-5 - 8$
13. $21 - (-37)$
14. $-60 - 32$
15. $57 - 63$

Evaluate each expression if $k = -3$, $p = 6$, $n = 1$, and $d = -8$.

16. $55 - k$
17. $p - 7$
18. $d - 15$
19. $n - 12$
20. $-51 - d$
21. $k - 21$
22. $n - k$
23. $-99 - k$
24. $p - k$
25. $d - (-1)$
26. $k - d$
27. $n - d$

Lesson 2-6

Pages 107–111

Multiply.

1. $5(-2)$
2. $6(-4)$
3. $4(21)$
4. $-11(-5)$
5. $-6(5)$
6. $-50(0)$
7. $-5(-5)$
8. $-4(8)$
9. $3(-13)$
10. $12(-5)$
11. $-9(-12)$
12. $15(-8)$
13. $(-6)^2$
14. $(-2)^2$
15. $(-4)^3$
16. $(-5)^3$

Evaluate each expression if $a = -5$, $b = 2$, $c = -3$, and $d = 4$.

17. $-2d$
18. $6a$
19. $3ab$
20. $-12d$
21. $-4b^2$
22. $-5cd$
23. a^2
24. $13ab$

Lesson 2-7

Pages 112–113

Solve using the *look for a pattern* strategy.

1. **NUMBERS** Determine the next three numbers in the pattern below.

   15, 21, 27, 33, 39, …

2. **TIME** Determine the next two times in the pattern below.

   2:30 A.M., 2:50 A.M., 3:10 A.M., 3:30 A.M., …

3. **MONEY** The table shows Abigail's savings. If the pattern continues, what will be the total amount in week 6?

Week	Total ($)
1	$400
2	$800
3	$1,200
4	$1,600
5	$2,000
6	▨

4. **SCIENCE** A single rotation of Earth takes about 24 hours. Copy and complete the table to determine the number of hours in a week.

Number of Days	Number of Hours
1	24
2	48
3	72
4	▨
5	▨
6	▨
7	▨

Lesson 2-8

Pages 114–118

Divide.

1. $4 \div (-2)$
2. $16 \div (-8)$
3. $-14 \div (-2)$
4. $\frac{32}{8}$
5. $18 \div (-3)$
6. $-18 \div 3$
7. $8 \div (-8)$
8. $0 \div (-1)$
9. $-25 \div 5$
10. $\frac{-14}{-7}$
11. $-32 \div 8$
12. $-56 \div (-8)$
13. $-81 \div 9$
14. $-42 \div (-7)$
15. $121 \div (-11)$
16. $-81 \div (-9)$
17. $18 \div (-2)$
18. $\frac{-55}{11}$
19. $\frac{25}{-5}$
20. $-21 \div 3$

Evaluate each expression if $a = -2$, $b = -7$, $x = 8$, and $y = -4$.

21. $-64 \div x$
22. $\frac{16}{y}$
23. $x \div 2$
24. $\frac{a}{2}$
25. $ax \div y$
26. $\frac{bx}{y}$
27. $2y \div 1$
28. $\frac{x}{ay}$
29. $-y \div a$
30. $x^2 \div y$
31. $\frac{ab}{1}$
32. $\frac{xy}{a}$

Write each phrase as an algebraic expression.

1. six less than p
2. twenty more than c
3. the quotient of a and b
4. Juana's age plus 6
5. x increased by twelve
6. $1{,}000 divided by z
7. 3 divided into y
8. the product of 7 and m
9. the difference of f and 9
10. twenty-six less q
11. 19 decreased by z
12. two less than x

Write each sentence as an algebraic equation.

13. Three times a number less four is 17.
14. The sum of a number and 6 is 5.
15. Twenty more than twice a number is -30.
16. The quotient of a number and -2 is -42.
17. Four plus three times a number is 18.
18. Five times a number minus 15 is 92.
19. Eight times a number plus twelve is 36.
20. The difference of a number and 24 is -30.

Lesson 3-2

Solve each equation. Check your solution.

1. $r - 3 = 14$
2. $t + 3 = 21$
3. $s + 10 = 23$
4. $7 + a = -10$
5. $14 + m = 24$
6. $-9 + n = 13$
7. $s - 2 = -6$
8. $6 + f = 71$
9. $x + 27 = 30$
10. $a - 7 = 23$
11. $-4 + b = -5$
12. $w + 18 = -4$
13. $k - 9 = -3$
14. $j + 12 = 11$
15. $-42 + v = -42$
16. $s + 1.3 = 18$
17. $x + 7.4 = 23.5$
18. $p + 3.1 = 18$
19. $w - 3.7 = 4.63$
20. $m - 4.8 = 7.4$
21. $x - 1.3 = 12$
22. $y + 3.4 = 18$
23. $7.2 + g = 9.1$
24. $z - 12.1 = 14$
25. $v - 18 = 13.7$
26. $w - 0.1 = 0.32$
27. $r + 6.7 = 1.2$

Lesson 3-3

Solve each equation. Check your solution.

1. $2m = 18$
2. $-42 = 6n$
3. $72 = 8k$
4. $-20r = 20$
5. $420 = 5s$
6. $325 = 25t$
7. $-14 = -2p$
8. $18q = 36$
9. $40 = 10a$
10. $100 = 20b$
11. $416 = 4c$
12. $45 = 9d$
13. $0.5m = 3.5$
14. $1.8 = 0.6x$
15. $0.4y = 2$
16. $1.86 = 6.2z$
17. $-8x = 24$
18. $8.34 = 2r$
19. $1.67t = 10.02$
20. $243 = 27a$
21. $0.9x = 4.5$
22. $4.08 = 1.2y$
23. $8d = 112$
24. $5f = 180.5$
25. $59.66 = 3.14m$
26. $98.4 = 8p$
27. $208 = 26k$

Lesson 4-3

Pages 190–191

Use the *make a list* strategy to solve each problem.

1. **MEASUREMENT** Gabriel has to make deliveries to three neighbors. He lives at house *b* on the map. Find the shortest route to make the deliveries and return home?

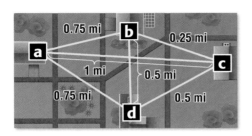

2. **FOOD** Daniel is making a peanut butter and jelly sandwich. His choices are creamy or crunchy peanut butter, white or wheat bread, and grape, apple, or strawberry jelly. How many different types of sandwiches can Daniel make?

3. **GAMES** On the game board, you plan to move two spaces away from square A. You can move horizontally, vertically, or diagonally. How many different moves can you make from square A? List them.

Lesson 4-4

Pages 192–195

Write each fraction in simplest form.

1. $\frac{14}{28}$
2. $\frac{15}{25}$
3. $\frac{100}{130}$
4. $\frac{14}{35}$
5. $\frac{9}{51}$

6. $\frac{54}{56}$
7. $\frac{75}{90}$
8. $\frac{24}{40}$
9. $\frac{180}{270}$
10. $\frac{312}{390}$

11. $\frac{240}{448}$
12. $\frac{71}{82}$
13. $\frac{333}{900}$
14. $\frac{85}{255}$
15. $\frac{84}{128}$

16. $\frac{64}{96}$
17. $\frac{99}{99}$
18. $\frac{3}{99}$
19. $\frac{44}{55}$
20. $\frac{57}{69}$

21. $\frac{15}{37}$
22. $\frac{204}{408}$
23. $\frac{5}{125}$
24. $\frac{144}{216}$
25. $\frac{15}{75}$

Lesson 4-5

Pages 196–200

Write each fraction or mixed number as a decimal. Use bar notation if the decimal is a repeating decimal.

1. $\frac{16}{20}$
2. $\frac{30}{120}$
3. $1\frac{7}{8}$
4. $\frac{1}{6}$

5. $\frac{11}{40}$
6. $5\frac{13}{50}$
7. $\frac{55}{300}$
8. $1\frac{1}{2}$

9. $\frac{5}{9}$
10. $2\frac{3}{4}$
11. $\frac{9}{11}$
12. $4\frac{1}{9}$

Write each decimal as a fraction or mixed number in simplest form.

13. 0.26
14. 0.75
15. 0.4
16. 0.1

17. 4.48
18. 9.8
19. 0.91
20. 11.15

Lesson 4-6

Pages 202–205

Write each ratio as a percent.

1. 39 out of 100
2. $\frac{23}{100}$
3. 17:100
4. 72 per 100
5. 4 to 100
6. 98 in 100

Write each fraction as a percent.

7. $\frac{1}{2}$
8. $\frac{2}{5}$
9. $\frac{60}{100}$
10. $\frac{17}{20}$
11. $\frac{7}{25}$
12. $\frac{1}{20}$
13. $\frac{8}{100}$
14. $\frac{7}{7}$
15. $\frac{9}{10}$
16. $\frac{1}{100}$
17. $\frac{50}{50}$
18. $\frac{49}{50}$

Write each percent as a fraction in simplest form.

19. 12%
20. 23%
21. 1%
22. 94%
23. 36%
24. 4%
25. 72%
26. 100%
27. 65%
28. 47%
29. 15%
30. 48%

Lesson 4-7

Pages 206–210

Write each percent as a decimal.

1. 42%
2. 100%
3. 8%
4. 20%
5. 35%
6. 3%
7. 62%
8. 50%
9. 28%
10. 87%
11. 7.5%
12. 87.5%
13. 1.8%
14. 99.9%
15. $85\frac{1}{4}\%$
16. $24\frac{1}{2}\%$
17. $64\frac{4}{5}\%$
18. $36\frac{3}{4}\%$
19. $1\frac{1}{5}\%$
20. $2\frac{1}{2}\%$

Write each decimal as a percent.

21. 0.16
22. 0.1
23. 0.5
24. 0.98
25. 0.31
26. 0.76
27. 0.07
28. 0.8
29. 0.07
30. 0.10
31. 0.90
32. 1.00
33. 0.666
34. 0.725
35. 0.138
36. 0.899
37. 0.256
38. 0.038
39. 0.0525
40. 0.017

Lesson 4-8

Pages 211–214

Find the LCM of each set of numbers.

1. 4, 9
2. 6, 16
3. 24, 36
4. 48, 84
5. 8, 9
6. 49, 56
7. 42, 66
8. 15, 39
9. 56, 64
10. 24, 42
11. 80, 250
12. 16, 24
13. 13, 14
14. 36, 48
15. 10, 100
16. 25, 200
17. 1, 2, 5
18. 2, 3, 7
19. 1, 9, 27
20. 2, 24, 36
21. 7, 21, 35
22. 12, 18, 28
23. 32, 80, 96
24. 5, 18, 45
25. 11, 22, 33
26. 35, 70, 140
27. 25, 200, 400
28. 100, 200, 300

Lesson 4-9
Pages 215–220

Replace each ● with < , >, or = to make a true sentence.

1. $-\dfrac{1}{5}$ ● $-\dfrac{3}{5}$

2. $-\dfrac{7}{8}$ ● $-\dfrac{5}{8}$

3. $-\dfrac{1}{6}$ ● $-\dfrac{5}{6}$

4. $-\dfrac{3}{4}$ ● $-\dfrac{1}{4}$

5. $-2\dfrac{1}{4}$ ● $-2\dfrac{2}{8}$

6. $-4\dfrac{3}{7}$ ● $-4\dfrac{2}{7}$

7. $-1\dfrac{4}{9}$ ● $-1\dfrac{8}{9}$

8. $-3\dfrac{4}{5}$ ● $-3\dfrac{2}{5}$

9. $\dfrac{7}{9}$ ● $\dfrac{3}{5}$

10. $\dfrac{14}{25}$ ● $\dfrac{3}{4}$

11. $\dfrac{8}{24}$ ● $\dfrac{20}{60}$

12. $\dfrac{5}{12}$ ● $\dfrac{4}{9}$

13. $\dfrac{18}{24}$ ● $\dfrac{10}{18}$

14. $\dfrac{4}{6}$ ● $\dfrac{5}{9}$

15. $\dfrac{11}{49}$ ● $\dfrac{12}{42}$

16. $\dfrac{5}{14}$ ● $\dfrac{2}{6}$

Order each set of numbers from least to greatest.

17. $70\%, 0.6, \dfrac{2}{3}$

18. $0.8, \dfrac{17}{20}, 17\%$

19. $\dfrac{61}{100}, 0.65, 61.5\%$

20. $0.\overline{42}, \dfrac{3}{7}, 42\%$

21. $2.15, 2.105, 2\dfrac{7}{50}$

22. $7\dfrac{1}{8}, 7.81, 7.18$

Lesson 5-1
Pages 230–235

Estimate.

1. $\dfrac{3}{7} + \dfrac{6}{8}$

2. $\dfrac{3}{9} + \dfrac{7}{8}$

3. $\dfrac{1}{8} + \dfrac{8}{9}$

4. $3\dfrac{1}{8} + 7\dfrac{6}{7}$

5. $4\dfrac{2}{3} + 6\dfrac{7}{8}$

6. $3\dfrac{2}{3} \times 2\dfrac{1}{3}$

7. $\dfrac{4}{5} \cdot 3$

8. $9\dfrac{7}{8} - 6\dfrac{2}{3}$

9. $\dfrac{3}{7} - \dfrac{1}{15}$

10. $\dfrac{3}{4} \cdot \dfrac{7}{8}$

11. $7\dfrac{1}{4} \div \dfrac{2}{3}$

12. $\dfrac{5}{6} \div \dfrac{2}{3}$

13. $9\dfrac{3}{5} + 3\dfrac{1}{8}$

14. $5\dfrac{1}{3} - 2\dfrac{3}{4}$

15. $13\dfrac{7}{8} - 2\dfrac{1}{3}$

16. $\dfrac{13}{15} \cdot \dfrac{3}{8}$

17. $\dfrac{1}{9} \div 2$

18. $\dfrac{5}{8} - \dfrac{1}{16}$

19. $9\dfrac{2}{3} + 4\dfrac{7}{8}$

20. $\dfrac{1}{2} \cdot 25$

21. $35\dfrac{1}{3} \div 6\dfrac{3}{4}$

Lesson 5-2
Pages 236–241

Add or subtract. Write in simplest form.

1. $\dfrac{5}{11} + \dfrac{9}{11}$

2. $\dfrac{5}{8} - \dfrac{1}{8}$

3. $\dfrac{7}{10} + \dfrac{7}{10}$

4. $\dfrac{9}{12} - \dfrac{5}{12}$

5. $\dfrac{2}{9} + \dfrac{1}{3}$

6. $\dfrac{1}{2} + \dfrac{3}{4}$

7. $\dfrac{1}{4} - \dfrac{3}{12}$

8. $\dfrac{3}{7} + \dfrac{6}{14}$

9. $\dfrac{1}{4} + \dfrac{3}{5}$

10. $\dfrac{4}{9} + \dfrac{1}{2}$

11. $\dfrac{5}{7} - \dfrac{4}{6}$

12. $\dfrac{3}{4} - \dfrac{1}{6}$

13. $\dfrac{3}{5} + \dfrac{3}{4}$

14. $\dfrac{2}{3} - \dfrac{1}{8}$

15. $\dfrac{9}{10} + \dfrac{1}{3}$

Evaluate each expression if $a = \dfrac{2}{3}$ and $b = \dfrac{7}{12}$.

16. $\dfrac{1}{5} + a$

17. $a - \dfrac{1}{2}$

18. $b + \dfrac{7}{8}$

19. $\dfrac{7}{8} - a$

20. $a + b$

21. $a - b$

Lesson 5-3

Pages 242–246

Add or subtract. Write in simplest form.

1. $2\frac{1}{3} + 1\frac{1}{3}$
2. $5\frac{2}{7} - 2\frac{3}{7}$
3. $6\frac{3}{8} + 7\frac{1}{8}$
4. $2\frac{3}{4} - 1\frac{1}{4}$
5. $5\frac{1}{2} - 3\frac{1}{4}$
6. $2\frac{2}{3} + 4\frac{1}{9}$
7. $7\frac{4}{5} + 9\frac{3}{10}$
8. $3\frac{3}{4} + 5\frac{5}{8}$
9. $10\frac{2}{3} + 5\frac{6}{7}$
10. $17\frac{2}{9} - 12\frac{1}{3}$
11. $6\frac{5}{12} + 12\frac{5}{12}$
12. $7\frac{1}{4} + 15\frac{5}{6}$
13. $6\frac{1}{8} + 4\frac{2}{3}$
14. $7 - 6\frac{4}{9}$
15. $8\frac{1}{12} + 12\frac{6}{11}$
16. $7\frac{2}{3} + 8\frac{1}{4}$
17. $12\frac{3}{11} + 14\frac{3}{13}$
18. $21\frac{1}{3} + 15\frac{3}{8}$
19. $19\frac{1}{7} + 6\frac{1}{4}$
20. $9\frac{2}{5} - 8\frac{1}{3}$
21. $18\frac{1}{4} - 3\frac{3}{8}$
22. $1\frac{1}{8} + 2\frac{1}{12}$
23. $2\frac{1}{12} - 1\frac{1}{8}$
24. $10 - \frac{2}{3}$

Lesson 5-4

Pages 247–248

Eliminate possibilities to solve each problem.

1. **MEASUREMENT** Guillermo has a 3-gallon cooler with $1\frac{3}{4}$ gallons of juice in it. If he wants the cooler full for his soccer game, how much juice should he add?

   A 4 gallons   C $1\frac{1}{4}$ gallons

   B $3\frac{1}{4}$ gallons   D $\frac{1}{4}$ gallon

2. **ELEPHANTS** An elephant in a zoo eats 58 cabbages in a week. About how many cabbages does an elephant eat in one year?

   F 7   H 1,500

   G 700   J 3,000

3. **TRAVEL** Mr. Rollins drove 780 miles on a 5-day trip. He rented a car for $23 per day plus $0.15 per mile after 500 free miles. About how much did the rental car cost?

   A $100

   B $130

   C $160

   D $180

Lesson 5-5

Pages 256–257

Multiply. Write in simplest form.

1. $\frac{2}{3} \times \frac{3}{5}$
2. $\frac{1}{6} \times \frac{2}{5}$
3. $\frac{4}{9} \times \frac{3}{7}$
4. $\frac{5}{12} \times \frac{6}{11}$
5. $\frac{3}{8} \times \frac{8}{9}$
6. $\frac{2}{5} \times \frac{5}{8}$
7. $\frac{7}{15} \times \frac{3}{21}$
8. $\frac{5}{6} \times \frac{15}{16}$
9. $\frac{2}{3} \times \frac{3}{13}$
10. $\frac{4}{9} \times \frac{1}{6}$
11. $3 \times \frac{1}{9}$
12. $5 \times \frac{6}{7}$
13. $\frac{3}{5} \times 15$
14. $3\frac{1}{2} \times 4\frac{1}{3}$
15. $\frac{4}{5} \times 2\frac{3}{4}$
16. $6\frac{1}{8} \times 5\frac{1}{7}$
17. $2\frac{2}{3} \times 2\frac{1}{4}$
18. $\frac{7}{8} \times 16$
19. $5\frac{1}{5} \times 2\frac{1}{2}$
20. $7 \times \frac{1}{14}$
21. $22 \times \frac{3}{11}$
22. $8\frac{2}{3} \times 1\frac{1}{2}$
23. $4 \times 6\frac{1}{2}$
24. $\frac{1}{2} \times 10\frac{2}{3}$
25. $\frac{2}{3} \times 21\frac{1}{3}$
26. $\frac{7}{8} \times \frac{8}{7}$
27. $21 \times \frac{1}{2}$
28. $11 \times \frac{1}{4}$

Lesson 6-5

Complete. Round to the nearest hundredth if necessary.

1. 400 mm = ▨ cm
2. 4 km = ▨ m
3. 660 cm = ▨ m
4. 0.3 km = ▨ m
5. 30 mm = ▨ cm
6. 84.5 m = ▨ km
7. ▨ m = 54 cm
8. 18 km = ▨ cm
9. ▨ mm = 45 cm
10. 4 kg = ▨ g
11. 632 mg = ▨ g
12. 4,497 g = ▨ kg
13. ▨ mg = 0.51 kg
14. 0.63 kg = ▨ g
15. ▨ kg = 563 g
16. 662 m = ▨ km
17. 5,283 mL = ▨ L
18. 0.24 cm = ▨ mm
19. 380 kL = ▨ L
20. 10.8 g = ▨ mg
21. 83,000 mL = ▨ L
22. 56 in. ≈ ▨ cm
23. 32.8 ft. ≈ ▨ m
24. 609 yd ≈ ▨ m
25. 21.78 mi ≈ ▨ km
26. 48 lb ≈ ▨ g
27. 2.3 T ≈ ▨ kg
28. 8.5 c ≈ ▨ mL
29. 33 gal ≈ ▨ L
30. 1.8 qt ≈ ▨ mL

Lesson 6-6

Determine if the quantities in each pair of ratios are proportional. Explain.

1. **MONEY** 2 coins for every 3 bills and 6 coins for every 9 bills

2. **SCALE** 3 feet for every 1 in and 15 feet for every 6 in

3. **FAMILY** 2 children for every 1 adult and 8 children for every 3 adults

Solve each proportion.

4. $\dfrac{u}{72} = \dfrac{2}{4}$
5. $\dfrac{12}{m} = \dfrac{15}{10}$
6. $\dfrac{36}{90} = \dfrac{16}{t}$
7. $\dfrac{g}{32} = \dfrac{8}{64}$
8. $\dfrac{5}{14} = \dfrac{10}{a}$
9. $\dfrac{k}{18} = \dfrac{5}{3}$
10. $\dfrac{15}{w} = \dfrac{60}{4}$
11. $\dfrac{81}{90} = \dfrac{y}{20}$
12. $\dfrac{45}{8} = \dfrac{36}{d}$
13. $\dfrac{125}{v} = \dfrac{20}{5}$
14. $\dfrac{4}{5} = \dfrac{x}{3}$
15. $\dfrac{45}{75} = \dfrac{j}{3}$

Lesson 6-7

Use the *draw a diagram* strategy to solve the following problems.

1. **TESTS** The scores on a test are found by adding or subtracting points as shown below. If Salazar's score on a 15-question test was 86 points, how many of his answers were correct, incorrect, and blank?

Answer	Points
Correct	+8
Incorrect	−4
No answer	−2

2. **GAMES** Six members of a video game club are having a tournament. In the first round, every player will play a video game against every other player. How many games will be in the first round of the tournament?

3. **FAMILY** At Latrice's family reunion, $\dfrac{4}{5}$ of the people are 18 years of age or older. Half of the remaining people are under 12 years old. If 20 children are under 12 years old, how many people are at the reunion?

Lesson 6-8

Pages 320–326

On a map, the scale is 1 inch = 50 miles. For each map distance, find the actual distance.

1. 5 inches
2. 12 inches
3. $2\frac{3}{8}$ inches
4. $\frac{4}{5}$ inch
5. $2\frac{5}{6}$ inches
6. 3.25 inches
7. 4.75 inches
8. 5.25 inches

On a scale drawing, the scale is $\frac{1}{2}$ inch = 2 feet. Find the dimensions of each room in the scale drawing.

9. 14 feet by 18 feet
10. 32 feet by 6 feet
11. 3 feet by 5 feet
12. 20 feet by 30 feet

Lesson 6-9

Pages 328–332

Write each percent as a fraction in simplest form.

1. 32%
2. 89%
3. 72%
4. 11%
5. 1%
6. 28%
7. 55%
8. 18.5%
9. 22.75%
10. 25.2%
11. 75.5%
12. 48.25%
13. 6.5%
14. 1.25%
15. 88.9%
16. $52\frac{1}{4}$%
17. 895%
18. 480%
19. 0.78%
20. 0.3%

Write each fraction as a percent. Round to the nearest hundredth if necessary.

21. $\frac{14}{25}$
22. $\frac{28}{50}$
23. $\frac{14}{20}$
24. $\frac{7}{10}$
25. $\frac{17}{17}$
26. $\frac{80}{125}$
27. $\frac{9}{12}$
28. $\frac{4}{6}$
29. $\frac{11}{12}$
30. $\frac{9}{16}$
31. $\frac{8}{9}$
32. $\frac{3}{16}$
33. $\frac{5}{32}$
34. $\frac{1}{16}$
35. $\frac{8}{15}$
36. $\frac{9}{11}$
37. $\frac{1}{250}$
38. $\frac{1}{500}$
39. $12\frac{1}{2}$
40. $18\frac{2}{5}$

Lesson 7-1

Pages 344–348

Find each number. Round to the nearest tenth if necessary.

1. 5% of 40
2. 10% of 120
3. 12% of 150
4. 12.5% of 40
5. 75% of 200
6. 13% of 25.3
7. 250% of 44
8. 0.5% of 13.7
9. 600% of 7
10. 1.5% of $25
11. 81% of 134
12. 43% of 110
13. 61% of 524
14. 100% of 3.5
15. 20% of 58.5
16. 45% of 125.5
17. 23% of 500
18. 80% of 8
19. 90% of 72
20. 32% of 54

Lesson 7-2

Pages 353–354

Find each number. Round to the nearest tenth if necessary.

1. What number is 25% of 280?
2. 38 is what percent of 50?
3. 54 is 25% of what number?
4. 24.5% of what number is 15?
5. What number is 80% of 500?
6. 12% of 120 is what number?
7. Find 68% of 50.
8. What percent of 240 is 32?
9. 99 is what percent of 150?
10. Find 75% of 1.
11. What number is $33\frac{1}{3}$% of 66?
12. 50% of 350 is what number?
13. What percent of 450 is 50?
14. What number is $37\frac{1}{2}$% of 32?
15. 95% of 40 is what number?
16. Find 30% of 26.
17. 9 is what percent of 30?
18. 52% of what number is 109.2?
19. What number is 65% of 200?
20. What number is 15.5% of 45?

Lesson 7-3

Pages 355–360

Estimate by using fractions.

1. 28% of 48
2. 99% of 65
3. 445% of 20
4. 9% of 81
5. 73% of 240
6. 65.5% of 75
7. 48.2% of 93
8. 39.45% of 51
9. 287% of 122
10. 53% of 80
11. 414% of 72
12. 59% of 105

Estimate by using 10%.

13. 30% of 42
14. 70% of 104
15. 90% of 152
16. 67% of 70
17. 78% of 92
18. 12% of 183
19. 51% of 221
20. 23% of 504
21. 81% of 390
22. 41% of 60
23. 59% of 178
24. 22% of 450

Estimate.

25. 50% of 37
26. 18% of 90
27. 300% of 245
28. 1% of 48
29. 70% of 300
30. 35% of 35
31. 60.5% of 60
32. $5\frac{1}{2}$% of 100
33. 40.01% of 16
34. 80% of 62
35. 45% of 119
36. 14.81% of 986

Lesson 7-4

Pages 361–365

Write an equation for each problem. Then solve. Round to the nearest tenth if necessary.

1. Find 45% of 50.
2. 75 is what percent of 300?
3. 16% of what number is 2?
4. 75% of 80 is what number?
5. 5% of what number is 12?
6. Find 60% of 45.
7. 90 is what percent of 95?
8. $28\frac{1}{2}$% of 64 is what number?
9. Find 46.5% of 75.
10. What number is 55.5% of 70?
11. 80.5% of what number is 80.5?
12. $66\frac{2}{3}$% of what number is 40?
13. Find 122.5% of 80.
14. 250% of what number is 75?

Solve each problem using the *reasonable answers* strategy.

1. **SKIING** Benito skied for 13.5 hours and estimated that he spent 30% of his time on the ski lift. Did he spend about 4, 6, or 8 hours on the ski lift?

2. **CLASS TRIP** The class trip at Wilson Middle School costs $145 per student. A fundraiser earns 38% of this cost. Will each student have to pay about $70, $80, or $90?

3. **GAS MILEAGE** Miguel's car gets 38 miles per gallon and has 2.5 gallons of gasoline left in the tank. Can he drive for 85, 95, or 105 more miles before he runs out of gas?

4. **DINING** At a restaurant, the total cost of a meal is $87.50. Nadia wants to leave a 20% tip. Should she leave a total of $95, $105, or $115?

Lesson 7-6

Find each percent of change. Round to the nearest whole percent if necessary. State whether the percent of change is an *increase* or *decrease*.

1. 450 centimeters to 675 centimeters

2. 77 million to 200.2 million

3. 500 albums to 100 albums

4. 350 yards to 420 yards

5. 3.25 meters to 2.95 meters

6. $65 to $75

7. 180 dishes to 160 dishes

8. 450 pieces to 445.5 pieces

9. 700 grams 910 grams

10. 55 women to 11 women

11. 412 children to 1,339 children

12. 464 kilograms to 20 kilograms

13. 24 hours to 86 hours

14. 16 minutes to 24 minutes

Lesson 7-7

Find the total cost or sale price to the nearest cent.

1. $45 sweater; 6% tax

2. $18.99 CD; 15% discount

3. $199 ring; 10% discount

4. $29 shirt; 7% tax

5. $19 purse; 25% discount

6. $145 coat; 6.25% tax

7. $12 meal; 4.5% tax

8. $899 computer; 20% discount

9. $105 skateboard; $7\frac{1}{2}$% tax

10. $599 TV; 12% discount

11. $12,500 car; $3\frac{3}{4}$% tax

12. $49.95 gloves; $5\frac{1}{4}$% tax

Find the percent of discount to the nearest percent.

13. sneakers: regular price, $72 sale price, $60

14. dress shirt: regular price, $90 sale price, $22.50

15. portable game player: regular price, $125 sale price, $100

16. car: regular price, $25,000 sale price, $22,000

17. hiking boots: regular price, $139 sale price, $113.98

18. airline tickets: regular price, $556 sale price, $500.40

19. CD: regular price, $15 sale price, $9

20. computer: regular price, $600 sale price, $450

Lesson 7-8

Pages 379–382

Find the simple interest earned to the nearest cent for each principal, interest rate, and time.

1. $2,000, 8%, 5 years
2. $500, 10%, 8 months
3. $750, 5%, 1 year
4. $175.50, $6\frac{1}{2}$%, 18 months
5. $236.20, 9%, 16 months
6. $89, $7\frac{1}{2}$%, 6 months
7. $800, 5.75%, 3 years
8. $225, $1\frac{1}{2}$%, 2 years
9. $12,000, $4\frac{1}{2}$%, 40 months

Find the simple interest paid to the nearest cent for each loan, interest rate, and time.

10. $750, 18%, 2 years
11. $1,500, 19%, 16 months
12. $300, 9%, 1 year
13. $4,750, 19.5%, 30 months
14. $2,345, 17%, 9 months
15. $689, 12%, 2 years
16. $390, 18.75%, 15 months
17. $1,250, 22%, 8 months
18. $3,240, 18%, 14 months

Lesson 8-1

Pages 395–400

Display each set of data in a line plot. Identify any clusters, gaps, or outliers.

1.

Number of Pets in the Home				
0	1	3	4	0
2	1	0	1	1
10	0	1	5	2

2.

High Temperatures for 18 Days (°F)					
75	81	75	65	76	81
77	80	65	65	80	80
76	85	66	75	80	75

3.

Number of Stories for Buildings in Denver				
56	43	36	42	29
54	42	32	34	
52	40	32	32	

Source: *The World Almanac and Book of Facts*

4.

Ages of Children at Sunny Day Care (years)					
4	1	6	4	5	3
4	5	1	2	5	4
3	2	4	1	3	3

Lesson 8-2

Pages 401–407

Find the mean, median, and mode for each set of data.

1. 1, 5, 9, 1, 2, 6, 8, 2
2. 2, 5, 8, 9, 7, 6, 3, 5, 1, 4
3. 82, 79, 93, 91, 95, 95, 81
4. 117, 103, 108, 120
5. 256, 265, 247, 256
6. 47, 54, 66, 54, 46, 66

7.

8.

Number of Absences	Tally	Frequency
0	IIII	4
1	JHT IIII	9
2	JHT I	6
3	JHT	5

Lesson 8-3

Pages 409–413

Display each set of data in a stem-and-leaf plot.

1. 23, 15, 39, 68, 57, 42, 51, 52, 41, 18, 29

2. 189, 182, 196, 184, 197, 183, 196, 194, 184

3.

Average Monthly High Temperatures in Albany, NY (°F)			
21	46	72	50
24	58	70	40
34	67	61	27

Source: *The World Almanac and Book of Facts*

4.

Super Bowl Winning Scores 1987–2004					
39	55	52	27	34	20
42	20	30	35	23	48
20	37	49	31	34	32

Source: *The World Almanac and Book of Facts*

Lesson 8-4

Pages 414–420

Select the appropriate graph to display each set of data: bar graph or histogram. Then display the data in the appropriate graph.

1.

Longest Snakes	
Snake Name	**Length (ft)**
Royal python	35
Anaconda	28
Indian python	25
Diamond python	21
King cobra	19
Boa constrictor	16

Source: *The Top 10 of Everything*

2.

Least Densely Populated States	
State	**People Per Square Mile**
Alaska	1
Wyoming	5
Montana	6
North Dakota	9
South Dakota	10
New Mexico	15

Source: *The Top 10 of Everything*

3.

Cost of a Movie Ticket at Selected Theaters			
$5.25	$6.50	$3.50	$3.75
$7.50	$9.25	$10.40	$4.75
$10.00	$4.50	$8.75	$7.25
$3.50	$6.70	$4.20	$7.50

4.

Highest Recorded Wind Speeds For Selected U.S. Cities (mph)					
52	55	81	46	73	57
75	54	58	76	46	58
60	91	53	53	51	56
80	60	73	46	49	47

Source: *The World Almanac and Book of Facts*

Lesson 8-5

Pages 424–425

WEATHER For Exercises 1–3, solve by using the graph.

1. In which month is the average high temperature about twice as high as the average low temperature for January?

2. What is the approximate difference between the average high temperature and the average low temperature each month?

3. Predict the high and low temperatures for June based on the data given on the graph.

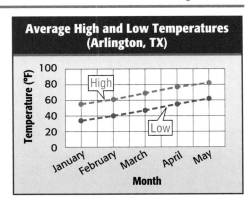

Average High and Low Temperatures (Arlington, TX)

Lesson 8-6

Pages 426–431

For Exercises 1–3, refer to the graph at the right which shows Rachel's quiz scores for six quizzes.

1. Describe the trend in Rachel's quiz scores.

2. If the trend continues, predict Rachel's score on the seventh quiz.

3. If the trend continues, predict Rachel's score on the tenth quiz.

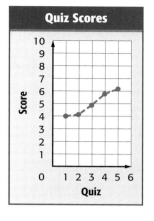

Quiz Scores

For Exercises 4–6, use the table which shows the average price paid to farmers per 100 pounds of sheep they sold.

4. Make a scatter plot of the data.

5. Describe the relationship, if any, between the two sets of data.

6. Predict the price per 100 pounds for 2010. Explain.

Year	Price Per 100 Pounds ($)
1940	4
1950	12
1960	6
1970	8
1980	21
1990	23
2000	34

Source: *The World Almanac and Book of Facts*

Lesson 8-7

Pages 435–437

1. **SURVEYS** The table shows the results of a survey of students' favorite cookies. Predict how many of the 424 students at Scobey High School prefer chocolate chip cookies.

Cookie	Number
chocolate chip	49
peanut butter	12
oatmeal	10
sugar	8
raisin	3

2. **VACATION** The circle graph shows the results of a survey of teens and where they would prefer to spend a family vacation. Predict how many of 4,000 teens would prefer to go to an amusement park.

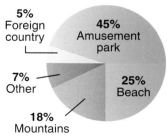

Vacation Survey

5% Foreign country

45% Amusement park

7% Other

25% Beach

18% Mountains

3. **TRAVEL** In 2000, about 29% of the foreign visitors to the U.S. were from Canada. If a particular hotel had 150,000 foreign guests in one year, how many would you predict were from Canada?

Lesson 8-8

Pages 438–443

Determine whether each conclusion is valid. Justify your answer.

1. To determine whether most students participate in after school activities, the principal of Humberson Middle School randomly surveyed 75 students from each grade level. Of these, 34% said they participate in after school activities. The principal concluded that about a third of the students at Humberson Middle School participate in after school activities.

2. To evaluate their product, the manager of an assembly line inspected the first 100 watches produced on Monday. Of these, 2 were defective. The manager concluded that about 2% of all watches produced are defective.

3. A television program asked its viewers to dial one of two phone numbers indicating their preference for one of two brands of shampoo. Of those that responded, 76% said they prefer Brand A. The program concluded that Brand A was the most popular brand of shampoo.

Lesson 8-9

Pages 444–449

Which graph could be misleading? Explain your reasoning.

1. Both graphs show pounds of grapes sold to Westview School in one week.

2. Both graphs show commissions made by Mr. Turner for a four-week pay period.

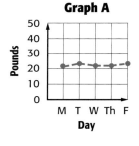

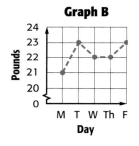

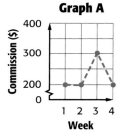

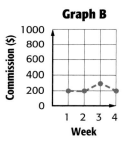

Lesson 9-1

Pages 460–464

Use the spinner at the right to find each probability. Write as a fraction in simplest form.

1. P(even number)
2. P(prime number)
3. P(factor of 12)
4. P(composite number)
5. P(greater than 10)
6. P(neither prime nor composite)

A package of balloons contains 5 green, 3 yellow, 4 red, and 8 pink balloons. Suppose you reach in the package and choose one balloon at random. Find the probability of each event. Write as a fraction in simplest form.

7. P(red balloon)
8. P(yellow balloon)
9. P(pink balloon)
10. P(orange balloon)
11. P(red or yellow balloon)
12. P(*not* green balloon)

Lesson 9-2

Pages 465–470

For each situation, find the sample space using a tree diagram.

1. rolling 2 number cubes

2. choosing an ice cream cone from waffle, plain, or sugar and a flavor of ice cream from chocolate, vanilla, or strawberry

3. making a sandwich from white, wheat, or rye bread, cheddar or Swiss cheese and ham, turkey, or roast beef

4. tossing a penny twice

5. choosing one math class from Algebra and Geometry and one foreign language class from French, Spanish, or Latin

Lesson 9-3

Pages 471–474

Use the Fundamental Counting Principle to find the total number of outcomes in each situation.

1. choosing a local phone number if the exchange is 398 and each of the four remaining digits is different

2. choosing a way to drive from Millville to Westwood if there are 5 roads that lead from Millville to Miamisburg, 3 roads that connect Miamisburg to Hathaway, and 4 highways that connect Hathaway to Westwood

3. tossing a quarter, rolling a number cube, and tossing a dime

4. spinning the spinners shown below

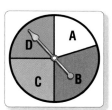 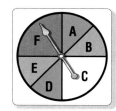 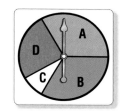

Lesson 9-4

Pages 475–478

1. **RACES** Eight runners are competing in a 100-meter sprint. In how many ways can the gold, silver, and bronze medals be awarded?

2. **LOCKERS** Five-digit locker combinations are assigned using the digits 1–9. In how many ways can the combinations be formed if no digit can be repeated?

3. **SCHEDULES** In how many ways can the classes math, language arts, science, and social studies be ordered on student schedules as the first four classes of their day?

4. **TOYS** At a teddy bear workshop, customers can select from black, brown, golden, white, blue, or pink for their bear's color. If a father randomly selects two bear colors, what is the probability that he will select a white bear for his son and a pink bear for his daughter? The father cannot pick the same color for both bears.

5. **WRITING** If you randomly select three of your last seven writing assignments to submit to an essay contest, what is the probability that you will select your first, fourth, and sixth essays in that order?

Extra Practice

1. **EXERCISE** How many ways can you choose to exercise three days of a week?

2. **BOOKS** In how many ways can six books be selected from a collection of 12?

3. **REPORTS** In how many ways can you select three report topics from a total of 8 topics?

4. **GROUPS** How many ways can four students be chosen from a class of 26?

5. **ROLLER COASTERS** In how many ways can you ride five out of nine roller coasters if you don't care in what order you ride them?

Use the *act it out* strategy to solve each problem.

1. **STAIRS** Lynnette lives on a certain floor of her apartment building. She goes up two flights of stairs to put a load of laundry in a washing machine on that floor. Then she goes down five flights to borrow a book from a friend. Next, she goes up 8 flights to visit another friend who is ill. How many flights up or down does Lynette now have to go to take her laundry out of the washing machine?

2. **LOGIC PUZZLE** Suppose you are on the west side of a river with a fox, a duck, and a bag of corn. You want to take all three to the other side of the river, but…

   - your boat is only large enough to carry you and either the fox, duck, or bag of corn.
   - you cannot leave the fox alone with the duck.
   - you cannot leave the duck alone with the corn.
   - you cannot leave the corn alone on the east side of the river because some wild birds will eat it.
   - the wild birds are afraid of the fox.
   - you cannot leave the fox, duck, and the corn alone.
   - you can bring something across the river more than once.

     If there is no other way to cross the river, how do you get everything to the other side?

The frequency table shows the results of a fair number cube rolled 40 times.

1. Find the experimental probability of rolling a 4.

2. Find the theoretical probability of *not* rolling a 4.

3. Find the theoretical probability of rolling a 2.

4. Find the experimental probability of *not* rolling a 6.

5. Suppose the number cube was rolled 500 times. About how many times would it land on 5?

Face	Frequency
1	5
2	9
3	2
4	8
5	12
6	4

Lesson 9-8
Pages 492–497

1. **COINS** Two evenly balanced nickels are tossed. Find the probability that one head and one tail result.

2. **MONEY** A wallet contains four $5 bills, two $10 bills, and eight $1 bills. A bill is randomly selected. Find $P(\$5 \text{ or } \$1)$.

3. **PROBABILITY** Two chips are selected from a box containing 6 blue chips, 4 red chips, and 3 green chips. The first chip selected is replaced before the second is drawn. Find $P(\text{red, green})$.

4. **PROBABILITY** A bag contains 7 blue, 4 orange, 8 red, and 5 purple marbles. Suppose one marble is chosen and not replaced. A second marble is then chosen. Find $P(\text{purple, red})$.

Lesson 10-1
Pages 510–513

Classify each angle as *acute, right, obtuse,* or *straight.*

1.

2.

3.

4.

5. Identify a pair of vertical angles in the diagram at the right.

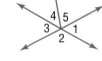

6. Identify a pair of adjacent angles in the diagram at the right.

Lesson 10-2
Pages 514–517

Classify each pair of angles as *complementary, supplementary,* or *neither.*

1.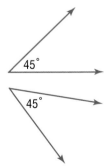

2.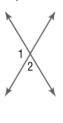

3.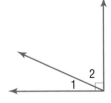

Find the value of *x* in each figure.

4.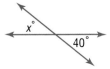

5.

6.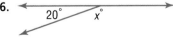

Lesson 10-3

Pages 518–523

Display each set of data in a circle graph.

1.

Car Sales	
Style	**Percent**
sedan	45%
SUV	22%
pickup truck	9%
sports car	13%
compact car	11%

2.

Favorite Flavor of Ice Cream	
Flavor	**Number**
vanilla	11
chocolate	15
strawberry	8
mint chip	5
cookie dough	3

Lesson 10-4

Pages 524–529

Find the value of x.

1.

2.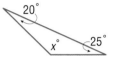

3.

Classify each triangle by its angles and by its sides.

4.

5.

6.

Lesson 10-5

Pages 530–531

Use the *logical reasoning* strategy to solve each problem.

1. **GEOMETRY** Draw several isosceles triangles and measure their angles. What do you notice about the measures of the angles of an isosceles triangle.

2. **BASKETBALL** Placido, Dexter, and Scott play guard, forward, and center on a team, but not necessarily in that order. Placido and the center drove Scott to practice on Saturday. Placido does not play guard. Who is the guard?

Lesson 10-6

Pages 533–539

Classify each quadrilateral using the name that *best* describes it.

1.

2.

3.

Find the missing angle measure in each quadrilateral.

4.

5.

6.

Lesson 10-7

Pages 540–545

Find the value of *x* in each pair of similar figures.

1,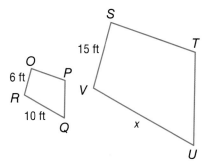

2.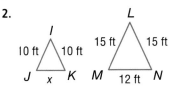

3.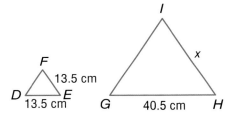

4.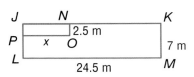

5.

6.

Lesson 10-8

Pages 546–551

Determine whether each figure is a polygon. If it is, classify the polygon and state whether it is regular. If it is *not* a polygon, explain why.

1.

2.

3.

4.

5.

6.

Find the measure of an angle in each polygon if polygon is regular. Round to the nearest tenth of a degree if necessary.

7. triangle   **8.** 30-gon   **9.** 18-gon   **10.** 14-gon

11. hexagon   **12.** nonagon   **13.** 27-gon   **14.** octagon

Lesson 10-9

Pages 553–557

1. Translate △*ABC* 2 units right and 1 unit down.

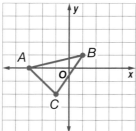

2. Translate quadrilateral *RSTU* 4 units left and 3 units down.

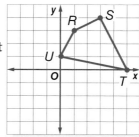

Triangle *TRI* has vertices *T*(1, 1), *R*(4, −2), and *I*(−2, −1). Find the vertices of *T′R′I′* after each translation. Then graph the figure and its translated image.

3. 2 units right, 1 unit down

4. 5 units left, 1 unit up

5. 3 units right

6. 2 units up

Lesson 10-10

Pages 558–562

Determine whether each figure has line symmetry. Write *yes* or *no*. If so, copy the figure and draw all lines of symmetry.

1.

2.

3.

4.

5.

6.

Graph each figure and its reflection over the *x*-axis. Then find the coordinates of the vertices of the reflected image.

7. quadrilateral *QUAD* with vertices *Q*(−1, 4), *U*(2, 2), *A*(1, 1), and *D*(−2, 2)

8. triangle △*ABC* with vertices *A*(0, −1), *B*(4, −3), and *C*(−4, −5)

Graph each figure and its reflection over the *y*-axis. Then find the coordinates of the vertices of the reflected image.

9. parallelogram *PARL* with vertices *P*(3, 5), *A*(5, 4), *R*(5, 1), and *L*(3, 2)

10. pentagon *PENTA* with vertices *P*(−1, 3), *E*(1, 1), *N*(0, −2), *T*(−2, −2), and *A*(−3, 1)

Lesson 11-1

Pages 570–574

Find the area of each parallelogram. Round to the nearest tenth if necessary.

1.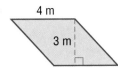
4 m
3 m

2.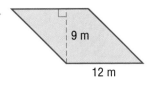
9 m
12 m

3.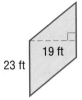
19 ft
23 ft

4. base = 19 m
height = 6 m

5. base = 135 in.
height = 15 in.

6. base = 8.2 cm
height = 5.5 cm

7. base = 29.3 m
height = 10.1 m

Lesson 11-2

Pages 576–580

Find the area of each figure. Round to the nearest tenth if necessary.

1.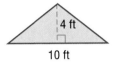
4 ft
10 ft

2.
6 cm
5 cm
3 cm

3.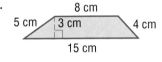
8 cm
5 cm   3 cm   4 cm
15 cm

4. triangle: base = 5 in., height = 9 in.

5. trapezoid: bases = 3 cm and 8 cm, height = 12 cm

6. trapezoid: bases = 10 ft and 15 ft, height = 12 ft

7. triangle: base = 12 cm, height = 8 cm

8. trapezoid: bases = 82.6 cm and 72.2 cm, height = 44.5 cm

9. triangle: base = 500.5 ft, height = 254.5 ft

Lesson 11-3

Pages 582–586

Find the circumference of each circle. Use 3.14 or $\frac{22}{7}$ for π. Round to the nearest tenth if necessary.

1.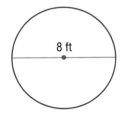
8 ft

2.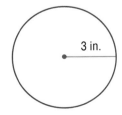
3 in.

3.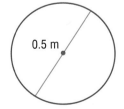
0.5 m

4.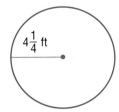
$4\frac{1}{4}$ ft

5. $r = 1$ m

6. $d = 2$ yd

7. $d = 5{,}280$ ft

8. $r = 0.5$ cm

9. $d = 6.4$ m

10. $r = 10.7$ km

11. $d = \frac{3}{16}$ in.

12. $r = 5\frac{1}{2}$ mi

13. $d = 42\frac{3}{4}$ ft

Lesson 11-4

Pages 587–591

Find the area of each circle. Round to the nearest tenth.

1.

2.

3.

4. radius = 8 in.

5. diameter = 5 ft

6. radius = 24 cm

7. diameter = 2.3 m

8. diameter = 82 ft

9. radius = 68 cm

10. radius = 9.8 mi

11. diameter = 25.6 m

12. diameter = 6.75 in.

13. radius = $1\frac{1}{4}$ ft

14. diameter = $5\frac{2}{3}$ yd

15. diameter = $45\frac{1}{2}$ mi

Lesson 11-5

Pages 592–593

Use the *solve a simpler problem* strategy to solve each problem.

1. **EARNINGS** Cedric makes $51,876 each year. If he is paid once every two weeks and actually takes home about 67% of his wages after taxes, how much does he take home each paycheck? Round to the nearest cent if necessary.

2. **CARS** Jorge plans to decorate the rims on his tires by putting a strip of shiny metal around the outside edge on each rim. The diameter of each tire is 17 inches, and each rim is 2.75 inches from the outside edge of each tire. If he plans to cut the four individual pieces for each tire from the same strip of metal, how long of a strip should he buy? Round to the nearest tenth.

3. **SAVINGS** Erin's aunt invested a total of $1,500 into three different savings accounts. She invested $450 into a savings account with an annual interest rate of 3.25% and $600 into a savings account with an annual interest rate of 4.75%. The third savings account had an annual interest rate of 4.375%. After 3 years, how much money will Erin's aunt have in the three accounts altogether if she made no more additional deposits or withdrawals? Round to the nearest cent.

BIOLOGY For Exercises 4–6, use the following information.
About five quarts of blood are pumped through the average human heart in one minute.

4. At this rate, how many quarts of blood are pumped through the average human heart in one year? (Use 365 days = 1 year)

5. If the average heart beats 72 times per minute, how many quarts of blood are pumped with each beat? Round to the nearest tenth.

6. About how many total gallons of blood are pumped through the average human heart in one week?

7. **LAND** A rectangular plot of land measures 1,450 feet by 850 feet. A contractor wishes to section off a portion of this land to build an apartment complex. If the complex is 425 feet by 550 feet, how many square feet of land will not be sectioned off to build it?

Lesson 11-6

Find the area of each figure. Round to the nearest tenth if necessary.

1.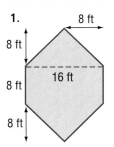
8 ft
8 ft
8 ft
8 ft
16 ft

2.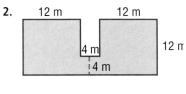
12 m   12 m
4 m
4 m
12 m

3.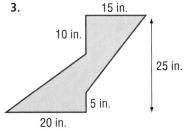
15 in.
10 in.
25 in.
5 in.
20 in.

4.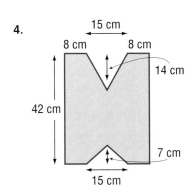
15 cm
8 cm   8 cm
14 cm
42 cm
7 cm
15 cm

5.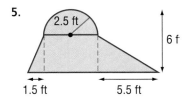
2.5 ft
6 ft
1.5 ft   5.5 ft

6.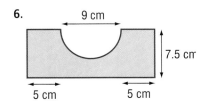
9 cm
7.5 cm
5 cm   5 cm

7.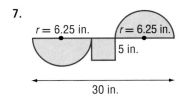
r = 6.25 in.   r = 6.25 in.
5 in.
30 in.

8.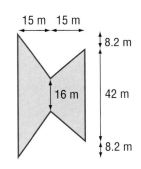
15 m   15 m
8.2 m
16 m   42 m
8.2 m

9.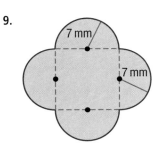
7 mm
7 mm

Lesson 11-7

For each figure, identify the shape of the base(s). Then classify the figure.

1.

2.

3.

4.

5.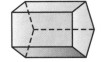

6.

7. **SOUP** Classify the shape of a soup can as a three-dimensional figure.

8. **APPLIANCES** Classify the shape of a microwave oven as a three-dimensional figure.

Lesson 11-8

Draw a top, a side, and a front view of each solid.

1.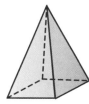

2.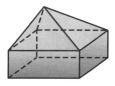

3.

Draw each solid using the top, side, and front views shown. Use isometric dot paper.

4.   top   side   front

5.   top

      side   front

Lesson 11-9

Find the volume of each prism. Round to the nearest tenth if necessary.

1.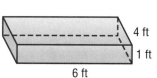
6 ft, 4 ft, 1 ft

2.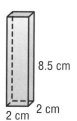
8.5 cm, 2 cm, 2 cm

3.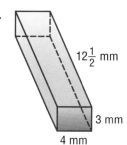
$12\frac{1}{2}$ mm, 3 mm, 4 mm

4.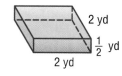
2 yd, $\frac{1}{2}$ yd, 2 yd

5.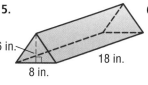
6 in., 18 in., 8 in.

6.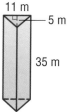
11 m, 5 m, 35 m

7.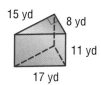
12 mm, 5 mm, 8 mm

8. 15 yd, 8 yd, 11 yd, 17 yd

Find the volume of each rectangular prism. Round to the nearest tenth if necessary.

9. length = 3 ft
width = 10 ft
height = 2 ft

10. length = 18 cm
width = 23 cm
height = 15 cm

11. length = 25 mm
width = 32 mm
height = 10 mm

12. length = 1.5 in.
width = 3 in.
height = 6 in.

13. length = 4.5 cm
width = 6.75 cm
height = 2 cm

14. length = 16 mm
width = 0.7 mm
height = 12 mm

15. length = $3\frac{1}{2}$ ft
width = 10 ft
height = 6 ft

16. length = $5\frac{1}{2}$ in.
width = 12 in.
height = $3\frac{3}{8}$ in.

17. Find the volume of a rectangular prism with a length of 3 yards, a width of 5 feet, and a height of 12 feet.

18. Find the volume of a triangular prism whose base has an area of 416 square feet and whose height is 22 feet.

Find the volume of each cylinder. Round to the nearest tenth. Use 3.14 for π.

1.
2 cm
4 cm

2.
3 yd
6.5 yd

3.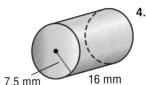
7.5 mm
16 mm

4.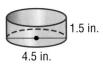
1.5 in.
4.5 in.

5. radius = 6 in.
   height = 3 in.

6. radius = 8 ft
   height = 10 ft

7. radius = 6 km
   height = 12 km

8. radius = 8.5 cm
   height = 3 cm

9. diameter = 16 yd
   height = 4.5 yd

10. diameter = 3.5 mm
    height = 2.5 mm

11. diameter = 12 m
    height = 4.75 m

12. diameter = $\frac{5}{8}$ in.
    height = 4 in.

13. diameter = 100 ft
    height = 35 ft

14. radius = 40.5 m
    height = 65.1 m

15. radius = 0.5 cm
    height = 1.6 cm

16. diameter = $8\frac{3}{4}$ in.
    height = $5\frac{1}{2}$ in.

17. Find the volume of a cylinder whose diameter is 6 inches and height is 2 feet. Round to the nearest tenth.

18. How tall is a cylinder that has a volume of 2,123 cubic meters and a radius of 13 meters? Round to the nearest tenth.

19. A cylinder has a volume of 310.2 cubic yards and a radius of 2.9 yards. What is the height of the cylinder? Round to the nearest tenth.

20. Find the height of a cylinder whose diameter is 25 centimeters and volume is 8,838 cubic centimeters. Round to the nearest tenth.

Estimate each square root to the nearest whole number.

1. $\sqrt{27}$
2. $\sqrt{112}$
3. $\sqrt{249}$
4. $\sqrt{88}$
5. $\sqrt{1,500}$
6. $\sqrt{612}$
7. $\sqrt{340}$
8. $\sqrt{495}$
9. $\sqrt{264}$
10. $\sqrt{350}$
11. $\sqrt{834}$
12. $\sqrt{3,700}$
13. $\sqrt{298}$
14. $\sqrt{101}$
15. $\sqrt{800}$

Graph each square root on a number line.

16. $\sqrt{58}$
17. $\sqrt{750}$
18. $\sqrt{1,200}$
19. $\sqrt{1,000}$
20. $\sqrt{5,900}$
21. $\sqrt{999}$
22. $\sqrt{374}$
23. $\sqrt{512}$
24. $\sqrt{3,750}$
25. $\sqrt{255}$
26. $\sqrt{83}$
27. $\sqrt{845}$
28. $\sqrt{200}$
29. $\sqrt{500}$
30. $\sqrt{10,001}$

31. **ALGEBRA** Evaluate $\sqrt{a - b}$ to the nearest tenth if $a = 16$ and $b = 4$.

32. **ALGEBRA** Estimate the value of $\sqrt{x + y}$ to the nearest whole number if $x = 64$ and $y = 25$.

Extra Practice

Extra Practice

Find the missing measure of each triangle. Round to the nearest tenth if necessary.

1.
4 ft
x ft
6 ft

2.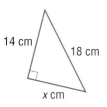
14 cm
18 cm
x cm

3.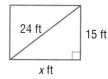
24 ft
15 ft
x ft

4. $a = 12$ cm, $b = 25$ cm

5. $a = 5$ yd, $c = 10$ yd

6. $b = 12$ mi, $c = 20$ mi

7. $a = 15$ yd, $b = 24$ yd

8. $a = 4$ m, $c = 12$ m

9. $a = 8$ mm, $b = 11$ mm

10. $a = 1$ mi, $c = 3$ mi

11. $a = 5$ yd, $b = 8$ yd

12. $b = 7$ in., $c = 19$ in.

13. $a = 50$ km, $c = 75$ km

14. $b = 82$ ft, $c = 100$ ft

15. $a = 100$ m, $b = 200$ m

Use the *make a model* strategy to solve each problem.

1. **ARCHITECTURE** An architect is designing a large skyscraper for a local firm. The skyscraper is to be 1,200 feet tall, 500 feet long, and 400 feet wide. If his model has a scale of 80 feet = 1 inch, find the volume of the model.

2. **STACKING BOXES** Box A has twice the volume of Box B. Box B has a height of 10 centimeters and a length of 5 centimeters. Box A has a width of 20 centimeters, a length of 10 centimeters, and a width of 5 centimeters. What is the width of Box B?

3. **TRAVEL** On Monday, Mara drove 400 miles as part of her journey to see her sister. She drove 60% of this distance on Tuesday. If the distance she drove on Tuesday represents one third of her total journey, how many more miles does she still need to drive?

4. **PIZZA** On Monday, there was a whole pizza in the refrigerator. On Tuesday, Enrico ate $\frac{1}{3}$ of the pizza. On Wednesday, he ate $\frac{1}{3}$ of what was left. On Thursday, he ate $\frac{1}{2}$ of what remained. What fractional part of the pizza is left?

5. **GARDENS** Mr. Blackwell has a circular garden in his backyard. He wants to build a curved brick pathway around the entire garden. The garden has a radius of 18 feet. The distance from the center of the garden to the outside edge of the brick pathway will be 21.5 feet. Find the area of the brick pathway. Round to the nearest tenth.

Lesson 12-4

Pages 647–651

Find the surface area of each rectangular prism. Round to the nearest tenth if necessary.

1.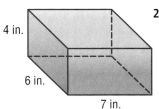
4 in.
6 in.
7 in.

2.
15 cm
4 cm
4 cm

3.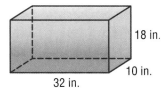
18 in.
10 in.
32 in.

4.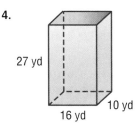
27 yd
16 yd
10 yd

5. length = 10 m
   width = 6 m
   height = 7 m

6. length = 20 mm
   width = 15 mm
   height = 25 mm

7. length = 16 ft
   width = 20 ft
   height = 12 ft

8. length = 52 cm
   width = 48 cm
   height = 45 cm

9. length = 8 ft
   width = 6.5 ft
   height = 7 ft

10. length = 9.4 m
    width = 2 m
    height = 5.2 m

11. length = 20.4 cm
    width = 15.5 cm
    height = 8.8 cm

12. length = 8.5 mi
    width = 3 mi
    height = 5.8 mi

13. length = $7\frac{1}{4}$ ft
    width = 5 ft
    height = $6\frac{1}{2}$ ft

14. length = $15\frac{2}{3}$ yd
    width = $7\frac{1}{3}$ yd
    height = 9 yd

15. length = $4\frac{1}{2}$ in.
    width = 10 in.
    height = $8\frac{3}{4}$ in.

16. length = 12.2 mm
    width = 7.4 mm
    height = 7.4 mm

17. Find the surface area of an open-top box with a length of 18 yards, a width of 11 yards, and a height of 14 yards.

18. Find the surface area of a rectangular prism with a length of 1 yard, a width of 7 feet, and a height of 2 yards.

Lesson 12-5

Pages 654–657

Find the surface area of each cylinder. Round to the nearest tenth.

1.
3 in.
7 in.

2.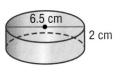
6.5 cm
2 cm

3.
1.5 m
6 m

4.
$\frac{1}{2}$ ft
$5\frac{3}{4}$ ft

5. height = 6 cm
   radius = 3.5 cm

6. height = 16.5 mm
   diameter = 18 mm

7. height = 22 yd
   radius = 10.5 yd

8. height = 6 ft
   radius = 18.5 ft

9. height = 10.2 mi
   diameter = 4 mi

10. height = 8.6 cm
    diameter = 8.2 cm

11. height = 5.8 km
    diameter = 3.6 km

12. height = 32.7 m
    radius = 21.5 m

13. height = $2\frac{2}{3}$ yd
    diameter = 6 yd

14. height = $12\frac{3}{4}$ ft
    radius = $7\frac{1}{4}$ ft

15. height = $5\frac{1}{5}$ mi
    radius = $18\frac{1}{3}$ mi

16. height = $5\frac{1}{2}$ in.
    diameter = 3 in.

Mixed Problem Solving

Chapter 1 Introduction to Algebra and Functions

1. **HISTORY** In 1932, Amelia Earhart flew 2,026 miles in 14 hours 56 minutes. To the nearest mile, what was her speed in miles per minute? (Lesson 1-1)

2. **LIGHT** The speed of light is about 67^3 kilometers per second. How many kilometers per second is this? (Lesson 1-2)

3. **FARMING** Find the length of one side of a square field with an area of 180,625 square feet. (Lesson 1-3)

4. **SALES** A department store is having a back-to-school sale. The table shows the prices of three popular items.

Item	Price ($)
Jeans	37.99
Sweatshirt	19.88
Polo Shirt	22.50

Latonia wants to buy 2 pairs of jeans, 3 sweatshirts, and 1 polo shirt. Write and evaluate a numerical expression that represents the total cost of all three items. (Lesson 1-4)

5. **MONEY** Mateo has $2.58 in coins. If he has quarters, dimes, nickels, and pennies, how many of each coin does he have? Use the *guess and check* strategy. (Lesson 1-5)

6. **FITNESS** You can estimate how fast you walk in miles per hour by evaluating the expression $\frac{n}{30}$, where n is the number of steps you take in one minute. Find your speed in miles per hour if you take 96 steps in one minute. (Lesson 1-6)

7. **BASEBALL** Last year, Scott attended 13 Minnesota Twins baseball games. This year, he attended 24. Solve $13 + n = 24$ to find how many more games he attended this year than last. (Lesson 1-7)

8. **HOT AIR BALLOONS** Miyoki paid $140 for a four-hour hot air balloon ride over the Bridger Mountains. Solve $4h = 140$ to find the cost per hour of the ride. (Lesson 1-7)

ENTERTAINMENT For Exercises 9 and 10, use the following information.
The five members of the Wolff family went to an amusement park. They each purchased an all-day ride pass and a water park pass, as shown below. (Lesson 1-8)

Item	Price ($)
All-Day Ride Pass	14.95
Water Park Pass	6.50

9. Use the Distributive Property to write two different expressions that represent the total cost for the family.

10. Find the total cost of the passes.

11. **NUMBER THEORY** Numbers that can be represented by a square arrangement of dots are called *square numbers*. The first four square numbers are shown below. (Lesson 1-9)

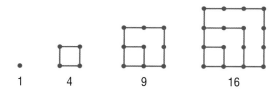

Write a sequence formed by the area of the first eight square numbers.

12. **TIME** Copy and complete the function table showing how many days y there are in various number of weeks x. Then identify the domain and range. (Lesson 1-10)

x	7x	y
1		
2		
3		
4		

1. **AIR CONDITIONING** Jacob turned on the air conditioning and the temperature in his apartment decreased 8 degrees. Write an integer to represent the change in temperature. (Lesson 2-1)

EARTH SCIENCE For Exercises 2 and 3, use the table below. It describes the deepest land depressions in the world in feet below sea level.

Depth (ft)			
220	436	511	282
383	505	235	230

Source: *The Top 10 Everything*

2. Write an integer to represent each depth. (Lesson 2-1)

3. Order the integers from greatest depth to least depth. (Lesson 2-2)

ENTERTAINMENT For Exercises 4–8, use the diagram below. It shows the locations of several rides at the Outlook Amusement Park. (Lesson 2-3)

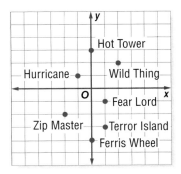

4. Which ride(s) is located in quadrant III?

5. Which ride(s) is located on the *y*-axis? Name the coordinates.

6. Which ride(s) has coordinates in which the *x*-coordinates and *y*-coordinates are equal?

7. In which quadrant is the Hurricane located?

8. A new ride is built with a location on the *x*-axis and 5 units left of the origin. Name the coordinates of this point.

9. **CAVERNS** Adriana is 52 feet underground touring the Lewis and Clark Caverns. She climbs a ladder up 15 feet. What is her new location? (Lesson 2-4)

10. **RECORDS** The lowest temperature recorded in Verkhoyansk, Russia, was about −90°F. The highest temperature was about 99°F. What is the difference between these temperatures? (Lesson 2-5)

11. **EARTH SCIENCE** The highest and lowest points in California are shown in the table. What is the difference in elevations? (Lesson 2-5)

Location	Elevation
Mount Whitney	14,494 ft above sea level
Death Valley	282 ft below sea level

Source: *The World Almanac of the U.S.A.*

12. **RIDES** A glider ride over the Crazy Mountains has a maximum altitude of 12,000 feet. It is descending at a rate of about 300 feet per minute. At what altitude will the glider be 20 minutes later? (Lesson 2-6)

13. **SPORTS** Every 12 times at bat, Simon hits the ball 3 times. About how many times will he hit the ball after 20 times at bat? 40? 84? Use the *look for a pattern* strategy. (Lesson 2-7)

Times at Bat	Number of Hits
12	3
20	
40	
84	

14. **TEMPERATURE** A temperature of −89°C was recorded in Antarctica. Use the expression $\frac{9C}{5} + 32$, where C is the temperature in degrees Celsius, to find the temperature in degrees Fahrenheit. (Lesson 2-8)

1. **TOURISM** The Statue of Liberty in New York, New York, and the Eiffel Tower in Paris, France, were designed by the same person. The Statue of Liberty is 152 feet tall. It is 732 feet shorter than the Eiffel Tower, x. Write an equation that models this situation. (Lesson 3-1)

ELECTIONS For Exercises 2 and 3, use the table below and the following information. New York has one more electoral vote than Texas. Pennsylvania has 9 fewer electoral votes than Texas. (Lesson 3-2)

Number of Electoral Votes 2000	
California	54
New York	33
Texas	
Florida	25
Pennsylvania	23

2. Write two different equations to find the number of electoral votes in Texas, n.

3. Find the number of electoral votes

4. **ROLLER COASTERS** The track length of a popular roller coaster is 5,106 feet. The roller coaster has an average speed of about 2,000 feet per minute. At that speed, how long will it take to travel its length of 5,106 feet? Use the formula $d = rt$. (Lesson 3-3)

5. **NUMBERS** A number is halved. Then three is subtracted from the quotient, and 5 is multiplied by the difference. Finally, 1 is added to the product. If the ending number is 26, what was the beginning number? Use the *work backward* strategy. (Lesson 3-4)

6. **BUSINESS** Carla's Catering charges a $25 fee to serve 15 or fewer people. In addition to that fee, they charge $10 per appetizer. You are having a party for 12 people and can spend a total of $85. How many appetizers can you order from Carla's Catering? (Lesson 3-5)

CHESS For Exercises 7–10, use the chess board below. (Lesson 3-6)

12 in.

12 in.

7. What is the perimeter of the chess board?

8. What is the area of the chess board?

9. What is the area of each small square?

10. A travel chess board has half the length and width of the board shown. What is the perimeter and area?

11. **FENCING** Mr. Hernandez will build a fence to enclose a rectangular yard for his horse. If the area of the yard to be enclosed is 1,944 square feet, and the length of the yard is 54 feet, how much fencing is needed? (Lesson 3-6)

12. **GEOMETRY** The formula for the perimeter of a square is $P = 4s$, where P is the perimeter and s is the length of a side. Graph the equation. (Lesson 3-7)

AGES For Exercises 13–16, use the table below. It shows how Jared's age and his sister Emily's age are related. (Lesson 3-7)

Jared's age (yr)	1	2	3	4	5
Emily's age (yr)	7	8	9	10	11

13. Write a verbal expression to describe how the ages are related.

14. Write an equation for the verbal expression. Let x represent Jared's age and y represent Emily's age.

15. Predict how old Emily will be when Jared is 10 years old.

16. Graph the equation.

LAND **For Exercises 1–3, use the information below.**

A section of land is one mile long and one mile wide. (Lesson 4-1)

1. Write the prime factorization of 5,280.

2. Find the area of the section of land in square feet. (*Hint*: 1 mile = 5,280 feet)

3. Write the prime factorization of the area that you found in Exercise 2.

DECORATIONS **For Exercises 4 and 5, use the information below.**

Benito is cutting streamers from crepe paper for a party. He has a red roll of crepe paper 144 inches long, a white roll 192 inches long, and a blue roll 360 inches long. (Lesson 4-2)

4. If he wants to have all colors of streamers the same length, what is the longest length that he can cut?

5. If he cuts the longest possible length, how many streamers can he cut?

6. **PRIZES** By reaching into a bag that has the letters A, B, and C, George will select three winners in order. How many possible combinations are there of the people who could win? Use the *make an organized list* strategy. (Lesson 4-3)

OLYMPICS **For Exercises 7 and 8, refer to the table below. It shows the medals won by the top three countries in the 2000 Summer Olympics.**

Country	Medals		
	Gold	Silver	Bronze
United States	40	24	33
Russia	32	28	28
China	28	16	15

Source: *The World Almanac*

7. Write the number of gold medals that Russia won as a fraction of the total number that Russia won in simplest form. (Lesson 4-4)

8. Write the fraction that you wrote in Exercise 7 as a decimal. (Lesson 4-5)

9. **SPORTS** At Belgrade Intermediate School, 75 out of every 100 students participate in sports. What percent of students do *not* participate in sports? (Lesson 4-6)

ADVERTISING **For Exercises 10–13, use the table below. It shows the results of a survey in which teens were asked which types of advertising they pay attention to.**

Type of Advertising	Percent of Teens
Television	80%
Magazine	62%
Product in a Movie	48%
Ad in an E-Mail	24%

Source: *E-Poll*

Write each percent as a fraction in simplest form. (Lesson 4-6)

10. television

11. magazine

12. product in a movie

13. ad in an e-mail

GEOMETRY **For Exercises 14–16, refer to the grid at the right.** (Lesson 4-7)

14. Write a decimal and a percent to represent the "T" shaded area.

15. Write a decimal and a percent to represent the area shaded pink.

16. What percent of the grid is *not* shaded?

17. **FLOWERS** Roses can be ordered in bunches of 6 and carnations in bunches of 15. If Ingrid wants to have the same number of roses as carnations for parent night, what is the least number of each flower that she must order? (Lesson 4-8)

18. **WATER** The table at the right shows the fraction of each state that is water. Order the states from least to greatest fraction of water. (Lesson 4-9)

What Part is Water?	
State	Fraction
Alaska	$\frac{3}{41}$
Michigan	$\frac{40}{97}$
Wisconsin	$\frac{1}{6}$

Source: *The World Almanac of the U.S.A*

Mixed Problem Solving

Mixed Problem Solving

1. **MEALS** A box of instant potatoes contains 20 cups of flakes. A family-sized bowl of potatoes uses $3\frac{2}{3}$ cups of the flakes. Estimate how many family-sized bowls can be made from one box. (Lesson 5-1)

2. **BAKING** A recipe calls for $2\frac{1}{3}$ cups of flour. Theo wants to make six batches of this recipe. About how much flour should he have available to use? (Lesson 5-1)

3. **CRAFTS** Kyle bought $\frac{5}{6}$ yard of fabric to make a craft item. He used $\frac{3}{4}$ yard in making the item. How much fabric was left over? (Lesson 5-2)

RAINFALL For Exercises 4 and 5, use the table. It shows the average annual precipitation for three of the driest locations on Earth. (Lesson 5-2)

Location	Precipitation (in.)
Arica, Chile	$\frac{3}{100}$
Iquique, Chile	$\frac{1}{5}$
Callao, Peru	$\frac{12}{25}$

Source: *The Top 10 Everything*

4. How much more rain does Iquique get per year than Arica?

5. How much more annual rain does Callao get than Iquique?

6. **INTERIOR DESIGN** A living room wall is $16\frac{1}{4}$ feet long. A window runs from the floor to the ceiling and has a length along the floor of $6\frac{3}{8}$ feet. How long is the wall without the window? (Lesson 5-3)

7. **HEALTH** The human body is about $\frac{7}{10}$ water. About how much would a person weigh if they had 70 pounds of water weight? Use the *eliminate possibilities* strategy. (Lesson 5-4)

   A 200 pounds   C 150 pounds
   B 100 pounds   D 70 pounds

8. **FOOD** The table below shows the carry-out menu for a Benito's Restaurant.

Take-out	Price ($)
Main Dish	5.00
Side Dishes	1.00
Dessert	2.00

A family of four spent $24.00 dollars for a take-home meal. What combination is possible for their meal? Use the *eliminate possibilities* strategy. (Lesson 5-4)

   F 3 main dishes and 2 side dishes
   G 4 main dishes and 3 side dishes
   H 3 main dishes, 3 side dishes, and 3 desserts
   J 4 main dishes and 4 desserts

9. **STARS** The star Sirius is about $8\frac{7}{10}$ light years from Earth. Alpha Centauri is half this distance from Earth. How far is Alpha Centauri from Earth? (Lesson 5-5)

10. **LIFE SCIENCE** Use the table below. It shows the average growth per month of hair and fingernails. Solve $3 = \frac{1}{2}t$ to find how long it takes hair to grow 3 inches. (Lesson 5-6)

Average Monthly Growth	
Hair	$\frac{1}{2}$ in.
Fingernails	$\frac{2}{25}$ in.

11. **SEWING** Jocelyn has nine yards of fabric to make table napkins for a senior citizens' center. She needs $\frac{3}{8}$ yard for each napkin. Use $\frac{3}{8}c = 9$ to find the number of napkins that she can make with this amount of fabric. (Lesson 5-6)

12. **WHALES** During the first year, a baby whale gains about $27\frac{3}{5}$ tons. What is the average weight gain per month? (Lesson 5-7)

1. **SCHOOLS** In a recent year, Oregon had 924 public elementary schools and 264 public high schools. Write a ratio in simplest form comparing the number of public high schools to elementary schools. (Lesson 6-1)

2. **MONTHS** Write a ratio in simplest form comparing the number of months that begin with the letter J to the total number of months in a year. (Lesson 6-1)

3. **EXERCISE** A person jumps rope 14 times in 10 seconds. What is the unit rate in jumps per second? (Lesson 6-2)

4. **FOOD** A 16-ounce box of cereal costs $3.95. Find the unit price to the nearest cent. (Lesson 6-2)

5. **MARKERS** The table below shows the number of markers per box. Graph the data. Then find the slope of the line. Explain what the slope represents. (Lesson 6-3)

Markers	8	16	24	32
Boxes	1	2	3	4

6. **TEMPERATURE** At 2:00, the temperature is 78°F. At 3:00, the temperature is 81°F. What is the rate of change? (Lesson 6-3)

7. **LIFE SCIENCE** An adult has about 5 quarts of blood. If a person donates 1 pint of blood, how many pints are left? (Lesson 6-4)

8. **COFFEE** In Switzerland, the average amount of coffee consumed per year is 1,089 cups per person. How many pints is this? (Lesson 6-4)

9. **BUILDINGS** A skyscraper is 0.484 kilometers tall. What is the height of the skyscraper in meters? (Lesson 6-5)

10. **WATER** A bottle contains 1,065 milliliters of water. About how many cups of water does the bottle hold? (Lesson 6-5)

11. **PHOTOGRAPHS** Mandy is enlarging a photograph that is 3 inches wide and 4.5 inches long. If she wants the width of the enlargement to be 10 inches, what will be the length? (Lesson 6-6)

12. **TILES** A kitchen is 10 feet long and 8 feet wide. If kitchen floor tiles are $2\frac{1}{2}$ inches by 3 inches, how many tiles are needed for the kitchen? Use the *draw a diagram* strategy. (Lesson 6-7)

13. **MAPS** Washington, D.C., and Baltimore, Maryland, are $2\frac{7}{8}$ inches apart on a map. If the scale is $\frac{1}{2}$ inch : 6 miles, what is the actual distance between the cities? (Lesson 6-8)

14. **MODELS** Ian is making a miniature bed for his daughter's doll house. The actual bed is $6\frac{3}{4}$ feet long. If he uses the scale $\frac{1}{2}$ inch = $1\frac{1}{2}$ feet, what will be the length of the miniature bed? (Lesson 6-8)

15. **POPULATION** According to the U.S. Census Bureau, 6.6% of all people living in Florida are 10–14 years old. What fraction is this? Write in simplest form. (Lesson 6-9)

COINS For Exercises 16 and 17, use the table below. It shows the fraction of a quarter that is made up of the metals nickel and copper. Write each fraction as a percent. Round to the nearest hundredth if necessary. (Lesson 6-9)

Metal	Fraction of Quarter
Nickel	$\frac{1}{12}$
Copper	$\frac{11}{12}$

16. nickel

17. copper

1. **SEEDS** A packet of beans guarantees that 95% of its 200 seeds will germinate. How many seeds are expected to germinate? (Lesson 7-1)

2. **SKIS** Toshiro spent $520 on new twin-tip skis. This was 40% of the money he earned at his summer job. How much did he earn at his summer job? (Lesson 7-2)

3. **GEOGRAPHY** In Washington, about 5.7% of the total area is water. If the total area of Washington is 70,637 square miles, estimate the number of square miles of water by using 10%. (Lesson 7-3)

4. **GOVERNMENT** Of the 435 members in the U.S. House of Representatives, 53 are from California and 13 are from North Carolina. To the nearest whole percent, what percent of the representatives are from California? from North Carolina? (Lesson 7-4)

FOOD For Exercises 5 and 6, use the graph below. It shows the results of a survey in which 1,200 people were asked how they determine how long food has been in their freezer. (Lesson 7-4)

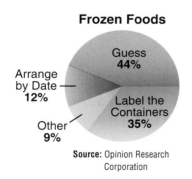

Frozen Foods

Guess 44%

Arrange by Date 12%

Other 9%

Label the Containers 35%

Source: Opinion Research Corporation

5. How many of the 1,200 surveyed guess to determine how long food has been in their freezer?

6. How many of the 1,200 surveyed label their freezer containers?

7. **DVDs** A store has 1,504 DVDs in stock. The store sold 19.8% of the DVDs last month. About how many DVDs did they sell last month? Use the *reasonable answers* strategy. (Lesson 7-5)

SPORTS For Exercises 8 and 9, use the table below. It shows the number of participants ages 7 to 17 in the sports listed. (Lesson 7-6)

Sport	Number (millions)	
	1990	2000
In-Line Skating	3.6	21.8
Snowboarding	1.5	4.3
Roller Hockey	1.5	2.2
Golf	23.0	26.4

Source: *National Sporting Goods Association*

8. What is the percent of change in in-line skaters 7 to 17 years old from 1990 to 2000? Round to the nearest percent and state whether the percent of change is an *increase* or *decrease*.

9. Find the percent of change from 1990 to 2000 in the number of children and teens who played roller hockey. Round to the nearest percent.

COMPUTERS For Exercises 10 and 11, use the following information.
The Wares want to buy a new computer with a regular price of $1,049. (Lesson 7-7)

10. If the store is offering a 20% discount, what will be the sale price of the computer?

11. If the sales tax on the computer is 5.25%, what will be the total cost with the discount?

BANKING For Exercises 12–15, complete the table below. The interest earned is simple interest. (Lesson 7-8)

	Principal	Rate	Time (yr)	Interest Earned
12.	$1,525.00	5%	$2\frac{1}{2}$	
13.	$2,250.00	4%		$337.50
14.		3.5%	4	$498.40
15.	$5,080.00		3	$952.50

Chapter 10 Geometry: Polygons

ART For Exercises 1 and 2, use the diagram of the Native American artifact.

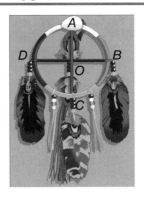

1. Name a right angle and a straight angle. (Lesson 10-1)

2. If $m\angle AOB = 90°$, what is $m\angle DOC$? (Lesson 10-2)

TELEVISION For Exercises 3 and 4, use the survey results shown in the table below. (Lesson 10-3)

Channels Families Watch	
Number	Percent
5 or fewer	30%
6–12	33%
13–25	19%
26 or more	14%

3. The fifth category in the survey is *no TV or no opinion*. What percent of the people surveyed were in this category?

4. Make a circle graph of the data.

5. **ART** Victor drew a right triangle so that one of the acute angles measures 55°. Without measuring, describe how Victor can determine the measure of the other acute angle in the triangle. Then find the angle measure. (Lesson 10-4)

6. **GARDENING** Mr. Sanchez has a flower bed with a length of 10 meters and a width of 5 meters. If he can only change the width of the flower bed, describe what he can do to increase the perimeter by 12 meters. Use the *logical reasoning* strategy. (Lesson 10-5)

7. **RUNNING** Four friends are entered in a race. Deirdre finishes directly ahead of Carlos. Mitchell finishes three places ahead of Tramaine and directly ahead of Deirdre. If Tramaine finishes fourth, place the runners in order from first to last. Use logical reasoning. (Lesson 10-5)

For Exercises 8 and 9, use the figure below.

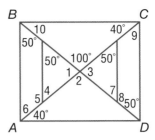

8. Find the measure of each angle numbered from 1–10. (Lesson 10-4)

9. Find the *best* name to classify quadrilateral *ABCD*. Explain your reasoning. (Lesson 10-6)

10. **CRAFTS** Priscilla makes porcelain dolls that are proportional to a real child. If Jody is $4\frac{2}{3}$ feet tall with a 23-inch waist, what should be the waist measure of a doll that is 13 inches tall? Round to the nearest inch. (Lesson 10-7)

11. **ART** Draw a tessellation using two of the polygons listed at the right. Identify the polygons and explain why the tessellation works. (Lesson 10-8)

regular triangles
quadrilaterals
pentagons
hexagons
octagons

For Exercises 12 and 13, use the quadrilateral *MOVE* shown below.

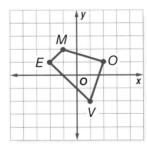

12. Describe the translation that will move *M* to the point at $(2, -2)$. Then graph quadrilateral *M'O'V'E'* using this translation. (Lesson 10-9)

13. Find the coordinates of the vertices of quadrilateral *MOVE* after a reflection over the *y*-axis. Then graph the reflection. (Lesson 10-10)

Mixed Problem Solving **713**

1. **CRAFTS** A quilt pattern uses 25 parallelogram-shaped pieces of fabric, each with a base of 4 inches and a height of $2\frac{1}{2}$ inches. How much fabric is used to make the 25 pieces? (Lesson 11-1)

2. **FURNITURE** A corner table is in the shape of a right triangle. If the side lengths of the tabletop are 3.5 feet, 3.5 feet, and 4.9 feet, what is the area? Round to the nearest tenth if necessary. (Lesson 11-2)

3. **PUZZLES** Find the area of each small and large shaded triangle. Round to the nearest whole. (Lesson 11-2)

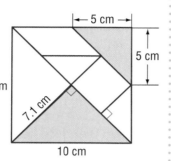

4. **EARTH SCIENCE** Earth has a diameter of 7,926 miles. Use the formula for the circumference of a circle to estimate the circumference of Earth at its equator. (Lesson 11-3)

5. **COOKIES** In New Zealand, a giant circular chocolate chip cookie was baked with a diameter of 81 feet 8 inches. To the nearest square foot, what was the area of the cookie? (Lesson 11-4)

6. **SPORTS PROFIT** A stadium seats 1,001,800 people. 22% of the tickets cost $134.87 each. 45% of the tickets cost $67.99 each. The remaining 33% cost only $35.87 each. About how much revenue is made from one game when each seat is sold out? Use the *solve a simpler problem* strategy. (Lesson 11-5)

7. **LANDSCAPING** Find the area of the flower garden shown in the diagram at the right. Round to the nearest square foot. (Lesson 11-6)

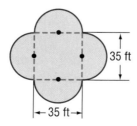

8. **GEOMETRY** A certain three-dimensional figure has four triangular faces and one square face. Classify this figure. (Lesson 11-7)

9. **RECORDS** According to the *Guinness Book of World Records*, the tallest hotel in the world is the 1,053-foot sail-shaped Burj Al Arab in Dubai, United Arab Emirates.

Draw possible sketches of the top, side, and front views of the hotel. (Lesson 11-8)

OCEANS For Exercises 10 and 11, use the following information.
The Atlantic Ocean has an area of about 33,420,000 square miles. Its average depth is 11,730 feet. (Lesson 11-9)

10. To the nearest hundredth, what is the average depth of the Atlantic Ocean in miles? (*Hint*: 1 mi = 5,280 ft)

11. What is the approximate volume of the Atlantic Ocean in cubic miles?

WATER For Exercises 12–13, use the cylinder-shaped water tank. (Lesson 11-10)

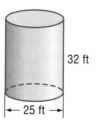

12. Find the volume of the tank. Round to the nearest cubic foot.

13. One cubic foot is approximately 7.48 gallons. Find the approximate volume of the water tank to the nearest gallon.

1. **OMELETS** In Japan, a gigantic omelet was made with an area of 1,383 square feet. If the omelet was a square, what would be its side lengths? Round to the nearest tenth. (Lesson 12-1)

2. **SOFTBALL** A softball diamond is a square measuring 60 feet on each side.

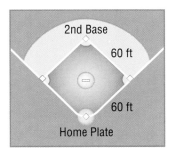

   How far does a player on second base throw when she throws from second base to home? Round to the nearest tenth. (Lesson 12-2)

3. **BANDS** Mr. Garcia is planning a band formation at a football game. The diagram shows the dimensions of the field.

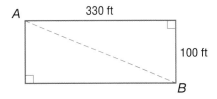

   To the nearest foot, what is the distance from *A* to *B*? (Lesson 12-2)

4. **GEOMETRY** Two right triangles are side by side such that they form a larger isosceles triangle. If the two right triangles are congruent and each have angle measures of 90°, 45°, and 45°, what type of triangle will the new isosceles triangle be? Use the *make a model* strategy. (Lesson 12-3)

5. **CUBES** A rectangular prism is formed from 48 centimeter cubes such that the height of the prism is one half of the width and one third of the length of the prism. Find the dimensions of the rectangular prism. Use the *make a model* strategy. (Lesson 12-3)

For Exercises 6 and 7, use the following information.
Liz is designing some gift boxes. The small box is 6 inches long, 4 inches wide, and 2.5 inches high. The medium box has dimensions that are each 3 times the dimensions of the small box. (Lesson 12-4)

6. Find the surface area of the small box.

7. What are the dimensions of the medium box? Then find the surface area of the medium box.

STORAGE For Exercises 8–10, use the following information.
The two canisters shown below each have a volume of about 628.3 cubic inches. (Lesson 12-5)

8. What is the radius of the blue canister? Round to the nearest tenth.

9. What is the height of the yellow canister? Round to the nearest tenth.

10. What is the difference between the surface areas of the two canisters?

HATS For Exercises 11–13, use the following information. (Lesson 12-5)
A certain cylinder-shaped hat box has a height of 9 inches and a radius of 5.5 inches. Its lid is also shaped as a cylinder, with a slightly larger diameter so that the lid fits over the box.

11. How many square inches of material are needed to make the hat box, not including the lid? Round to the nearest tenth.

12. If the lid has a height of 3.5 inches and a diameter of 11.8 inches, how many square inches of material are needed to make the lid? Round to the nearest tenth.

13. How many times more material is needed to make the hat box than the lid? Round to the nearest tenth.

Mixed Problem Solving

Preparing for Standardized Tests

Throughout the school year, you may be required to take several standardized tests, and you may have many questions about them. Here are some answers to help you get ready.

How Should I Study?

The good news is that you've been studying all along—a little bit every day. Here are some of the ways your textbook has been preparing you.

- **Every Day** Each lesson had multiple-choice practice questions.

- **Every Week** The Mid-Chapter Quiz and Practice Test also had several practice questions.

- **Every Month** The Test Practice pages at the end of each chapter had even more questions, including short-response/grid-in and extended-response questions.

Are There Other Ways to Review?

Absolutely! The following pages contain even more practice for standardized tests.

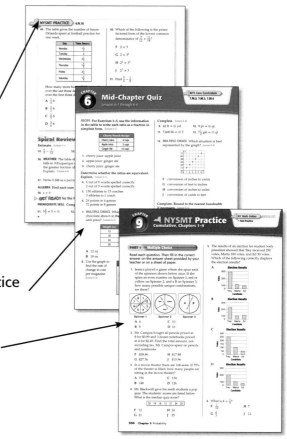

Tips for SUCCESS

Prepare
- Go to bed early the night before the test. You will think more clearly after a good night's rest.
- Become familiar with common formulas and when they should be used.
- Think positively.

During the Test
- Read each problem carefully. Underline key words and think about different ways to solve the problem.
- Watch for key words like *not*. Also look for order words like *least, greatest, first,* and *last*.
- Answer questions you are sure about first. If you do not know the answer to a question, skip it and go back to that question later.
- Check your answer to make sure it is reasonable.
- Make sure that the number of the question on the answer sheet matches the number of the question on which you are working in your test booklet.

Whatever you do...
- Don't try to do it all in your head. If no figure is provided, draw one.
- Don't rush. Try to work at a steady pace.
- Don't give up. Some problems may seem hard to you, but you may be able to figure out what to do if you read each question carefully or try another strategy.

RELAX!
Just do your best.

Multiple-Choice Questions

Multiple-choice questions are the most common type of question on standardized tests. These questions are sometimes called *selected-response questions*. You are asked to choose the best answer from four or five possible answers.

To record a multiple-choice answer, you may be asked to shade in a bubble that is a circle or an oval or just to write the letter of your choice. Always make sure that your shading is dark enough and completely covers the bubble.

The answer to a multiple-choice question may not stand out from the choices. However, you may be able to eliminate some of the choices. Another answer choice might be that the correct answer is not given.

Incomplete shading
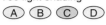

Too light shading
Ⓐ Ⓑ ⊘ Ⓓ

Correct shading
Ⓐ Ⓑ ● Ⓓ

NYSMT EXAMPLE

Notice that the problem asks for the expression that *cannot* represent the situation.

1. **Mrs. Hon's seventh grade students are purchasing stuffed animals to donate to a charity. They bought 3 boxes containing eight animals each and 2 boxes containing twelve animals each. Which expression *cannot* be used to find the total number of animals they bought to give to the charity?**

   A $8 + 8 + 8 + 12 + 12$

   B $3 \times 8 + 2 \times 12$

   C $3(8) + 2(12)$

   D $5 \times (8 + 12)$

Read the problem carefully and locate the important information. There are 3 boxes that have eight animals, so that is 3×8, or 24 animals. There are 2 boxes of twelve animals, so that is 2×12, or 24 animals. The total number of animals is $24 + 24$, or 48.

You know from reading the problem that you are looking for the expression that *does not* simplify to 48. Simplify each expression to find the answer.

A $8 + 8 + 8 + 12 + 12 = (8 + 8 + 8) + (12 + 12)$
$$= 24 + 24$$
$$= 48$$

B $3 \times 8 + 2 \times 12 = 24 + 24$
$$= 48$$

C $3(8) + 2(12) = 24 + 24$
$$= 48$$

D $5 \times (8 + 12) = 5 \times 20$
$$= 100$$

The only expression that *does not* simplify to 48 is D. The correct choice is D.

Some problems are easier to solve if you draw a diagram. If you cannot write in the test booklet, draw a diagram on scratch paper.

 NYSMT EXAMPLE

STRATEGY

Diagrams
Draw a diagram for the situation.

② **On a hiking trip, Grace and Alicia traveled 10 miles south and 4 miles west. If they take the shortest return route, how far will the hike be back to their starting point? Round to the nearest tenth of a mile.**

A 6.0 mi   **B** 9.2 mi   **C** 10.8 mi   **D** 14.0 mi

To solve this problem, you need to draw a diagram of the situation. Label the directions and the important information from the problem.

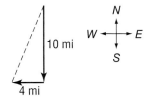

Use the Pythagorean Theorem to find the distance that they will hike back to their starting point.

$c^2 = a^2 + b^2$ Pythagorean Theorem

$c^2 = 4^2 + 10^2$ Replace a with 4 and b with 10.

$c^2 = 16 + 100$ Simplify.

$c^2 = 116$ Add.

$\sqrt{c^2} = \sqrt{116}$ Take the square root of each side.

$c \approx 10.8$ Use a calculator to simplify.

Round the answer to the correct decimal place.

The hike back will be about 10.8 miles. The correct choice is C.

Some problems give you more information than you need to solve the problem. Read the question carefully to determine the information you need.

NYSMT EXAMPLE

STRATEGY

Formulas
Use the reference sheet to find the correct formula.

③ **One of the biggest pieces of cheese ever produced was made in 1866 in Ingersoll, Canada. It weighed 7,300 pounds. It was shaped as a cylinder with a diameter of 7 feet and a height of 3 feet. To the nearest cubic foot, what was the volume of the cheese? Use 3.14 for π.**

A 462 ft³   **B** 143 ft³   **C** 115 ft³   **D** 63 ft³

You need to use the formula for the volume of a cylinder. The diameter is 7 feet, so the radius is $\frac{7}{2}$ or 3.5 feet. The height is 3 feet.

$V = \pi r^2 h$ Volume of a cylinder

$V \approx (3.14)(3.5)^2(3)$ Replace π with 3.14, r with 3.5, and h with 3.

$V \approx 115.395$ Simplify.

The volume of the cheese is about 115 cubic feet. The correct choice is C.

Multiple-Choice Practice

Choose the best answer.

Number and Operations

1. The world's smallest fruit is the fruit of a wolffia plant, which measures about 0.01 inch in length. Another small fruit is the eye of a sewing needle, with a length of 0.20 inch. How many times longer is the eye of a sewing needle fruit than a wolffia fruit?

   A 0.002   **C** 2

   B 0.005   **D** 20

2. The table shows what types of trash fill landfills in the United States. What fraction of the trash in landfills is plastic?

Type of Trash	Percent in Landfills
metal	8%
plastic	24%
food, yard waste	11%
rubber, leather	6%
paper	21%
other trash	30%

   Source: *The World Almanac for Kids*

   F $\dfrac{1}{100}$ **H** $\dfrac{1}{6}$

   G $\dfrac{1}{24}$ **J** $\dfrac{6}{25}$

3. A recent movie earned 317 million dollars in ticket sales. What is this value in scientific notation?

   A 3.17

   B 317×10^{6}

   C 3.17×10^{8}

   D 3.17×10^{11}

4. Mercury orbits the sun at 29.75 miles per second. Earth orbits the sun at 18.51 miles per second. How many more miles does Mercury travel in one minute than Earth?

   F 11.24 mi   **H** 674.4 mi

   G 269.76 mi   **J** 40,464 mi

Algebra

5. The table shows the population growth of a certain bacteria. How many bacteria will there be after 5 hours?

Hours	0	1	2	3	4	5
Number of Bacteria	32	48	72	104	144	?

   A 243   **B** 200   **C** 192   **D** 178

6. Which function rule describes the relationship between distance from home y and hours traveled x?

Time (h)	Distance from Home (mi)
x	y
0	0
1	65
2	130
3	195

   F $65y = x$ **H** $y = x + 65$

   G $y = 65 \div x$ **J** $y = 65x$

7. For a family portrait, a photographer charges a sitting fee and an amount of money per portrait ordered. Which function rule describes the relationship between the total cost y and number of portraits x?

   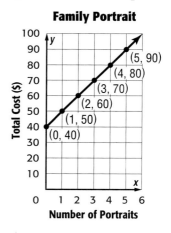

   Family Portrait

   A $y = 10 + 40x$ **C** $y = 40 + 10x$

   B $y = 40x + 10x$ **D** $y = 10x$

Geometry

8. The three towns on the map form a triangle. Which term *best* describes the angle with vertex at Worthington?

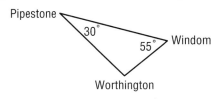

    F obtuse   **H** acute

    G right   **J** straight

9. Pedro is using a triangle in a computer graphics design. On a coordinate plane, the vertices of the triangle are $A(-1, 1)$, $B(0, 2)$, and $C(5, -1)$. If Pedro translates the triangle 4 units left and 3 units down, what will be the coordinates of B'?

    A $(-1, -4)$ **C** $(4, 5)$

    B $(4, -1)$ **D** $(-4, -1)$

10. Jasmine is using this polygon on a poster she is making for the basketball team. Which term *best* describes the polygon?

    F quadrilateral   **H** hexagon

    G decagon   **J** heptagon

Measurement

11. Crispy Crackers are packaged in a box that measures 6 inches by 2.5 inches by 10 inches. Which dimensions are of a prism that has the same volume as the Crispy Crackers box?

    A 6.5 in. by 2 in. by 10 in.

    B 6.25 in. by 3 in. by 8 in.

    C 7.25 in. by 2 in. by 9.5 in.

    D 4 in. by 7.5 in. by 6 in.

TEST-TAKING TIP

Question 11 Most standardized tests will include any commonly used formulas at the front of the test booklet. Quickly review the list before you begin so that you know what formulas are available.

12. The Crab nebula is a cloud of gas and dust particles in space that is expanding at a rate of 930 miles per second. What is its rate of expansion in miles per hour?

    F 22,320 mph   **H** 1,339,200 mph

    G 55,800 mph   **J** 3,348,000 mph

13. Super Toys makes two sizes of building blocks shaped as cubes. The large block has side length four times the length of the small block. What is the ratio of the surface area of the small block to the surface area of the large block?

    A 1 to 4   **C** 1 to 16

    B 1 to 6   **D** 1 to 32

Data Analysis and Probability

14. The table shows the number of students playing each sport at Wilson Junior High. Find the mean of the data.

Sport	Number of Students
baseball/softball	49
basketball	74
soccer	82
swimming	21
track and field	115
volleyball	25

    F 23   **G** 49   **H** 61   **J** 94

15. The spinner is divided into four equal-sized sections. If you spin the spinner 62 times, which is the *best* estimate for the number of times you will land on 2?

    A 15   **B** 30   **C** 40   **D** 50

Preparing for Standardized Tests

Gridded-Response Questions

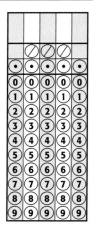

Gridded-response questions are another type of question on standardized tests. These questions are sometimes called *student-produced response* or *grid in*.

For gridded response, you must mark your answer on a grid printed on an answer sheet. The grid contains a row of four or five boxes at the top, two rows of ovals or circles with decimal and fraction symbols, and four or five columns of ovals, numbered 0–9. An example of a grid from an answer sheet is shown.

TEST EXAMPLE

1 **Mr. Byrd builds and sells storage buildings. The dimensions of his most popular model are shown in the diagram. What is the volume of the building in cubic feet?**

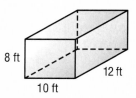

8 ft

12 ft

10 ft

STRATEGY

Formulas
Use the reference sheet to find the formula you need.

What do you need to find?

You need to find the volume of a rectangular prism. Use the formula and the dimensions given in the diagram.

$V = \ell w h$ \qquad Volume of a rectangular prism

$V = 12 \cdot 10 \cdot 8$ \quad Replace ℓ with 12, *w* with 10, and *h* with 8.

$V = 960$ \qquad Multiply.

The volume is 960 cubic feet.

How do you fill in the grid for the answer?

- Write your answer in the answer boxes.

- Write only one digit or symbol in each answer box.

- Do not write any digits or symbols outside the answer boxes.

- You may write your answer with the first digit in the left answer box, or with the last digit in the right answer box. You may leave blank any boxes you do not need on the right or the left side of your answer.

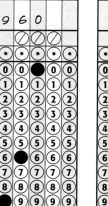

- Fill in only one bubble for every answer box that you have written in. Be sure not to fill in a bubble under a blank answer box.

Many gridded-response questions result in an answer that is a fraction or a decimal. These values can also be filled in on the grid.

TEST EXAMPLE

2. A prize box contains 12 glitter pencils, 5 fluorescent pens, and 13 mechanical pencils. If Alex randomly selects a prize, what is the probability that he will choose a glitter pencil?

$$P(\text{glitter pencil}) = \frac{\text{number of favorable outcomes}}{\text{number of possible outcomes}}$$

$$= \frac{12}{12 + 5 + 13} = \frac{12}{30} \text{ or } \frac{2}{5}$$

You can either grid the fraction $\frac{12}{30}$ or $\frac{2}{5}$. You also can rewrite the fraction as a decimal and grid 0.4. Be sure to write the decimal point or fraction bar in the answer box. The following are acceptable answers.

> Any equivalent fraction that fits the grid will be counted as correct.

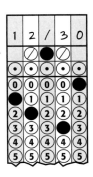 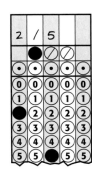 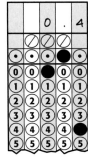 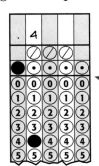

> Do not leave a blank answer box in the middle of an answer.

If the answer is a mixed number, change it to an equivalent improper fraction or decimal.

TEST EXAMPLE

3. A hummingbird measures $2\frac{1}{2}$ inches in length. A sandpiper, measures $4\frac{1}{2}$ inches in length. How many times as long is the sandpiper as the hummingbird?

Divide the length of the sandpiper by the length of the hummingbird.

$$4\frac{1}{2} \div 2\frac{1}{2} = \frac{9}{2} \div \frac{5}{2} \quad \text{Rename } 2\frac{1}{2} \text{ as } \frac{5}{2}.$$

$$= \frac{9}{2} \cdot \frac{2}{5} \quad \text{Multiply by the reciprocal of } \frac{5}{2}, \text{ which is } \frac{2}{5}.$$

$$= \frac{9}{5}$$

You can either grid the improper fraction $\frac{9}{5}$, or rewrite it as 1.8 and grid the decimal. Do not enter 14/5, as this will be interpreted as $\frac{14}{5}$.

Gridded-Response Practice

Solve each problem. Then copy and complete a grid like the one shown on page 722.

Number and Operations

1. The seventh-grade class at Willow Creek Middle School is planning a class trip. Each student will need to pay $4.50 for the bus ride, $8.00 for a ticket to the museum, and $5.25 for lunch. If there are 52 students in the class, what will be the total cost in dollars of the trip?

2. The highest point in Louisiana is Driskill Mountain at 585 feet. The lowest point is −8 feet in New Orleans. What is the difference in feet between the highest and lowest elevation points?

3. People in the United States own about 204 million cars. If the number of cars is written in scientific notation, what is the exponent of the 10 in the expression?

4. Ashlee has $5\frac{1}{4}$ cups of cocoa powder. If each batch of chocolate cookies uses $\frac{1}{2}$ cup of cocoa powder, how many batches could she make?

Algebra

5. The number of televisions per 1,000 people in France is 598. The number of televisions per 1,000 people in the United States is 208 greater than the number in France. How many televisions are there in the United States per 1,000 people?

6. The table shows the number of white beads that Carmen uses in each row for a particular pattern in a necklace that she designed. How many white beads will there be in the sixth row?

Row	1	2	3	4	5
Beads	1	2	4	8	16

7. The table shows the cost of renting a booth at the week-long Fall Festival. There is an initial charge for reserving a booth and a fee per day. What is the cost in dollars of renting a booth for the 7 days of the festival?

Days	0	1	2	3
Cost ($)	50	90	130	170

8. The graph shows the cost to rent a power paint sprayer. Let x be the number of hours the sprayer is rented and y be the total cost of the rental. Suppose an equation of the form $y = ax$ represents the data in the graph. What is the value of a?

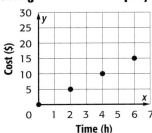

Renting a Power Paint Sprayer

9. The temperature on a January morning is −18°F. The temperature is expected to rise at a rate of 5° each hour for the next several hours. In how many hours will the temperature be 7°F?

Geometry

10. Triangle ABC is similar to triangle XYZ. What is the measure of $\angle Z$ in degrees?

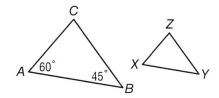

11. Triangle DEF has vertices $D(1, 1)$, $E(4, -2)$, and $F(0, -3)$. Find the y-coordinate of point E after the triangle is reflected over the x-axis.

12. In the diagram, ∠1 and ∠2 are supplementary. Find the measure of ∠3 in degrees.

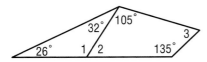

13. Rectangle *B* has width and length 3 times the width and length of Rectangle *A*. As a fraction, what is the ratio of the area of Rectangle *A* to the area of Rectangle *B*?

Measurement

14. In the United States, the average amount of meat eaten per person is 261 pounds per year. If there are 365 days in a year, what is the average number of ounces of meat eaten per day? Round to the nearest ounce.

TEST-TAKING TIP

Question 14 The units of measure given in a question may not be the same as the units of measure asked for in the answer. Check that your solution is in the correct unit.

15. A rectangle has an area of 318 square centimeters and a width of 12 centimeters. What is the length of the rectangle in centimeters?

16. The circle graph shows the results of a survey in Ms. Chen's fifth period math class. What is the exact degree measure of the section of the circle graph representing soccer?

**Favorite Sports of
Ms. Chen's Students**

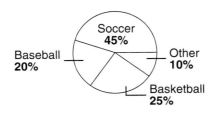

17. One acre of land is 43,560 square feet. How many square yards are in one acre?

18. The Browns plan to paint the top, bottom, and sides of the cylinder-shaped barrel with a waterproof coating. What is the surface area in square feet of the barrel? Round to the nearest square foot.

Data Analysis and Probability

19. The table shows the world's highest dams and the countries in which they are located. Find the range of the data.

Country	Height (ft)
Tajikistan	984
Switzerland	935
Georgia	892
Italy	859
Mexico	856

Source: *Scholastic Book of World Records*

20. The table shows the prices in dollars that Mike spent on his college textbooks. What is the mean of the data?

66.00	51.25	50.00	63.00
56.25	9.00	82.00	50.50

21. Kaya has the following scores for the first four tests in her science class: 82%, 95%, 100%, and 90%. She wants the mean of her first five tests to be 93%. What score as a percent must she earn on the fifth test?

22. In a carnival game, the probability of winning is 0.005. As a percent, what is the probability of losing?

23. A beverage cooler contains 6 regular colas, 5 orange drinks, 7 iced teas, and 7 diet colas. James reaches into the cooler and randomly takes two drinks, one after the other. Find the probability that he will choose a regular cola and then an orange drink.

Short-Response Questions

Short-response questions require you to provide a solution to the problem as well as any method, explanation, and/or justification you used to arrive at the solution. These are sometimes called *constructed-response, open-response, open-ended, free-response,* or *student-produced questions.*

The following is a sample **rubric**, or scoring guide.

Credit	Score	Criteria
Full	2	Full credit: The answer is correct and a full explanation is provided that shows each step in arriving at the final answer.
Partial	1	Partial credit: There are two different ways to receive partial credit. • The answer is correct, but the explanation provided is incomplete or incorrect. • The answer is incorrect, but the explanation and method of solving the problem is correct.
None	0	No credit: Either an answer is not provided or the answer does not make sense.

On some standardized tests, no credit is given for a correct answer if your work is not shown.

NYSMT EXAMPLE

1. **Hanna is deciding between a desktop and a laptop that are on sale. The desktop costs $609.00 with a 10% discount. The laptop costs $725.00 with a 25% discount. There is also a 6.75% sales tax on all purchases. Which computer is less expensive? What will be the total cost of the computer including discount and sales tax?**

STRATEGY

Reread the Problem
Look for the important information in the problem.

Full Credit Solution

Since there are two computers to compare, I will first find the discounted price of each computer. I will change each percent to a decimal to make the calculations.

desktop	laptop
$609.00 \times 0.10 = 60.90$	$725.00 \times 0.25 = 181.25$
$609.00 - 60.90 = 548.10$	$725.00 - 181.25 = 543.75$

The laptop is less expensive after the discount.

I still need to find the cost of the laptop computer with tax.

$$543.75 \times 0.0675 \qquad 6.75\% = 0.0675$$
$$= 36.703125 \qquad \text{Use a calculator.}$$
$$\approx 36.70$$

Now I will add the sales tax to the cost of the laptop.

$$543.75 + 36.70 = 580.45$$

The laptop will cost Hanna $580.45, including sales tax.

The steps, calculations, and reasoning are clearly stated.

Partial Credit Solution

In this sample solution, the calculations are correct and the answer is correct. However, there is no explanation for any of the calculations.

Notice that the student multiplies the discounted price by 1.0675 since the cost is 1 and the tax is 0.0675. So, the total is then 1.0675.

$609.00 \times 0.10 = 60.90$

$609.00 - 60.90 = 548.10$

$548.10 \times 1.0675 = 585.09675$

$725.00 \times 0.25 = 181.25$

$725.00 - 181.25 = 543.75$

$543.75 \times 1.0675 = 580.453125$

The laptop is cheaper for $580.45.

Partial Credit Solution

In this sample solution, the answer is partially incorrect because the student does not add the sales tax.

I will find the discount price for each set.

Desktop: Since the current price is 100% and the discount is 10%, the sale price will be

$100 - 10 = 90\%$ or 0.9.

$609.00 \times 0.9 = 548.10$

Laptop: Since the current price is 100% and the discount is 25%, the sale price will be

$100 - 25 = 75\%$ or 0.75.

$725.00 \times 0.75 = 543.75$

Hanna should get the laptop for $543.75.

The student does not add the cost of the tax.

No Credit Solution

In this sample solution, the student does not understand how to find discounted prices and the sales tax. There are just some calculations using the numbers in the problem.

$609.00 - 10\% = 602.91$

$725.00 - 25\% = 625.00$

$602.91 + 6.75\% = 609.66$

$625.00 + 6.75 = 631.75$

I think Hanna should buy the desktop for $609.66.

Short-Response Practice

Solve each problem. Show all your work.

Number and Operations

1. A main unit of currency in Egypt is the pound. One U.S. dollar is equal to $3\frac{4}{5}$ pounds. How many pounds are equivalent to $10.00 in the U.S.?

2. The average daytime temperature on Venus is 870°F. The average temperature on Jupiter is −160°F. What is the difference between the average temperatures on Venus and Jupiter?

3. A bag of chocolate candies has a nutrition label stating that each serving contains 20% of the recommended daily amount of fat. A serving has 13 grams of fat. Using this information, what is the total recommended daily amount of fat in grams?

4. The Montana Department of Fish, Wildlife, and Parks raised the price of a tag to catch a paddlefish from $2.50 to $5.00 for residents and from $7.50 to $15.00 for nonresidents. Which percent of increase is greater, the increase for residents or for nonresidents?

5. A recent article in the newspaper said that there were 75 cell phones for every 100 people in Finland. The number of cell phones in Finland was given to be 3,893,000. Estimate the population of Finland using this information.

Algebra

6. Florida has 8,426 miles of shoreline. Alaska has 25,478 more miles of shoreline than Florida. Write and solve an equation to find the number of miles of shoreline for Alaska.

7. Juana is saving money to buy a skateboard that costs $95. She has $25 and plans to save $5 per week. In how many weeks will she have enough money for the skateboard?

8. Tyler delivers televisions for Electronics Depot. The graph shows the amount Tyler charges for delivery based on distance. Name the slope and y-intercept of the graph and describe what they mean in this situation.

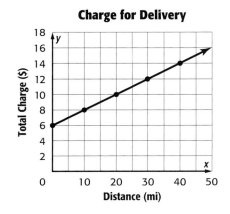

9. Solve $\frac{2}{3}b = \frac{8}{7}$.

Geometry

10. The formula for the area of a trapezoid is $A = \frac{1}{2}h(b_1 + b_2)$. Find the area of the trapezoid.

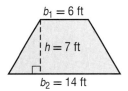

11. Angles MNP and PNO are supplementary. Find $m\angle PNO$.

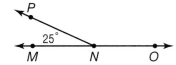

12. What is the value of x?

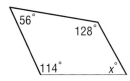

13. Colby is planning to use △CDE for a design by using transformations. He plans to reflect it over the *y*-axis. Find the coordinates of △CDE after this reflection. Graph the reflected image of the triangle.

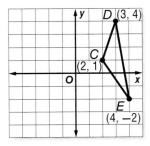

Measurement

14. Diego ran in a 10-kilometer race. What is the distance of this race in meters?

15. What is the surface area of the box?

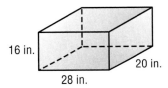

16. A certain aircraft uses 350 gallons of fuel per hour. If the plane used 1,925 gallons of fuel on a flight between two cities, how many hours long was the flight?

17. Curtis is painting the four columns for the set of a school play. What is the surface area that needs to be painted? Assume that the tops and bottoms of the columns do *not* need to be painted.

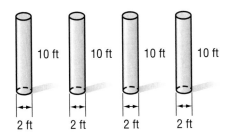

18. The record for women running the Boston Marathon was set in 2002 when Margaret Okayo of Kenya ran the 26.2-mile course in approximately 2 hours and 21 minutes. What was her average speed in miles per hour?

TEST-TAKING TIP

Questions 17 and 18 After finding the solution, always go back and read the problem again to make sure your solution answers what the problem is asking.

Data Analysis and Probability

19. The stem-and-leaf plot shows the scores on the last math test in Mr. Hill's class. What is the range of the data?

Stem	Leaf
9	1 2 3 5 6 9
8	0 2 5 6 6 7 8 8
7	1 2 3 5 7 9 9
6	0 5 8 9 9 8|2 = 82 points

20. The table shows the heights of the world's largest flightless birds. Make a bar graph of the data.

Bird	Height (in.)
Ostrich	96
Emu	60
Cassowary	60
Rhea	54
Emperor Penguin	45

21. The table shows the average precipitation in inches for each month in Syracuse, New York. Make a scatter plot of the data. Use the months on the horizontal axis and the precipitation on the vertical axis. Describe the graph.

Month	Precipitation (in.)	Month	Precipitation (in.)
Jan	2.34	July	3.81
Feb	2.15	Aug	3.51
Mar	2.77	Sep	3.79
Apr	3.33	Oct	3.24
May	3.28	Nov	3.72
June	3.79	Dec	3.2

22. Two number cubes each marked with 1, 2, 3, 4, 5, 6 on their faces are rolled. List all the possible outcomes.

Extended-Response Questions

Extended-response questions are often called *open-ended* or *constructed-response questions*. Most extended-response questions have multiple parts. You must answer all parts to receive full credit.

Extended-response questions are similar to short-response questions in that you must show all of your work in solving the problem and a rubric is used to determine whether you receive full, partial, or no credit. The following is a sample rubric for scoring extended-response questions.

Credit	Score	Criteria
Full	4	Full credit: A correct solution is given that is supported by well-developed, accurate explanations.
Partial	3, 2, 1	Partial credit: A generally correct solution is given that may contain minor flaws in reasoning or computation, or an incomplete solution is given. The more correct the solution, the greater the score.
None	0	No credit: An incorrect solution is given indicating no mathematical understanding of the concept, or no solution is given.

On some standardized tests, no credit is given for a correct answer if your work is not shown.

Make sure that when the problem says to *Show your work*, you show every aspect of your solution including figures, sketches of graphing calculator screens, or the reasoning behind computations.

NYSMT EXAMPLE

1. **Each fall, the community pool is drained. The pool contains 100,000 gallons of water. The pool has two drains that together drain the pool at a rate of 10,000 gallons per hour.**

   a. Make a function table for this situation. Let x represent the time in hours the pool drains. Let y represent the gallons of water left in the pool.

   b. Make a graph of the data in the function table.

   c. Predict how many hours it will take for the pool to drain.

Full Credit Solution

Part a A complete table includes labeled columns.

Every 2 hours there will be 20,000 gallons less water.

Hours (x)	Water Left (y)	(x, y)
2	80,000	(2, 80,000)
4	60,000	(4, 60,000)
6	40,000	(6, 40,000)
8	20,000	(8, 20,000)
10	0	(10, 0)

Part b   A complete graph includes a title for the graph, appropriate scales and labels for the axes, and correctly graphed points.

To make a graph, I graphed the ordered pairs from my table and decided to connect them with a line.

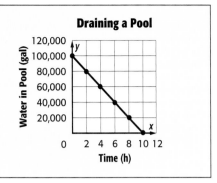

Draining a Pool

Part c

The pool is drained when the y-value is 0. That is at 10 hours. It will take 10 hours to drain the pool.

Partial Credit Solution

Part a   This sample answer has an incomplete table.

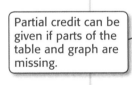

1	90,000
2	80,000
3	70,000

Part b   Partial credit is given because the points are graphed correctly, but the scale on the *y*-axis jumps from 0 to 70,000 with no break.

Water in the Pool

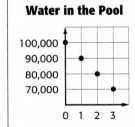

Part c

I wrote a function rule to find how long it will take to drain the pool. I let $y = 0$ to find the number of hours.

$$y = 100,000 - 10,000x$$
$$0 = 100,000 - 10,000x$$
$$10,000x = 100,000$$
$$x = 10$$ It will take 10 hours to drain the pool.

No Credit Solution

If the student demonstrates no understanding of making a table or graph or making a prediction, then no credit is given.

Extended-Response Practice

Solve each problem. Show all your work.

Number and Operations

1. The table shows the land area and water area of five states.

Land and Water Area of Five States		
State	**Land Area (mi²)**	**Water Area (mi²)**
Florida	53,937	5991
Colorado	103,729	371
Alaska	570,374	44,856
Iowa	55,875	401
Rhode Island	1,045	186

a. Find the percent of each state that is water. The entire state area is the sum of the land and water areas.

b. Order the states from the state with the least percent water to the greatest percent water.

c. Suppose a state had land and water areas such that the water area was 20% of the total area. Give a possible land and water area, with land area greater than 1,000 square miles, such that this is true.

Algebra

2. The table shows the rental rates for a crane rental service. There is an initial fee to reserve the crane and a daily fee.

Days	Cost	Days	Cost
0	$100	4	$280
1	$145	5	$325
2	$190	6	$370
3	$235		

a. Graph the data. Let x = number of days and y = cost. Connect the points.

b. What is the slope of the line? What does the slope represent?

c. What would be the charge for renting the crane for 10 days?

3. A long-distance phone company charges 40¢ per call plus 4¢ per minute.

a. Write and solve an equation to find the number of minutes used for a call that costs $2.80.

b. Another long-distance company charges 6¢ per minute and no connection fee. What is the cost of a 60-minute call?

c. Which service would charge less for a 90-minute call?

Geometry

4. Ava and Bryn are making rectangular fleece blankets. Ava is making her blanket 40 inches by 60 inches. Bryn says she wants to make her blanket "twice as big."

a. What is the area of Ava's blanket?

b. Bryn makes her blanket such that both the length and width are each twice the length and width of Ava's blanket. What will be the area of Bryn's blanket?

c. What is the ratio of the area of Bryn's blanket in Part b to the area of Ava's blanket in Part a?

d. Ava tells Bryn that a blanket is "twice as big" if the area is twice the area of the other. Give a possible length and width for a blanket twice as big as Ava's using this idea of "twice as big."

5. The map shows five towns and the angle measures for roads connecting them.

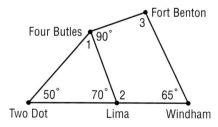

a. Find the measure of ∠1. Explain.

b. Find the measures of ∠2 and ∠3. Explain.

6. City planners are considering enlarging the circular manholes in their streets. Currently, the diameter of a manhole cover is 24 inches. The new size being considered is 2 inches greater all the way around.

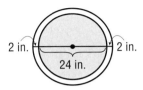

a. Find the area of a current manhole cover with diameter of 24 inches. Round to the nearest square inch.

b. Find the area of a new manhole cover. Round to the nearest square inch.

c. What would be the percent increase in the area of a manhole cover if the new covers are made?

TEST-TAKING TIP

Question 6 Be sure to completely and carefully read the problem before beginning any calculations. If you read too quickly, you may miss a key piece of information.

7. A company that sells beads for craft projects has the two containers shown for packaging the beads.

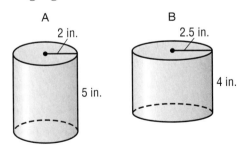

a. Find the volume of each container. Round to the nearest tenth of a cubic inch.

b. Find the ratio of the volume of container A to the volume of container B.

c. The company wants to make a container shaped as a rectangular prism. Find a possible length, width, and height for the container such that its volume is twice the volume of container B.

8. The table shows the normal monthly temperatures in degrees Fahrenheit for Honolulu, Hawaii, and Minneapolis, Minnesota.

Month	Honolulu	Minneapolis
Jan	73	13
Feb	73	20
Mar	74	32
Apr	76	47
May	77	59
Jun	80	68
Jul	81	73
Aug	82	71
Sep	82	61
Oct	80	49
Nov	78	33
Dec	75	19

Source: *The World Almanac*

a. Make a scatter plot of the data for Honolulu. Place the months on the x-axis and the temperatures on the y-axis. Describe the graph.

b. Make a scatter plot of the data for Minneapolis. Place the months on the x-axis and the temperatures on the y-axis. Describe the graph.

c. Compare the range for each city's temperature data. Explain a possible reason for the difference in the data.

9. Two 4-sided number cubes are rolled for a game, one white and one black. Each number cube has faces numbered 1 through 4.

a. List the sample space for rolling the two number pieces.

b. What is the theoretical probability that a player will roll a sum of 8?

c. If you played this game 50 times, how many times would you expect to get a sum of 4?

Concepts and Skills Bank

1 Divisibility Patterns

In $54 \div 6 = 9$, the quotient, 9, is a whole number. So, we say that 54 is **divisible** by 6. You can use the following rules to determine whether a number is by 2, 3, 4, 5, 6, 9, and 10.

A number is divisible by:

- 2 if the ones digit is divisible by 2.
- 3 if the sum of the digits is divisible by 3.
- 4 if the number formed by the last two digits is divisible by 4.
- 5 if the ones digit is 0 or 5.
- 6 if the number is divisible by both 2 and 3.
- 9 if the sum of the digits is divisible by 9.
- 10 if the ones digit is 0.

EXAMPLE Use Divisibility Rules

Determine whether 972 is divisible by 2, 3, 4, 5, 6, 9, or 10.

2: Yes; the ones digit, 2, is divisible by 2.
3: Yes; the sum of the digits, $9 + 7 + 2 = 18$, is divisible by 3.
4: Yes; the number formed by the last two digits, 72, is divisible by 4.
5: No; the ones digit is not 0 or 5.
6: Yes; the number is divisible by 2 and 3.
9: Yes; the sum of the digits, 18, is divisible by 9.
10: No; the ones digit is not 0.

Exercises

Use divisibility rules to determine whether the first number is divisible by the second number.

1. 447; 3
2. 135; 6
3. 240; 4
4. 419; 3
5. 831; 3
6. 4,408; 4
7. 7,110; 5
8. 1,287; 9
9. 2,984; 9
10. 7,026; 6
11. 1,260; 10
12. 8,903; 6

Determine whether each number is divisible by 2, 3, 4, 5, 6, 9, or 10.

13. 712
14. 1,035
15. 8,901
16. 462
17. 270
18. 1,005
19. 32,221
20. 8,340
21. 920
22. 50,319
23. 64,042
24. 3,498

25. **MEASUREMENT** Jordan has 5,280 feet of rope. Can he cut the rope into 9-foot pieces and use all of the rope? Explain.

Estimating with Decimals

Estimation can be used to provide quick answers when an exact answer is not necessary. It is also an excellent way to check whether your answer is reasonable. One method of estimating is to use rounding. Round numbers to any place value that makes estimation easier.

EXAMPLES Estimate by Rounding

Estimate by rounding.

1. $23.485 - 9.757$

$$
\begin{array}{rcl}
23.485 & \rightarrow & 23 \\
-\ 9.757 & \rightarrow & -\ 10 \\
\hline
& & 13
\end{array}
$$
Round to the nearest whole numbers.

The difference is about 13.

2. $6.43 + 2.17 + 9.1 + 4.87$

$$
\begin{array}{rcl}
6.43 & \rightarrow & 6 \\
2.17 & \rightarrow & 2 \\
9.1 & \rightarrow & 9 \\
+\ 4.87 & \rightarrow & +\ 5 \\
\hline
& & 22
\end{array}
$$
Round to the nearest whole numbers.

The sum is about 22.

Another way to estimate sums is to use **clustering**. This strategy is used when all the numbers are close to a common value.

EXAMPLE Estimate by Clustering

3. Estimate $9.775 + 9.862 + 9.475 + 9.724$ by clustering.

All of the numbers are clustered around 10. There are four numbers.

So, the sum is about 4×10 or 40.

Exercises

Estimate by rounding.

1. $8.56 + 5.34$
2. $34.84 - 17.69$
3. $6.8 + 2.4$
4. $40.79 - 6.8$
5. $6.9 + 5.2$
6. $23.84 + 12.13$
7. $34.3 - 18.9$
8. $7.5 + 8.4$
9. $65.48 - 9.3$
10. $26.3 + 9.7$
11. $33.21 - 8.23$
12. $67.86 - 24.35$
13. $8.99 - 2.6$
14. $121.5 + 487.8$
15. $32.5 + 81.4$

Estimate by clustering.

16. $18.4 + 22.5 + 20.7$
17. $56.9 + 63.2 + 59.3 + 61.1$
18. $42.3 + 41.5 + 39.8 + 40.4$
19. $77.8 + 75.6 + 81.2 + 79.9$
20. $239.8 + 242.43 + 236.20 + 240.77$
21. $9.9 + 10.0 + 10.3 + 11.1 + 9.8 + 11.2$
22. $50.4 + 51.1 + 48.9 + 49.5 + 50.8$
23. $100.5 + 97.8 + 101.6 + 100.2 + 99.3$

Multiplying Decimals

To multiply decimals, multiply as with whole numbers. The product has the same number of decimal places as the sum of the decimal places of the factors. Use estimation to determine whether your answers are reasonable.

EXAMPLES Multiply Decimals

Multiply.

1 **1.3 × 0.9** Estimate $1 \times 1 = 1$

$$
\begin{array}{r}
1.3 \quad \leftarrow \text{1 decimal place} \\
\times 0.9 \quad \leftarrow \text{1 decimal place} \\
\hline
1.17 \quad \leftarrow \text{2 decimal places}
\end{array}
$$

The product is reasonable.

2 **0.054 × 1.6** Estimate $0 \times 2 = 0$

$$
\begin{array}{r}
0.054 \quad \leftarrow \text{3 decimal places} \\
\times 1.6 \quad \leftarrow \text{1 decimal place} \\
\hline
324 \\
540 \\
\hline
0.0864
\end{array}
$$
Annex a zero on the left so the answer has four decimal places. Compare to the estimate.

Exercises

Place the decimal point in each product. Add zeros if necessary.

1. $1.32 \times 4 = 528$
2. $0.07 \times 1.1 = 77$
3. $0.4 \times 0.7 = 28$
4. $1.9 \times 0.6 = 114$
5. $1.4 \times 0.09 = 126$
6. $5.48 \times 3.6 = 19728$
7. $4.5 \times 0.34 = 153$
8. $0.45 \times 0.02 = 9$
9. $150.2 \times 32.75 = 4919050$

Multiply.

| 10. $\begin{array}{r} 0.2 \\ \times\, 6 \\ \hline \end{array}$ | 11. $\begin{array}{r} 0.3 \\ \times\, 0.9 \\ \hline \end{array}$ | 12. $\begin{array}{r} 0.45 \\ \times\, 0.12 \\ \hline \end{array}$ | 13. $\begin{array}{r} 0.0023 \\ \times\, 32 \\ \hline \end{array}$ | 14. $\begin{array}{r} 1.5 \\ \times\, 2.7 \\ \hline \end{array}$ |

| 15. $\begin{array}{r} 10.1 \\ \times\, 9 \\ \hline \end{array}$ | 16. $\begin{array}{r} 2 \\ \times\, 0.3 \\ \hline \end{array}$ | 17. $\begin{array}{r} 6.78 \\ \times\, 1.3 \\ \hline \end{array}$ | 18. $\begin{array}{r} 200 \\ \times\, 0.004 \\ \hline \end{array}$ | 19. $\begin{array}{r} 0.0023 \\ \times\, 0.35 \\ \hline \end{array}$ |

20. 15.8×11
21. 88×2.5
22. 33×0.03
23. 36×0.46
24. 0.003×482
25. 1.88×1.11
26. 0.6×2
27. 38.3×29.1
28. 0.7×18
29. 8×0.3
30. 12.2×12.4
31. 380×1.25
32. 42×0.17
33. 0.4×16
34. 0.23×0.2
35. 0.44×0.5
36. 0.44×55
37. 44×0.55

38. **JOBS** Antonia earns $10.75 per hour. What are her total weekly earnings if she works 34.5 hours? Round to the nearest cent.

 Concepts and Skills Bank

Powers of Ten

You can use a pattern to mentally find the product of any number and a power of 10 that is greater than 1. Count the number of zeros in the power of 10 or use the exponent. Then move the decimal point that number of places to the right.

Decimal Power of 10	Product
19.7×10^1 (or 10)	= 197
19.7×10^2 (or 100)	= 1,970
19.7×10^3 (or 1,000)	= 19,700
19.7×10^4 (or 10,000)	= 197,000

EXAMPLES **Use Mental Math to Multiply**

Multiply mentally.

1 12.562×100

$12.562 \times 100 = 12.562$ Move the decimal point two places
$\qquad\qquad\qquad = 1,256.2$ to the right, since 100 has two zeros.

2 0.59×10^4

$0.59 \times 10^4 = 0.5900$ Move the decimal point four places
$\qquad\qquad\quad = 5,900$ to the right, since the exponent is 4.

To mentally multiply by a power of ten that is less than 1, count the number of decimal places. Or, if the power is written as a fraction, use the exponent in the denominator. Then move the decimal point that number of places *to the left*.

Decimal Power of 10	Product
$19.7 \times 0.1 \left(\text{or } \frac{1}{10^1}\right)$	= 1.97
$19.7 \times 0.01 \left(\text{or } \frac{1}{10^2}\right)$	= 0.197
$19.7 \times 0.001 \left(\text{or } \frac{1}{10^3}\right)$	= 0.0197

EXAMPLES **Use Mental Math to Multiply**

Multiply mentally.

3 10.5×0.01

$10.5 \times 0.01 = 10.5$ Move the decimal point
$\qquad\qquad\quad = 0.105$ two places to the left.

4 $5,284 \times 0.00001$

$5,284 \times 0.00001 = 05284$ Move the decimal point
$\qquad\qquad\qquad\quad = 0.05284$ five places to the left.

Exercises

Multiply mentally.

1. 12.53×10
2. 4.6×10^3
3. 78.4×0.01
4. 0.05×100
5. 4.527×100
6. $2.78 \times 1,000$
7. 13.58×0.01
8. 5.49×10^3
9. 0.1×0.8
10. 0.925×10
11. 99.44×10^2
12. 0.01×16
13. 1.32×10^3
14. $0.56 \times 10,000$
15. 1.4×0.001
16. 11.23×10^5
17. 68.94×0.01
18. 0.8×10^4
19. 28.1×0.01
20. 9.3×10^7
21. $625,799 \times 0.0001$

Dividing Decimals

To divide two decimals, use the following steps.

- If necessary, change the divisor to a whole number by moving the decimal point to the right. You are multiplying the divisor by a power of ten.
- Move the decimal point in the dividend the same number of places to the right. You are multiplying the dividend by the same power of ten.
- Divide as with whole numbers.

EXAMPLES **Divide Decimals**

Divide.

1 **25.8 ÷ 2** **Estimate** 26 ÷ 2 = 13

$$\begin{array}{r} 12.9 \\ 2\overline{)25.8} \\ -2 \\ \hline 5 \\ -4 \\ \hline 18 \\ -18 \\ \hline 0 \end{array}$$

The divisor, 2, is already a whole number, so you do not need to move the decimal point. Divide as with whole numbers. Then place the decimal directly above the decimal point in the dividend.

Compared to the estimate, the quotient, 12.9, is reasonable.

2 **199.68 ÷ 9.6** **Estimate** 200 ÷ 10 = 20

$$\begin{array}{r} 20.8 \\ 9.6\overline{)199.68} \\ -192 \\ \hline 7\ 68 \\ -7\ 68 \\ \hline 0 \end{array}$$

Move each decimal point one place to the right.

Compare the answer to the estimate.

Exercises

Divide.

1. $0.3\overline{)9.81}$ 2. $12\overline{)0.12}$ 3. $3.2\overline{)5.76}$ 4. $0.22\overline{)0.0132}$ 5. $0.04\overline{)0.008}$

6. $3.18\overline{)0.636}$ 7. $0.2\overline{)8.24}$ 8. $82.3\overline{)823}$ 9. $12.02\overline{)24.04}$ 10. $0.5\overline{)85}$

11. $74.9\overline{)5.992}$ 12. $19.2\overline{)4.416}$ 13. $1.9\overline{)38.57}$ 14. $13.8\overline{)131.1}$ 15. $6.48\overline{)259.2}$

16. 812 ÷ 0.4   17. 0.34 ÷ 0.2   18. 14.4 ÷ 0.12   19. 90.175 ÷ 2.5

20. 39.95 ÷ 799   21. 88.8 ÷ 444   22. 613.8 ÷ 66   23. 2,445.3 ÷ 33

24. 20.24 ÷ 2.3   25. 45 ÷ 0.09   26. 2.475 ÷ 0.03   27. 4.6848 ÷ 0.366

28. 180 ÷ 0.36   29. 97.812 ÷ 1.1   30. 23 ÷ 0.023   31. 1,680.042 ÷ 44.2

32. **OLYMPICS** In the 2000 Olympics, Michael Johnson of the U.S. ran the 400-meter run in 43.84 seconds. To the nearest hundredth, find his speed in meters per second.

33. **SCIENCE** It takes Pluto 247.69 Earth years to revolve once around the Sun. It takes Jupiter 11.86 Earth years to revolve once around the Sun. About how many times longer does it take Pluto than Jupiter to revolve once around the Sun?

Converting Currencies

NYSCC 7.M.7
Convert money between different currencies with the use of an exchange rate table and a calculator

Countries around the world use different types of currency. By using an exchange rate table, proportions, and a calculator, you can convert from one type of currency to another.

Examine the following exchange rate table. All values in the same column are equivalent to each other.

Echange Rates					
United States Dollars	1	2.020590	0.999950	1.407465	0.864699
British Pounds	0.494905	1	0.494880	0.696562	0.427944
Canadian Dollars	1.000050	2.020691	1	1.407536	0.864742
Euros	0.710497	1.435623	0.710461	1	0.614366
Australian Dollars	1.156472	2.336756	1.156414	1.627694	1

*Exchange rates as of September 21, 2007

You can use proportions to convert any amount of money from one currency to another.

EXAMPLE

Julian has 16 United States dollars to exchange for Euros. How many Euros will he receive?

U.S. dollar ⟶ $\dfrac{1}{0.710497} = \dfrac{16}{x}$ Set up a proportion.
euros ⟶

$1(x) = 16(0.710497)$ Cross multiply.

$x = 11.367952$ Simplify.

He will receive about 11.37 euros in exchange for 16 United States dollars.

Exercises

Fill in the blanks by converting each currency. Round to the nearest hundredth.

1. 26.41 Canadian dollars = _____ euros

2. 46.82 British pounds = _____ United States dollars

3. 4.7 Australian dollars = _____ euros

4. 285.31 United States dollars = _____ British pounds

5. 567.31 Canadian dollars = _____ Australian dollars

6. 91.67 euros = _____ Canadian dollars

7. 6 Australian dollars = _____ British pounds

8. 37.24 euros = _____ United States dollars

9. 64.28 British pounds = _____ Canadian dollars

10. 74.28 United States dollars = _____ Australian dollars

Concepts and Skills Bank

Solving Inequalities

You have already solved one- and two-step equations. You can apply what you learned about equations to solve one- and two-step inequalities. **Inequalities** are sentences that compare quantities that are not equal. The symbols used are $<, >, \leq, \geq,$ and $\neq$.

 **NYSCC** **7.A.5** Solve one-step inequalities (positive coefficients only) **7.G.10** Graph the solution set of an inequality (positive coefficients only) on a number line

Inequality Symbols					Key Concept
Symbols	<	>	≤	≥	≠
Words	• less than • fewer than	• greater than • more than	• less than or equal to • no more than • at most	• greater than or equal to • no less than • at least	• not equal to

EXAMPLE **Solve a One-Step Inequality**

① Solve $s + \frac{3}{4} \leq 1\frac{1}{4}$.

$$s + \frac{3}{4} \leq 1\frac{1}{4}$$ Write the inequality.

$$s + \frac{3}{4} - \frac{3}{4} \leq 1\frac{1}{4} - \frac{3}{4}$$ Subtract $\frac{3}{4}$ from each side.

$$s + \frac{3}{4} - \frac{3}{4} \leq \frac{5}{4} - \frac{3}{4}$$ Change $1\frac{1}{4}$ to an improper fraction, $\frac{5}{4}$.

$$s \leq \frac{2}{4} \text{ or } \frac{1}{2}$$ Subtract.

The solution is $s \leq \frac{1}{2}$.

EXAMPLE **Solve a Two-Step Inequality**

② Solve $7p + 21 \leq 49$.

$$7p + 21 \leq 49$$ Write the inequality.

$$7p + 21 - 21 \leq 49 - 21$$ Subtract 21 from each side.

$$7p \leq 28$$ Simplify.

$$p \leq \frac{28}{7}$$ Divide each side by 7.

$$p \leq 4$$ Simplify.

The solution is $p \leq 4$.

Unlike the solution to an equation, the solution $p \leq 4$ reflects a set of numbers. One way to show this is in a number line graph.

Concepts and Skills Bank

Rotations

A **rotation** occurs when a figure is rotated around a point. A rotation does not change the size or shape of the figure. The rotations shown below are clockwise around the origin.

90° Rotation	**180° Rotation**	**270° Rotation**
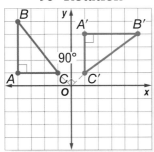	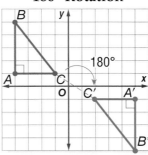	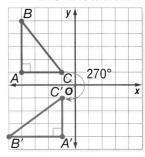

EXAMPLE **Rotate a Figure About the Origin**

1. Triangle *ABC* has vertices *A*(−4, 4), *B*(−1, 2), and *C*(−3, 1). Graph the figure and its rotated image after a clockwise rotation of 90° about the origin. Then give the coordinates of the vertices for △*A′B′C′*.

STEP 1 Graph △*ABC* on a coordinate plane.

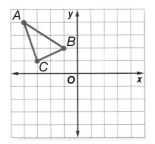

STEP 2 Sketch segment $\overline{BO}$ connecting point *B* to the origin. Sketch another segment, $\overline{BO}$ so that the angle between point *B*, *O*, and *B′* measures 90° and the segment is congruent to $\overline{BO}$.

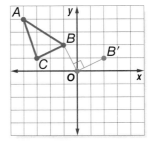

STEP 3 Repeat Step 2 for points *A* and *C*. Then connect the vertices to form △*A′B′C′*.

So, the coordinates of the vertices of △*A′B′C′* are *A′*(4, 4), *B′*(2, 1), and *C′*(1, 3).

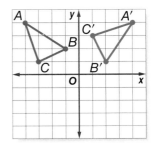

2 **Triangle *LMN* has vertices *L*(5, 4), *M*(5, 7), and *N*(8, 7). Graph the figure and its rotated image after a counterclockwise rotation of 180° about vertex *L*. Then give the coordinates of the vertices for △*L′M′N′*.**

STEP 1 Graph the original triangle.

STEP 2 Graph the rotated image.

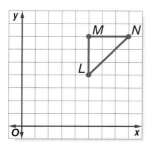

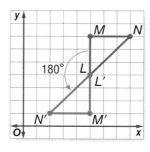

So, the coordinates of the vertices of △*L′M′N′* are *L′*(5, 4), *M′*(5, 1), and *N′*(2, 1).

Exercises

Graph △*XYZ* and its rotated image after each rotation. Then give the coordinates of the vertices for △*X′Y′Z′*.

1. 180° clockwise about the origin

2. 270° counterclockwise about vertex *X*

3. 90° counterclockwise about the origin

4. 270° clockwise about vertex *Y*

5. 180° counterclockwise about vertex *Z*

6. 90° clockwise about the origin

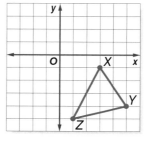

Graph quadrilateral *ABCD* and its rotated image after each rotation. Then give the coordinates of the vertices for quadrilateral *A′B′C′D′*.

7. 90° counterclockwise about the origin

8. 90° clockwise about vertex *A*

9. 180° counterclockwise about vertex *D*

10. 270° clockwise about the origin

11. 90° clockwise about the origin

12. 180° clockwise about vertex *B*

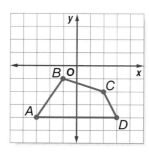

10 Cross Sections of Three-Dimensional Figures

The intersection of a solid and a plane is called a **cross section** of the solid.

Inequality Symbols		Key Concept
Vertical Slice	**Angled Slice**	**Horizontal Slice**
The cross section is a rectangle.	The cross section is an oval.	The cross section is a circle.

EXAMPLE Describe Cross Sections

1. Draw and describe the shape resulting from a vertical, angled, and horizontal cross section of a square pyramid.

	Slice	Drawing	Description
Vertical			The cross section is a triangle.
Angled			The cross section is a trapezoid.
Horizontal			The cross section is a square.

Exercises

Draw and describe the shape resulting from each cross section.

1. cone

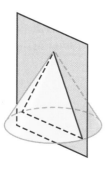

2. triangular pyramid

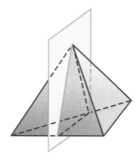

3. rectangular prism

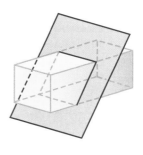

4. cone

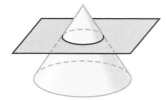

5. rectangular prism

6. cone

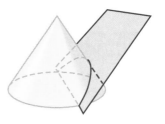

7. triangular pyramid

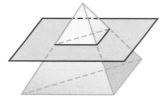

8. rectangular prism

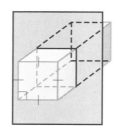

9. triangular prism

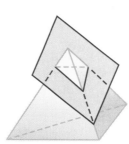

SPORTS For Exercises 10 and 11, use the following information.

A standard basketball is shaped like a sphere.

10. Draw a basketball with a vertical, angled, and horizontal slice.

11. Draw and describe the cross section made by each slice.

SPORTS For Exercises 12 and 13, use the following information.

A soup can is shaped like a cylinder.

12. Draw a soup can with a vertical, angled, and horizontal slice.

13. Draw and describe the cross section made by each slice.

Converting Between Measurement Systems

Dimensional analysis is the process of including units of measurement as factors when you compute. You can use ratios to convert measurements between the two common measurement systems. For example, since 1 in. ≈ 2.54 cm, the ratio $\frac{1 \text{ in.}}{2.54 \text{ cm}} \approx 1$. So, you can multiply a measurement by this ratio without changing its value.

Conversion Factors for Length	
1 in. ≈ 2.54 cm	1 yd ≈ 0.914 m
1 ft ≈ 0.305 m	1 mi ≈ 1.609 km
Conversion Factors for Capacity and Mass or Weight	
1 fl oz ≈ 29.574 mL	1 qt ≈ 0.946 L
1 pt ≈ 0.473 L	1 gal ≈ 3.785 L
1 oz ≈ 28.35 g	1 lb ≈ 0.454 kg

EXAMPLES Convert Between Systems

1 **15 feet to meters**

Use 1 ft ≈ 0.305 meters.

$15 \text{ ft} \approx 15 \text{ ft} \cdot \dfrac{0.305 \text{ m}}{1 \text{ ft}}$ Since 1 ft ≈ 0.305 m, multiply by $\dfrac{0.305 \text{ m}}{1 \text{ ft}}$.

$\approx 15 \,\cancel{\text{ft}} \cdot \dfrac{0.305 \text{ m}}{1 \,\cancel{\text{ft}}}$ Divide out common units, leaving the desired unit, meter.

$\approx 15 \text{ ft} \cdot 0.305 \text{ or } 4.58 \text{ m}$ Multiply.

So, 15 feet is approximately 4.58 meters.

2 **22 kilograms to pounds.**

Use 1 lb ≈ 0.454 kg.   Since 1 qt ≈ 0.946 L, multiply by $\dfrac{1 \text{ lb}}{0.454 \text{ kg}}$.

$22 \text{ kg} \approx 22 \text{ kg} \cdot \dfrac{1 \text{ lb}}{0.454 \text{ kg}}$ Divide out common units, leaving the desired unit, pound.

$\approx 22 \,\cancel{\text{kg}} \cdot \dfrac{1 \text{ lb}}{0.454 \,\cancel{\text{kg}}}$

$\approx \dfrac{22}{0.454} \text{ or } 48.46 \text{ pounds.}$ Divide.

So, 22 kilograms is approximately 48.46 pounds.

Exercises

Complete each conversion. Round to the nearest hundredth.

1. 17 in. ≈ ■ cm
2. 13 L ≈ ■ gal
3. 15 oz ≈ ■ g
4. 12 m ≈ ■ ft

5. 10 L ≈ ■ pt
6. 150 cm ≈ ■ in.
7. 18 fl oz ≈ ■ mL
8. 15 km ≈ ■ mi

9. 8 yd ≈ ■ m
10. 25 mi ≈ ■ km
11. 200 mL ≈ ■ fl oz
12. 12 L ≈ ■ qt

13. 32 ft ≈ ■ m
14. 12 gal ≈ ■ L
15. 575 g ≈ ■ oz
16. 55 lb ≈ ■ kg

17. **ANIMALS** A peregrine falcon can reach a speed of 250 kilometers per hour. How many miles is this per hour?

18. **FOOD** The heaviest apple ever picked weighed about 4 pounds. About how many kilograms did the apple weigh?

12 Volume and Surface Area of Composite Figures

Volume is the measure of space occupied by a three-dimensional figure. The volume of a composite figure can be found by separating the figure into solids whose volumes you know how to find.

EXAMPLE **Find the Volume of a Composite Figure**

1. **Find the volume of the toy house shown at the right.**

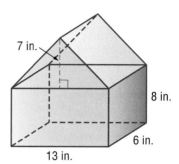

The toy house is made of one rectangular prism and one triangular prism. Find the volume of each prism.

Rectangular Prism 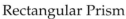 Triangular Prism

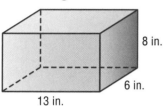

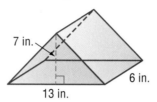

$V = Bh$

$V = (13 \cdot 6)\, 8$ or 624

$V = Bh$

$V = (\frac{1}{2} \cdot 13 \cdot 7)\, 6$ or 273

The volume of the toy house is $624 + 273$ or 897 cubic inches.

EXAMPLE **Find the Surface Area of a Composite Figure**

2. **Max is painting the mailbox shown. What is the area of the surface that is to be painted? Round to the nearest tenth.**

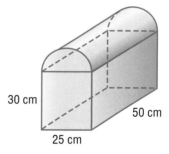

The mail box is made of one half of one cylinder and one rectangular prism.

Cylinder 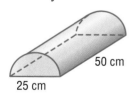 Rectangular Prism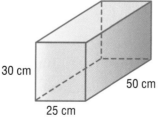

> There is only one base needed. The top is covered by the half cylinder

$S = \dfrac{2\pi r^2 + 2\pi rh}{2}$

$S = \dfrac{2\pi(12.5)^2 + 2\pi(12.5)(50)}{2}$

$S \approx 2{,}454.4$

$S = 2\ell w + 2\ell h + wh$

$S = 2 \cdot 25 \cdot 30 + 2 \cdot 25 \cdot 50 + 30 \cdot 50$

$S = 5{,}500$

So, the area to be painted is equal to $2{,}454.4 + 5{,}500$ or $7{,}954.4$ square centimeters.

Exercises

Find the volume of each figure. Round to the nearest tenth if necessary.

1.

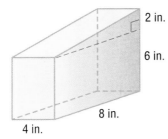

2 in.
6 in.
8 in.
4 in.

2.

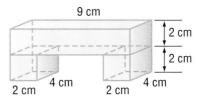

9 cm
2 cm
2 cm
2 cm   4 cm   2 cm   4 cm

3.

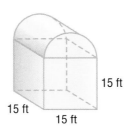

15 ft
15 ft
15 ft

4.

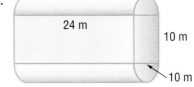

24 m
10 m
10 m

Find the surface area of each figure. Round to the nearest tenth if necessary.

5.

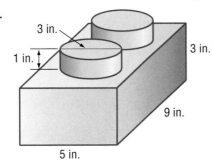

3 in.
3 in.
1 in.
9 in.
5 in.

6.

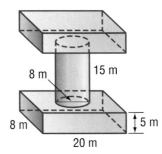

8 m
15 m
8 m
5 m
20 m

7.

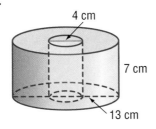

4 cm
7 cm
13 cm

8.

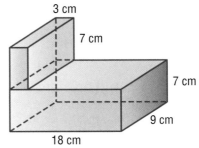

3 cm
7 cm
7 cm
9 cm
18 cm

Relative Error and Magnitude

NYSCC 7.M.10 Identify the relationships between relative error and magnitude when dealing with large numbers (e.g., money, population)

Significant error can occur when estimating or making predictions using large numbers. By comparing estimates to exact calculations, you can identify the amount of error that occurred as a result of rounding or estimating.

The **absolute error** is the magnitude of the difference between the estimate and the actual answer. **Relative error** is the absolute error in estimating a quantity divided by the magnitude of the exact value.

EXAMPLE

Estimate the population increase for New York City from 2003 to 2006. Then find the exact population increase and compare.

Estimate: $8,200,000 - 8,100,000 = 100,000$

Exact: $8,214,426 - 8,085,742 = 128,684$

New York City	
Year	**Population**
2006	8,214,426
2005	8,143,197
2004	8,104,079
2003	8,085,742

The absolute error is the difference between the estimate and the exact answer: $128,684 - 100,000$, or 28,684 people.

The relative error is $\frac{28,684}{128,684}$ or approximately 0.22 or 22%.

Exercises

Solve each problem. Compare your estimate to the exact answer by finding the relative error.

1. Find the area of a circle with a radius of 14 centimeters. First calculate the area using 3 as an estimate for π. Then use 3.14 to find the area.

2. Jennifer is planning a wedding reception for 300 guests. The table shows the cost per person for each part of the meal. She needs to know the total approximate cost of the food for the reception.

Cost (per person)	
Appetizer	$2.75
Entrée	$6.75
Beverage	$1.55
Cake	$2.90

3. Antonio is driving 2,452 miles from Albany, New York, to Los Angeles, California. He drives about 68 miles per hour for most of the trip. Antonio estimates how long his trip will take by rounding 68 to 70 and 2,452 to 2,500.

3. As part of their grand opening, a new grocery store decides to give a free gallon of milk to their first 1,500 customers. Normally the milk sells for $3.49 per gallon. How much will the grand-opening special cost the grocery store?

Photo Credits

Cover Richard Hutchings/Digital Light Source, shot on location courtesy of Busch Gardens/Adventure Island, Tampa Florida; **iv** File Photo; **v** Digital Vision/Getty Images, **vi** Aaron Haupt; **vii** (l to r)Aaron Haupt, (bl)File Photo; **viii-ix** Jose Fuste Raga/CORBIS; **xix** David Muench/CORBIS; **xx-xxi** David Frazier/CORBIS; **xxii-xxiii** Robert Glusic/Photodisc/Getty Images; **xxiv-xxv** Brand X Pictures/PunchStock; **xxvi-xxvii** Eastcott Momatiuk/Photodisc/Getty Images; **xxviii-xxix** Michael Ventura/Alamy Images, **xxx-xxxi** Kim Karpeles/Alamy Images; **xxxii-xxxiii** W.A.Hamilton/Alamy; **xxxiv** (t c)CMCD/Getty Images, (b)Comstock Select/CORBIS; **NY26** Digital Vision/Getty Images; **1** Digital Vision/PunchStock; **3** colinspics/Alamy Images; **4** Glen Allison/Getty Images; **5** Stockbyte/Getty Images; **6** Tomasso DeRosa/CORBIS; **7** Jeff Greenberg/Alamy Images; **8** Jonathan King; **9** Jonathan Nourok/PhotoEdit; **10** Lee Snider/Photo Images/CORBIS; **11** Erik Freeland/CORBIS; **12** Digital Vision/Alamy Images; **13** Simon Jonathan Webb/Alamy Images; **14** Comstock/Punchstock; **15** David R. Frazier Photolibrary, Inc./Alamy Images; **16** Marie Read/Animals Animals; **17** (t)Ross M Horowitz/Getty Images, (bl)ACE STOCK LIMITED/Alamy Images, (br)Philip Scalia/Alamy Images; **18** Dennis MacDonald/Alamy Images; **19** (t)Royalty-Free/CORBIS, (b)photo courtesy of Enchanted Forest/Water Safari, Old Forge, NY; **20-21** JG Photography/Alamy Images; **22** Andre Jenny/Alamy Images; **26** Jose Luis Pelaez, Inc./Blend Images/Getty Images; **32** Stefano Bianchetti/CORBIS; **35** Lawrence Manning/CORBIS; **39** The McGraw-Hill Companies Inc.; **41** (l)Michael Newman/PhotoEdit, (r)David Young-Wolff/PhotoEdit; **42** Kevin Peterson/Photodisc/Getty Images; **45** SW Productions/Photodisc/Getty Images; **49** Photo courtesy of BGSU Photo Services; **52** (l)David Young-Wolff/PhotoEdit, (r)Tony Freeman/PhotoEdit; **54** Tim de Waele/CORBIS; **64** (l)Noel Hendrickson/Digital Vision/Getty Images, (r)Eric and David Hosking/CORBIS; **66** Herbert Kehrer/zefa/CORBIS; **78** SW Productions/Brand X/CORBIS; **80** Chase Jarvis/CORBIS; **90** Kevin Schafer/Photographer's Choice RF/Getty Images; **99** (l)Myrleen Ferguson Cate/PhotoEdit, (r)David Young-Wolff/PhotoEdit; **104** NASA/Photodisc/Getty Images; **106** (l)Cleve Bryant/PhotoEdit, (r)David Young-Wolff/PhotoEdit; **109** Ralph White/CORBIS; **112** John Evans; **116** Lothar Lenz/zefa/CORBIS; **126** Sarah Hadley/Alamy Images; **128** JPL/NASA; **129** Photo by Brad Kuntz; **132** (t)David Siccardi/CORBIS, (bl)Comstock/CORBIS, (br)Dann Tardif/LWA/Blend Images/Getty Images; **137** Stephen Frink/zefa/CORBIS; **139** Eddie Adams/Sygma/CORBIS; **140** Streeter Lecka/Stringer/Getty Images; **143** Jose Luis Pelaez, Inc./Blend Images/Getty Images; **146** (l)Ryan McVay/Photodisc/Getty Images, (r)Comstock/CORBIS; **148** John Evans; **153** John Eder/Stone/Getty Images; **154** Robert Glusic/Photodisc/Getty Images; **160** David Young-Wolff/PhotoEdit; **165** Erich Schlegel/NewSport/CORBIS; **176-177** Brand X Pictures/PunchStock; **178** Robert Marien/CORBIS; **183** Zigmund Leszcynski/Animals Animals; **184** Mark Duffy/Alamy Images; **185** Raymond K Gehman/National Geographic/Getty Images; **190** Matt Meadows; **193** Jose Luis Pelaez Inc/Blend Images/Getty Images; **195** (l)Seth Kushner/Taxi/Getty Images, (r)Image Source/SuperStock; **198** Comstock/SuperStock; **203** Bäumle Studios/Alamy Images; **208** Steve Bloom/Taxi/Getty Images; **210** (l)RubberBall/SuperStock, (r)RubberBall/SuperStock; **212** Roy Ooms/Masterfile; **216** John Wilkes/SuperStock; **218** United States coin images from the United States Mint; **219** Yiorgos Karahalis/Reuters/CORBIS; **228** Eclipse Studios; **230** Eastcott Momatiuk/Photodisc/Getty Images; **232** Michael Newman/PhotoEdit; **234** Russell Kightley/Photo Researchers; **239** David R. Frazier Photolibrary, Inc./Alamy Images; **240** (l)George Doyle/Stockbyte/Getty Images, (r)Michael Newman/PhotoEdit; **242** Diane Macdonald/Stockbyte/Getty Images; **244** Stockbyte/Getty Images; **248** Getty Images; **254** Garry Black/Masterfile; **256** Rob Lewine/CORBIS; **262** Jeff Greenberg/Alamy Images; **268** Phyllis Greenberg/Animals Animals; **270** (l)Jack Hollingsworth/CORBIS, (r)Barbara Penoyar/Photodisc/Getty Images; **278-279** Christie's Images/CORBIS; **280** Alan Schein/zefa/CORBIS; **283** Brian Snyder/Reuters/CORBIS; **285** (l)Tom Brakefield/CORBIS, (r)Judith Collins/Alamy Images; **286** (l)Donna Day/ImageState, (r)JupiterImages/Thinkstock/Alamy Images; **287** David Young-Wolff/PhotoEdit; **289** Doug Menuez/Photodisc/Getty Images; **290** Ryan McVay/Stone+/Getty Images; **291** Don Emmert/AFP/Getty Images; **295** Thomas Allen/Digital Vision/Getty Images; **298** John Lambert/Brand X/CORBIS; **300** Norbert Wu/Minden Pictures; **305** age fotostock/SuperStock; **307** Lawrence Lawry/Photodisc/Getty Images; **308** (t)Bruce Edwards/America 24-7/Getty Images, (bl)Jonathan Nourok/PhotoEdit, (br)Michael Newman/PhotoEdit; **312** CORBIS; **314** Alley Cat Productions/Brand X/CORBIS; **318** KS Studios; **319** Michael Newman/PhotoEdit; **325** Bilderbuch/Design Pics/CORBIS; **331** Franz Marc Frei/CORBIS; **340** Masterfile; **344** DLILLC/CORBIS; **347** G. Biss/Masterfile; **350** Ingemar Edfalk/Pixonnet.com/Alamy Images; **352** Donna Ikenberry/Animals Animals; **354** Digital Vision/Getty Images; **356** Big Cheese Photo/SuperStock; **358** D. Robert & Lorri Franz/CORBIS; **359** (l)FIRST LIGHT ASSOCIATED PHOTOGRAPHERS, (r)Flying Colours Ltd/Digital Vision/Getty Images; **363** Alaskastock; **366** Adrian Peacock/Digital Vision/Getty Images; **374** (l)RubberBall/ImageState, (r)Barbara Penoyar/Photodisc/Getty Images; **376** Gregor Schuster/Photographer's Choice RF/Getty Images; **380** Don Mason/Blend Images/Getty Images; **392-393** Matthias Kulka/zefa/CORBIS; **394** Michael Ventura/Alamy Images; **399** Harald Sund/Photographer's Choice/Getty Images; **400** (l)MedioImages/Alamy Images, (r)Tony Freeman/PhotoEdit; **406** CORBIS; **410** Barbara Peacock/Taxi/Getty Images; **411** JUPITERIMAGES/Liquid Library Value/Alamy; **413** Brandon D. Cole/CORBIS; **419** Daryl Balfour/The Image Bank/Getty Images; **424** Gabe Palmer/Alamy Images; **431** Bob Daemmrich/Stock Boston; **438** Paul Costello/Getty Images; **441** photocuisine/CORBIS; **446** Andrew J.G. Bell; Eye Ubiquitous/CORBIS; **458** Robert Michael/CORBIS; **461** Bettmann/ CORBIS; **468** Ariel Skelley/CORBIS; **469** Tom Hauck/Getty Images; **473** O'Brien Productions/CORBIS; **476** Shenval/Alamy Images; **477** Tetra Images/Alamy Images; **481** Mitchell Layton/NewSport/CORBIS; **483** (l)George Doyle/Stockbyte Platinum/Alamy Images, (r)Jose Luis Pelaez Inc/Blend Images/Getty Images; **484** Laura Sifferlin; **488** Zefa RF/Alamy Images; **492** United States coin images from the United States Mint; **506-507** Oberto Gill/Beateworks/CORBIS; **508** Dennis MacDonald/Alamy Images; **510** Kelly-Mooney Photography/CORBIS; **519** Stephen Dalton/Animals Animals; **526** Jeff Greenberg/PhotoEdit; **528** Damir Frkovic/Masterfile; **530** Ryan McVay/Photodisc/Getty Images; **537** (l)Michael Newman/PhotoEdit, (r)Michelle Pedone/zefa/CORBIS; **544** Martin Shields/Alamy Images; **548** (t)Andriy Doriy Textures/Alamy Images, (b)Somos/Veer/Getty Images; **550** (l)Jeff Greenberg/PhotoEdit, (r)Todd Gipstein/National Geographic/Getty Images; **556** Artwork courtesy of Marjorie Rice; **560** Robert Spoenlein/zefa/CORBIS; **561** (t)Jose Fuste Raga/CORBIS, (b)Allen Wallace/Photonica/Getty Images; **570** Kim Karpeles/Alamy Images; **573** Don Farrall/Photodisc/Getty Images; **575** Jules Frazier/Photodisc/Getty Images; **580** Byron Schumaker/Alamy; **583** Getty Images; **584** Iconotec/Alamy Images; **587** (t)Studio Wartenberg/zefa/CORBIS, (c)Klaus Leidorf/zefa/CORBIS, (bl)BananaStock/PictureQuest/Jupiter Images, (br)BananaStock/PunchStock; **590** United States coin images from the United States Mint; **592** (bl)MedioImages/CORBIS, (br)Michael Newman/PhotoEdit; **594** Jim Esposito Photography LLC/Photodisc/Getty Images; **604** Judith Collins/Alamy images; **605** (t)C Squared Studios/Photodisc/Getty Images, (c)Glencoe, (b)PhotoLink/Photodisc/Getty Images; **608** Phil Degginger/Alamy Images; **610** Chris Bell/Taxi/Getty Images; **611** (tl)Mark Karrass/CORBIS, (tr)Ryan McVay/Photodisc/Getty Images, (b)Lew Robertson/CORBIS; **614** Charles Gupton/CORBIS; **617** Keate/Masterfile; **620** Tony Freeman/PhotoEdit; **621** Duncan Usher/Foto Natura/Minden Pictures; **629** Gregor Schuster/zefa/CORBIS; **634** Horizons Companies; **638** Myrleen Ferguson Cate/PhotoEdit; **641** Steve Bly/Alamy Images; **645** (l)Comstock/SuperStock, (r)Stockbyte/SuperStock; **646** Michael Newman/PhotoEdit; **652** Daly & Newton/Getty Images; **657** W.A.Hamilton/Alamy; **666** Eclipse Studios; **LA0-LA1** Brand X Pictures/Punchstock; **LA2** Charles O'Rear/CORBIS; **LA3** Chris A Crumley/Alamy Images; **LA9** Doug Menuez/Getty Images; **LA13** allOver photography/Alamy Images; **LA14** Brand X Pictures/PunchStock; **LA17** Royalty-Free-CORBIS; **LA18** dbphots/Alamy Images; **LA19** Brandon Haist, Flagstaff, Arizona; **LA21** Merlin D. Tuttle, Bat Conservation International; **LA24** Royalty-Free-CORBIS

Glossary/Glosario

NY Math Online A mathematics multilingual glossary is available at glencoe.com.

The glossary includes the following languages.

Arabic	Cantonese	Korean	Tagalog
Bengali	English	Russian	Urdu
Brazilian Portuguese	Haitian Creole	Spanish	Vietnamese
	Hmong		

Cómo usar el glosario en español:

1. Busca el término en inglés que desees encontrar.
2. El término en español, junto con la definición, se encuentran en la columna de la derecha.

English / Español

absolute value (p. 81)   The distance the number is from zero on a number line.

valor absoluto   Distancia a la que se encuentra un número de cero en la recta numérica.

acute angle (p. 511)   An angle with a measure greater than 0° and less than 90°.

ángulo agudo   Ángulo que mide más de 0° y menos de 90°.

acute triangle (p. 525)   A triangle having three acute angles.

triángulo acutángulo   Triángulo con tres ángulos agudos.

Addition Property of Equality (p. 138) If you add the same number to each side of an equation, the two sides remain equal.

propiedad de adición de la igualdad Si sumas el mismo número a ambos lados de una ecuación, los dos lados permanecen iguales.

additive inverse (p. 96)   The opposite of an integer. The sum of an integer and its additive inverse is zero.

inverso aditivo   El opuesto de un entero. La suma de un entero y su inverso aditivo es cero.

adjacent angles (p. 511)   Angles that have the same vertex, share a common side, and do not overlap.

ángulos adyacentes   Ángulos que comparten el mismo vértice y un común lado, pero no se sobreponen.

algebra (p. 44)   The branch of mathematics that involves expressions with variables.

álgebra   Rama de las matemáticas que involucra expresiones con variables.

algebraic expression (p. 44)   A combination of variables, numbers, and at least one operation.

expresión algebraica   Combinación de variables, números y por lo menos una operación.

analyze (p. 397)   To describe, summarize, and compare data.

analizar   Describir, resumir o comparar datos.

angle (p. 510)   Two rays with a common endpoint form an angle. The rays and vertex are used to name the angle.

∠ABC, ∠CBA, or ∠B

ángulo   Dos rayos con un extremo común forman un ángulo. Los rayos y el vértice se usan para nombrar el ángulo.

∠ABC, ∠CBA o ∠B

area (p. 157)   The number of square units needed to cover a surface enclosed by a geometric figure.

área   El número de unidades cuadradas necesarias para cubrir una superficie cerrada por una figura geométrica.

arithmetic sequence (p. 57)   A sequence in which each term is found by adding the same number to the previous term.

sucesión aritmética   Sucesión en que cada término se encuentra sumando el mismo número al término anterior.

Associative Property (p. 54)   The way in which three numbers are grouped when they are added or multiplied does not change their sum or product.

propiedad asociativa   La manera de agrupar tres números al sumarlos o multiplicarlos no cambia su suma o producto.

average (p. 402)   The mean of a set of data.

promedio   La media de un conjunto de datos.

B

bar graph (p. 415)   A graphic form using bars to make comparisons of statistics.

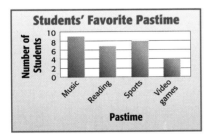

gráfica de barras   Forma gráfica que usa barras para hacer comparaciones estadísticas.

bar notation (p. 197)   In repeating decimals, the line or bar placed over the digits that repeat. For example, $2.\overline{63}$ indicates that the digits 63 repeat.

notación de barra   Línea o barra que se coloca sobre los dígitos que se repiten en decimales periódicos. Por ejemplo, $2.\overline{63}$ indica que los dígitos 63 se repiten.

base (p. 30)   In a power, the number used as a factor. In 10^3, the base is 10. That is, $10^3 = 10 \times 10 \times 10$.

base   En una potencia, el número usado como factor. En 10^3, la base es 10. Es decir, $10^3 = 10 \times 10 \times 10$.

base (p. 572)   The base of a parallelogram or triangle is any side of the figure. The bases of a trapezoid are the parallel sides.

base   La base de un paralelogramo o triángulo es el lado de la figura. Las bases de un trapecio son los lados paralelos.

base (p. 603)   The top or bottom face of a three-dimensional figure.

base   La cara inferior o superior de una figura tridimensional.

biased sample (p. 439)   A sample drawn in such a way that one or more parts of the population are favored over others.

muestra sesgada   Muestra en que se favorece una o más partes de una población.

center (p. 584)   The given point from which all points on a circle or sphere are the same distance.

centro   Un punto dado del cual equidistan todos los puntos de un círculo o de una esfera.

circle (p. 584)   The set of all points in a plane that are the same distance from a given point called the center.

círculo   Conjunto de todos los puntos en un plano que equidistan de un punto dado llamado centro.

circle graph (p. 518)   A type of statistical graph used to compare parts of a whole.

gráfica circular   Tipo de gráfica estadística que se usa para comparar las partes de un todo.

Area of Oceans

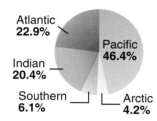

Área de superficie de los océanos

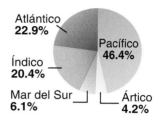

circumference (p. 584)   The distance around a circle.

circunferencia   La distancia alrededor de un círculo.

cluster (p. 397)   Data that are grouped closely together.

agrupamiento   Datos estrechamente agrupados.

coefficient (p. 45)   The numerical factor of a term that contains a variable.

coeficiente   El factor numérico de un término que contiene una variable.

combination (p. 480)   An arrangement, or listing, of objects in which order is not important.

combinación   Arreglo o lista de objetos donde el orden no es importante.

common denominator (p. 215)   A common multiple of the denominators of two or more fractions. 24 is a common denominator for $\frac{1}{3}$, $\frac{5}{8}$, and $\frac{3}{4}$ because 24 is the LCM of 3, 8, and 4.

común denominador   El múltiplo común de los denominadores de dos o más fracciones. 24 es un denominador común para $\frac{1}{3}$, $\frac{5}{8}$ y $\frac{3}{4}$ porque 24 es el mcm de 3, 8 y 4.

Commutative Property (p. 54)   The order in which two numbers are added or multiplied does not change their sum or product.

propiedad conmutativa   El orden en que se suman o multiplican dos números no afecta su suma o producto.

complementary angles (p. 514)   Two angles are complementary if the sum of their measures is 90°.

ángulos complementarios   Dos ángulos son complementarios si la suma de sus medidas es 90°.

∠1 and ∠2 are complementary angles.

∠1 y ∠2 son complementarios.

complementary events (p. 462)   The events of one outcome happening and that outcome not happening are complementary events. The sum of the probabilities of complementary events is 1.

complex figure (p. 596)   A figure made of circles, rectangles, squares, and other two-dimensional figures.

composite number (p. 181)   A whole number greater than 1 that has more than two factors.

compound event (p. 492)   An event consisting of two or more simple events.

cone (p. 604)   A three-dimensional figure with a curved surface and a circular base.

congruent angles (p. 511)   Angles that have the same measure.

∠1 and ∠2 are congruent angles.

congruent figures (p. 554)   Figures with equal corresponding sides of equal length and corresponding angles of equal measure.

congruent segments (p. 525)   Segments having the same measure.

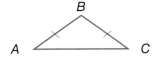

Side $\overline{AB}$ is congruent to side $\overline{BC}$.

convenience sample (p. 439)   A sample which includes members of the population that are easily accessed.

coordinate plane (p. 88)   A plane in which a horizontal number line and a vertical number line intersect at their zero points. Also called a coordinate grid.

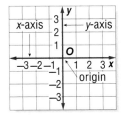

eventos complementarios   Se dice de los eventos de un resultado que ocurren y el resultado que no ocurre. La suma de las probabilidades de eventos complementarios es 1.

figura compleja   Una figura compuesta por círculos, rectángulos, cuadrados y otras dos figuras bidimencionales.

número compuesto   Un número entero mayor que 1 que tiene más de dos factores.

evento compuesto   Un evento que consiste en dos o más eventos simples.

cono   Figura tridimensional con una superficie curva y una base circular.

ángulos congruentes   Ángulos que tienen la misma medida.

∠1 y ∠2 son congruentes.

figuras congruentes   Figuras cuyos lados y ángulos correspondientes son iguales.

segmentos congruentes   Segmentos que tienen la misma medida.

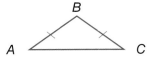

$\overline{AB}$ es congruente a $\overline{BC}$.

muestra de conveniencia   Muestra que incluye miembros de una población fácilmente accesibles.

plano de coordenadas   Plano en el cual se han trazado dos rectas numéricas, una horizontal y una vertical, que se intersecan en sus puntos cero. También conocido como sistema de coordenadas.

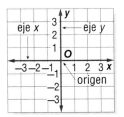

corresponding angles (p. 540)   Congruent angles of similar figures.

corresponding sides (p. 540)   Congruent or proportional sides of similar figures.

counterexample (p. 56)   An example showing that a statement is not true.

cross product (p. 310)   In a proportion, a cross product is the product of the numerator of one ratio and the denominator of the other ratio.

cubed (p. 30)   The product in which a number is a factor three times. Two cubed is 8 because $2 \times 2 \times 2 = 8$.

cylinder (p. 604)   A three-dimensional figure with two parallel congruent circular bases.

ángulos correspondientes   Ángulos iguales de figuras semejantes.

lados correspondientes   Lados iguales o proporcionales de figuras semejantes.

contraejemplo   Ejemplo que demuestra que un enunciado no es verdadero.

productos cruzados   En una proporción, un producto cruzado es el producto del numerador de una razón y el denominador de la otra razón.

al cubo   El producto de un número por sí mismo, tres veces. Dos al cubo es 8 porque $2 \times 2 \times 2 = 8$.

cilindro   Figura tridimensional que tiene dos bases circulares congruentes y paralelas.

D

data (p. 376)   Pieces of information, which are often numercical.

decagon (p. 546)   A polygon having ten sides.

defining the variable (p. 50)   Choosing a variable to represent an unknown value in a problem, and using it to write an expression or equation to solve the problem.

degrees (p. 510)   The most common unit of measure for angles. If a circle were divided into 360 equal-sized parts, each part would have an angle measure of 1 degree.

dependent events (p. 493)   Two or more events in which the outcome of one event affects the outcome of the other event(s).

diameter (p. 584)   The distance across a circle through its center.

disjoint events (p. 494)   Events that cannot happen at the same time.

datos   Información, la cual a menudo se presenta de manera numérica.

decágono   Un polígono con diez lados.

definir una variable   El elegir una variable para representar un valor desconocido en un problema y usarla para escribir una expresión o ecuación para resolver el problema.

grados   La unidad más común para medir ángulos. Si un círculo se divide en 360 partes iguales, cada parte tiene una medida angular de 1 grado.

eventos dependientes   Dos o más eventos en que el resultado de un evento afecta el resultado de otro u otros eventos.

diámetro   La distancia a través de un círculo pasando por el centro.

eventos disjuntos   Eventos que no pueden ocurrir al mismo tiempo.

Distributive Property (p. 53)   To multiply a sum by a number, multiply each addend of the sum by the number outside the parentheses.

propiedad distributiva   Para multiplicar una suma por un número, multiplica cada sumando de la suma por el número fuera del paréntesis.

Division Property of Equality (p. 142)   If you divide each side of an equation by the same nonzero number, the two sides remain equal.

propiedad de igualdad de la división   Si divides ambos lados de una ecuación entre el mismo número no nulo, los lados permanecen iguales.

domain (p. 63)   The set of input values for a function.

dominio   El conjunto de valores de entrada de una función.

E

edge (p. 603)   The segment formed by intersecting faces of a three-dimensional figure.

arista   Segmento de recta formado por la intersección de las caras en una figura tridimensional.

equation (p. 49)   A mathematical sentence that contains an equals sign, =.

ecuación   Enunciado matemático que contiene un signo de igualdad, =.

equilateral triangle (p. 525)   A triangle having three congruent sides.

triángulo equilátero   Triángulo con tres lados congruentes.

equivalent expressions (p. 53)   Expressions that have the same value.

expresiones equivalentes   Expresiones que tienen el mismo valor.

equivalent fractions (p. 192)   Fractions that have the same value. $\frac{2}{3}$ and $\frac{4}{6}$ are equivalent fractions.

fracciones equivalentes   Fracciones que tienen el mismo valor. $\frac{2}{3}$ y $\frac{4}{6}$ son fracciones equivalentes.

equivalent ratios (p. 288)   Two ratios that have the same value.

razones equivalentes   Dos razones que tienen el mismo valor.

evaluate (p. 31)   To find the value of an expression.

evaluar   Calcular el valor de una expresión. probabilidad experimental

experimental probability (p. 486)   An estimated probability based on the relative frequency of positive outcomes occurring during an experiment.

probabilidad experimental   Estimado de una probabilidad que se basa en la frecuencia relativa de los resultados positivos que ocurren durante un experimento.

exponent (p. 30)   In a power, the number that tells how many times the base is used as a factor. In 5^3, the exponent is 3. That is, $5^3 = 5 \times 5 \times 5$.

exponente   En una potencia, el número que indica las veces que la base se usa como factor. En 5^3, el exponente es 3. Es decir, $5^3 = 5 \times 5 \times 5$.

exponential form (p. 31)   Numbers written with exponents.

forma exponencial   Números escritos usando exponentes.

F

face (p. 603)   The flat surface of a three-dimensional figure.

cara   Superficies planas de una figura tridimensional.

factors (p. 30)   Two or more numbers that are multiplied together to form a product.

factores   Dos o más números que se multiplican entre sí para formar un producto.

factor tree (p. 182)   A diagram showing the prime factorization of a number. The factors branch out from the previous factors until all of the factors are prime numbers.

diagrama de árbol   Diagrama que muestra la factorización prima de un número. Los factores se ramifican a partir de los factores previos hasta que todos los factores son números primos.

formula (p. 144)   An equation that shows the relationship among certain quantities.

fórmula   Ecuación que muestra la relación entre ciertas cantidades.

function (p. 63)   A relation in which each element of the input is paired with exactly one element of the output according to a specified rule.

función   Relación en que cada elemento de entrada es apareado con un único elemento de salida, según una regla específica.

function rule (p. 63)   The operation performed on the input of a function.

regla de función   Operación que se efectúa en el valor de entrada.

function table (p. 63)   A table used to organize the input numbers, output numbers, and the function rule.

tabla de funciones   Tabla que organiza las entradas, la regla y las salidas de una función.

Fundamental Counting Principle (p. 471)   Uses multiplication of the number of ways each event in an experiment can occur to find the number of possible outcomes in a sample space.

Principio Fundamental de Contar   Este principio usa la multiplicación del número de veces que puede ocurrir cada evento en un experimento para calcular el número de posibles resultados en un espacio muestral.

G

gram (p. 304)   A unit of mass in the metric system equivalent to 0.001 kilogram.

gramo   Unidad de masa del sistema métrico. Un gramo equivale a 0.001 de kilogramo.

graph (p. 80)   The process of placing a point on a number line at its proper location.

graficar   Proceso de dibujar o trazar un punto en una recta numérica en su ubicación correcta.

greatest common factor (GCF) (p. 186)   The greatest of the common factors of two or more numbers. The GCF of 18 and 24 is 6.

máximo común divisor (MCD)   El mayor factor común de dos o más números. El MCD de 18 y 24 es 6.

H

height (p. 572)   The length of the segment perpendicular to the base with endpoints on opposite sides. In a triangle, the distance from a base to the opposite vertex.

altura   Longitud del segmento perpendicular a la base y con extremos en lados opuestos. En un triángulo, es la distancia desde una base al vértice opuesto.

heptagon (p. 546)   A polygon having seven sides.

heptágono   Polígono con siete lados.

hexagon (p. 546)   A polygon having six sides.

hexágono   Polígono con seis lados.

histogram (p. 416)   A special kind of bar graph in which the bars are used to represent the frequency of numerical data that have been organized in intervals.

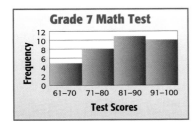

histograma   Tipo especial de gráfica de barras que usa barras para representar la frecuencia de los datos numéricos, los cuales han sido organizados en intervalos iguales.

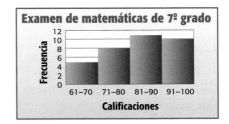

hypotenuse (p. 640)   The side opposite the right angle in a right triangle.

hipotenusa   El lado opuesto al ángulo recto en un triángulo rectángulo.

I

Identity Property (p. 54)   The sum of an addend and 0 is the addend. The product of a factor and 1 is the factor.

propiedad de identidad   La suma de un sumando y 0 es el sumando mismo. El producto de un factor y 1 es el factor mismo.

independent events (p. 492)   Two or more events in which the outcome of one event does not affect the outcome of the other event(s).

eventos independientes   Dos o más eventos en los cuales el resultado de uno de ellos no afecta el resultado de los otros eventos.

indirect measurement (p. 542)   Finding a measurement by using similar triangles and writing a proportion.

medida indirecta   Técnica que se usa para calcular una medida a partir de triángulos semejantes y proporciones.

integer (p. 80)   Any number from the set $\{... , -4, -3, -2, -1, 0, 1, 2, 3, 4, ...\}$

entero   Todo número del conjunto $\{... , -4, -3, -2, -1, 0, 1, 2, 3, 4, ...\}$

inverse operations (p. 136)   Operations that "undo" each other. Addition and subtraction are inverse operations.

operaciones inversas   Operaciones que se "anulan" mutuamente. La adición y la sustracción son operaciones inversas.

irrational number (p. 637)   A number that cannot be expressed as the quotient of two integers.

número irracional   Número que no se puede expresar como el cociente de dos enteros.

isosceles triangle (p. 525)   A triangle having at least two congruent sides.

triángulo isósceles   Triángulo que tiene por lo menos dos lados congruentes.

kilogram (p. 304)   The base unit of mass in the metric system equivalent to 1,000 grams.

kilogramo   Unidad fundamental de masa del sistema métrico. Un kilogramo equivale a mil gramos.

L

lateral face (p. 603)   One side of a three-dimesional figure.

cara lateral   Lado de una figura tridimensional.

leaf (p. 410)   The second greatest place value of data in a stem-and-leaf plot.

hoja   El segundo valor de posición mayor en un diagrama de tallo y hojas.

least common denominator (LCD) (p. 215)   The least common multiple of the denominators of two or more fractions.

mínimo común denominador (mcd)   El menor múltiplo común de los denominadores de dos o más fracciones.

least common multiple (LCM) (p. 211)   The least of the common multiples of two or more numbers. The LCM of 2 and 3 is 6.

mínimo común múltiplo (mcm)   El menor múltiplo común de dos o más números. El mcm de 2 y 3 es 6.

leg (p. 640)   Either of the two sides that form the right angle of a right triangle.

cateto   Cualquiera de los lados que forman el ángulo recto de un triángulo rectángulo.

like fractions (p. 236)   Fractions that have the same denominator.

fracciones semejantes   Fracciones con el mismo denominador.

line graph (p. 426)   A type of statistical graph using lines to show how values change over a period of time.

gráfica lineal   Tipo de gráfica estadística que usa segmentos de recta para mostrar cómo cambian los valores durante un período de tiempo.

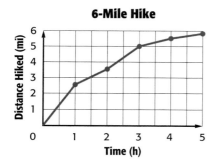

line of reflection (p. 559)   The line over which a figure is reflected.

eje de reflexión   La línea sobre la cual se refleja una figura.

line of symmetry (p. 558)   A line that divides a figure into two halves that are reflections of each other.

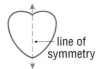

line plot (p. 396)   A diagram that shows the frequency of data. An × is placed above a number on a number line each time that number occurs in a set of data.

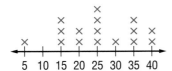

line symmetry (p. 558)   Figures that match exactly when folded in half have line symmetry.

linear equation (p. 164)   An equation for which the graph is a straight line.

liter (p. 304)   The base unit of capacity in the metric system. A liter is a little more than a quart.

eje de simetría   Recta que divide una figura en dos mitades que son reflexiones entre sí.

esquema lineal   Grafica que muestra la frecuencia de datos. Se coloca una × sobre la recta numérica, cada vez que el número aparece en un conjunto de datos.

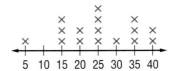

simetría lineal   Exhiben simetría lineal las figuras que coinciden exactamente al doblarse una sobre otra.

ecuación lineal   Ecuación cuya gráfica es una recta.

litro   Unidad básica de capacidad del sistema métrico. Un litro es un poco más de un cuarto de galón.

mean (p. 402)   The sum of the data divided by the number of items in the data set.

measures of central tendency (p. 402)   Numbers that are used to describe the center of a set of data. These measures include the mean, median, and mode.

median (p. 403)   The middle number in a set of data when the data are ordered from least to greatest. If the data has an even number of items, the median is the mean of the two numbers closer to the middle.

meter (p. 304)   The base unit of length in the metric system.

metric system (p. 304)   A base-ten system of measurement using the base units: meter for length, kilogram for mass, and liter for capacity.

media   La suma de los datos dividida entre el número total de artículos en el conjunto de datos.

medidas de tendencia central   Números que se usan para describir el centro de un conjunto de datos. Estas medidas incluyen la media, la mediana y la moda.

mediana   El número del medio en un conjunto de datos cuando los datos se ordenan de menor a mayor. Si los datos tienen un número par de artículos, la mediana es la media de los dos números más cercanos al medio.

metro   Unidad fundamental de longitud del sistema métrico.

sistema métrico   Sistema de medidas de base diez que usa las unidades fundamentales: metro para longitud, kilogramo para masa y litro para capacidad.

mode (p. 403)   The number or numbers that appear most often in a set of data. If there are two or more numbers that occur most often, all of them are modes.

moda   El número o números que aparece con más frecuencia en un conjunto de datos. Si hay dos o más números que ocurren con más frecuencia, todosellos son modas.

multiple (p. 211)   The product of a number and any whole number.

múltiplo   El producto de un número y cualquier número entero.

Multiplication Property of Equality (p. 259) If you multiply each side of an equation by the same nonzero number, the two sides remain equal.

propiedad de multiplicación de la igualdad Si multiplicas ambos lados de una ecuación por el mismo número no nulo, lo lados permanecen iguales.

multiplicative inverse (p. 258)   The product of a number and its multiplicative inverse is 1.

The multiplicative inverse of $\frac{2}{3}$ is $\frac{3}{2}$.

inverso multiplicativo   El producto de un número y su inverso multiplicativo es 1.

El inverso multiplicativo de $\frac{2}{3}$ es $\frac{3}{2}$.

negative integer (p. 80)   An integer that is less than zero.

entero negativo   Un entero menor que cero.

net (p. 600)   A two-dimensional figure that can be used to build a three-dimensional figure.

red   Figura bidimensional que sirve para hacer una figura tridimensional.

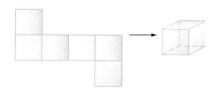

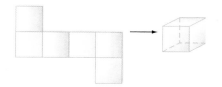

nonagon (p. 546)   A polygon having nine sides.

enágono   Polígono que tiene nueve lados.

numerical expression (p. 38)   A combination of numbers and operations.

expresión numérica   Combinación de números y operaciones.

obtuse angle (p. 511)   Any angle that measures greater than 90° but less than 180°.

ángulo obtuso   Cualquier ángulo que mide más de 90° pero menos de 180°.

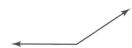

obtuse triangle (p. 525)   A triangle having one obtuse angle.

triángulo obtusángulo   Triángulo que tiene un ángulo obtuso.

octagon (p. 546)   A polygon having eight sides.

octágono   Polígono que tiene ocho lados.

opposites (p. 96)   Two integers are opposites if they are represented on the number line by points that are the same distance from zero, but on opposite sides of zero. The sum of two opposites is zero.

opuestos   Dos enteros son opuestos si, en la recta numérica, están representados por puntos que equidistan de cero, pero en direcciones opuestas. La suma de dos opuestos es cero.

order of operations (p. 38)   The rules to follow when more than one operation is used in a numerical expression.
1. Evaluate the expressions inside grouping symbols
2. Evaluate all powers
3. Multiply and divide in order from left to right.
4. Add and subtract in order from left to right.

orden de operaciones   Reglas a seguir cuando se usa más de una operación en una expresión numérica.
1. Primero ejecuta todas las operaciones dentro de los símbolos de agrupamiento
2. Evalúa todas las potencias antes que las otras operaciones.
3. Multiplica y divide en orden de izquierda a derecha.
4. Suma y resta en orden de izquierda a derecha.

ordered pair (p. 88)   A pair of numbers used to locate a point in the coordinate plane. An ordered pair is written in the form (x-coordinate, y-coordinate).

par ordenado   Par de números que se utiliza para ubicar un punto en un plano de coordenadas. Se escribe de la siguiente forma: (coordenada x, coordenada y).

origin (p. 88)   The point at which the x-axis and the y-axis intersect in a coordinate plane.

origen   Punto en que el eje x y el eje y se intersecan en un plano de coordenadas.

outcome (p. 460)   One possible result of a probability event. For example, 4 is an outcome when a number cube is rolled.

resultado   Uno de los resultados posibles de un evento probabilístico. Por ejemplo, 4 es un resultado posible cuando se lanza un dado.

outlier (p. 397)   A piece of data that is quite separated from the rest of the data.

valor atípico   Dato que se encuentra muy separado del resto de los datos.

parallel lines (p. 533)   Lines in a plane that do not intersect.

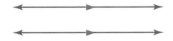

líneas paralelas   Rectas situadas en un mismo plano y que no se intersecan.

parallelogram (p. 533)   A quadrilateral with opposite sides parallel and opposite sides congruent.

paralelogramo   Cuadrilátero cuyos lados opuestos son paralelos y congruentes.

part (p. 350)   In a percent proportion, the number that is compared to the whole quantity.

parte   En una proporción porcentual, el número que se compara con la cantidad total.

pentagon (p. 546)   A polygon having five sides.

percent (p. 202)   A ratio that compares a number to 100.

percent equation (p. 361)   An equation that describes the relationship between the part, whole, and percent.
part = percent · whole

percent of change (p. 369)   A ratio that compares the change in a quantity to the original amount.

percent of decrease (p. 369)   A percent of change when the original quantity decreased.

percent of increase (p. 369)   A percent of change when the original quantity increased.

percent proportion (p. 350)   Compares part of a quantity to the whole quantity using a percent.

$$\frac{\text{part}}{\text{whole}} = \frac{\text{percent}}{100}$$

perfect squares (p. 34)   Numbers with square roots that are whole numbers. 25 is a perfect square because the square root of 25 is 5.

perimeter (p. 156)   The distance around a closed geometric figure.

permutation (p. 475)   An arrangement, or listing, of objects in which order is important.

perpendicular lines (p. 512)   Lines that meet to form right angles.

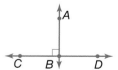

pi (π) (p. 584)   The ratio of the circumference of a circle to its diameter. An approximation often used for π is 3.14.

polygon (p. 546)   A simple closed figure in a plane formed by three or more line segments.

pentágono   Polígono que tiene cinco lados.

por ciento   Razón que compara un número con 100.

ecuación porcentual   Ecuación que describe la relación entre la parte, el todo y el por ciento. parte = por ciento · todo.

porcentaje de cambio   Razón que compara el cambio en una cantidad, con la cantidad original.

porcentaje de disminución   Porcentaje de cambio cuando disminuye la cantidad original.

porcentaje de aumento   Porcentaje de cambio cuando aumenta la cantidad original.

proporción porcentual   Comparar partes de una cantidad, a la cantidad entera, usando un porcentaje.

$$\frac{\text{parte}}{\text{todo}} = \frac{\text{porcentaje}}{100}$$

cuadrados perfectos   Números cuya raíz cuadrada es un número entero. 25 es un cuadrado perfecto porque la raíz cuadrada de 25 es 5.

perímetro   La distancia alrededor de una figura geométrica cerrada.

permutación   Arreglo o lista en que el orden es importante.

rectas perpendiculares   Rectas que al encontrarse forman ángulos rectos.

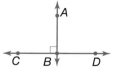

pi (π)   Razón entre la circunferencia de un círculo y su diámetro. A menudo, se usa 3.14 como aproximación del valor de π.

polígono   Figura simple cerrada en un plano, formada por tres o más segmentos de recta.

population (p. 434)   The entire group of items or individuals from which the samples under consideration are taken.

positive integer (p. 80)   An integer that is greater than zero.

powers (p. 30)   Numbers expressed using exponents. The power 3^2 is read *three to the second power, or three squared.*

prime factorization (p. 182)   Expressing a composite number as a product of prime numbers. For example, the prime factorization of 63 is $3 \times 3 \times 7$.

prime number (p. 181)   A whole number greater than 1 that has exactly two factors, 1 and itself.

principal (p. 379)   The amount of money deposited or invested.

prism (p. 601)   A three-dimensional figure with at least three rectangular lateral faces and top and bottom faces parallel.

probability (p. 460)   The chance that some event will happen. It is the ratio of the number of ways a certain event can occur to the number of possible outcomes.

properties (p. 54)   Statements that are true for any number or variable.

proportion (p. 310)   An equation that shows that two ratios are equivalent.

proportional (p. 310)   The relationship between two ratios with a constant rate or ratio.

protractor (p. 680)   An instrument used to measure angles.

pyramid (p. 603)   A three-dimensional figure with at least three lateral faces that are triangles and only one base.

Pythagorean Theorem (p. 640)   In a right triangle, the square of the length of the hypotenuse is equal to the sum of the squares of the lengths of the legs. $c^2 = a^2 + b^2$

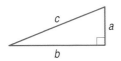

población   El grupo total de individuos o de artículos del cual se toman las muestras bajo estudio.

entero positivo   Un entero mayor que cero.

potencias   Números que se expresan usando exponentes. La potencia 3^2 se lee *tres a la segunda potencia o tres al cuadrado.*

factorización prima   Escritura de un número compuesto como el producto de números primos. La factorización prima de 63 es $3 \times 3 \times 7$.

número primo   Número entero mayor que 1 que sólo tiene dos factores, 1 y sí mismo.

capital   La cantidad de dinero depositada o invertida.

prisma   Figura tridimensional que tiene por lo menos tres caras laterales rectangulares y caras paralelas superior e inferior.

probabilidad   La posibilidad de que suceda un evento. Es la razón del número de maneras en que puede ocurrir un evento al número total de resultados posibles.

propiedades   Enunciados que se cumplen para cualquier número o variable.

proporción   Ecuación que muestra que dos razones son equivalentes.

proporcional   Relación entre dos razones con una tasa o razón constante.

transportador   Instrumento que sirve para medir ángulos.

pirámide   Figura tridimensional que tiene por lo menos tres caras laterales triangulares que son triángulos y una sola base.

Teorema de Pitágoras   En un triángulo rectángulo, el cuadrado de la longitud de la hipotenusa es igual a la suma de los cuadrados de las longitudes de los catetos. $c^2 = a^2 + b^2$

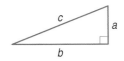

Q

quadrant (p. 88)   One of the four regions into which the two perpendicular number lines of the coordinate plane separate the plane.

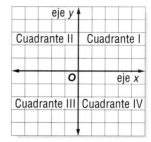

quadrilateral (p. 533)   A closed figure having four sides and four angles.

cuadrante   Una de las cuatro regiones en que dos rectas numéricas perpendiculares dividen el plano de coordenadas.

cuadrilátero   Figura cerrada que tiene cuatro lados y cuatro ángulos.

R

radical sign (p. 35)   The symbol used to indicate a nonnegative square root, $\sqrt{\ }$.

radius (p. 584)   The distance from the center of a circle to any point on the circle.

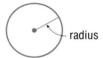

random (p. 461)   Outcomes occur at random if each outcome is equally likely to occur.

range (p. 63)   The set of output values for a function.

range (p. 397)   The difference between the greatest and least numbers in a data set.

rate (p. 287)   A ratio that compares two quantities with different kinds of units.

rate of change (p. 293)   A ratio that shows a change in one quantity with respect to a change in another quantity.

ratio (p. 202)   A comparison of two numbers by division. The ratio of 2 to 3 can be written as 2 out of 3, 2 to 3, 2 : 3, or $\frac{2}{3}$.

signo radical   Símbolo que se usa para indicar una raíz cuadrada no negativa, $\sqrt{\ }$.

radio   Distancia desde el centro de un círculo hasta cualquier punto del mismo.

aleatorio   Un resultado ocurre al azar si la posibilidad de ocurrir de cada resultado es equiprobable.

rango   Conjunto de los valores de salida de una función.

rango   La diferencia entre el número mayor y el menor en un conjunto de datos.

tasa   Razón que compara dos cantidades que tienen distintas unidades de medida.

tasa de cambio   Razón que representa el cambio en una cantidad con respecto al cambio en otra cantidad.

razón   Comparación de dos números mediante división. La razón de 2 a 3 puede escribirse como 2 de cada 3, 2 a 3, 2:3 ó $\frac{2}{3}$.

rational number (p. 216) A number that can be expressed as a fraction.

número racional Número que puede expresarse como fracción.

reciprocal (p. 258) The multiplicative inverse of a number.

recíproco El inverso multiplicativo de un número.

rectangle (p. 533) A parallelogram having four right angles.

rectángulo Paralelogramo con cuatro ángulos rectos.

rectangular prism (p. 611) A solid figure that has two parallel and congruent bases that are rectangles.

prisma rectangular Figura sólida con dos bases paralelas y congruentes que son rectángulos.

reflection (p. 559) A type of transformation in which a figure is flipped over a line of symmetry.

reflexión Tipo de transformación en el que se da vuelta a una figura sobre un eje de simetría.

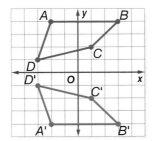

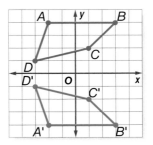

regular polygon (p. 546) A polygon that has all sides congruent and all angles congruent.

polígono regular Polígono con todos los lados y todos los ángulos congruentes.

repeating decimals (p. 197) A decimal whose digits repeat in groups of one or more. Examples are 0.181818... and 0.83333... .

decimales periódicos Decimal cuyos dígitos se repiten en grupos de uno o más. Por ejemplo: 0.181818... y 0.83333... .

rhombus (p. 533) A parallelogram having four congruent sides.

rombo Paralelogramo que tiene cuatro lados congruentes.

right angle (p. 511) An angle that measures 90°.

ángulo rect Ángulo que mide exactamente 90°.

right triangle (p. 525)   A triangle having one right angle.

triángulo rectángulo   Triángulo que tiene un ángulo recto.

S

sample (p. 438)   A randomly selected group chosen for the purpose of collecting data.

muestra   Grupo escogido al azar o aleatoriamente que se usa con el propósito de recoger datos.

sample space (p. 465)   The set of all possible outcomes of a probability experiment.

espacio muestral   Conjunto de todos los resultados posibles de un experimento probabilístico.

sampling (p. 312)   A practical method used to survey a representative group.

muestreo   Método conveniente que facilita la elección y el estudio de un grupo representativo.

scale (p. 320)   On a map, intervals used representing the ratio of distance on the map to the actual distance.

escala   En un mapa, los intervalos que se usan para representar la razón de las distancias en el mapa a las distancias verdaderas.

scale drawing (p. 320)   A drawing that is similar but either larger or smaller than the actual object.

dibujo a escala   Dibujo que es semejante, pero más grande o más pequeño que el objeto real.

scale factor (p. 322)   A scale written as a ratio in simplest form.

factor de escala   Escala escrita como una tasa en forma reducida.

scale model (p. 320)   A model used to represent something that is too large or too small for an actual-size model.

modelo a escala   Réplica de un objeto real, el cual es demasiado grande o demasiado pequeño como para construirlo de tamaño natural.

scalene triangle (p. 525)   A triangle having no congruent sides.

triángulo escaleno   Triángulo sin lados congruentes.

scatter plot (p. 427)   In a scatter plot, two sets of related data are plotted as ordered pairs on the same graph.

diagrama de dispersión   Diagrama en que dos conjuntos de datos relacionados aparecen graficados como pares ordenados en la misma gráfica.

School Commute

Commute Time (min) vs Distance From School (mi)

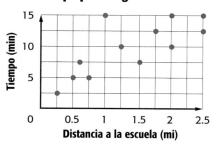

Tiempo para llegar a la escuela

Tiempo (min) vs Distancia a la escuela (mi)

semicircle (p. 696)   Half of a circle.

sequence (p. 57)   A list of numbers in a certain order, such as 0, 1, 2, 3, or 2, 4, 6, 8.

similar figures (p. 540)   Figures that have the same shape but not necessarily the same size.

simple event (p. 460)   One outcome or a collection of outcomes.

simple interest (p. 425)   The amount paid or earned for the use of money. The formula for simple interest is $I = prt$.

simple random sample (p. 438)   A sample where each item or person in the population is as likely to be chosen as any other.

simplest form (p. 192)   A fraction is in simplest form when the GCF of the numerator and the denominator is 1.

simulate (p. 491)   A way of acting out or modeling a problem situation.

slope (p. 293)   The rate of change between any two points on a line. The ratio of vertical change to horizontal change.

solution (p. 49)   A value for the variable that makes an equation true. The solution of $12 = x + 7$ is 5.

solving an equation (p. 49)   The process of finding a solution to an equation.

sphere (p. 604)   A three-dimensional figure in which all points are equal distance from the center.

square (p. 34)   The product of a number and itself. 36 is the square of 6.

square (p. 533)   A parallelogram having four right angles and four congruent sides.

square root (p. 35)   One of the two equal factors of a number. The square root of 9 is 3.

standard form (p. 31)   Numbers written without exponents.

semicírculo   Mitad de un círculo con el mismo diámetro.

sucesión   Lista de números en cierto orden, tales como 0, 1, 2, 3 ó 2, 4, 6, 8.

figuras semejantes   Figuras que tienen la misma forma, pero no necesariamente el mismo tamaño.

eventos simples   Un resultado o una colección de resultados.

interés simple   Cantidad que se paga o que se gana por el uso del dinero. La fórmula para calcular el interés simple es $I = prt$.

muestra aleatoria simple   Muestra de una población que tiene la misma probabilidad de escogerse que cualquier otra.

forma reducida   Una fracción está escrita en forma reducida si el MCD de su numerador y denominador es 1.

simulación   Manera de modelar o representar un problema.

pendiente   Razón de cambio entre cualquier par de puntos en una recta. La razón del cambio vertical al cambio horizontal.

solución   Valor de la variable de una ecuación que hace verdadera la ecuación. La solución de $12 = x + 7$ es 5.

resolver una ecuación   Proceso de encontrar el número o números que satisfagan una ecuación.

esfera   Figura tridimensional en que todos los puntos están equidistantes del centro.

cuadrado   El producto de un número por sí mismo. 36 es el cuadrado de 6.

cuadrado   Paralelogramo con cuatro ángulos rectos y cuatro lados congruentes.

raíz cuadrada   Uno de dos factores iguales de un número. La raíz cuadrada de 9 es 3.

forma estándar   Números escritos sin exponentes.

statistics (p. 396)   The branch of mathematics that deals with collecting, organizing, and interpreting data.

estadística   Rama de las matemáticas cuyo objetivo primordial es la recopilación, organización e interpretación de datos.

stem (p. 410)   The greatest place value common to all the data values is used for the stem of a stem-and-leaf plot.

tallo   El mayor valor de posición común a todos los datos es el que se usa como tallo en un diagrama de tallo y hojas.

stem-and-leaf plot (p. 410)   A system used to condense a set of data where the greatest place value of the data forms the stem and the next greatest place value forms the leaves.

diagrama de tallo y hojas   Sistema que se usa para condensar un conjunto de datos y en el cual el mayor valor de posición de los datos forma el tallo y el segundo mayor valor de posición de los datos forma las hojas.

straight angle (p. 511)   An angle that measures exactly 180°.

ángulo llano   Ángulo que mide exactamente 180°.

Subtraction Property of Equality (p. 136)   If you subtract the same number from each side of an equation, the two sides remain equal.

propiedad de sustracción de la igualdad   Si restas el mismo número de ambos lados de una ecuación, los dos lados permanecen iguales.

supplementary angles (p. 514)   Two angles are supplementary if the sum of their measures is 180°.

ángulos suplementarios   Dos ángulos son suplementarios si la suma de sus medidas es 180°.

∠1 and ∠2 are supplementary angles.

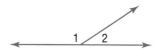

∠1 y ∠2 son suplementarios.

surface area (p. 649)   The sum of the areas of all the surfaces (faces) of a three-dimensional figure.

área de superficie   La suma de las áreas de todas las superficies (caras) de una figura tridimensional.

survey (p. 434)   A question or set of questions designed to collect data about a specific group of people.

encuesta   Pregunta o conjunto de preguntas diseñadas para recoger datos sobre un grupo específico de peronas.

term (p. 57)   Each number in a sequence.

término   Cada número en una sucesión.

terminating decimals (p. 197)   A decimal whose digits end. Every terminating decimal can be written as a fraction with a denominator of 10, 100, 1,000, and so on.

decimales terminales   Decimal cuyos dígitos terminan. Todo decimal terminal puede escribirse como una fracción con un denominador de 10, 100, 1,000, etc.

tessellation (p. 548)   A repetitive pattern of polygons that fit together with no holes or gaps.

teselado   Un patrón repetitivo de polígonos que coinciden perfectamente, sin dejar huecos o espacios.

theoretical probability (p. 486)   The ratio of the number of ways an event can occur to the number of possible outcomes.

probabilidad teórica   La razón del número de maneras en que puede ocurrir un evento al número total de resultados posibles.

three-dimensional figures (p. 603)   A figure with length, width, and depth (or height).

figuras tridimensionales   Figuras que poseen largo, ancho y profundidad (o altura).

transformation (p. 553)   A movement of a geometric figure.

transformación   Movimientos de figuras geométricas.

translation (p. 553)   One type of transformation where a geometric figure is slid horizontally, vertically, or both.

traslación   Tipo de transformación en que una figura se desliza horizontal o verticalmente o de ambas maneras.

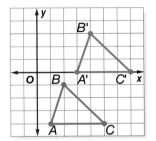

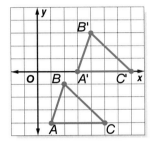

trapezoid (p. 533)   A quadrilateral with one pair of parallel sides.

trapecio   Cuadrilátero con un único par de lados paralelos.

tree diagram (p. 466)   A diagram used to show the total number of possible outcomes in a probability experiment.

diagrama de árbol   Diagrama que se usa para mostrar el número total de resultados posibles en experimento probabilístico.

triangle (p. 524)   A polygon that has three sides and three angles.

triángulo   Polígono que posee tres lados y tres ángulos.

triangular prism (p. 615)   A prism that has bases that are triangles.

prisma triangular   Prisma cuyas bases son triángulos.

two-step equation (p. 151)   An equation having two different operations.

ecuación de dos pasos   Ecuación que contiene dos operaciones distintas.

unbiased sample (p. 438)   A sample representative of the entire population.

muestra no sesgada   Muestra que se selecciona de modo que se representativa de la población entera.

unit rate (p. 287)   A rate with denominator of 1.

tasa unitaria   Una tasa con un denominador de 1.

unit ratio (p. 298)   A unit rate where the denominator is one unit.

razón unitaria   Tasa unitaria en que el denominador es la unidad.

unlike fractions (p. 237)   Fractions with different denominators.

fracciones con distinto denominador   Fracciones cuyos denominadores son diferentes.

variable (p. 44)   A placeholder, usually a letter, used to represent an unspecified value in mathematical expressions or sentences. In $3 + a = 6$, a is a variable.

variable   Marcador de posición, por lo general, una letra, que se usa para representar un valor desconocido en expresiones o enunciados matemáticos. En $3 + a = 6$, a es una variable.

Venn diagram (p. 186)   A diagram that uses overlapping circles to show how elements among sets of numbers or objects are related.

diagrama de   Venn Diagrama que usa círculos para mostrar la relación entre los elementos en un conjunto de números u objetos

vertex (p. 510)   A vertex of an angle is the common endpoint of the rays forming the angle.

vértice   El vértice de un ángulo es el extremo común de los rayos que lo forman.

vertex (p. 603)   The point where the edges of a three dimensional figure intersect.

vértice   Punto donde se intersecan las aristas de una figura tridimensional.

vertical angles (p. 511)   Opposite angles formed by the intersection of two lines.

ángulos opuestos por el vértice   Ángulos opuestos que se forman de la intersección de dos rectas.

∠1 and ∠2 are vertical angles.

∠1 y ∠2 son ángulos opuestos por el vértice.

volume (p. 613)   The number of cubic units needed to fill the space occupied by a solid.

volumen   Número de unidades cúbicas que se requieren para llenar el espacio que ocupa un sólido.

voluntary response sample (p. 439)   A sample which involves only those who want to participate in the sampling.

muestra de respuesta voluntaria   Muestra que involucra sólo aquellos que quieren participar en el muestreo.

whole (p. 350)   In a percent proportion, the number to which the part is being compared.

todo   En una proporción porcentual, el número con que se compara la parte.

x-axis (p. 88)   The horizontal number line in a coordinate plane.

eje x   La recta numérica horizontal en el plano de coordenadas.

x-coordinate (p. 88)   The first number of an ordered pair. It corresponds to a number on the x-axis.

coordenada x   El primer número de un par ordenado. Corresponde a un número en el eje x.

y-axis (p. 88)   The vertical number line in a coordinate plane.

eje y   La recta numérica vertical en el plano de coordenadas.

y-coordinate (p. 88)   The second number of an ordered pair. It corresponds to a number on the y-axis.

coordenada y   El segundo número de un par ordenado. Corresponde a un número en el eje y.

zero pair (p. 93)   The result when one positive counter is paired with one negative counter.

par nulo   Resultado que se obtiene cuando una ficha positiva se aparea con una ficha negativa.

Selected Answers

Chapter 1 Introduction to Algebra and Functions

Page 23 **Chapter 1** **Getting Ready**
1. 105.8 **3.** 60.64 **5.** $72.94 **7.** 2.5 **9.** 6.2 **11.** 29.4
13. 10.2 **15.** 5.3 **17.** 4.46

Pages 27–29 **Lesson 1-1**
1. Sample answer: 4 times; $16{,}000 \div 4{,}000 = 4$
3. 3,000 **5.** $763.75
7.

9. every 45 minutes **11.** 5:10 P.M. **15.** 13 wk
17. Sample answer: For the school bake sale, Samantha bakes 79 cookies and 42 brownies. If two other students baked the same amount of cookies and brownies, how many items did they bake altogether? **19.** C **21.** 100 **23.** 625

Pages 31–33 **Lesson 1-2**
1. $9 \cdot 9 \cdot 9$ **3.** $8 \cdot 8 \cdot 8 \cdot 8 \cdot 8$ **5.** 49 **7.** Sample answer: 9,765,625 people **9.** 1^4 **11.** $1 \cdot 1 \cdot 1 \cdot 1 \cdot 1$
13. $3 \cdot 3 \cdot 3 \cdot 3 \cdot 3 \cdot 3 \cdot 3 \cdot 3$ **15.** $9 \cdot 9 \cdot 9$ **17.** 64
19. 2,401 **21.** 1 **23.** 1,000,000 **25.** 3^2 **27.** 1^8
29. $4 \cdot 4 \cdot 4 \cdot 4 \cdot 4$ **31.** 1,296 **33.** 3^3 **35.** $5^4 \cdot 4^3$
37. $1^{14}, 17^3, 6^5, 4^{10}$ **39.** $7^2, 5^3, 2^{11}, 4^6$ **41.** $8^2 = 64$ and $4^3 = 64$ **43.** Sample answer: The pattern is that each successive term is $\frac{1}{2}$ of the previous one, so $2^0 = 1$ and $2^{-1} = \frac{1}{2}$. **45.** 7 wins **47.** 4 **49.** 25

Pages 36–37 **Lesson 1-3**
1. 36 **3.** 289 **5.** 3 **7.** 11 **9.** 24 in. by 24 in. **11.** 1
13. 121 **15.** 400 **17.** 1,156 **19.** 4 **21.** 10 **23.** 16
25. 25 **27.** 40 ft **29.** 361 **31.** 53,824 mi^2
33.

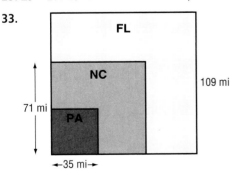

35. 41m^2 greater; The area of a 7 by 7 square has an area of 49 m^2. **37.** Yes; for example, a pen that measures 10 feet by 10 feet has the same perimeter, but its area is 100 square feet, which is greater than 84 square feet. **39.** Sample answer: It is called squaring the number because the area of a square is found by multiplying the two side lengths together. **41.** J
43. $8 \cdot 8 \cdot 8 \cdot 8 \cdot 8$ **45.** $2 \cdot 2 \cdot 2 \cdot 2 \cdot 2 \cdot 2$ **47.** 21 **49.** 30

Pages 40–41 **Lesson 1-4**
1. 11; Sample answer: Subtract first since $5 - 2$ is in parentheses. Then add 8. **3.** 11; Sample answer: Multiply 2 by 6 first since multiplication comes before addition or subtraction. Then subtract and add in order from left to right. **5.** 400; Sample answer: Evaluate 10^2 first since it is a power. Then multiply by 4. **7.** 11; Sample answer: Subtract first since $6 - 3$ is in parentheses. Then multiply the difference by 2 and multiply 3 by 4 since multiplication comes before addition or subtraction. Finally, add $17 + 6$ and subtract 12 in order from left to right. **9.** $3(0.05) + 2(0.25) + 2(0.10) + 7(0.01)$; $0.92 **11.** 3; Sample answer: Add first since $3 + 4$ is in parentheses. Then subtract. **13.** 1; Sample answer: Subtract first since $11 - 2$ is in parentheses. Then divide. **15.** 8; Sample answer: Divide first since division comes before addition or subtraction. Then subtract 1 and add 7 in order from left to right. **17.** 5; Sample answer: Multiply first since multiplication comes before addition or subtraction. Then subtract the product from 118 and add 5. **19.** 30,000; Sample answer: Evaluate 10^4 first since it is a power. Then multiply by 3. **21.** 386; Sample answer: Evaluate 7^2 first since it is a power. Then multiply by 8 since multiplication comes before subtraction. Finally, subtract. **23.** 75; Sample answer: Evaluate 9^2 since it is a power. Then divide 14 by 7 and multiply the quotient by 3 since multiplication and division occur from left to right. Finally, subtract. **25.** 22; Sample answer: Add first since $6 + 5$ is in parentheses. Then subtract 6 from 8 since $8 - 6$ is in parentheses. Finally, multiply.
27. 23; Sample answer: Add first since $4 + 7$ is in parentheses. Then multiply the sum by 3. Next, multiply 5 by 4 and divide the product by 2 since multiplication and division occur in order from left to right. Finally, subtract. **29.** $3.25 **31.** 19; Sample answer: Evaluate 3^3 first since it is a power. Then add 8. Next subtract 6 from 10 since $10 - 6$ is in parentheses. Then square the difference since the power is 2. Finally subtract 16 from 35. **33.** 64; Sample answer: Subtract first since $4 - 3.2$ is in parentheses. Multiply 7 by 9 next since multiplication occurs from left to right. Then subtract 0.8 from the product, 63, and add 1.8 since addition and subtraction occur from left to right. **35.** Peggy; the first step is to do the division $24 \div 6$. Phoung incorrectly multiplied 6 and 2 first. **37.** Yes; For example, a pen that measures 10 feet by 10 feet has the same perimeter, but its area is 100 square feet, which is greater than 84 square feet. **39.** H **41.** 8 **43.** 28 **45.** 968

Pages 42–43 **Lesson 1-5**
1. Sample answer: You need to keep track of what numbers you have already guessed, so that you do not make the same guess twice. You also need to know what numbers produce answers that are too large or

too small, so you can make better guesses.   **3.** 125 adult tickets and 250 student tickets   **5.** birthday, holiday, vacation   **7.** The circumference of Earth is 24,900 miles long at the Equator.   **9.** 512 and 1,024   **11.** 324.8 in. of snow.   **13.** 2, 3, 6

Pages 46–47   **Lesson 1-6**
1. 10   **3.** 2   **5.** 32   **7.** 22   **9.** 7   **11.** 3   **13.** 4   **15.** 7   **17.** 5   **19.** 12   **21.** 48   **23.** 4   **25.** 18   **27.** 36   **29.** 5   **31.** 4 quarts   **33.** 5.1   **35.** $19.99d + 0.17m$ **37.** 64 ft   **39.** Sample answer: $5x - 37$ if $x = 8$ **41.** Sample answer: Sometimes; $x - 3$ and $y - 3$ represent the same value only when $x = y$ **43.** G   **45.** 28   **47.** 56   **49.** 19   **51.** false

Pages 51–52   **Lesson 1-7**
1. 3   **3.** 54   **5.** $2.25   **7.** 7   **9.** 28   **11.** 46   **13.** 33   **15.** 64   **17.** 7   **19.** 11   **21.** $m =$ the number of miles Derrick walked on Monday; $m + 2.5 = 6.3$; 3.8 mi   **23.** 5.4   **25.** 4.4   **27.** 1.2   **29.** $1.75 + c = 6.25; $c =$ $4.50   **31.** Antonio; $105 - 35 = 70$ is a true statement. $35 - 35 \neq 70$ **33.** A   **35.** J   **37.** 31   **39.** 19   **41.** 28   **43.** 150

Pages 55–56   **Lesson 1-8**
1. $7(4) + 7(3)$; 49   **3.** $3(9 + 6)$; 45   **5.** $4(12 + 5)$; $68; Sample answer: The expression $12 + 5$ represents the cost of one ticket and one hot dog. The expression $4(12 + 5)$ represents the cost of four tickets and four hot dogs. Since $4 \times 12 = 48$ and $4 \times 5 = 20$, find $48 + 20$, or 68, to find the total cost of four tickets and four hot dogs.   **7.** Sample answer: Rewrite $44 + (23 + 16)$ as $44 + (16 + 23)$ using the Commutative Property of Addition. Rewrite $44 + (16 + 23)$ as $(44 + 16) + 23$ using the Associative Property of Addition. Find $44 + 16$, or 60, mentally. Then find $60 + 23$, or 83, mentally.   **9.** $2(6) + 2(7)$; 26   **11.** $4(3 + 8)$; 44   **13.** Sample answer: Rewrite $(8 + 27) + 52$ as $(27 + 8) + 52$ using the Commutative Property of Addition. Rewrite $(27 + 8) + 52$ as $27 + (8 + 52)$ using the Associative Property of Addition. Find $8 + 52$, or 60, mentally. Then find $60 + 27$, or 87, mentally.   **15.** Sample answer: Rewrite $91 + (15 + 9)$ as $91 + (9 + 15)$ using the Commutative Property of Addition. Rewrite $91 + (9 + 15)$ as $(91 + 9) + 15$ using the Associative Property of Addition. Find $91 + 9$, or 100, mentally. Then find $100 + 15$, or 115, mentally.   **17.** Sample answer: Rewrite $(4 \cdot 18) \cdot 25$ as $(18 \cdot 4) \cdot 25$ using the Commutative Property of Multiplication. Rewrite $(18 \cdot 4) \cdot 25$ as $18 \cdot (4 \cdot 25)$ using the Associative Property of Multiplication. Find $4 \cdot 25$, or 100, mentally. Then find $100 \cdot 18$, or 1,800, mentally.   **19.** Sample answer: Rewrite $15 \cdot (8 \cdot 2)$ as $15 \cdot (2 \cdot 8)$ using the Commutative Property of Multiplication. Rewrite $15 \cdot (2 \cdot 8)$ as $(15 \cdot 2) \cdot 8$ using the Associative Property of Multiplication. Find $15 \cdot 2$, or 30, mentally. Then find $30 \cdot 8$, or 240, mentally.   **21.** Sample answer: Rewrite $5 \cdot (30 \cdot 12)$ as $5 \cdot (12 \cdot 30)$ using the Commutative Property of Multiplication. Rewrite $5 \cdot (12 \cdot 30)$ as $(5 \cdot 12) \cdot 30$ using the Associative Property

of Multiplication. Find $5 \cdot 12$, or 60, mentally. Then find $60 \cdot 30$, or 1,800, mentally.   **23.** $5(20 + 7)$; 135 million   **25.** $7(9 - 3)$; 42   **27.** $9(7 - 3)$; 36   **29.** $y + 5$ **31.** $32b$ **33.** $2x + 6$ **35.** $6c + 6$ **37.** $55 + 184 = 184 + 55$ **39.** Sample answer: $(5 + z) + 9 = 5 + (z + 9)$ **41.** Sample answer: Since $24 \div (12 \div 2) = 4$ and $(24 \div 12) \div 2 = 1$, $24 \div (12 \div 2) \neq (24 \div 12) \div 2$.   **43.** B   **45.** 11.3   **47.** 75 yr   **49.** 10   **51.** 4.6

Pages 59–61   **Lesson 1-9**
1. 9 is added to each term; 36, 45, 54   **3.** 0.1 is added to each term; 1.4, 1.5, 1.6   **5.** $3n$; 36 in.   **7.** 6 is added to each term; 25, 31, 37   **9.** 12 is added to each term; 67, 79, 91   **11.** 5 is added to each term; 53, 58, 63   **13.** 0.8 is added to each term; 5.6, 6.4, 7.2   **15.** 1.5 is added to each term; 10.5, 12.0, 13.5   **17.** 4 is added to each term; 20.6, 24.6, 28.6   **19.** $7n$; 42 laps   **21.** 25 is added to each term; 120, 145, 170   **23.** 256, 1,024, 4,096   **25.** 324, 972, 2,916   **27.** 1,200   **29.** 4,950   **31.** The Fibonacci sequence is 1, 1, 2, 3, 5, 8, 13, …. In this sequence, each term after the second term is the sum of the two terms before it. Fibonacci numbers occur in many areas of nature, including pine cones, shell spirals, and branching plants.   **33.** $+ 2, + 4, + 6, + 8, …$; 30, 42, 56   **35.** Sample answer: paper/pencil; Write the equation that represents this situation, $15n$. Since 2 years = 24 months, evaluate the expression when n is 24. $15(24) = 360$. So, after 2 years, $360 will be saved.   **37.** D   **39.** 48; Sample answer: Rewrite $(23 + 18) + 7$ as $(18 + 23) + 7$ using the Communicative Property of Addition. Rewrite $(18 + 23) + 7$ as $18 + (23 + 7)$ using the Associative Property of Addition. Find $23 + 7$, or 30, mentally. Then find $18 + 30$, or 48, mentally.   **41.** 29   **43.** 20   **45.** 8   **47.** 3

Pages 65–67   **Lesson 1-10**

1.

x	3x	y
1	3 · 1	3
2	3 · 2	6
3	3 · 3	9
4	3 · 4	12

; domain: {1, 2, 3, 4}; range: {3, 6, 9, 12}

3.

x	8x	y
1	8 · 1	8
2	8 · 2	16
3	8 · 3	24
4	8 · 4	32

; domain: {1, 2, 3, 4}; range: {8, 16, 24, 32}

5. 693 mi; Sample answer: Replace h with 3 in the equation $m = 231h$ to find the distance in miles the race car travels in 3 hours.

7.

x	6x	y
1	6 · 1	6
2	6 · 2	12
3	6 · 3	18
4	6 · 4	24

; domain: {1, 2, 3, 4}; range: {6, 12, 18, 24}

9.

x	25x	y
1	25 · 1	25
2	25 · 2	50
3	25 · 3	75
4	25 · 4	100

; domain: {1, 2, 3, 4}; range: {25, 50, 75, 100}

11. $c = 40m$ **13.** $t = 35m$

15.

x	x − 1	y
1	1 − 1	0
2	2 − 1	1
3	3 − 1	2
4	4 − 1	3

; domain: {1, 2, 3, 4}; range: {0, 1, 2, 3}

17.

x	x + 0.25	y
0	0 + 0.25	0.25
1	1 + 0.25	1.25
2	2 + 0.25	2.25
3	3 + 0.25	3.25

; domain: {0, 1, 2, 3}; range: {0.25, 1.25, 2.25, 3.25}

19.

Width (units)	6w	Area (sq units)
2	6 · 2	12
3	6 · 3	18
4	6 · 4	24
5	6 · 5	30

21. $m = 8s$ **23.** 480 mi; 1,140 mi; Sample answer: Replace s with 60 in the equation $m = 8s$ and in the equation $m = 19s$ to find the number of miles Jupiter travels in 1 minute the number of miles Earth travels in 1 minute, respectively. **25.** $y = 3x$ **27.** Sample answer: Sam charges $3 for each dog that he walks. In the equation $y = 3x$, x represents the number of dogs and y represents the total amount of money earned. **29.** C **31.** 63, 72, 81 **33.** (12)4 +(4)4; 64 **35.** 10(6 − 5); 10 **37.** 7 **39.** 7

Pages 70–74 Chapter 1 Study Guide and Review

1. false, equivalent expressions **3.** false, domain **5.** true **7.** false, square **9.** 60 ft **11.** 3 · 3 · 3 · 3 **13.** 5 **15.** 5 · 5 · 5 · 5 **17.** 40,353,607 **19.** 324 **21.** 100 **23.** 8 **25.** 169 **27.** 18 **29.** 25 **31.** 33 **33.** 36 ÷ 4 + 12 ÷ 3; 13 **35.** 22 **37.** 5 **39.** 48 **41.** 8 **43.** 108 **45.** 9 + x = 15; 6 tickets **47.** 68; Sample answer: Rewrite 14 + (38 + 16) as 14 + (16 + 38) using the Commutative Property of Addition. Rewrite 14 + (16 + 38) as (14 + 16) + 38 using the Associative Property of Addition. Find 14 + 16, or 30, mentally. Then find 30 + 38, or 68, mentally. **49.** $2(15 + 12); $54; Sample answer: The expression 15 + 12 represents the total number of roses Wesley sold. The expression $2(15 + 12) represents the total amount of money Wesley earned. Since $2 × 15 = $30 and $2 × 12 = $24, find $30 + $24, or $54, to find the total amount Wesley earned. **51.** Each term is found by adding 0.8 to the previous term; 6.6, 7.4, 8.2 **53.** $4.50n

55.

x	4x	y
5	4(5)	20
6	4(6)	24
7	4(7)	28
8	4(8)	32

; domain: {5, 6, 7, 8}; range: {20, 24, 28, 32}

Chapter 2 Integers

Page 79 **Chapter 2** **Getting Ready**

1. < **3.** < **5.** Garrett **7.** 20 **9.** 9 **11.** 216 **13.** 29 **15.** 1,900 mi

Pages 82–83 **Lesson 2-1**

1. −11 **3.** 16 **5.** −15

7.

9. 8 **11.** 9 **13.** −53 **15.** −2 **17.** 12 **19.** −7

21.
−3 −2 −1 0 1

23.
−10 −8 −6 −4 −2 0 2 4 6 8 10

25. 10 **27.** 2 **29.** 14 **31.** 25 **33.** 5 **35.** 17 positive charges: 17; 25 negative charges: −25 **37.** false; 0 **39.** C

41.

x	x − 4	y
4	4 − 4	0
5	5 − 4	1
6	6 − 4	2
7	7 − 4	3

; domain: {4, 5, 6, 7}; range: {0, 1, 2, 3}.

43.

x	5x + 1	y
1	5 · 1 + 1	6
2	5 · 2 + 1	11
3	5 · 3 + 1	16
4	5 · 4 + 1	21

; domain: {1, 2, 3, 4}; range: {6, 11, 16, 21}

45. > **47.** >

Pages 85–87 **Lesson 2-2**

1. > **3.** > **5.** {−18, −16, −10, 12, 19} **7.** < **9.** > **11.** > **13.** < **15.** {−8, −5, −3, 6, 11} **17.** {−7, −6, −4, 1, 3, 5} **19.** Sunlight, Twilight, Midnight, Abyssal, Hadal **21.** < **23.** < **25.** −5, −2, 5, 10, 20 **27.** 5° with a 10-mile-per-hour wind **29.** true **31.** true **33.** −1 **35.** C **37.** −9 **39.** $r = 6t$

41–44.
10
8
6
4
2
0
−2
−4
−6
−8
−10

Pages 90–92   Lesson 2-3
1. (−2, −4), III   **3.** (0, 3), *y*-axis

5–8.
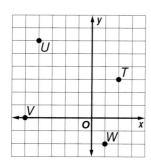

9. Sea Cliffs
11. (−2, 2), II
13. (−5, 0), *x*-axis
15. (2, −2), IV
17. (−4, −1), III
19. (2, 2), I
21. (0, 4); *y*-axis

23–34.
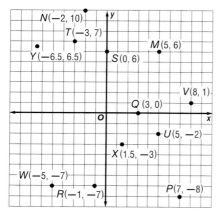

35. Africa   **37.** South America

39–41.
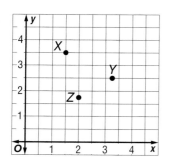

43. Sample answer: Rene Descarte is often credited with inventing the coordinate plane and so the coordinate plane is sometimes called the Cartesian plane, in his honor. The myth is that while a child, Descarte lay in bed one day watching a fly crawling around on the ceiling. In wondering how to tell someone else where the fly was, he realized that he could describe its position by its distance from the walls of the room.   **45.** Sample answer: Sometimes; both (0, −2) and (0, 2) lie on the *y*-axis.

47. Sample answer: Using the graphic, you can see that an ordered pair such as (−5, −4) is in Quadrant III.

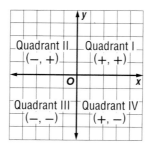

49. Sample answer: Point *A* is 1 unit to the right and 2 units down from the origin, in quadrant IV. Point *B* is 2 units to the left and 1 unit up from the origin, in quadrant II.   **51.** G   **53.** >   **55.** <   **57.** 41 mi   **59.** 1,326   **61.** 11,737

Pages 98–99   Lesson 2-4
1. −14   **3.** 7   **5.** −4   **7.** 0   **9.** −$25 + $18; −$7; Camilia still owes her brother $7.   **11.** −25   **13.** 28   **15.** −6   **17.** −13   **19.** −2   **21.** 2   **23.** 0   **25.** 22   **27.** −19   **29.** 60 + (−60); 0; The pelican is now at sea level.   **31.** −5 + (−15) + 12; The team has lost a total of 8 yards.   **33.** 4   **35.** −3   **39.** *a*   **41.** *m* + (−15)   **43.** Sample answer: Look at the signs. If the numbers being added are both positive, the sum is positive. If the numbers being added are both negative, the sum is negative. If the numbers being added have different signs, subtract their absolute values and give the sum the sign of the number with the greatest absolute value. If the numbers being added are opposites, the sum is zero.   **45.** B   **47.** (−2, 4); 11   **49.** (−3, −1); 111   **51.** −8, −4, −3, 0, 1, 4, 6   **53.** 103   **55.** 3,109

Pages 105–106   Lesson 2-5
1. −3   **3.** −12   **5.** 24   **7.** −2   **9.** −21   **11.** 22   **13.** −10   **15.** −14   **17.** 23   **19.** 31   **21.** −14   **23.** −30   **25.** 104   **27.** 6   **29.** 0   **31.** 0   **33.** 15   **35.** 11   **37.** 8,757 ft   **39.** 5,066 ft   **41.** −31   **43.** 23   **45.** Mei; Alicia did not add the additive inverse of −18.   **47.** Sample answer: To subtract an integer, add its additive inverse.   **49.** J   **51.** −11   **53.** −14   **55.** 8   **57.** −33   **59.** −24

Pages 109–111   Lesson 2-6
1. −60   **3.** −28   **5.** 45   **7.** 64   **9.** −12   **11.** 100(−3) = −300; Tamera's investment is now worth $300 less than it was before the price of the stock dropped.   **13.** 70   **15.** −220   **17.** −70   **19.** −50   **21.** 80   **23.** −125   **25.** 81   **27.** −45   **29.** 49   **31.** −24   **33.** 24   **35.** 64   **37.** −160   **39.** 5(−650) = −3,250; Ethan burns 3,250 Calories each week.   **41.** −243   **43.** 88

47.
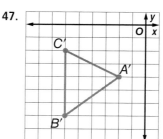
Sample answer: new triangle *A′B′C′* is on the other side of the origin (in quadrant III) from original triangle *ABC* (which is in Quadrant I).

49. Sample answer: −6 × 3 = −18   **51.** Sample answer: 1; (−1)(−1) = 1. Since there are 50 ÷ 2 = 25 pairs of (−1) factors, (−1)^{50} = 1^{25}. One raised to any power is still 1.   **53.** Sample answer: The product of three integers is positive when exactly two integers are negative or all three integers are positive.   **55.** J   **57.** 8   **59.** −21   **61.** −6   **63.** −9   **65.** 18 + *x* = 51; *x* = 33

Pages 112–113   Lesson 2-7

1. Sample answer: Use the look for a pattern strategy when there is a data table, a series of numbers, or a geometric pattern as part of the problem.   **3.** Sample answer: Amanda has $2 in change in her bank. If she adds $0.50 each week for 7 weeks, how much money will be in her bank? $5.50   **5.** 8 months   **7.** Sample answer: 3 quarters, 2 nickels, and 1 penny   **9.** 2 letters and 10 postcards   **11.** 8,914,113   **13.** 84 cm

Pages 116–118   Lesson 2-8

1. −4   **3.** −6   **5.** 5   **7.** −3   **9.** −48.3°C   **11.** −7
13. −9   **15.** 10   **17.** −7   **19.** −9   **21.** 9   **23.** 5
25. −12   **27.** −3   **29.** 2   **31.** −1   **33.** −10°F   **35.** 8
37. 1   **39.** −53 ft   **41.** −32 ÷ (−4) has a positive quotient; the others have negative quotients
43. −20, −10, −5, −4, −2, −1, 1, 2, 4, 5, 10, 20   **45.** A
47.   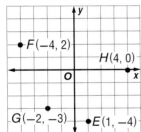   **49.** 60   **51.** 81   **53.** −10   **55.** 30

Pages 119–122   Chapter 2   Study Guide and Review

1. false, negative   **3.** false, opposite
5. false, y-coordinate   **7.** true   **9.** false, positive
11. 350 ft   **13.** −12°   **15.** 32   **17.** −48 mL
19. >   **21.** <   **23.** <   **25.** {−32, −23, −21, 14, 19, 25}
27. −10, −6, −5, 0, 2, 5, 10, 12, 20, 25

28–31.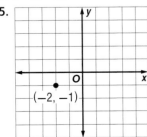

33. 2   **35.** −5
37. 123 ft   **39.** −3
41. 4   **43.** −12
45. 35   **47.** 28
49. −35
51. 17,280
53. 576 ft
55. −3   **57.** 9

Chapter 3 Algebra: Linear Equations and Functions

Page 127   Chapter 3   Getting Ready

1. 4   **3.** −11
5.

(graph with point (−2, −1))

7. −8   **9.** 4
11. −11   **13.** 14
15. 2   **17.** −2

Pages 131–133   Lesson 3-1

1. $n + 8$ **3.** $n − 9 = 24$ **5.** $2m = 18$ **7.** $x − 1 = 34.3$
9. $15 + t$ **11.** $n − 10$ **13.** $8r$ **15.** $\frac{a}{3}$ **17.** $n + 4 = −8$
19. $5d = −20$ **21.** $h − 10 = 26$ **23.** $c + 3.5 = 5.5$
25. the length is 4 times the width   **27.** the width is 5

less than the length   **29.** $2b + 2$ **31.** $3(a − 43)$
33. $13k^2$ **35.** American toad   **37.** 3 less than a number is 6.   **39.** $x + 2$; $x − 2$ **41.** C   **43.** −7   **45.** 15
47. 25   **49.** 49   **51.** Plan A   **53.** −11   **55.** −8

Pages 139–141   Lesson 3-2

1. 2   **3.** −2   **5.** $d + 120 = 364$; 244 ft   **7.** 5   **9.** 7   **11.** 7
13. −3   **15.** −9   **17.** 17   **19.** 7   **21.** $7 = w + 2$; 5
23. $15 = t − 3$; 18 years old   **25.** 61   **27.** −5   **29.** −12
31. 18.4   **33.** 6.4   **35.** −0.68   **37.** $d − 5 = 18$; $23
39. $35 + 45 + x = 180$; 100°   **41.** $1 − 1 + s = −5$; −5
43. $s − 65 = 13$; 78 mph   **45.** $n − 13 = 163$; 176 ft
47. The value of y decreases by 2.   **49.** C   **51.** $p + 180$
53. 46 pages   **55.** 2.6   **57.** 1.52

Pages 144–146   Lesson 3-3

1. 3   **3.** −3   **5.** 8 h   **7.** 7   **9.** −3   **11.** 7   **13.** −9
15. 4   **17.** −8   **19.** $15w = 300$; 20 weeks   **21.** $3h$
23. 23   **25.** 18   **27.** 4.7   **29.** Sample answer: Evelyn Ashford has the faster average speed. Her race is half the distance as Sanya Richards, and it took her less than half the time as Sanya Richards to complete the race.   **31.** $20.88h = 145$; $h ≈ 6.94$ **33.** Steve; the variable is multiplied by −6. To solve for x, you need to divide each side of the equation by the entire coefficient, −6.   **35.** Sample answer: Billie has twice as many cards as Tyree. If Billie has 16 cards, how many does Tyree have?   **37.** Sample answer: If it takes a scuba diver 4 seconds to swim 8 meters below the surface of the water, what is the rate of descent?
39. 8   **41.** 5   **43.** −3y

45.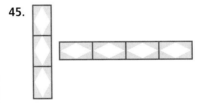

Pages 148–149   Lesson 3-4

1. When you are given the final result and asked to find an earlier amount.   **3.** Sample answer: In the first four games, Hannah scored a total of 83 points. In the fourth game she scored 19 points. In game three, she scored 27 points and in the second game she scored 22 points. How many points did she score in the first game? To solve, first subtract 19 from 83, which is 64. Then subtract 27 from 64 to get 37. Finally, subtract 22 from 37. So, Hannah scored 15 points in her first game.
5. 4   **7.** 19,200 tennis balls   **9.** Brie is 14 years old.
11.   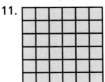   **13.** Raquels's car gets 1,774,074 more inches per gallon than an aircraft carrier.
15. 1-$10 bill, 2-$5 bills, and 7-$1 bills

Pages 153–155   Lesson 3-5

1. 2   **3.** 3   **5.** 3   **7.** $14c + 23 = 65$; 3 CDs   **9.** 3
11. −4   **13.** 4   **15.** 9   **17.** 8   **19.** 11   **21.** $2c + 10 = 14$;

2 cups   **23.** 2.25   **25.** 2.1   **27.** 28   **29.** $8 + 47.6d = 960$;
20 days   **31.** 92.2°F   **33.** $\frac{1}{2}(20x) - 18 = 200$; $x = 21.8$;
They must sell at least 22 subscriptions.   **35.** Sample
answer: A flower shop charges \$2 for each flower in a
vase and \$5 for the vase. How many flowers can you
place in a vase if you have \$15 to spend?   **37.** G   **39.** 7
41. 41   **43.** 213 ft   **45.** 14   **47.** 16

Pages 158–161   **Lesson 3-6**
1. 18 yd   **3.** 7 in.   **5.** 26.25 ft²   **7.** 36 ft   **9.** 14.8 mm
11. 11 ft   **13.** 18 in.   **15.** 78 ft²   **17.** 183.6 m²
19. 6.5 in²   **21.** 5 squares   **23.** $\ell = 33$ ft   **25.** 3,200 yd²
27. perimeter; 4 mi   **29.** area; 13 ft   **31.** 270 m²

35. 9;

1 yd
1 yd
1 yd²

3 ft
3 ft
9 ft²

=

37. 144;

1 ft
1 ft
1 ft²

12 in.
12 in.
144 in²

=

39. When the width of a rectangle is doubled, the
perimeter becomes $2\ell + 4w$ and the area becomes $2\ell w$,
or in other words, the area is doubled.   **41.** $P = 2(3w + 1) + 2w$ or $P = 8w + 2$ **43.** A   **45.** −2   **47.** 3.5
49. $5x = 11.25$; \$2.25   **51.** −27   **53.** 11 years old

55.
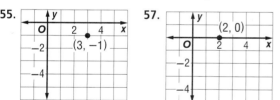

(3, −1)

57.
(2, 0)

Pages 166–167   **Lesson 3-7**
1.

Total Cost of Baseballs

Total Cost (\$)

Number of Baseballs

3.
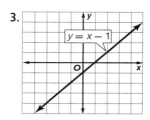

$y = x - 1$

5.
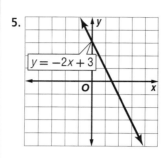

$y = -2x + 3$

7.
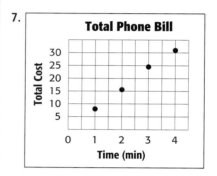

Total Phone Bill

Total Cost

Time (min)

9.

$y = x + 1$

11.

$y = x$

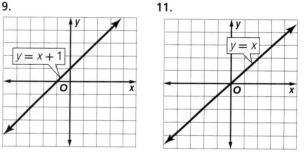

13.

$y = 2x + 3$

15.

Distance (mi)

Gasoline (gal)

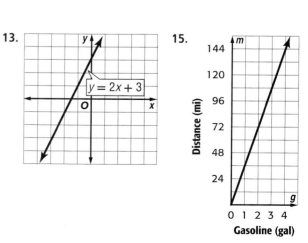

17.

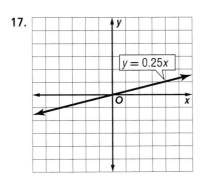

19.

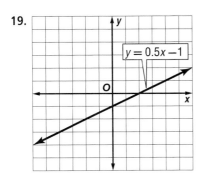

23.

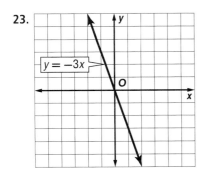

25. C   **27.** −3   **29.** 4   **31.** 15

Pages 169–172   Chapter 3   Study Guide and Review

1. true   **3.** true   **5.** true   **7.** False; subtract 3 from each.
9. true   **11.** false; perimeter   **13.** $x + 5$ **15.** $2a$
17. $n − 4 = 19$ **19.** $f + \$8.75$ **21.** −6   **23.** 23
25. 37   **27.** 4   **29.** −9   **31.** $14w = 98$; 7 weeks
33. \$146.70   **35.** 4   **37.** −2   **39.** 5   **41.** 53.4 in.;
142.82 in^2 **43.** 7 mi   **45.** 16 yd   **47.** 12 ft

49.

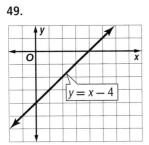

51.

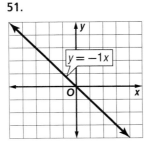

53.

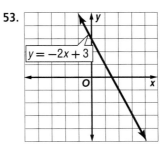

55. $y = 9x$

Chapter 4 Fractions, Decimals, and Percents

Page 179   Chapter 4   Getting Ready
1. 0.61   **3.** 0.33   **5.** Kirsten   **7.** 2, 3, 6   **9.** Yes; the sum
of the digits, 6, is divisible by 3.   **11.** 6   **13.** 0.75
15. $5 \times 5 \times 5 \times 5 \times 5$ **17.** $9 \times 9 \times 9 \times 9$

Pages 183–184   Lesson 4-1
1. prime   **3.** prime   **5.** 2×17 **7.** $2^2 \times 3$
9. $2 \cdot 2 \cdot 2 \cdot 2 \cdot x \cdot x$ **11.** composite   **13.** prime
15. composite   **17.** prime   **19.** $2^5 \times 3$ **21.** $3^2 \times 11$
23. $2 \times 3 \times 5 \times 7$ **25.** $2 \times 3^2 \times 7$ **27.** $2 \cdot 5^2$
29. $3 \cdot 5 \cdot m \cdot n$ **31.** $2 \cdot 17 \cdot j \cdot k \cdot m$ **33.** $2 \cdot 2 \cdot 13 \cdot g \cdot h \cdot h$ **35.** 7   **37.** 5^2 **39.** $2 \times 5^2 \times 29$ **41.** prime
43. 81: 9×9, 3×27; 225: 9×25, 5×45; 441: 9×49;
7×63 **45.** 36   **47.** If $n = 1$, $2n$ is a prime number. If $n > 1$, $2n$ is a composite number with at least three
factors.   **49.** H   **51.** 36 ft; 65 ft^2 **53.** −4   **55.** 0
57. 2, 5, 10   **59.** 3, 9

Pages 188–189   Lesson 4-2
1. 6   **3.** 10   **5.** 4   **7.** 9   **9.** 6   **11.** 5   **13.** 24   **15.** 8
17. 6   **19.** 7   **21.** 8 students   **23.** 6 care packages
25. 25¢   **27.** $6a$ **29.** $5y$ **31.** Sample answer: 24 and
36   **33.** Sample answer: 60 and 90   **35.** Sample answer:
The first prism is 8 in. high, 3 in. long, and 4 in. wide. The second prism is 8 in. high, 6 in. long, and 5 in. wide. The third prism is 8 in. high, 5 in. long, and 5 in. wide.   **37.** always   **39.** Sample answer: 4 and 12 are
factors of 24. The greatest common factor of 4, 12, and
24 is 4.   **41.** G

43.

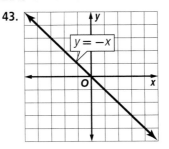

45.

$y = 2x - 1$

47. 20 songs

Pages 190–191   Lesson 4-3

1. Sample answer: By making an organized list, you can show all the possible pizza combinations. Count the number of pizzas to determine the answer. **3.** 6 outfits **5.** 6 ways **7.** 27, 29 **9.** Paul: 9; Angelina: 5; Bret: 9; Jill: 1 **11.** 7 in. **13.** 0.5m

Pages 194–195   Lesson 4-4

1. $\frac{1}{3}$ **3.** $\frac{2}{5}$ **5.** $\frac{2}{3}$ **7.** $\frac{5}{7}$ **9.** $\frac{7}{10}$ **11.** $\frac{4}{7}$ **13.** $\frac{6}{7}$ **15.** 1 **17.** $\frac{5}{6}$ **19.** $\frac{5}{7}$ **21.** $\frac{2}{5}$ **23.** $\frac{1}{5}$ **25.** $\frac{3}{4}$ **27.** $\frac{1}{3}$ **31.** No, because both the numerator and denominator can be divided by 2. **33.** Sample answer: A fraction is in simplest form if the GCF of the numerator and denominator is 1. **35.** F **37.** 9 **39.** 5 **41.** 0.5 **43.** 0.7

Pages 199–200   Lesson 4-5

1. 0.4 **3.** 7.5 **5.** 0.125 **7.** $0.\overline{5}$ **9.** $\frac{11}{50}$ **11.** $4\frac{3}{5}$ **13.** 0.8 **15.** 4.16 **17.** 0.3125 **19.** 0.66 **21.** 5.875 **23.** $0.\overline{4}$ **25.** $0.\overline{16}$ **27.** $5.\overline{3}$ **29.** $\frac{1}{5}$ **31.** $\frac{11}{20}$ **33.** $5\frac{24}{25}$ **35.** $30\frac{1}{2}$ cm **39.** $\frac{22}{3}$ **41.** $\frac{-16}{5}$ **43.** Felisa; Sample answer: Felisa's batting average is about 0.286. Harmony's batting average is about 0.263. 0.286 > 0.263, so Felisa's average is better than Harmony's. **45.** Sample answer: $3\frac{1}{7} \approx 3.14286\ldots$ and $3\frac{10}{71} \approx 3.14085$; Since 3.1415927… is between $3\frac{1}{7}$ and $3\frac{10}{71}$, Archimedes was correct. **47.** B **49.** $\frac{5}{12}$ **51.** $\frac{2}{7}$ **53.** 9 pizzas **55.** $\frac{3}{9}$ or $\frac{1}{3}$ **57.** $\frac{4}{12}$ or $\frac{1}{3}$

Pages 204–205   Lesson 4-6

1. 57% **3.** 25% **5.** 85% **7.** $\frac{9}{10}$ **9.** $\frac{11}{50}$ **11.** 42% **13.** 99.9% **15.** $66\frac{2}{3}\%$ **17.** 80% **19.** 26% **21.** 60% **23.** 100% **25.** 30% **27.** $\frac{3}{10}$ **29.** $\frac{22}{25}$ **31.** $\frac{13}{100}$ **33.** 3 **35.** < **37.** = **39.** > **41.** 64% **43.** 5%; the other ratios equal 25% **45.** C **47.** $\frac{3}{5}$ **49.** $2\frac{4}{5}$ **51.** 3 **53.** 4 **55.** 71 **57.** 0.791

Pages 208–210   Lesson 4-7

1. 0.68 **3.** 0.276 **5.** 9% **7.** 73% **9.** 0.27 **11.** 0.06 **13.** 0.185 **15.** 0.022 **17.** 0.277 **19.** 0.3025 **21.** 0.686 **23.** 70% **25.** 580% **27.** 95% **29.** 17% **31.** 67.5% **33.** 1.2% **35.** 34.7% **37.** 33.9% **39.** < **41.** > **43.** < **45.** 0.01389; 0.04167; 0.02083; 0.05; 0.03125 **47.** 1.2 ft **49.** Sample answer: 0.25, $\frac{1}{4}$, 25%

51. 37.5% **53.** 3.125% **55.** D **57.** 72% **59.** 3.1% **61.** −125 **63.** $2 \times 5 \times 5$ **65.** $2 \times 2 \times 19$

Pages 213–214   Lesson 4-8

1. 28 **3.** 60 **5.** 60 **7.** 2020 **9.** 72 **11.** 72 **13.** 315 **15.** 72 **17.** 420 **19.** 144 **21.** 6:00 P.M. **23.** 50¢ **25.** Sample answer: 5, 7 **27.** Sample answer: 10, 35 **29.** 4 packages of juice boxes and 5 packages of oatmeal snack bars **31.** $2^2 \cdot 3 \cdot 5$, or 60 **33.** Sample answer: 3, 10, 15 **35.** C **37.** 0.55 **39.** 0.0025 **41.** $\frac{17}{25}$ **43.** $2s + 7$ **45.** <

Pages 218–220   Lesson 4-9

1. > **3.** < **5.** Eliot; 3 out of 4 has an average of 0.75; 7 out of 11 has an average of 0.64 **7.** C **9.** < **11.** = **13.** < **15.** > **17.** > **19.** < **21.** > **23.** = **25.** Jim; $\frac{10}{16} > \frac{4}{15}$ **27.** $\frac{8}{10}$, 0.805, 81% **29.** −1.4, −1.25, $-1\frac{1}{25}$ **31.** 3.47, $3\frac{4}{7}$, $3\frac{3}{5}$ **33.** > **35.** < **37.** 6c, $6\frac{1}{3}$c, 6.5c **39.** $\frac{1}{5}$g, 1.5g, 5g **41.** Eastern Chipmunk **43.** Bustos: 0.346; Kretschman: 0.333; Nuveman: 0.313; Watley: 0.400; Watley **45.** 0.08; 0.08 equals 8% and the other ratios equal 80%. **47.** Sample answer: Gwen needs $\frac{2}{5}$ yard of fabric and $\frac{3}{8}$ yard of ribbon to make a pillow. Which item does she need more, fabric or ribbon? Answer: fabric **49.** H **51.** 42 **53.** 48 **55.** 0.06 **57.** 18 **59.** 4 **61.** 12

Pages 221–224   Chapter 4   Study Guide and Review

1. false; division **3.** true **5.** false; least common multiple **7.** true **9.** 2^7 **11.** 5×19 **13.** $65 = 5 \cdot 13$ **15.** 6 **17.** 21 **19.** $12 **21.** 15 different plans **23.** $\frac{7}{12}$ **25.** $\frac{2}{9}$ **27.** $\frac{1}{2}$ **29.** 0.75 **31.** $0.\overline{5}$ **33.** 6.4 **35.** $\frac{7}{10}$ **37.** $\frac{1}{20}$ **39.** $\frac{27}{50}$ **41.** $5.1\overline{3}$ min **43.** 44% **45.** 40% **47.** $\frac{19}{20}$ **49.** $\frac{4}{25}$ **51.** 0.48 **53.** 0.125 **55.** 61% **57.** 19% **59.** 0.12 **61.** 8 **63.** 24 **65.** 120 **67.** < **69.** > **71.** English

Chapter 5   Applying Fractions

Page 229   Chapter 5   Getting Ready

1. 35 **3.** 30 **5.** 21.6 **7.** 0.83 **9.** 4 **11.** $2\frac{6}{5}$ **13.** $5\frac{5}{4}$ **15.** $\frac{5}{3}$ or $1\frac{2}{3}$ c

Pages 233–235   Lesson 5-1

1–41. Sample answers are given. **1.** $8 + 2 = 10$ **3.** $6 \times 3 = 18$ **5.** $0 + \frac{1}{2} = \frac{1}{2}$ **7.** $\frac{1}{2} \times 1 = \frac{1}{2}$ **9.** $\frac{1}{4} \cdot 16 = 4$ **11.** $\frac{2}{3}$ of 6 ≈ 4ft **13.** $1 + 6 = 7$ **15.** $4 - 2 = 2$ **17.** $2 \cdot 3 = 6$ **19.** $9 \div 3 = 3$ **21.** $\frac{1}{2} + \frac{1}{2} = 1$ **23.** $1 - \frac{1}{2} = \frac{1}{2}$ **25.** $\frac{1}{2} \cdot 1 = \frac{1}{2}$ **27.** $0 \div 1 = 0$ **29.** 16 + 1 or 17 in. **31.** $\frac{1}{6} \times 36 = 6$ **33.** $25 \div 5 = 5$ **35.** $24 \div 2 = 12$ **37.** 39 **39.** 24 **41.** $\frac{1}{2} \times 200$ or 100

43. Sample answer: $\frac{1}{4} \times 60$ inches or 15 inches
45. Sample answer: $\frac{11}{12}$ and $\frac{7}{15}$; $\frac{11}{12} - \frac{7}{15} \approx 1 - \frac{1}{2}$ or $\frac{1}{2}$
and $\frac{11}{12} \cdot \frac{7}{15} \approx 1 \cdot \frac{1}{2}$ or $\frac{1}{2}$ **47.** Estimation; Dion doesn't
need an exact answer. Sample answer: $3\frac{1}{4} + 1\frac{2}{3} + 1\frac{2}{3}$ is
about halfway between 6 and 7 cups. Since this is
more than 6 cups, Dion cannot use this bowl to mix the
ingredients.   **49.** 38 cups   **51.** >   **53.** <   **55.** 5 pkg. of
necklaces and 3 pkg. of bracelets **57.** 37.5%   **59.** 1.9%
61. 10   **63.** 15

Pages 238–241   **Lesson 5-2**
1. $\frac{2}{3}$ **3.** $\frac{1}{4}$ **5.** $\frac{13}{24}$ **7.** $\frac{1}{4}$ **9.** addition; Sample answer:
To find how much smaller the total height of the photo
is now, add $\frac{5}{16}$ and $\frac{3}{8}$; $\frac{11}{16}$ in.   **11.** $\frac{4}{7}$ **13.** $\frac{2}{3}$ **15.** $\frac{2}{3}$
17. $1\frac{13}{24}$ **19.** $\frac{4}{9}$ **21.** $\frac{14}{45}$ **23.** addition; Sample answer:
To find the smallest width to make the shelf, add $\frac{4}{5}$
and $\frac{3}{4}$; $1\frac{11}{20}$ ft   **25.** subtraction; Sample answer: To find
how much more turkey Makayla bought, subtract $\frac{1}{4}$
from $\frac{5}{8}$; $\frac{3}{8}$ **27.** $\frac{23}{28}$ **29.** $\frac{7}{12}$ **31.** $1\frac{1}{4}$ **33.** $2\frac{2}{3}$ **35.** $\frac{2}{15}$
37. $1\frac{1}{4}$ **39.** $\frac{1}{12}$ **41.** $\frac{2}{3} - \left(\frac{1}{6} + \frac{1}{4}\right) = \frac{1}{4}$ **43.** $\frac{3}{4}$
45. Sample answer: The sum of two unit fractions $\frac{1}{a}$
and $\frac{1}{b}$, where a and b are not 0, is $\frac{a + b}{ab}$.

$\frac{1}{a} + \frac{1}{b} = \frac{1 \cdot b}{a \cdot b} + \frac{1 \cdot a}{b \cdot a}$ Rename each fraction using
                  the LCD, ab.

$= \frac{b}{ab} + \frac{a}{ab}$ Simplify.

$= \frac{b + a}{ab}$ Add the numerators.

$= \frac{a + b}{ab}$ Commutative Property

So, $\frac{1}{99} + \frac{1}{100} = \frac{99 + 100}{99 \cdot 100}$ or $\frac{199}{9,900}$ **47.** Lourdes;
Meagan did not rename the fractions using the LCD.
49. A   **51.** D   **53.** Sample answer: $4 + 4 = 8$
55. Sample answer: $6 \cdot 3 = 18$ **57.** 24.8%   **59.** −5
61. $\frac{2}{3}$ **63.** 5

Pages 244–246   **Lesson 5-3**
1. $9\frac{6}{7}$ **3.** $4\frac{2}{3}$ **5.** $1\frac{1}{2}$ **7.** $4\frac{5}{8}$ **9.** $3\frac{3}{20}$ gal   **11.** $7\frac{5}{7}$ **13.** $2\frac{1}{7}$
15. $7\frac{5}{12}$ **17.** $18\frac{17}{24}$ **19.** $3\frac{1}{2}$ **21.** $2\frac{11}{20}$ **23.** $5\frac{7}{8}$ **25.** $7\frac{1}{6}$
27. addition; Sample answer: To find the length of the
necklace, add $7\frac{1}{4}$ in. and $10\frac{5}{8}$ in.; $17\frac{7}{8}$ in.
29. subtraction; Sample answer: To find how many
inches Alameda had cut, subtract $6\frac{1}{2}$ from $9\frac{3}{4}$; $3\frac{1}{4}$ in.
31. $15\frac{1}{4}$ **33.** $1\frac{3}{4}$ **35.** $7\frac{1}{8}$ yd   **37.** Estimation; You do
not need an exact answer; Less than; $7 + 1 < 2 + 7$
39. Sample answer: Since the garden is a rectangle, the
length of one side added to the length of the other side
would equal half the length of the perimeter. If one

side of the garden is $2\frac{5}{12}$ ft long, find 6 ft $- 2\frac{5}{12}$ ft,
or $3\frac{7}{12}$ ft.   **41.** H   **43.** Sample answer: $1 \div 1 = 1$
45. Sample answer: $9 \times 7 = 63$ **47.** 15 ft

Pages 248–249   **Lesson 5-4**
1. Sample answer: Use estimation, look for a pattern,
work backward.   **3.** Sample answer: A fishbowl holds
$1\frac{1}{2}$ gallons of water. If there is $\frac{1}{3}$ gallon of water in the
bowl, how many more gallons are needed to fill the
bowl; $\frac{2}{3}$ gallons, $1\frac{1}{6}$ gallons, or $2\frac{1}{6}$ gallons. Answer:
$\frac{2}{3}$ gallon is not enough because that would make
exactly 1 gallon of water in the tank. $2\frac{1}{6}$ gallons is too
much, since the tank only holds $1\frac{1}{2}$ gallons.
The answer is $1\frac{1}{6}$ gallons.   **5.** J   **7.** $\frac{5}{8}$ in.
9.

11. $\frac{1}{49}$ inch   **13.** $\frac{4}{15}$

Pages 255–257   **Lesson 5-5**
1. $\frac{2}{9}$ **3.** $\frac{2}{3}$ **5.** $1\frac{1}{2}$ **7.** 32 pounds   **9.** $\frac{4}{15}$ **11.** $4\frac{4}{5}$ **13.** $\frac{1}{9}$
15. $\frac{1}{20}$ **17.** $\frac{3}{8}$ **19.** $\frac{3}{4}$ **21.** $\frac{3}{16}$ **23.** 14 c   **25.** $1\frac{9}{16}$
27. 30   **29.** $31\frac{1}{3}$ **31.** 20   **33.** $7\frac{1}{20}$ mi   **35.** $\frac{8}{21}$ **37.** $\frac{11}{48}$
39. $16\frac{1}{3}$ yd; $10\frac{5}{6}$ yd^2 **41.** one pint   **43.** one centimeter
45. $15\frac{3}{4}$ **47.** $28\frac{3}{4}$ **49.** broccoli: $1\frac{7}{8}$ c, pasta: $5\frac{5}{8}$ c, salad
dressing: 1 c, cheese: 2 c; Multiply each amount by $1\frac{1}{2}$.
51. $\frac{2}{5} \times \frac{3}{5}$; $\frac{6}{25}$; Sample answer: The model shows that $\frac{3}{5}$
of the rectangle is 15 sections out of 25 sections. Since $\frac{2}{5}$
of 15 sections is six sections, $\frac{2}{5}$ of $\frac{3}{5}$ is $\frac{6}{25}$.   **53.** Always;
Sample answer: Improper fractions are always greater
than 1.   **55.** Sample answer: The model shows that $\frac{3}{4}$
of one rectangle is 3 sections out of 4 sections shaded
since 2 sets of rectangles with 3 sections out of
4 sections is $\frac{6}{4}$, 2 of $\frac{3}{4}$ is $\frac{6}{4}$, or $1\frac{1}{2}$.   **57.** H   **59.** $2\frac{1}{10}$ in.
61. >   **63.** $4.95 + 0.6x = 22.95$; 300 min   **65.** 3

Pages 261–263   **Lesson 5-6**
1. $\frac{5}{8}$ **3.** $\frac{5}{29}$ **5.** 32   **7.** 20.5   **9.** $\frac{4}{5}$ **11.** 32   **13.** $\frac{6}{5}$ or
$1\frac{1}{5}$ **15.** $\frac{6}{1}$ or 6   **17.** $\frac{1}{3}$ **19.** $\frac{8}{41}$ **21.** 36   **23.** 14.4
25. 2.88   **27.** $\frac{6}{5}$ **29.** $\frac{20}{21}$ **31.** $6\frac{2}{3}$ **33.** 195 mi
35. −75   **37.** 41.4   **39.** $1\frac{1}{2}$ **41.** $x =$ elevation change of
the Wild Cave Tour; $140 = \frac{7}{15}x$; 300ft   **43.** $x =$ number
of servings; $\frac{3}{4}x = 16\frac{1}{2}$; 22 servings   **45.** $x =$ the
person's age; $\frac{1}{3}x = 26$; 78 years old

47. 20; Sample answer: By solving $8 = \frac{m}{4}$, you find that $m = 32$. So, replace m with 32 to find $32 - 12 = 20$.
49. Sample answer: Multiply each side by 2. Then divide each side by $(b_1 + b_2)$. So, $\frac{2A}{b_1 + b_2} = h$. **51.** A
53. $\frac{1}{6}$ **55.** $\frac{2}{5}$ **57.** $3\frac{5}{12}$ c **59.** 0.08 **61.** 1.23
63. -1 **65.** -8 **67.** Sample answer: $18 \div 3 = 6$
69. Sample answer: $0 \div 1 = 0$

Pages 267–270 **Lesson 5-7**
1. $\frac{3}{8}$ **3.** $3\frac{1}{2}$ **5.** $\frac{1}{15}$ **7.** $1\frac{1}{5}$ **9.** 56 segments **11.** $\frac{7}{16}$
13. $1\frac{1}{3}$ **15.** 12 **17.** $\frac{2}{3}$ **19.** 12 **21.** $\frac{4}{15}$ **23.** $\frac{2}{3}$ **25.** $2\frac{17}{20}$
27. $7\frac{4}{5}$ **29.** 36 **31.** $3\frac{34}{35}$

33. $1\frac{1}{4}$;

35. $2\frac{3}{4}$;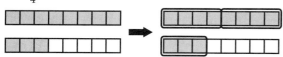

37. 25 **39.** $\frac{1}{22}$ **41.** $\frac{3}{11}$ **43.** 4 **45.** $7\frac{9}{10}$ times as many
47. 6:30 P.M.; Sample answer: $105 \div 35 = 3$, The storm will travel 105 miles in 3 sets of $\frac{1}{2}$ hour, or $1\frac{1}{2}$ hours. Adding $1\frac{1}{2}$ hours to 5:00 P.M. will make it 6:30 P.M.
49. Sample answer; You could use paper and pencil; $\frac{3}{4}$ of what number is $1\frac{1}{2}$? $1\frac{1}{2}$ feet $\div \frac{3}{4} = \frac{3}{2}$ feet $\cdot \frac{4}{3} = 2$ feet.
51. Yes. If the first proper fraction is larger than the second proper fraction then the resulting quotient will be a whole number or mixed number. **53.** H
55. $\frac{13}{4}$ or $3\frac{1}{4}$ **57.** $\frac{4}{21}$ **59.** 4.5 ft

Pages 271–274 **Chapter 5** **Study Guide and Review**
1. numerators **3.** reciprocal **5.** unlike **7.** $1\frac{11}{7}$ **9.** 3
11. reciprocal **13.** Sample answer: $3 \div 1 = 3$
15. Sample answer: $1 \times 0 = 0$ **17.** Sample answer: $\frac{1}{2} \times 26 = 13$ **19.** Sample answer: 19×10 or about 190 ft^2 **21.** $\frac{1}{6}$ **23.** $\frac{2}{3}$ **25.** $\frac{5}{24}$ **27.** $\frac{5}{8}$ in. **29.** $9\frac{11}{15}$
31. $10\frac{1}{7}$ **33.** $2\frac{4}{15}$ **35.** $15\frac{1}{8}$ **37.** $7\frac{11}{12}$ h **39.** G **41.** $\frac{5}{27}$
43. $2\frac{3}{5}$ **45.** $9\frac{3}{8}$ **47.** $\frac{12}{7}$ or $1\frac{5}{7}$ **49.** $\frac{3}{10}$ **51.** 15 **53.** 0.75
55. $\frac{7}{10}$ **57.** $3\frac{3}{10}$ **59.** $1\frac{21}{22}$ **61.** 54

Chapter 6 Ratios and Proportions

Page 281 **Chapter 6** **Getting Ready**
1. 48.1 **3.** 7.4 **5.** $\frac{1}{5}$ **7.** $\frac{19}{23}$ **9.** $\frac{39}{50}$ **11.** $\frac{3}{50}$ **13.** 450
15. 2,200

Pages 284–286 **Lesson 6-1**
1. $\frac{2}{15}$ **3.** $\frac{1}{51}$ **5.** yes. $\frac{12}{20} = \frac{3}{5}$ and $\frac{6}{10} = \frac{3}{5}$ **7.** no;
Sample answer: $\frac{2 \text{ boxes}}{\$5} \neq \frac{6 \text{ boxes}}{\$20}$, since $2 \cdot 3 = 6$, but $5 \cdot 3 \neq 20$ **9.** $\frac{3}{2}$ **11.** $\frac{1}{3}$ **13.** $\frac{2}{5}$ **15.** $\frac{21}{1,600}$ **17.** $\frac{21}{32}$
19. yes; $\frac{4}{16} = \frac{1}{4}$ and $\frac{10}{40} = \frac{1}{4}$, so $\frac{4}{16} = \frac{10}{40}$ **21.** no; $\frac{8}{6} = \frac{4}{3}$
and $\frac{12}{10} = \frac{6}{5}$, so $\frac{8}{6} \neq \frac{12}{10}$ **23.** no; Sample answer: $\frac{12 \text{ in.}}{3 \text{ in.}}$
$= \frac{4}{1}$ and $\frac{6 \text{ in.}}{1 \text{ in.}} = \frac{6}{1}$, so $\frac{12 \text{ in.}}{3 \text{ in.}} \neq \frac{6 \text{ in.}}{1 \text{ in.}}$ **25.** no; $\frac{71}{42} \neq \frac{16}{9}$
27. 9 lb **29.** yes; $\frac{330}{396} = \frac{5}{6}$; **31.** Areas A and C; both
ratios simplify to a growth-to-removal ratio of $\frac{11}{30}$.
33. 80; Sample answer: $440 + 80 = 520$; $\frac{520}{1,200} = \frac{13}{30}$ which
is the same ratio as area B **35.** 2,400; The denominator of the ratios increases by 1. $\frac{20}{40} = \frac{1}{2}, \frac{40}{120} = \frac{1}{3}, \frac{120}{480} = \frac{1}{4}$
37. B **39.** $\frac{6}{7}$ **41.** $1\frac{2}{3}$ **43.** $5\frac{2}{3}$ **45.** 4.9 **47.** $0.31

Pages 289–292 **Lesson 6-2**
1. 6 mi per gal **3.** $0.50 per lb **5.** C **7.** 60 mi/h
9. 30.4 people per class **11.** 3.5 m/s **13.** $0.14/oz
15. about $0.50 per pair **17.** Susanna; 1.78 m/s > 1.66 m/s > 1.23 m/s **19.** Soft drinks A and B have about 3 grams of sodium per ounce and Soft drink C has about 6 grams per ounce. **21.** 510 words
23. Sample answer: about 60 **25.** 182.7 **27.** 38¢ per lb; $1.89 $\div$ 5 $\approx$ $1.90 $\div$ 5 or $0.38 **29.** 3 c
31. about 1 h 29 min 49 s **33.** The bear's heart beats 120 times in 2 minutes when it is active. **35.** the bear's heart rate in beats per minute **37.** when it is active; Sample answer: The active line increases faster than the hibernating line when read from left to right.
41. Always; every rate is a ratio, because it is a comparison of two quantities by division. **43.** Sample answer: a; $\frac{30 \text{ ft}}{2 \text{ min}} = 15$ ft/min, $\frac{40 \text{ ft}}{2 \text{ min}} = 20$ ft/min
45. C **47.** $\frac{2}{9}$ **49.** $\frac{9}{14}$ **51.** 8 subs **53.** $\frac{1}{2}$ **55.** $\frac{1}{4}$

Pages 295–297 **Lesson 6-3**
1. 1.5°F every hour **3.** slope: $\frac{8}{1}$ or 8; There are 8 packs of fruit snacks in each box.

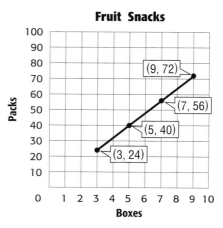

Fruit Snacks

5. $9 per h **7.** $9 per shirt **9.** 0.45mi/min

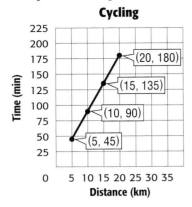

Cycling

(20, 180)
(15, 135)
(10, 90)
(5, 45)

Time (min) / *Distance (km)*

11. $\frac{1}{6}$ inch per minute

13.

Length of Poster (ft)	Ribbon Needed (in.)
3	18
6	36
9	54
12	72

15. A **17.** $11.90 **19.** $\frac{18}{1}$ **21.** 50 **23.** 6.5

Pages 301–303   **Lesson 6-4**
1. 48 **3.** 52 **5.** 3 **7.** $7\frac{1}{2}$ **9.** 0.0075 mi/s **11.** $4\frac{1}{2}$
13. 16 **15.** 24 **17.** 6,600 **19.** $6\frac{1}{2}$ **21.** 6,750
23. 3,520 ft **25.** $17\frac{1}{2}$ qt **27.** 1,056,000 ft/h **29.** No;
15 in. $+ 4\frac{1}{2}$ in. $+ 6\frac{3}{4}$ in. $= 26\frac{1}{4}$ in. and $26\frac{1}{4} \div 12 = 2\frac{3}{16}$.
So, it snowed a total of $2\frac{3}{16}$ ft or about 2 ft, not $2\frac{1}{2}$ ft.
31. $\frac{1}{2}$ **33.** about $8\frac{1}{2}$ mi **35.** $\frac{1}{4}$ **37.** 3 gal; Sample
answer: the graph increases by 1 gallon for every 4
quarts. **39.** < **41.** = **43.** 720 in²; Square feet mean
a unit of feet × feet. To divide out each unit, you must
multiply by two conversion factors that have feet in
the denominator and inches in the numerator.
5 ft² · $\frac{12 \text{ in.}}{1 \text{ ft}}$ · $\frac{12 \text{ in.}}{1 \text{ ft}}$ = 720 in² **45.** G **47.** $9 per h
49. 8 ft **51.** $2v$ **53.** 118.9 **55.** 142.127

Pages 307–309   **Lesson 6-5**
1. 370 **3.** 1.46 **5.** 8.52 **7.** 128.17 **9.** about
5,333.33 ft **11.** 0.98 **13.** 3.0 **15.** 0.08 **17.** 130,500
19. 106.17 **21.** 36.01 **23.** 15.75 **25.** 0.51 kg
27. about 35,420 meters per hour **29.** 16,582.42
31. 403,704 **33.** 0.06 L, 660 mL, 6.6 kL **35.** 130 cm,
2650 mm, 5 m **37.** 50 cm **39.** Jake; Gerardo divided
3.25 by 1,000; he should have multiplied. **41.** about
621,118.01 mi **43.** There are a greater number of
smaller units. **45.** H

47. slope: $\frac{12}{1}$ or 12; There are 12 inches in 1 foot.

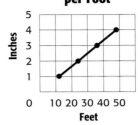

Number of Inches per Foot

Inches / *Feet*

49. $6\frac{1}{5}$ **51.** $5\frac{19}{20}$ **53.** 10 **55.** 3.2

Pages 313–315   **Lesson 6-6**
1. No; sample answer: $\frac{10 \text{ children}}{2 \text{ adults}} = \frac{5 \text{ children}}{1 \text{ adult}}$ and
$\frac{12 \text{ children}}{3 \text{ adults}} = \frac{4 \text{ children}}{1 \text{ adult}}$. The unit rates are not equal.
3. Yes; sample answer: The cross products of the ratios
$\frac{8}{21}$ and $\frac{12}{31.5}$, 8 × 31.5 and 21 × 12, are both equal to 252.
5. 15 **7.** 29.4 **9.** 20 **11.** $8.33 **13.** No; sample answer:
The cross products of the ratios $\frac{20}{6}$ and $\frac{16}{5}$, 20 × 5 or
100 and 6 × 16 or 96, are not equal. **15.** Yes; sample
answer: The cross products of the ratios $\frac{16}{200}$ and $\frac{28}{350}$,
16 × 350 and 200 × 28, are both equal to 5,600.
17. No; sample answer: The cross products of the ratios
$\frac{1.4 \text{ T}}{18 \text{ days}}$ and $\frac{10.5 \text{ T}}{60 \text{ days}}$, 1.4 × 60 or 84 and 18 × 10.5 or
189, are not equal. **19.** No; $\frac{45 \text{ min}}{25 \text{ pages}} = \frac{1.8 \text{ min}}{1 \text{ page}}$ and
$\frac{60 \text{ min}}{30 \text{ pages}} = \frac{2 \text{ min}}{1 \text{ page}}$. These rates are not equivalent. **21.** 15
23. 3 **25.** 1.5 **27.** 63 **29.** 2.4 **31.** 13.5 **33.** 4.2 lb
35. 1.6 oz **37.** (3, 15) represents 3 pizzas cost $15.
(5, 25) represents 5 pizzas cost $25. Yes, as the number
of pizzas increases by 1, the cost increases by $5. **39.** 5,
with delivery fee; 5 no delivery fee; the slope represents
the cost per pizza **41.** $19.50 **43.** $\frac{325}{13} = \frac{100}{x}$; 4 **45.** $5.70
for 6 lb; Sample answer: This ratio has a unit rate of
0.95/lb, while all of the other ratios have a unit rate
of $0.90/lb. **47.** Sample answer: Mental math; $10 is 4
times more than $2.50, so the number of ears is 4 times
more than a dozen, which is 4 dozen or 48 ears of corn.
49. B **51.** about 2.27 kg **53.** 56 **55.** $20\frac{9}{16}$ **57.** $17\frac{5}{32}$

Pages 318–319   **Lesson 6-7**
1. $\frac{3}{4}$ feet

8 ft
$5\frac{1}{3}$ ft
$3\frac{5}{9}$ ft
$2\frac{10}{27}$ ft
12 ft

1   2   3   4

3. 30 miles **5.** 125 minutes **7.** 28 games
9. Mary, Isabela, Anna, Pilar

Pages 323–326   **Lesson 6-8**

1. 50 km **3.** 130 km **5.** 14 feet **7.** $16\frac{2}{3}$ in. **9.** $\frac{1}{48}$
11. $\frac{1}{63,360}$ **13.** 81 mi **15.** 40 mi **17.** 12 ft by 9 ft
19. $11\frac{3}{5}$ in.; $\frac{1}{90}$ **21.** 6 in.; $\frac{1}{720}$ **23.** 26 mi **25.** $\frac{1}{3}$; about
6.33 feet **29.** $\frac{5,000}{1}$ **31a.** A; 0.5 cm is larger than 1 mm.
If 0.5 cm on the model is equal to 1 mm on the actual
figure, then model A must be larger than the actual
figure. **31b.** B; 1.5 mm is smaller than 4 cm. If 1.5 mm
on the model is equal to 4 cm on the actual figure, then
model B must be smaller than the actual figure. **31c.** C;
0.25 cm is equal to 2.5 mm. If 0.25 cm is equal to 2.5 mm
on the actual figure, then model C must be the same
size as the actual figure. **33.** Sample answer: Using the
scale given on the map, look at the distance between the
two cities on the map and then estimate the actual
distance based on the distance given in the scale.
35. F **37.** 200 people **39.** 15 **41.** Sample answer:
$7 + 9 + 10 + 12 + 7$ or 45 mi **43.** 24 **45.** 30 **47.** 60
49. $\frac{13}{30}$ **51.** $\frac{7}{8}$

Pages 331–332   **Lesson 6-9**

1. $\frac{27}{100}$ **3.** $\frac{3}{40}$ **5.** $\frac{5}{8}$ **7.** 0.16% **9.** 11.11% **11.** $\frac{5}{8}$
13. 0.13% **15.** $\frac{1}{3}$ **17.** $\frac{15}{16}$ **19.** $\frac{1}{1000}$ **21.** 555%
23. 375% **25.** 96.67% **27.** 71.43% **29.** 0.13%
31. 0.42% **33.** 140% **35.** < **37.** > **39.** $\frac{1}{2}$%, $\frac{2}{5}$, 0.48,
0.5 **43.** Less than; $\frac{26}{125} = 20.8\%$ **45.** Sample answer:
Since a percent is a ratio that compares a number to
100, 80% is the ratio $\frac{80}{100}$. The ratio $\frac{80}{100}$ can be read as
eighty-hundredths or written as a decimal, 0.80 or 0.8.
The ratio $\frac{80}{100}$ also simplifies to $\frac{4}{5}$ if you divide the
numerator and denominator by the same number, 20.
47. G **49.** 2 **51.** $\frac{9}{20}$ **53.** $1\frac{3}{4}$

Pages 333–336   **Chapter 6**   **Study Guide and Review**

1. ratio **3.** rate **5.** scale drawing **7.** scale factor
9. unit rate **11.** $\frac{4}{3}$ **13.** $\frac{2}{3}$ **15.** no; $\frac{18}{24} = \frac{3}{4}$, $\frac{5}{20} = \frac{1}{4}$,
and $\frac{3}{4} \neq \frac{1}{4}$ **17.** 90 mi per day
19.

José's Savings

slope: $\frac{30}{1}$ or 30; José saved $30 every week.
21. 24 **23.** $4\frac{1}{2}$ **25.** 50 bushels **27.** 51,528.96

29. 18.43 **31.** about 345 lb **33.** 54 **35.** 9 **37.** $2\frac{1}{3}$
39. 27 feet **41.** 24 mi **43.** $\frac{11}{40}$ **45.** $\frac{181}{400}$ **47.** 91.67%

Chapter 7   Applying Percents

Page 341   **Chapter 7**   **Getting Ready**

1. 48 **3.** 1,512 **5.** $54.75 **7.** 0.34 **9.** 0.75 **11.** 75
13. 38.9 **15.** 0.17 **17.** 1.57 **19.** 0.075

Pages 346–348   **Lesson 7-1**

1. 4 **3.** 110.5 **5.** 23 **7.** $3.25 **9.** $194.40 **11.** 45.9
13. 14.7 **15.** 17.5 **17.** 62.5 **19.** $290 **21.** 3.5
23. 97.8 **25.** 92.5 **27.** about 19.5 million **29.** 3.3
31. 990 **33.** 520 **35.** 0.24 **37.** $9.75 **39.** $297
41. about 316 **43.** Sample answer: 53% of 60 → 50%
$\times$ 60 or $\frac{1}{2} \times 60 = 30$ **45.** Sample answer: 75% of 19 →
75% $\times$ 20 or $\frac{3}{4} \times 20 = 15$ **47.** 80 **49.** Sample answer:
24% $\times$ 250 or 60 people prefer cherries. So, 250 − 60
or 190 people did not prefer cherries. **51.** 15
53. Sample answer: Determine the number of
questions answered correctly on a test and find how
much to tip a restaurant server. **55.** Less than the
original number; you are subtracting 10% of a greater
number. **57.** C **59.** 71% **61.** $\frac{3}{5}$ **63.** $\frac{17}{24}$ **65.** 1267.81

Pages 353–354   **Lesson 7-2**

1. 36% **3.** 0.7 **5.** 75 **7.** 3 c **9.** 7.5% **11.** 8.6
13. 375 **15.** 24 **17.** 40% **19.** 4.1 **21.** 192
23. 0.2% **25.** $8 **27.** about 3.4% **29.** about 6,378 km
31. 20% of 500, 20% of 100, 5% of 100; If the percent is
the same but the base is greater, then the part is
greater. If the base is the same but the percent is
greater, then the part is greater. **33.** B **35.** 30
37. 2.75 **39.** 3 **41.** 30 **43.** 18

Pages 357–360   **Lesson 7-3**

1. 5; $\frac{1}{2} \cdot 10 = 5$; $0.1 \cdot 10 = 1$ and $5 \cdot 1 = 5$ **3.** 24; $\frac{2}{5} \cdot 60$
$= 24$; $0.1 \cdot 60 = 6$ and $4 \cdot 6 = 24$ **5.** $(1 \cdot 70) + \left(\frac{1}{2} \cdot 70\right)$
$= 105$ **7.** about $50; $\frac{1}{4} \cdot \$200 = \$50$; $0.1 \cdot 200 = 20$ and
$2.5 \cdot 20 = 50$ **9.** about 160,000 acres; $0.1 \cdot 20,000,000$
$= 200,000$ and $\frac{4}{5}$ of $200,000 = 160,000$ **11.** 18; $\frac{1}{5} \cdot 90 =$
18; $0.1 \cdot 90 = 9$ and $2 \cdot 9 = 18$ **13.** 135; $\frac{3}{4} \cdot 180 = 135$;
$0.1 \cdot 180 = 18$ and $7.5 \cdot 18 = 135$ **15.** 90; $\frac{9}{10} \cdot 100 = 90$;
$0.1 \cdot 100 = 10$ and $9 \cdot 10 = 90$ **17.** 36; $\frac{3}{10} \cdot 120 = 36$;
$0.1 \cdot 120 = 12$ and $3 \cdot 12 = 36$ **19.** 90; $\frac{3}{5} \cdot 150 = 90$; 0.1
$\cdot 150 = 15$ and $6 \cdot 15 = 90$ **21.** 168; $\frac{7}{10} \cdot 240 = 168$; 0.1
$\cdot 240 = 24$ and $7 \cdot 24 = 168$ **23.** about 12 muscles;
$\frac{3}{10} \cdot 40 = 12$ **25.** $(2 \cdot 300) + \left(\frac{1}{4} \cdot 300\right) = 675$ **27.** 0.01 $\cdot$
$200 = 2$ and $\frac{3}{4} \cdot 2 = 1.5$ **29.** $0.01 \cdot 70 = 0.7$ **31.** about
2,700 birds; $0.01 \cdot 450,000 = 4,500$ and $\frac{3}{5} \cdot 4,500 = 2,700$
33. $\frac{1}{2} \cdot 80 = 40$ **35.** $\frac{1}{10} \cdot 240 = 24$ **37.** $1 \cdot 45 = 45$
39. about 3 hours; Method 1: $\frac{1}{3} \cdot 24 = 8$ and $\frac{1}{5} \cdot 24 \approx 5$;

$8 - 5 = 3$ **41.** 1,200   **43.** Sample answer: about 400,000 people; .08 • 5,000,000 = 400,000   **45.** 3.6 ounces   **47.** Sample answer: Find 1% of $800, then multiply by $\frac{3}{8}$.   **49.** Sometimes; sample answer: one estimate for 37% of 60 is $\frac{2}{5} \times 60 = 24$. This is greater than the actual answer because $\frac{2}{5}$ is greater than 37%. Another estimate is $\frac{1}{3} \times 60 = 20$. This is less than the actual answer because $\frac{1}{3}$ is less than 37%.   **51.** B   **53.** D   **55.** 64.8   **57.** 157.1   **59.** Sample answer: $1 + 0 = 1$ **61.** Sample answer: $1 - \frac{1}{2} = \frac{1}{2}$ **63.** 50   **65.** 357.1   **67.** 0.7   **69.** 0.75

Pages 363–365   Lesson 7-4

1. $p = 88 \cdot 300$; 264   **3.** $75 = n \cdot 150$; 50%   **5.** $3 = 0.12 \cdot w$; 25   **7.** 39 loaves   **9.** $p = 0.39 \cdot 65$; 25.4   **11.** $p = 0.53 \cdot 470$; 249.1   **13.** $26 = n \cdot 96$; 27.1%   **15.** $30 = n \cdot 64$; 46.9%   **17.** $84 = 0.75 \cdot w$; 112   **19.** $64 = 0.8 \cdot w$; 80   **21.** 4,400 games   **23.** 1   **25.** $p = 0.004 \cdot 82.1$; 0.3   **27.** $230 = n \cdot 200$; 115%   **29.** about 42%   **31.** about 27%   **33.** Sample answer: 30 is 125% of what number?; 24   **35.** Sample answer: It may be easier if the percent and the base are known because after writing the percent as a decimal or fraction, the only step is to multiply. When using the percent proportion, you must first find the cross products and then divide.   **37.** H   **39.** $1.95   **41.** 170.73%   **43.** 37 yr

Pages 366–367   Lesson 7-5

1. Sample answer: Look for a rule or pattern in the data or number facts, estimation, guess and check, make an organized list, or work backward.   **3.** Sample answer; $10 \cdot 12 = $120   **5.** 500; 60% • 830 ≈ 500   **7.** 2 quarters, 1 dime, 4 nickels, and 3 pennies   **9.** $30   **11.** $3.50   **13.** $15 \cdot 18 + 18 \cdot 20 = 630$ ft³. Then convert 630 ft³ to square yards; 630 ÷ (3 • 3) = 70 yd²

Pages 372–374   Lesson 7-6

1. 20% decrease   **3.** 19% increase   **5.** C   **7.** 40% increase   **9.** 71% decrease   **11.** 25% decrease   **13.** 5% decrease   **15.** 13% increase   **17.** 38% decrease   **19.** 2% increase   **21.** 75% decrease   **23.** 150% increase   **25.** 400%   **29.** about $31 billion   **31.** 23%   **33.** 2005 to 2006; 50%   **35.** The $60 sound system since 10 is a greater part of 60 than of 90.   **37.** No; after a 10% increase, the quantity is greater then the original quantity. Decreasing a larger number by the same percent results in a greater change.   **39.** D   **41.** Sample answer: 0.5 • 800 or 400 students   **43.** $n = 0.21 \cdot 62$; 13.0   **45.** 0.065   **47.** 0.0825

Pages 377–378   Lesson 7-7

1. $3.10   **3.** $1,338.75   **5.** $98.90   **7.** $1,605   **9.** $4.90   **11.** $7.99   **13.** $96.26   **15.** $7.50   **17.** $180.00   **19.** $35.79   **21.** $333.60   **23.** $25   **25.** $50, $25; The percent of discount is 50%. All of the other pairs have a discount of 25%.   **27.** C   **29.** C   **31.** 9% decrease   **33.** about 500 × 0.7 or 350 mi   **35.** $\frac{1}{18}$

Pages 381–382   Lesson 7-8

1. $38.40   **3.** $5.80   **5.** $1,417.50   **7.** $1,219.00   **9.** $21.38   **11.** $123.75   **13.** $45.31   **15.** $14.06   **17.** $1,353.13   **19.** $116.25   **21.** Yes, he would have $5.208.   **23.** $825.60, $852.02, $879.28   **25.** C   **27.** $21.39   **29.** 37% decrease   **31.** $1\frac{1}{5}$ **33.** $2\frac{2}{15}$

Pages 384–388   Chapter 7   Study Guide and Review

1. true   **3.** true   **5.** false; 10%   **7.** true   **9.** false; original   **11.** 39   **13.** 135   **15.** 14 games   **17.** 0.3   **19.** $27.49   **21.** Sample answer: 40; $\frac{1}{3} \cdot 120 = 40$ **23.** Sample answer: 20; $\frac{1}{5} \cdot 100 = 20$ **25.** Sample answer: 360; 0.1 • 400 = 40; 9 • 40 = 360   **27.** $32 = p \cdot 50$; 64%   **29.** $n = 42 \cdot 300$; 126   **31.** $108 = 0.12 \cdot w$; 900   **33.** 333   **35.** Sample answer: $700 × 0.4 = $280   **37.** 93% increase   **39.** 10% increase   **41.** $26.75   **43.** $8,440   **45.** 14%   **47.** $3.51   **49.** $1,500   **51.** $101.25   **53.** $311.85

Chapter 8 Statistics: Analyzing Data

Page 395   Chapter 8   Getting Ready

1. 95.89, 96.02, 96.2   **3.** 22, 22.012, 22.02   **5.** 74.7, 74.67, 74.65   **7.** 3.340, 3.304, 3.04   **9.** 2.32

Pages 398–401   Lesson 8-1

1.
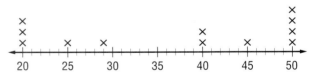

3. clusters: 6–12; gaps: 12–20; outlier: 20; range: 20 − 4, or 16   **5.** 1 or 2 glasses per day   **7.** 5 glasses

9.
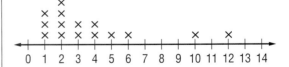

11.
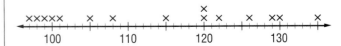

13. 34°   **15.** Sample answer: cluster 104°F–122°F; gaps 100°F–104°F, 122°F–125°F, 125°F–128°F, and 128°F–134°F; outliers 100°F and 134°F   **17.** 36 − 10 or 26   **19.** 15   **21.** Sometimes; the range will only change if the new data value lies above or below the greatest and least points, respectively, of the original data set.   **23.** 10   **25.** about 72%   **27.** Sample answer: range 28; cluster 1–19; gap 19–29; outlier 29

29.

The House of Representatives, Southwestern States

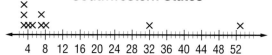

Sample answer: The data are more clustered between 4 and 25 with a range of 35 and a gap from 25 to 39. Most frequently occurring is 7, 8, and 9. 39 is an outlier. **31.** Sample answer: The range of a data set excluding the outlier(s) is a lesser value than the range of the same data set including the outlier(s). This is because any outliers will lie below and above the least and greatest values, respectively, of the data set and the range will change to include these values. **33.** Sample answer: A frequency table shows the number of times data occur by using tally marks. A line plot shows the number of times data occur by using ×s. A frequency table usually shows intervals of data and is useful when a summary is needed. A line plot shows individual data points and is useful when you need to see how all the data points are spread out. **35.** C **37.** $60 **39.** $4.46 **41.** 47 **43.** 12.6

Pages 405–408 **Lesson 8-2**

1. 52.3; 57; 59 **3.** 26.5, 25.5, 23, and 25 **5.** Sample answer: Either the mean, median, or mode could be used to represent the data. The mean is slightly less than most of the data items, and therefore is a less accurate description of the data. **7.** 87; 90; 80 and 93 **9.** $12; $9; $6 **11.** 46, 45, 44 **13.** Sample answer: The mean, 15.6, is higher than most of the data. The median, 1, or mode, 1, best represents the data. **15.** $3.50, $3.50, $3.50 **17.** Always; Sample answer: Any value that is added which is greater than the maximum value, 23, will increase the average, or mean, of the values. **19.** Sometimes; Sample answer: The mean of the data set is currently 14.5. If a value that is greater than 14.5 is added, the mean will increase. If a value that is less than 14.5 is added, the mean will decrease. If a value that is exactly 14.5 is added, the mean will remain unchanged. **21.** Sample answer: A length of 960 inches is much greater than the other pieces of data. So, if this piece of data is added, the mean will increase. **23.** 10 points; Sample answer: The sum of the points for the first thirteen games is 158. In order for the average number of points to be 12, the total number of points for fourteen games would need to be 12 × 14 or 168. So, 168 − 158 or 10 points need to be scored during the last game. **25.** Sample answer: 4, 5, 2, 2, 3, 3, 1, 0, 1, 2, 4, 68, 5; the mean, 7.7, is not the best representation since it is greater than all the data items except one. **27.** Sometimes; if there is an odd number of items, the median is the middle number. If there is an even number of items, the median is the mean of the two middle numbers. **29.** Mean; a mode must be a member of the data set, and it is impossible to have 2.59 family members. **31.** H

33.

High Temperatures for July in Kentucky

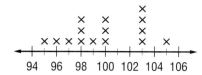

35. $10.43 **37.** Commutative (×) **39.** ones **41.** tenths

Pages 412–414 **Lesson 8-3**

1.

Height of Trees(ft)

Stem	Leaf
0	8 8
1	0 2 5 5 6 8
2	0 5

$2|0 = 20\ ft$

3. 5 **5.** mean

7.

Low Temperatures (°F)

Stem	Leaf
1	3 3 5
2	0 4 8
3	0 1 2 2 5 6 8 8 8

$1|3 = 13°F$

9.

School Play Attendance

Stem	Leaf
22	5 7 9
23	0
24	3 6
25	
26	7 9 9
27	8 8 8

$26|7 = 267\ people$

11. 4; 1 **13.** $45 **15.** mean **17.** 26 **19.** Sample answer: Yes; Thirty-six of the 56 signers were 30–49 years. Since 36 out of 56 is greater than half, you can say that the majority of the signers were 30–49 years old.

21.

Average Length (ft) of Crocodiles

Stem	Leaf
6	3
7	
8	1
9	8 8
10	
11	4
12	
13	6
14	
15	
16	0 3 3 3 3
17	
18	
19	5

$13|6 = 13.6\ ft$

Sample answer: A reasonable length for an average crocodile is about 16 feet.

23. Diana; three out of the six or 50% of the pieces of ribbon are 20–30 inches in length.

25. Fiber in Cereal (g)

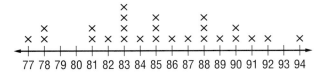

Stem	Leaf
0	0 1 1 1 1 1 1 1
	2 3 3 3 4 5 5

$0|1 = 1$ gram

Both representations show the frequency of data occurring. The line plot gives a good picture of the spread of the data. The stem-and-leaf plot shows individual grams of fiber as in the line plot, as well as intervals. See students' favorites and reasons. **27.** C **29.** 48.7; 50; 55

31. Test Scores

33. $\frac{3}{8}$ **35.** Sample answer: 20; 20–120

Pages 418–421 **Lesson 8-4**

1. histogram

State Sales Tax Rates

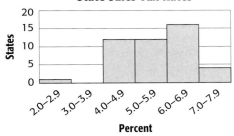

3. health

5. bar graph

Most Threatened Reptiles

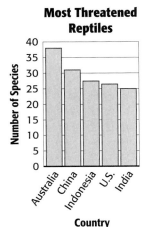

7. histogram

Major U.S. Rivers

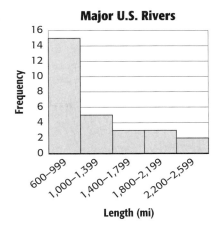

9. 13 students **11.** The 60–69 interval is three times larger than the 70–79 interval. **13.** 43 **15.** Sample answer: The number of zoos with 0.0–0.9 million visitors is about 4 times the number of zoos with 3.0–3.9 million visitors. **17.** a **19.** b **21.** about 15.8% **23.** Sample answer: It is easier to compare two sets of data. **25.** tennis **27.** Yes, 20% of girls prefer basketball and 35% of boys prefer basketball. **29.** Sample answer: Each interval represents a portion of the data set. The number of items in each interval is indicated by the frequency, typically shown along the vertical scale. By adding the frequencies for each interval, you can determine the number of values in the data set. **31.** C

33. Number of Wins

Stem	Leaf
1	5 7 9
2	3 5 6
3	0 1 2 6
4	0 0 1 3 4 5 6 7 7
5	0 0 0 1 2 3 3 5 6 6

$5|3 = 53$ wins

35. Mental math; the numbers are easy to computer mentally. Sample answer: $\frac{1}{5}$ of $50 is $10 and $50 − $10 = $40. So, he will need an additional $40.

Pages 424–425 **Lesson 8-5**

1. Sample answer: Graphs provide a visual representation of a situation involving comparisons. A graphical model can sometimes show conclusively what is often difficult to interpret from looking at lists alone.

3.

Temperature Conversions

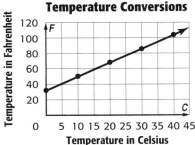

5. Taylor **7.** Wednesday **9.** 27, 29 **11.** 2

Pages 428–431   **Lesson 8-6**
1. Sample answer: The graph shows a positive relationship. That is, as the years pass, the population increases.   **3.** about 155–160 people   **5.** about 400   **7.** about 95 min   **9.** Sample answer: As sleep decreases, the math test score decreases.

11.   **Free Throws Made out of Attempts**

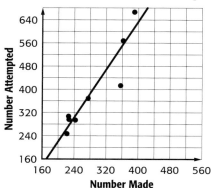

13. There is a positive relationship. That is, as the number of attempts increases, the number of free throws made increases.   **15.** about $55,000   **17.** Sample answer: Both lines have a positive trend. The population of Phoenix, Arizona, is now more than San Diego, California.   **19.** mode; the other three are ways to display data.   **21.** Graphs often show trends over time. If you continue the pattern, you can use it to make a prediction.   **23.** J

25.   **Favorite Color**   **27.** 83°; 83°

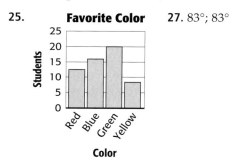

Pages 435–437   **Lesson 8-7**
1. 19,800   **3.** about 59   **5.** 1,419 people   **7.** 62,280 teens   **9.** b.   **11.** a.   **13.** about 38 people   **15.** about 1,040   **19.** D   **21.** B   **23.** It takes Dale a little bit longer to run each successive mile of the 5-mile run.
25. −24   **27.** 54   **29.** $\frac{3}{10}$ or 0.3   **31.** $\frac{9}{2}$ or 4.5

Pages 440–443   **Lesson 8-8**
1. The conclusion is invalid. This is a biased sample, since people in other states might have more umbrellas than those in Arizona. The sample is a convenience sample since all the people are from the same state.
3. This is an unbiased random survey, so the sample is valid; about 102 students.   **5.** The conclusion is invalid. This is a biased, convenience sample.

7. The conclusion is valid. This is an unbiased random sample.   **9.** The conclusion is invalid. This is a biased, convenience sample.   **11.** This is an unbiased random sample, so the results are valid; about 132 boxes.
13. This sample is a voluntary response sample. Therefore, no valid conclusion can be made.
15. about 550   **17.** Sample answer: This is an unbiased random sample. The time a student spends on the Internet during this week may not be typical of other weeks.   **19.** Sample answer: This is a convenience sample. The softball team may not represent the entire student population.   **21.** Not necessarily; Sample answer: Because you may be surveying different people in each sample, you may get different results.
23. Sample answer: If the entire population is too large to survey, it will be less time-consuming and easier to use a sample.   **25.** Yes; Sample answer: Every 10th person at a basketball game is asked whether they prefer basketball or baseball. This survey is a convenience sample because people attending a basketball game probably prefer basketball.
27. C   **29.** about 292 students   **31.** $58.44   **33.** $\frac{5}{8}$
35. true

Pages 447–449   **Lesson 8-9**
1. Graph B; From the length of the bars, it appears that it appears that Cy Young had about 3 times as many wins as Jim Galvin. However, Jim Galvin had 365 wins and Cy Young had 511 wins. So, the conclusion is not valid.   **3.** Sample answer: The mean is 8,638 and the median is 8,941. Since the median is greater than the mean, use the median to emphasize the average length.   **5.** The sample is a biased convenience sample. Mr. Kessler's first period class may not be representative of all his students. The display is biased because the data used to create the display came from a biased sample.   **7.** The median or the mode because they are much closer in value to most of the pieces of data.   **9.** The sample is a biased convenience sample. The first 100 batteries produced may not be representative of all the batteries produced. The display is biased because the data used to create the display came from a biased sample.

11.   **Monthly Cost to Rent an Apartment**

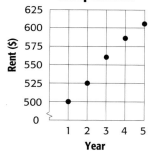

13. Sample answer: Outliers may distort measures of central tendency; data shown in graphs may be exaggerated or minimized by manipulating scales and intervals.   **15.** G   **17.** about 391 teens

Selected Answers

1. true   **3.** true   **5.** false; outlier   **7.** true

9.

Temperatures

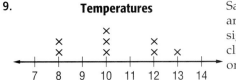

Sample answer: no significant clusters, gaps, or outliers.

11.

Number of Calories

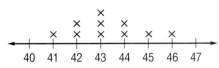

Sample answer: cluster 42–44; no gaps or outliers.

13. 84.3°; 86°; none

15.

Hours Worked

Stem	Leaf
2	1 3 6 9
3	1 2 7 8
4	6
5	4

3|2 = 32 hours

17.

Birthdates

Stem	Leaf
0	3 5 7 9
1	0 1 2 4 6 8
2	1 4

1|2 = 12

19. 10.75 million   **21.** *Crazy Horse*   **23.** Sample answer: The graph shows a positive relationship. That is, as the number of people in a family increases, the number of telephone calls per week increases.   **25.** about 208   **27.** 162 teens   **29.** The conclusion is valid. This is an unbiased random sample.   **31.** The graph does not include the hot summer months or the cold winter months, which could have higher electricity bills.

Chapter 9 Probability

1. 105   **3.** 52   **5.** 160   **7.** 210   **9.** 360   **11.** 5,040
13. $315   **15.** $\frac{1}{6}$ **17.** $\frac{1}{3}$ **19.** 5   **21.** 4

1. $\frac{1}{8}$ **3.** $\frac{1}{8}$ **5.** $\frac{3}{5}$ **7.** $\frac{23}{30}$ **9.** 1   **11.** $\frac{1}{20}$ **13.** $\frac{3}{10}$ **15.** $\frac{19}{20}$
17. $\frac{5}{8}$ **19.** $\frac{7}{20}$ **21.** 1   **23.** $\frac{3}{4}$ **25.** 60%   **27.** $\frac{14}{33}$

29. Sample answer: The complementary event is the chance of no rain. Its probability is 63%   **31a.** 1; Sample answer: Since 2032 is a leap year, there will be 29 days in February, making this event certain to happen;   **31b.** 0; Sample answer: Since 2058 is not a

leap year, there will only be 28 days in February, making this event impossible to happen.   **33.** 0.33, 0.44 are probabilities that are not complementary because $0.33 + 0.44 \neq 1$. The other sets of probability are complementary.   **35.** C   **37.** There are no labels on the vertical scale.   **39.** $\frac{1}{3}$ **41.** $\frac{1}{2}$ **43.** $\frac{9}{16}$

1. Sample answer:

	Outcomes					
	1	**2**	**3**	**4**	**5**	**6**
1	1, 1	1, 2	1, 3	1, 4	1, 5	1, 6
2	2, 1	2, 2	2, 3	2, 4	2, 5	2, 6
3	3, 1	3, 2	3, 3	3, 4	3, 5	3, 6
4	4, 1	4, 2	4, 3	4, 4	4, 5	4, 6
5	5, 1	5, 2	5, 3	5, 4	5, 5	5, 6
6	6, 1	6, 2	6, 3	6, 4	6, 5	6, 6

3. C   **5.** Sample answer:

Shape	Number	Sample Space
heads	1	heads, 1
	2	heads, 2
	3	heads, 3
	4	heads, 4
	5	heads, 5
tails	1	tails, 1
	2	tails, 2
	3	tails, 3
	4	tails, 4
	5	tails, 5

7. Sample answer:

Coin	Number Cube	Sample Space
H	1	H, 1
	2	H, 2
	3	H, 3
	4	H, 4
	5	H, 5
	6	H, 6
T	1	T, 1
	2	T, 2
	3	T, 3
	4	T, 4
	5	T, 5
	6	T, 6

9. Sample answer:

Color	Speeds	Sample Space
purple	10	purple, 10-speed
	18	purple, 18-speed
	21	purple, 21-speed
	24	purple, 24-speed
green	10	green, 10-speed
	18	green, 18-speed
	21	green, 21-speed
	24	green, 24-speed
black	10	black, 10-speed
	18	black, 18-speed
	21	black, 21-speed
	24	black, 24-speed
silver	10	silver, 10-speed
	18	silver, 18-speed
	21	silver, 21-speed
	24	silver, 24-speed

11. Sample answer:

Outcomes		
Short Sleeve	Gray	Small
Short Sleeve	Gray	Medium
Short Sleeve	Gray	Large
Short Sleeve	White	Small
Short Sleeve	White	Medium
Short Sleeve	White	Large
Long Sleeve	Gray	Small
Long Sleeve	Gray	Medium
Long Sleeve	Gray	Large
Long Sleeve	White	Small
Long Sleeve	White	Medium
Long Sleeve	White	Large

13. Sample answer:

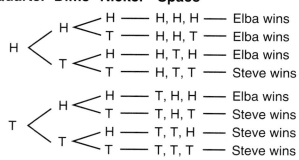

Quarter	Dime	Nickel	Sample Space	
H	H	H	H, H, H	Elba wins
		T	H, H, T	Elba wins
	T	H	H, T, H	Elba wins
		T	H, T, T	Steve wins
T	H	H	T, H, H	Elba wins
		T	T, H, T	Steve wins
	T	H	T, T, H	Steve wins
		T	T, T, T	Steve wins

There are 8 equally likely outcomes with 4 favoring Elba. So, the probability that Elba wins is $\frac{1}{2}$. **15.** $\frac{1}{8}$
17. $\frac{3}{8}$ **19.** $\frac{1}{8}$ **21.** 16 **25.** Sample answer: Mei can draw a model of the situation using a tree diagram to

show the sample space. Then she can determine the probability. The probability of guessing correctly is $\frac{1}{4}$.

Right/ Wrong	Right/ Wrong	Sample Space
R	R	R, R
	W	R, W
W	R	W, R
	W	W, W

27. Sample game: Each player tosses a coin 10 times. If it comes up heads, player 1 receives 1 point. If it comes up tails, player 2 receives 1 point. The player with the most points at the end of 20 tosses wins. **29.** $\frac{2}{5}$
31. $\frac{13}{20}$ **33.** $\frac{3}{10}$ **35.** The cost is increasing over time. In about a year's time, the cost has more than tripled.
37. $3,876 **39.** 15.6 **41.** 154 **43.** 460

Pages 472–474   Lesson 9-3

1. 8 **3.** 24 **5.** 12 **7.** 84 **9.** 24 **11.** 6 possible routes; $\frac{1}{6}$ **13.** 7,776 **15.** No; the number of selections is 32 • 11 or 352, which is less than 365. **17.** 2; 4; 8; 2^n; Sample answer: I used a pattern to determine the number of outcomes for n coins. One coin: 2^1 outcomes, two coins: $2 • 2$ or 2^2 outcomes, three coins: $2 • 2 • 2$ or 2^3 outcomes, n coins: 2^n outcomes.
19. Sample answer: When there are multiple events, the Fundamental Counting Principle is a much faster method of obtaining the total number of outcomes than drawing a tree diagram. The Fundamental Counting Principle also saves paper space and can often be done mentally. When you need to see what the specific outcomes are, make a tree diagram since the Fundamental Counting Principle only gives the number of outcomes. **21.** 6 **23.** $\frac{1}{2}$ **25.** $\frac{1}{5}$, 0.22, 27%, 20.1 **27.** 6 **29.** 120

Pages 476–478   Lesson 9-4

1. 5,040 **3.** $\frac{1}{20}$ **5.** 24 **7.** 720 **9.** $\frac{1}{90}$ **11.** $\frac{1}{12}$
13. Sample answer: Calculator; An exact answer is required. $24 \times 23 \times 22 \times 21 = 255,024$ **15.** $\frac{1}{4}$
17. Sample answer: The number of ways you can order 3 books on a shelf is $3 • 2 • 1$ or 6. **19.** H

21.

Meat	Cheese	Outcomes
turkey	cheddar	turkey, cheddar
	Swiss	turkey, Swiss
ham	cheddar	ham, cheddar
	Swiss	ham, Swiss
salami	cheddar	salami, cheddar
	Swiss	salami, Swiss

23. $1\frac{13}{15}$ **25.** $4\frac{11}{48}$ **27.** 56 **29.** 3

Pages 482–483   Lesson 9-5
1. 21 ways   **3.** 6   **5.** 210   **7.** 924   **9.** $\frac{1}{20}$ **11.** 10; $\frac{1}{10}$
13. permutation; 90   **15.** 15   **17.** The number of ways you can choose three CDs from a collection of ten CDs is 120 ways.   **19.** H   **21.** 72   **23.** Sample answer: $\frac{1}{2} - 0 = \frac{1}{2}$ **25.** $\frac{1}{5}$

Pages 484–485   Lesson 9-6
1. Sample answer: Results would vary slightly.
3. No; Sample answer: the experiment produces about 1–2 correct answers, so using a spinner with 4 sections is not a good way to answer a 5-question multiple-choice quiz.   **5.** 30   **7.** 65 is a reasonable answer because 40% of 160 is 64.   **9.** No, the 6th row should have the numbers 1, 5, 10, 10, 5, 1.   **11.** No; Sample answer: the experiment produces about 2–3 correct answers, so using a number cube is not a good way to answer a 5-question true-false quiz.   **13.** $195

Pages 488–490   Lesson 9-7
1. $\frac{14}{25}$ **3.** The experimental probability, $\frac{14}{25}$ or 56%, is close to its theoretical probability of $\frac{1}{2}$ or 50%.
5. 22 people   **7.** $\frac{9}{10}$; the experimental probability, $\frac{9}{10}$ or 90%, is close to its theoretical probability of $\frac{5}{6}$, or about 83%   **9.** $\frac{3}{35}$ **11.** 134   **13.** $\frac{6}{25}$; $\frac{13}{50}$ **15.** $\frac{13}{25}$ **17.** No; The experimental probability that a mother will receive jewelry is $\frac{17}{100}$. Out of 750 mothers that receive gifts, only about 128 can expect to receive jewelry not 250.   **19.** Yes; Sample answer: If there are 40 unsharpened pencils in the box, then there are twice as many unsharpened pencils as there are sharpened pencils. If there are five sharpened pencils in the sample that was taken out, then there should be ten unsharpened pencils which would give a total of 15 pencils in the sample, which was the size of the sample. It is important to note that this was only one sample. To be sure that the sample represents the population, other samples should be taken and compared to the first sample.   **21.** B   **23.** 6   **25.** 60%   **27.** 55%   **29.** 78%

Pages 495–497   Lesson 9-8
1. $\frac{1}{30}$ **3.** $\frac{1}{10}$ **5.** $\frac{2}{9}$ **7.** $\frac{1}{90}$ **9.** $\frac{2}{3}$ **11.** $\frac{1}{2}$ **13.** $\frac{1}{6}$ **15.** $\frac{3}{25}$
17. 0   **19.** $\frac{3}{16}$ **21.** $\frac{15}{92}$ **23.** $\frac{3}{7}$ **25.** $\frac{4}{7}$ **27.** $\frac{3}{20}$ **29.** $\frac{1}{10}$
31. $\frac{1}{1,024}$ **33.** 21%

35.

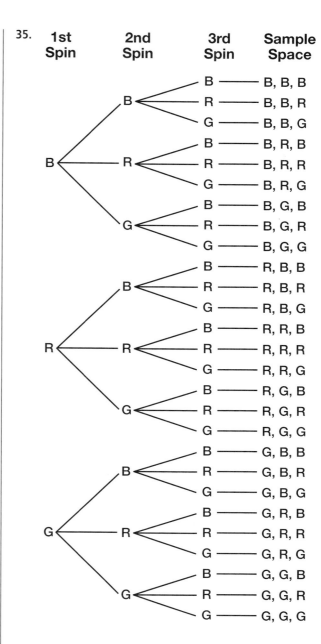

1st Spin	2nd Spin	3rd Spin	Sample Space
B	B	B	B, B, B
		R	B, B, R
		G	B, B, G
	R	B	B, R, B
		R	B, R, R
		G	B, R, G
	G	B	B, G, B
		R	B, G, R
		G	B, G, G
R	B	B	R, B, B
		R	R, B, R
		G	R, B, G
	R	B	R, R, B
		R	R, R, R
		G	R, R, G
	G	B	R, G, B
		R	R, G, R
		G	R, G, G
G	B	B	G, B, B
		R	G, B, R
		G	G, B, G
	R	B	G, R, B
		R	G, R, R
		G	G, R, G
	G	B	G, G, B
		R	G, G, R
		G	G, G, G

37. These events are dependent, not independent. Selecting one book and not returning it to the shelf limits your choices for the next pick to 2 books, not 3, as on the first pick.   **39.** G   **41.** 6   **43.** −48   **45.** 80

Pages 498–502   Chapter 9   Study Guide and Review
1. true   **3.** false; probability   **5.** false; independent events   **7.** false; disjoint   **9.** $\frac{1}{6}$ **11.** $\frac{13}{18}$ **13.** 8%

15.

Outcomes	
Pepperoni	Water
Pepperoni	Milk
Pepperoni	Juice
Mushroom	Water
Mushroom	Milk
Mushroom	Juice
Cheese	Water
Cheese	Milk
Cheese	Juice

17. 36   **19.** $\frac{1}{12}$ **21.** 40,320
23. 105   **25.** 364   **27.** No; Sample answer: the experiment produces about 3 correct answers, so tossing a coin is not a good way to answer a 6-question true-false quiz.   **29.** 8   **31.** $\frac{5}{16}$
33. $\frac{9}{32}$ **35.** $\frac{23}{32}$ **37.** $\frac{3}{64}$; $\frac{1}{20}$
39. $\frac{1}{16}$; $\frac{7}{120}$ **41.** $\frac{1}{3}$

Chapter 10 Geometry: Polygons

Page 509   Chapter 10   Getting Ready

1. 306   **3.** 0.15   **5.** 0.11   **7.** 44   **9.** 105   **11.** 14   **13.** 36

Pages 512–513   Lesson 10-1

1. ∠MNP, ∠PNM, ∠N, ∠1; obtuse   **3.** ∠1 and ∠3; Sample answer: Since ∠1 and ∠3 are opposite angles formed by the intersection of two lines, they are vertical angles.   **5.** ∠DEF, ∠FED, ∠E, ∠5; right   **7.** ∠MNP, ∠PNM, ∠N, ∠7; straight   **9.** ∠RTS, ∠STR, ∠T, ∠9; acute   **11.** neither   **13.** adjacent   **15.** vertical   **17.** ∠1 and ∠2; Sample answer: Since ∠1 and ∠2 share a common vertex, a common side, and do not overlap, they are adjacent angles.
19. True; Sample answer:

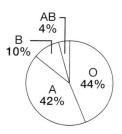

21. A   **23.** $\frac{1}{24}$ **25.** $\frac{5}{24}$ **27.** 46   **29.** 54

Pages 516–517   Lesson 10-2

1. supplementary   **3.** 135   **5.** supplementary
7. supplementary   **9.** neither   **11.** 65°   **13.** 137°
15. Sample answer: ∠CGK, ∠KGJ   **17.** adjacent; adjacent; vertical   **19.** m∠1 = 180° − m∠2; m∠3 = 180° − m∠2; Sample answer: m∠1 and m∠3 both equal the same expression.   **21.** m∠E = 39°, m∠F = 51° **23.** B
25. ∠P, ∠I, ∠RPQ, ∠QPR; acute   **27.** 223.2   **29.** 0.12

Pages 520–523   Lesson 10-3

1.   **Blood Types in U.S.**

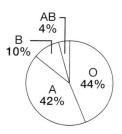

3. blue

5.   **U.S. Steel Roller Coasters**

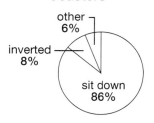

7. Endangered Species in U.S.

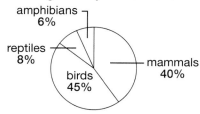

9. paper   **11.** 48 million tons   **13.** about 12 million
15. 15
17. bar graph

Birthplaces of Presidents

19.   **Sizes of U.S Great Lakes**

21. Lake Ontario is one third the size of Lake Michigan.   **23.** No; a 50% increase in 126 students is 189, and 189 is not equal to 366. So, it is not reasonable to say that 50% more students said they could make a difference. Since 300% of 126 is 378, it is reasonable to say that 300% more students said they could make a difference than those who said they cannot make a difference.   **25.** 12.5%; English is half of the circle. Since Science is half of English, Science is half of 50% or 25%. Math is half of Science, which is half of 25% or 12.5%.   **27.** Sample answer: No; the sum of the percents is greater than 100. The people surveyed must have been able to choose more than one fruit juice. **29.** Sample answer: ∠1 and ∠3 **31.** 68 **33.** 101

Pages 527–529 **Lesson 10-4**
1. 44 **3.** 45 **5.** C **7.** right, scalene
9.

11. 118 **13.** 27 **15.** 90 **17.** 53° **19.** acute, equilateral
21. obtuse, isosceles **23.** obtuse, isosceles **25.** right
27.

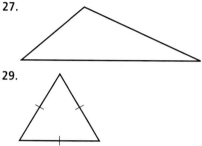

29.

31. 79.5 **33.** 79.2 **35.** 21.3 **37.** 80° **39.** 60 **41.** 77.5
43.

Sample answer: All of the angles are acute and there are no congruent sides. **45.** Never; Sample answer: The sum of the three angles in a triangle is 180°. If a triangle has two obtuse angles, the sum of these two angles, not including the third angle, would already be greater than 180°. **47.** B **49.** about 101°
51. supplementary **53.** about $9

Pages 530–531 **Lesson 10-5**
1. Sample answer: They used inductive reasoning because they made a rule after seeing four examples.
3. See students' work; none of the angles are congruent.
5. Julio: mango, Rashanda: banana, Perry: orange.
7. $3.75 **9.** 196 sq units **11.** 3 packs **13.** Sample answer: addition and division; Find the score that when added to the others, the total sum divided by the number of scores, 5, is equal to 82. The answer is 79.

Pages 535–538 **Lesson 10-6**
1. rectangle **3.** parallelogram **5.** 120° **7.** 64°
9. square **11.** quadrilateral **13.** trapezoid **15.** 56°
17. 67° **19.** 90° **21.** 116° **23.** Bricks A, B, and D are rectangles and brick C is a square. **25.** 129.1°
27. 131.8° **29.** trapezoids, squares, scalene triangles, equilateral triangles **31.** right isosceles triangles, squares, trapezoids **33.** No; a quadrilateral with three right angles will have both pairs of opposite sides parallel. So, it cannot be a trapezoid. **35.** 45 **37.** 80
39. Property B states that there are 4 right angles; Sample answer: A rectangle has 4 right angles in addition to property A. So, property B must state that

there are 4 right angles. **41.** Never; Sample answer: A trapezoid has only one pair of parallel sides. A parallelogram has 2 pairs of parallel sides.
43. Sometimes; Sample answer; A rhombus is only a square if all 4 angles are right angles. **45.** Since a square has all the properties of a rectangle and a rhombus, the diagonals of a square must be congruent and perpendicular. Nothing can be concluded about the diagonals of a parallelogram unless more information is provided. If a quadrilateral is a parallelogram, it is not necessarily a rectangle or a rhombus. So, it would not necessarily have the properties of a rectangle or a rhombus. **47.** J
49. Neva: hamster, Sophie: turtle, Seth: dog
51. right, scalene **53.** 720 **55.** $3.45 **57.** 3 **59.** 7

Pages 543–545 **Lesson 10-7**
1. rectangle *PQRS* **3.** 45 mm **5.** triangle *CAB*
7. 25 mi **9.** 7.2 in. **11.** 6 ft **13.** 12 m **15.** 1,207 feet
17. 120 m **19.** 1:16 **21.** B **23.** C **25.** trapezoid
27. 69° **29.** 90 **31.** 120

Pages 549–551 **Lesson 10-8**
1. decagon; not regular **3.** hexagon; regular **5.** 128.6°
7. not a polygon; the figure is not simple **9.** isosceles right triangle; not regular **11.** hexagon; not regular
13. 144° **15.** 90° **17.** No; the figure is a decagon. Each angle of a decagon measures 144°; since 144° does not divided evenly into 360°, a decagon cannot make a tessellation. **19.** hexagon and triangle
21. octagon and square **23.** $36\frac{1}{4}$ yd **25.** No; the stop sign is an octagon in shape. An octagon cannot make a tessellation. So, there will be some steel that is wasted after the nine signs are cut from the sheet.
27. trapezoid
29. square **31.** pentagon   hexagon

the Pentagon   section of honeycomb
33. Sample answer; A copy of the parallelogram can fit next to it since 45° + 135° = 180° and above or below it since 135° + 45° = 180°. **35.** B **37.** trapezoid
39. 1.5 m per s **41.** $3\frac{1}{6}$ **43.** $4\frac{37}{40}$

44–47.

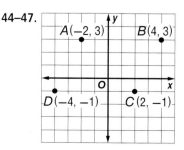

1.

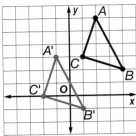

3. $D'(7, 0)$, $E'(4, -2)$, $F'(8, 4)$, $G'(12, -3)$

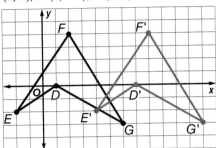

5.

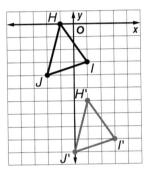

7. $P'(6, 5)$, $Q'(11, 3)$, $R'(3, 11)$

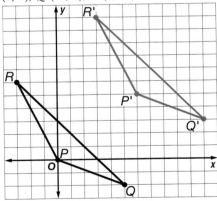

9. $P'(-3, 0)$, $Q'(2, -2)$, $R'(-6, 6)$

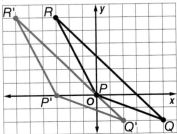

11. 1 unit right and 2 units up: (1, 2)

13.

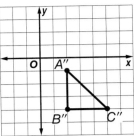

15. Sample answer: There are two main images, the small fish and the large fish. The fish are translated to different parts of the picture. These translations allow for the tessellation of the fish.

17. $F'\left(8\frac{1}{2}, 2\frac{1}{2}\right)$, $G'\left(4\frac{1}{2}, \frac{1}{2}\right)$, $H'\left(2\frac{1}{2}, 1\frac{1}{2}\right)$

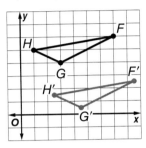

19. 5 units left and 3 units down; $(-5, -3)$ **21.** 5 units right and 4 units up; (5, 4)   **23.** Transformation A is not a translation; the others are translations.   **25.** B
27. octagon   **29.** $2 \times 4 \times 3$ or 24 dinners   **31.** The range would be 62 instead of 55.   **33.** The range would be 8 instead of 6.   **35.** Sample answer: 0.567   **37.** Sample answer: 1.026   **39.** no   **41.** yes

1. no   **3.** yes   **5.** $A'(5, -8)$, $B'(1, -2)$, and $C'(6, -4)$

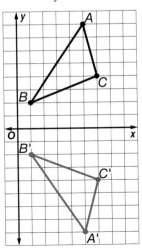

7. $Q'(-2, -5)$, $R'(-4, -5)$, and $S'(-2, 3)$

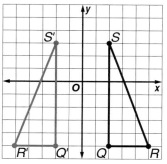

9. yes

11. yes

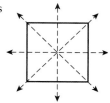

13. yes

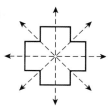

15. $T'(-6, 1)$, $U'(-2, 3)$, and $V'(5, 4)$

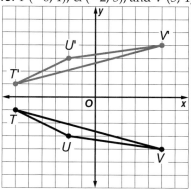

17. $A'(2, -4)$, $B'(-2, -4)$, $C'(2, -8)$, $D'(-2, -8)$

19. $R'(5, 3)$, $S'(4, -2)$, $T'(2, 3)$

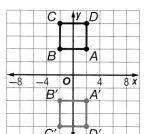 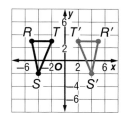

21. $H'(1, 3)$, $I'(1, -1)$, $J'(-2, -2)$, and $K'(-2, 2)$

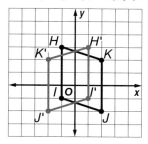

23. There is a line of symmetry vertically down the center of the picture.   **25.** 1   **27.** figures A and C
29. Sample answer; A reflection over the y-axis followed by a reflection over the x-axis.

31.

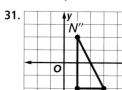

33. x-axis   **35.** y-axis   **37.** $J''(7, -4)$, $K''(-7, -1)$, and $L''(-2, 2)$ **39.** C
41. $F'(1, 6)$, $G'(3, 4)$, $H'(2, 1)$

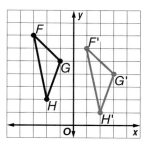

43. Sample answer: $\frac{1}{2} + 8 = 8\frac{1}{2}$ **45.** Sample answer: $12 \div 6 = 2$

Pages 563–566   Chapter 10   Study Guide and Review
1. false; supplementary angles   **3.** false; acute angle
5. false; $(-2, 1)$ **7.** $\angle 1$ and $\angle 4$; Sample answer: Since $\angle 1$ and $\angle 4$ are opposite angles formed by the intersection of two lines, they are vertical   **9.** neither
11.

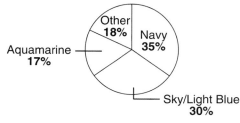

Favorite Shades of Blue

13. 45   **15.** right, isosceles   **17.** parallelogram
19. 10 cm   **21.** 8m   **23.** nonagon; regular

25. $P'(-2, 1)$, $Q'(-8, 0)$, and $R'(-7, 9)$

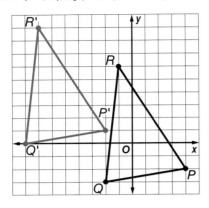

27. $P'(1, -2)$, $Q'(-5, -3)$, and $R'(-4, 6)$

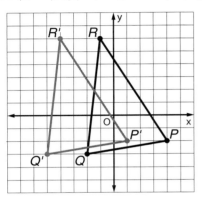

29. $R'(-1, -3)$, $S'(2, -6)$, and $T'(6, -1)$

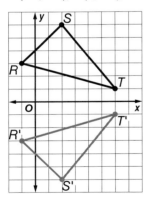

31. $E'(4, -2)$, $F'(-2, -2)$, $G'(-2, 5)$, $H'(4, -5)$

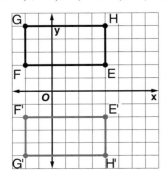

Chapter 11 Measurement: Two- and Three-Dimensional Figures

Page 571   Chapter 11   Getting Ready
1. 136   **3.** 1,248   **5.** 77   **7.** $41.67   **9.** 121   **11.** 36
13. 12.6   **15.** 31.4   **17.** 254.3

Pages 574–576   Lesson 11-1
1. 135 cm² **3.** 17.5 in² **5.** 32 in² **7.** 60 cm²
9. 0.2 cm² **11.** 49.5 yd² **13.** 190.625 m²
15. 525 mm² **17.** 216 in² or 1.5 ft² **19.** 972 in² or
0.8 yd² **21.** 42,000 mi² **23.** 15 in. **25.** $11\frac{2}{3}$ ft²
27. False; if the base and height are each doubled,
then the area is $2b \cdot 2h = 4bh$, or 4 times greater.
29. D   **31.** $A'(2, 5)$, $B'(1, 2)$, $C'(4, 1)$

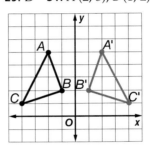

33. Sample answer: The sum of the measures of an
octagon is 1,080°. Each angle measure is 1,080° ÷ 8 or
135°. Since 135 does not go into 360 evenly, an octagon
does not tessellate the plane.   **35.** 84   **37.** 19.5

Pages 580–582   Lesson 11-2
1. 6 in² **3.** 90.4 ft² **5.** 147 in² **7.** 4.5 cm²
9. 183.7 in² **11.** about 95,000 mi² **13.** 125 ft
15. Sample answer: 10 cm² **17.** Sample answer: 7 ft²

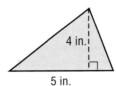

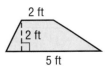

19. 6,500 ft² **21.** Sample answer: the ratio of the area
is the square of the ratio of the bases.   **23.** The area of
a triangle is half the area of a parallelogram with the
same base and height, because two of these triangles
make up the parallelogram.   **25.** 11,000 ft²
27. $J'(-1, 4)$, $L'(1, -1)$, $K'(3, 2)$

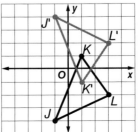

29. 25%   **31.** 156.3   **33.** 91.1   **35.** 72.4

Pages 586–588   Lesson 11-3

1. 31.4 ft   **3.** 44 m   **5.** 67.2 cm   **7.** 50.2 ft   **9.** 22.6 cm
11. 66 ft   **13.** 33 in.   **15.** 33.9 km   **17.** $38\frac{1}{2}$ mi
19. $28\frac{2}{7}$ in.   **21.** 94.2 ft   **23.** 11.0 in.   **29.** 8.0 ft
31. 4.8 yd   **33.** Sample answer: Using the
circumference formula, you find that the unicycle
travels 3.14 • 10 • 2 or 62.8 inches on one revolution.
So, in 5 revolutions, it will travel 62.8 • 5 or 314 inches.
To change into feet, divide 314 by 12. So, the unicycle
will travel $26\frac{1}{6}$ feet in 5 revolutions.   **37.** Elsa; Logan
incorrectly applied the formula that uses the diameter.
39. Both will be doubled. If the value of x is doubled,
the diameter will be $2x$ instead of x. The circumference
will increase from $2\pi x$ to $2\pi(2x)$ to $4\pi x$.   **41.** C   **43.** B
45. 3.7 ft^2 **47.** $\frac{15}{27}$ or $\frac{5}{9}$ **49.** 153.9   **51.** 63.6

Pages 591–593   Lesson 11-4

1. 78.5 cm^2 **3.** 201.1 m^2 **5.** B   **7.** 28.3 in^2
9. 227.0 cm^2 **11.** 32.2 mm^2 **13.** 124.7 cm^2 **15.** 44.2 ft^2
17. 338.2 yd^2 **19.** 2,827.4 ft^2 **21.** Sample answer:
$3 \cdot 6^2 = 108$ ft^2 **23.** 32; 60   **25.** about 50.3 cm^2
27. 29.0 m^2 **29.** 52,276.1 km^2 • 0.1 = 1,582.9 yd^2
31. 62.8 m^2 **33.** 103.5 cm^2 **35.** Sample answer: If the
radius of a circular garden is 6 feet, how much room is
there for gardening? about 113.1 ft^2 **37.** J   **39.** 50.2 yd
41. 120 in^2 **43.** 100.1 m^2 **45.** 113.04   **47.** 150.5

Pages 594–595   Lesson 11-5

1. Sample answer: Finding the areas of the separate
geometric figures and then adding is easier than trying
to find the area of the irregular figure as a whole.
3. Sample answer: Find the area of the wall below to
determine how much paint to buy. To solve, find the
area of the triangle and the area of the rectangle, then
add. The answer is 77 ft^2.

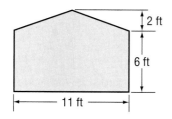

5. Sample answer: Asia, 17,251,712.4 mi^2; Africa,
11,616,153.02; N. America, 9,488,441.82 mi^2;
S. America, 6,900,684.96 m^2; Antarctica, 5,118,008.012
mi^2; Europe, 3,852,882.436 mi^2; Australia/Oceania,
3,047,802,524 mi^2 **7.** B   **9.** 175.84 ft^2

Pages 597–599   Lesson 11-6

1. 112 m^2 **3.** 145 m^2 **5.** 195 ft^2 **7.** 58.6 in^2

9. 257.1 mm^2 **11.** 66.2 yd^2 **13.** approximately
847.2 ft^2 **15.** 196.1 in^2 **17.** $x^2 + \frac{1}{2}(6x)$ or $x^2 + 3x$
19. $467.4 \div 350 \approx 1.34$; Since only whole gallons of
paint can be purchased, you will need 2 gallons of
paint. At \$20 each, the cost will be 2 • \$20 or \$40.
21. Sample answer: Add the areas of a small rectangle
and a trapezoid. Area of a rectangle: $3 \times 1 = 3$; Area of
trapezoid: $\left(\frac{1}{2} \times 6.5 \,(4 + 5)\right) = 29.25$; $3 + 29.25 = 32.25$.
So, an approximate area is 32.25 = 2,400 or 77,400 mi^2.
23. B   **25.** Sample answer: about 30% of \$500 or \$150
27. 452.2 in^2
29.   **31.**

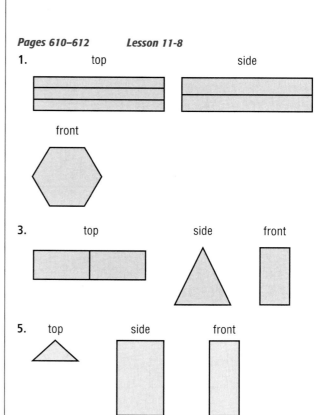

Pages 605–606   Lesson 11-7

1. square; square pyramid   **3.** circle; cylinder
5. triangle; triangular pyramid   **7.** rectangle;
rectangular pyramid   **9.** cone   **11.** trapezoid;
trapezoidal prism   **13.** octagon; octagonal prism
15. triangular prism and rectangular prism
17. rectangular prism   **19.** Sample answer: A cone
has only one base that is a circle. A pyramid also has
only one base, but its base is a polygon. They both
have only one vertex. A cone does not have any lateral
faces and a pyramid has at least three lateral faces.
21. F   **23.** 102.0 m^2 **25.** 53°   **27.** square   **29.** circle

Pages 610–612   Lesson 11-8

1.   top   side

          front

3.   top   side   front

5.   top   side   front

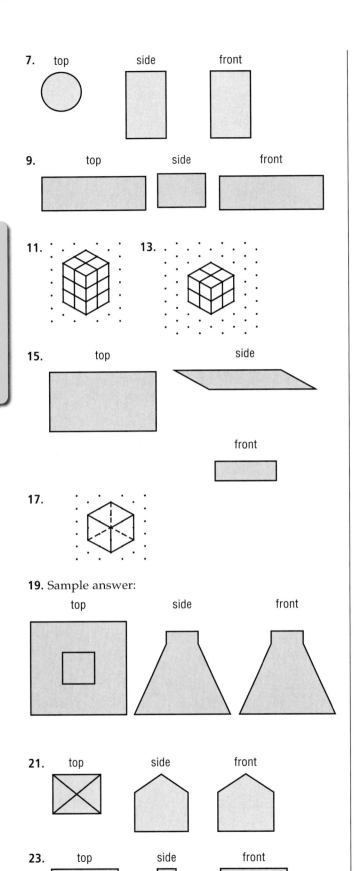

7.
top   side   front

9.
top   side   front

11.   **13.**

15.
top   side

front

17.

19. Sample answer:
top   side   front

21.
top   side   front

23.
top   side   front

25. Sample answer:

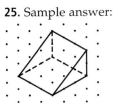

29. B   **31.** sphere   **33.** 96 ft^2 **35.** 112.8 in^2 **37.** $77.39
39. 22   **41.** 68

Pages 616–618   **Lesson 11-9**
1. 220 in^3 **3.** 63 yd^3 **5.** 37.5 ft^3 < 63 ft^3; second
cabinet   **7.** 90 ft^3 **9.** 236.3 cm^3 **11.** 108 m^3
13. 20.4 mm^3 **15.** 40 ft^3 > 36 ft^3, so too much was
bought   **17.** $166\frac{1}{4}$ yd^3 **19.** 2,157,165 ft^3 **21.** 306.52 =
19.4h; 15.8 m **23.** Sample answer: 5 × 4 × 2 or 40 ft^3
25. 6 ft   **27.** No; the area of Prism A is 80 in^2, and the
area of Prism B is 640 in^2, which is eight times
greater.   **29.** D

31.
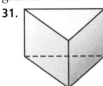
    33. rectangle; rectangular prism
    35. 50 mph
    37. Sample answer: 5 · 3^2 or 45
    39. Sample answer: 3 · 2^2 · 2 or 24

Pages 620–623   **Lesson 11-10**
1. 141.4 in^3 **3.** 617.7 ft^3 **5.** about 603.2 cm^3
7. 4,071.5 ft^3 **9.** 2,770.9 yd^3 **11.** 35.6 m^3 **13.** 103.4 m^3
15. 288.6 in^3 **17.** about 226.2 in^3 **19.** 124,642.7 m^3
21. d.   **23.** a.   **25.** 2,375 cm^3 **27.** 8 in.   **29.** The volume
is 8 times greater than the previous cylinder.
31. Sample answer:

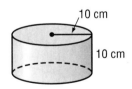

10 cm

10 cm

33. 1 to 4   **35.** D   **37.** 152.9 m^3
39. Sample answer:

41. $\frac{1}{4}$ **43.** $\frac{5}{12}$

Pages 626–630   **Chapter 11**   **Study Guide and Review**
1. rectangular prism   **3.** trapezoid   **5.** cylinder
7. circle   **9.** 2,520 in^2 **11.** 36 ft^2 **13.** 527.3 yd^2
15. 75.4 in.   **17.** $26\frac{2}{5}$ ft   **19.** 29.8 ft   **21.** 1,520.5 cm^2
23. 800 cakes   **25.** $105   **27.** 67.8 yd^2 **29.** triangle;
triangular prism   **31.** cylinder   **33.** rectangular pyramid

35.

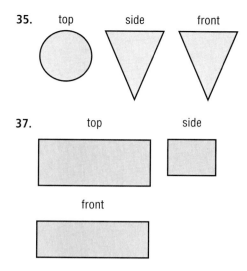

37.

top

side

front

39. 639.6 in^3 **41.** 1,330 ft^3 **43.** 3,728.7 mm^3
45. 942 in.^{3}

Chapter 12 Geometry and Measurement

Page 635   Chapter 12   Getting Ready

1. 16   **3.** 169   **5.** 89   **7.** 225   **9.** 130 yr   **11.** 226
13. 42.1   **15.** 527.8

Pages 637–639   Lesson 12-1

1. 6   **3.** 9   **5.** 2.6   **7.** 7.1   **9.** 10 ft   **11.** 4   **13.** 8
15. 11   **17.** 20   **19.** 2.8   **21.** 9.4   **23.** 23.9   **25.** 52.9
27. 6   **29.** 50   **31.** 81   **33.** 0.7   **35.** 1.7   **37.** 3.8
39. 3.17 cm   **43.** Sample answer: 17, 18, 19   **45.** 2
47. 5   **49.** It cannot be written as a fraction.   **51.** G
53. 28.3 in^3 **55.** acute   **57.** obtuse   **59.** 74   **61.** 16

Pages 643–645   Lesson 12-2

1. 26 mm   **3.** 18.5 cm   **5.** 4.0 ft   **7.** 16.1 m   **9.** 14.1 m
11. 5.4 ft   **13.** 2.8 yd   **15.** 25 in.   **17.** 5.6 mi   **19.** 72.1 in.
21. Sample answer: The plank will not fit horizontally
or vertically. However, the diagonal of the doorway
measures about 18 feet. So, the plank will fit through
the doorway if you tilt it diagonally.   **23.** about
10.4 in.   **27.** G   **29.** 1,413 in^3 **31.** 0.08   **33.** 2.65

Pages 646–647   Lesson 12-3

1. Sample answer: making a model helps you see
what is happening in the problem to help solve it.
3. 8 cars and 4 motorcycles   **5.** 47 in.   **7.** $5\frac{1}{3}$ min
9. Sample answer: The boxes could be arranged on the
shelf 4 boxes wide and 5 boxes deep.   **11.** $125\frac{7}{16}$ ft^2
13. 6 DVDs

Pages 651–653   Lesson 12-4

1. 108 ft^2 **3.** Yes; the surface area of the box is
252 in^2. The surface area of the paper is 288 in^2.
Since 252 in^2 < 288 in^2, she has enough paper.
5. 314 cm^2 **7.** 833.1 mm^2 **9.** 125.4 in^2 or $124\frac{3}{8}$ in^2
11. 234 in^2 **13.** 243.8 ft^2 **15.** 64.5 in^2 **17.** $s = 6x^2$
19. Sample answer: Not all sides of the rectangular
prism will be painted. The top will not be painted.
The number of square feet to be painted is (18)(12) +
2(18)(6) + 2(12)(6).   **21.** Surface area measures the area
of the faces, and area is measured in square units.
23. G   **25.** 12.4 ft   **27.** 804.2 ft^2 **29.** 145.3 yd^2

Pages 657–659   Lesson 12-5

1. 88.0 mm^2 **3.** about 471.2 m^2 **5.** 1,215.8 m^2
7. 272.0 mm^2 **9.** 1,120.0 in^2 or $1,119\frac{44}{45}$ in^2
11. 61.3 cm^2 **13.** Sample answer: $2 \cdot 3 \cdot 4^2 + 2 \cdot 3 \cdot 4 \cdot$
4 or 192 m^2 **15.** 205.0 in^2 **17.** No; the surface area of
the side of the cylinder will double, but the area of the
bases will not.   **19.** A cylinder with radius 6 cm and
height 3 cm has a greater surface area than a cylinder
with height 6 cm and radius 3 cm; Sample answer: The
first cylinder has a surface area of 339.3 cm^2 while the
second cylinder has a surface area of 169.6 cm^2.
21. G   **23.** 112 cm^2 **25.** 12.8 in.   **27.** 7.2 cm

Pages 660–662   Chapter 12   Study Guide and Review

1. false; right triangle   **3.** true   **5.** true   **7.** false;
positive and negative   **9.** false; cylinder   **11.** 2
13. 7   **15.** 4   **17.** 7.8   **19.** 21.1   **21.** 11 ft   **23.** 26.7 in.
25. 29.8 ft   **27.** 875 in^2 **29.** 202 yd^2 **31.** 43 ft^2
33. 2,261.9 in^2

Index

Index

Key Concept

Index

602, 648

Millimeter, 304

Mini Lab
adding fractions on a ruler, 236
angle relationships, 514
area of a circle, 589
area of a parallelogram, 572
area of a triangle, 578
converting length units, 304
finding a pattern, 44
finding the mean, 402
folding triangles, 524
making a scale drawing, 320
making predictions, 426
modeling division by a fraction, 265
modeling division of integers, 114
modeling multiples, 211
modeling multiplicative inverses, 258
modeling percent of change, 369
modeling primes, 181
modeling subtraction of integers, 128
patterns and sequences, 57
percents, 329
perimeter and area, 34
permutations, 475
probability game, 465, 486
Pythagorean Theorem, 640
rates, 287
similar figures, 540
solving equations using models, 142
square roots, 636
squares, 34
tessellations, 553
volume of cylinders, 619

Misleading statistics, 444–449

Mixed numbers
adding and subtracting, 242, 243
dividing, 266
improper fractions and, 677
multiplying, 253

Mixed Problem Solving, 704–715

Mode, 403, 404

Models
concrete, 34, 57, 62, 93–94, 101–102, 107, 114, 134, 135, 142, 151, 180, 211, 250, 265, 287, 316, 320, 369, 426, 475, 486, 491, 518, 524, 532, 540, 552, 553, 583, 589, 600, 607, 613, 619, 646–647, 649, 656
objects, 34, 57, 62, 93–94, 101–102, 107, 114, 134, 135,

142, 151, 180, 211, 250, 265, 287, 316, 320, 369, 426, 475, 486, 491, 518, 524, 532, 540, 552, 553, 583, 589, 600, 607, 613, 619, 646–647, 649, 656
pictorial, 32, 33, 34, 37, 48, 59, 118, 134–135, 136, 141, 142, 147, 151, 174, 175, 181, 192, 196, 200, 202, 203, 205, 216, 225, 226, 231, 236, 237, 241, 243, 250, 251, 252, 256, 257, 258, 266, 272, 275, 276, 333, 336, 342–343, 475, 486, 505, 510, 532, 552, 572, 600–601, 608–612, 640, 654–655, 656, 664–665
number line, 80, 81, 83, 84, 85, 95, 96, 99, 100, 103, 106, 108, 120, 121, 124, 125, 239, 231
verbal, 39, 50, 64, 128, 129, 137, 138, 143, 153, 196, 232, 238, 254, 266, 312, 350, 351, 363, 458, 376, 534

Monomials, 742

Multiple, 211

Multiple Choice. *See* Test Practice

Multiple representations, 57–61, 62, 156–161, 162, 163–167, 168, 304–309, 572–576, 577, 578–582, 583, 589–593, 613–618, 619–623, 624, 654–655

Multiplication
Associative Property of, 54
Commutative Property of, 54
fractions, 252, 253
Identity Property of, 54
integers, 107, 108
Inverse Property of, 258
mixed numbers, 253
phrases indicating, 128
solving equations, 259
Property of Equality, 259
using dot for, 30

Multiplicative Inverse Property, 258

Mutually exclusive events, 494

Negative integers, 80

Nets, 600, 601

Nonagon, 546

Nonexamples, 546–547, 566, 604. *See also* Which One Doesn't Belong?

Note taking. *See* Foldables® Study Organizer

Number line
absolute value, 81
adding integers, 95–96
graphing integers, 80
graphing irrational numbers, 637

Number Sense. *See* H.O.T. Problems

Numbers
comparing and ordering rational numbers, 216
compatible, 232
composite, 181
exponential form, 31
integers, 80
irrational, 637
percent, 342, 344
perfect squares, 35
prime, 181
squares, 34, 35
standard form, 31

Numerical expressions, 38

Obtuse angle, 511

Obtuse triangle, 525, 678

Octagon, 546

Open Ended. *See* H.O.T. Problems

Opposites, 96–97

Ordered pairs, 88
graphing, 89

Order of operations, 38

Origin, 88

Ounce, 298

Outcomes. *See* Probability

Outliers, 397

Output. *See* Functions

Parallel lines, 533
transversals, LA10

Parallelograms, 533
area, 572, 573
base, 572
height, 572

Part. *See* Percents

Index

less than (<), is, 84
not equal to, is, 49
pi, 584
similar to, is, 541

Symmetry, lines of, 558

Tally. *See* Frequency tables

Technology. *See* Calculators, Graphing Calculator Lab, Internet Connections, Spreadsheet Lab

Temperature, 116

Term, 45

Terminating. *See* Decimals

Terms, of a sequence, 57

Tessellations, 548, 552

Test Practice, 76–77, 124–125, 174–175, 226–227, 276–277, 338–339, 390–391, 456–457, 504–505, 568–569, 632–633, 664–665

   Extended Response, 77, 125, 175, 227, 277, 339, 391, 457, 505, 569, 633, 665

   Mid-Chapter Quiz, 48, 100, 147, 201, 247, 317, 368, 423, 479, 539, 602, 648

   Multiple Choice, 29, 33, 37, 41, 47, 52, 56, 61, 67, 83, 85, 87, 92, 99, 106, 111, 118, 305, 131, 133, 141, 146, 147, 155, 161, 167, 184, 189, 200, 205, 214, 220, 235, 241, 246, 257, 260, 262, 270, 286, 292, 297, 303, 309, 315, 326, 332, 348, 354, 360, 365, 364, 460, 372, 374, 378, 382, 401, 405, 408, 414, 421, 431, 449, 464, 467, 470, 474, 478, 483, 490, 517, 523, 529, 538, 545, 551, 557, 562, 576, 582, 588, 592, 594, 599, 606, 611, 612, 617, 618, 622, 639, 645, 653, 659

   Practice Test, 75, 123, 173, 225, 275, 337, 389, 455, 503, 567, 631, 663

   Preparing for Standardized Tests, 716–733

   Short Response/Grid In, 52, 77, 99, 125, 146, 175, 214, 227, 235, 277, 297, 339, 348, 391, 421, 457, 474, 505, 551, 569, 582, 633, 639, 665

   Worked-Out Example, 50, 85, 130, 217, 260, 288, 371, 396, 417, 404, 466, 525, 590, 642

Test-Taking Tip, 77, 125, 175, 227, 260, 277, 339, 391, 457, 505, 569, 633, 665

   alternative method, 288
   back solving, 50
   check the results, 371, 525
   comparing measures, 404
   educated guess, 466
   eliminating answer choices, 85
   reading choices, 217
   verify your answer, 260

Theoretical probability, 486, 487

Three-dimensional figures
   building, 607
   classifying, 604
   cross-sectional views, 745
   cylinder, 619
   drawing, 608
   edges, 603
   effect of changing dimensions, 622, 654–655
   face, 603–604
   lateral faces, 603
   nets, 600–601, 656, 659, 664–665
   pyramid, 601
   rectangular prism, 613–614
   similar solids, 654–655
   sketch top, side, front, and corner views of, 607–612, 618, 623, 631
   solids, 608
   surface area of, 649–659
   triangular prism, 615
   vertices, 603–605
   volume of, 613–614

Transformations, 553
   reflections, 558
   translations, 553, 554

Translations, 553, 554
   coordinates, 554

Transversal, LA10

Trapezoid, 533
   area, 579
   base, 579
   height, 579
   Tree diagrams, 466
   Triangle, 524, 525
   acute, 525
   area, 578
   base, 578
   classifying, 525
   equilateral, 525
   height, 578
   isosceles, 525
   obtuse, 525
   right, 525
      hypotenuse, 640

leg, 640
scalene, 525
similar, 540, 542
Two-dimensional figures. *See also*
   congruent
   decagon,
   heptagon,
   hexagon, 54
   nets, 600–601,
   octagon, 546
   parallelogram, 5
   pentagon, 546   65
   polygons, 546
   quadrilateral, 533–53
   rectangle, 533
   regular polygon, 546
   similar, 540–545
   square, 533
   tessellation, 548
   transformations of, 553–561
   triangle, 524–529

Two-step equations, 151, 153

Unbiased sample, 438

Unit cost, 289, 290, 291, 292. *See also* Unit rates

Units,
   customary, 298
   converting between, 298–303
   metric, 304
   converting between, 304, 305

Unit rates, 287
   best buy, 289

Units, metric, 304
   converting between, 304, 305

Variables, 44
   defining, 50

Venn diagram, 186, 532

Verbal,
   model for problem solving, 39, 50, 64, 128, 129, 137, 138, 143, 153, 196, 232, 238, 254, 266, 312, 350, 351, 363, 458, 376, 534
   phrases as algebraic expressions, 128

Vertex/vertices, 510, 603